OPERATIONAL PHYSICS

OPERATIONAL PHYSICS

Modern physics increasingly replaces the object-based world-view of the old physics with an operation-based world-view. This process is supported by a similar one that is taking place in mathematics at the same time. *The International Journal of Theoretical Physics* dedicates the series "Operational Physics" to this process. Its goal is to provide scientists and researchers with textbooks, monographs, and reference works to address the growing need for information.

Prospective authors are encouraged to correspond with the Editor-in-Chief in advance of submitting a manuscript. Submission of manuscripts should be made to the Editor-in-Chief.

Editor-in-Chief

David Finkelstein
Georgia Institute of Technology
School of Physics
837 State Street
Atlanta, GA 30332-0430

Volumes Published in This Series:

Operational Quantum Theory I—Nonrelativistic Structures
by Saller, H. 2006

Operational Quantum Theory II—Relativistic Structures
by Saller, H. 2006

Heinrich Saller

Operational Quantum Theory II

Relativistic Structures

Springer

Heinrich Saller
Werner-Heisenberg Institut
Max-Planck Institut für Physik
München, Germany 80805
hms@mpp.mu.mpg.de

Library of Congress Control Number: 2006920923

ISBN-10: 0-387-29776-6

Printed on acid-free paper.

Printed in the United States of America. (TB/MVY)

9 8 7 6 5 4 3 2 1

springer.com

I would like to thank three people from three generations without whom this book could not have been written.

First, the late Werner Heisenberg, who implanted in me the conviction that symmetries with their operations are appropriate basic concepts for understanding physical interactions and objects.

Second, David Finkelstein, who gave me the feeling, in our fruitful collaboration and work over the decades, of not being alone in giving priority to the operational approach.

Finally, I learned a lot from my first son, Christian, who is a much better mathematician than I. He taught me that many of the mathematical concepts denigrated as esoteric and academic by those physicists who have a direct pipeline to God are basic and exactly the right tools for the physical structures to be formalized. He also helped me very much by knowing and recommending the appropriate advanced mathematical literature.

Contents

INTRODUCTION

Quantum theory is connected especially with the names Planck, Bohr, Heisenberg, Pauli, and Dirac. The quantum revolution describes our deepest insight, so far, into the physical structure of nature. It is comparable only with the Copernican revolution, switching from a finally oriented anthropocentric description of physical phenomena to one using general laws with initial or boundary conditions, connected with the names Kepler, Galileo, and Newton, or with the change from tangible mass points as basic structures to Faraday's and Maxwell's field concepts and, shortly before quantum theory, with the relativization of space and time by the lonely genius Einstein.

In retrospect, the label "quantum" or, as adjective, "quantal," is too weak to characterize the extent of the revolution involved in abandoning the classical theory as a basic epistemological framework for physics. The word "quantal" – in contrast to the assumed classical "continuous" ("natura non facit saltus") – was motivated by the finite jumps and the discreteness as seen, for example, in the photoelectric effect or in the spectral lines for atoms or in the discrete split of atomic rays in Stern-Gerlach experiments.

One has to distinguish in quantum theory between two kinds of "jumps": First, the quantum structure relies on the noncommutativity of operations, e.g., of the not commuting position-momentum operator pair $[i\mathbf{p}, \mathbf{x}] = \hbar$, with a nontrivial quantum $\hbar$ (Planck's constant) or of the not anticommuting conjugate operator pair of an electron-positron field $\{\overline{\mathbf{\Psi}}(\vec{y}), \mathbf{\Psi}(\vec{x})\} = \hbar\gamma^0\delta(\vec{x}-\vec{y})$. Second, there are the jumps, characterized by integers. These jumps, as seen in the atomic spectral lines, were the starting point of quantum theory. However, after the dust has settled, they cannot be addressed as the revolutionary characteristics of quantum theory: Integers characterize compact operation groups. Take a circle, say a closed rubber string, cut it, wind it around your wrist, and glue both ends together again; the number of possible windings is always an integer. Does rubber band winding characterize quantum theory? The rubber band stands for the circle, parametrizing the compact Lie group $\mathbf{U}(1) = \exp i\mathbb{R}$ or the isomorphic group $\mathbf{SO}(2)$ with the rotations around one space axis. The irreducible representations of the circle (1-dimensional torus), as realized by the different rubber band windings and thus of all compact Lie groups involving higher-dimensional tori, come with integer winding numbers, "quantum numbers" in the narrow sense. Since bound waves in quantum mechanics are related to compact representations of the noncompact time translation group $\mathbb{R}$, they give rise to integer-related discrete (rational) quantum jumps. The same situation occurs for spin, which is related to the 3-dimensional position rotations, parametrizable by the compact volume of a sphere. However, in addition to these discrete jumps (integer winding numbers $z \in \mathbb{Z}$) continuous quantum numbers can also occur, e.g., real energies $E \in \mathbb{R}$ or momenta $\vec{q} \in \mathbb{R}^3$, or, apparently, the particle masses $m^2 \in \mathbb{R}_+$ from a continuous spectrum as eigenvalues or invariants for representations of time and space translations. Continuous numbers require operations with noncompact action groups, whereas compact groups come with rational ("quantum") numbers.

At the core of quantum theory is the relativization of the ontic structures in contrast to the absolute ontology in classical theories, e.g., of the position of mass points or of the spin direction of particles. The appropriate characterization "quantum relativity" alludes to the relativity of time and space. A quantum description starts from practic stuctures, e.g., from translations or rotations. Quantum theory describes operations with the dynamics itself an operation. Quantum theory is operation theory. A classical ontology requires a projection of the nonabelian operational framework to an abelian substructure. In a classical description, objects are primary with interactions between them as a secondary structure. In a quantum description the hierarchy is reversed: objects arise as eigenvectors of operations.

Appropriate questions in quantum theory ask for operations: What is the operational meaning of spin and mass of a particle? Invariants for rotations and spacetime translations. What is the operational meaning of a Coulomb and Yukawa potential? Representation distributions, 2-sphere spreads of position translations. What is the operational meaning of a gauge coupling constant? The relative normalization of the gauge–transformation–inducing operational Lie algebra in the Lorentz Lie algebra. What is the operational meaning of a Feynman propagator? Matrix elements of spacetime translation representations, unitary for on-shell contributions.

And one may ask even about quite specific structures: What is the operational meaning of cosines and exponentials, of Bessel and Macdonald functions, or of Laguerre polynomials, etc.? Representation coefficients of specific operations. With respect to a formulation of physics by special functions arising as solutions of "special differential equations," e.g., equations of motion in time and space, there is a unified view, initiated by Wigner and elaborated in exhaustive encyclopedic detail by Vilenkin, who writes in the introduction of his subject–related book, "a really unified view on the theory of the basic classes of special functions ... was established by employing the considerations that belong to a field of mathematics seemingly quite far from the subject under consideration, the theory of representations of Lie groups." Essentially all physically relevant special functions arise as coefficients of Lie group representations. Therefore in the following, Lie operations are of paramount importance.

Weyl was the first to connect with each other, basically and in a systematic form, "The theory of groups and quantum mechanics" in his like-named book. Wigner especially proceeded to extend the group–theoretic method in mathematical detail to relativistic quantum theory.

There was always a symmetry strain in physical theories: The Greeks started with the association of the five Platonic solids with the four basic elements: fire, water, earth, and air; the fifth polyhedron, the dodecahedron, with pentagonal sides, called quintessence, was taken as the all-encompassing cosmos. The idea was revived by Kepler in his Mysterium Cosmographicum to understand the six planets known in his time as regularly circling on the simultaneous in- and out-spheres between the five Platonic solids, nested one within the other. It is fascinating to realize how Kepler's fantastic ideas, completely wrong and without any reasonable contact with any physical dynamics, hit upon an apparently immensely important basic structure in nature: The five Platonic solids have as their sides regular triangles, squares, and pentagons. Exactly these two-dimensional symmetric Euclidean polygons characterize the symmetry operations related to simple Lie groups as classified by E. Cartan. The four main series of symmetry operations can be related, via the characterizing weight and root diagrams, to regular squares and triangles lumped together in higher and higher dimensions (details in the chapters "Simple Lie Operations" and "Rational Quantum Numbers"). All the semisimple symmetries we use in fundamental theories of particles and their interactions can be associated with those operational structures. Every particle physics student today knows the quark triangles as weight diagrams for the color operations $\mathbf{SU}(3)$. The squares as weight diagrams for orthogonal symmetries show up, for instance, in the electron occupation numbers (twice a square) of the atomic shells, $2 = 2\times 1^2$, $8 = 2\times 2^2$, $18 = 2\times 3^2$, etc., originating in the nonrelativistic framework from the orthogonal group $\mathbf{SO}(4)$ desribing rotation and perihelion conservation.

The main mistake of Kepler (forgive me) was, with our knowledge today, to look for the symmetry of the objects, not for the symmetries of the dynamical law; he was no quantum theorist. The possibility in quantum field theories to have less symmetric state vectors or objects as a result of operations with a larger symmetry plays an important role in reconciling the asymetry of the world as we see it with basic symmetric operations.

The quantum concepts as a unifying picture for the basic physical laws, at least without any experimental contradiction thus far, are not "anschaulich." Particles have no positions in the naive classical sense. To call them basically "pointlike" does not make sense. All this makes our physical intuition very difficult. The classical physical concepts dissolve like Dali's clock in the desert. Let me quote from the last public talk of Heisenberg in Munich, 1975 (my translation):

"It is unavoidable that we use a language originating from classical philosophy. We ask, What does the proton consist of? Is the quantum of light elementary or composite? etc. However, all these questions are incorrectly posed since the words "divide" and "consist of" have lost almost all their meaning. Therefore it should be our task to adjust our language, our thinking, i.e., our scientific philosophy to this new situation that has been created by experiments. Unfortunately, that is very difficult. Therefore, there creep into particle physics, again and again, wrong questions and wrong conceptions....

We have to come to terms with the fact that experimental knowledge from very small and very large distances no longer provides us with an "anschauliches Bild," and we have to learn to live there without "Anschauung." In this case we realize that the antinomy of the infinitely small for the elementary particles is resolved in a very subtle way, in a way which neither Immanuel Kant nor the Greek philosophers could have imagined, the word "to divide" loses its sense.

If one wants to compare the insights of today's particle physics with any earlier philosophy, it could be only the philosophy of Plato, since the particles of today's physics are representations of symmetry groups – that is what quantum theory teaches us – and hence the particles resemble the regular Platonic polyhedra."

Physical properties are registered in experiments, i.e., they describe a relation with an observer. They are mathematically formulated as eigenvalues of operations, e.g., energy and momentum or the spin in the direction of a magnetic field. Different ontic (asymptotic) structures as projections of one practic structure (interaction) are determined by an experimental setup that distinguishes one of possibly many eigenvector bases for the operations under consideration. Behind different setups there are the characterizing invariants, e.g., the mass of a particle for the Lorentz transformation-dependent energy-momenta, as measured in different spacetime frames, or its spin as measured in one space direction, which is determined, e.g., in a Stern-Gerlach experiment by the spatial inhomogeneity of a magnetic field. An experimental setup is related, mathematically, to a diagonalization of a set of operations. Since a set of diagonalizable matrices is simultaneously diagonalizable if and only if its elements commute with each other, an ontic interpretation of a set of operations depends on the experimenter's decision, concretized in the chosen apparatus, to distinguish a subset of simultaneously diagonalizable matrices. In general, there exist many different inequivalent diagonalizable subsets. Mathematically, this is a relatively simple theorem; its physical interpretation and coordination with our daily life experience, relying on an absolute ontic description existing and remaining without an ongoing measurement, is difficult and counterintuitive. An operator is not exhaustively described by the property (eigenvalue) of one object (particle, bound state vector, eigenvector) and even more for a set with more than one operator.

A transition from operations to particle- or state-related experimental numbers has to do with a maximal diagonalization of linear transformations as introduced for the characterization of Lie groups by E. Cartan. In this sense, an experimental test of quantum operations can be maximal, but because of the basic noncommutativity, it is never complete.

A vector is not a collection (row or column) of some numbers; this is a representation of the vector in a chosen basis, physically implemented by a given experimental apparatus. Not only for a mathematician, perhaps even more for a physicist, the distinction and choice of a basis has to be justified and the imposed restrictions have to be discussed carefully.

The ontic interpretation of one operator, e.g., acting on a two-dimensional vector space and diagonalizable as the matrix $l^3 \cong \begin{pmatrix} 1 & 0 \\ 0 & -1 \end{pmatrix}$ displaying its orthonormal eigenvectors by the two columns with their property (eigenvalues) $\{\pm 1\}$, may prevent the ontic interpretation of a second operator via simultaneous eigenvalues, e.g., of $l^1 \cong \begin{pmatrix} 0 & 1 \\ 1 & 0 \end{pmatrix}$; one eigenvector basis may not be usable twice. However, that's not all. Since there exist nondiagonalizable operators, in the simplest case of an operator on a complex two-dimensional vector space with a basis representation $n \cong \begin{pmatrix} 0 & 1 \\ 0 & 0 \end{pmatrix}$ (triangular Jordan structures), quantum theory involves even operations without ontic particle interpretation at all, i.e., without eigenvector bases. Such nondiagonalizable operations really occur, e.g., in connection with the quantum gauge field describing the Coulomb interaction as one degree of freedom in the four-component electromagnetic field (potential) that has only two degrees of freedom with an ontic particle interpretation, the left and right circularly polarized photons.

In quantum theories a clear distinction has to be made between the full operational interaction language and the restricted projections to objects. Physical objects, e.g., bound state vectors or elementary particles, as seen in experiments are eigenvectors with respect to transformation groups. Particles are eigenvectors with respect to space and time translations, rotations, and electromagnetic transformations that are formalized with the real Lie groups $\mathbb{R}$, $\mathbf{SU}(2)$, and $\mathbf{U}(1)$ and give rise to the properties mass, spin, and electromagnetic charge number and, at least until now, nothing more. The bound waves of the nonrelativistic hydrogen atom are eigenvectors for the operation groups $\mathbb{R}$ and $\mathbf{SO}(4)$ with the time translations and the space rotations with perihelial transformations respectively, giving rise to the properties energy and the space rotations-related quantum numbers. Interactions are characterizable by groups that in general are larger than the asymptotic symmetry groups that determine the object's properties. Elementary interactions implement internal ("chargelike") transformation groups as used in the standard model, i.e., hypercharge $\mathbf{U}(1)$, isospin $\mathbf{SU}(2)$, and color $\mathbf{SU}(3)$, in addition to the external spacetime translations $\mathbb{R}^4$ and the orthochronous Lorentz group $\mathbf{SO}_0(1,3)$ or, more precisely, its twofold cover $\mathbf{SL}(\mathbb{C}^2)$. The projective transition from the operations characterizing the interactions to those for the objects involves a dramatic operation group reduction, e.g., in the standard model for electroweak and strong interactions

$$\text{for interactions } \Big[\underbrace{\mathbf{SL}(\mathbb{C}^2) \vec{\times} \mathbb{R}^4}_{\text{external: Poincaré}}\Big] \times \Big[\underbrace{\mathbf{U}(1) \circ \big(\mathbf{SU}(2) \times \mathbf{SU}(3)\big)}_{\text{internal: hyperisospin-color}}\Big]$$

$$\to [\mathbf{SU}(2) \times \mathbb{R}^4] \times \mathbf{U}(1) \text{ for massive particles}$$

The interaction operation groups, e.g., isospin $\mathbf{SU}(2)$ for the nuclear interactions, which vanish as symmetries for asymptotic objects, e.g., for proton and neutron with different masses, may leave their traces in multiplicities, e.g., in the two nucleons arising from an isospin doublet. Sometimes not only the symmetries may vanish, but even the related nontrivial multiplicities, as proposed for the color $\mathbf{SU}(3)$ interaction symmetry leaving asymptotically only $\mathbf{SU}(3)$-singlets (color confinement, not proved yet).

There was a development in geometry culminating in the "Erlanger Programm" (1872) of Felix Klein that can serve as an analogue for the operational point of view to characterize quantum physics. A geometry, according to Klein, can be characterized by a Lie group G acting on an analytic manifold M, in the irreducible case on the equivalence classes in the homogeneous space G/H with a subgroup $H \subseteq G$ as fixgroup ("little group") or on a vector space. An example is the spherical geometry with the rotation group $\mathbf{SO}(3)$ acting on the 2-sphere $\Omega^2 \cong \mathbf{SO}(3)/\mathbf{SO}(2)$ that parametrizes the axial rotation subgroups, or the Euclidean geometry $\mathbf{SO}(3) \vec{\times} \mathbb{R}^3$ with the rotation group acting on 3-space or the pseudo-Euclidean Poincaré geometry $\mathbf{SO}_0(1,3) \vec{\times} \mathbb{R}^4$ with the Lorentz group acting on spacetime where the Minkowski translations $\mathbb{R}^4$ can be looked on as the tangent space of the homogeneous space $\mathbf{GL}(\mathbb{C}^2)/\mathbf{U}(2)$, or the special and general linear affine geometries $\mathbf{SL}(\mathbb{R}^n) \vec{\times} \mathbb{R}^n$ and $\mathbf{GL}(\mathbb{R}^n) \vec{\times} \mathbb{R}^n$. In a Klein space $G \bullet M$ only concepts compatible with or even invariant under the operation group G make sense. For example, for general linear geometry, the invariant concepts "parallelity" and "dimension", in addition "volume" for the special linear geometry, in addition the concepts "causal order" and "length" for Poincaré geometry, and in addition "angle" and "distance" for orthogonal geometry. The decreasing group chain $G_1 \supset G_2 \supset \cdots$ is reflected in the increasing number of invariants for the space acted on: To characterize smaller subgroups one has to invoke more and more properties. In a physical interpretation of Klein's program the acting groups are the interaction governing groups like $\mathbf{SO}(4)$ for the periodic system of the atoms in nonrelativistic mechanics or the internal hypercharge-isospin-color group $\mathbf{U}(1) \circ [\mathbf{SU}(2) \times \mathbf{SU}(3)]$ for interactions in the standard model. The vector spaces with the interaction group representations, characterized by invariants, e.g., mass and spin or hyperisospin and electromagnetic charge number, contain, after symmetry reduction, the bound state vectors or the particles.

It is not the purpose of this book to teach quantum theory to the beginner; it is not an introduction, but intended for the graduate student with a good knowledge of, on the one hand, the conventional presentations of nonrelativistic quantum mechanics and canonical quantum field theory, and, on the other hand, some knowledge of groups and Lie algebras, their algebraic and topological structures and their representations. Parts of it have been used for lectures on "Algebraic Methods in Quantum Mechanics," on "Introduction to Quantum Field Theory," on the "Standard Model of Strong and Electroweak Interactions," and on "Time, Space, and Spacetime in Quantum Mechanics and Quantum Field Theory." My motivation and aim is to understand and to explain quantum physics as far as possible by operational structures: why we apply them, which structures are unavoidable, which ones are immanent already in the mathematical framework used, and which structures seem artifically complicated and should be looked at with some suspicion. I work with the prejudice that fundamental physical structures are simple, not trivial, to understand and to formulate and esthetically beautiful, in some sense definable not only by personal taste. Relevant questions, worked with, but not necessarily satisfactorily answered, are of the kind, What follows from the real Lie structure of the complex represented operations? For example, the Hilbert space formulation with probability amplitudes. Is there a connection between the causal order of time and spacetime and the probability interpretation of quantum experiments and the positivity of energy? What is the operational origin of the Yukawa and Coulomb interaction? Which transformations are represented by a Feynman propagator, by its "on-shell" and its "off-shell" contributions? Are the divergences of the canonical quantum field theories related to a misrepresentation of the operations involved? What causes the dichotomy between internal compact and external spacetime-related operations that are also noncompact? Where does the gauge structure come from?

And the deepest question is, What is the common conceptual basic root branching into the phenomenological concepts interaction, spacetime, and matter? Wigner's classification of particles as unitary representations of the Poincaré group can be taken as an indication that it is impossible to think about spacetime and matter separately. One step to further this program is to show that scattering states and interaction-bound states arise from operation group representations.

Mathematically elegant formulations in physics may leave us with an empty taste: Answers to all the questions above are physically satisfactory only if they lead also to experimentally testable numbers. Mathematics alone is not enough: The richness of mathematical forms, even esthetically appealing simple structures, seems to be inexhaustible. To paraphrase a word of Kant: Physical theories without experimental numbers are empty. The determination of one number, e.g., of a gauge coupling constant, may justify a huge theoretical building. However, also this is true: Numbers without a theoretical understanding are blind; think of numerologists. To take up the first sentence of this paragraph: Mathematically ugly formulations in physics leave us with a bad taste.

The mathematical level is not undergraduate; I have tried to use the best mathematical tools at my disposal. A. Knapp, one of the mathematical experts in the field of "Representation theory of semisimple Lie groups", writes in the preface of the like-named textbook (about 800 pages), "The subject of semisimple Lie groups is especially troublesome in this respect" (learning by logical progression). "It has a reputation for being both beautiful and difficult, and many mathematicians seem to want to know something about it. But it seems impossible to penetrate. A thorough logical-progression approach might require ten thousand pages." The application of these beautiful tools in physics would presuppose their understanding, although, I hope at least, not with the completeness and depth necessary for mathematicians. I shall try to assist this understanding by sections with mathematical tools. In the beginning, it is not necessary to master all the concepts mentioned there. The pragmatic "battle tested physical approach to mathematics" carries rather far. But in the end, a pedestrian mathematical attitude with some knowledge of the rotation group is not enough. Mathematical simplicity does not coincide with conceptual triviality. The relevant simple concepts are, in most cases, very deep.

In the historical development of physics the causal equations of motion, introduced by Newton for time development, were derived later with extremal and variational principles from Lagrangians and Hamiltonians, which, in turn, could be characterized, for important cases, by their invariance or transformation properties with respect to operation groups. In this book I will go the historical route in the opposite direction: In contrast to the familiar procedure starting with equations of motion, I start with operational structures. The equations of motion do not play the basic role. They are a Lie parameter-related formulation of the local behavior with respect to the operation group involved as expressed for a Lie group by the action of its Lie algebra (tangent space translations). Time and space for the interpretation of a physical dynamics with the conventional equations of motion are a very important, but from the operational point of view only one example of, tangent space-related structures. Therefore the time and spacetime dependence of operators or eigenvectors and equations of motion reflects properties of acting groups and Lie algebras or, to include also semigroups and symmetric spaces with their tangent translations, of acting Lie operations. Equations of motion are a powerful method to diagonalize, to find eigenvalues and invariants of the operations involved.

To illustrate this reversed procedure in the simple example of a harmonic oscillator, time operations or causality as the starting point is formalized, qualitatively and quantitatively, by the additive ordered group $\mathbb{R}$. The Lie group $\mathbb{R}$ has its irreducible complex representations in the compact group $\mathbf{U}(1)$ acting on 1-dimensional vector spaces. The represented time translations define time orbits in the representation space, especially the irreducible orbits of a dual eigenvector basis $(\mathrm{u}(t), \mathrm{u}^\star(t))$ for the two $\mathbb{C}$-isomorphic dual representation spaces with imaginary time action eigenvalues $\pm i\omega \in i\mathbb{R}$:

$$\mathbb{R} \ni t \longmapsto e^{\pm i\omega t} \in \mathbf{U}(1) \Rightarrow \left\{ \begin{array}{rcl} \mathrm{u}(t) & = & e^{i\omega t} \ \mathrm{u}(0), \\ \mathrm{u}^\star(t) & = & e^{-i\omega t} \mathrm{u}^\star(0). \end{array} \right.$$

The Lie algebra (time translation) action can be expressed by first-order differential equations for the representation orbits

$$(\tfrac{d}{dt} \mp i\omega)(\mathrm{u}, \mathrm{u}^\star)(t) = 0.$$

The Lagrangian L yields another formulation of the time translation action on dual eigenvectors

$$iL = iL_0 - iH_0 = \mathrm{u}^\star \tfrac{d}{dt}\mathrm{u} - i\omega \mathrm{u}\mathrm{u}^\star$$

with the kinetic term L_0 implementing the duality of the basic pair $(\mathrm{u}, \mathrm{u}^\star)$ and the Hamiltonian H_0 as product of the basic space identity $\mathrm{u}\mathrm{u}^\star$ and eigenvalue (frequency) ω the represented time translation (Lie algebra) basis. The dual irreducible representation characteristic invariant $|\omega|$ sets the intrinsic time unit.

The representation connected $\mathbf{U}(1)$-conjugation of the irreducible complex spaces with the time orbits implements the time reflection $t \overset{\mathrm{T}}{\leftrightarrow} -t$ and $\mathrm{u} \overset{\mathrm{T}}{\leftrightarrow} \mathrm{u}^\star$ and allows the definition of Hermitian orbits, called position-momentum $(\mathbf{x}, \mathbf{p})$. It thus becomes possible to interpret the time orbits in position and momentum space, e.g., by an oscillating spring or a pendulum. The position-momentum orbits arise from real self-dual representations of the time operations in the group $\mathbf{SO}(2)$, as Lie group isomorphic to $\mathbf{U}(1)$:

$$\begin{array}{rl}\mathbf{x} &= \ell\frac{\mathrm{u}+\mathrm{u}^\star}{\sqrt{2}}\\ i\mathbf{p} &= \frac{\mathrm{u}-\mathrm{u}^\star}{\ell\sqrt{2}}\end{array} \Rightarrow \begin{pmatrix}\mathbf{x}(t)\\ \mathbf{p}(t)\end{pmatrix} = \begin{pmatrix}\cos\omega t & \ell^2\sin\omega t\\ -\frac{\sin\omega t}{\ell^2} & \cos\omega t\end{pmatrix}\begin{pmatrix}\mathbf{x}(0)\\ \mathbf{p}(0)\end{pmatrix},$$
$$L = \mathbf{p}\tfrac{d}{dt}\mathbf{x} - \omega(\ell^2\tfrac{\mathbf{p}^2}{2} + \tfrac{1}{\ell^2}\tfrac{\mathbf{x}^2}{2}),$$

where ℓ is the characteristic length in the dual position-momentum pair, defined by the $\mathbf{SO}(2)$-metric $\begin{pmatrix}\ell^2 & 0\\ 0 & \frac{1}{\ell^2}\end{pmatrix}$ and defining together with the frequency ω two phenomenological units, the inert mass $M = \frac{1}{\omega\ell^2}$ and the spring constant $k = \frac{\omega}{\ell^2}$. The usual starting point, the classical Lagrangian $L = \mathbf{p}\frac{d}{dt}\mathbf{x} - (\frac{\mathbf{p}^2}{2M} + k\frac{\mathbf{x}^2}{2})$, encapsulating the self-dual irreducible real representations of the time operations, comes at the end of the procedure.

In quantum mechanics, much more in quantum field theory, the definition of an operator Lagrangian with explicit spacetime derivatives is in general rather difficult, if not impossible. The dual pair structure, classically encoded in the kinetic Lagrangian, e.g., in $iL_0 = \mathrm{u}^\star\frac{d}{dt}\mathrm{u}$, formulates the quantization $[\mathrm{u}^\star, \mathrm{u}] = 1$ or, for a Hermitian-anti-Hermitian pair $iL_0 = i\mathbf{p}\frac{d}{dt}\mathbf{x}$, the Born-Heisenberg relation $[i\mathbf{p}, \mathbf{x}] = 1$. The time translations are realized by the adjoint action (quantum commutator) with a Hamiltonian

$$\mathbf{H}_0 = \omega\frac{\{\mathrm{u},\mathrm{u}^\star\}}{2} = \frac{\mathbf{p}^2}{2M} + k\frac{\mathbf{x}^2}{2} \Rightarrow \left\{\begin{array}{llll}[i\mathbf{H}_0, \mathrm{u}] &= \frac{d}{dt}\mathrm{u}, & [i\mathbf{H}_0, \mathrm{u}^\star] &= \frac{d}{dt}\mathrm{u}^\star,\\ [i\mathbf{H}_0, \mathbf{x}] &= \frac{d}{dt}\mathbf{x}, & [i\mathbf{H}_0, \mathbf{p}] &= \frac{d}{dt}\mathbf{p}.\end{array}\right.$$

The time derivative $\frac{d}{dt}$ can be considered to be a shorthand notation, familiar from the classical derivative, for the adjoint-action-induced Lie algebra transformation. From this point of view the first-order time differential equations for dual pairs, e.g., for position-momentum $(\mathbf{x}, \mathbf{p})$, or the second-order equations for one Hermitian combination, e.g., for position $\mathbf{x}$, are a consequence of the quantum-implemented linear Lie algebra action, i.e., of $\frac{d}{dt} = [i\mathbf{H},\ ]$.

The conjugation group $\mathbf{U}(1)$ with the represented time operation by phase transformations $\mathbb{R} \ni t \longmapsto e^{\pm i\omega t} \in \mathbf{U}(1)$ endows the one-dimensional representation with a scalar product and a Hilbert space structure that allows Born's "probability amplitudes" for the ontological interpretation of the operations via experiments. The spectrum of the position operator $x \in \mathrm{spec}\,\mathbf{x}$ is used for Schrödinger wave functions $x \longmapsto \psi(x)$, which are orbits (representation coefficients) of position translations.

Also, for quantum field theory the classically oriented approach relying on differential equations of first and second orders, e.g., Dirac and Klein-Gordon equations, will not be in the foreground. Representations for external spacetime and internal unitary groups and their actions as seen, for example, in the standard model are more basic for the understanding as their projections to asymptotic particle state vectors, as used for experimental tests. An illustration of the method used in this book may be given, for instance, by a Dirac field $\boldsymbol{\Psi}$ for a massive spinor particle. Here the unitarily represented group is the

Poincaré group $\mathbf{SL}(\mathbb{C}^2) \vec{\times} \mathbb{R}^4$, induced by representations of a direct product subgroup $\mathbf{SU}(2) \times \mathbb{R}^4$ involving spin $\mathbf{SU}(2)$ as double cover of position rotations $\mathbf{SO}(3)$ and spacetime translations to define the embedded particle, e.g., the electron-positron, with its spin invariant $\frac{1}{2}$ from a rational spectrum and its mass m^2 from a continuous spectrum. The Fock expectation value $\langle \ldots \rangle$ of the commutator, with Dirac matrices $\{\gamma^k\}_{k=0,1,2,3}$,

$$\langle [\overline{\mathbf{\Psi}}(y), \mathbf{\Psi}(x+y)] \rangle = \langle [\overline{\mathbf{\Psi}}, \mathbf{\Psi}] \rangle(x) = \int \frac{d^4q}{(2\pi)^3} (\gamma^k q_k + m)\delta(m^2 - q^2)e^{iqx},$$

is a matrix element of a Hilbert representation of spacetime translations. The projection to time translation representation matrix elements $e^{\pm imx_0}$ can be obtained by position integration

$$\mathbb{R} \ni x_0 \longmapsto \int d^3x \; \gamma_0 \langle [\overline{\mathbf{\Psi}}, \mathbf{\Psi}] \rangle(x) = \mathbf{1}_2 \otimes \begin{pmatrix} i \sin mx_0 & \cos mx_0 \\ \cos mx_0 & i \sin mx_0 \end{pmatrix}.$$

The corresponding position projections by time integration is trivial:

$$\int dx_0 \; \langle [\overline{\mathbf{\Psi}}, \mathbf{\Psi}] \rangle(x) = 0.$$

This is in contrast to the position projection of the time-ordered quantization anticommutator arising in the Feynman propagator. Here one obtains a Yukawa potential and force as noncompact representation coefficients $e^{-m|z|}$ of position translations, distributed with the Kepler factor $\frac{1}{r}$ on the 2-spheres in 3-dimensional position space

$$\begin{aligned} \mathbb{R}^3 \ni \vec{x} \longmapsto \int dx_0 \; \epsilon(x_0)\gamma_0 \{\overline{\mathbf{\Psi}}, \mathbf{\Psi}\}(x) &= \begin{pmatrix} \frac{\vec{\sigma}\vec{x}}{r} \frac{1+mr}{r} & m\mathbf{1}_2 \\ m\mathbf{1}_2 & -\frac{\vec{\sigma}\vec{x}}{r} \frac{1+mr}{r} \end{pmatrix} \frac{e^{-mr}}{2\pi r}, \\ \int dxdy \; \frac{e^{-mr}}{2\pi r} &= \frac{e^{-m|z|}}{m}. \end{aligned}$$

Spacetime cannot be thought of without interactions. Spacetime is perceived by its operational representations, which are given by and act on what we call quantum fields, which may or may not have particles as projections in a Hilbert space.

A customary approach to quantum structures uses ad hoc Hilbert spaces with square integrable position space functions at a very early stage. The operational approach puts the Hilbert spaces in a representational perspective. As each Lie group defines its representations, so each Lie group with real operations defines its complex Hilbert spaces on which it acts. The Hilbert spaces of nonrelativistic quantum mechanics are defined, as shown in the Stone-von Neumann theorem, by the Heisenberg Lie algebra, whose three real operations are characterized by the Lie bracket $[\mathbf{x}, \mathbf{p}] = \mathbf{I}$. Those historically first Hilbert spaces in quantum theory are not appropriate for all operation groups. They are not suited for fermionic quantum structures and not used in quantum field theory. Already quantum–mechanical scattering theory is formulated more appropriately in the Hilbert spaces defined by the Euclidean group $\mathbf{SO}(3) \vec{\times} \mathbb{R}^3$ of rotations acting on position translations. The Hilbert space for a free relativistic particle is defined, as shown by Wigner, by a representation of the

Poincaré group $\mathbf{SO}_0(1,3) \vec{\times} \mathbb{R}^4$. Or there are Hilbert spaces for the Lorentz groups $\mathbf{SO}_0(1,2)$ and $\mathbf{SO}_0(1,3)$ for two or three position dimensions whose elements cannot be formulated with square integrable functions, as shown by Bargmann and Gel'fand and Naimark.

To understand the strength and appropriateness of the operational point of view it is useful to learn, to test, and to apply it in the well-established areas of nonrelativistic quantum mechanics and relativistic quantum field theory. Therefore, the first volume of the book deals essentially, after an introductory presentation of time and space translations, with the time and space-related finite-dimensional representation structures, with compact Lie operations, and, as a nonrelativistic application, with an operationally oriented formulation of the always fascinating Kepler problem.

Here arise already continuous eigenvalues and invariants for noncompact operations, which, in the context of relativistic quantum field theory with the noncompact nonabelian Lorentz group, are looked at more closely in the first part of the second volume. The representation structure of free particle fields, massive and massless, and its implementation in the familiar formalism are given. This part ends with an application of those structures to the standard model of elementary particles. Perturbation theory with its normalization-regularization procedure will not be discussed.

The second part of the second volume works with the – mathematically rather demanding – harmonic analysis of noncompact nonabelian Lie groups and their homogeneous spaces, e.g., the Lorentz and Poincaré group or the causal spacetime cone, to understand the spacetime representations in Feynman propagators and their shortcomings. One has to face the question whether the concepts of "virtual particles" ("off-shell") with the so–called energy-time uncertainty and the virtual particle-exchange in an "anschauliche" description of interactions, as suggested by Feynman diagrams, are not of the same dangerous quality as the point-particle and position-orbit concepts for electrons inside atoms to understand their spectral lines.

In the end, an attempt is made to proceed from the Wigner classification of the particles as vectors acted on with irreducible unitary Poincaré group representations, i.e., from a classification of tangent structures, to the constitution of these tangent structures. An operational spacetime model is proposed in the form of a nonlinear symmetric space whose spectrum includes as invariants particle masses and, especially, gauge coupling constants as normalization of its irreducible representations. Since this is an extremely difficult problem, such an attempt should be seen not as a solution, but as one proposal for a direction on the way to a solution.

Perhaps it is necessary to mention that essentially up to parts of the last two chapters in the second volume, the material in the following is general as concerns the results. I do not propose new theories. The aim is, on an operational basis, to understand more deeply what we are working with in quantum theory. The appropriate language and the conceptional presentation may not be so familiar.

MATHEMATICAL TOOLS

The basic mathematics used in the following is strongly influenced by the Bourbaki school. The concepts, the notation, and the names I use may be unfamiliar to many physicists. They are the usual ones in the mathematical literature and, as I found after getting used to them, also appropriate for physics. Sometimes the abstract structural concepts of mathematics are easier to probe more deeply than the ad hoc coined concepts in physics.

The structural formulation helps, as far as possible, to separate the specific problems in physics from the mathematical-logical ones. With respect to the structure of Lie groups and their representations, especially for the noncompact and nonabelian operations, I have learned much, especially from the books of Folland, Gel'fand, Helgason, Kirillov, Knapp, and Vilenkin, which are highly recommended.

In general, each chapter starts with the more physically oriented sections, which, after a summary, are followed (not always) by more mathematically oriented ones dealing with the concepts used before. Sometimes, especially in later chapters, a distinction between "mathematical" and "physical" would look too arbitrary.

Presumably, one cannot learn the mathematics only from what is given in the mathematical sections: they may already require much mathematical experience. As I know from personal experience, there is "no free mathematical lunch." The mathematical sections are intended to place the mathematical manipulations in physics in their structural context. They should define, introduce, and make familiar to some degree with or remind of the structures used, give a coarse orientation, and stimulate a deeper study of the mathematical literature, which is given with all important references, also in journals, in the books quoted above.

It is not the purpose of this book to prove mathematical theorems that can be found in mathematical textbooks. One "opens up" for the mathematical tools if one really needs them in physics. Then, many proofs become unnecessary if one dives deeply enough into the structures. The mathematical structures are treated eclectically, reflecting my personal taste and my limited abilities and avoiding cumbersome complications. Nevertheless, I am sure, that there will be mistakes I have overlooked and subtleties, even major ones, that I have not taken into account. The representation is by no means hierarchical and complete; some basic concepts are tacitly assumed as familiar and other basic concepts are briefly explained. Mathematical formulas are not always easy to read. Since, however, mathematics is the language of science, it will not be assumed to be necessary to express each formula before or after in everyday language.

The operation concept is clearly formalized in the language of *categories and functors*, which will be used only superficially, mnemotechnically, and for notational purposes. The notation **kat** denotes a category in which the objects are sets, e.g., the categories

sets:	$\underline{\mathbf{set}}$
monoids:	$\underline{\mathbf{mon}}$
groups:	$\underline{\mathbf{grp}}$
abelian groups:	$\underline{\mathbf{abgrp}}$
rings:	$\underline{\mathbf{rng}}$
modules over a ring R:	$\underline{\mathbf{mod}}_R$
vector spaces over K (abelian field):	$\underline{\mathbf{vec}}_K$
algebras over K:	$\underline{\mathbf{ag}}_K$
associative algebras over K:	$\underline{\mathbf{aag}}_K$
Lie algebras over K:	$\underline{\mathbf{lag}}_K$
differentiable manifolds over K:	$\underline{\mathbf{dif}}_K$
Lie groups over K:	$\underline{\mathbf{lgrp}}_K$
topological spaces:	$\underline{\mathbf{top}}$
measure spaces:	$\underline{\mathbf{mes}}$

Elements in the categories used are *morphisms* $(A \xrightarrow{f} B) \in \underline{\mathbf{kat}}$ (mappings, arrows). An object pair (A, B) of a category $\underline{\mathbf{kat}}$ has the set of morphisms $\mathbf{kat}(A, B) = \{A \longrightarrow B\}$, compatible with the category characterizing structure and associatively composable. Morphisms are called *endomorphisms* $\mathbf{kat}(A, A)$ for $A = B$, the *isomorphisms* $\overset{\circ}{\mathbf{kat}}(A, B)$ are called *automorphisms* $\overset{\circ}{\mathbf{kat}}(A, A)$ for $A = B$ with id_A the identity.

Isomorphies hold in a category; therefore they should be qualified, e.g., $L \overset{\underline{\mathbf{vec}}_{\mathbb{R}}}{\cong} \mathbb{R}^n$ for a real vector space isomorphy of a Lie algebra. For a simpler notation, such qualifications are omitted; they should be obvious from the context.

Starting with operations as basic structures, one may use the identity operation id_A as the neutral operation in the nontrivially acting ones ("constancy in change") to define an object A. The sloppy notation $f \in \underline{\mathbf{kat}}$ and $A \in \underline{\mathbf{kat}}$ is used for morphisms and objects.

The categories above can be arranged with the inclusion order

$$\begin{array}{ccccccccccccc}
\underline{\mathbf{top}} & \subset & \underline{\mathbf{mes}} & \subset & \underline{\mathbf{set}} & & & & & & & & \\
 & & \cup & & \cup & & & & & & & & \\
 & & \underline{\mathbf{dif}}_K & & \underline{\mathbf{mon}} & & & & & & & & \\
 & & \cup & & \cup & & & & & & & & \\
 & & \underline{\mathbf{lgrp}}_K & \subset & \underline{\mathbf{grp}} & \supset & \underline{\mathbf{abgrp}} & \supset & \underline{\mathbf{mod}}_R & \supset & \underline{\mathbf{vec}}_K & & \\
 & & & & & & & & & & \cup & & \\
 & & & & \cup & & & & & & \underline{\mathbf{ag}}_K & \supset & \underline{\mathbf{lag}}_K \\
 & & & & & & & & & & \cup & & \\
 & & & & \underline{\mathbf{rng}} & & & \supset & & & \underline{\mathbf{aag}}_K & &
\end{array}$$

Isomorphic objects of a category define classes as objects in the associated equivalence category. Categories may have additional properties. For example, they are *morphism stable* if the morphisms are objects of the same category

$\mathbf{kat}(A,B) \in \underline{\mathbf{kat}}$, e.g., $\mathbf{set}(S,T) \in \underline{\mathbf{set}}$, $\mathbf{vec}_K(V,W) \in \underline{\mathbf{vec}}_K$, linear mappings constitute a vector space.

As basic operational structures, set and vector space endomorphisms (arrow monoids and arrow algebras) as well as set and vector space automorphism groups (permutation groups and linear groups) deserve special symbols

$$\begin{array}{llll} \mathbf{set}(S,S) &= \mathbf{A}(S) &\in \underline{\mathbf{mon}}, & \overset{\circ}{\mathbf{set}}\,(S,S) = \mathbf{G}(S) \in \underline{\mathbf{grp}},\\ \mathbf{vec}_K(V,V) &= \mathbf{AL}(V) &\in \underline{\mathbf{aag}}_K, & \overset{\circ}{\mathbf{vec}}_K\,(V,V) = \mathbf{GL}(V) \in \underline{\mathbf{grp}}. \end{array}$$

Co- and contravariant functors are mappings for categories $\underline{\mathbf{kat}}^{1,2}$

$$\mathcal{F} : \underline{\mathbf{kat}}^1 \longrightarrow \underline{\mathbf{kat}}^2, \quad \begin{array}{c} A \\ f \big\downarrow \\ B \end{array} \longmapsto \begin{array}{c} \mathcal{F}(A) \\ \big\downarrow \mathcal{F}(f) \\ \mathcal{F}(B) \end{array} \quad \text{or} \longmapsto \begin{array}{c} \mathcal{F}(A) \\ \big\uparrow \mathcal{F}(f) \\ \mathcal{F}(B) \end{array}$$

$$\text{with } \mathrm{id}_{\mathcal{F}(A)} = \mathcal{F}(\mathrm{id}_A) \quad \begin{cases} \mathcal{F}(f\circ g) = \mathcal{F}(f)\circ\mathcal{F}(g), & \text{covariant},\\ \mathcal{F}(f\circ g) = \mathcal{F}(g)\circ\mathcal{F}(f), & \text{contravariant}. \end{cases}$$

For example, a Lie group G has a unique Lie algebra, denoted by $\log G$, with the covariant logarithm functor

$$\log : \underline{\mathbf{lgrp}}_{\mathbb{K}} \longrightarrow \underline{\mathbf{lag}}_{\mathbb{K}}, \quad G \longmapsto \log G.$$

A functor may have additional properties, e.g., additive if direct sums of vector spaces are involved $\mathcal{F}(V_1 \oplus V_2) = \mathcal{F}(V_1) \oplus \mathcal{F}(V_2)$ or exponential $\mathcal{F}$ for (tensor) products $\mathcal{F}(V_1 \oplus V_2) = \mathcal{F}(V_1) \otimes \mathcal{F}(V_2)$.

Mappings can inherit structures of their domains, e.g., a vector space can arise from a set with mappings into a field K as expressed in the covariant *free functor* (linear extension or span functor)

$$K^{(\ )} : \underline{\mathbf{set}} \longrightarrow \underline{\mathbf{vec}}_K, \quad \begin{array}{c} S \\ f \big\downarrow \\ T \end{array} \longmapsto \begin{array}{c} K^{(S)} \\ \big\downarrow K^{(f)} \\ K^{(T)} \end{array}$$

The vector space $K^{(S)} \cong \{\sum_{\text{finite}} \alpha_s s\}$ contains the finite linear K-combinations of set elements (or the mappings $\alpha : S \longrightarrow K$ with finite support); it has S as canonical basis. For $K^{(f)}$ the set mapping is linearily extended.

Important functors arise with universal extensions (structures): Given a structure expressed with the category $\underline{\mathbf{kat}}$ there may be objects with more structure in a subcategory $\underline{\mathbf{ukat}} \subset \underline{\mathbf{kat}}$, e.g., algebras in vector spaces $\underline{\mathbf{ag}}_K \subset \underline{\mathbf{vec}}_K$ or abelian groups in abelian semigroups with cancellation rule or complete Hausdorff spaces in uniform (e.g., metric) spaces.

A *universal extension functor* $\mathcal{E}$ from a category in a more structured subcategory

$$\mathcal{E} : \underline{\mathbf{kat}} \longrightarrow \underline{\mathbf{ukat}}$$

is the solution of a universal problem if for any $A \in \underline{\mathbf{kat}}$ there exists a more structured "universal" object $\mathcal{E}(A) \in \underline{\mathbf{ukat}}$ and a natural injection ι that factorizes any kat-morphisms f to a $\underline{\mathbf{ukat}}$-object U with a unique $\underline{\mathbf{ukat}}$-morphism $\tilde{f}$ as shown in the commutative diagram[1]

$$A, \iota, f \in \underline{\mathbf{kat}}, \qquad \begin{array}{ccc} A & \xrightarrow{\iota} & \mathcal{E}(A) \\ {\scriptstyle f}\downarrow & & \downarrow{\scriptstyle \tilde{f}} \\ U & \xrightarrow[\mathrm{id}_U]{} & U \end{array}, \qquad \mathcal{E}(A), U, \tilde{f} \in \underline{\mathbf{ukat}},$$

$$f = \tilde{f} \circ \iota, \quad \mathbf{kat}(A, U) \cong \mathbf{ukat}(\mathcal{E}(A), U).$$

If $\mathcal{E}$ exists, the object $\mathcal{E}(A)$ is unique up to $\underline{\mathbf{ukat}}$-isomorphisms. The induced functor $\mathcal{E}$ is covariant: take $U = \mathcal{E}(B)$ with $B \in \underline{\mathbf{kat}}$. With a unique $\tilde{f}$ the corresponding morphism sets are set-isomorphic (equal cardinality).

An example is the linear extension functor above,

$$S, \iota, f \in \underline{\mathbf{set}}, \qquad \begin{array}{ccc} S & \xrightarrow{\iota} & K^{(S)} \\ {\scriptstyle f}\downarrow & & \downarrow{\scriptstyle \tilde{f}} \\ V & \xrightarrow[\mathrm{id}_V]{} & V \end{array}, \qquad K^{(S)}, V, \tilde{f} \in \underline{\mathbf{vec}}_K,$$

or the *tensor algebra functor* (multilinear extension functor)

$$\bigotimes : \underline{\mathbf{vec}}_K \longrightarrow \underline{\mathbf{aag}}_K, \quad V \longmapsto \bigotimes V.$$

Also, the numbers, denoted by

natural: $\mathbb{N}_k = \{k, k+1, \dots\}$, $\mathbb{N} = \mathbb{N}_1 \supseteq \mathbb{N}_k$,
integer: $\mathbb{Z}$, rational: $\mathbb{Q}$, algebraic: $\mathbb{A}$,
number fields $\mathbb{K} \in \{\mathbb{R}, \mathbb{C}\}$ with real $\mathbb{R}$ and complex $\mathbb{C}$,
positive (negative): $\mathbb{Z}_\pm = \pm\mathbb{N}_0 = \pm|\mathbb{Z}|$, $\mathbb{R}_\pm = \pm|\mathbb{R}|$,

are examples of natural structures and basic operations. They start from an additive semigroup $\mathbb{N}$ with cancellation rule, extended to and embedded naturally into $\mathbb{Z}$, which formalizes binary operations on $\mathbb{N}$. Since $\mathbb{Z}$ forms an abelian multiplicative monoid with cancellation rule it is extendable, analogously, to $\mathbb{Q}$ formalizing binary $\mathbb{Q}$- or quartic $\mathbb{N}$-operations. $\mathbb{Q}$ allows the natural Cauchy completion to the reals $\mathbb{R}$, which formalizes approximation operations

$$\mathbb{N} \longmapsto \mathbb{Z} \longmapsto \mathbb{Q} \longmapsto \mathbb{R}.$$

Good guesses to look for universal extensions are self-relations in the set products, e.g., $\mathbb{N} \times \mathbb{N}$ for $\mathbb{Z}$ or $\mathbb{Z} \times \mathbb{Z}$ for $\mathbb{Q}$ or the countably infinite relations (Cauchy series) $\mathbb{Q}^{\aleph_0}$ for $\mathbb{R}$.

[1] If not stated otherwise, all such diagrams are commutative.

1

LORENTZ OPERATIONS

Spacetime translations are characterized by a causality (order) compatible "metric" with indefinite $(1,3)$-signature, defining or defined by Lorentz transformations (chapter "Spacetime Translations"). In the complex formulation of quantum structures the noncompact Lorentz group also has to be represented in a unitary group - because of the unbounded group volume necessarily indefinite unitary for finite-dimensional nontrivial representations.

If the rotation group $\mathbf{SO}(3)$ for position translations $\mathbb{S} \cong \mathbb{R}^3$ with the spin Lie algebra[1] $A_1^c \cong (i\mathbb{R})^3$ and its Lie group $\mathbf{SU}(2)$ is represented by actions on complex vector spaces with canonical conjugation (chapter "Antistructures: The Real in the Complex"), it is embedded into representations of the doubled Lie algebra $A_1^c \hookrightarrow A_1^c \oplus iA_1^c \cong A_{(1,1)} \cong \mathbb{R}^6$. This involves an embedding for the Cartan subalgebras $i\mathbb{R} \hookrightarrow i\mathbb{R} \oplus \mathbb{R} = \mathbb{C}_\mathbb{R}$ and their groups $\mathbf{SO}(2) \hookrightarrow \mathbf{SO}(2) \times \mathbf{SO}_0(1,1) = \mathbf{SO}(\mathbb{C}_\mathbb{R}^2)$. The subindex $\mathbb{R}$ in $\mathbb{C}_\mathbb{R}$ denotes a real structure represented in the complex, i.e., with a conjugation. For a less-cumbersome notation, it will be omitted in the following, only real Lie operations will be considered.

The doubled Lie algebra $A_{(1,1)}$ is the Lie algebra of the Lorentz group whose defining representation space gives a model for Minkowski spacetime $\mathbb{M} \cong \mathbb{R}^4$. The noncompact Lorentz structures arise by complexification of the compact spin structures $\mathbf{SO}(3) \hookrightarrow \mathbf{SO}(\mathbb{C}^3)$. The classes of the real 6-dimensional Lie group $\mathbf{SL}(\mathbb{C}^2) = \exp A_{(1,1)}$ with respect to its center constitute the orthochronous Lorentz group $\mathbf{SO}_0(1,3) \cong \mathbf{SL}(\mathbb{C}^2)/\mathbb{I}(2) \cong \mathbf{SO}(\mathbb{C}^3)$. The orientation manifold of spin groups in a Lorentz group is parametrizable by the real 3-dimensional noncompact symmetric boost space, the hyperboloid $\mathcal{Y}^3 \cong \mathbf{SL}(\mathbb{C}^2)/\mathbf{SU}(2) \cong \mathbf{SO}_0(1,3)/\mathbf{SO}(3)$. In such a complexification approach the causal order of Minkowski spacetime as $\mathbf{SL}(\mathbb{C}^2)$-representation space comes as a surprise. The connection between complexification and causality (order) is considered in more detail in the chapter "Spacetime as Unitary Operation Classes."

In this chapter all finite-dimensional irreducible $\mathbf{SL}(\mathbb{C}^2)$-representations are given. They arise by a doubling of the irreducible $\mathbf{SU}(2)$-representations starting from the Weyl doubling of the fundamental Pauli spinor representation. For those finite-dimensional representations, the integer winding numbers $\mathbb{Z}$

[1] In this chapter a Lie algebra structure of a vector space is defined up to linear equivalence.

as eigenvalues for compact spin $\mathbf{SU}(2)$ are paired with integers $\mathbb{Z}$ for the noncompact boosts $\mathbf{SL}(\mathbb{C}^2)/\mathbf{SU}(2)$. The relevant unitary group for the complex finite-dimensional $\mathbf{SL}(\mathbb{C}^2)$-representations is the indefinite anticonjugation group $\mathbf{U}(2,2)$.

Definite unitary, i.e., Hilbert space representations of the group $\mathbf{SL}(\mathbb{C}^2)$ are, if faithful, necessarily infinite-dimensional; the noncompact boosts have eigenvalues from a continuous spectrum. They will be discussed in the chapter "Harmonic Analysis."

1.1 Spacetime Lie Algebras

1.1.1 Lorentz Lie Algebra

The operational structure for spacetime translations can be introduced as canonical complexification of the spin operations for position translations.

The Lie algebra $A_1^c \oplus iA_1^c \cong \mathbb{R}^6$, doubling the spin Lie algebra A_1^c, has as Lie brackets in a doubled orthogonal basis

$$\text{basis of } A_1^c \oplus iA_1^c: \quad \{l^a, b^a = il^a \mid a = 1,2,3\}, \quad \begin{cases} [l^a, l^b] &= -\epsilon^{abc} l^c, \\ [l^a, b^b] &= -\epsilon^{abc} b^c, \\ [b^a, b^b] &= +\epsilon^{abc} l^c. \end{cases}$$

The *Lorentz Lie algebra* $A_{(1,1)} \cong \mathbb{R}^6$ is the, up to linear equivalence, unique Lie algebra with the neutral signature $(3,3)$ for the Killing form. It allows Cartan decompositions $A_{(1,1)} \cong A_1^c \oplus iA_1^c$ into a compact 3-dimensional Lie subalgebra and a noncompact 3-dimensional vector subspace. It is simple with rank 2, i.e., its eigenvectors are characterized by two eigenvalues. From the A_1^c-Casimir element $-\frac{\delta_{ab}}{2} l^a \otimes l^b$, the inverse definite Killing form for the angular momenta, the complexification leads to two invariant power-2 tensors, the inverses of the two signature $(3,3)$ invariant forms for the Lorentz Lie algebra:

$$\begin{aligned} I_+(A_{(1,1)}) &= -\tfrac{\delta_{ab}}{4}(l^a \otimes l^b - b^a \otimes b^b), \\ I_-(A_{(1,1)}) &= -\tfrac{\delta_{ab}}{2} l^a \otimes b^b = -\tfrac{\delta_{ab}}{4}(l_+^a \otimes l_+^b - l_-^a \otimes l_-^b), \quad l_\pm^a = \tfrac{l^a \pm b^a}{\sqrt{2}}, \end{aligned}$$

where, $I_+(A_{(1,1)})$ as the inverse Killing form of $A_{(1,1)}$, is called the *Killing-Casimir element*, $I_-(A_{(1,1)})$ the *chiral Casimir element*. They generate all invariants for Lorentz transformations, i.e., the center of the enveloping algebra $\mathbf{E}(A_{(1,1)})$.

As A_1^c is isomorphic to the angular momentum Lie algebra $\log \mathbf{SO}(3)$ of the rotation group, so $A_{(1,1)}$ is isomorphic to $\log \mathbf{SO}(\mathbb{C}^3)$ or to the Lorentz Lie algebra $\log \mathbf{SO}_0(1,3)$ with angular momenta and boosts, in orthogonal bases

$$\begin{array}{ccccc} A_{(1,1)} & \cong & \log \mathbf{SO}(\mathbb{C}^3) & \cong & \log \mathbf{SO}_0(1,3), \\ \varphi_a l^a + \psi_a b^a & \cong & \begin{pmatrix} 0 & -(\varphi_3 + i\psi_3) & \varphi_2 + i\varphi_2 \\ \varphi_3 + i\psi_3 & 0 & -(\varphi_1 + i\psi_1) \\ -(\varphi_2 + i\psi_2) & \varphi_1 + i\psi_1 & 0 \end{pmatrix} & \cong & \left(\begin{array}{c|ccc} 0 & \psi_1 & \psi_2 & \psi_3 \\ \hline \psi_1 & 0 & -\varphi_3 & \varphi_2 \\ \psi_2 & \varphi_3 & 0 & -\varphi_1 \\ \psi_3 & -\varphi_2 & \varphi_1 & 0 \end{array}\right), \end{array}$$

$$\begin{aligned} l^a &\cong \tfrac{\epsilon^{abc}}{2} \mathbf{1}^{bc}, \\ b^a &\cong \mathbf{1}^{a0} = -\mathbf{1}^{0a}, \\ a,b &= 1,2,3, \end{aligned} \quad \begin{cases} [l^a, l^b] &\cong [\frac{\epsilon^{aef}}{2}\mathbf{1}^{ef}, \frac{\epsilon^{bcd}}{2}\mathbf{1}^{cd}] = \mathbf{1}^{ba}, \\ [b^a, b^b] &\cong [\mathbf{1}^{a0}, \mathbf{1}^{b0}] = \mathbf{1}^{ab}, \\ [b^a, l^b] &\cong [\mathbf{1}^{a0}, \frac{\epsilon^{bcd}}{2}\mathbf{1}^{cd}] = \epsilon^{bcd}\frac{\eta^{ad}\mathbf{1}^{c0} - \eta^{ac}\mathbf{1}^{d0}}{2} = -\epsilon^{abc}\mathbf{1}^{c0}. \end{cases}$$

In a $\log \mathbf{SO}_0(1,3)$-basis, the Killing and chiral Casimir elements look as follows:

$$\begin{array}{ll} I_+(\log \mathbf{SO}_0(1,3)) & = -\frac{1}{8} g_{jm} g_{kn} \mathbf{l}^{mn} \otimes \mathbf{l}^{jk} = -\frac{1}{8} \mathbf{l}_{jk} \otimes \mathbf{l}^{jk}, \\ I_-(\log \mathbf{SO}_0(1,3)) & = \frac{1}{4} \epsilon_{jkmn} \mathbf{l}^{mn} \otimes \mathbf{l}^{jk}, \\ & \quad j,k,\dots = 0,1,2,3, \quad \epsilon_{0123} = -1, \end{array}$$

i.e., the chiral Casimir element is related to the 4-dimensional volume forms of Minkowski spacetime. Examples are the invariant for the square of the electrodynamic field strength tensor $F_{jk}F^{jk}$ and the product with its dual $\tilde{F}_{jk}F^{jk}$. The occurrence of the Lorentz "metric" g with signature $(1,3)$ in the framework of $A_{(1,1)}$ will be discussed below.

1.1.2 Poincaré Lie Algebra

The real 6-dimensional Lorentz group together with the four spacetime translations on which they act defines the real 10-dimensional Poincaré Lie algebra:

$$\begin{array}{l} \text{basis of } \log \mathbf{SO}_0(1,3) \ \vec{\oplus}\ \mathbb{R}^4 : \{\mathbf{l}^{jk}, \mathbf{p}^j \mid j,k = 0,1,2,3\}, \\ [\mathbf{l}^{jk}, \mathbf{l}^{nm}] = g^{jn}\mathbf{l}^{km} - g^{kn}\mathbf{l}^{jm} - g^{jm}\mathbf{l}^{kn} + g^{km}\mathbf{l}^{jn}, \quad [\mathbf{p}^j, \mathbf{p}^k] = 0, \\ [\mathbf{l}^{jk}, \mathbf{p}^n] = g^{jn}\mathbf{p}^k - g^{kn}\mathbf{p}^j \iff \left\{ \begin{array}{llll} [\mathbf{l}^a, \mathbf{p}^0] & = 0, & [\mathbf{l}^a, \mathbf{p}^b] & = -\epsilon^{abc}\mathbf{p}^c, \\ [\mathbf{b}^a, \mathbf{p}^0] & = -\mathbf{p}^a, & [\mathbf{b}^a, \mathbf{p}^b] & = -\delta^{ab}\mathbf{p}^0. \end{array} \right. \end{array}$$

It is useful to consider the Poincaré operations in a larger context with subgroups and supergroups:

$$\begin{array}{ccccc} & \nearrow & \mathbf{SO}(3)\ \vec{\times}\ \mathbb{R}^3 & \hookrightarrow & \mathbf{SO}_0(1,3)\ \vec{\times}\ \mathbb{R}^4 \\ \mathbf{SO}(3) & & \uparrow & & \uparrow \text{ contractions} \\ & \searrow & \mathbf{SO}_0(1,3) & \hookrightarrow & \mathbf{SO}_0(1,4), \mathbf{SO}_0(2,3) \end{array}$$

The nonsemisimple Euclidean and Poincaré Lie algebra and the simple Lie algebras of their group expansion have all real rank 2 (number of independent invariants),[2] all these noncompact operations embed the compact rank-1 angular momenta $\log \mathbf{SO}(3)$.

With the compact and noncompact factor in $\mathbf{SO}(\mathbb{C}^2) = \mathbf{SO}(2) \times \mathbf{SO}_0(1,1)$ the three Cartan subalgebra types of the de Sitter group $\mathbf{SO}_0(1,4)$ and anti-de Sitter group $\mathbf{SO}_0(2,3)$ come from $\mathbf{SO}(2) \times \mathbf{SO}(2)$, $\mathbf{SO}(2) \times \mathbf{SO}_0(1,1)$, and $\mathbf{SO}_0(1,1) \times \mathbf{SO}_0(1,1)$. Hence the Cartan subalgebras for the Poincaré Lie algebra

$$\left(\begin{array}{c||c} \log \mathbf{SO}(1,3) & \mathbb{R}^4 \\ \hline 0 & 0 \end{array} \right) \ni \left(\begin{array}{c|ccc||c} 0 & \psi_1 & \psi_2 & \psi_3 & x_0 \\ \hline \psi_1 & 0 & \varphi_3 & -\varphi_2 & x_1 \\ \psi_2 & -\varphi_3 & 0 & \varphi_1 & x_2 \\ \psi_3 & \varphi_2 & -\varphi_1 & 0 & x_3 \\ \hline 0 & 0 & 0 & 0 & 0 \end{array} \right)$$

[2] $\mathbf{SO}_0(p,q)$ with $p+q = 2r$ or $p+q = 1+2r$ has r independent invariants, related to the block-diagonalization with $\mathbf{SO}(2)^n \times \mathbf{SO}_0(1,1)^m$, $n+m = r$ (axial rotations or boosts), and, for odd $1+2r$, one additional 1.

can be characterized as follows: For trivial translations, i.e., for the Lorentz Lie algebra, the Cartan Lie algebra $\log \mathbf{SO}(\mathbb{C}^2)$ generates axial rotations around and boosts along one position axis, e.g., $\{\mathbf{l}^3, \mathbf{b}^3\}$. For nontrivial translations one Cartan subalgebra type comprises time translations and rotations around one position axis, e.g., $\{\mathbf{p}^0, \mathbf{l}^3\}$ and one type position translations and commuting boosts, e.g., $\{\mathbf{p}^1, \mathbf{b}^3\}$:

$$\begin{array}{ccll} \log \mathbf{SO}(2) & \oplus & \log \mathbf{SO}(1,1) & \text{for } \varphi_3, \psi_3 \neq 0, \\ \log \mathbf{SO}(2) & \oplus & \mathbb{R} & \text{for } \varphi_3, x_0 \neq 0, \\ \log \mathbf{SO}(1,1) & \oplus & \mathbb{R} & \text{for } \psi_3, x_1 \neq 0. \end{array}$$

The one rotation invariant in the angular momenta enveloping algebra gives rise to two invariants in the Poincaré enveloping algebra (notation e.g., $\vec{\mathbf{p}}^2 = \mathbf{p}^a \otimes \mathbf{p}^a$, $\vec{\mathbf{b}} \times \vec{\mathbf{p}} \cong \epsilon_{abc}\mathbf{b}^a \otimes \mathbf{p}^b$):

$$\begin{array}{ccccc} & & \boxed{\begin{array}{c}\mathbb{R}^3 \overleftarrow{\times} \mathbf{SO}(3) \\ -\vec{\mathbf{p}}^2,\ \vec{\mathbf{l}}\vec{\mathbf{p}}\end{array}} & \hookrightarrow & \boxed{\begin{array}{c}\mathbb{R}^4 \overleftarrow{\times} \mathbf{SO}_0(1,3) \\ \mathbf{p}^2 = \mathbf{p}_0^2 - \vec{\mathbf{p}}^2,\ \ \mathbf{S}^2\end{array}} \\ & \nearrow & & & \\ \boxed{\begin{array}{c}\mathbf{SO}(3) \\ \vec{\mathbf{l}}^2\end{array}} & & \uparrow & & \uparrow \\ & \searrow & & & \\ & & \boxed{\begin{array}{c}\mathbf{SO}_0(1,3) \\ (\vec{\mathbf{l}}^2 \pm i\vec{\mathbf{b}})^2 \Rightarrow \vec{\mathbf{l}}^2 - \vec{\mathbf{b}}^2,\ \vec{\mathbf{l}}\vec{\mathbf{b}}\end{array}} & \hookrightarrow & \mathbf{SO}_0(1,4), \mathbf{SO}_0(2,3) \end{array}$$

The 3-dimensional spin vector $\mathbf{l}^a = \frac{\epsilon^{abc}}{2}\mathbf{l}^{bc}$ is embedded into the *Pauli-Lubanski vector*

$$\begin{aligned} \mathbf{S}_j &= \tfrac{1}{2}\epsilon_{jmkl}\mathbf{l}^{kl}\mathbf{p}^m = (\vec{\mathbf{l}}\vec{\mathbf{p}}, \vec{\mathbf{l}}\mathbf{p}_0 + \vec{\mathbf{b}} \times \vec{\mathbf{p}}), \quad \mathbf{S}_j\mathbf{p}^j = 0, \\ -\mathbf{S}^2 &= \vec{\mathbf{l}}^2\mathbf{p}_0^2 - (\vec{\mathbf{l}}\vec{\mathbf{p}})^2 + \vec{\mathbf{b}}^2\vec{\mathbf{p}}^2 - (\vec{\mathbf{b}}\vec{\mathbf{p}})^2 + 2(\vec{\mathbf{l}} \times \vec{\mathbf{b}})\vec{\mathbf{p}}\mathbf{p}_0. \end{aligned}$$

The two Poincaré invariants are the power-2 translation-invariant $\mathbf{p}^2$ with the values for "mass" taken from a continuous spectrum and the power-4 rotation invariant $\mathbf{S}^2$ with the values for "spin" from a discrete spectrum for a nonnegative translation invariant. The corresponding eigenvalues are momentum and helicity, i.e., the spin component in the momentum direction (replacing $\mathbf{l}_3$ for $\mathbf{SO}(3)$)

$$(\mathbf{p}^2, \mathbf{S}^2) \text{ with } (\vec{\mathbf{p}}, \tfrac{\vec{\mathbf{l}}\vec{\mathbf{p}}}{|\vec{\mathbf{p}}|}).$$

1.2 Left- and Right-Handed Weyl Spinors

The embedding of the spin group into the Lorentz group by complexification goes with the embedding of the fundamental Pauli spinor representation into Weyl spinors and Dirac spinors that have nontrivial boost properties.

Without the antihermiticity restriction $l = -l^\star$ for the spin Lie algebra A_1^c, the Lorentz Lie algebra $A_{(1,1)}$ is definable by all traceless $\mathbb{C}^2$-endomorphisms

$$A_{(1,1)} \cong \{l \in \mathbf{AL}(\mathbb{C}^2) \bigm| \operatorname{tr} l = 0\} \cong \mathbb{R}^6,$$

considered, with a Euclidean conjugation $l \overset{\star}{\leftrightarrow} l^\star$, as a real Lie algebra. The Pauli representation of the complex $\mathbf{SL}(\mathbb{C}^2)$-Lie algebra $A_1 \cong \mathbb{C}^3$ on spinors

$\mathbb{C}^2$ gives rise to two representations $A_1^c \oplus (\pm i A_1^c)$, conjugate to each other. The *fundamental Weyl representations* act on a complex quartet of 2-dimensional vector spaces $(V, V^T, \overline{V}, \overline{V}^T) \cong \mathbb{C}^2$ (chapter "Antistructures: The Real in the Complex"). The following notation for dual bases and antibases is used (l stands for "left" and r for "right"):

$$\begin{array}{l} \mathrm{l}^A \in V, \ \ \mathrm{r}_A^\times \in V^T, \ \ \mathrm{l}^{\times A} \in \overline{V}, \ \ \mathrm{r}_A \in \overline{V}^T; \ \ A = 1,2, \\ \langle \mathrm{r}_B^\times, \mathrm{l}^A \rangle = \langle \mathrm{l}^{\times A}, \mathrm{r}_B \rangle = \delta_B^A. \end{array}$$

The *left-handed Weyl spinors*, self-dual with the $\mathbb{C}^2$-volume forms (spinor "metric"), are denoted by V-elements:

$$\begin{array}{l} A_{(1,1)} \longrightarrow \mathbf{AL}(V)_0, \ \ \mathcal{D}^{[1|0]}(l) = l, \ \ \left\{ \begin{array}{ll} \vec{l} & = \frac{i}{2}\vec{\sigma}_A^B \, \mathrm{l}^A \otimes \mathrm{r}_B^\times, \\ \vec{b} & = -\frac{1}{2}\vec{\sigma}_A^B \, \mathrm{l}^A \otimes \mathrm{r}_B^\times, \end{array} \right. \\ \text{dual representation:} \left\{ \begin{array}{l} -l^T = \epsilon^{[1|0]} \circ l \circ \epsilon^{[1|0]-1}, \quad -\vec{\sigma}_B^A = \epsilon^{AC}\vec{\sigma}_C^D \epsilon_{DB}, \\ \epsilon^{[1|0]} : V \longrightarrow V^T, \ \ \mathrm{l}^A \longmapsto \epsilon^{AB}\mathrm{r}_B^\times. \end{array} \right. \end{array}$$

The antirepresentation, also self-dual, acts on the *right-handed Weyl spinors* $\overline{V}^T$:

$$\begin{array}{l} A_{(1,1)} \longrightarrow \mathbf{AL}(\overline{V}^T)_0, \ \ \mathcal{D}^{[0|1]}(l) = \hat{l} = -l^\times, \ \ \left\{ \begin{array}{ll} \hat{\vec{l}} & = \frac{i}{2}\overline{\vec{\sigma}_A^B} \, \mathrm{r}_B \otimes \mathrm{l}^{\times A}, \\ \hat{\vec{b}} & = +\frac{1}{2}\overline{\vec{\sigma}_A^B} \, \mathrm{r}_B \otimes \mathrm{l}^{\times A}, \end{array} \right. \\ \text{dual representation:} \left\{ \begin{array}{l} -\hat{l}^T = \epsilon^{[0|1]} \circ \hat{l} \circ \epsilon^{[0|1]-1}, \\ \epsilon^{[0|1]} : \overline{V}^T \longrightarrow \overline{V}, \ \ \mathrm{r}_B \longmapsto \epsilon_{BA}\mathrm{l}^{\times A}. \end{array} \right. \end{array}$$

With respect to the Euclidean $\mathbf{U}(2)$-conjugation $\overline{\vec{\sigma}_A^B} = \delta^{BC}\vec{\sigma}_C^D\delta_{DA}$, i.e., $\vec{\sigma} = \vec{\sigma}^\star$, the following index notation is used:

$$\text{in } \overline{V}^T : \mathrm{r}^{\dot{A}} = \mathrm{r}_B\delta^{B\dot{A}}, \ \ \text{in } \overline{V} : \mathrm{l}_{\dot{A}}^\times = \mathrm{l}^{\times B}\delta_{B\dot{A}}, \ \ \langle \mathrm{l}_{\dot{B}}^\times, \mathrm{r}^{\dot{A}} \rangle = \delta_{\dot{B}}^{\dot{A}},$$

leading to the notation for the antirepresentation

$$\mathcal{D}^{[0|1]} : \ \ \left\{ \begin{array}{l} \hat{\vec{l}} = \frac{i}{2}\vec{\sigma}_{\dot{A}}^{\dot{B}} \, \mathrm{r}^{\dot{A}} \otimes \mathrm{l}_{\dot{B}}^\times, \ \ \hat{\vec{b}} = +\frac{1}{2}\vec{\sigma}_{\dot{A}}^{\dot{B}} \, \mathrm{r}^{\dot{A}} \otimes \mathrm{l}_{\dot{B}}^\times, \\ \epsilon^{[0|1]} : \overline{V}^T \longrightarrow \overline{V}, \ \ \mathrm{r}^{\dot{A}} \longmapsto \epsilon^{\dot{A}\dot{B}}\mathrm{l}_{\dot{B}}^\times. \end{array} \right.$$

This *Weyl notation with dotted and undotted indices* keeps track of the two types of fundamental representations also in representation products.

The anticonjugation connects left- and right-handed Weyl representations:

$$\begin{array}{lll} \times : V \longrightarrow \overline{V}, & \mathrm{l}^A \longmapsto \mathrm{l}^{\times A} = \delta^{A\dot{B}}\mathrm{l}_{\dot{B}}^\times, \\ \times : V^T \longrightarrow \overline{V}^T, & \mathrm{r}_A^\times \longmapsto \mathrm{r}_A = \delta_{A\dot{B}}\mathrm{r}^{\dot{B}}. \end{array}$$

The Killing and the chiral invariant have conjugated values for the left- and right-handed Weyl spinors

$$\begin{array}{ll} I_+^{[1|0]}(A_{(1,1)}) = \frac{3}{8}\,\mathrm{id}_V, & I_-^{[1|0]}(A_{(1,1)}) = +\frac{3}{8}i\,\mathrm{id}_V, \\ I_+^{[0|1]}(A_{(1,1)}) = \frac{3}{8}\,\mathrm{id}_{\overline{V}^T}, & I_-^{[0|1]}(A_{(1,1)}) = -\frac{3}{8}i\,\mathrm{id}_{\overline{V}^T}. \end{array}$$

The product of spin axis and boost direction $\frac{4i}{3}\vec{l}\,\vec{b} = \pm 1$ (parallel and antiparallel) motivates the names "left-handed" and "right-handed."

Exponentation of $A_{(1,1)}$ gives the Lie group of the special linear automorphisms

$$\exp A_{(1,1)} \cong \mathbf{SL}(\mathbb{C}^2) = \{\lambda \in \mathbf{GL}(\mathbb{C}^2) \mid \det \lambda = 1\} \in \underline{\mathbf{lgrp}}_{\mathbb{R}}$$

in both fundamental representations with a local Lie algebra parametrization

$$\begin{array}{l} \mathbf{SL}(\mathbb{C}^2) \longrightarrow \mathbf{SL}(V), \quad D^{[1|0]}(\lambda) = \lambda = \lambda(\vec{\alpha}, \vec{\beta})^B_A \, \mathrm{l}^A \otimes \mathrm{r}^\times_B \cong e^{i\vec{\alpha}-\vec{\beta}}, \\ \mathbf{SL}(\mathbb{C}^2) \longrightarrow \mathbf{SL}(\overline{V}^T), \ D^{[0|1]}(\lambda) = \hat{\lambda} = \hat{\lambda}(\vec{\alpha}, \vec{\beta})^{\dot{B}}_{\dot{A}} \, \mathrm{r}^{\dot{A}} \otimes \mathrm{l}^\times_{\dot{B}} \cong e^{i\vec{\alpha}+\vec{\beta}}; \end{array}$$

$\mathbb{R}^3$-vectors are written as Hermitian traceless (3×3) matrices, e.g., $\vec{\alpha} = \alpha_a \sigma^a$.

1.3 Finite-Dimensional Representations of the Lorentz Operations

Each *complex finite-dimensional* representation of the Lie algebra $A_{(1,1)}$ and the Lie group $\mathbf{SL}(\mathbb{C}^2)$ is semisimple. With rank 2 the finite-dimensional irreducible representations are characterized[3] by two natural numbers $2L, 2R = 0, 1, \ldots$ with $(1+2L)(1+2R)$ the dimensionality of the representation

$$\left.\begin{array}{llll} \mathbf{SL}(\mathbb{C}^2) & \longrightarrow \mathbf{SL}(W), & \lambda & \longmapsto D^{[2L|2R]}(\lambda), \\ A_{(1,1)} & \longrightarrow \mathbf{AL}(W)_0, & l & \longmapsto \mathcal{D}^{[2L|2R]}(l) \end{array}\right. \text{ on } W \cong \mathbb{C}^{(1+2L)(1+2R)}.$$

They can be obtained, up to equivalence, by the tensor product of totally symmetric tensor powers:

$$\begin{array}{ll} D^{[2L|2R]}(\lambda) & = \bigvee^{2L} \lambda \otimes \bigvee^{2R} \hat{\lambda}, \\ \mathcal{D}^{[2L|2R]}(l) & = \mathcal{D}^{2L}(l) \otimes \mathrm{id}_{W_R} + \mathrm{id}_{W_L} \otimes \mathcal{D}^{2R}(\hat{l}), \\ \text{on } W \cong W_L \otimes W_R, & W_L \cong \mathbb{C}^{1+2L}, \quad W_R \cong \mathbb{C}^{1+2R}, \end{array}$$

from the two fundamental Weyl representations

$$\begin{array}{llll} D^{[1|0]}(\lambda) = \lambda \cong e^{i\vec{\alpha}-\vec{\beta}}, & D^{[0|1]}(\lambda) & = \hat{\lambda} & = \lambda^{-1\star} \cong e^{i\vec{\alpha}+\vec{\beta}}, \\ \mathcal{D}^{[1|0]}(l) = l \cong i\vec{\alpha} - \vec{\beta}, & \mathcal{D}^{[0|1]}(l) & = \hat{l} & = -l^\star \cong i\vec{\alpha} + \vec{\beta}. \end{array}$$

$L = J_L, R = J_R$ may be called "left and right spin," respectively. There is, however, only one spin. Sometimes, as for spin $\mathbf{SU}(2)$, representations with (half)integer sum $L + R$ are called *(spinor) vector representations.*

With the $\mathbf{SU}(2)$-representations in the "left-" and "right-"handed part an $\mathbf{SL}(\mathbb{C}^2)$-representation shows its spin content in the decomposition into $\mathbf{SU}(2)$-representations

[3] Again, both notations, integer $[2L|2R]$ or possibly half-integer $[L|R]$, have advantages and disadvantages.

$$\mathbf{SL}(\mathbb{C}^2) \cong \bigoplus \mathbf{SU}(2): \quad D^{[2L|2R]} \cong \bigoplus_{J=|L-R|}^{L+R} D^{2J}.$$

(Spinor) vector representations contain only (half)integer spin J.

The canonical conjugation $\times$, restricted to the 2×2 matrices, is the transposition with canonical number conjugation and may be written as $\lambda^\times = \lambda^\star$. The difference between the two conjugations $\star$ (definite) and $\times$ (indefinite) can be formulated in $\mathbf{U}(2,2)$, i.e., in 4×4 matrices as used for Dirac matrices (below).

The irreducible representations are self-dual with the products of the spinor "metrics"

$$D^{[2L|2R]}(\lambda) \cong D^{[2L|2R]}(\lambda^{-1})^T, \quad \mathcal{D}^{[2L|2R]}(l) \cong -\mathcal{D}^{[2L|2R]}(l)^T$$
$$\text{with } \epsilon^{[2L|2R]} = \bigvee^{2L} \epsilon^{[1|0]} \otimes \bigvee^{2R} \epsilon^{[0|1]}.$$

Spinor (vector) representations are symplectic (orthogonal) self-dual:

$$D^{[2L|2R]}[\mathbf{SL}(\mathbb{C}^2)] \subseteq \begin{cases} \mathbf{Sp}(\mathbb{C}^{(1+2L)(1+2R)}), & L+R = \frac{1}{2}, \frac{3}{2}, \ldots, \\ \mathbf{SO}(\mathbb{C}^{(1+2L)(1+2R)}), & L+R = 0, 1, \ldots. \end{cases}$$

The represented Casimir elements are the sum and difference of the Casimir elements for the "left" and "right" $\mathbf{SU}(2)$-representation

$$I_+^{[2L|2R]}(A_{(1,1)}) = [\tbinom{1+L}{2} + \tbinom{1+R}{2}] \operatorname{id}_W, \quad I_-^{[2L|2R]}(A_{(1,1)}) = i[\tbinom{1+L}{2} - \tbinom{1+R}{2}] \operatorname{id}_W.$$

A tensor product of finite-dimensional irreducible $\mathbf{SL}(\mathbb{C}^2)$-representations has the Clebsch-Gordan decomposition

$$D^{[2L_1|2R_1]} \otimes D^{[2L_2|2R_2]} \cong \bigoplus_{L=|L_1-L_2|}^{L_1+L_2} \bigoplus_{R=|R_1-R_2|}^{R_1+R_2} D^{[2L|2R]}.$$

With the discrete group $\mathbb{I}(2) \cong \{\pm \mathbf{1}_2\}$ as its center, the adjoint group is isomorphic to the orthochronous Lorentz group

$$\operatorname{Int} \mathbf{SL}(\mathbb{C}^2) = \mathbf{SL}(\mathbb{C}^2)/\mathbb{I}(2) \cong \mathbf{SO}_0(1,3).$$

The centrality of the representation, called *twoality*, characterizes the representation of the center $\mathbb{I}(2)$:

$$\begin{aligned} (2L+2R)\text{mod}2 &= \begin{cases} 1 \text{ for } L+R = \frac{1}{2}, \frac{3}{2}, \ldots, \\ 0 \text{ for } L+R = 0, 1, \ldots, \end{cases} \\ D^{[2L|2R]}(-\mathbf{1}_2) &= (-\mathbf{1}_W)^{2(L+R)}. \end{aligned}$$

The equivalence classes of the complex finite-dimensional irreducible $A_{(1,1)}$- and $\mathbf{SL}(\mathbb{C}^2)$-representations define the discrete $\mathbf{SL}(\mathbb{C}^2)$*-representation cone* with abelian composition $\vee$ (the "largest" representation in the product representation) and the trivial representation as neutral element $[0|0]$:

$$\begin{aligned}\mathbf{irrep}_{\text{fin}}\mathbf{SL}(\mathbb{C}^2) &\cong \{[2L|2R]\}\\ &\cong \mathbf{irrep}\,\mathbf{SU}(2)\times\mathbf{irrep}\,\mathbf{SU}(2)\cong\mathbb{N}_0\times\mathbb{N}_0,\\ [2L_1|2R_1]\vee[2L_2|2R_2] &= [2(L_1+L_2)|2(R_1+R_2)].\end{aligned}$$

The two Weyl representations are a cone basis.

In the polar decomposition, the simply connected real 6-dimensional Lie group $\mathbf{SL}(\mathbb{C}^2)$ is parametrizable by the points of a full 3-sphere for $\mathbf{SU}(2)\cong\Omega^3$ times a 3-hyperboloid for the noncompact boost manifold $\mathcal{Y}^3\cong\mathbf{SL}(\mathbb{C}^2)/\mathbf{SU}(2)$:

$$\begin{aligned}A_{(1,1)} &\cong A_1^c\oplus iA_1^c\cong(i\mathbb{R})^3\oplus\mathbb{R}^3,\\ \mathbf{SL}(\mathbb{C}^2) &= \mathbf{SU}(2)\circ\mathbf{SO}_0(1,1)\circ\mathbf{SU}(2)\\ \mathbf{SL}(\mathbb{C}^2) &\cong\Omega^3\times\mathcal{Y}^3,\quad\left\{\begin{array}{ll}\Omega^3\cong & \{\vec{\alpha}\in\mathbb{R}^3\,\big|\,|\vec{\alpha}|\le\pi\},\\ \mathcal{Y}^3\cong & \{\vec{\beta}\in\mathbb{R}^3\}.\end{array}\right.\end{aligned}$$

Instead of the left-right spin oriented notation a compact-noncompact Cartan subgroup $\mathbf{SO}(2)\times\mathbf{SO}_0(1,1)$ notation can be chosen:

$$\mathbb{N}_0\times\mathbb{N}_0\ni[2L|2R]\cong(2J;2D)\in\mathbb{N}_0\times\mathbb{Z}\text{ with }\left\{\begin{array}{ll}J&=L+R,\\ D&=L-R,\end{array}\right.$$

where J is the maximal spin. The invariant D, related to the noncompact boosts, is integer-valued for the finite-dimensional representations and takes continuous values for the infinite-dimensional representations (chapter "Harmonic Analysis").

The representation cone carries the anticonjugation

$$\begin{aligned}&[2L|2R]=[2R|2L]^\times:\ D^{[2L|2R]}(\lambda)=D^{[2R|2L]}(\hat{\lambda}),\ \mathcal{D}^{[2L|2R]}(l)=\mathcal{D}^{[2R|2L]}(\hat{l}),\\ &(2J;2D)=(2J;-2D)^\times,\end{aligned}$$

leading to two types of *indefinite unitary complex finite-dimensional representations*, either irreducible or complex decomposable:

$$\begin{aligned}[2J|2J],&\quad J=0,\tfrac{1}{2},1,\dots,\\ [2L|2R]\oplus[2R|2L],&\quad L\neq R.\end{aligned}$$

The two types start nontrivially with the real 4-dimensional Minkowski representation $[1|1]$ with $\mathbf{SO}_0(1,3)\subset\mathbf{SU}(1,3)$ and the complex 4-dimensional Dirac representation $[1|0]\oplus[0|1]$ in $\mathbf{SU}(2,2)$.

The equivalence classes of the finite-dimensional real irreducible representations of the locally isomorphic orthochronous Lorentz group $\mathbf{SO}_0(1,3)\cong\mathbf{SO}(\mathbb{C}^3)\cong\mathbf{SL}(\mathbb{C}^2)/\mathbb{I}(2)$ are given by the "even" subcone of the $\mathbf{SL}(\mathbb{C}^2)$-representations. The complex irreducible vector representations $[2L|2R]$ with even sum $2L+2R\in2\mathbb{N}_0$ (integer spins) give rise to two types of real irreducible $\mathbf{SO}_0(1,3)$-representations

$$\begin{aligned}\mathbf{irrep}_{\text{fin}}\mathbf{SO}_0(1,3)\cong&\ \{[2J|2J]\,\big|\,2J=0,1,\dots\}\\ \uplus&\ \{[2L|2R]\oplus[2R|2L]\,\big|\,L+R=1,2,\dots,\ L\neq R\}.\end{aligned}$$

The representation spaces with complex dimension $(1+2J)^2$ and $2(1+2L)(1+2R)$ are decomposable into symmetric and antisymmetric real vector subspace, real isomorphic to each other. The two *fundamental* representations of the rank-2 group $\mathbf{SO}_0(1,3)$ are the defining *Minkowski representation* $[1|1]$ (real dimension 4) and the *adjoint representation* $[2|0] \oplus [0|2]$ with real dimension $6 = \binom{4}{2}$, which can be obtained as an antisymmetric product

$$[2|0] \oplus [0|2] = [1|1] \wedge [1|1].$$

A prominent example is the electromagnetic field strength $\{F^{jk}\}_{j,k=0}^3$ as antisymmetric derivative $F^{jk} = \partial^j A^k - \partial^k A^j$ of the potential $\{A^j\}_{j=0}^3$.

The $\mathbf{SO}_0(1,3)$-representations of type $[2J|2J]$ are isomorphic to totally symmetric products of Minkowski representations $[1|1]$. They have a trivial noncompact invariant

$$[2J|2J] \cong (4J;0).$$

The Lorentz group is represented in indefinite orthogonal groups of $\mathbb{R}^{(1+2J)^2}$ with the definite subspaces for even or odd angular momenta in the $\mathbf{SO}(3)$-subrepresentations

$$D^{[2J|2J]} \cong \bigoplus_{L=0}^{2J} D^{2L} = \bigoplus_{L=0,2,\dots} D^{2L} \oplus \bigoplus_{L=1,3,\dots} D^{2L}, \quad (1+2J)^2 = \binom{2+2J}{2} + \binom{1+2J}{2},$$

$$D^{[2J|2J]}[\mathbf{SO}_0(1,3)] \subseteq \begin{cases} \mathbf{SO}_0(\binom{2+2J}{2}, \binom{1+2J}{2}), & 2J = 0,2,\dots, \\ \mathbf{SO}_0(\binom{1+2J}{2}, \binom{2+2J}{2}), & 2J = 1,3,\dots . \end{cases}$$

The complex representation spaces have an eigenvector basis of a real 2-dimensional Cartan subalgebra, e.g., of $\mathbb{R}l^3 \oplus \mathbb{R}b^3$ with the Lie group

$$\mathbf{SO}(\mathbb{C}^2) \cong \mathbf{SO}(2) \times \mathbf{SO}_0(1,1) \cong \{e^{(i\alpha_3 - \beta_3)\sigma^3} \bigm| \alpha_3, \beta_3 \in \mathbb{R}\} \subset \mathbf{SL}(\mathbb{C}^2).$$

For the simple Lie group $\mathbf{SL}(\mathbb{C}^2)$ there arise only self-dual $\mathbf{U}(1) \times \mathbf{D}(1)$ representations in $\mathbf{SO}(2) \ni \begin{pmatrix} e^{2ji\alpha_3} & 0 \\ 0 & e^{-2ji\alpha_3} \end{pmatrix}$ and $\mathbf{SO}_0(1,1) \ni \begin{pmatrix} e^{-2d\beta_3} & 0 \\ 0 & e^{2d\beta_3} \end{pmatrix}$ with integer weights $(\pm 2j; \pm 2d) \in \mathbb{Z} \times \mathbb{Z}$.

The Cartan subalgebra eigenvalues (weights) of a representation can be given either with the "left-right" spin components (on an $\mathbb{R}^2$-rectangle),

$$\textbf{weights}\,[2L|2R] = \{[2l|2r] \bigm| l = -L, -L+1, \dots, L,\ r = -R, \dots, R\},$$

or with the $\mathbf{SO}(2) \times \mathbf{SO}_0(1,1)$-weights $(2j; 2d)$, arising from the spin pair by a rotation

$$[2l|2r] \cong (2j; 2d), \quad j = l + r,\ d = l - r,$$

e.g., for the two Weyl representations

$$\begin{aligned} \textbf{weights}\,[1|0] &= \underset{\begin{pmatrix} e^{i\alpha_3 - \beta_3} & 0 \\ 0 & e^{-i\alpha_3 + \beta_3} \end{pmatrix}}{\{[1|0],\ [-1|0]\}} \cong \{(1;1),\ (-1;-1)\}, \\ \textbf{weights}\,[0|1] &= \underset{\begin{pmatrix} e^{i\alpha_3 + \beta_3} & 0 \\ 0 & e^{-i\alpha_3 - \beta_3} \end{pmatrix}}{\{[0|1],\ [0|-1]\}} \cong \{(1;-1),\ (-1;1)\}, \end{aligned}$$

and for the 4-dimensional Minkowski representation

$$\textbf{weights}\,[1|1]\left\{\begin{array}{l} = \{[1|1],\ [1|-1],\ [-1|1],\ [-1|-1]\},\\ \cong \{(0;2),\ (2;0),\ (-2;0),\ (0;-2)\},\end{array}\right.$$

$$\begin{pmatrix} e^{-2\beta_3} & 0 & 0 & 0\\ 0 & e^{2i\alpha_3} & 0 & 0\\ 0 & 0 & e^{-2i\alpha_3} & 0\\ 0 & 0 & 0 & e^{2\beta_3}\end{pmatrix} \cong \begin{pmatrix} \cosh\psi_3 & 0 & 0 & \sinh\psi_3\\ 0 & \cos\varphi_3 & \sin\varphi_3 & 0\\ 0 & -\sin\varphi_3 & \cos\varphi_3 & 0\\ \sinh\psi_3 & 0 & 0 & \cosh\psi_3\end{pmatrix},\quad \begin{array}{ll}\vec{\varphi} & = 2\vec{\alpha},\\ \vec{\psi} & = 2\vec{\beta}.\end{array}$$

The discrete $\mathbf{SL}(\mathbb{C}^2)$-*weight module* shows two winding numbers,

$$\begin{aligned}\textbf{weights}_{\text{fin}}\,\mathbf{SL}(\mathbb{C}^2) &= \{[2l|2r]\} = \mathbb{Z}\times\mathbb{Z}\\ &= \textbf{weights}\,\mathbf{SU}(2)\times\textbf{weights}\,\mathbf{SU}(2)\\ &= \textbf{weights}\,\mathbf{U}(1)\times\textbf{weights}\,\mathbf{U}(1),\end{aligned}$$

with the conjugation

$$[2r\ 2l]^\times = [2l|2r] \cong (2j;2d) = (2j;-2d)^\times.$$

It contains the $\mathbf{SO}_0(1,3)$-*weight submodule*

$$\textbf{weights}_{\text{fin}}\mathbf{SO}_0(1,3) = \{[2l|2r] \mid l+r\in\mathbb{Z}\} \cong \mathbb{Z}\times\mathbb{Z}.$$

The representation cones are given by the dominant weights, i.e., they are the positive cones of the weight modules.

1.4 Spacetime Translations as Spinor Transformations

The sum of the two inequivalent Weyl representations of the Lorentz group acting on the antidoubling $V_{\text{doub}} = V \oplus \overline{V}^T \cong \mathbb{C}^4$ is the decomposable *Dirac representation* in the real 32-dimensional *Dirac algebra*

$$\mathbf{AL}(V_{\text{doub}}) = \begin{pmatrix} V\otimes V^T & V\otimes\overline{V}\\ \overline{V}^T\otimes V^T & \overline{V}^T\otimes\overline{V}\end{pmatrix} = \begin{pmatrix}\mathbf{AL}(V) & \mathbf{P}(V)\\ \mathbf{P}(V)^T & \mathbf{AL}(V)^\times\end{pmatrix} \cong \mathbb{C}^{16}.$$

The $\times$-unitary group of $\mathbf{AL}(\mathbb{C}^4)$ is $\mathbf{U}(2,2)$.

In addition to the self-dual antialgebras in the diagonal with the image of the two Weyl representations one has two vector spaces in the skew-diagonal, dual to each other and invariant with respect to the anticonjugation. They are stable under the adjoint $\mathbf{SL}(\mathbb{C}^2)$-action:

$$\begin{array}{rl} \mathbf{AL}(V_{\text{doub}}) \longrightarrow & \begin{pmatrix}\lambda & 0\\ 0 & \hat\lambda\end{pmatrix}\circ\mathbf{AL}(V_{\text{doub}})\circ\begin{pmatrix}\lambda & 0\\ 0 & \hat\lambda\end{pmatrix}^{-1},\\ \mathbf{SL}(\mathbb{C}^2)\longrightarrow\mathbf{SL}(V\otimes\overline{V}), & \left\{\begin{array}{l}\lambda\longmapsto D^{[1|1]}(\lambda) = \lambda\otimes\lambda^\star = \Lambda = \Lambda^\star,\\ \Lambda(1\otimes 1^\times) = \lambda(1)\otimes\lambda(1)^\times,\end{array}\right.\\ \mathbf{SL}(\mathbb{C}^2)\longrightarrow\mathbf{SL}(\overline{V}^T\otimes V^T), & \left\{\begin{array}{l}\lambda\longmapsto\check\Lambda = \Lambda^{-1T} = \hat\lambda\otimes\hat\lambda^\star,\\ \check\Lambda(\mathrm{r}\otimes\mathrm{r}^\times) = \hat\lambda(\mathrm{r})\otimes\hat\lambda(\mathrm{r})^\times,\end{array}\right.\end{array}$$

and contain real 4-dimensional vector spaces, symmetric for the anticonjugation and dual to each other:

$$\begin{array}{llll} \mathbf{P}(V) &= \mathbb{C}\otimes\mathbb{M}, & \mathbb{M} &= \{x\in\mathbf{P}(V) \mid x=x^\times\} \cong \mathbb{R}^4,\\ \mathbf{P}(V)^T &= \mathbb{C}\otimes\mathbb{M}^T, & \mathbb{M}^T &= \{p\in\mathbf{P}(V)^T \mid p=p^\times\} \cong \mathbb{R}^4. \end{array}$$

These real vector spaces are the *Cartan representation* of Minkowski spacetime and its dual energy-momentum space by Weyl spinor transformations

$$\mathbb{C}\otimes\mathbb{M}\cong\{\overline{V}^T\longrightarrow V\},\ \ \mathbb{C}\otimes\mathbb{M}^T\cong\{V\longrightarrow\overline{V}^T\}.$$

The Pauli representation $\vec{x}=\begin{pmatrix} x_3 & x_1-ix_2\\ x_1+ix_2 & -x_3\end{pmatrix}\in\mathbb{S}\cong\mathbb{R}^3$ of position translations is embedded into the Cartan representation $\begin{pmatrix} x_0+x_3 & x_1-ix_2\\ x_1+ix_2 & x_0-x_3\end{pmatrix}\in\mathbb{M}\cong\mathbb{R}^4$ of Minkowski spacetime: The decomposition of $\mathbb{M}$ with respect to spin A_1^c-images contains a trivial A_1^c-representation $x_0\mathbf{1}_2\in\mathbb{T}\cong\mathbb{R}$ with the time translations (Clebsch-Gordan projection with $\delta^A_{\dot D}$) in addition to the Pauli representation of the position translations (Clebsch-Gordan projection with $\vec{\sigma}^A_{\dot D}$). Explicitly: By the Fierz recoupling of the product of four spinors

$$(\mathbf{1}\otimes\mathbf{1})^{A\dot C}_{B\dot D}=\delta^A_B\,\delta^{\dot C}_{\dot D}\overset{\text{Fierz}}{=}\frac{\delta^{\dot C}_B\delta^A_{\dot D}+\vec{\sigma}^{\dot C}_B\vec{\sigma}^A_{\dot D}}{2}=\tfrac12(\sigma^j)^{\dot C}_B\,(\check\sigma_j)^A_{\dot D}=\tfrac12(\sigma^j\otimes\check\sigma_j)^{\dot CA}_{B\dot D},$$

one obtains two sets with four 2×2 matrices, dual to each other:

$$\textit{Sylvester-Weyl matrices:}\ \begin{cases}\sigma^j=(\mathbf{1}_2,\vec\sigma),\\ \check\sigma_j=(\mathbf{1}_2,\vec\sigma).\end{cases}$$

The Sylvester-Weyl matrices contain Clebsch-Gordan coefficients for the spin decomposition of Minkowski spacetime.

The Fierz recoupling describes a basis transformation from *bispinor bases* of the dual spaces $\mathbf{P}(V)$ (spacetime translations) and $\mathbf{P}(V)^T$ (energy-momenta) to *vector bases*

$$\mathrm{l}^A\otimes\mathrm{l}^\times_{\dot D}=\frac{(\check\sigma_j)^A_{\dot D}}{2}\mathbf{P}^j,\ \ \mathrm{r}^{\dot C}\otimes\mathrm{r}^\times_B=\frac{(\sigma^j)^{\dot C}_B}{2}\mathbf{X}_j.$$

They are symmetric with respect to the anticonjugation and dual to each other:

$$\begin{array}{rl}\text{vector basis of }\ \mathbb{C}\otimes\mathbb{M}: & \begin{cases}\{\mathbf{P}^j=(\sigma^j)^{\dot B}_A\mathrm{l}^A\otimes\mathrm{l}^\times_{\dot B}\mid j=0,1,2,3\},\\ \mathbf{P}^j:\overline{V}^T\longrightarrow V,\ \ \mathbf{P}^j=(\mathbf{P}^j)^\times\cong\begin{pmatrix}0&\sigma^j\\0&0\end{pmatrix},\end{cases}\\ \text{vector basis of }\ \mathbb{C}\otimes\mathbb{M}^T: & \begin{cases}\{\mathbf{X}_j=(\check\sigma_j)^B_{\dot A}\mathrm{r}^{\dot A}\otimes\mathrm{r}^\times_B\mid j=0,1,2,3\},\\ \mathbf{X}_j:V\longrightarrow\overline{V}^T,\ \ \mathbf{X}_j=\mathbf{X}_j^\times\cong\begin{pmatrix}0&0\\\check\sigma_j&0\end{pmatrix},\end{cases}\\ \text{dual:} & \langle\mathbf{X}_j,\mathbf{P}^k\rangle=\tfrac12\operatorname{tr}\mathbf{X}_j\circ\mathbf{P}^k=\delta^k_j=\tfrac12\operatorname{tr}\sigma^k\check\sigma_j.\end{array}$$

The Weyl matrices $\{\sigma^j\}_{j=0}^3$ and $\{\check\sigma_j\}_{j=0}^3$ are dual bases for spacetime translations and energy-momentum space

$$\begin{array}{llll}\overline{V}^T\longrightarrow V, & \mathbb{M}\ni x &= x_j\sigma^j=&\begin{pmatrix}x_0+x_3 & x_1-ix_2\\ x_1+ix_2 & x_0-x_3\end{pmatrix},\\ V\longrightarrow\overline{V}^T, & \mathbb{M}^T\ni\check p &= p^j\check\sigma_j=&\begin{pmatrix}p^0+p^3 & p^1-ip^2\\ p^1+ip^2 & p^0-p^3\end{pmatrix}.\end{array}$$

The $\mathbf{SO}(3)$-rotations of the position translations via the adjoint $\mathbf{SU}(2)$-action $\vec{\sigma} \longmapsto u\circ\vec{\sigma}\circ u^\star$ are embedded into the $\mathbf{SO}_0(1,3)$-Lorentz transformations of Minkowski spacetime:

$$\begin{array}{lll} \mathbf{SL}(\mathbb{C}^2)\times\mathbb{M} & \longrightarrow \mathbb{M}, \quad \sigma^k \longmapsto \lambda\sigma^k\lambda^\star & = \Lambda^k_j\sigma^j \Rightarrow \Lambda^k_j = \frac{1}{2}\operatorname{tr}\lambda\sigma^k\lambda^\star\check{\sigma}_j, \\ \mathbf{SL}(\mathbb{C}^2)\times\mathbb{M}^T & \longrightarrow \mathbb{M}^T, \quad \check{\sigma}_j \longmapsto \hat{\lambda}\check{\sigma}_j\hat{\lambda}^\star & = (\Lambda^{-1})^k_j\check{\sigma}_k. \end{array}$$

The components in the Cartan representation of Minkowski spacetime,

$$\begin{pmatrix} x_0+x_3 & x_1-ix_2 \\ x_1+ix_2 & x_0-x_3 \end{pmatrix} = (x_0+x_3)\pi^+ + (x_0-x_3)\pi^- + (x_1-ix_2)\sigma^+ + (x_1+ix_2)\sigma^-$$
$$\text{with } \pi^\pm = \frac{\mathbf{1}_2\pm\sigma^3}{2}, \quad \sigma^\pm = \frac{\sigma^1\pm i\sigma^2}{2},$$

are an $\mathbf{SO}(2)\times\mathbf{SO}_0(1,1)$-eigenvector basis

$$e^{(i\alpha_3+\beta_3)\sigma^3}\circ\begin{pmatrix} x_0+x_3 & x_1-ix_2 \\ x_1+ix_2 & x_0-x_3 \end{pmatrix}\circ e^{(-i\alpha_3+\beta_3)\sigma^3} = \begin{pmatrix} e^{2\beta_3}(x_0+x_3) & e^{2i\alpha_3}(x_1-ix_2) \\ e^{-2i\alpha_3}(x_1+ix_2) & e^{-2\beta_3}(x_0-x_3) \end{pmatrix}.$$

The conjugate adjoint group of the full linear group $\mathbf{GL}(\mathbb{C}^2)$ is given by the classes with respect to the unitary center subgroup $\mathbf{U}(1)$:

$$\operatorname{Int}_*\mathbf{GL}(\mathbb{C}^2) = \mathbf{GL}(\mathbb{C}^2)/\mathbf{U}(\mathbf{1}_2) = \mathbf{D}(\mathbf{1}_2)\times\mathbf{SL}(\mathbb{C}^2)/\mathbb{I}(2) \cong \mathbf{D}(1)\times\mathbf{SO}_0(1,3).$$

It involves a dilation group in addition to the Lorentz group. The spacetime translations are isomorphic to the symmetric vector subspace of the $\mathbf{GL}(\mathbb{C}^2)$-Lie algebra, whereas the antisymmetric subspace is the Lie algebra of the unitary group $\mathbf{U}(2)$ (chapter "Spacetime as Unitary Operation Classes"):

$$\begin{array}{rll} \mathbb{R}^8 \cong & \log\mathbf{GL}(\mathbb{C}^2) & = \log\mathbf{GL}(\mathbb{C}^2)_- \oplus \log\mathbf{GL}(\mathbb{C}^2)_+ \in \underline{\mathbf{lag}}_{\mathbb{R}}, \\ & z = i\frac{\gamma}{2}+x & = \frac{i}{2}\begin{pmatrix} \gamma_0+\gamma_3 & \gamma_1-i\gamma_2 \\ \gamma_1+i\gamma_2 & \gamma_0-\gamma_3 \end{pmatrix} + \begin{pmatrix} x_0+x_3 & x_1-ix_2 \\ x_1+ix_2 & x_0-x_3 \end{pmatrix}, \\ (i\mathbb{R})^4 \cong & \log\mathbf{GL}(\mathbb{C}^2)_- & = \log\mathbf{U}(1)\oplus\log\mathbf{SU}(2) \cong i\mathbb{R}\oplus A_1^c \in \underline{\mathbf{lag}}_{\mathbb{R}}, \\ \mathbb{R}^4 \cong & \log\mathbf{GL}(\mathbb{C}^2)_+ & \cong \mathbb{R}\oplus iA_1^c \cong \mathbb{T}\oplus\mathbb{S} \in \underline{\mathbf{vec}}_{\mathbb{R}}. \end{array}$$

The dilation Poincaré group is the *conjugate adjoint affine group* of the full linear group $\mathbf{GL}(\mathbb{C}^2)$, considered as real 8-dimensional Lie group

$$\operatorname{Int}_*\mathbf{GL}(\mathbb{C}^2)\ \vec{\times}\ \log\mathbf{GL}(\mathbb{C}^2)_+ = [\mathbf{D}(1)\times\mathbf{SO}_0(1,3)]\ \vec{\times}\ \mathbb{R}^4.$$

The 4×4-Dirac representation of the dilation-Lorentz group in $\mathbf{U}(2,2)$ acts on translations by inner automorphisms. It gives in the nontrivial submatrix the conjugate adjoint action on the translations in the Cartan 2×2 representation

$$g,\hat{g} = g^{-1\star}\in\mathbf{D}(\mathbf{1}_2)\times\mathbf{SL}(\mathbb{C}^2): \begin{pmatrix} g & 0 \\ 0 & \hat{g} \end{pmatrix}\begin{pmatrix} 0 & x \\ 0 & 0 \end{pmatrix}\begin{pmatrix} g & 0 \\ 0 & \hat{g} \end{pmatrix}^{-1} = \begin{pmatrix} 0 & g\circ x\circ g^\star \\ 0 & 0 \end{pmatrix}.$$

The adjoint action of the Lie algebra representation in $\log\mathbf{U}(2,2)$,

$$\left[\begin{pmatrix} i\vec{\alpha}-\beta_0\mathbf{1}_2-\vec{\beta} & 0 \\ 0 & i\vec{\alpha}+\beta_0\mathbf{1}_2+\vec{\beta} \end{pmatrix}, \begin{pmatrix} 0 & x \\ 0 & 0 \end{pmatrix}\right] = \begin{pmatrix} 0 & [i\vec{\alpha},x]-\{\beta_0\mathbf{1}_2+\vec{\beta},x\} \\ 0 & 0 \end{pmatrix},$$

has as nontrivial contributions the conjugate adjoint action with commutators and anticommutators for the $\mathbf{U}(2)$-anti-Hermitian and Hermitian (2×2) matrices which represent the three compact rotation and the four noncompact dilation and boost generators respectively.

The *Lorentz "metric"* with its characteristic signature $(1,3)$ is the product of both spinor "metrics":

$$g=\epsilon^{[1|1]}:\mathbb{M}\longrightarrow\mathbb{M}^T,\quad \epsilon^{[1|1]}(1\otimes 1^\times)=-\epsilon^{[0|1]-1}(1^\times)\otimes\epsilon^{[1|0]}(1).$$

It contains the negative definite product $-\mathbf{1}_3$ for the position translations

$$\left.\begin{array}{ll}\epsilon^{[1|1]}(\mathbf{P}^j) &= g^{jk}\mathbf{X}_k,\\ \epsilon^{AC}(\sigma^j)^{\dot D}_C\epsilon_{\dot D\dot B} &= g^{jk}(\check\sigma_k)^A_{\dot B},\end{array}\right\}\Rightarrow\quad \begin{array}{l} g\cong\eta=\begin{pmatrix}1&0\\0&-\mathbf{1}_3\end{pmatrix}\\ \text{for }\sigma^j=(\mathbf{1}_2,\vec\sigma),\end{array}$$

$$\epsilon^{\dot A\dot C}\epsilon_{DB}=\tfrac12(\sigma^k)^{\dot A}_B g_{kj}(\sigma^j)^{\dot C}_D,\quad \epsilon^{AC}\epsilon_{\dot D\dot B}=\tfrac12(\check\sigma_k)^A_{\dot B}g^{kj}(\check\sigma_j)^C_{\dot D}.$$

Hence $\Lambda=D^{[1|1]}(\lambda)\in\mathbf{SO}_0(1,3)$ is orthogonal self-dual.

The Sylvester-Weyl matrices $(\mathbf{1}_2,\vec\sigma)$ are a basis for the Cartan spacetime representation. All translation bases are related to each other by tetrads $\sigma^j\longmapsto h^j_k\sigma^k$ with $h\in\mathbf{GL}(\mathbb{R}^4)/\mathbf{O}(1,3)$ (orientation manifold of the Lorentz group). General Weyl bases of spacetime translations and energy-momenta $\mathbb{M},\mathbb{M}^T$ use Hermitian 2×2 matrices

$$\textit{general Weyl matrices:}\quad \{\sigma^j,\check\sigma_j\}_{j=0}^3\quad \left\{\begin{array}{l}\frac12\operatorname{tr}\sigma^j\check\sigma_k=\delta^j_k,\\ (\sigma^j)^\star=\sigma^j,\quad(\check\sigma_k)^\star=\check\sigma_k,\end{array}\right.$$

$$\text{with }\sigma_k=g_{kj}\sigma^j,\quad \check\sigma^j=g^{jk}\check\sigma_k\Rightarrow\left\{\begin{array}{l}\check\sigma^j\sigma^k+\check\sigma^k\sigma^j=2g^{jk}\mathbf{1}_2,\\ \sigma^j\check\sigma^k+\sigma^k\check\sigma^j=2g^{jk}\mathbf{1}_2,\end{array}\right.$$

e.g., the Sylvester-Weyl matrices above or the

$$\textit{Witt-Weyl matrices:}\left\{\begin{array}{l}\sigma^j\simeq(\frac{\mathbf{1}_2+\sigma^3}{\sqrt2},\sigma^{1,2},\frac{\mathbf{1}_2-\sigma^3}{\sqrt2}),\\ \check\sigma_j\simeq(\frac{\mathbf{1}_2+\sigma^3}{\sqrt2},\sigma^{1,2},\frac{\mathbf{1}_2-\sigma^3}{\sqrt2}),\end{array}\right.,\quad g\cong\iota=\begin{pmatrix}0&0&1\\0&-\mathbf{1}_2&0\\1&0&0\end{pmatrix}.$$

The lightlike subspaces are spanned by the projectors $\pi^\pm=\frac{\mathbf{1}_2\pm\sigma^3}{2}$.

The Lorentz square g of the translations is the determinant

$$g(x,x)=\det x,\quad g(x,y)=\tfrac{\det(x+y)-\det(x-y)}{4}.$$

The spectrum of a translation x defines its two *Cartan coordinates* $\xi_{1,2}$ by the sum and difference of time translations and the length of position translations

$$x=\begin{pmatrix}x_0+x_3&x_1-ix_2\\x_1+ix_2&x_0-x_3\end{pmatrix},\quad \det x=x_0^2-\vec x^2,\quad \operatorname{tr}x=2x_0,$$

$$\operatorname{spec}x=\{\xi\mid\det(x-\xi\,\mathrm{id}_V)=0\},\quad\left\{\begin{array}{l}\xi^2-\xi\operatorname{tr}x+\det x=0,\\ \xi_1\xi_2=\det x,\quad \xi_1+\xi_2=\operatorname{tr}x,\\ \xi_1,\xi_2=x_0\pm\sqrt{\vec x^2}.\end{array}\right.$$

The signs or the triviality of the Cartan coordinates determine the spacetime order with the properties "timelike," "lightlike" and "spacelike,"

$$\operatorname{sign}(\xi_1,\xi_2)=\left\{\begin{array}{ll}(+,+)\text{ and }(-,-) & \text{strictly future and past timelike,}\\ (+,0),(0,+)\text{ and }(-,0),(0,-) & \text{future and past lightlike,}\\ (0,0) & \text{trivial,}\\ (+,-),(-,+) & \text{strictly spacelike.}\end{array}\right.$$

The order is Lorentz-compatible:

$$\begin{array}{rl} \mathbf{AL}(\mathbb{C}^2) \ni x \succeq 0 & \left\{ \begin{array}{l} \iff x = x^* \text{ and } \operatorname{spec} x \subset \mathbb{R}_+ \\ \iff x = x^* \text{ and } \det x \geq 0, \ \operatorname{tr} x \geq 0 \\ \iff x = s \circ s^* \text{ with } s \in \mathbf{AL}(\mathbb{C}^2), \end{array} \right. \\ \lambda \in \mathbf{SL}(\mathbb{C}^2) & \Rightarrow \lambda \circ x \circ \lambda^* = z \circ z^* \succeq 0 \text{ with } z = \lambda \circ s. \end{array}$$

Here, the positivity of $s \circ s^\star$ for elements of the C*-algebra $\mathbf{AL}(\mathbb{C}^2)$ (chapter "Spacetime as Unitary Operation Classes") is used.

1.4.1 Weyl Spinors with Minkowski Notation

With the local isomorphism $\mathbf{SL}(\mathbb{C}^2) \sim \mathbf{SO}_0(1,3)$ the Weyl representations of the Lorentz symmetry can be written with two sets of six *Lorentz generators*, i.e., 2×2 matrices $\{\sigma^{jk}, \hat{\sigma}^{jk}\}_{j,k=0,1,2,3}$:

$$\begin{array}{l} \sigma^{jk} = -\sigma^{kj} = -\frac{\sigma^j \check{\sigma}^k - \sigma^k \check{\sigma}^j}{4}, \quad \hat{\sigma}^{jk} = -\hat{\sigma}^{kj} = -\frac{\check{\sigma}^j \sigma^k - \check{\sigma}^k \sigma^j}{4}, \\ \text{Sylvester basis:} \left\{ \begin{array}{l} \sigma^j = (\mathbf{1}_2, \vec{\sigma}) = \check{\sigma}_j, \\ \sigma^{ab} = \epsilon^{abc} \frac{i}{2} \sigma^c = \hat{\sigma}^{ab}, \\ \sigma^{a0} = -\frac{1}{2} \sigma^a = -\hat{\sigma}^{a0}, \end{array} \right. \end{array}$$

with the identities

$$\begin{array}{l} \frac{i}{2} \epsilon_{jklm} \sigma^{lm} = +\sigma_{jk}, \quad \frac{i}{2} \epsilon_{jklm} \hat{\sigma}^{lm} = -\hat{\sigma}_{jk}, \\ \epsilon^{0123} = 1 = -\epsilon_{0123}. \end{array}$$

In this notation the Lorentz Lie algebra elements are

$$\begin{array}{l} 2(i\alpha_a \mp \beta_a)\sigma^a = \alpha_a \epsilon^{abc}[\sigma^b, \sigma^c] \mp 2\beta_a \sigma^a \cong \omega_{jk} \left\{ \begin{array}{l} \sigma^{jk}, \\ \hat{\sigma}^{jk}, \end{array} \right. \\ \omega_{jk} = -\omega_{kj} \in \mathbb{R}, \text{ Sylvester basis:} \left\{ \begin{array}{l} \omega_{ab} = 2\epsilon^{abc}\alpha_c = \epsilon^{abc}\varphi_c \text{ (rotations)}, \\ \omega_{a0} = 2\beta_a = \psi_a \text{ (boosts)}, \end{array} \right. \end{array}$$

with the two Weyl spinor representations

$$\mathcal{D}^{[1|0]}: \left\{ \begin{array}{l} \mathrm{l}^{jk} = (\sigma^{jk})^B_A \, \mathrm{l}^A \otimes \mathrm{r}^\times_B, \quad \epsilon^{CA}(\sigma^{jk})^B_A \epsilon_{BD} = -(\sigma^{jk})^C_D, \\ [\sigma^{jk}, \sigma^{nm}] = g^{jn}\sigma^{km} - g^{kn}\sigma^{jm} - g^{jm}\sigma^{kn} + g^{km}\sigma^{jn}, \\ \lambda(\omega) = e^{\frac{1}{2}\omega_{jk}\mathrm{l}^{jk}} \cong e^{\frac{1}{2}\omega_{jk}\sigma^{jk}}, \end{array} \right.$$

$$\mathcal{D}^{[0|1]}: \left\{ \begin{array}{l} \hat{\mathrm{l}}^{jk} = (\hat{\sigma}^{jk})^{\dot{B}}_{\dot{A}} \, \mathrm{r}^{\dot{A}} \otimes \mathrm{l}^\times_{\dot{B}}, \quad \epsilon^{\dot{C}\dot{A}}(\hat{\sigma}^{jk})^{\dot{B}}_{\dot{A}} \epsilon_{\dot{B}\dot{D}} = -(\hat{\sigma}^{jk})^{\dot{C}}_{\dot{D}}, \\ [\hat{\sigma}^{jk}, \hat{\sigma}^{nm}] = g^{jn}\hat{\sigma}^{km} - g^{kn}\hat{\sigma}^{jm} - g^{jm}\hat{\sigma}^{kn} + g^{km}\hat{\sigma}^{jn}, \\ \hat{\lambda}(\omega) = e^{\frac{1}{2}\omega_{jk}\hat{\mathrm{l}}^{jk}} \cong e^{\frac{1}{2}\omega_{jk}\hat{\sigma}^{jk}}, \end{array} \right.$$

and the two Casimir elements

$$\begin{array}{lll} \sigma_{jk}\sigma^{jk} & = +3\frac{i}{4!}\epsilon_{jklm}\sigma^j\check{\sigma}^k\sigma^l\check{\sigma}^m & = 3 \cdot \mathbf{1}_2, \\ \hat{\sigma}_{jk}\hat{\sigma}^{jk} & = -3\frac{i}{4!}\epsilon_{jklm}\check{\sigma}^j\sigma^k\check{\sigma}^l\sigma^m & = 3 \cdot \mathbf{1}_2. \end{array}$$

The involutory Fierz recouplings of the Pauli spinor representations give the Fierz recouplings for the Weyl spinor representations with one Fierz-symmetric and one Fierz-antisymmetric linear combination:

$$\begin{pmatrix} \mathbf{1}_2 \otimes \mathbf{1}_2 \\ \sigma_{jk} \otimes \sigma^{jk} \end{pmatrix}_{AC}^{BD} = \begin{pmatrix} S \\ T \end{pmatrix}_{AC}^{BD} \stackrel{\text{Fierz}}{=} \begin{pmatrix} \frac{1}{2} & -\frac{1}{2} \\ -\frac{3}{2} & -\frac{1}{2} \end{pmatrix} \begin{pmatrix} S \\ T \end{pmatrix}_{AC}^{DB}, \quad \begin{pmatrix} S - \frac{T}{3} \\ S + T \end{pmatrix} \stackrel{\text{Fierz}}{\longleftrightarrow} \begin{pmatrix} S - \frac{T}{3} \\ -(S + T) \end{pmatrix}.$$

The recoupling matrix $\begin{pmatrix} \frac{1}{2} & -\frac{1}{2} \\ -\frac{3}{2} & -\frac{1}{2} \end{pmatrix}$ with square $\mathbf{1}_2$ has eigenvalues $\{\pm 1\}$.

1.5 Minkowski Clifford Algebras

The Clifford algebra (chapter "Quantum Algebras") for Minkowski spacetime $\mathbb{M} \cong \mathbb{R}^4$ is constituted by the equivalence classes in the tensor algebra $\bigotimes \mathbb{M}$ with respect to the equality of the tensor square of a spacetime translation to its Lorentz square g. It has real dimension $2^4 = 16$:

$$\text{in CLIFF}(1,3) \cong \mathbb{R}^{16} : \begin{cases} x \otimes x &= g(x,x), \quad x \in \mathbb{M}, \\ \frac{1}{2}\{e^j, e^k\} &= g^{jk} \cong \begin{pmatrix} 1 & 0 \\ 0 & -\mathbf{1}_3 \end{pmatrix} \text{ for } \mathbb{M}\text{-basis } \{e^j\}_{j=0}^3. \end{cases}$$

The Minkowski Clifford algebra comes with an adjoint action of the Lorentz Lie algebra $\text{CLAG}(\mathbb{M}, g) \cong \mathbb{R}^6$ with basis $\{l^{jk}\}_{j,k=0,1,2,3}$ given by the translation commutators

$$l^{jk} = -\frac{[e^j, e^k]}{4} : \quad \begin{cases} [l^{jk}, e^n] &= g^{jn} e^k - g^{kn} e^j, \\ [l^{jk}, l^{nm}] &= g^{jn} l^{km} - g^{kn} l^{jm} - g^{jm} l^{kn} + g^{km} l^{jn}. \end{cases}$$

The ten elements $\{l^{jk}, e^j\}$ with the Clifford product induced commutator do not represent the Poincaré Lie algebra since the commutators $[e^j, e^k]$ do not vanish.

The even subalgebra of the Clifford algebra contains Lorentz scalars and 2-tensors. Geometrically: numbers, 2-dimensional areas and 4-dimensional volumes; the odd subspace contains Lorentz vectors. Geometrically: vectors and 3-dimensional volumes (axial vectors):

$$\begin{aligned} &\text{CLIFF}(1,3) = \text{CLIFF}(1,3)_0 \oplus \text{CLIFF}(1,3)_1 \cong \mathbb{R}^{1+6+1} \oplus \mathbb{R}^{4+4}, \\ &\text{basis of } \text{CLIFF}(1,3)_0 : \{1, \ l^{jk}, \ e_5 = \tfrac{\epsilon_{jkmn}}{4!} e^j \otimes e^k \otimes e^m \otimes e^n\}, \\ &\text{basis of } \text{CLIFF}(1,3)_1 : \{e^j, \ e^j \otimes e_5\}. \end{aligned}$$

With the imaginary roots of the minimal polynomial $p_{e_5}(X) = X^2 + 1$, i.e., $e_5 \otimes e_5 = -1$, the *chiral projectors*

$$\mathcal{P}_\pm = \frac{1 \pm i e_5}{2},$$

can be constructed only in the complexified Clifford algebra $\mathbb{C} \otimes \text{CLIFF}(1,3) \cong \mathbb{C}^{16}$.

The two possible Minkowski Clifford algebras are isomorphic to endomorphism algebras, 4×4 matrices with real entries

$$\text{CLIFF}(3,1) \cong \mathbb{R}(4 \times 4)$$

and 2×2 matrices with entries from the quaternionic field

$$\mathrm{CLIFF}(1,3) \cong \mathbb{H}(2 \times 2), \quad \mathbb{H} \cong \{\alpha_0 \mathbf{1}_2 + i\vec{\alpha}\} \cong \mathbb{R}^4$$

representable by endomorphisms in the Dirac algebra (next section)

$$\begin{aligned} &\mathrm{CLIFF}(1,3) \longrightarrow \mathbb{C}(4 \times 4), \quad e^j \longmapsto \gamma^j \\ &\Rightarrow (1, e_5, l^{jk}, e^j, e^j \otimes e_5) \longmapsto (\mathbf{1}_4, \gamma_5, \gamma^{jk}, \gamma^j, \gamma^j\gamma_5), \\ &\qquad \{\gamma^j, \gamma^k\} = 2g^{jk}\mathbf{1}_4. \end{aligned}$$

1.6 Dirac Spinors and Dirac Algebra

The direct sum of both Weyl spinor representations acting on the antidoubling V_{doub} defines the complex 4-dimensional *Dirac representation* $[1|0] \oplus [0|1]$ of $\mathbf{SL}(\mathbb{C}^2)$ with the anticonjugation $\times$ induced $\mathbf{U}(2,2)$-unitarity. Dirac spinors have as dual bases

$$\begin{aligned} &\text{basis of } V_{\text{doub}} = V \oplus \overline{V}^T \cong \mathbb{C}^4: \quad \{\psi^K \mid \psi^{1,2} = (\mathrm{l}^{1,2}, 0), \quad \psi^{3,4} = (0, \mathrm{r}^{1,2})\}, \\ &\text{basis of } V_{\text{doub}}^T = V^T \oplus \overline{V} \cong \mathbb{C}^4: \quad \{\overline{\psi}_K \mid \overline{\psi}_{1,2} = \begin{pmatrix} \mathrm{r}^{\times}_{1,2} \\ 0 \end{pmatrix}, \quad \overline{\psi}_{3,4} = \begin{pmatrix} 0 \\ \mathrm{l}^{\times}_{1,2} \end{pmatrix}\}, \\ &\text{dual: } \langle \overline{\psi}_K, \psi^M \rangle = \delta_K^M, \\ &\times : V_{\text{doub}} \longrightarrow V_{\text{doub}}^T, \quad \psi^K = (\mathrm{l}^A, \mathrm{r}^{\dot{A}}) \longmapsto \begin{pmatrix} 0 & \delta^{A\dot{B}} \\ \delta^{\dot{A}B} & 0 \end{pmatrix} \begin{pmatrix} \mathrm{r}^{\times}_B \\ \mathrm{l}^{\times}_{\dot{B}} \end{pmatrix}. \end{aligned}$$

The Euclidean $\mathbf{U}(2)$-conjugation for 2×2 matrices $\lambda \leftrightarrow \lambda^\star$, used in the complex quartet with the chiral basis, is part of the indefinite $\mathbf{U}(2,2)$-conjugation (anticonjugation) $\times$ for 4×4 matrices. It combines the $\mathbf{U}(4)$-conjugation $\star$ with the linear transformation $\begin{pmatrix} 0 & \mathbf{1}_2 \\ \mathbf{1}_2 & 0 \end{pmatrix}$:

$$\text{chiral basis: } \begin{pmatrix} \lambda & \mu \\ \rho & \tau \end{pmatrix}^\times = \begin{pmatrix} 0 & \mathbf{1}_2 \\ \mathbf{1}_2 & 0 \end{pmatrix} \begin{pmatrix} \lambda & \mu \\ \rho & \tau \end{pmatrix}^\star \begin{pmatrix} 0 & \mathbf{1}_2 \\ \mathbf{1}_2 & 0 \end{pmatrix} = \begin{pmatrix} \tau^\star & \mu^\star \\ \rho^\star & \lambda^\star \end{pmatrix}.$$

The Lorentz cover group $\mathbf{SL}(\mathbb{C}^2)$ is represented as an $\mathbf{SU}(2,2)$-subgroup in the Dirac algebra $\mathbf{AL}(V_{\text{doub}}) \cong \mathbb{C}^{16}$:

$$\log \mathbf{SO}_0(1,3) \cong \log \mathbf{SL}(\mathbb{C}^2) \longrightarrow \log \mathbf{SU}(2,2) \cong \log \mathbf{SO}_0(2,4).$$

Explicitly,

$$\begin{aligned} \lambda \longmapsto \lambda_{\text{doub}} = \begin{pmatrix} \lambda & 0 \\ 0 & \hat{\lambda} \end{pmatrix} &= (\mathrm{l}^A, \mathrm{r}^{\dot{A}}) \otimes \begin{pmatrix} (e^{\frac{1}{2}\omega_{jk}\sigma^{jk}})_A^B & 0 \\ 0 & (e^{\frac{1}{2}\omega_{jk}\hat{\sigma}^{jk}})_{\dot{A}}^{\dot{B}} \end{pmatrix} \begin{pmatrix} \mathrm{r}^{\times}_B \\ \mathrm{l}^{\times}_{\dot{B}} \end{pmatrix} \\ &= (\lambda_{\text{doub}})_K^M \; \psi^K \otimes \overline{\psi}_M = e^{\frac{1}{2}\omega_{jk}\mathbf{L}^{jk}} \cong e^{\frac{1}{2}\omega_{jk}\gamma^{jk}}, \\ \lambda_{\text{doub}}^\times &= \begin{pmatrix} 0 & \mathbf{1}_2 \\ \mathbf{1}_2 & 0 \end{pmatrix} \lambda_{\text{doub}}^\star \begin{pmatrix} 0 & \mathbf{1}_2 \\ \mathbf{1}_2 & 0 \end{pmatrix} = \begin{pmatrix} \hat{\lambda}^\star & 0 \\ 0 & \lambda^\star \end{pmatrix} = \lambda_{\text{doub}}^{-1}. \end{aligned}$$

The $\times$-antisymmetric basis $\{\mathbf{L}^{jk}\}_{j,k=0,1,2,3}$ for the Lorentz Lie algebra

$$\begin{aligned} &\mathbf{L}^{jk} : V_{\text{doub}} \longrightarrow V_{\text{doub}}, \quad \mathbf{L}^{jk} = \mathbf{l}^{jk} \oplus \hat{\mathbf{l}}^{jk} = (\gamma^{jk})_K^M \psi^K \otimes \overline{\psi}_M, \\ &\gamma^{jk} = \begin{pmatrix} \sigma^{jk} & 0 \\ 0 & \hat{\sigma}^{jk} \end{pmatrix} = -\tfrac{1}{4}[\gamma^j, \gamma^k], \\ &[\gamma^{jk}, \gamma^{nm}] = g^{jn}\gamma^{km} - g^{kn}\gamma^{jm} - g^{jm}\gamma^{kn} + g^{km}\gamma^{jn}, \end{aligned}$$

is combinable by *Dirac matrices* as a sum of the Weyl matrices:

$$\mathbf{V}^j : V_{\mathrm{doub}} \longrightarrow V_{\mathrm{doub}}, \quad \mathbf{V}^j = i(\mathbf{P}^j \oplus \mathbf{X}^j) = i(\gamma^j)^M_K \psi^K \otimes \overline{\psi}_M,$$
$$\gamma^j = \begin{pmatrix} 0 & \sigma^j \\ \check{\sigma}^j & 0 \end{pmatrix}, \quad \begin{cases} \{\gamma^j, \gamma^k\} = 2g^{jk}\mathbf{1}_4, \quad \operatorname{tr}\gamma^j = 0, \quad \operatorname{tr}\gamma^j\gamma^k = 4g^{jk}, \\ [\gamma^{jk}, \gamma^n] = g^{jn}\gamma^k - g^{kn}\gamma^j. \end{cases}$$

The Lorentz action for the Dirac matrices arises by inner $\mathbf{SU}(2,2)$-automorphisms

$$\gamma^k \longmapsto \lambda_{\mathrm{doub}}\gamma^k\lambda^\times_{\mathrm{doub}} = \Lambda^k_j\gamma^j.$$

Inner automorphisms $\gamma^k \longmapsto g \circ \gamma^k \circ g^{-1}$ with $g \in \mathbf{GL}(\mathbb{C}^4)$ lead from the *chiral representation of Dirac matrices* above to other forms, e.g., to the *time-diagonal representation* $(\gamma^0_U, \vec{\gamma}_U) = (\beta, \beta\vec{\alpha})$ with diagonal γ^0_U:

$$\underset{\text{chiral}}{(\gamma^0, \vec{\gamma}) = \left(\begin{pmatrix} 0 & \mathbf{1}_2 \\ \mathbf{1}_2 & 0 \end{pmatrix}, \begin{pmatrix} 0 & \vec{\sigma} \\ -\vec{\sigma} & 0 \end{pmatrix}\right)} \longmapsto \underset{\text{time-diagonal}}{(\gamma^0_U, \vec{\gamma}_U) = \left(\begin{pmatrix} \mathbf{1}_2 & 0 \\ 0 & -\mathbf{1}_2 \end{pmatrix}, \begin{pmatrix} 0 & \vec{\sigma} \\ -\vec{\sigma} & 0 \end{pmatrix}\right)},$$
$$\gamma^k_U = w_U \circ \gamma^k \circ w_U^{-1} \text{ with } w_U = \frac{1}{\sqrt{2}}\begin{pmatrix} \mathbf{1}_2 & \mathbf{1}_2 \\ -\mathbf{1}_2 & \mathbf{1}_2 \end{pmatrix}.$$

The time-diagonal representation was originally introduced by Dirac.

The time-diagonal-associated $\mathbf{U}(2,2)$-conjugation combines the Euclidean $\mathbf{U}(4)$-conjugation with a signature $(2,2)$ matrix:

$$\text{time-diagonal basis: } \begin{pmatrix} \lambda & \mu \\ \rho & \tau \end{pmatrix}^\times = \begin{pmatrix} \mathbf{1}_2 & 0 \\ 0 & -\mathbf{1}_2 \end{pmatrix}\begin{pmatrix} \lambda & \mu \\ \rho & \tau \end{pmatrix}^\star\begin{pmatrix} \mathbf{1}_2 & 0 \\ 0 & -\mathbf{1}_2 \end{pmatrix} = \begin{pmatrix} \lambda^\star & -\rho^\star \\ -\mu^\star & \tau^\star \end{pmatrix}.$$

In the Dirac representation the Killing-Casimir element is proportional to the unit matrix $\mathbf{1}_4$,

$$\mathbf{L}_{jk} \circ \mathbf{L}^{jk} = -3\mathbf{I}, \quad \gamma_{jk}\gamma^{jk} = -3\mathbf{1}_4,$$
$$\mathbf{I} : V_{\mathrm{doub}} \longrightarrow V_{\mathrm{doub}}, \quad \mathbf{I} = \mathrm{id}_{V_{\mathrm{doub}}} = \mathrm{id}_V \oplus \mathrm{id}_{\overline{V}^T} = \psi^K \otimes \overline{\psi}_K,$$

and the chiral Casimir element to the volume element γ_5, diagonal for chiral Dirac matrices

$$\mathbf{J} : V_{\mathrm{doub}} \longrightarrow V_{\mathrm{doub}}, \quad \mathbf{J} = \mathrm{id}_V \oplus (-\mathrm{id}_{\overline{V}^T}) = i(\gamma_5)^M_K\ \psi^K \otimes \overline{\psi}_M,$$
$$i\gamma_5 = \frac{i}{4!}\epsilon_{jkmn}\gamma^j\gamma^k\gamma^m\gamma^n, \quad \begin{cases} \gamma_5^2 = -\mathbf{1}_4, \\ \{\gamma^j, \gamma_5\} = 0, \quad [\gamma^{jk}, \gamma_5] = 0, \\ \frac{1}{2}\epsilon_{jkmn}\gamma^{mn} = \gamma_5\gamma_{jk}, \quad \operatorname{tr}\gamma_5 = 0, \end{cases}$$
$$\text{chiral } i\gamma_5 = \begin{pmatrix} \mathbf{1}_2 & 0 \\ 0 & -\mathbf{1}_2 \end{pmatrix} \longmapsto i\gamma_{5U} = w_U \circ i\gamma_5 \circ w_U^{-1} = \begin{pmatrix} 0 & -\mathbf{1}_2 \\ -\mathbf{1}_2 & 0 \end{pmatrix} \text{ time-diagonal}.$$

Hence one can construct the projectors from the Dirac representation to the two irreducible Weyl representations

$$\mathcal{P}_+ = \mathcal{P} = \frac{\mathbf{1}_4 + i\gamma_5}{2}, \quad \mathcal{P}_- = \mathcal{P}^\times = \frac{\mathbf{1}_4 - i\gamma_5}{2}, \quad \mathcal{P}\mathcal{P}^\times = 0, \ \mathcal{P}^2 = \mathcal{P},$$
$$\mathcal{P}\overline{\psi}_K = \begin{cases} \begin{pmatrix} \mathrm{r}^\times_K \\ 0 \end{pmatrix}, & K = 1,2, \\ 0, & K = 3,4, \end{cases} \quad \mathcal{P}^\times\overline{\psi}_K = \begin{cases} 0, & K = 1,2, \\ \begin{pmatrix} 0 \\ \mathrm{l}^\times_{K-2} \end{pmatrix}, & K = 3,4, \end{cases}$$
$$\frac{\mathbf{1}_4 + i\gamma_5}{2}\gamma_j = \begin{pmatrix} 0 & \sigma_j \\ 0 & 0 \end{pmatrix}, \quad \frac{\mathbf{1}_4 - i\gamma_5}{2}\gamma_j = \begin{pmatrix} 0 & 0 \\ \check{\sigma}_j & 0 \end{pmatrix}.$$

The difference of the Weyl matrices gives the axial vectors

$$\begin{array}{l}\mathbf{A}^j : V_{\text{doub}} \longrightarrow V_{\text{doub}}, \quad \mathbf{A}^j = i\mathbf{P}^j \oplus (-i\mathbf{X}^j) = (\gamma^j\gamma_5)^M_K \psi^K \otimes \overline{\psi}_M, \\ i\gamma^j\gamma_5 = \begin{pmatrix} 0 & -\sigma^j \\ \check{\sigma}^j & 0 \end{pmatrix}, \quad \{\gamma_5, \gamma^j\} = 0, \quad \operatorname{tr} \gamma^j\gamma_5 = 0.\end{array}$$

A Lorentz group representation adapted basis of the complex 16-dimensional Dirac algebra $\mathbf{AL}(V_{\text{doub}})$ or of the real 16-dimensional Lie algebra $\log \mathbf{U}(2,2)$ is given by the decomposition of the tensor product of two Dirac representations into two scalar representations, the adjoint tensor representation and two vector representations, $16 = 1 + 1 + 6 + 4 + 4$:

$$\begin{array}{cccccc} \bigotimes^2 \left(\mathcal{D}^{[1|0]} \oplus \mathcal{D}^{[0|1]}\right) & = 2 \times \mathcal{D}^{[0|0]} & \oplus & \left[\mathcal{D}^{[2|0]} \oplus \mathcal{D}^{[0|2]}\right] & \oplus & 2 \times \mathcal{D}^{[1|1]}, \\ \log \mathbf{U}(2,2)\text{-basis} & \{i\mathbf{I}, \mathbf{J}\} & \uplus & \{\mathbf{L}^{jk}\} & \uplus & \{\mathbf{V}^j, \mathbf{A}^j\}. \end{array}$$

The Fierz recoupling has two Fierz-symmetric and three Fierz-antisymmetric linear combinations as eigenvectors:

$$\begin{pmatrix} \mathbf{1}_4 \otimes \mathbf{1}_4 \\ \gamma_5 \otimes \gamma_5 \\ \gamma_{jk} \otimes \gamma^{jk} \\ \gamma_j \otimes \gamma^j \\ \gamma_5\gamma_j \otimes \gamma_5\gamma^j \end{pmatrix}^{MN}_{KL} = \begin{pmatrix} s \\ p \\ t \\ v \\ a \end{pmatrix}^{MN}_{KL} \stackrel{\text{Fierz}}{=} \begin{pmatrix} \frac{1}{4} & -\frac{1}{4} & -\frac{1}{8} & \frac{1}{4} & \frac{1}{4} \\ -\frac{1}{4} & \frac{1}{4} & \frac{1}{8} & \frac{1}{4} & \frac{1}{4} \\ -3 & 3 & -\frac{1}{2} & 0 & 0 \\ 1 & 1 & 0 & -\frac{1}{2} & \frac{1}{2} \\ 1 & 1 & 0 & \frac{1}{2} & -\frac{1}{2} \end{pmatrix} \begin{pmatrix} s \\ p \\ t \\ v \\ a \end{pmatrix}^{NM}_{KL}$$

$$\Rightarrow \begin{pmatrix} s - p - \frac{t}{6} \\ s - p + \frac{t}{2} \\ s + p + \frac{v+a}{2} \\ s + p - \frac{v+a}{2} \\ v - a \end{pmatrix} \stackrel{\text{Fierz}}{\longleftrightarrow} \begin{pmatrix} s - p - \frac{t}{6} \\ -(s - p + \frac{t}{2}) \\ s + p + \frac{v+a}{2} \\ -(s + p - \frac{v+a}{2}) \\ -(v - a) \end{pmatrix}.$$

As a reflection the recoupling matrix has square $\mathbf{1}_5$ and eigenvalues $\{\pm 1\}$.

1.7 Reflections for Position and Time

Time and position translations have linear reflections (chapter "Spacetime Translations")

$$\begin{array}{ll} \mathbf{O}(1) \cong \mathbf{I}(2), \quad \mathbf{O}(3) = \mathbb{I}(2) \times \mathbf{SO}(3) & \mathbb{I}(2) = \{\pm \mathbf{1}_3\}. \\ t \stackrel{\mathsf{T}}{\leftrightarrow} -t & \vec{x} \stackrel{\mathsf{P}}{\leftrightarrow} -\vec{x} \end{array}$$

embedded in a spacetime reflection Klein group in the Lorentz group

$$\mathbf{O}(1,3) = \mathbb{I}(2) \vec{\times} \mathbf{SO}(1,3) = \mathbb{I}(2) \vec{\times} [\mathbb{I}(2) \times \mathbf{SO}_0(1,3)].$$

The central spacetime reflection $x \stackrel{-\mathbf{1}_4}{\leftrightarrow} -x$ is compatible with $\mathbf{O}(1,3)$. There are as many position reflections P with associated time reflection T as there are decompositions into time and position translations. $\{\mathsf{P}, \mathsf{T}\}$ are only rotation, not boost invariant:

$$\mathbf{O}(1,3)/\mathbf{SO}_0(1,3) \cong \mathbb{I}(2) \times \mathbb{I}(2) \cong \{\mathbf{1}_4, \mathsf{P}\} \times \{\pm \mathbf{1}_4\} = \{\pm \mathbf{1}_4, \mathsf{P}, \mathsf{T}\},$$

$$\begin{array}{l} \mathsf{P} \in \mathbf{O}(1,3)/\mathbf{SO}(1,3), \\ \mathsf{T} = -\mathsf{P}, \quad \det \mathsf{P} = -1 = \det \mathsf{T}, \\ [\mathsf{P}, \mathbf{SO}_0(1,3)] \neq \{0\}, [\mathsf{P}, \mathbf{O}(3)] = \{0\}. \end{array}$$

$\mathbf{1}_4$	$-\mathbf{1}_4$	T	P
$-\mathbf{1}_4$	$\mathbf{1}_4$	P	T
T	P	$\mathbf{1}_4$	$-\mathbf{1}_4$
P	T	$-\mathbf{1}_4$	$\mathbf{1}_4$

Spinors are acted on with the simply connected covering groups $\mathbf{SU}(2)$ (rotations) and $\mathbf{SL}(\mathbb{C}^2)$ (Lorentz transformations)

$$\mathbf{SO}(3) \cong \mathbf{SU}(2)/\mathbb{I}(2), \quad \mathbf{SO}_0(1,3) \cong \mathbf{SL}(\mathbb{C}^2)/\mathbb{I}(2).$$

The time and position reflections are not contained in the orthochronous Lorentz group. Therefore, the Klein groups $\mathbb{I}(2) \times \mathbb{I}(2)$ in $\mathbf{O}(1,3)$ used for spacetime reflections have to be implemented, on the spinor level, differently. The twoality group $\mathbb{I}(2) = \{\pm \mathbf{1}_2\}$ for the $\mathbf{SO}(3)$-classes in $\mathbf{SU}(2)$ and the $\mathbf{SO}_0(1,3)$-classes in $\mathbf{SL}(\mathbb{C}^2)$ has nothing to do with the reflection Klein group.

For spacetime translations in the Cartan representation $x_0 + \vec{x}$ (with x_0 for $x_0\mathbf{1}_2$) as Weyl spinor product $[1|1] = [1|0] \otimes [0|1]$ the reflections can be constructed as products of spinor reflections, i.e., from bi- and sesquilinear forms of spinor spaces.

The general structure: A vector space reflection (chapter "Spacetime Translations"), which can be linear or antilinear for complex spaces, is a representation of the reflection group $\mathbb{I}(2) = \{\pm 1\}$ with an involution $-1 \longmapsto \mathtt{R}$:

$$\mathbb{I}(2) \times V \longrightarrow V, \quad -1 \bullet v = \mathtt{R}.v = v^{\mathtt{R}}, \quad \mathtt{R}^2.v = v^{\mathtt{RR}} = v.$$

It is either faithful or trivial. The vector space is decomposed into the reflection invariants V_+, the eigenspace with $\mathtt{R}$-eigenvalue $\mathtt{r} = +1$ for a linear reflection and with the Hermitian elements for an antilinear $\mathtt{R} = *$, and the eigenvectors V_- with eigenvalue $\mathtt{r} = -1$ and the anti-Hermitian vectors $V_- = iV_+$ for $\mathtt{R} = *$:

$$V = V_+ \oplus V_-, \quad V_\pm = \{v = \pm v^{\mathtt{R}} \mid v \in V\} = \{v \pm v^{\mathtt{R}} \mid v \in V\}.$$

An $\mathbb{I}(2)$-action on two vector spaces defines the reflection for the tensor product:

$$\mathbb{I}(2) \times (V_1 \otimes V_2) \longrightarrow V_1 \otimes V_2, \quad \mathtt{R}_{1\otimes 2}(v_1 \otimes v_2) = (\mathtt{R}_1.v_1) \otimes (\mathtt{R}_2.v_2).$$

1.7.1 Linear Spinor Reflections

All spinors (Pauli, Weyl, and Dirac) are linear self-dual with the $\mathbb{C}^2$-volume form (spinor "metric").

The defining $\mathbf{SU}(2)$-representation [1] acts on Pauli spinors $V \cong \mathbb{C}^2$. They have an, up to a scalar factor unique, $\mathbf{SL}(\mathbb{C}^2)$-invariant antisymmetric bilinear form that defines an isomorphism with the dual space V^T, in this context called *Pauli spinor reflection*:

$$\begin{array}{ccc} V & \xrightarrow{u} & V \\ \epsilon \downarrow & & \downarrow \epsilon \\ V^T & \xrightarrow[\check{u}]{} & V^T \end{array}, \quad \begin{aligned} u &= e^{i\vec{\alpha}}, \quad \check{u} = u^{-1T} = e^{-i\vec{\alpha}^T}, \\ u &= \epsilon^{-1} \circ \check{u} \circ \epsilon, \\ -\vec{\sigma} &= \epsilon^{-1} \circ \vec{\sigma}^T \circ \epsilon, \end{aligned}$$

$$[1] \stackrel{\epsilon}{\longleftrightarrow} [1], \quad V \stackrel{\epsilon}{\longleftrightarrow} V^T, \quad \mathrm{u}^A \leftrightarrow \epsilon^{AB}\mathrm{u}^\star_B.$$

Equally, the two fundamental $\mathbf{SU}(2)$-embedding representations of $\mathbf{SL}(\mathbb{C}^2)$, the left- and right-handed Weyl representations $[1|0]$, $[0|1]$ acting on $V_L, V_R \cong \mathbb{C}^2$ with the dual representations on $V^T_{L,R}$, are self-dual with the Lorentz compatible *Weyl spinor reflections*

$$\begin{array}{ccc} V_L & \xrightarrow{\lambda} & V_L \\ \epsilon_L \downarrow & & \downarrow \epsilon_L \\ V_L^T & \xrightarrow[\check\lambda]{} & V_L^T \end{array}, \quad \begin{array}{ccc} V_R & \xrightarrow{\hat\lambda} & V_R \\ \epsilon_R \downarrow & & \downarrow \epsilon_R \\ V_R^T & \xrightarrow[\overline\lambda]{} & V_R^T \end{array},$$

$$\begin{array}{rl} \lambda & = e^{i\vec\alpha-\vec\beta}, \\ \check\lambda = \lambda^{-1T} & = e^{(-i\vec\alpha+\vec\beta)^T}, \end{array} , \quad \begin{array}{rll} \hat\lambda & = \lambda^{-1\star} & = e^{i\vec\alpha+\vec\beta}, \\ \overline\lambda & = \lambda^{T\star} & = e^{(-i\vec\alpha-\vec\beta)^T}, \end{array}$$

$$\begin{array}{l} [1|0] \overset{\epsilon_L}{\leftrightarrow} [1|0], \quad V_L \overset{\epsilon_L}{\longleftrightarrow} V_L^T, \quad \mathrm{l}^A \leftrightarrow \epsilon^{AB}\mathrm{r}^\times_B, \\ [0|1] \overset{\epsilon_R}{\leftrightarrow} [0|1], \quad V_R \overset{\epsilon_R}{\longleftrightarrow} V_R^T, \quad \mathrm{r}^{\dot A} \leftrightarrow \epsilon^{\dot A\dot B}\mathrm{l}^\times_{\dot B}. \end{array}$$

1.7.2 Spinor Conjugations

All spinors (Pauli, Weyl, Dirac) are complex vectors, acted on by a real group, defined by a conjugation.

The definition of $\mathbf{SU}(2)$ requires a *Euclidean conjugation* $\star$, for Pauli spinors

$$\begin{array}{ccc} V & \xrightarrow{u} & V \\ \star \downarrow & & \downarrow \star \\ V^T & \xrightarrow[\check u]{} & V^T \end{array}, \quad \begin{array}{rl} u & = \hat u = u^{-1\star}, \\ \vec\sigma & = \vec\sigma^\star, \end{array}$$

$$[1] \overset{\star}{\leftrightarrow} [1], \quad V \overset{\star}{\longleftrightarrow} V^T, \quad \mathrm{u}^A \leftrightarrow \delta^{AB}\mathrm{u}^\star_B.$$

The $\mathbf{SL}(\mathbb{C}^2)$-embeddings define antilinear isomorphisms between the left- and right-handed Weyl spinors, compatible only with the spin group action, not with the boosts $\mathbf{SL}(\mathbb{C}^2)/\mathbf{SU}(2)$:

$$\begin{array}{ccc} V_L & \xrightarrow{u_L} & V_L \\ \star \downarrow & & \downarrow \star \\ V_L^T & \xrightarrow[\check u_L]{} & V_L^T \end{array}, \quad u_L = \lambda|_{\vec\beta=0} = u_R = \hat\lambda|_{\vec\beta=0} = e^{i\vec\alpha} \in \mathbf{SU}(2),$$

$$[1|0] \overset{\star}{\leftrightarrow} [0|1], \quad \begin{array}{l} V_L \overset{\star}{\longleftrightarrow} V_L^T, \quad \mathrm{l}^A \leftrightarrow \delta^{AB}\mathrm{r}^\times_B, \\ V_R \overset{\star}{\longleftrightarrow} V_R^T, \quad \mathrm{r}^{\dot A} \leftrightarrow \delta^{\dot A\dot B}\mathrm{l}^\times_{\dot B}. \end{array}$$

For Weyl spinors there is the *anticonjugation* $\times$ in the complex quartet; it is Lorentz compatible:

$$\begin{array}{ccc} V_L & \xrightarrow{\lambda} & V_L \\ \times\downarrow & & \downarrow\times \\ \overline{V}_L & \xrightarrow[\overline{\lambda}]{} & \overline{V}_L \end{array}, \quad \overline{V}_L = V_R^T,$$

$$[1|0] \stackrel{\times}{\longleftrightarrow} [0|1], \quad \begin{array}{ll} V_L \stackrel{\times}{\longleftrightarrow} V_R^T, & \mathrm{l}^A \leftrightarrow \delta^{A\dot{A}} \mathrm{l}^\times_{\dot{A}}, \\ V_R \stackrel{\times}{\longleftrightarrow} V_L^T, & \mathrm{r}^{\dot{A}} \leftrightarrow \delta^{\dot{A}A} \mathrm{r}^\times_A. \end{array}$$

The difference between positive $\mathbf{U}(2)$-conjugation $\star$ and indefinite $\mathbf{U}(2,2)$-conjugation $\times$ is visible in the Dirac algebra, e.g., with the (4×4)-Dirac matrices in the chiral representation where $\begin{pmatrix} \lambda & \mu \\ \rho & \tau \end{pmatrix}^\times = \begin{pmatrix} \tau^\star & \mu^\star \\ \rho^\star & \lambda^\star \end{pmatrix}$:

$$i\gamma_5 = \begin{pmatrix} \mathbf{1}_2 & 0 \\ 0 & -\mathbf{1}_2 \end{pmatrix}, \ \gamma^j = \begin{pmatrix} 0 & \sigma^j \\ \check{\sigma}^j & 0 \end{pmatrix}, \ i\gamma_5\gamma^j = \begin{pmatrix} 0 & \sigma^j \\ -\check{\sigma}^j & 0 \end{pmatrix}, \ \gamma^{jk} = \begin{pmatrix} \sigma^{jk} & 0 \\ 0 & \check{\sigma}^{jk} \end{pmatrix},$$
$$(\mathbf{1}_4, i\gamma_5, \gamma^j, i\gamma_5\gamma^j, \gamma^{jk})^\times = (\mathbf{1}_4, -i\gamma_5, \gamma^j, i\gamma_5\gamma^j, -\gamma^{jk}).$$

1.7.3 C, P, T from Spinor Reflections

The three spinor reflections (dual isomorphisms) – linear reflections ϵ, Euclidean $\mathbf{U}(2)$-conjugation $\star$. and $\mathbf{U}(2,2)$-anticonjugation $\times$ – implement, in product representations, the reflections of spacetime translations.

In the following table the reflection properties of spinors and their scalar and vector products are given, first for Pauli spinors

dual reflection	for spinors	for products	compatible
$\stackrel{\epsilon}{\leftrightarrow}$ linear	$\mathrm{u} \leftrightarrow \epsilon\mathrm{u}^\star$	$\mathrm{uu}^\star \leftrightarrow \mathrm{uu}^\star$ $\mathrm{u}\vec{\sigma}\mathrm{u}^\star \leftrightarrow -\mathrm{u}\vec{\sigma}\mathrm{u}^\star$	$\mathbf{SU}(2)$
$\stackrel{\star}{\leftrightarrow}$ antilinear	$\mathrm{u} \leftrightarrow \mathrm{u}^\star$	$\mathrm{uu}^\star \leftrightarrow \mathrm{uu}^\star$ $\mathrm{u}\vec{\sigma}\mathrm{u}^\star \leftrightarrow \mathrm{u}\vec{\sigma}\mathrm{u}^\star$	$\mathbf{SU}(2)$

and then for Weyl spinors

reflection	for spinors	for products	compatible
$\stackrel{\epsilon_{L,R}}{\longleftrightarrow}$ linear, dual	$\mathrm{l} \leftrightarrow \epsilon\mathrm{r}^\times$ $\mathrm{r} \leftrightarrow \epsilon\mathrm{l}^\times$	$\mathrm{lr}^\times \leftrightarrow \mathrm{lr}^\times$ $\mathrm{l}\sigma^j\mathrm{l}^\times \leftrightarrow \mathrm{r}\check{\sigma}^j\mathrm{r}^\times$	$\mathbf{SL}(\mathbb{C}^2)$
$\stackrel{\times}{\longleftrightarrow}$ antilinear, anti	$\mathrm{l} \leftrightarrow \mathrm{l}^\times$ $\mathrm{r} \leftrightarrow \mathrm{r}^\times$	$\mathrm{lr}^\times \leftrightarrow \mathrm{rl}^\times$ $\mathrm{l}\sigma^j\mathrm{l}^\times \leftrightarrow \mathrm{l}\sigma^j\mathrm{l}^\times$ $\mathrm{r}\check{\sigma}^j\mathrm{r}^\times \leftrightarrow \mathrm{r}\check{\sigma}^j\mathrm{r}^\times$	$\begin{pmatrix} \mathbf{SL}(\mathbb{C}^2) & 0 \\ 0 & \mathbf{SL}(\mathbb{C}^2) \end{pmatrix} \subset \mathbf{SU}(2,2)$
$\stackrel{\star}{\longleftrightarrow}$ antilinear, dual	$\mathrm{l} \leftrightarrow \mathrm{r}^\times$ $\mathrm{r} \leftrightarrow \mathrm{l}^\times$	$\mathrm{lr}^\times \leftrightarrow \mathrm{lr}^\times$ $\mathrm{l}(\mathbf{1}_2, \vec{\sigma})\mathrm{l}^\times \leftrightarrow \mathrm{r}(\mathbf{1}_2, \vec{\sigma})\mathrm{r}^\times$	$\mathbf{SU}(2)$
$\stackrel{\times\circ\star}{\longleftrightarrow}$ linear, antidual	$\mathrm{l} \leftrightarrow \mathrm{r}$ $\mathrm{l}^\times \leftrightarrow \mathrm{r}^\times$	$\mathrm{lr}^\times \leftrightarrow \mathrm{rl}^\times$ $\mathrm{l}(\mathbf{1}_2, \vec{\sigma})\mathrm{l}^\times \leftrightarrow \mathrm{r}(\mathbf{1}_2, \vec{\sigma})\mathrm{r}^\times$	$\mathbf{SU}(2)$

From the Weyl spinors one obtains the reflections for Dirac spinors $\psi \in V_{\text{doub}} = V_L \oplus V_R \cong \mathbb{C}^4$ - in the chiral representation:

$$\left[\begin{pmatrix} \epsilon_{\text{doub}} \\ \times \\ \star \end{pmatrix}\right] : \quad V_{\text{doub}} \longrightarrow V_{\text{doub}}^T, \quad \psi^K = (\mathrm{l}^A, \mathrm{r}^{\dot{A}}) \longmapsto \left[\begin{matrix} \begin{pmatrix} \epsilon^{AB} & 0 \\ 0 & \epsilon^{\dot{A}\dot{B}} \end{pmatrix} \\ \begin{pmatrix} 0 & \delta^{A\dot{B}} \\ \delta^{\dot{A}B} & 0 \end{pmatrix} \\ \begin{pmatrix} \delta^{AB} & 0 \\ 0 & \delta^{\dot{A}\dot{B}} \end{pmatrix} \end{matrix}\right] \begin{pmatrix} \mathrm{r}^\times_B \\ \mathrm{l}^\times_{\dot{B}} \end{pmatrix},$$

$$\times \circ \star : \quad V_{\text{doub}} \longrightarrow V_{\text{doub}}, \quad \psi^K = (\mathrm{l}^A, \mathrm{r}^{\dot{A}}) \longmapsto (\mathrm{l}^B, \mathrm{r}^{\dot{B}}) \begin{pmatrix} 0 & \delta^{\dot{A}}_B \\ \delta^A_{\dot{B}} & 0 \end{pmatrix},$$
$$\psi \longmapsto \psi\gamma^0.$$

The linear self-duality $\psi = (\mathrm{l}, \mathrm{r}) \longmapsto (\epsilon \mathrm{r}^\times, \epsilon \mathrm{l}^\times) = \epsilon_{\text{doub}}\psi^\times$ defines the Majorana reflected Dirac spinor. A Majorana spinor identifies both Dirac spinors $\psi = \epsilon_{\text{doub}}\psi^\times$. The reflections in the Dirac algebra are

γ	$\mathbf{1}_4$	$i\gamma_5$	γ^j	$i\gamma_5\gamma^j$	γ^{jk}
$\epsilon_{\text{doub}} \circ \gamma^T \circ \epsilon_{\text{doub}}^{-1} = \gamma^{\mathtt{C}}$	$\mathbf{1}_4$	$i\gamma_5$	γ^j	$-i\gamma_5\gamma^j$	$-\gamma^{jk}$
$\gamma^\times = \gamma^{\mathtt{T}\circ\mathtt{P}}$	$\mathbf{1}_4$	$-i\gamma_5$	γ^j	$i\gamma_5\gamma^j$	$-\gamma^{jk}$
$\gamma^\star = \gamma^{\mathtt{T}}$	$\mathbf{1}_4$	$i\gamma_5$	$\begin{pmatrix} \gamma^0 \\ -\gamma^a \end{pmatrix}$	$i\gamma_5 \begin{pmatrix} -\gamma^0 \\ \gamma^a \end{pmatrix}$	$\begin{pmatrix} \gamma^{0a} \\ -\gamma^{ab} \end{pmatrix}$
$\gamma^{\times\circ\star} = \gamma^{\mathtt{P}}$	$\mathbf{1}_4$	$-i\gamma_5$	$\begin{pmatrix} \gamma^0 \\ -\gamma^a \end{pmatrix}$	$i\gamma_5 \begin{pmatrix} -\gamma^0 \\ \gamma^a \end{pmatrix}$	$\begin{pmatrix} -\gamma^{0a} \\ \gamma^{ab} \end{pmatrix}$

The ϵ-reflection for angular momentum vectors $i\vec{\sigma}$ for Pauli spinors is embedded in the $\epsilon_{L,R}$-reflection for the boosts and angular momenta $(\sigma^{jk}, \hat{\sigma}^{jk})$ for Weyl spinors.

Altogether, the antilinear $\mathbf{U}(2)$-conjugation of Weyl spinors implements the *reflection of time translations* $\mathtt{T}$, the linear product of $\mathbf{U}(2)$-Euclidean and $\mathbf{U}(2,2)$-anticonjugation the *reflection of position translations* $\mathtt{P}$ and the volume-form-induced linear spinor "metric" the *particle-antiparticle reflection* $\mathtt{C}$. One has a group $\mathbb{I}(2) \times \mathbb{I}(2) \times \mathbb{I}(2)$, generated by three reflections, wherein the two conjugations implement a Klein reflection group

$\mathbf{1}$	$\times \cong \mathtt{T}\circ\mathtt{P}$	$\star \cong \mathtt{T}$	$\times\circ\star \cong \mathtt{P}$
$\times \cong \mathtt{T}\circ\mathtt{P}$	$\mathbf{1}$	$\mathtt{P}$	$\mathtt{T}$
$\star \cong \mathtt{T}$	$\mathtt{P}$	$\mathbf{1}$	$\mathtt{T}\circ\mathtt{P}$
$\times\circ\star \cong \mathtt{P}$	$\mathtt{T}$	$\mathtt{T}\circ\mathtt{P}$	$\mathbf{1}$

$\mathbb{I}(2) \times \mathbb{I}(2)$

and

$\mathbf{1}$	$\epsilon_{L,R} \cong \mathtt{C}$
$\epsilon_{L,R} \cong \mathtt{C}$	$\mathbf{1}$

,

$\mathbb{I}(2)$

$$[\mathtt{P}, \mathtt{T}] = 0, \quad [\mathtt{C}, \mathtt{P}] = 0, \quad [\mathtt{C}, \mathtt{T}] = 0.$$

The reflections have the invariances

$$[\times \text{ and } \mathtt{C}, \mathbf{SL}(\mathbb{C}^2)] = \{0\}, \quad \begin{aligned} [\mathtt{T} \text{ and } \mathtt{P}, \mathbf{SL}(\mathbb{C}^2)] &\neq \{0\}, \\ [\mathtt{T} \text{ and } \mathtt{P}, \mathbf{SU}(2)] &= \{0\}. \end{aligned}$$

The relation between angular momentum invariant and position reflection eigenvalue in (pseudo)scalars and (axial)vectors is generalized to define polar and axial rotation group representations

$$\mathbf{O}(3)\text{-representation with } L = 0, 1, 2, \cdots : \ \mathrm{p} = \left\{ \begin{array}{ll} (-1)^L, & \text{polar}, \\ (-1)^{1+L}, & \text{axial}. \end{array} \right.$$

1.8 Dirac Equation

The Dirac equation describes the energy-momentum parametrization for the spinor representations of the boost hyperboloid $\mathbf{SL}(\mathbb{C}^2)/\mathbf{SU}(2)$.

With the spacetime translations $\mathbb{R}^4$ related to the tangent space of the boost classes $\mathbf{D}(1) \times \mathbf{SO}_0(1,3)/\mathbf{SO}(3)$, the momenta can be chosen as Lie parameters for the boosts. The connection of the three noncompact Lie parameters $\vec{\beta}$ of $\mathbf{SL}(\mathbb{C}^2)/\mathbf{SU}(2)$ with the characteristic momentum (velocity) of a boost $\mathbf{SO}_0(1,3)/\mathbf{SO}(3)$,

$$\tfrac{q^2}{m^2} = \tfrac{q_0^2 - \vec{q}^2}{m^2} = 1, \quad 2\vec{\beta} = \vec{\psi} = \tfrac{\vec{q}}{|\vec{q}|} \operatorname{artanh} \tfrac{|\vec{q}|}{q_0}, \quad \tfrac{\vec{q}}{q_0} = \tfrac{\vec{v}}{c}, \quad q_0 = \sqrt{m^2 + \vec{q}^2},$$

allows the parametrization of the Weyl representations ($m > 0$)

$$\left. \begin{array}{ll} s(\tfrac{q}{m}) & = e^{-\vec{\beta}} \\ \hat{s}(\tfrac{q}{m}) = s^{-1\star}(\tfrac{q}{m}) & = e^{\vec{\beta}} \end{array} \right\}, \ e^{\mp\vec{\beta}} = \mathbf{1}_2 \cosh\beta \mp \tfrac{\vec{\beta}}{\beta} \sinh\beta = \sqrt{\tfrac{q_0+m}{2m}} \left[\mathbf{1}_2 \mp \tfrac{\vec{q}}{q_0+m} \right],$$

involving the hyperbolic functions with $\tanh 2\beta = \frac{|\vec{q}|}{q_0}$:

$$\beta^2 = \vec{\beta}^2, \quad \cosh\beta = \sqrt{\tfrac{q_0+m}{2m}}, \quad \sinh\beta = \sqrt{\tfrac{q_0-m}{2m}}, \quad \tanh\beta = \tfrac{|\vec{q}|}{q_0+m}.$$

The boost representatives are determined up to $\mathbf{SU}(2)$, they are chosen to be Hermitian $s(\frac{q}{m}) = s^\star(\frac{q}{m}) \in \mathbf{SL}(\mathbb{C}^2)/\mathbf{SU}(2)$, i.e., the absolute values in the polar decomposition of $\mathbf{SL}(\mathbb{C}^2)$.

The combination of the Minkowski representation by the two Weyl spinor representations $[1|1] \cong [1|0] \otimes [0|1]$,

$$s(\tfrac{q}{m})\sigma^k \hat{s}^{-1}(\tfrac{q}{m}) = \Lambda(\tfrac{q}{m})^k_j \sigma^j, \quad \hat{s}(\tfrac{q}{m})\check{\sigma}^k s^{-1}(\tfrac{q}{m}) = \Lambda(\tfrac{q}{m})^k_j \check{\sigma}^j,$$

gives the explicit momentum parametrization of the Lorentz boosts

$$\exp\beta \begin{pmatrix} 0 & \frac{q_a}{|\vec{q}|} \\ \frac{q_b}{|\vec{q}|} & 0 \end{pmatrix} = \frac{1}{m} \begin{pmatrix} q_0 & q_a \\ q_b & m\delta_{ab} + \frac{q_a q_b}{q_0+m} \end{pmatrix} = \Lambda(\tfrac{q}{m}) \in \mathbf{SO}_0(1,3)/\mathbf{SO}(3).$$

From the relation for the time component

$$\left. \begin{array}{l} \Lambda(\frac{q}{m})^j_0 = \frac{q^j}{m}, \\ \sigma_0 = \mathbf{1}_2 = \check{\sigma}_0, \end{array} \right\} \Rightarrow \left\{ \begin{array}{ll} s(\frac{q}{m})\hat{s}^{-1}(\frac{q}{m}) & = \frac{\sigma_j q^j}{m} = \frac{q}{m}, \\ \hat{s}(\frac{q}{m})s^{-1}(\frac{q}{m}) & = \frac{\check{\sigma}_j q^j}{m} = \frac{\check{q}}{m}, \end{array} \right\} \iff \left\{ \begin{array}{ll} \frac{q}{m}\hat{s}(\frac{q}{m}) & = s(\frac{q}{m}), \\ \frac{\check{q}}{m}s(\frac{q}{m}) & = \hat{s}(\frac{q}{m}), \end{array} \right.$$

one obtains the *Dirac equation*, which expresses the momentum dependence of the Weyl representations of the Lorentz boosts

$$\begin{aligned} \frac{q^j}{m}\begin{pmatrix} 0 & \sigma_j \\ \check{\sigma}_j & 0 \end{pmatrix}\begin{pmatrix} s(\frac{q}{m}) & 0 \\ 0 & \hat{s}(\frac{q}{m}) \end{pmatrix} &= \begin{pmatrix} s(\frac{q}{m}) & 0 \\ 0 & \hat{s}(\frac{q}{m}) \end{pmatrix}, \\ (\frac{\gamma_j q^j}{m} - \mathbf{1}_4) s_{\text{doub}}(\tfrac{q}{m}) &= 0. \end{aligned}$$

With a Fourier transformation, it can be written in the more familiar conventional form

$$\begin{aligned} (i\gamma_j\partial^j + m)\mathbf{\Psi}(x) = 0, \quad \mathbf{\Psi}(x) &= \int \frac{d^3q}{2q_0} e^{iqx} s_{\text{doub}}(\tfrac{q}{m})\Psi(q)\big|_{q_0=\sqrt{\vec{q}^2+m^2}} \\ &= \int d^4q\ \vartheta(q_0)\delta(q^2-m^2)e^{iqx} s_{\text{doub}}(\tfrac{q}{m})\Psi(q). \end{aligned}$$

The (four columns of the) Dirac representation $s_{\text{doub}}(\frac{q}{m})$ of the Lorentz boosts are the *solutions of the Dirac equation*, given in chiral and time-diagonal Dirac matrices

$$\begin{aligned} s_{\text{doub}}(\tfrac{q}{m}) &= s^\star_{\text{doub}}(\tfrac{q}{m}) &&= \sqrt{\frac{q_0+m}{2m}}\begin{pmatrix} \mathbf{1}_2 - \frac{\vec{q}}{q_0+m} & 0 \\ 0 & \mathbf{1}_2 + \frac{\vec{q}}{q_0+m} \end{pmatrix} \in \mathbf{SU}(2,2), \\ s_U(\tfrac{q}{m}) &= s^\star_U(\tfrac{q}{m}) &&= w_U \circ s_{\text{doub}}(\tfrac{q}{m}) \circ w_U^{-1} \\ & &&= \sqrt{\frac{q_0+m}{2m}}\begin{pmatrix} \mathbf{1}_2 & \frac{\vec{q}}{q_0+m} \\ \frac{\vec{q}}{q_0+m} & \mathbf{1}_2 \end{pmatrix} \in \mathbf{SU}(2,2). \end{aligned}$$

The $\mathbf{U}(2,2)$-unitarity, in a chiral and time-diagonal basis, of the Dirac equation solutions expresses their indefinite orthonormality and their completeness:

$$\begin{aligned} s_{\text{doub}}(\tfrac{q}{m}) s^\times_{\text{doub}}(\tfrac{q}{m}) &= \mathbf{1}_4 \text{ with } & s^\times_{\text{doub}}(\tfrac{q}{m}) &= \begin{pmatrix} 0 & \mathbf{1}_2 \\ \mathbf{1}_2 & 0 \end{pmatrix} s^\star_{\text{doub}}(\tfrac{q}{m}) \begin{pmatrix} 0 & \mathbf{1}_2 \\ \mathbf{1}_2 & 0 \end{pmatrix}, \\ s_U(\tfrac{q}{m}) s^\times_U(\tfrac{q}{m}) &= \mathbf{1}_4 \text{ with } & s^\times_U(\tfrac{q}{m}) &= \begin{pmatrix} \mathbf{1}_2 & 0 \\ 0 & -\mathbf{1}_2 \end{pmatrix} s^\star_U(\tfrac{q}{m}) \begin{pmatrix} \mathbf{1}_2 & 0 \\ 0 & -\mathbf{1}_2 \end{pmatrix}. \end{aligned}$$

A basis $\{\mathrm{u}^A, \mathrm{u}^{\times A}\}_{A=1,2}$ of the representation space $V_{\text{doub}} \cong \mathbb{C}^4$ with time-diagonal Dirac matrices is related by w_U^{-1} to a basis $\{\mathrm{l}^A, \mathrm{r}^A\}_{A=1,2}$ with chiral Dirac matrices

$$\begin{pmatrix} \mathrm{l}^A \\ \mathrm{r}^A \end{pmatrix} = \frac{1}{\sqrt{2}}\begin{pmatrix} \mathbf{1}_2 & -\mathbf{1}_2 \\ \mathbf{1}_2 & \mathbf{1}_2 \end{pmatrix}\begin{pmatrix} \mathrm{u}^A \\ \mathrm{u}^{\times A} \end{pmatrix} = \begin{pmatrix} \frac{\mathrm{u}^A - \mathrm{u}^{\times A}}{\sqrt{2}} \\ \frac{\mathrm{u}^A + \mathrm{u}^{\times A}}{\sqrt{2}} \end{pmatrix}.$$

These structures are relevant for relativistic Dirac particle quantum fields (chapter "Massive Particle Quantum Fields").

1.9 Polynomials with Lorentz Group Action

The polynomials in the vectors of any finite-dimensional representation space carry the Lorentz Lie algebra action via derivatives (chapter "Spin, Rotations, and Position"). The fundamental Weyl and Minkowski representations are of special interest.

1.9.1 Weyl Spinor Polynomials

All finite-dimensional irreducible representations $\mathcal{D}^{[2L|2R]}$ for the Lie algebra $A_{(1,1)}$ can be realized by derivations of *Weyl spinor polynomials* in two complex indeterminates with conjugation from a basis $\{\mathrm{l}^A, \mathrm{l}^\times_A\}_{A=1,2}$ of the spinor spaces $V, \overline{V} \cong \mathbb{C}^2$:

$$\mathbb{C}[\mathrm{l}, \mathrm{l}^\times] \cong \bigvee \mathbb{C}^2 \otimes (\bigvee \mathbb{C}^2)^\times.$$

The dual-product-induced derivations

$$\begin{array}{lll} \langle \mathrm{r}^\times_B, \mathrm{l}^A \rangle & = \delta^A_B = \partial_B \mathrm{l}^A & \text{with } \partial_B & = \frac{\partial}{\partial \mathrm{l}^B}, \\ \langle \mathrm{l}^\times_B, \mathrm{r}^A \rangle & = \delta^A_B = \partial^{\times A} \mathrm{l}^\times_B & \text{with } \partial^{\times A} & = \frac{\partial}{\partial \mathrm{l}^\times_A}, \end{array}$$

give the Lie algebra representing derivations

$$A_{(1,1)} = A^c_1 \oplus iA^c_1 \longrightarrow \operatorname{der} \mathbb{C}[\mathrm{l}, \mathrm{l}^\times], \quad \begin{cases} \vec{l} \longmapsto \vec{l}_{\text{der}} & = \frac{i}{2} \vec{\sigma}^B_A [\mathrm{l}^A \partial_B + \mathrm{l}^\times_B \partial^{\times A}], \\ \vec{b} \longmapsto \vec{b}_{\text{der}} & = \frac{1}{2} \vec{\sigma}^B_A [-\mathrm{l}^A \partial_B + \mathrm{l}^\times_B \partial^{\times A}], \end{cases}$$

with a represented Cartan subalgebra $\log \mathbf{SO}(2) \oplus \log \mathbf{SO}(1,1)$:

$$\begin{array}{ll} h_{\text{der}} = (\sigma_0)^B_A [\mathrm{l}^A \partial_B + \mathrm{l}^\times_B \partial^{\times A}], & h \bullet (\mathrm{l}^1, \mathrm{l}^2, \mathrm{l}^\times_1, \mathrm{l}^\times_2) = (+\mathrm{l}^1, -\mathrm{l}^2, -\mathrm{l}^\times_1, +\mathrm{l}^\times_2), \\ d_{\text{der}} = (\sigma_0)^B_A [\mathrm{l}^A \partial_B - \mathrm{l}^\times_B \partial^{\times A}], & d \bullet (\mathrm{l}^1, \mathrm{l}^2, \mathrm{l}^\times_1, \mathrm{l}^\times_2) = (+\mathrm{l}^1, -\mathrm{l}^2, +\mathrm{l}^\times_1, -\mathrm{l}^\times_2). \end{array}$$

The totally symmetric tensor powers of the Weyl spinors as irreducible representation spaces are isomorphic to the polynomials, *homogeneous* of degree $[2L|2R]$ in l and $\mathrm{l}^\times$:

$$\begin{array}{l} \text{basis of } \mathbb{C}[\mathrm{l}, \mathrm{l}^\times]^{[2L|2R]} \cong \bigvee^{2L} \mathbb{C}^2 \otimes \bigvee^{2R} \mathbb{C}^2 \cong \mathbb{C}^{(1+2L)(1+2R)}, \\ \{(\mathrm{l}^1)^{L+l} (\mathrm{l}^2)^{L-l} (\mathrm{l}^\times_1)^{R+r} (\mathrm{l}^\times_2)^{R-r} \,\big|\, l = -L, \dots, L, \quad r = -R, \dots, R\}. \end{array}$$

The monomials in the given basis are Cartan subalgebra eigenvectors with weights $(2j; 2d) = (2(l + r); 2(l - r))$.

1.9.2 Harmonic Spacetime Translation Polynomials

All finite-dimensional irreducible $\mathbf{SO}_0(1,3)$-representations of indefinite unitary type $\mathcal{D}^{[2J|2J]}$ for the Lie algebra $A_{(1,1)}$ can be realized by derivations of complex *Minkowski spacetime polynomials*, i.e., by derivations of polynomials in four indeterminates $\{x^k\}^3_{k=0}$ from a basis of $V \otimes \overline{V} = \mathbb{C} \otimes \mathbb{M} \cong \mathbb{C}^4$:

$$\mathbb{C}[x] \cong \bigvee \mathbb{C}^4 \cong \mathbb{C} \otimes \mathbb{R}[x].$$

They contain the real polynomials of the spacetime translations. A Cartan basis of the complexified Minkowski space $\mathbb{C} \otimes \mathbb{M}$ is given by

$$\begin{pmatrix} x_0 + x_3 & x_1 - ix_2 \\ x_1 + ix_2 & x_0 - x_3 \end{pmatrix} = \begin{pmatrix} \xi_+ & x_- \\ x_+ & \xi_- \end{pmatrix} = \begin{pmatrix} x_0 + r\cos\theta & re^{-i\varphi}\sin\theta \\ re^{i\varphi}\sin\theta & x_0 - r\cos\theta \end{pmatrix},$$

with Lebesque measure of the spacetime translations

$$\int d^4x = \int dx_0 \int d^3x, \quad \int d^3x = \int_0^\infty r^2 dr \int d^2\omega, \quad \int d^2\omega = \int_0^{2\pi} d\varphi \int_{-1}^1 d\cos\theta.$$

Since the Lorentz-invariant degree-2 polynomial, i.e., the translation square

$$\det x = x^2 = g_{jk}x^j x^k = \xi_+\xi_- - x_+x_- = (x^0)^2 - \vec{x}^2,$$

vanishes for lightlike translations, the position directions $\frac{\vec{x}}{\sqrt{\vec{x}^2}} \in \Omega^2$ for the $\mathbf{SO}(3)$-spherical harmonics have no analogue in the noncompact group $\mathbf{SO}_0(1,3)$.

Polar-hyperbolic coordinates can be used only for nonlightlike translations $|x^2| > 0$, in the general case with $s = 1, 2, \dots$ position dimensions

$$\mathbf{SO}_0(1,s) \vec{\times} \mathbb{R}^{1+s} : \begin{cases} x_0 + \vec{x} = \begin{cases} \tau[\epsilon(x_0)\cosh\psi + \frac{\vec{x}}{r}\sinh\psi] \\ \text{timelike } x^2 = \tau^2 > 0, \\ \\ \rho(\sinh\psi + \frac{\vec{x}}{r}\cosh\psi) \\ \text{spacelike } x^2 = -\rho^2 < 0, \end{cases} \\ \\ \int \vartheta(\pm x_0)\vartheta(x^2) d^{1+s}x = \pm\int_0^{\pm\infty} \tau^s d\tau \int_0^\infty \sinh^{s-1}\psi d\psi \int d^{s-1}\omega, \\ \int \vartheta(x^2) d^{1+s}x = \int \tau^s d\tau \int_0^\infty \sinh^{s-1}\psi d\psi \int d^{s-1}\omega, \\ \int \vartheta(-x^2) d^{1+s}x = \int_0^\infty \rho^s d\rho \int_{-\infty}^\infty \cosh^{s-1}\psi d\psi \int d^{s-1}\omega. \end{cases}$$

The integration over the future and past lightcone can be parametrized as follows:

$$\text{lightlike } x^2 = 0 : \quad x_0 + \vec{x} = \vartheta(\pm x_0)|\vec{x}| + \vec{x}, \\ \int \vartheta(\pm x_0)\delta(x^2) d^{1+s}x = \int \frac{d^s x}{2|\vec{x}|} = \frac{1}{2}\int_0^\infty r^{s-2} dr \int d^{s-1}\omega.$$

In the representation of the Lorentz Lie algebra by derivations, using the duality $\frac{\partial}{\partial x^i} x^j = \delta_i^j$,

$$A_{(1,1)} \cong \log \mathbf{SO}_0(1,3) \longrightarrow \operatorname{der} \mathbb{C}[x], \; l^{jk} \longmapsto g^{ji}x_i \frac{\partial}{\partial x^k} - g^{ki}x_i \frac{\partial}{\partial x^j},$$

a Cartan subalgebra $\log \mathbf{SO}(2) \oplus \log \mathbf{SO}_0(1,1)$ is spanned by

$$\begin{aligned} h_{\text{der}} &= x_+ \frac{\partial}{\partial x_+} - x_- \frac{\partial}{\partial x_-} && \Rightarrow h \bullet (\xi_\pm, x_\pm) = (0, \pm 1), \\ d_{\text{der}} &= \xi_+ \frac{\partial}{\partial \xi_+} - \xi_- \frac{\partial}{\partial \xi_-} && \Rightarrow d \bullet (\xi_\pm, x_\pm) = (\pm 1, 0). \end{aligned}$$

The spacetime polynomials with the Lorentz action are decomposable into homogeneous polynomials of degree $N = 0, 1, \dots$:

$$\bigvee^N \mathcal{D}^{[1|1]} : A_{(1,1)} \longrightarrow \mathbf{AL}(\mathbb{C}[x]^N), \quad \mathbb{C}[x]^N \cong \bigvee^N \mathbb{C}^4 \cong \mathbb{C}^{\binom{3+N}{3}}.$$

A Cartan subalgebra eigenvector basis of the degree-N polynomials with weights $(2j; 2d)$ is given by

$$\{x_+^{J+j} x_-^{J-j} \xi_+^{D+d} \xi_-^{D-d} \mid N = 2(J+D), \;\; 2J, 2D \in \mathbb{N}_0, \;\; |j| \le J, \;\; |d| \le D\}.$$

Only the complex polynomials have a basis with eigenvectors of a Cartan subalgebra.

Homogeneity of spacetime polynomials does not entail irreducibility: The degree-N polynomials for $x^2 \neq 0$ are decomposable into irreducible representation spaces:

$$\bigvee^N \mathcal{D}^{[1|1]} = \begin{cases} \bigoplus_{2J=0,2,\dots,N} \mathcal{D}^{[2J|2J]}, & \binom{3+N}{3} = \sum_{2J=0,2,\dots,N} (1+2J)^2, \quad N = 0,2,\dots, \\ \bigoplus_{2J=1,3,\dots,N} \mathcal{D}^{[2J|2J]}, & \binom{3+N}{3} = \sum_{2J=1,3,\dots,N} (1+2J)^2, \quad N = 1,3,\dots, \end{cases}$$

with $x^{2n} = (x^2)^n$ powers as factors. They have even or odd integers $2J$ for even and odd degree N respectively:

$$\begin{aligned} \mathbb{C}[x]^0 &= \mathbb{C}, \quad \mathbb{C}[x]^1 = \mathbb{C} \otimes \mathbb{M} \cong \mathbb{C}^4, \\ \mathbb{C}[x]^N &= x^2\mathbb{C}[x]^{N-2} \oplus \mathbb{C}^{(1+N)^2} \cong \mathbb{C}^{\binom{3+N}{3}}, \quad N = 2,3,\dots, \\ &\cong \begin{cases} x^N\mathbb{C} \oplus x^{N-2}\mathbb{C}^9 \oplus \dots \oplus \mathbb{C}^{(1+N)^2}, & N = 0,2,\dots, \\ x^{N-1}\mathbb{C}^4 \oplus x^{N-3}\mathbb{C}^{16} \oplus \dots \oplus \mathbb{C}^{(1+N)^2}, & N = 1,3,\dots. \end{cases} \end{aligned}$$

Spacetime translation polynomial bases for the irreducible representation spaces are given with the Lorentz "metric" as follows:

$$\begin{aligned} &N = 0 : \{1\}, \\ &N = 1 : \{x^j \mid j = 0,1,2,3\}, \\ &N = 2 : \{x^j x^k\} \cong \{x^2\} \oplus \{x^j x^k - \tfrac{x^2}{4} g^{jk}\}, \\ &N = 3 : \{x^j x^k x^l\} \cong \{x^2 x^j\} \oplus \{x^j x^k x^l - \tfrac{x^2}{4}(g^{jk}x^l + g^{jl}x^k + g^{kl}x^j)\}, \\ &\dots \end{aligned}$$

The highest-order polynomials are the *harmonic* $\mathbf{SO}_0(1,3)$*-polynomials*

$$\partial^2 P^N(x) = 0 \text{ for } \begin{cases} P^0(x) = 1, \\ P^1(x) \in \{x^j\}, \\ P^2(x) \in \{x^j x^k - \tfrac{x^2}{4} g^{jk}\}, \\ \dots \end{cases}$$

1.10 Summary

By compact-noncompact doubling (canonical complexification with the anticonjugation $\mathbf{U}(2,2)$) the representations of the spin Lie algebra $A_1^c \cong \mathbb{R}^3$ are embedded into representations of the simple real rank-2 Lie algebra $A_{(1,1)} \cong A_1^c \oplus iA_1^c$. Its real 6-dimensional simply connected Lie group $\mathbf{SL}(\mathbb{C}^2)$, embedding the spin group $\mathbf{SU}(2)$, covers twofold the orthochronous Lorentz group $\mathbf{SO}_0(1,3) \cong \mathbf{SL}(\mathbb{C}^2)/\mathbb{I}(2) \cong \mathbf{SO}(\mathbb{C}^3)$.

All finite-dimensional representations of $A_{(1,1)}$ and $\mathbf{SL}(\mathbb{C}^2)$ arise by doubling from spin representations and are equivalent to products of right- and left-handed Weyl spinors that are the symplectically self-dual fundamental representations, antidual to each other and doubling the Pauli spin representation. The defining vector representation of $A_{(1,1)}$ leads to the Cartan representation of Minkowski spacetime $\mathbb{M} \cong \mathbb{R}^4$ as 2×2 Weyl spinor transformations with the orthogonal Lorentz "metric" as product of the symplectic spinor "metrics." The dilation Poincaré group $[\mathbf{D}(1) \times \mathbf{SO}_0(1,3)] \vec{\times} \mathbb{R}^4$ is the conjugate adjoint affine group of the full group $\mathbf{GL}(\mathbb{C}^2)$. Minkowski spacetime is the real noncompact part $\log \mathbf{GL}(\mathbb{C}^2)_+ \cong \mathbb{R}^4$ of the full Lie algebra $\log \mathbf{GL}(\mathbb{C}^2) \cong \mathbb{R}^8$. The three reflections for Weyl spinors – the linear spinor "metric" ϵ and the antilinear $\mathbf{U}(2,2)$-anticonjugation $\times$, both $\mathbf{SL}(\mathbb{C}^2)$-compatible, together with the only $\mathbf{SU}(2)$-compatible antilinear Euclidean $\mathbf{U}(2)$-conjugation $\star$ – define the particle-antiparticle reflection $\mathtt{C} \cong \epsilon$ (linear) and the reflections for time and space translations, $\mathtt{T} \cong \star$ (antilinear) and $\mathtt{P} \cong \times \circ \star$ (linear).

	$\mathbf{SL}(\mathbb{C}^2) = \exp A_{(1,1)}$	$\mathbf{SO}_0(1,3) \cong \mathbf{SL}(\mathbb{C}^2)/\mathbb{I}(2)$
weight module	$\mathbf{weights}_{\text{fin}}\mathbf{SL}(\mathbb{C}^2) = \mathbb{Z} \times \mathbb{Z}$	$\mathbf{weights}_{\text{fin}}\mathbf{SO}_0(1,3) \ni [2L\|2R]$ for $l + r \in \mathbb{Z}$
representation cone	$\mathbf{irrep}_{\text{fin}}\mathbf{SL}(\mathbb{C}^2) \cong \mathbb{N}_0 \times \mathbb{N}_0$	$\mathbf{irrep}_{\text{fin}}\mathbf{SO}_0(1,3) \ni [2L\|2R]$ for $L + R = 0, 1, \ldots$
representations	$D^{[2L\|2R]}(\lambda) \in \begin{cases} \mathbf{Sp}(\mathbb{C}^{(1+2L)(1+2R)}) \\ J = \frac{1}{2}, \frac{3}{2}, \ldots \\ \mathbf{SO}(\mathbb{C}^{(1+2L)(1+2R)}) \\ J = 0, 1, \ldots \end{cases}$, $J = L + R$, $D = L - R$; $D^{[2L\|2R]}(\lambda) \cong \bigvee^{2L} \lambda \otimes \bigvee^{2R} \hat{\lambda}$	either $[2J\|2J]$ with $2J = 0, 1, \ldots$ or $[2L\|2R] \oplus [2R\|2L]$ with $L \neq R,\ L + R = 1, 2, \ldots$
Lie algebra	$A_{(1,1)} = \{i\vec{\alpha} - \vec{\beta}\} \cong \mathbb{R}^6$	$\log \mathbf{SO}_0(1,3) \ni b + l$; $b + l = 2\begin{pmatrix} 0 & \beta_a \\ \beta_b & -\epsilon^{abc}\alpha_a \end{pmatrix}$
fundamental representations	left Weyl $\lambda = e^{i\vec{\alpha} - \vec{\beta}}$, $[1\|0]$; right Weyl $\hat{\lambda} = \lambda^{-1\star}$, $[0\|1]$; Dirac spinors $\lambda \oplus \hat{\lambda}$, $[1\|0] \oplus [0\|1]$	$\Lambda \cong \lambda \otimes \lambda^\star$, $[1\|1]$ Minkowski; $\Lambda \wedge \Lambda$, $[2\|0] \oplus [0\|2]$ adjoint
fundamental "metric"	$\epsilon \cong \begin{pmatrix} 0 & 1 \\ -1 & 0 \end{pmatrix}$ spinor "metric"	$\epsilon \otimes \epsilon = \eta \cong \begin{pmatrix} 1 & 0 \\ 0 & -\mathbf{1}_3 \end{pmatrix}$ Lorentz "metric"; $\eta \wedge \eta = \kappa \cong \begin{pmatrix} \mathbf{1}_3 & 0 \\ 0 & -\mathbf{1}_3 \end{pmatrix}$ Killing form

finite-dimensional $\mathbf{SL}(\mathbb{C}^2)$-representations and weights

$$\begin{array}{ccc}
\boxed{\begin{array}{c} \mathrm{l}^A \in V \\ i\vec{\alpha} - \vec{\beta} \end{array}} & \stackrel{\times = \mathtt{P}\circ\mathtt{T}}{\longleftrightarrow} & \boxed{\begin{array}{c} \mathrm{l}^{\times}_{\dot{A}} \in \overline{V} \\ -(i\vec{\alpha} + \vec{\beta})^T \end{array}} \\
\epsilon = \mathtt{C} \updownarrow \star = \mathtt{T} & & \star = \mathtt{T} \updownarrow \epsilon = \mathtt{C} \\
\boxed{\begin{array}{c} \mathrm{r}^{\times}_A \in V^T \\ -(i\vec{\alpha} - \vec{\beta})^T \end{array}} & \stackrel{\times = \mathtt{P}\circ\mathtt{T}}{\longleftrightarrow} & \boxed{\begin{array}{c} \mathrm{r}^{\dot{A}} \in \overline{V}^T \\ i\vec{\alpha} + \vec{\beta} \end{array}}
\end{array}$$

quartet of Weyl spinors
with $\log \mathbf{SL}(\mathbb{C}^2)$-representations

MATHEMATICAL TOOLS

1.11 Doubled Lie Algebra

The *doubling (canonical complexification)* $L_{\rm doub} = L_{\mathbb{R}} \oplus iL_{\mathbb{R}} = \begin{pmatrix} L_{\mathbb{R}} \\ L_{\mathbb{R}} \end{pmatrix} \in \underline{\mathbf{lag}}_{\mathbb{R}}$ *of a real Lie algebra* $L_{\mathbb{R}}$ (chapter "Simple Lie Operations") has as Lie bracket, suggested by $i^2 = -1$,

$$\begin{aligned}
&[l_1 + ik_1, l_2 + ik_2] = \Big([l_1, l_2] - [k_1, k_2]\Big) + i\Big([k_1, l_2] + [l_1, k_2]\Big),\\
&\left[\begin{pmatrix} l_1 \\ k_1 \end{pmatrix}, \begin{pmatrix} l_2 \\ k_2 \end{pmatrix}\right] = \begin{pmatrix} [l_1, l_2] - [k_1, k_2] \\ [l_2, k_1] + [l_1, k_2] \end{pmatrix},\\
&[L_{\mathbb{R}}, L_{\mathbb{R}}] \subseteq L_{\mathbb{R}}, \quad [iL_{\mathbb{R}}, iL_{\mathbb{R}}] \subseteq L_{\mathbb{R}}, \quad [L_{\mathbb{R}}, iL_{\mathbb{R}}] \subseteq iL_{\mathbb{R}}.
\end{aligned}$$

The *linear* reflection is nontrivial only for $L_{\mathbb{R}}$:

$$* : L_{\rm doub} \longrightarrow L_{\rm doub}, \quad (l + ik)^* = -l + ik, \quad [l, k]^* = [k^*, l^*],$$

$L_{\rm doub}$ is a real twin vector space with exchange vector space isomorphism

$$L_{\mathbb{R}} \cong iL_{\mathbb{R}}, \quad l \cong il, \quad \text{bases:}\{l^a\}_a \cong \{il^a\}_a \begin{cases} [l^a, l^b] = \epsilon^{ab}_c l^c, \\ [il^a, il^b] = -\epsilon^{ab}_c l^c, \\ [il^a, l^b] = \epsilon^{ab}_c il^c. \end{cases}$$

The Killing form κ of $L_{\mathbb{R}}$ can be doubled in two ways, even or odd with respect to the linear reflection

$$\begin{aligned}
\kappa_\pm(\ ,\ ) : L_{\rm doub} \times L_{\rm doub} &\longrightarrow \mathbb{R},\\
\kappa_+ \cong \begin{pmatrix} 2\kappa & 0 \\ 0 & -2\kappa \end{pmatrix}, &\begin{cases} \kappa_+(l, k) = 2\kappa(l, k) = -\kappa_+(il, ik), \\ \kappa_+(l, ik) = 0, \\ \kappa_+(l_1 + ik_1, l_2 + ik_2) = \kappa_+(-l_1 + ik_1, -l_2 + ik_2), \end{cases}\\
\kappa_- \cong \begin{pmatrix} 0 & \kappa \\ \kappa & 0 \end{pmatrix}, &\begin{cases} \kappa_-(l, k) = 0 = \kappa_-(il, ik), \\ \kappa_-(l, ik) = \kappa_-(il, k) = \kappa(l, k), \\ \kappa_-(l_1 + ik_1, l_2 + ik_2) = -\kappa_-(-l_1 + ik_1, -l_2 + ik_2). \end{cases}
\end{aligned}$$

For nondegenerate κ there are the invariant tensors in the enveloping algebra:

$$\begin{aligned}
\text{in } \mathbf{E}(L_{\mathbb{R}}) :&\quad \kappa_{ab} l^a \otimes l^b,\\
\text{in } \mathbf{E}(L_{\rm doub}) :&\quad \tfrac{\kappa_{ab}}{2}(l^a \otimes l^b - il^a \otimes il^b), \quad \kappa_{ab} l^a \otimes il^b.
\end{aligned}$$

Each complex representation of the real Lie algebra $\mathcal{D} : L_{\mathbb{R}} \longrightarrow \mathbf{AL}(V)$ can be extended to two representations of its doubling $L_{\rm doub}$:

$$\mathcal{D}_\pm : L_{\rm doub} \longrightarrow \mathbf{AL}(V), \quad \mathcal{D}_\pm(l + ik) = \mathcal{D}(l) \pm i\mathcal{D}(k),$$

$$\begin{array}{ccc} L_{\rm doub} & \stackrel{\mathcal{D}_+}{\longrightarrow} & \mathbf{AL}(V) \\ * \downarrow & & \downarrow \times \\ L_{\rm doub} & \underset{\mathcal{D}_-}{\longrightarrow} & \mathbf{AL}(\overline{V}^T) \end{array}, \quad \begin{array}{ccc} l + ik & \longmapsto & \mathcal{D}(l) + i\mathcal{D}(k) \\ \updownarrow & & \updownarrow \\ -l + ik & \longmapsto & \mathcal{D}(l)^\times - i\mathcal{D}(k)^\times \end{array}.$$

1.12 Conjugate-Adjoint Representations

A group G with reflection (conjugation) $*$ defines the *doubled group*

$$G_{\rm doub} = G \times \hat{G} = \{(g_1, \hat{g}_2) \mid g_{1,2} \in G\} \in \underline{\mathbf{grp}}, \quad \hat{g} = g^{-1*}.$$

The restriction of inner $G_{\rm doub}$-automorphisms leads to the group realization by *conjugate bijections*,

$$\begin{array}{l} \mathrm{Int}_* : G \longrightarrow \mathbf{G}(G), \quad \mathrm{Int}_* g : G \longrightarrow G, \quad \mathrm{Int}_* g(a) = ga\hat{g}^{-1} = gag^*, \\ \mathrm{Int}_* g_1 \circ \mathrm{Int}_* g_2 = \mathrm{Int}_* g_1 g_2, \quad (\mathrm{Int}_* g)^{-1} = \mathrm{Int}_* \hat{g}. \end{array}$$

For a unitary element $u \in U(G,*) = \{u \in G \mid \hat{u} = u\} \in \underline{\mathbf{grp}}$ the conjugate bijection is an inner automorphism, $\mathrm{Int}_* u = \mathrm{Int}\, u$. The kernel, a normal subgroup, contains the unitary subgroup of the center

$$\begin{array}{l} U(\mathrm{centr}\, G, *) \subseteq \mathrm{kern}\ \mathrm{Int}_* = \{h \mid hgh^* = g \text{ for all } g \in G\}, \\ \mathrm{Int}_* G = G/\,\mathrm{kern}\ \mathrm{Int}_*. \end{array}$$

The symmetric domain

$$D(G,*) = \{d \in G \mid d^* = d\} \in \underline{\mathbf{set}}$$

is stable and defines the *conjugate adjoint symmetric space*

$$G \,\vec{\times}\, D(G,*) = \mathrm{Int}_* G \,\vec{\times}\, D(G,*), \quad (g,d) \longmapsto gdg^* = (gdg^*)^*.$$

Analogous structures hold for the conjugate bijections of a unital algebra with conjugation

$$\begin{array}{l} 1 \in A \in *\underline{\mathbf{aag}}_{\mathbb{K}}, \quad \mathrm{Ad}_* : A^\circ \longrightarrow \mathbf{G}(A), \\ \mathrm{Ad}_* g : A \longrightarrow A, \quad a \longmapsto gag^*. \end{array}$$

A complex Lie group G and its Lie algebra with conjugation have a unitary real Lie subgroup $U(G,*) \in \underline{\mathbf{lgrp}}_{\mathbb{R}}$ with antisymmetric real Lie subalgebra and a symmetric real submanifold $D(G,*) \in \underline{\mathbf{dif}}_{\mathbb{R}}$ with symmetric real tangent space:

$$\begin{array}{ll} \log G & = \log G_- \oplus \log G_+, \\ \log G_- & = \log U(G,*) = \{l \in \log G \mid l^* = -l\} \in \underline{\mathbf{lag}}_{\mathbb{R}}, \\ \log G_+ & = i \log G_- = \{x \in \log G \mid x^* = +x\} \in \underline{\mathbf{vec}}_{\mathbb{R}}. \end{array}$$

If the conjugate adjoint representation of the group G can be defined also on its Lie algebra (this is not always the case, e.g., not for $G = \mathbf{SL}(\mathbb{C}^2_{\mathbb{R}})$ with unitary $\mathbf{SU}(2)$-subgroup and symmetric boost domain $\mathbf{SL}(\mathbb{C}^2_{\mathbb{R}})/\mathbf{SU}(2)$),

$$G \times \log G \longrightarrow \log G, \quad \mathrm{Ad}_*(g)(l) = g \circ l \circ g^*,$$

then one can define the *conjugate adjoint affine group and Lie algebra* on the real tangent space of the symmetric domain

$$\mathrm{Ad}_* G \,\vec{\times}\, \log G_+, \quad \log \mathrm{Ad}_* G \,\vec{\oplus}\, \log G_+.$$

2

SPACETIME AS UNITARY OPERATION CLASSES

In quantum theory, time and position are really parametrized operations acting on complex vector spaces. The causal homogeneous manifolds that will be discussed in this chapter are n^2-dimensional generalizations of 1-dimensional time and 4-dimensional spacetime. They are constituted by classes of compact unitary transformations in complex linear ones as suggested by the Cartan presentation of Minkowski spacetime by Hermitian (2×2) matrices (chapter "Lorentz Operations"). The description of these causal manifolds with real rank n clarifies the structures of 4-dimensional spacetime with real rank 2 as the physically most important case.

From a mathematical point of view, the first sections of this chapter contain, in physical terms, a reformulation of familiar structures of the stellar algebras (C*-algebras) $\mathbf{AL}(\mathbb{C}^n)$ with $n \times n$ matrices acting on $\mathbb{C}^n$-isomorphic Hilbert spaces (chapter "Quantum Probability"). If spacetime translations constitute a real vector subspace in complex linear transformations $x : \mathbb{C}^n \longrightarrow \mathbb{C}^n$, $x = x^\star$, they are, from the outset, recognizable as binary relations in the sense of Leibniz, for Minkowski spacetime $n = 2$ as binary spinor relations.

The polar decomposition of the full linear group $\mathbf{GL}(\mathbb{C}^n)$ into the group $\mathbf{U}(n)$ with the phases and the symmetric space $\mathbf{D}(n)$ with the absolute values – the uniquely defined positive cone in the stellar algebra of all complex $(n \times n)$ matrices – is proposed to establish the dichotomy of compact internal ("chargelike") and noncompact external ("spacetimelike") degrees of freedom respectively as used in quantum field theories for $n = 2$ leading to the compact hyperisospin group $\mathbf{U}(2)$ (chapter "Gauge Interactions") and the noncompact nonlinear spacetime $\mathbf{D}(2)$ with tangent Minkowski translations $\mathbb{R}^4$.

2.1 Spacetime Translations

Cartan's parametrization of the spacetime translations (real 4-dimensional Minkowski vector space) uses the Hermitian complex 2×2 matrices (chapter

"Lorentz Operations"). Together with the time translations (real numbers)

$$x = \begin{cases} t = \overline{t} & \in \mathbb{C}_{\mathbb{R}}(1) = i\mathbb{R} \oplus \mathbb{R}, \quad n = 1, \\ \begin{pmatrix} x_0 + x_3 & x_1 - ix_2 \\ x_1 + ix_2 & x_0 - x_3 \end{pmatrix} = x^\star & \in \mathbb{C}_{\mathbb{R}}(2) = (i\mathbb{R})^4 \oplus \mathbb{R}^4, \quad n = 2, \end{cases}$$

they should be used as illustrations for general n.

The complex $n \times n$ matrices $z \in \mathbb{C}(n) = \mathbf{AL}(\mathbb{C}^n)$ constitute, with the $\mathbf{U}(n)$-conjugation $\star$, a *stellar algebra* (C*-algebra) as endomorphisms of a $\mathbb{C}^n$-isomorphic Hilbert space. They are decomposable into two isomorphic vector spaces of real dimension n^2:

$$z = \tfrac{i}{2}\gamma + x \in \mathbb{C}_{\mathbb{R}}(n) = i\mathbb{R}(n) \oplus \mathbb{R}(n) \cong \mathbb{R}^{2n^2}.$$

The vector subspace $\mathbb{R}(n)$ is called the *matrix parametrization of the spacetime translations* with $n \in \mathbb{N}$ the *real rank of spacetime.*

A basis for $\mathbb{C}_{\mathbb{R}}(n)$ is given by generalized Weyl matrices

$$z = z_j \sigma(n)^j \cong z_A^{\dot{A}}, \quad A, \dot{A} = 1, \ldots, n \text{ with } \{\sigma(n)^j\}_{j=0}^{n^2-1} = \{\mathbf{1}_n, \sigma(n)^a\}_{a=1}^{n^2-1}$$

where $\mathbf{1}_n$ is the unit matrix and $\sigma(n)^a$ for $n \geq 2$ are (n^2-1) generalized Hermitian traceless Pauli matrices, i.e., three Pauli matrices $\vec{\sigma}$ for $n = 2$, eight Gell-Mann matrices $\sigma(3)^a = \lambda^a$ for $n = 3$, etc. (chapter "Simple Lie Operations").

The determinant defines the *abelian projection* on the complex numbers ($\star$ monoid morphism):

$$\det : \mathbb{C}_{\mathbb{R}}(n) \longrightarrow \mathbb{C}_{\mathbb{R}}, \quad \det z^\star = \overline{\det z}.$$

By polarization, i.e., by an appropriate combination of $(z_1 \pm z_2 \pm \cdots \pm z_n)^n$, one obtains a totally symmetric $\star$-compatible multilinear form, generalizing the well-known bilinear form of the Minkowski translations $\mathbb{R}(2)$:

$$\begin{aligned} \eta: &\quad \mathbb{C}_{\mathbb{R}}(n) \times \cdots \times \mathbb{C}_{\mathbb{R}}(n) \longrightarrow \mathbb{C}, \\ &\quad (z_1, \ldots, z_n) \longmapsto \eta(z_1, \ldots, z_n) = \epsilon^{A_1 \ldots A_n} \epsilon_{\dot{A}_1 \ldots \dot{A}_n} (z_1)_{A_1}^{\dot{A}_1} \cdots (z_n)_{A_n}^{\dot{A}_n}, \\ n = 1: &\quad \eta(z) = \det z = z, \\ n = 2: &\quad \eta(z_1, z_2) = \tfrac{(z_1+z_2)^2 - (z_1-z_2)^2}{4}, \quad \operatorname{sign} \eta|_{\mathbb{R}(2)} = (1, 3). \end{aligned}$$

The trace and the traceless parts of a translation are called a time translation and a position translation respectively. The position translation $\vec{x}$ denotes a traceless $(n \times n)$ matrix and, in this connection, $x_0 \cong x_0 \mathbf{1}_n$:

$$\begin{aligned} \operatorname{tr} \mathbb{C}_{\mathbb{R}}(n) &= i \operatorname{tr} \mathbb{R}(n) \oplus \operatorname{tr} \mathbb{R}(n), \quad \operatorname{tr} \mathbb{R}(n) \cong \mathbb{R}, \\ \mathbb{C}_{\mathbb{R}}(n)_0 &= \{z \in \mathbb{C}_{\mathbb{R}}(n) \mid \operatorname{tr} z = 0\} = i\mathbb{R}(n)_0 \oplus \mathbb{R}(n)_0, \quad \mathbb{R}(n)_0 \cong \mathbb{R}^{n^2-1}, \\ x &= x_j \sigma(n)^j = x_0 \mathbf{1}_n + x_a \sigma(n)^a = x_0 + \vec{x}. \end{aligned}$$

A spacetime decomposition into time and position translation subspaces is incompatible with the determinant, since in general, $\det(x+y) \neq \det x + \det y$ for $n \geq 2$.

$z \in \mathbb{C}_{\mathbb{R}}(n)$ is unitarily diagonalizable if and only if it is normal. Then functions of z are defined via its spectrum:

$$z \circ z^\star = z^\star \circ z \Rightarrow f(z) = u(z) \circ f(\operatorname{diag} z) \circ u(z)^\star, \ \ u(z) \in \mathbf{SU}(n).$$

Both the Hermitian translations $x \in \mathbb{R}(n)$ and the anti-Hermitian vectors $\frac{i}{2}\gamma \in i\mathbb{R}(n)$ are diagonalizable, not, however, each matrix $z \in \mathbb{C}_{\mathbb{R}}(n)$ for $n \geq 2$. All spacetime translations are unitarily equivalent to a real diagonal matrix:

$$\begin{array}{lll} x \in \mathbb{R}(n) & \Rightarrow \operatorname{spec} x = \{\xi \mid \det(x - \xi \mathbf{1}_n) = 0\} \subset \mathbb{R}, \\ n = 1: & x \in \mathbb{R}(1), \ \ \operatorname{spec} x = \{t\}, \\ n = 2: & x \in \mathbb{R}(2), \ \ \operatorname{spec} x = \{x_0 \pm r\}. \end{array}$$

The n real spectral values $\operatorname{diag} x = \begin{pmatrix} \xi_1 & 0 & \dots \\ 0 & \xi_2 & \dots \\ & \dots & \\ \dots & 0 & \xi_n \end{pmatrix}$ are called *Cartan spacetime coordinates*. The unique stellar algebra *order* uses the spectrum

$$\begin{array}{rl} x \succeq 0 \iff & x = x^\star, \text{ i.e., } x \in \mathbb{R}(n), \text{ and } \operatorname{spec} x \geq 0 \\ \iff & \text{There exists } z \in \mathbb{C}_{\mathbb{R}}(n) \text{ with } x = z^\star \circ z. \end{array}$$

The Cartan coordinates can change under dilation-Lorentz transformations $x \longmapsto g \circ x \circ g^\star$ with $g \in \mathbf{GL}(\mathbb{C}^n_{\mathbb{R}})$. Positivity or triviality of spectral values is invariant, $g \circ z^\star \circ z \circ g^\star = z_g^\star \circ z_g$ with $z_g = z \circ g^\star$. The stellar norm topology via the largest eigenvalue of the square $z^\star \circ z$ (spectral radius),

$$\| z \| = \max\{|\xi| \mid \xi^2 \in \operatorname{spec} z^\star \circ z\},$$

is the order topology on the spacetime translations $\mathbb{R}(n)$ with a "diamond" (double cone) basis for $n = 2$ (chapter "Spacetime Translations").

The stellar order generalizes the familiar natural orders of the time and of the spacetime translations. With a nontrivial positive causal vector $c \succ 0$ positivity is expressible by positive c-projected products (n causal projections)

$$x \in \mathbb{R}(n): \ \ x \succeq 0 \iff x_c^r = \eta(\underbrace{x, \dots, x}_{r}, \underbrace{c, \dots, c}_{n-r}) \geq 0 \text{ for } r = 1, \dots, n.$$

The idempotent characteristic function of a causal translation uses the spectral values

$$\begin{array}{rl} x \in \mathbb{R}(n): \ \ \vartheta(x) & = \prod_{r=1}^{n} \vartheta(\xi_r) = \prod_{r=1}^{n} \vartheta(x_c^r), \ \ c \succ 0, \\ \epsilon(x) & = \vartheta(x) - \vartheta(-x), \ \ \epsilon(x)^2 = \vartheta(x) + \vartheta(-x), \\ x \succeq 0 & \iff x = \epsilon(x) x, \end{array}$$

with the Minkowski translations as example:

$$x \in \mathbb{R}(2): \ \ \left\{ \begin{array}{ll} \vartheta(x) & = \vartheta(x_0 + r)\vartheta(x_0 - r) = \vartheta(x_0)\vartheta(x^2), \\ \epsilon(x) & = \epsilon(x_0)\vartheta(x^2), \ \ \epsilon(x)^2 = \vartheta(x^2), \\ x \succeq 0 & \iff x = \epsilon(x_0)\vartheta(x^2) x. \end{array} \right.$$

The vector space of all spacetime translations $\mathbb{R}(n)$ is the union of the positive and of the negative *causal cone* and the *spacelike* submanifold:

$$\begin{array}{rlrl} \mathbb{R}(n) & = \mathbb{R}(n)_{\text{caus}} \cup \mathbb{R}(n)_{\text{position}}, & \mathbb{R}(n)_{\text{caus}} \cap \mathbb{R}(n)_{\text{position}} & = \{0\}, \\ \mathbb{R}(n)_{\text{caus}} & = \mathbb{R}(n)^+_{\text{caus}} \cup \mathbb{R}(n)^-_{\text{caus}}, & \mathbb{R}(n)^+_{\text{caus}} \cap \mathbb{R}(n)^-_{\text{caus}} & = \{0\}, \\ \mathbb{R}(n)^+_{\text{caus}} & = \{x \in \mathbb{R}(n) \mid \operatorname{spec} x \geq 0\} = -\mathbb{R}(n)^-_{\text{caus}}. \end{array}$$

The positive causal cone is the disjoint union

$$\mathbb{R}(n)^+_{\text{caus}} = \{0\} \uplus \mathbb{R}(n)^+_{\text{time}} \uplus \mathbb{R}(n)^+_{\text{light}}$$

of the trivial translation (vertex of the cone), the strictly positive *timelike* translations (open cone), whose spectrum does not contain 0,

$$\mathbb{R}(n)^+_{\text{time}} = \{x \in \mathbb{R}(n)^+_{\text{caus}} \mid 0 \notin \operatorname{spec} x\},$$

and the strictly positive *lightlike* translations (skin of the tipless cone), where 0 is a spectral value

$$\mathbb{R}(n)^+_{\text{light}} = \{x \in \mathbb{R}(n)^+_{\text{caus}} \mid x \neq 0, \quad 0 \in \operatorname{spec} x\}.$$

Spacetime translations have a *positive causal and a causal projection*:

$$\begin{array}{lll} \mathbb{R}(n) \longrightarrow \mathbb{R}(n)^+_{\text{caus}}, & x \longmapsto \vartheta(x)x & = \begin{cases} \vartheta(t)t & \text{for } n = 1, \\ \vartheta(x_0)\vartheta(x^2)x & \text{for } n = 2, \end{cases} \\ \mathbb{R}(n) \longrightarrow \mathbb{R}(n)_{\text{caus}}, & x \longmapsto \epsilon(x)^2 x & = \begin{cases} t & \text{for } n = 1, \\ \vartheta(x^2)x & \text{for } n = 2. \end{cases} \end{array}$$

The causal projection coincides for the abelian case $n = 1$ with the *eigentime projection* that is the causal projection on the real numbers $\mathbb{R}(1) = \mathbb{R}$:

$$\mathbb{R}(n) \longrightarrow \mathbb{R}, \quad x \longmapsto \epsilon(x) \Big| \det x^{\frac{1}{n}} \Big| = \begin{cases} t & \text{for } n = 1, \\ \epsilon(x_0)\vartheta(x^2)|\sqrt{x^2}| & \text{for } n = 2. \end{cases}$$

All translations can be written as a sum of a strictly positive and a strictly negative timelike translation

$$\mathbb{R}(n) = \{x_+ + x_- \mid x_+, -x_- \in \mathbb{R}(n)^+_{\text{time}}\}.$$

In the case of the 1-dimensional time translations $\mathbb{R}(1) = \mathbb{R}$ the position translations are trivial, $\mathbb{R}(1)_{\text{position}} = \{0\}$. The nontrivial spacelike manifold for $n \geq 2$ is the disjoint union of $(n-1)$ manifolds with m strictly positive and $n-m$ strictly negative Cartan coordinates $\begin{pmatrix} \mathbf{1}_m & 0 \\ 0 & -\mathbf{1}_{n-m} \end{pmatrix} \in \mathbb{R}(m, n-m)_{\text{position}}$:

$$n \geq 2: \quad \mathbb{R}(n)_{\text{position}} \setminus \{0\} = \biguplus_{m=1}^{n-1} \mathbb{R}(m, n-m)_{\text{position}}.$$

In the 1-dimensional case there is no light, $\mathbb{R}(1)^{\pm}_{\text{light}} = \emptyset$. For $n \geq 2$ the strictly positive (strictly negative) lightlike manifold is the disjoint union of $(n-1)$ manifolds with exactly m strictly positive (negative) and $n-m$ trivial Cartan coordinates $\begin{pmatrix} \pm\mathbf{1}_m & 0 \\ 0 & \mathbf{0}_{n-m} \end{pmatrix} \in \mathbb{R}(m,0)^{\pm}_{\text{light}}$:

$$n \geq 2: \quad \mathbb{R}(n)^{\pm}_{\text{light}} = \biguplus_{m=1}^{n-1} \mathbb{R}(m,0)^{\pm}_{\text{light}}.$$

The linear forms $\mathbb{R}(n)^T$ of the spacetime translations are called the *frequency (energy) space* for $n = 1$ and the *energy-momentum space* for $n \geq 2$. The double trace with one "open slot" describes an isomorphism between translations and energy-momenta:

$$\begin{array}{l}
\mathbb{R}(n) \longrightarrow \mathbb{R}(n)^T, \quad q \longmapsto \check{q} = \operatorname{tr} q \circ \cdots, \\
\text{dual product: } \mathbb{R}(n)^T \times \mathbb{R}(n) \longrightarrow \mathbb{R}, \quad \langle \check{q}, x \rangle = \operatorname{tr} q \circ x, \\
\mathbb{R}(n)^T\text{-basis: } \{\check{\sigma}(n)_j\}_{j=0}^{n^2-1} = \{\tfrac{2}{n}\mathbf{1}_n, \sigma(n)^a\}_{a=1}^{n^2-1}, \\
\text{dual: } \tfrac{1}{2} \operatorname{tr} \sigma(n)^j \check{\sigma}(n)_k = \delta^j_k.
\end{array}$$

2.2 Nonlinear Spacetime

The complex $(n \times n)$ matrices $\mathbb{C}(n) = \mathbf{AL}(\mathbb{C}^n)$ with commutator define the complex rank-n Lie algebra of the Lie group $\mathbf{GL}(\mathbb{C}^n)$ and, with $\mathbf{U}(n)$-conjugation as real $2n^2$-dimensional vector space $\mathbb{C}_{\mathbb{R}}(n)$, the rank-$2n$ Lie algebra of the real group $\mathbf{GL}(\mathbb{C}^n_{\mathbb{R}})$. The antisymmetric vector subspace $i\mathbb{R}(n)$ in $\mathbb{C}_{\mathbb{R}}(n)$ defines the imaginary rank-n Lie algebra of the unitary group $\mathbf{U}(n)$:

$$\begin{array}{rlrl}
\mathbb{C}(n) & = \log \mathbf{GL}(\mathbb{C}^n), & \mathbf{GL}(\mathbb{C}^n) & = \exp \mathbb{C}(n), \\
i\mathbb{R}(n) \oplus \mathbb{R}(n) & = \log \mathbf{GL}(\mathbb{C}^n_{\mathbb{R}}), & \mathbf{GL}(\mathbb{C}^n_{\mathbb{R}}) & = \exp[i\mathbb{R}(n) \oplus \mathbb{R}(n)], \\
i\mathbb{R}(n) & = \log \mathbf{U}(n), & \mathbf{U}(n) & = \exp i\mathbb{R}(n), \\
i\mathbb{R}(n)_0 & = \log \mathbf{SU}(n), & \mathbf{SU}(n) & = \exp i\mathbb{R}(n)_0.
\end{array}$$

From now on in this chapter, the subindex in $\mathbb{C}_{\mathbb{R}}$ is omitted for notational convenience, in the following $\mathbb{C} = \mathbb{R} \oplus i\mathbb{R}$.

The vector space with the spacetime translations $\mathbb{R}(n)$ is isomorphic to the classes of the unitary Lie algebra in the full real Lie algebra. Its exponent is isomorphic to the corresponding homogeneous space, the real n^2-dimensional manifold with the right orbits $\mathbf{U}(n)g$ of the unitary group $\mathbf{U}(n)$ in the full group $\mathbf{GL}(\mathbb{C}^n)$:

$$\begin{array}{rlrl}
\mathbb{R}(n) & \cong \log \mathbf{GL}(\mathbb{C}^n)/\log \mathbf{U}(n), & \mathbf{D}(n) & = \exp \mathbb{R}(n) \cong \mathbf{U}(n)\backslash\mathbf{GL}(\mathbb{C}^n), \\
\mathbb{R}(n)_0 & \cong \log \mathbf{SL}(\mathbb{C}^n)/\log \mathbf{SU}(n), & \mathbf{SD}(n) & = \exp \mathbb{R}(n)_0 \cong \mathbf{SU}(n)\backslash\mathbf{SL}(\mathbb{C}^n).
\end{array}$$

One could equally take the left orbits with corresponding changes, e.g., for the action of the external group.

$\mathbf{D}(n)$ is the n-bein manifold $\mathbf{GL}(\mathbb{C}^n)$, all linear automorphisms, up to the positive unitary operations $\mathbf{U}(n)$, it will be called *nonlinear spacetime (manifold)*. It has the direct manifold factor $\mathbf{SD}(n)$, the *nonlinear position (Sylvester or boost manifold)*, trivial only for the abelian case $n = 1$. The *causal* manifold $\mathbf{D}(n)$ is the strictly positive cone of the uniquely ordered C*-algebra $\mathbb{C}(n)$.

Also the name *scalar product or* $\mathbf{U}(n)$ *(orientation) manifold* is justified for $\mathbf{D}(n)$: It parametrizes all possible scalar products of a complex $\mathbb{C}^n$-isomorphic Hilbert space, e.g., with basis $\{\Phi^A\}_{A=1}^n$:

$$\mathbf{D}(n) = \{d = g^\star \circ g \cong \begin{pmatrix} \langle\Phi^1|\Phi^1\rangle & \dots & \langle\Phi^1|\Phi^n\rangle \\ \dots & \dots & \dots \\ \langle\Phi^n|\Phi^1\rangle & \dots & \langle\Phi^n|\Phi^n\rangle \end{pmatrix} \bigm| g \in \mathbf{GL}(\mathbb{C}^n)\}.$$

The groups involved have the centrum, the phase correlations, and the adjoint groups

$$\begin{aligned} \text{centr}\,\mathbf{GL}(\mathbb{C}^n) &\cong \mathbb{C}^\diamond, & \text{centr}\,\mathbf{U}(n) &\cong \mathbf{U}(1), \\ \mathbb{C}^\diamond \cap \mathbf{SL}(\mathbb{C}^n) &\cong \mathbb{I}(n), & \mathbf{U}(\mathbf{1}_n) \cap \mathbf{SU}(n) &\cong \mathbb{I}(n), \\ \text{Int}\,\mathbf{GL}(\mathbb{C}^n) &\cong \mathbf{SL}(\mathbb{C}^n)/\mathbb{I}(n), & \text{Int}\,\mathbf{U}(n) &\cong \mathbf{SU}(n)/\mathbb{I}(n). \end{aligned}$$

There is another chain of causal spacetime manifolds, characterized by the orthogonal structures

$$s = 0: \quad \mathbf{D}(1), \quad s \geq 1: \quad \mathbf{D}(1) \times \mathbf{SO}(s) \backslash \mathbf{SO}_0(1, s)$$

with $s \geq 0$ the position dimension. This chain of orthogonal groups acting on real spaces with dimensions $1 + s$ meets the $\mathbf{D}(n)$-chain only for the two spacetime dimensions $1 + s = n^2 = 1, 4$. The orthogonal structures have an invariant bilinear form for all dimensions with $s \geq 1$. $\mathbf{D}(1)$ has real rank 1; all the other causal orthogonal manifolds with $s \geq 1$ have real rank 2.

For $n = 2$ one has as isomorphisms with the orthochronous Lorentz group $\mathbf{SO}_0(1, 3)$ and the rotation group $\mathbf{SO}(3)$:

$$\begin{aligned} \mathbf{GL}(\mathbb{C}^2)/\mathbb{C}^\diamond &\cong \mathbf{SL}(\mathbb{C}^2)/\mathbb{I}(2) &&\cong \mathbf{SO}_0(1, 3), \\ \mathbf{U}(2)/\mathbf{U}(1) &\cong \mathbf{SU}(2)/\mathbb{I}(2) &&\cong \mathbf{SO}(3), \\ \mathbf{D}(2) &\cong \mathbf{U}(2) \backslash \mathbf{GL}(\mathbb{C}^2) &&\cong \mathbf{D}(\mathbf{1}_2) \times \mathbf{SO}(3) \backslash \mathbf{SO}_0(1, 3). \end{aligned}$$

If one visualizes real 4-dimensional nonlinear spacetime $\mathbf{D}(2)$ as the open future cone $\mathbb{R}(2)^+_{\text{time}}$ in the Minkowski translations $\mathbb{R}(2)$, this cone can be foliated[1] with the hyperboloids $\mathcal{Y}^3$ as nonlinear positions, each isomorphic to $\mathbf{SD}(2) \cong \mathbf{SO}(3) \backslash \mathbf{SO}_0(1, 3)$. The $\mathbf{D}(1)$ causal group acts on the manifold by a "hyperbolic hopping," whereas the orthochronous group $\mathbf{SO}_0(1, 1)$-action on the individual hyperboloids can be described as a "hyperbolic stretching." The total semiorder is the "foliation order" of the future hyperboloids. The Minkowski translations $\mathbb{R}(2)$ as tangent structure of the spacetime manifold $\mathbf{D}(2)$ can be visualized by means of a 3-dimensional tangent space of a timelike hyperboloid $\mathbf{SO}(3) \backslash \mathbf{SO}_0(1, 3)$ and the tangent line of "blowing up" or "shrinking" this hyperboloid with $\mathbf{D}(1)$.

[1]Take the 3-dimensional projection with hyperboloids $\mathbf{SO}(2)\backslash\mathbf{SO}_0(1, 2)$ and 2-dimensional tangent planes.

2.3 Spacetime and Hyperisospin

The elements of the real $2n^2$-dimensional group $\mathbf{GL}(\mathbb{C}^n)$ have a unique *stellar polar decomposition* into unitary phase (first factor) and strictly positive absolute value (second factor) and, for the elements of the stellar algebra $\mathbb{C}(n) = \mathbf{AL}(\mathbb{C}^n)$, into an anti-Hermitian Lie algebra element and a Hermitian spacetime translation (both with generalized Weyl matrices as bases)

$$\begin{array}{rlll} g \in \mathbf{GL}(\mathbb{C}^n) \Rightarrow & g = U(g) \circ |g|, & U^\star(g) = U^{-1}(g), & |g| = \sqrt{g^\star \circ g} \succ 0, \\ & g = e^{\frac{i}{2}\gamma} \circ e^{\psi}, & \gamma = \gamma_j \tau(n)^j, & \psi = \psi_j \sigma(n)^j, \\ z \in \mathbb{C}(n) \Rightarrow & z = \frac{i}{2}\gamma + x, & \gamma^\star = \gamma, & x = x^\star = \frac{z^\star + z}{2}. \end{array}$$

There is the corresponding "left" decomposition (below) with exchanged order: absolute value space $\mathbf{D}(n)$ times phase group $\mathbf{U}(n)$. Given a matrix g, the Hermitian product $g^\star \circ g$ can by unitarily diagonalized, which gives, with the positive square roots of the diagonal elements, the absolute value matrix $|g|$:

$$\begin{array}{rl} g^\star \circ g = e^{2\psi} = u \circ \operatorname{diag}(g^\star \circ g) \circ u^\star & \Rightarrow |g| = u \circ \sqrt{\operatorname{diag}(g^\star \circ g)} \circ u \\ & \Rightarrow U(g) = g \circ |g|^{-1}. \end{array}$$

The factorization of the group into compact *internal group* $\mathbf{U}(n)$ and noncompact *external symmetric spacetime* $\mathbf{D}(n)$ and the associated Lie algebra decomposition into unitary Lie algebra and tangent spacetime translations

$$\begin{array}{llll} \mathbf{GL}(\mathbb{C}^n) & = \mathbf{U}(n) \circ \mathbf{D}(n), & \mathbb{C}(n) & = i\mathbb{R}(n) \oplus \mathbb{R}(n), \\ \mathbf{SL}(\mathbb{C}^n) & = \mathbf{SU}(n) \circ \mathbf{SD}(n), & \mathbb{C}(n)_0 & = i\mathbb{R}(n)_0 \oplus \mathbb{R}(n)_0, \end{array}$$

induces also a unique factorization of the adjoint group, the generalized *Lorentz group*, into compact adjoint subgroup, the *rotation group* for $n = 2$, and position (boost) manifold

$$\begin{array}{rl} \mathbf{SL}(\mathbb{C}^n)/\mathbb{I}(n) \ni \Lambda & = O_\Lambda \circ |\Lambda| \in \mathbf{SU}(n)/\mathbb{I}(n) \circ \mathbf{SD}(n), \\ |\Lambda| = |\Lambda|^T & = \sqrt{\Lambda^T \circ \Lambda} = R_{|\Lambda|} \circ \sqrt{\operatorname{diag}(\Lambda^T \circ \Lambda)} \circ R^T_{|\Lambda|}, \; R_{|\Lambda|} \in \mathbf{SU}(n)/\mathbb{I}(n), \\ \text{e.g., } \mathbf{SO}_0(1,3) & = \mathbf{SO}(3) \circ \mathbf{SD}(2). \end{array}$$

In general, for $n \geq 2$, the polar decomposition of conjugated group elements have a different absolute value with equal eigenvalues:

$$\begin{array}{rl} g = U \circ |g|, \;\; g^\star = U^\star \circ |g^\star| & \Rightarrow |g^\star| = U \circ |g| \circ U^\star \\ & \Rightarrow \operatorname{spec}|g| = \operatorname{spec}|g^\star|. \end{array}$$

Hence the right and left orbit decompositions have equal phase, but, in general, different absolute values:

$$\begin{array}{llll} \mathbf{GL}(\mathbb{C}^n) = \mathbf{U}(n) \circ \mathbf{D}(n), & g = U \circ |g_R|, & |g_R| = |g| & = |g|^\star \;\; = \sqrt{g^\star \circ g}, \\ \mathbf{GL}(\mathbb{C}^n) = \mathbf{D}(n) \circ \mathbf{U}(n), & g = |g_L| \circ U, & |g_L| = |g^\star| & = |g^\star|^\star = \sqrt{g \circ g^\star}, \\ & |g_L| = e^{\psi_L} = e^{\frac{i}{2}\gamma} \circ e^{\psi_R} \circ e^{-\frac{i}{2}\gamma}. & & \end{array}$$

In addition to the stellar polar decomposition there is the *Cartan polar decomposition (diagonalization)* for both factors, nontrivial for $n \geq 2$: The unitary diagonalization transformations – different, in general, for internal and external factors – are determined up to diagonal phases[2]

$$\begin{array}{rl} & |g| \in \mathbf{D}(n),\ x \in \mathbb{R}(n),\ U \in \mathbf{U}(n),\ \frac{i}{2}\gamma \in \log \mathbf{U}(n) \text{ are diagonalizable,}\\ \text{e.g.,} & |g| = u_{|g|} \circ \operatorname{diag} |g| \circ u^\star_{|g|}\\ \text{with} & u_{|g|} \cong u_x,\ u_U \cong u_{\frac{i}{2}\gamma} \Subset \mathbf{SU}(n)/\mathbf{SO}(2)^{n-1}. \end{array}$$

The diagonal matrix for a spacetime element contains n strictly positive spectral values $\{e^{\psi_r}\}_{r=1}^n$:

$$|g| \in \mathbf{D}(n): \quad \operatorname{diag} |g| = \begin{pmatrix} e^{\psi_1} & 0 & \dots & 0 \\ 0 & e^{\psi_2} & \dots & 0 \\ \dots & & \dots & \\ 0 & 0 & \dots & e^{\psi_n} \end{pmatrix}.$$

In a nontrivial position (boost) manifold $\mathbf{SD}(n)$ the group $\mathbf{D}(1)$ comes in self-dual decomposable representations, isomorphic to the orthochronous group $\mathbf{SO}_0(1,1)$:

$$\mathbf{SO}_0(1,1) \ni \begin{pmatrix} \cosh\psi & \sinh\psi \\ \sinh\psi & \cosh\psi \end{pmatrix} \cong \begin{pmatrix} e^{\psi} & 0 \\ 0 & e^{-\psi} \end{pmatrix}.$$

The Cartan subgroups of $\mathbf{GL}(\mathbb{C}^n)$ and $\mathbf{SL}(\mathbb{C}^n)$ with rank $2n$ and $2(n-1)$ are isomorphic to $(\mathbb{C}^\diamond)^n$ and $(\mathbb{C}^\diamond)^{n-1}$ respectively. The real ranks n and $n-1$ of the manifolds $\mathbf{D}(n)$ and $\mathbf{SD}(n)$ with the Cartan subgroups above has to be seen in analogy to the imaginary ranks n and $n-1$ for $\mathbf{U}(n)$ and $\mathbf{SU}(n)$ in the Cartan factorizations:

$$\begin{array}{ll} \mathbf{GL}(\mathbb{C}^n) \cong \mathbf{D}(1) \supset \mathbf{U}(1) \circ \mathbf{SU}(n) \circ \mathbf{SO}_0(1,1)^{n-1} \circ \mathbf{SU}(n), & \\ \mathbf{U}(n) \cong \mathbf{U}(1) \circ \mathbf{SU}(n), & \mathbf{SU}(n) \cong \mathbf{SO}(2)^{n-1} \circ \mathbf{SU}(n)/\mathbf{SO}(2)^{n-1},\\ \mathbf{D}(n) \cong \mathbf{D}(1) \times \mathbf{SD}(n), & \mathbf{SD}(n) \cong \mathbf{SO}_0(1,1)^{n-1} \circ \mathbf{SU}(n)/\mathbf{SO}(2)^{n-1}. \end{array}$$

For $\mathbf{U}(n)$ there is the central correlation for phase group and special group $\mathbf{U}(\mathbf{1}_n) \cap \mathbf{SU}(n) \cong \mathbb{I}(n)$.

In the tangent space (Lie algebra for $\mathbf{U}(n)$), one has the corresponding *Cartan translations*

$$\begin{pmatrix} \log \mathbf{D}(n) \\ \log \mathbf{U}(n) \end{pmatrix} = \begin{pmatrix} \mathbb{R}(n) \\ i\mathbb{R}(n) \end{pmatrix} \cong \begin{pmatrix} \mathbb{R}^n \\ (i\mathbb{R})^n \end{pmatrix} \circ \mathbf{SU}(n)/\mathbf{SO}(2)^{n-1}.$$

The relativistic case $\mathbf{GL}(\mathbb{C}^2) = \mathbf{U}(2) \circ \mathbf{D}(2)$ uses two polar coordinates for the 2-sphere $\mathbf{SU}(2)/\mathbf{SO}(2)$ in addition to two Cartan coordinates, e.g., for spacetime translations $x \in \mathbb{R}(2)$:

$$\begin{array}{l} \begin{pmatrix} x_0 + x_3 & x_1 - ix_2 \\ x_1 + ix_2 & x_0 - x_3 \end{pmatrix} = u(\frac{\vec{x}}{r}) \circ \operatorname{diag} x \circ u^\star(\frac{\vec{x}}{r}), \quad \operatorname{diag} x = \begin{pmatrix} x_0 + r & 0 \\ 0 & x_0 - r \end{pmatrix},\\ u(\frac{\vec{x}}{r}) = \begin{pmatrix} \cos\frac{\theta}{2} & -e^{-i\varphi}\sin\frac{\theta}{2} \\ e^{i\varphi}\sin\frac{\theta}{2} & \cos\frac{\theta}{2} \end{pmatrix} = \sqrt{\frac{r+x_3}{2r}} \begin{pmatrix} 1 & -\frac{x_1 - ix_2}{r+x_3} \\ \frac{x_1+ix_2}{r+x_3} & 1 \end{pmatrix} \Subset \mathbf{SU}(2)/\mathbf{SO}(2). \end{array}$$

[2]The double-element symbol denotes a class representative $u \Subset G/H \iff u \in gH \in G/H$.

The two Cartan coordinates in $\operatorname{diag} x$ are expressible by a $\mathbf{D}(1)$-factor multiplying a boost $\mathbf{SO}_0(1,1)$ for time- and spacelike translations:

$$\begin{aligned}
x^2 > 0,\quad x \succ 0 \;&\Rightarrow\; \operatorname{diag} x = \sqrt{x^2}\; e^{\sigma^3|\vec{\psi}|}\mathbf{1}_2,\quad e^{|\vec{\psi}|} = \sqrt{\tfrac{x_0+r}{x_0-r}},\quad \tanh|\vec{\psi}| = \tfrac{r}{x_0},\\
x^2 < 0,\quad &\Rightarrow\; \operatorname{diag} x = \sqrt{-x^2} e^{\sigma^3|\vec{\psi}|}\sigma^3,\quad e^{|\vec{\psi}|} = \sqrt{\tfrac{r+x_0}{r-x_0}},\quad \tanh|\vec{\psi}| = \tfrac{x_0}{r},\\
x^2 = 0,\quad x \succ 0 \;&\Rightarrow\; \operatorname{diag} x = \begin{pmatrix} 2r & 0\\ 0 & 0\end{pmatrix}.
\end{aligned}$$

2.3.1 Internal and External Action Groups

The polar decomposition of the automorphisms $\mathbf{GL}(\mathbb{C}^n) = \mathbf{U}(n) \circ \mathbf{D}(n)$ into internal group $\mathbf{U}(n)$ (phases) and external spacetime $\mathbf{D}(n)$ (absolute values, right $\mathbf{U}(n)$-orbits, no group for $n \geq 2$) comes with corresponding groups, called *internal and external action groups.* The internal-external doubling of the action group has its origin in the independent left and right group multiplication $G \ni k \longmapsto g_1 k g_2^{-1}$. This two-sided regular realization of the doubled group $G \times G$ is extensively used in the theory of group representations (chapter "Harmonic Analysis").

The internal action group is $\mathbf{U}(n)$, for $n = 2$ called the *hyperisospin group.* The adjoint action on its Lie algebra $i\mathbb{R}(n)$,

$$\mathbf{U}(n) \times \log \mathbf{U}(n) \longrightarrow \log \mathbf{U}(n),\quad \tfrac{i}{2}\gamma \longmapsto u \circ \tfrac{i}{2}\gamma \circ u^\star,$$

is faithful for $\operatorname{Int} \mathbf{U}(n) = \mathbf{U}(n)/\mathbf{U}(\mathbf{1}_n) \cong \mathbf{SU}(n)/\mathbb{I}(n)$. For example, for $n = 2$, the hyperisospin action on the hyperisospin gauge fields (chapter "Gauge Interactions") defines the adjoint semidirect group $\operatorname{Int} \mathbf{U}(2) \;\vec{\times}\; \log \mathbf{U}(2) \cong \mathbf{SO}(3) \;\vec{\times}\; \mathbb{R}^4$ with a trivial adjoint action of the hypercharge group $\mathbf{U}(1)$. The internal group action respects the Lie algebra decomposition $i\mathbb{R}(n) \ni \frac{i}{2}\gamma = \frac{i}{2}(\gamma_0 + \vec{\gamma})$ into abelian and simple parts and can be exponentiated for the inner automorphisms of $\mathbf{U}(n)$:

$$\mathbf{U}(n) \times \mathbf{U}(n) \longrightarrow \mathbf{U}(n),\quad \exp \tfrac{i}{2}\gamma \longmapsto u \circ \exp \tfrac{i}{2}\gamma \circ u^\star = \exp(u \circ \tfrac{i}{2}\gamma \circ u^\star).$$

The spacetime manifold $\mathbf{D}(n)$ has as external action group the full linear group $\mathbf{GL}(\mathbb{C}^n)$, via right conjugate multiplication

$$\mathbf{D}(n) \times \mathbf{GL}(\mathbb{C}^n) \longrightarrow \mathbf{D}(n),\quad d = |d| \longmapsto |d \circ g^\star|.$$

The external group is the product of the *causal group with a phase group and the generalized Lorentz covering group*

$$\mathbf{GL}(\mathbb{C}^n) = \mathbf{D}(\mathbf{1}_n) \times [\mathbf{U}(\mathbf{1}_n) \circ \mathbf{SL}(\mathbb{C}^n)].$$

The spacetime translations are acted on by the generalized Poincaré group

$$\mathbf{GL}(\mathbb{C}^n)/\mathbf{U}(\mathbf{1}_n) \;\vec{\times}\; \mathbb{R}(n) \cong \begin{cases} \mathbf{D}(1) \;\vec{\times}\; \mathbb{R}, & \text{for } n = 1,\\ [\mathbf{D}(\mathbf{1}_4) \times \mathbf{SO}_0(1,3)] \;\vec{\times}\; \mathbb{R}^4, & \text{for } n = 2,\end{cases}$$

with the defining representation of the generalized *Lorentz group* $\mathbf{SL}(\mathbb{C}^n)/\mathbb{I}(n)$:

$$g, \hat{g} = g^{-1\star} \in \mathbf{GL}(\mathbb{C}^n): \quad \begin{cases} \mathbb{R}(n) \longrightarrow \mathbb{R}(n), & x \longmapsto g \circ x \circ g^\star, \\ \mathbb{R}(n)^T \longrightarrow \mathbb{R}(n)^T, & q \longmapsto \hat{g} \circ q \circ \hat{g}^\star. \end{cases}$$

For time, $n = 1$, the causal transformations (dilations) act on the time translations. In the relativistic case, $n = 2$, one has the action of the causal (dilation) Lorentz group. In contrast to the corresponding transition from internal Lie algebra $\log \mathbf{U}(n)$ to group $\mathbf{U}(n)$, the exponentiation from linear spacetime to nonlinear spacetime (for $n \geq 2$ no group) is incompatible with the action of the group $\mathbf{GL}(\mathbb{C}^n)$:

$$g \in \mathbf{GL}(\mathbb{C}^n): \quad |d| = e^\psi \longmapsto |e^\psi \circ g^\star|, \text{ but } e^x \longmapsto e^{g \circ x \circ g^\star}.$$

The exponentiation does not respect the decomposition $\mathbb{R}(n) \ni x = x_0 + \vec{x}$ into time and position translations, in contrast to the adjoint action for the internal $\mathbf{U}(n)$-degrees of freedom. Nonlinear spacetime can be parametrized by the translations of the open forward cone:

$$\begin{array}{lll} \mathbf{D}(n) & \cong \mathbb{R}(n)^+_{\text{time}} & = \{x \in \mathbb{R}(n) \mid \operatorname{spec} x > 0\} = \{\vartheta(x)x \mid x \in \mathbb{R}(n)\}, \\ \mathbf{D}(2) & \cong \mathbb{R}(2)^+_{\text{time}} & = \{\vartheta(x_0)\vartheta(x^2)x \mid x \in \mathbb{R}(2)\}. \end{array}$$

Then the action of the external dilation-Lorentz group on nonlinear spacetime can be written as an action on the translations:

$$\mathbf{GL}(\mathbb{C}^n) \times \mathbf{D}(n) \longrightarrow \mathbf{D}(n), \quad (g, \vartheta(x)x) \longmapsto g \circ \vartheta(x)x \circ g^\star.$$

The future cone parametrization with translation parameters is related to an exponential Lie parametrization, i.e., an orbit parametrization as follows:

$$\begin{array}{rrl} & \vartheta(x)x = \vartheta(x)(x_0 + \vec{x}) & = e^\psi = e^{\psi_0} e^{\vec{\psi}} = 1 + \psi_0 + \vec{\psi} + \cdots, \\ \text{for } n = 2: & \vartheta(x_0)\vartheta(x^2)(x_0 + \vec{x}) & = e^{\psi_0}(\cosh|\vec{\psi}| + \frac{\vec{\psi}}{|\vec{\psi}|}\sinh|\vec{\psi}|) \\ & = u(\frac{\vec{x}}{r}) \circ (x_0 + \sigma^3 r) \circ u(\frac{\vec{x}}{r})^\star & = u(\frac{\vec{\psi}}{|\vec{\psi}|}) \circ e^{\psi_0 + \sigma^3|\vec{\psi}|} \circ u(\frac{\vec{\psi}}{|\vec{\psi}|})^\star \\ & \text{with } e^{2\psi_0} = x_0^2 - r^2, & \tanh^2|\vec{\psi}| = \frac{r^2}{x_0^2}, \quad \frac{\vec{\psi}}{|\vec{\psi}|} = \frac{\vec{x}}{r}. \end{array}$$

The $\mathbf{D}(1)$-parameter ψ_0 gives eigentime.

2.4 Orbits and Fixgroups of Hyperisospin

Orbits and fixgroups ("little groups") of hyperisopin $\mathbf{U}(2)$ and the related transmutators from hyperisospin to electromagnetic transformations are relevant for the symmetry reduction from hyperisospin fields to charged particles in the electroweak standard model (chapter "Gauge Interactions").

2.4.1 Higgs Hilbert Space and Goldstone Manifold

For the internal group $\mathbf{U}(n)$-action, $n \geq 2$, the defining representation acts on the *Higgs Hilbert space* H. There is only one fixgroup type for all nontrivial vectors:

$$\begin{aligned} \Phi \in H \cong \mathbb{C}^n: \quad & \mathbf{U}(n)_\Phi = \{u \in \mathbf{U}(n) \mid u(\Phi) = \Phi\}; \\ \langle \Phi | \Phi \rangle = \sum_{A=1}^{n} |\Phi^A|^2 = |\Phi|^2 > 0 \Rightarrow \quad & \mathbf{U}(n)_\Phi \cong \mathbf{U}(n-1). \end{aligned}$$

The nontrivial vector can be used in a basis:

$$\begin{aligned} \Phi = e^n \quad & \Rightarrow \mathbf{U}(n)_{e^n} = \left(\begin{array}{c|c} \mathbf{U}(n-1) & 0 \\ \hline 0 & 1 \end{array} \right), \\ \text{e.g., } n=2: \quad & \mathbf{U}(2) = \{ e^{\frac{i}{2}(\gamma_0 + \vec{\gamma})} \mid \text{Pauli matrices } \vec{\gamma} = \gamma_a \tau^a, a = 1,2,3 \}, \\ & \mathbf{U}(2)_{\begin{pmatrix} 0 \\ 1 \end{pmatrix}} = \{ e^{i\gamma_0 \frac{\mathbf{1}_2 + \tau^3}{2}} = \begin{pmatrix} e^{i\gamma_0} & 0 \\ 0 & 1 \end{pmatrix} \} = \mathbf{U}(1)_+. \end{aligned}$$

The fixgroup for $n = 2$ is a factor of a Cartan torus $\mathbf{U}(1)_+ \times \mathbf{U}(1)_-$, not the phase group $\mathbf{U}(\mathbf{1}_2) \subset \mathbf{U}(2)$ or $\mathbf{U}(1)_3 \subset \mathbf{SU}(2)$. $\mathbf{U}(1)_+$ is called an *electromagnetic group* in the hyperisospin group (one could equivalently take $\mathbf{U}(1)_-$). A fixgroup $\mathbf{U}(n-1)$ is a stabilgroup for an orthogonal Higgs space decomposition $H \cong \mathbb{C}e^n \perp \mathbb{C}^{n-1}$ with a 1-dimensional subspace.

With the fixgroup classes, the internal group $\mathbf{U}(n) = \mathbf{U}(\mathbf{1}_n) \circ \mathbf{SU}(n)$ gives as fixgroup orientations the compact *Goldstone manifold*, for $n = 2$ the *orientation manifold of the electromagnetic group in the hyperisospin group*:

$$\begin{aligned} \mathcal{G}^{2n-1} & \cong \mathbf{U}(n)/\mathbf{U}(n-1), \quad \dim_{\mathbb{R}} \mathcal{G}^{2n-1} = 2n-1, \\ \mathcal{G}^3 & \cong \mathbf{U}(2)/\mathbf{U}(1)_+. \end{aligned}$$

The Higgs vectors are, as a real manifold, isomorphic to the product of their absolute values $\mathbb{R}_+ \ni |\Phi|$ with the orientation manifold

$$H \cong \mathbb{C}^n \cong \mathbb{R}^{2n} \cong \mathbb{R}_+ \times \mathcal{G}^{2n-1}.$$

This decomposition is the unitary analogue to orthogonal polar coordinates $\mathbb{R}^n \cong \mathbb{R}_+ \times \Omega^{n-1}$ with the sphere $\Omega^{n-1} \cong \mathbf{SO}(n)/\mathbf{SO}(n-1)$. The Goldstone manifold is parametrizable by the orbit of a nontrivial Higgs vector

$$\mathbf{U}(n)/\mathbf{U}(n-1) \cong \mathbf{U}(n).\Phi \text{ for } \Phi \neq 0.$$

A Higgs vector can be used for a parametrization of the fundamental representation of the fixgroup classes as acting on the Higgs space

$$\begin{aligned} \mathbf{U}(n)/\mathbf{U}(n-1) \cong \mathcal{G}^{2n-1} \ni \tfrac{\Phi}{|\Phi|} & \longmapsto v(\tfrac{\Phi}{|\Phi|}) \in \mathbf{U}(n)/\mathbf{U}(n-1) \\ \text{with } v(\tfrac{\Phi}{|\Phi|}) \begin{pmatrix} 0 \\ \cdots \\ 0 \\ |\Phi| \end{pmatrix} & = \begin{pmatrix} \Phi^1 \\ \cdots \\ \Phi^{n-1} \\ \Phi^n \end{pmatrix}, \quad v(e^n) = \mathbf{1}_n. \end{aligned}$$

Given the Higgs vector Φ as nth column, the columns of the matrix $v(\frac{\Phi}{|\Phi|})$ are a $\mathbf{U}(n)$-orthonormal basis of the Higgs space. For the start vector e^n, there arises the unit matrix. This is analogous to a (3×3) matrix representation of axial rotation classes $\mathbf{SO}(3)/\mathbf{SO}(2)$ with $\frac{\vec{\varphi}}{|\vec{\varphi}|}$ in the third column (chapter "Spacetime Translations"). For example, the $\mathbf{U}(2)$ matrices, determined up to an electromagnetic $\mathbf{U}(1)_+$-transformation, constitute the *fundamental Higgs representation of the Goldstone manifold* $\mathcal{G}^3$, it is parametrizable with the two complex Higgs vector components

$$\begin{aligned}\begin{pmatrix}\Phi^1\\ \Phi^2\end{pmatrix} = v(\tfrac{\Phi}{|\Phi|})\begin{pmatrix}0\\ |\Phi|\end{pmatrix} \Rightarrow v(\tfrac{\Phi}{|\Phi|}) &= \tfrac{1}{|\Phi|}\begin{pmatrix}\Phi_2^\star & \Phi^1\\ -\Phi_1^\star & \Phi^2\end{pmatrix} \in \mathbf{U}(2)/\mathbf{U}(1)_+,\\ v(e^2) &= \mathbf{1}_2.\end{aligned}$$

The *Goldstone translations* are representatives of the fix-Lie algebra classes in the tangent space $\log \mathbf{U}(n)$, e.g.,

$$\begin{aligned} i\begin{pmatrix}\gamma_0 & 0\\ 0 & 0\end{pmatrix} \in \log \mathbf{U}(2)_{e^2} &= \log \mathbf{U}(1)_+,\\ \tfrac{i}{2}\gamma_\perp = \tfrac{i}{2}\begin{pmatrix}0 & \gamma_1 - i\gamma_2\\ \gamma_1 + i\gamma_2 & -2\gamma_3\end{pmatrix} &= i\tfrac{-\gamma_3 + \vec{\gamma}}{2} \quad \in \log \mathbf{U}(2)/\log \mathbf{U}(1)_+,\\ e^{\frac{i}{2}\gamma_\perp} = \tfrac{e^{-\frac{i}{2}\gamma_3}}{\sqrt{1+\vec{\vartheta}^2}}\begin{pmatrix}1 + i\vartheta_3 & \vartheta_2 + i\vartheta_1\\ -\vartheta_2 + i\vartheta_1 & 1 - i\vartheta_3\end{pmatrix} &= v(\tfrac{\Phi}{|\Phi|}) \quad \in \mathbf{U}(2)/\mathbf{U}(1)_+, \quad \vec{\vartheta} = \tfrac{\vec{\gamma}}{|\vec{\gamma}|}\tan\tfrac{|\vec{\gamma}|}{2}.\end{aligned}$$

2.4.2 Electromagnetism-Hyperisospin Transmutators

A representation of a symmetric space has a typical hybrid transformation behavior - here: A left $\mathbf{U}(n)$ action on the Goldstone manifold representation $v(\frac{\Phi}{|\Phi|}) \in \mathbf{U}(n)/\mathbf{U}(n-1)$ gives the representation with the $\mathbf{U}(n)$-transformed Higgs vector $u.\Phi$ up to a right action with a *Wigner element* from the fixgroup $\mathbf{U}(n-1)$. The fixgroup action "goes through":

$$\begin{aligned} u \in \mathbf{U}(n) \;\Rightarrow\; u \circ v(\tfrac{\Phi}{|\Phi|}) &= v(u.\tfrac{\Phi}{|\Phi|}) \circ t(u, \tfrac{\Phi}{|\Phi|}) \text{ with } t \in \mathbf{U}(n-1),\\ t \in \mathbf{U}(n-1) \;\Rightarrow\; t \circ v(\tfrac{\Phi}{|\Phi|}) &= v(t.\tfrac{\Phi}{|\Phi|}) \circ t.\end{aligned}$$

e.g., for the electromagnetic orientation manifold $\mathcal{G}^3$

$$\begin{aligned} u = e^{\frac{i}{2}(\gamma_0 + \vec{\gamma})} \in \mathbf{U}(2) \Rightarrow u \circ v(\tfrac{\Phi}{|\Phi|}) &= v(u.\tfrac{\Phi}{|\Phi|}) \circ t(u)\\ &\qquad \text{with } t(u) = e^{i\gamma_0 \frac{\mathbf{1}_2 + \tau^3}{2}} \in \mathbf{U}(1)_+,\\ t = \begin{pmatrix}e^{i\gamma_0} & 0\\ 0 & 1\end{pmatrix} \in \mathbf{U}(1)_- \Rightarrow \quad v(t.\tfrac{\Phi}{|\Phi|}) &= \tfrac{1}{|\Phi|}\begin{pmatrix}\Phi_2^\star & e^{i\gamma_0}\Phi^1\\ -e^{-i\gamma_0}\Phi_1^\star & \Phi^2\end{pmatrix} = t \circ v(\tfrac{\Phi}{|\Phi|}) \circ t^\star.\end{aligned}$$

The two columns in the defining representation of the Goldstone manifold $\mathbf{U}(2)/\mathbf{U}(1)_+ \cong \mathcal{G}^3$ for the electromagnetic group orientation on $H \cong \mathbb{C}^2$ are acted on from the left with hyperisospin $\mathbf{U}(2)$-transformations and from the right with a $\mathbf{U}(1)_+$-transformation. Therefore they are called *transmutators between the electromagnetic and hyperisospin groups*:

$$\begin{aligned} v(\tfrac{\Phi}{|\Phi|}) = \tfrac{1}{|\Phi|}\left(\begin{array}{c|c}\Phi_2^\star & \Phi^1\\ -\Phi_1^\star & \Phi^2\end{array}\right) &= \tfrac{1}{|\Phi|}(\tilde{\Phi}, \Phi) \text{ with } \tilde{\Phi}^\alpha = \epsilon^{\alpha\beta}\Phi_\beta^\star,\\ u \circ v(\tfrac{\Phi}{|\Phi|})^1 = v(u.\tfrac{\Phi}{|\Phi|})^1 \circ e^{i\gamma_0}, &\quad u \circ v(\tfrac{\Phi}{|\Phi|})^2 = v(u.\tfrac{\Phi}{|\Phi|})^2.\end{aligned}$$

For a hyperisospinor field, embedding a charged particle, e.g., for the lepton field of the standard model, embedding the electron (chapter "Gauge Interactions"), this transformation property may be expressed as follows: A hyperisopin $\mathbf{U}(2)$-transformation of the lepton field where the electron's charge group is defined by $\mathbf{U}(1)_+ \subset \mathbf{U}(2)$ gives a $\mathbf{U}(2)$-transformed lepton field with the embedded electron's electromagnetic $\mathbf{U}(1)_+$-property, in general, "rotated" with respect to the original one.

Transmutators are an important tool in the theory of inducing group G-representations from a representation of a subgroup $H \subseteq G$ (chapter "Harmonic Analysis").

The general Goldstone manifold case: A representation of the Goldstone manifold in the automorphisms of a finite-dimensional vector space V,

$$\begin{aligned} \mathbf{U}(n)/\mathbf{U}(n-1) \cong \mathcal{G}^{2n-1} \ni \tfrac{\Phi}{|\Phi|} &\longmapsto D(v(\tfrac{\Phi}{|\Phi|})) \in \mathbf{U}(V), \\ \text{where } D : \mathbf{U}(n) \ni v &\longmapsto D(v) \in \mathbf{U}(V). \end{aligned}$$

The decomposition of V with respect to fixgroup-stable irreducible subspaces and representations (square block matrices in $\mathbf{U}(W^\iota)$)

$$\begin{aligned} &\mathbf{U}(n) \cong \bigoplus \mathbf{U}(n-1), \quad V \overset{\mathbf{U}(n-1)}{\cong} \bigoplus_{\iota=1}^{N} W^\iota, \\ &\mathbf{U}(n-1) \ni t \longmapsto D(t) \cong \bigoplus_{\iota=1}^{N} D^\iota(t) \cong \begin{pmatrix} D^1(t) & 0 & \cdots & 0 \\ 0 & D^2(t) & \cdots & 0 \\ & & \cdots & \\ 0 & 0 & \cdots & D^N(t) \end{pmatrix}, \end{aligned}$$

gives a corresponding decomposition of the $\mathcal{G}^{2n-1}$-representation into irreducible transmutators from $\mathbf{U}(n)$ to $\mathbf{U}(n-1)$. They are rectangular matrices in $W^\iota \otimes V^T$,

$$\begin{aligned} &D(v(\tfrac{\Phi}{|\Phi|})) \cong \bigoplus_{\iota=1}^{N} d^\iota(v(\tfrac{\Phi}{|\Phi|})) \cong \begin{pmatrix} d^1(v\tfrac{\Phi}{|\Phi|})) \,\Big|\, d^2(v(\tfrac{\Phi}{|\Phi|})) \,\Big|\, \cdots \,\Big|\, d^N(v(\tfrac{\Phi}{|\Phi|})) \end{pmatrix}, \\ &d^\iota(v(\tfrac{\Phi}{|\Phi|})) : V \longrightarrow W^\iota, \end{aligned}$$

and have the characteristic hybrid transformation behavior involving the "large" group $\mathbf{U}(n)$ and the fixgroup ("little group") $\mathbf{U}(n-1)$

$$\begin{aligned} u \in \mathbf{U}(n) \Rightarrow \quad & D(u) \circ d^\iota(v(\tfrac{\Phi}{|\Phi|})) = d^\iota(v(u.\tfrac{\Phi}{|\Phi|})) \circ D^\iota(t(u, \tfrac{\Phi}{|\Phi|})) \\ & \text{with } t(u, \tfrac{\Phi}{|\Phi|}) \in \mathbf{U}(n-1) \text{ Wigner transformation}, \\ t \in \mathbf{U}(n-1) \Rightarrow \quad & D(t) \circ d^\iota(v(\tfrac{\Phi}{|\Phi|})) = d^\iota(v(t.\tfrac{\Phi}{|\Phi|})) \circ D^\iota(t). \end{aligned}$$

Back to the special case $n = 2$; The irreducible representations of hyperisospin $\mathbf{U}(2)$ with central correlation $\mathbf{SU}(2) \cap \mathbf{U}(\mathbf{1}_2) = \{\pm \mathbf{1}_2\}$ (chapter "Rational Quantum Numbers") can be constructed by products of the defining representations $[y|2T] = [\pm\frac{1}{2}|1]$ with hypercharge $y = \frac{1}{2}$ and isospin $T = \frac{1}{2}$:

$$\begin{aligned} \mathbf{irrep}\,\mathbf{U}(2) \ni [\pm n + T|2T] &\cong [\pm 1|0]^n \otimes \bigvee^{2T} [\tfrac{1}{2}|1], \\ [\pm 1|0] &\cong [\pm\tfrac{1}{2}|1] \wedge [\pm\tfrac{1}{2}|1]. \end{aligned}$$

They show the hypercharge-isospin correlation $y = T \pm n$ with natural n, i.e., (y,T) either both integer or both half-integer.

The decomposition of a $\mathbf{U}(2)$-representation with respect to irreducible representations of the electromagnetic group $\mathbf{U}(1)_+ \ni e^{i\gamma_0} \longmapsto e^{zi\gamma_0}$ is characterized by integer charge numbers $z \in \mathbb{Z}$:

$$\mathbf{U}(2) \cong \bigoplus \mathbf{U}(1)_+ : \quad [\pm n + T|2T] \cong \bigoplus_{z=\pm n}^{\pm n+2T} [z],$$
$$\text{e.g.,} \quad \begin{cases} [\pm\frac{1}{2}|1] \cong [0] \oplus [\pm 1], \\ [0|2] \cong [-1] \oplus [0] \oplus [1]. \end{cases}$$

These $\mathbf{U}(2)$-representations have to be used for Higgs parametrized representations of the orientation manifold $\mathcal{G}^3$ of the electromagnetic group. One obtains products of the defining representation and its conjugate,

$$\begin{array}{l} [\frac{1}{2}|1](\frac{\Phi}{|\Phi|}) = v(\frac{\Phi}{|\Phi|}), \quad [-\frac{1}{2}|1](\frac{\Phi}{|\Phi|}) = v^\star(\frac{\Phi}{|\Phi|}), \\ \mathcal{G}^3 \longrightarrow \mathbf{U}(1+2T), \quad \frac{\Phi}{|\Phi|} \longmapsto [\pm n + T|2T](\frac{\Phi}{|\Phi|}), \end{array}$$

with the examples for the $\mathbf{U}(1)_+$ to $\mathbf{U}(2)$ transmutation on $\mathbb{C}$ with hypercharge $\mathbf{U}(1)$ nontrivial isospin $\mathbf{SU}(2)$-singlets,

$$\begin{array}{rll} [1|0](\frac{\Phi}{|\Phi}) & = \frac{\Phi^\alpha \epsilon_{\alpha\beta}\Phi^\beta}{|\Phi|^2} \in \mathbf{U}(1) \text{ with } [1|0] & \overset{\mathbf{U}(1)_+}{\cong} [1], \\ [-1|0](\frac{\Phi}{|\Phi}) & = \frac{\Phi^\star_\alpha \epsilon^{\alpha\beta}\Phi^\star_\beta}{|\Phi|^2} \in \mathbf{U}(1) \text{ with } [-1|0] & \overset{\mathbf{U}(1)_+}{\cong} [-1], \end{array}$$

and on $\mathbb{C}^3$ with hypercharge $\mathbf{U}(1)$ trivial isospin $\mathbf{SU}(2)$-triplets. Here the columns define three transmutators for charge $z = -1, 0, 1$:

$$\begin{array}{rl} [0|2](\frac{\Phi}{|\Phi|}) & = \frac{1}{2} \operatorname{tr} \tau^a v(\frac{\Phi}{|\Phi|}) \tau^b v^\star(\frac{\Phi}{|\Phi|}) \\ & = \left(\frac{\Phi^\star \vec\tau \tilde\Phi + \tilde\Phi^\star \vec\tau \Phi}{2|\Phi|^2} \,\Big|\, -i\frac{\Phi^\star \vec\tau \tilde\Phi - \tilde\Phi^\star \vec\tau \Phi}{2|\Phi|^2} \,\Big|\, \frac{\tilde\Phi^\star \vec\tau \tilde\Phi - \Phi^\star \vec\tau \Phi}{2|\Phi|^2} \right) \in \mathbf{SO}(3) \\ \text{with} & [0|2] \overset{\mathbf{U}(1)_+}{\cong} [-1] \oplus [0] \oplus [1]. \end{array}$$

2.5 Orbits and Fixgroups in Spacetime

Orbits and fixgroups ("little groups") of the Lorentz group $\mathbf{SO}_0(1,3)$ and the related transformations from the Lorentz group to rotation groups are relevant for the embedding of particles into fields (chapters "Massive Particle Quantum Fields" and "Massless Quantum Fields").

2.5.1 Fixgroups of Spacetime Translations

The action of the noncompact external group on the spacetime translations defines different fixgroup types. The fixgroups

$$x \in \mathbb{R}(n): \quad \mathbf{SL}(\mathbb{C}^n)_x = \{s \in \mathbf{SL}(\mathbb{C}^n) \,\big|\, x = s \circ x \circ s^*\}$$

are easily determined for diagonal translations $x = \operatorname{diag} x$: They are invariance groups of sesquilinear forms, for the nondegenerate case $\mathbb{I}(n)$-classes of unitary $\mathbf{SL}(\mathbb{C}^n)$-subgroups.

First the fixgroups and stabilgroups of the Minkowski translations $\mathbb{R}(2)$: The fixgroup ("little group") with respect to the action of the Lorentz group

$$x \in \mathbb{R}(2): \ \mathbf{SO}_0(1,3)_x = \{\Lambda(s) \mid s \in \mathbf{SL}(\mathbb{C}^2), \quad s \circ x \circ s^\star = x\} \cong \mathbf{SL}(\mathbb{C}^2)_x/\mathbb{I}(2)$$

has the fix-Lie algebra

$$\log \mathbf{SL}(\mathbb{C}^2)_x = \{l = i\vec{\alpha} + \vec{\beta} \in \log \mathbf{SL}(\mathbb{C}^2) \mid lx + xl^\star = 0\}.$$

It will be given in the Minkowski representation for a Sylvester basis with a decomposition into time and position and a Cartan basis with $\mathbf{SO}(2) \times \mathbf{SO}_0(1,1)$-eigenvectors

$$\log \mathbf{SO}_0(1,3) \ni \begin{pmatrix} 0 & \psi_1 & \psi_2 & \psi_3 \\ \psi_1 & 0 & \varphi_3 & -\varphi_2 \\ \psi_2 & -\varphi_3 & 0 & \varphi_1 \\ \psi_3 & \varphi_2 & -\varphi_1 & 0 \end{pmatrix} \quad \text{for } \begin{pmatrix} x_0 \\ \vec{x} \end{pmatrix}$$

$$\text{with } \vec{\varphi} = 2\vec{\alpha}, \quad \vec{\psi} = 2\vec{\beta}$$

$$\sim \begin{pmatrix} \psi_3 & \gamma_+ & \overline{\gamma_+} & 0 \\ \overline{\gamma_-} & i\varphi_3 & 0 & \overline{\gamma_+} \\ \gamma_- & 0 & -i\varphi_3 & \gamma_+ \\ 0 & \gamma_- & \overline{\gamma_-} & -\psi_3 \end{pmatrix} \quad \text{for } \begin{pmatrix} x_0 + x_3 \\ x_1 - ix_2 \\ x_1 + ix_2 \\ x_0 - x_3 \end{pmatrix}$$

$$\text{with } \gamma_\pm = \frac{\psi_1 + i\psi_2 \pm (\varphi_2 - i\varphi_1)}{2}.$$

The fixgroups are isomorphic for all translations of a Lorentz group orbit and also for those translations that arise by transformation with the centralizer of $\mathbf{SO}_0(1,3)$ in $\mathbf{GL}(\mathbb{R}^4)$, given by the causal (dilation) and the reflection group

$$\{g \in \mathbf{GL}(\mathbb{R}^4) \mid g \circ \Lambda \circ g^{-1} = \Lambda \text{ for all } \Lambda \in \mathbf{SO}_0(1,3)\} \cong \mathbf{D}(1) \times \mathbb{I}(2).$$

Any inner automorphism with $g \in \mathbf{GL}(\mathbb{R}^4)$ gives a Lorentz group in the tetrad manifold $\mathbf{GL}(\mathbb{R}^4)/\mathbf{SO}_0(1,3)$. To stay with the same Lorentz group, g has to be an element from the centralizer.

Therefore there exist, in addition to the full fixgroup $\mathbf{SL}(\mathbb{C}^2)$ for $x = 0$, three proper fixgroup types for the nontrivial translations

$$\begin{aligned}
&\text{timelike:} && \det x = x^2 > 0 \ \Rightarrow \ x \cong \begin{pmatrix} 1 & 0 \\ 0 & 1 \end{pmatrix} = \mathbf{1}_2 \in \mathbb{R}(2)^+_{\text{time}}, \\
&\text{spacelike:} && x^2 < 0 \ \Rightarrow \ x \cong \begin{pmatrix} 1 & 0 \\ 0 & -1 \end{pmatrix} = \sigma^3 \in \mathbb{R}(2)_{\text{position}}, \\
&\text{lightlike:} && x^2 = 0, \ x \neq 0 \ \Rightarrow \ x \cong \begin{cases} \begin{pmatrix} 1 & 0 \\ 0 & 0 \end{pmatrix} = \frac{\mathbf{1}_2 + \sigma^3}{2} = \pi^+ \in \mathbb{R}(2)^+_{\text{light}}, \\ \begin{pmatrix} 0 & 0 \\ 0 & 1 \end{pmatrix} = \frac{\mathbf{1}_2 - \sigma^3}{2} = \pi^- \in \mathbb{R}(2)^+_{\text{light}}. \end{cases}
\end{aligned}$$

The *fixgroup of a nontrivial timelike translation*, e.g., of $\mathbf{1}_2$, is the compact rotation group

$$x^2 > 0 \Rightarrow \begin{cases} \{s \in \mathbf{SL}(\mathbb{C}^2) \mid s \circ s^\star = \mathbf{1}_2\} = \mathbf{SU}(2) \ni e^{i\vec{\alpha}}, \\ \mathbf{SO}_0(1,3)_{\mathbf{1}_2} \cong \mathbf{SO}(3), \\ \log \mathbf{SO}(3) \ni \begin{pmatrix} 0 & 0 & 0 & 0 \\ 0 & 0 & \varphi_3 & -\varphi_2 \\ 0 & -\varphi_3 & 0 & \varphi_1 \\ 0 & \varphi_2 & -\varphi_1 & 0 \end{pmatrix}. \end{cases}$$

The rotation group is the fixgroup of all associated time translations $\mathbb{T} = \mathbb{R}\mathbf{1}_2$ and the stabilgroup of the corresponding orthogonal Sylvester decomposition $\mathbb{R}(2) \cong \mathbb{T} \perp \mathbb{S}^3$ into time and position translations. It distinguishes in the Poincaré group the direct product subgroups with timelike translations

$$\mathbf{SO}_0(1,3) \vec{\times} \mathbb{R}^4 \supset \mathbf{SO}(3) \times \mathbb{R} \ni (O,x) \text{ with } x^2 > 0.$$

The *fixgroup of a spacelike translation*, e.g., of σ^3, is the Lorentz group for two position dimensions

$$x^2 < 0 \Rightarrow \begin{cases} \{s \in \mathbf{SL}(\mathbb{C}^2) \mid s \circ \sigma^3 \circ s^\star = \sigma^3\} = \mathbf{SU}(1,1) \ni e^{i\alpha_3\sigma^3 + \beta_1\sigma^1 + \beta_2\sigma^2}, \\ \mathbf{SO}_0(1,3)_{\sigma^3} \cong \mathbf{SO}_0(1,2), \\ \log \mathbf{SO}(1,2) \ni \begin{pmatrix} 0 & \psi_1 & \psi_2 & 0 \\ \psi_1 & 0 & \varphi_3 & 0 \\ \psi_2 & -\varphi_3 & 0 & 0 \\ 0 & 0 & 0 & 0 \end{pmatrix}. \end{cases}$$

This $(1,2)$-Lorentz group is the stabilgroup of the corresponding orthogonal decomposition $\mathbb{R}(2) \cong \mathbb{S}^1 \perp \mathbb{M}^{1+2}$ with 1-dimensional position translations $\mathbb{S}^1 = \mathbb{R}\sigma^3$ and an $\mathbf{SO}_0(1,2)$-spacetime. It distinguishes the direct product subgroups with spacelike translations

$$\mathbf{SO}_0(1,3) \vec{\times} \mathbb{R}^4 \supset \mathbf{SO}_0(1,2) \times \mathbb{R} \ni (\Lambda,x) \text{ with } x^2 < 0.$$

For the lightlike translations $x^2 = 0$, $x \neq 0$ the situation is more complicated, since the sesquilinear form, e.g., for $\pi^\pm$, is degenerate. The *fixgroup of one lightlike translation*, e.g., of π^+, is the noncompact semidirect Euclidean group in two dimensions where the $\mathbb{R}^2$-translations arise from the noncompact boosts

$$x^2 = 0,\ x \neq 0 \Rightarrow \begin{cases} \{s \in \mathbf{SL}(\mathbb{C}^2) \mid s \circ \pi^+ \circ s^\star = \pi^+\} \cong \mathbf{SO}(2) \vec{\times} \mathbb{R}^2, \\ \ni e^{(\psi_1 + i\psi_2)\sigma_+} e^{i\alpha_3\sigma^3} = \begin{pmatrix} e^{i\alpha_3} & e^{-i\alpha_3}(\psi_1 + i\psi_2) \\ 0 & e^{-i\alpha_3} \end{pmatrix}, \\ \mathbf{SO}_0(1,3)_{\pi^+} \cong \mathbf{SO}(2) \vec{\times} \mathbb{R}^2, \\ \log[\mathbf{SO}(2) \vec{\times} \mathbb{R}^2] \ni \begin{pmatrix} 0 & \psi_1 & \psi_2 & 0 \\ \psi_1 & 0 & \varphi_3 & -\psi_1 \\ \psi_2 & -\varphi_3 & 0 & -\psi_2 \\ 0 & \psi_1 & \psi_2 & 0 \end{pmatrix} \\ \sim \begin{pmatrix} 0 & \psi & \overline{\psi} & 0 \\ 0 & i\varphi_3 & 0 & \overline{\psi} \\ 0 & 0 & -i\varphi_3 & \psi \\ 0 & 0 & 0 & 0 \end{pmatrix}, \ \psi = \psi_1 + i\psi_2, \end{cases}$$

The fixgroup for lightlike translations $\mathbb{L}_+ = \mathbb{R}\pi^+$ is not a stabilgroup for a translation decomposition, since no direct $\mathbb{L}_+$-complement in $\mathbb{R}(2)$ is stable. It distinguishes the direct product subgroups with lightlike translations

$$\mathbf{SO}_0(1,3) \vec{\times} \mathbb{R}^4 \supset [\mathbf{SO}(2) \vec{\times} \mathbb{R}^2] \times \mathbb{R} \ni (L,x) \text{ with } x^2 = 0,\ x \neq 0.$$

The fixgroup of two linearly independent lightlike translations, e.g., of $\pi^\pm$, i.e., the *fixgroup of all lightlike translations* in a decomposition $\mathbb{L}^2 = \mathbb{L}_+ \oplus \mathbb{L}_-$, is the compact axial rotation group

$$x_\pm^2 = 0,\ x_\pm \neq 0 \Rightarrow \begin{cases} \{s \in \mathbf{SL}(\mathbb{C}^2) \mid & s \circ \pi^\pm \circ s^\star = \pi^\pm\} = \mathbf{SO}(2) \ni e^{i\alpha_3\sigma^3}, \\ \mathbf{SO}_0(1,3)_{\mathbb{L}^2} & = \mathbf{SO}(3) \cap \mathbf{SO}_0(1,2) \cong \mathbf{SO}(2), \\ \log \mathbf{SO}(2) & \ni \begin{pmatrix} 0 & 0 & 0 & 0 \\ 0 & i\varphi_3 & 0 & 0 \\ 0 & 0 & -i\varphi_3 & 0 \\ 0 & 0 & 0 & 0 \end{pmatrix}. \end{cases}$$

This fixgroup is the intersection of the time- and spacelike fixgroups. It is a fixgroup in the fixgroup $\mathbf{SO}(2) \vec{\times} \mathbb{R}^2$, i.e., the fixgroup for the action of the axial rotations on trivial boosts.

Hence the manifolds with all nontrivial time-, light-, and spacelike translations are isomorphic to symmetric spaces with the characteristic fixgroups as equivalences

translations	fixgroup in $\mathbf{SO}_0(1,3)$	isomorphic manifold
$\mathbb{R}(2)^\pm_{\text{time}}$	$\mathbf{SU}(2)/\mathbb{I}(2) \cong \mathbf{SO}(3)$	$\mathbf{GL}(\mathbb{C}^2)/\mathbf{U}(2)$ $\cong \mathbf{D}(1) \times \mathbf{SO}_0(1,3)/\mathbf{SO}(3)$ $\cong \mathbf{D}(1) \times \mathcal{Y}^3$
$\mathbb{R}(2)^\pm_{\text{light}}$	$\mathbf{SO}(2) \vec{\times} \mathbb{R}^2$	$\mathbf{SO}_0(1,3)/\mathbf{SO}(2) \vec{\times} \mathbb{R}^2$ $\cong \mathbf{D}(1) \times \Omega^2$
$\mathbb{R}(2)_{\text{position}} \setminus \{0\}$	$\mathbf{SU}(1,1)/\mathbb{I}(2) \cong \mathbf{SO}_0(1,2)$	$\mathbf{GL}(\mathbb{C}^2)/\mathbf{U}(1,1)$ $\cong \mathbf{D}(1) \times \mathbf{SO}_0(1,3)/\mathbf{SO}_0(1,2)$ $\cong \mathbf{D}(1) \times \mathcal{Y}^{(1,2)}$

For the general case $n \geq 2$ the distribution of the spectral values $\{\pm 1, 0\}$ in a normalized diagonal translation $\operatorname{diag} x$ and its invariance group $s \circ \operatorname{diag} x \circ s^\star = \operatorname{diag} x$ characterizes the fixgroup. The disjoint decompositions of the spacelike and lightlike manifold above correspond to the $(n-1)$ different fixgroups

translations	fixgroup in $\mathbf{SL}(\mathbb{C}^n)/\mathbb{I}(n)$	isomorphic manifold
$\pm\mathbf{1}_n \in \mathbb{R}(n)^\pm_{\text{time}}$	$\mathbf{SU}(n)/\mathbb{I}(n)$	$\mathbf{GL}(\mathbb{C}^n)/\mathbf{U}(n)$ $\mathbb{R}$-dimension: n^2
$\begin{pmatrix} \pm\mathbf{1}_m & 0 \\ 0 & \mathbf{0}_{n-m} \end{pmatrix} \in \mathbb{R}(m,0)^\pm_{\text{light}}$ $m = 1, \dots, n-1$	$[\mathbf{U}(m) \vec{\times} \mathbb{C}^{m(n-m)}] \circ \mathbf{SL}(\mathbb{C}^{n-m})$ fix-fixgroup: $\mathbf{SU}(m)/\mathbb{I}(m)$	$\dfrac{\mathbf{SL}(\mathbb{C}^n)}{[\mathbf{U}(m) \vec{\times} \mathbb{C}^{m(n-m)}] \circ \mathbf{SL}(\mathbb{C}^{n-m})}$ $\mathbb{R}$-dimension: $m(2n-m)$
$\begin{pmatrix} \mathbf{1}_m & 0 \\ 0 & -\mathbf{1}_{n-m} \end{pmatrix} \in \mathbb{R}(m,n-m)_{\text{position}}$ $m = 1, \dots, n-1$	$\mathbf{SU}(m,n-m)/\mathbb{I}(n)$	$\mathbf{GL}(\mathbb{C}^n)/\mathbf{U}(m,n-m)$ $\mathbb{R}$-dimension: n^2

For the orthogonal chain $\mathbf{SO}_0(1,s) \vec{\times} \mathbb{R}^{1+s}$ with $s \geq 1$ one has the following fixgroups: $\mathbf{SO}(s)$ for timelike translations, and $\mathbf{SO}_0(1,s-1)$ for spacelike ones. The fixgroup for lightlike translations is trivial $\{1\}$ for $s = 1$; for $s \geq 2$, it is the Euclidean group $\mathbf{SO}(s-1) \vec{\times} \mathbb{R}^{s-1}$ with the forward lightcone V^s as homogeneous space

$$s \geq 2: \quad \mathbf{SO}_0(1,s)/\mathbf{SO}(s-1) \vec{\times} \mathbb{R}^{s-1} \cong \mathbf{D}(1) \times \Omega^{s-1} \cong \mathrm{V}^s.$$

2.5.2 Transmutators for the Lorentz Group

The action of the Lorentz group $\mathbf{SO}_0(1,3) \cong \mathbf{SL}(\mathbb{C}^2)/\mathbb{I}(2)$ on spacetime translations $\mathbb{R}(2)$ and, equivalently, on energy-momenta $\mathbb{R}(2)^T$ defines the compact stabilgroups $\mathbf{SO}(3) \cong \mathbf{SU}(2)/\mathbb{I}(2)$ and $\mathbf{SO}(2)$ for Sylvester and Witt decompositions respectively:

$$\begin{array}{ccccc} \left\{ \begin{array}{c} \mathbb{L}_+ \oplus \mathbb{L}_- \perp \mathbb{S}^2 \\ \cong \mathbb{T} \perp \mathbb{S}^1 \perp \mathbb{S}^2 \end{array} \right\} & \longrightarrow & \mathbb{T} \perp \mathbb{S}^3 & \longrightarrow & \mathbb{R}(2), \\ \mathbf{SO}(2) & \hookrightarrow & \mathbf{SO}(3) & \hookrightarrow & \mathbf{SO}_0(1,3). \end{array}$$

The orientation manifold of the rotation groups in a Lorentz group (3-hyperboloid) can be parametrized by an energylike energy-momentum (nontrivial mass) and the axial rotation groups in a rotation group (2-sphere) by a nontrivial lightlike one (trivial mass)

$$\begin{array}{llll} \mathbf{SO}_0(1,3)/\mathbf{SO}(3) & \cong \mathcal{Y}^3 & = \mathbf{SO}_0(1,3).q & \text{with } q^2 = m^2 > 0, \\ \mathbf{SO}(3)/\mathbf{SO}(2) & \cong \Omega^2 & = \mathbf{SO}(3).\vec{q} & \text{with } q^2 = 0,\ q = (|\vec{q}|, \vec{q}) \neq 0. \end{array}$$

The representations of those symmetric spaces give the transmutators from Lorentz group to rotation groups and from rotation group to axial rotation groups.

2.5.3 Rotation Groups in a Lorentz Group

A nontrivial mass $q^2 = m^2 > 0$ induces via a rest system a Sylvester decomposition into time and position or energy and momenta. The rotation group $\mathbf{SO}(3)$ as stabilgroup is the, up to isomorphy unique, maximal compact subgroup of the orthochronous Lorentz group $\mathbf{SO}_0(1,3)$.

The representations of the *orientation manifold of the rotation groups in a Lorentz group* (special relativity) can be parametrized with the orbit, e.g., of the energylike vector $e^0 = m\mathbf{1}_2$, $m > 0$, for a rest system

$$\begin{array}{l} s(\frac{q}{m})\begin{pmatrix} m & 0 \\ 0 & m \end{pmatrix} s^\star(\frac{q}{m}) = q = \begin{pmatrix} q_0 + q^3 & q^1 - iq^2 \\ q^1 + iq^2 & q_0 - q^3 \end{pmatrix}, \quad \left\{ \begin{array}{l} q = (q_0, \vec{q}), \\ q_0 = \sqrt{m^2 + \vec{q}^2}, \end{array} \right. \\ \hat{s}(\frac{q}{m}) = s^{-1\star}(\frac{q}{m}), \quad s(1,0,0,0) = \mathbf{1}_2, \\ s(\frac{q}{m}) = e^{\vec{\beta}} \text{ with } 2\vec{\beta} = \frac{\vec{q}}{|\vec{q}|} \operatorname{artanh} \frac{|\vec{q}|}{q_0}. \end{array}$$

The three noncompact momenta $\frac{\vec{q}}{m}$ parametrize the *fundamental Weyl representations of the manifold $\mathcal{Y}^3$ for special relaitivity* (chapter "Lorentz Operations")

$$\begin{array}{lll} s(\frac{q}{m}) & = & \sqrt{\frac{m+q_0}{2m}}[\mathbf{1}_2 + \frac{\vec{q}}{m+q_0}] \\ & = & \frac{1}{\sqrt{2m(m+q_0)}} \begin{pmatrix} m + q_0 + q^3 & q^1 - iq^2 \\ q^1 + iq^2 & m + q_0 - q^3 \end{pmatrix} \in \mathbf{SL}(\mathbb{C}^2)/\mathbf{SU}(2), \\ \hat{s}(\frac{q}{m}) & = & \sqrt{\frac{m+q_0}{2m}}[\mathbf{1}_2 - \frac{\vec{q}}{m+q_0}] \in \mathbf{SL}(\mathbb{C}^2)/\mathbf{SU}(2). \end{array}$$

Being $\mathbf{SU}(2)$-irreducible, they are also the fundamental transmutators from rotation to Lorentz group.

A "left" Lorentz action on a transmutator $s(\frac{q}{m})$ gives the transmutator for the Lorentz-transformed energy-momenta up to a "right" action with the fixgroup, in this context called *Wigner rotation*. The rotations "go through":

$$\begin{array}{rrl}
\lambda \in \mathbf{SL}(\mathbb{C}^2) \Rightarrow & \lambda \circ s(\frac{q}{m}) & = s(\lambda \circ \frac{q}{m} \circ \lambda^\star) \circ u(\lambda, \frac{q}{m}), \quad u(\lambda, \frac{q}{m}) \in \mathbf{SU}(2), \\
& s(\lambda \circ \frac{q}{m} \circ \lambda^\star) & = \sqrt{\lambda \circ s(\frac{q}{m})^2 \circ \lambda^\star}, \quad u(\lambda, \frac{q}{m}) = \frac{\lambda \circ s(\frac{q}{m})}{\sqrt{\lambda \circ s(\frac{q}{m})^2 \circ \lambda^\star}}, \\
\text{for} & q = (m, 0): & \lambda = s(\lambda \circ \lambda^\star) \circ u(\lambda) \text{ (polar decomposition)}, \\
u \in \mathbf{SU}(2) \Rightarrow & u \circ s(\frac{q}{m}) & = s(u \circ \frac{q}{m} \circ u^\star) \circ u.
\end{array}$$

If used in the context of relativistic particle fields (chapter "Massive Particle Quantum Fields"), this cooperation of transformations can be expressed as follows: The Lorentz transformation of a spinning particle with energy-momentum q leads to the particle with Lorentz transformed energy-momentum and spinning around the Wigner rotated direction.

All 3-hyperboloid representations (boost representations), i.e., all Lorentz to rotation transmutators, can be built from the Weyl representations (transmutators)

$$\begin{array}{rcl}
\mathcal{Y}^3 \cong \mathbf{SL}(\mathbb{C}^2)/\mathbf{SU}(2) & \longrightarrow & \mathbf{SL}(\mathbb{C}^{(1+2L)(1+2R)}), \\
\frac{q}{m} & \longmapsto & [2L|2R](\frac{q}{m}) = \bigvee^{2L} s(\frac{q}{m}) \otimes \bigvee^{2R} \hat{s}(\frac{q}{m}), \\
{[2L|2R]} & \stackrel{\mathbf{SU}(2)}{\cong} & \bigoplus_{J=|L-R|}^{L+R} [2J].
\end{array}$$

The vector representation $[1|1]$, e.g., acting on the energy-momentum space itself, gives two irreducible transmutators from Lorentz group to rotation groups, the first column for spin 0 and the three remaining columns for spin 1 with $a, b = 1, 2, 3$:

$$\begin{array}{rl}
[1|1](\frac{q}{m}) = \Lambda(\frac{q}{m})^j_k & \cong \frac{1}{2} \operatorname{tr} s(\frac{q}{m}) \sigma^j s^\star(\frac{q}{m}) \check{\sigma}_k \\
& = \frac{1}{m} \left(\begin{array}{c|c} q_0 & q_a \\ q_b & \delta_{ab} m + \frac{q_a q_b}{m + q_0} \end{array} \right) \in \mathbf{SO}_0(1,3)/\mathbf{SO}(3), \\
\Lambda(1,0,0,0) & = \mathbf{1}_4, \quad \Lambda(\frac{q}{m}) \begin{pmatrix} m \\ 0 \end{pmatrix} = \begin{pmatrix} q_0 \\ q_b \end{pmatrix}, \quad \Lambda(\frac{q}{m})^i_0 = \frac{q^i}{m}.
\end{array}$$

The four columns of the matrix $\Lambda(\frac{q}{m})^j_{0,a}$ are a general Sylvester basis in the distinguished Sylvester basis that arises for $\vec{q} = 0$ (rest system). Therefore the follwing relations hold for the metric tensors of $\mathbf{SO}_0(1,3)$ and $\mathbf{SO}(3)$:

$$\Lambda(\tfrac{q}{m})^i_{0,a} \eta_{ij} \Lambda(\tfrac{q}{m})^j_{0,b} = \left(\begin{array}{c|c} 1 & 0 \\ \hline 0 & -\delta_{ab} \end{array} \right), \quad \Lambda(\tfrac{q}{m})^i_a \delta^{ab} \Lambda(\tfrac{q}{m})^j_b = -\eta^{ij} + \frac{q^i q^j}{m^2}.$$

2.5.4 Axial Rotation Groups in a Rotation Group

There is no rest system for a trivial mass $q^2 = m^2 = 0$, $q \neq 0$. Here a Witt decomposition into two fixed 1-dimensional lightlike directions and 2-dimensional position translations, or, equivalently, into time and position translations with one fixed axis, is appropriate. To parametrize the symmetric space representations and the transmutators associated with a Witt decomposition into three subspaces where axial position rotations $\mathbf{SO}(2)$ remain as stabilgroup, one can use an energy- and a lightlike vector. The symmetric space representations are constructible in two stages from $\mathbf{SO}_0(1,3)$ to $\mathbf{SO}(3)$ and from $\mathbf{SO}(3)$ to $\mathbf{SO}(2)$, in the fundamental Weyl and Pauli representations:

$$\left.\begin{array}{l} p^2 = m^2 > 0: \ s(\frac{p}{m}) \in \mathbf{SL}(\mathbb{C}^2)/\mathbf{SU}(2) \\ q^2 = 0, \ q \neq 0: \ u(\frac{\vec{q}}{|\vec{q}|}) \in \mathbf{SU}(2)/\mathbf{SO}(2) \end{array}\right\} \Rightarrow s(\tfrac{p}{m}) \circ u(\tfrac{\vec{q}}{|\vec{q}|}) \in \mathbf{SL}(\mathbb{C}^2)/\mathbf{SO}(2).$$

Therefore the real 5-dimensional Witt manifold $\mathbf{SL}(\mathbb{C}^2)/\mathbf{SO}(2)$ is parametrized by three noncompact parameters $\frac{\vec{p}}{m}$ and two compact ones $\frac{\vec{q}}{|\vec{q}|}$.

A basis of a fixed light space $\mathbb{L}_+ \oplus \mathbb{L}_-$ is given with two distinguished vectors $(1,0,0,\pm 1)$ (components in a Sylvester basis) from the orbit $q^2 = 0$, $q \neq 0$. Any other $\mathbf{SO}(3)$-equivalent basis is defined by two lightlike vectors, parametrizable by two vectors from the same orbit

$$\tfrac{1}{q_0}(q_0, \pm\vec{q}), \quad q_0 = |\vec{q}|.$$

In the Cartan representation lightlike vectors are projectors

$$\begin{aligned} \pi^\pm &= \tfrac{\mathbf{1}_2 \pm \sigma^3}{2}, \\ p^\pm(\tfrac{\vec{q}}{|\vec{q}|}) = p^\pm(\tfrac{\vec{q}}{|\vec{q}|})^\star &= \tfrac{\mathbf{1}_2 \pm \frac{\vec{q}}{|\vec{q}|}}{2} = \tfrac{1}{2|\vec{q}|}\begin{pmatrix} |\vec{q}| \pm q^3 & \pm(q^1 - iq^2) \\ \pm(q^1 + iq^2) & |\vec{q}| \mp q^3 \end{pmatrix}. \end{aligned}$$

These fundamental *axial rotation projectors* can be obtained also with the dilation factor $\sqrt{m}$ as limit from the Weyl transmutators

$$\lim_{m\to 0} \sqrt{m}(s, \hat{s})(\tfrac{q}{m}) = \sqrt{2|\vec{q}|}\; p^\pm(\tfrac{\vec{q}}{|\vec{q}|}).$$

With the axial rotation fixgroup of the nontrivial position translations, e.g., of σ^3,

$$\mathbf{SU}(2)_{\sigma^3} = \{r \in \mathbf{SU}(2) \mid r \circ \sigma^3 \circ r^\star = \sigma^3\} = \mathbf{SO}(2) \ni e^{i\alpha_3\sigma^3},$$

the *fundamental Pauli representation* $u(\frac{\vec{q}}{|\vec{q}|})$ *of the 2-sphere* with the momentum orientations $\frac{\vec{q}}{|\vec{q}|} = \vec{\omega} \in \Omega^2$ is defined by the condition to transform the distinguished third momentum axis into the general momentum direction (chapter "Spin, Rotations, and Position")

$$\begin{aligned} u(\tfrac{\vec{q}}{|\vec{q}|}) \circ |\vec{q}|\sigma^3 \circ u^\star(\tfrac{\vec{q}}{|\vec{q}|}) &= \vec{q}, \quad u(0,0,1) = \mathbf{1}_2 \\ \Rightarrow u(\tfrac{\vec{q}}{|\vec{q}|}) \circ \pi^\pm \circ u^\star(\tfrac{\vec{q}}{|\vec{q}|}) &= p^\pm(\tfrac{\vec{q}}{|\vec{q}|}). \end{aligned}$$

The Lie parameters $\vec{\alpha}$ for this rotation are determined by the momenta, where $\vec{\alpha}_\perp$ has to be orthogonal to both σ^3 and $\vec{q}$:

$$\begin{aligned} u(\tfrac{\vec{q}}{|\vec{q}|}) &= e^{i\vec{\alpha}_\perp} \text{ with } 2\vec{\alpha}_\perp = \tfrac{\vec{q}_\perp}{|\vec{q}_\perp|} \arctan \tfrac{|\vec{q}_\perp|}{|\vec{q}|}, \quad i\vec{q}_\perp = \begin{pmatrix} 0 & -q^1 + iq^2 \\ q^1 + iq^2 & 0 \end{pmatrix}, \\ u(\tfrac{\vec{q}}{|\vec{q}|}) &= \sqrt{\tfrac{|\vec{q}|+q^3}{2|\vec{q}|}}[\mathbf{1}_2 + i\tfrac{\vec{q}_\perp}{|\vec{q}|+q^3}] \\ &= \tfrac{1}{\sqrt{2|\vec{q}|(|\vec{q}|+q^3)}}\left(\begin{array}{c|c} |\vec{q}| + q^3 & -q^1 + iq^2 \\ q^1 + iq^2 & |\vec{q}| + q^3 \end{array}\right) \in \mathbf{SU}(2)/\mathbf{SO}(2). \end{aligned}$$

The two columns of the matrix are the two Pauli transmutators from the rotation spin group $\mathbf{SU}(2)$ to the axial rotation groups $\mathbf{U}(1) \cong \mathbf{SO}(2)$

$$u(\tfrac{\vec{q}}{|\vec{q}|}) = u(\tfrac{\vec{q}}{|\vec{q}|})_+ \oplus u(\tfrac{\vec{q}}{|\vec{q}|})_-, \quad u(\tfrac{\vec{q}}{|\vec{q}|})_\pm = u(\tfrac{\vec{q}}{|\vec{q}|}) \circ \pi^\pm \in \mathbf{SU}(2)/\mathbf{U}(1).$$

A "left" action of the rotation group $\mathbf{SU}(2)$ on a transmutator $u(\frac{\vec{q}}{|\vec{q}|})_\pm$ is correlated with the "right" action of the axial group $\mathbf{SO}(2)$. The axial group "goes through":

$$\begin{array}{ll} r \in \mathbf{SU}(2) \Rightarrow & r \circ u(\frac{\vec{q}}{|\vec{q}|}) = u(r \circ \frac{\vec{q}}{|\vec{q}|} \circ r^\star) \circ o(r, \frac{\vec{q}}{|\vec{q}|}) \text{ with } o(r, \frac{\vec{q}}{|\vec{q}|}) \in \mathbf{SO}(2), \\ o \in \mathbf{SO}(2) \Rightarrow & o \circ u(\frac{\vec{q}}{|\vec{q}|}) = u(o \circ \frac{\vec{q}}{|\vec{q}|} \circ o^\star) \circ o. \end{array}$$

All 2-sphere representations, i.e., all transmutators from a rotation group to its axial rotation subgroups, arise from products of Pauli representations

$$\begin{array}{l} \Omega^2 \cong \mathbf{SU}(2)/\mathbf{SO}(2) \longrightarrow \mathbf{SU}(1+2J), \ \frac{\vec{q}}{|\vec{q}|} \longmapsto [2J](\frac{\vec{q}}{|\vec{q}|}) = \bigvee^{2J} u(\frac{\vec{q}}{|\vec{q}|}), \\ [2J] \overset{\mathbf{SO}(2)}{\cong} \bigoplus_{|z|} [\pm z], \quad |z| = \left\{ \begin{array}{ll} 0, 2, \ldots, 2J, & J = 0, 1, \ldots, \\ 1, \ldots, 2J, & J = \frac{1}{2}, \frac{3}{2}, \ldots, \end{array} \right. \end{array}$$

e.g., the rotation in momentum space $\mathbb{R}^3$ with $a, b = 1, 2, 3$ and $\alpha, \beta = 1, 2$:

$$\begin{aligned} [2](\tfrac{\vec{q}}{|\vec{q}|}) \cong O(\tfrac{\vec{q}}{|\vec{q}|})^b_a &= \tfrac{1}{2} \operatorname{tr} u(\tfrac{\vec{q}}{|\vec{q}|}) \sigma^b u^\star(\tfrac{\vec{q}}{|\vec{q}|}) \sigma^a \\ &= \tfrac{1}{|\vec{q}|} \left(\begin{array}{c|c} \delta^{\alpha\beta}|\vec{q}| - \frac{q^\alpha q^\beta}{|\vec{q}|+q^3} & q^\alpha \\ -q^\beta & q^3 \end{array} \right) \in \mathbf{SO}(3)/\mathbf{SO}(2), \\ O(0,0,1) = \mathbf{1}_3, \quad O(\tfrac{\vec{q}}{|\vec{q}|}) \begin{pmatrix} 0 \\ 0 \\ |\vec{q}| \end{pmatrix} &= \vec{q}, \quad O(\tfrac{\vec{q}}{|\vec{q}|})^a_3 = \tfrac{q^a}{|\vec{q}|}, \end{aligned}$$

with the relations for the $\mathbf{SO}(3)$ and $\mathbf{SO}(2)$ metric tensors

$$O(\tfrac{\vec{q}}{|\vec{q}|})^a_{\alpha,3} \delta_{ab} O(\tfrac{\vec{q}}{|\vec{q}|})^b_{\beta,3} = \left(\begin{array}{c|c} \delta_{\alpha\beta} & 0 \\ \hline 0 & 1 \end{array} \right), \quad O(\tfrac{\vec{q}}{|\vec{q}|})^a_\alpha \delta^{\alpha\beta} O(\tfrac{\vec{q}}{|\vec{q}|})^b_\beta = \delta^{ab} - \tfrac{q^a q^b}{\vec{q}^2}.$$

2.6 Summary

Spacetime can be represented by the operations of the real $2n^2$-dimensional Lie group $\mathbf{GL}(\mathbb{C}^n)$, time for $n = 1$, relativistic spacetime for $n = 2$. n is the real rank, n^2 the real dimension of spacetime.

$\mathbf{GL}(\mathbb{C}^n)$ acting on a $\mathbb{C}^n$-isomorphic Hilbert space is the regular group of the stellar algebra $\mathbb{C}(n) = \mathbf{AL}(\mathbb{C}^n)$ whose unique spectral order defines a causal structure and topology. $\mathbf{GL}(\mathbb{C}^n)$ is the polar product of the unitary internal group $\mathbf{U}(n)$ (compact) and the strictly positive external causal spacetime manifold $\mathbf{D}(n) \cong \mathbf{GL}(\mathbb{C}^n)/\mathbf{U}(n)$ (noncompact). Correspondingly, the $\mathbb{R}^{2n^2}$-isomorphic tangent space is decomposable into the internal Lie algebra $\log \mathbf{U}(n) = i\mathbb{R}(n)$ and the spacetime translations $\mathbb{R}(n) \cong \log \mathbf{GL}(\mathbb{C}^n)/\log \mathbf{U}(n)$.

The symmetric space $\mathbf{D}(n)$, isomorphic to the unitary operation classes (unitary relativity), i.e., for $n = 2$ the orientation manifold of the hyperisospin group $\mathbf{U}(2)$ in the general linear group $\mathbf{GL}(\mathbb{C}^2)$, is taken as model for nonlinear spacetime. It can be parametrized by the strict future $\mathbb{R}(n)^+_{\rm time}$ (open causal cone) in the spacetime translations.

Spacetime $\mathbf{D}(n)$ is acted on by the group $\mathbf{GL}(\mathbb{C}^n) = \mathbf{D}(\mathbf{1}_n) \times [\mathbf{U}(\mathbf{1}_n) \circ \mathbf{SL}(\mathbb{C}^n)]$ (extended Lorentz group for $n = 2$).

The defining representation of $\mathbf{U}(n)$ on a Higgs Hilbert space $H \cong \mathbb{C}^n$ gives as proper fixgroup type $\mathbf{U}(n-1)$, for $n = 2$ called the electromagnetic group $\mathbf{U}(1)_+$. It defines the Goldstone manifold $\mathbf{U}(2)/\mathbf{U}(1)_+$, the orientation manifold of the electromagnetic group in the hyperisospin group (electromagnetic relativity).

The proper fixgroup types of the spacetime translations (energy-momenta) with the Lorentz group action for $n = 2$ are $\mathbf{SO}(3)$, $\mathbf{SO}_0(1,2)$, and $\mathbf{SO}(2) \vec{\times} \mathbb{R}^2$. They decompose the nontrivial spacetime translations into disjoint strata, the nontrivial timelike, spacelike, and lightlike translations. Those manifolds are isomorphic to the symmetric spaces $\mathbf{D}(1) \times \mathcal{Y}^3$, $\mathbf{D}(1) \times \mathcal{Y}^{(1,2)}$, and $\mathbf{D}(1) \times \Omega^2$.

MATHEMATICAL TOOLS

2.7 Fixgroups of Representations

A vector space V with group action $G \subseteq \mathbf{GL}(V)$ is decomposable into disjoint G-orbits characterized by fixgroups (chapter "Time Representations"):

$$V \setminus \{0\} = \biguplus_{\text{repr } v_r} G \bullet v_r \cong \biguplus_{\text{repr } v_r} G/G_{v_r}.$$

The trivial translation is an orbit with fixgroup $G_0 = G$. The strata decomposition collects orbits with isomorphic fixgroups

$$V \setminus \{0\} = \biguplus_{\text{repr } v_R} [G \bullet v_R] \cong \biguplus_{\text{repr } v_R} [G/G_{v_R}]$$

with less representatives $\{v_R\} \subseteq \{v_r\}$.

In this way one can associate to a group G the different fixgroup types $\{H_R = G_{v_R}\}$, i.e., the different irreducible group realizations $\{G/H_R\}$, which are composed in the linear representation space.

2.8 Orbits with Signatures

An orthogonal $\mathbf{O}(p,q)$ bilinear form of $V \cong \mathbb{R}^n$, $n \geq 1$, and a unitary $\mathbf{U}(p,q)$ sesquilinear form of $V \cong \mathbb{C}^n$ gives rise to *four orbit types $\mathbf{O}(p,q) \bullet v$ and $\mathbf{U}(p,q) \bullet v$ with definite signature*:

$$\textit{trivial: } v = 0; \quad \text{strictly} \left\{ \begin{array}{l} \textit{positive: } \zeta(v,v) > 0, \\ \textit{negative: } \zeta(v,v) < 0, \end{array} \right. \qquad \textit{singular: } \zeta(v,v) = 0, \ v \neq 0.$$

For $pq = 0$ there are only the trivial and strictly positive (negative) orbits.

Sylvester decompositions have two definite orthogonal vector subspaces

$$V \cong \mathbb{R}^p \bot \mathbb{R}^q, \quad V \cong \mathbb{C}^p \bot \mathbb{C}^q, \ \zeta \cong \begin{pmatrix} \mathbf{1}_p & 0 \\ 0 & -\mathbf{1}_q \end{pmatrix}$$

with their stabilgroups *maximal compact subgroups* in $\mathbf{O}(p,q)$ and $\mathbf{U}(p,q)$ respectively and the decomposition of the group dimension $d = d_c + d_{nc}$ into compact and noncompact dimension:

$$\begin{array}{lllcll} \mathbf{O}(p,q) & \supset \mathbf{O}(p) \times \mathbf{O}(q), & \binom{p+q}{2} & = & [\binom{p}{2} + \binom{q}{2}] & + \quad pq, \\ \mathbf{U}(p,q) & \supset \mathbf{U}(p) \times \mathbf{U}(q), & (p+q)^2 & = & [p^2 + q^2] & + \quad 2pq. \end{array}$$

The distinction of singular vectors leads to *Witt decompositions*, being direct sums of two or three subspaces for $p = q \geq 1$ and $p > q \geq 1$ respectively. The sum of the two singular subspaces is orthogonal to the definite space. One has skew-diagonal metrical matrices $\mathbf{z}$:

$$\begin{array}{lll} V \cong [\mathbb{R}^q \oplus \mathbb{R}^q] \bot \mathbb{R}^{p-q}, & \mathbf{O}(p,q) & \supset \mathbf{O}(q,q) \times \mathbf{O}(p-q), \\ V \cong [\mathbb{C}^q \oplus \mathbb{C}^q] \bot \mathbb{C}^{p-q}, & \mathbf{U}(p,q) & \supset \mathbf{U}(q,q) \times \mathbf{U}(p-q), \end{array}$$

$$\zeta \cong \begin{pmatrix} 0 & \mathbf{z}_q & 0 \\ \mathbf{z}_q & 0 & 0 \\ 0 & 0 & \mathbf{1}_{p-q} \end{pmatrix}.$$

2.9 Fix- and Stabil-Lie Algebras

Fixgroup and stabilgroup corresponding concepts can also be given for Lie algebras.

The *fix-Lie algebra* (also invariance Lie algebra) of a vector subspace $U \subseteq V$ with action of a Lie algebra $L \times V \longrightarrow V$ is defined by the trivially acting Lie algebra elements

$$U \subseteq V: \quad L_U = \{l \in L \mid l \bullet u = 0 \text{ for all } u \in U\} \in \underline{\mathbf{lag}}_{\mathbb{K}}.$$

The *Lie-centralizer* of a vector subspace W of a Lie algebra L is its invariance Lie algebra with respect to the adjoint action

$$W \subseteq L: \quad L_W^{\text{ad}} = \{l \in L \mid [l,w] = 0 \text{ for all } w \in W\} \in \underline{\mathbf{lag}}_{\mathbb{K}}.$$

The centralizer of the whole Lie algebra is the centrum of L.

The *stabil-Lie algebra* of a vector subspace $U \subseteq V$ consists of those Lie algebra elements that keep U stable

$$U \subseteq V: \quad L_{\{U\}} = \{l \in L \mid l \bullet U = U\} \in \underline{\mathbf{lag}}_{\mathbb{K}}, \quad L_U \subseteq L_{\{U\}}.$$

The *Lie-normalizer* of a vector subspace W of a Lie algebra L is its stabil-Lie algebra with respect to the adjoint action:

$$W \subseteq L: \quad L^{\rm ad}_{\{W\}} = \{l \in L \mid [l,W] \subseteq W\} \in \underline{\mathbf{lag}}_{\mathbb{K}}.$$

For an ideal W the full Lie algebra L is the normalizer.

If the vector space V carries an L-invariant quadratic form

$$\zeta : V \times V \longrightarrow \mathbb{K}, \quad \begin{cases} \zeta(v,w) = \overline{\zeta(w,v)}, \\ \zeta(l \bullet v, w) - \zeta(v, l \bullet w) = 0, \end{cases}$$

then each vector $v \in V$ induces a quadratic form of the Lie algebra

$$\begin{array}{c} v \in V, \quad \zeta_v : L \times L \longrightarrow \mathbb{K}, \quad \zeta_v(l,m) = \zeta(l \bullet v, m \bullet v) = \overline{\zeta_v(m,l)} \\ \zeta \text{ bilinear } \Rightarrow \zeta_v \text{ bilinear}, \\ V \in \underline{\mathbf{vec}}_{\mathbb{C}} \text{ and } \zeta \text{ sesquilinear } \Rightarrow \begin{cases} \zeta_v \text{ bilinear for } L \in \underline{\mathbf{lag}}_{\mathbb{R}}, \\ \zeta_v \text{ sesquilinear for } L \in \underline{\mathbf{lag}}_{\mathbb{C}}. \end{cases} \end{array}$$

The quadratic form ζ_v is trivial for the fix-Lie algebra of $\mathbb{K}v$ and defines a quadratic form $\overline{\zeta}_v$ of the quotient

$$\begin{array}{l} \zeta_v(l,m) = 0 \text{ for } l \text{ or } m \in L_{\mathbb{K}v}, \\ \overline{\zeta}_v : L/L_{\mathbb{K}v} \times L/L_{\mathbb{K}v} \longrightarrow \mathbb{K}. \end{array}$$

2.10 Transmutators as Coset Representations

For a subgroup $H \subseteq G$ there are class representatives

$$\begin{array}{l} G/H \longrightarrow (G/H)_{\rm repr} \subseteq G, \quad kH \longmapsto k_r \text{ for } kH = k_r H, \\ G = \biguplus_{k_r} k_r H = (G/H)_{\rm repr} \circ H. \end{array}$$

A natural choice may be given by a polar decomposition. In general, there is no natural choice. For an exponential parametrization with Lie algebra coefficients

$$\begin{array}{rl} \mathbb{K}^d \cong \log G & = \log H \oplus W, \quad W \cong \log G / \log H, \\ \log G \ni l & = \alpha_a l^a + \beta_k b^k, \quad g(\alpha,\beta) = e^l \in G, \end{array}$$

a representative is given with trivial H-parameters:

$$G = \{g(\alpha,\beta) \mid \alpha_a, \beta_k \in \mathbb{K}\} = \{g(0,\beta)H \mid \beta_k \in \mathbb{K}\}$$

with the examples

$$\begin{array}{rll} \mathbf{SL}(\mathbb{C}^2) & = \{e^{i\vec{\alpha}+\vec{\beta}} \mid \vec{\alpha},\vec{\beta} \in \mathbb{R}^3\} & = \{e^{\vec{\beta}}\, \mathbf{SU}(2)\}, \\ \mathbf{SU}(2) & = \{e^{i\vec{\alpha}} \mid \vec{\alpha} \in \mathbb{R}^3\} & = \{e^{i(\alpha_1\sigma^1+\alpha_2\sigma^2)}\, \mathbf{SO}(2)\}, \\ \mathbf{U}(2) & = \{e^{i(\alpha_0+\vec{\alpha})} \mid \alpha_j \in \mathbb{R}\} & = \{e^{i(-\alpha_3+\vec{\alpha})}\, \mathbf{U}(1)\}. \end{array}$$

Since in general a group subset $\{g(\alpha,\beta) \mid \alpha_a \in \mathbb{K}\}$ with fixed β is not an H-coset, $g(0,\beta)$ cannot be thought of as its representative, e.g., $e^{\vec{\beta}}\, \mathbf{SU}(2) \neq \{e^{i\vec{\alpha}+\vec{\beta}} \mid \vec{\alpha} \in \mathbb{R}^3\}$.

The action of the group $g \in G$ from left on representatives k_r is complicated: It gives the representative $(gk)_r$ of the class gk_rH for the product up to a right multiplication with a *Wigner fixgroup element* which depends on the choice of the representatives:

$$\begin{array}{rl} G \times (G/H)_{\rm repr} \longrightarrow (G/H)_{\rm repr}, & gk_r = (gk)_r h(k_r, g)^{-1}, \\ \text{since } gk_rH = (gk)_rH, & \text{Wigner element } h(k_r, g) \in H. \end{array}$$

The representatives can be chosen is such a way that the fixgroup H-action is described by *inner H-automorphisms*

$$H \times (G/H)_{\rm repr} \longrightarrow (G/H)_{\rm repr}, \quad hk_r = hk_rh^{-1}.$$

Then the representatives are the disjoint union of H-orbits

$$(G/H)_{\rm repr} = \biguplus_R \text{Int}\, H(k_R).$$

A group representation $D : G \longrightarrow \mathbf{GL}(V)$ defines a linear representation of the classes G/H:

$$\begin{array}{rl} G/H & \longrightarrow D[G]/D[H], \\ gH & \longmapsto D[gH] = D(g) \circ D[H] = \{D(g) \circ D(h) \bigm| h \in H\}. \end{array}$$

The fixgroup $D[H]$ of a vector $v \in V$ is isomorphic - as $D[G]$-set - to the G-orbit of v:

$$D[G]_v = D[H] \Rightarrow D[G]/D[H] \cong G \bullet v.$$

The $D[H]$-classes can be parametrized with the orbit parameters, i.e., with the components of the orbit vectors

$$D[kH] \cong D(kH \bullet v),$$

e.g., $\mathbf{SO}_0(1,3)/\mathbf{SO}(3)$ by the Minkowski vectors of a timelike hyperboloid or $\mathbf{SO}(3)/\mathbf{SO}(2)$ by the direction vectors of a 2-sphere.

A G-representation gives a linear representation of the representatives

$$\begin{array}{rl} & (G/H)_{\rm repr} \longrightarrow \mathbf{GL}(V) \quad k_r \longmapsto D(k_r), \\ g \in G \Rightarrow & D(gk_r) = D((gk)_r) \circ D(h^{-1}) \text{ with Wigner element } h(k_r, g). \end{array}$$

The set $\{D(k_r) \bigm| \text{representatives}\}$ cannot be used as a group representation, since, in general, the product k_rk_s is no class representative.

A finite-dimensional group representation with $(n \times n)$ matrices

$$G \ni k \longmapsto D(k) \in \mathbf{GL}(V), \quad D(k)_{l=1,\dots,n}^{j=1,\dots,n},$$

is decomposable into subgroup H-representations with square $(m_\iota \times m_\iota)$ matrices from $\mathbf{GL}(W^\iota)$:

$$V \overset{H}{\cong} \bigoplus_{\iota=1}^N W^\iota, \quad W^\iota \cong \mathbb{K}^{m_\iota}, \quad \sum_{\iota=1}^N m_\iota = n,$$

$$H \ni h \longmapsto D(h) = \bigoplus_{\iota=1}^N d^\iota(h) = \left(\begin{array}{c|c|c|c} d^1(h) & 0 & \cdots & 0 \\ \hline 0 & d^2(h) & \cdots & 0 \\ \hline & & \cdots & \\ \hline 0 & 0 & \cdots & d^N(h) \end{array}\right),$$

$$d^\iota(h)_{a=1,\dots,m_\iota}^{b=1,\dots,m_\iota}.$$

A corresponding *representation of the symmetric space* G/H,

$$(G/H)_{\text{repr}} \ni k_r \longmapsto D(k_r) \in \mathbf{GL}(V),$$

is decomposed correspondingly into *transmutators between* $H \bullet W^\iota$ *and* $G \bullet V$ with rectangular $(m_\iota \times n)$ matrices from $W^\iota \otimes V^T$:

$$D(k_r) = \bigoplus_{\iota=1}^{N} D^\iota(k_r) = \left(\begin{array}{c|c|c|c} m_1 \text{ columns} & m_2 \text{ columns} & \ldots & m_N \text{ columns} \\ \hline & & & \\ D^1(k_r) & D^2(k_r) & \ldots & D^N(k_r) \\ & & & \end{array} \right),$$
$$D^\iota(k_r)^{j=1,\ldots,n}_{a=1,\ldots,m_\iota}.$$

The transmutators have a $G \times H$-transformation behavior

$$D^\iota(gk_r h^{-1})^j_a = D(g)^j_l D^\iota(k_r)^l_b d^\iota(h^{-1})^b_a.$$

3

PROPAGATORS

Feynman propagators characterize the spacetime behavior of particles. They will be introduced as Lorentz compatible relativistic distributions of matrix elements of time representations. The particle interpretation is discussed in the chapters "Massive Particle Quantum Fields" and "Massless Quantum Fields."

Representations of the causal group $\mathbf{D}(1) \cong \exp \mathbb{R}$, generated by and isomorphic to the time translations $\mathbb{R}$, can be embedded, by position distribution, into a Lorentz-action-compatible framework. The invariant time operation eigenvalues (energies, frequencies) are distributed by energy-momentum $(q_0, \vec{q})$-measures (generalized functions) supported by the Lorentz invariant mass hyperboloid $q^2 = m^2$. As special relativistic supplement for the compact time representation matrix elements $e^{\pm i|q_0|t} \in \mathbf{U}(1) \cong \mathbf{SO}(2)$, there arise $r = 0$-regular spherical waves $\frac{\sin |\vec{q}|r}{r}$, $|\vec{x}| = r$, which are representation coefficients of the Euclidean group $\mathbf{SO}(3) \vec{\times} \mathbb{R}^3$ (chapter "The Kepler Factor"). The causal time representations $e^{\pm i|q_0 t|}$ are supplemented by $r = 0$-singular Yukawa potentials $\frac{e^{-|\vec{Q}|r}}{r}$.

The relation of relativistic distributions of time representations to representations of the Poincaré group $\mathbf{SO}_0(1,3) \vec{\times} \mathbb{R}^4$ is discussed in the chapters "Harmonic Analysis" and "Residual Spacetime Representations."

3.1 Point Measures for Energies

To prepare the relativistic embedding, the time representation matrix elements are formulated as Fourier transformed energy measures. The continuous eigenvalues of the irreducible unitary time representations can be embedded as the real axis $m \in \mathbb{R}$ into the complex energy plane. The Dirac distributions of the energies define *point supported measures of the complex energy plane.* They can also be written as a loop integration around an energy pole:

$$1 = \int dE\ \delta(m - E) = \oint \frac{dE}{2i\pi} \frac{1}{E-m} \text{ for } m \in \mathbb{R}.$$

Here the following notation with Lebesgue measure dE is used:

$\int dE$ for $\int_{-\infty}^{\infty} dE = \int_{\mathbb{R}} dE$ on the real axis,
$\oint dE$ for a positive (counterclockwise) loop around all poles.

All distributions (generalized functions) used for propagators are tempered $\mathcal{S}'(\mathbb{R}^d)$ with the Fourier isomorphism $\mathcal{S}'(\check{\mathbb{R}}^d) \cong \mathcal{S}'(\mathbb{R}^d)$.

The Dirac point measure, equivalent to a *residue* $\int dE\ \delta(E-m)f(E) = \oint \frac{dE}{2i\pi}\frac{f(E)}{E-m}$, is the real part of a complex generalized function where the principal value function, denoted by the subscript P, comes as imaginary part:

$$a \in \mathbb{R},\ \ \pm\frac{1}{i\pi}\frac{1}{a\mp io} = \delta(a) \pm \frac{1}{i\pi}\frac{1}{a_{\rm P}} \iff \left\{\begin{array}{lll} \delta(a) & = \frac{1}{2i\pi}\left[\frac{1}{a-io} - \frac{1}{a+io}\right] & = \frac{1}{\pi}\frac{o}{a^2+o^2}, \\ \frac{1}{a_{\rm P}} & = \frac{1}{2}\left[\frac{1}{a-io} + \frac{1}{a+io}\right] & = \frac{a}{a^2+o^2}. \end{array}\right.$$

The symbol o in the generalized function prescribes a pole with a *real positive* $o > 0$, an integration on the real axis and, afterward, the limit $o \to 0$.

The complex point measures with a pole in the energy plane are Fourier transforms of the *advanced and retarded* time representations

$$\vartheta(\pm t)e^{imt} = \pm\int \frac{dE}{2i\pi}\frac{1}{E\mp io-m}e^{iEt}.$$

The distributional imaginary part determines the time direction, the upper half-plane pole for $E - io$ leads to support by the future, the lower half-plane pole $E + io$ to support by the past.

With those measures and functions time representation matrix elements can be written in different forms, with a closed loop integration, with a Dirac measure, or with a time-ordered principal value integration:

$$\begin{array}{rl} \mathbb{R} \longrightarrow \mathbf{U}(1) \ni e^{imt} & = \oint \frac{dE}{2i\pi}\frac{1}{E-m}e^{iEt} = \int dE\ \delta(m-E)e^{iEt} \\ & = \epsilon(t)\int \frac{dE}{i\pi}\frac{1}{E_{\rm P}-m}e^{iEt}. \end{array}$$

The self-dual time representations with the trigonometric functions use an energy measure self-dually supported by $\pm m$:

$$\begin{array}{rl} \mathbb{R} \longrightarrow \mathbf{SO}(2) & \ni \begin{pmatrix} \cos mt & i\sin mt \\ i\sin mt & \cos mt \end{pmatrix}, \\ \text{with } \begin{pmatrix} \cos mt \\ i\sin mt \end{pmatrix} & = \int dE\ \epsilon(m)\begin{pmatrix} m \\ E \end{pmatrix}\delta(m^2-E^2)e^{iEt} \\ & = \int dE\ \epsilon(E)\begin{pmatrix} E \\ m \end{pmatrix}\delta(m^2-E^2)e^{iEt} \\ & = \oint \frac{dE}{i\pi}\frac{1}{E^2-m^2}\begin{pmatrix} E \\ m \end{pmatrix}e^{iEt} = \epsilon(t)\int \frac{dE}{i\pi}\frac{1}{E_{\rm P}^2-m^2}\begin{pmatrix} E \\ m \end{pmatrix}e^{iEt}. \end{array}$$

The causal time representations have as energy measures

$$\begin{pmatrix} 1 \\ \epsilon(t) \end{pmatrix}e^{\pm i|mt|} = \begin{pmatrix} 1 \\ \epsilon(t) \end{pmatrix}(\cos mt \pm \epsilon(mt) i\sin mt) = \int \frac{dE}{i\pi}\begin{pmatrix} \pm|m| \\ E \end{pmatrix}\frac{1}{E^2\mp io-m^2}e^{iEt}.$$

Representations with finite closed integration contours in the complex plane like e^{imt} for the group $\mathbb{R}$ obey homogeneous differential $\frac{d}{dt}$ equations, those with infinite unclosed contours like $\sin|mt|$ for $\mathbb{R} = \mathbb{R}_+ \uplus \mathbb{R}_-$ as ordered double cone inhomogeneous ones.

3.2 Relativistically Distributed Time Representations

For time $\mathbb{R}$-representations the eigenvalue energy E coincides with the representation invariant. For Minkowski spacetime $\mathbb{R}^4$ the Lorentz invariant q^2 involves the square $q_0^2 - \vec{q}^2$ of the eigenvalues $(q_j)_{j=0}^3$ for all four translation subgroups $\mathbb{R}$. Hence two different relativistic distributions of time representations are possible: Since the frequency (energy) comes in an energy-momentum (vector)

$$\check{\mathbb{R}} \ni E \hookrightarrow (q_j)_{j=0}^3 \in \check{\mathbb{R}}^4,$$

one can distribute time representation cosine and sine either by a Lorentz-"scalar" or a Lorentz-"vector" cosine paired with a Lorentz-"vector" and Lorentz-"scalar" sine, respectively:

$$\begin{aligned}
\begin{pmatrix} \cos mt \\ i\sin mt \end{pmatrix} &= \int dE\ \epsilon(m) \begin{pmatrix} m \\ E \end{pmatrix} \delta(m^2 - E^2) e^{iEt} \hookrightarrow \begin{pmatrix} \mathbf{C}(m|x) \\ i\mathbf{S}_j(m|x) \end{pmatrix}, \\
&= \int dE\ \epsilon(E) \begin{pmatrix} E \\ m \end{pmatrix} \delta(m^2 - E^2) e^{iEt} \hookrightarrow \begin{pmatrix} \mathbf{c}_j(m|x) \\ i\mathbf{s}(m|x) \end{pmatrix}, \\
d_t \begin{pmatrix} \cos mt \\ i\sin mt \end{pmatrix} &= im \begin{pmatrix} i\sin mt \\ \cos mt \end{pmatrix} \hookrightarrow \begin{cases} \partial^j \begin{pmatrix} \mathbf{C}(m|x) \\ i\mathbf{S}_j(m|x) \end{pmatrix} &= im \begin{pmatrix} i\mathbf{S}^j(m|x) \\ \mathbf{C}(m|x) \end{pmatrix}, \\ \partial^j \begin{pmatrix} \mathbf{c}_j(m|x) \\ i\mathbf{s}(m|x) \end{pmatrix} &= im \begin{pmatrix} i\mathbf{s}(m|x) \\ \mathbf{c}^j(m|x) \end{pmatrix}. \end{cases}
\end{aligned}$$

The distribution with a Dirac energy-momentum measure on the mass shell for a mass $m \in \mathbb{R}$,

$$\begin{pmatrix} \mathbf{C}(m|x) \\ i\mathbf{S}_j(m|x) \end{pmatrix} = \epsilon(m) \int \frac{d^4q}{(2\pi)^3} \begin{pmatrix} m \\ q_j \end{pmatrix} \delta(m^2 - q^2) e^{iqx},$$

defines functions that will occur as Fock state functions for relativistic particle fields. Their sum can be given as an "exponential" with (4×4) Dirac matrices in the Dirac algebra $\mathbf{AL}(\mathbb{C}^4)$:

$$\begin{aligned}
e^{imt} \hookrightarrow \mathbf{EXP}(im|x) &= \mathbf{1}_4 \mathbf{C}(m|x) + i\gamma^j \mathbf{S}_j(m|x) \\
&= \epsilon(m) \int \frac{d^4q}{(2\pi)^3} \delta(\gamma q - m) e^{iqx}, \\
\text{with} \quad & \delta(\gamma q - m) = (\gamma q + m)\delta(q^2 - m^2).
\end{aligned}$$

The distribution with an ordered Dirac energy-momentum measure

$$\begin{pmatrix} \mathbf{c}_j(m|x) \\ i\mathbf{s}(m|x) \end{pmatrix} = \int \frac{d^4q}{(2\pi)^3} \epsilon(q_0) \begin{pmatrix} q_j \\ m \end{pmatrix} \delta(m^2 - q^2) e^{iqx} = \epsilon(x_0) \int \frac{d^4q}{i\pi(2\pi)^3} \begin{pmatrix} q_j \\ m \end{pmatrix} \frac{1}{q_{\mathrm{P}}^2 - m^2} e^{iqx}$$

defines the distributions that will occur for the relativistic field quantization. The ordered Lebesgue measure $d^4q\epsilon(q_0)\vartheta(q^2)$ leads to *causal support*

$$\begin{pmatrix} \mathbf{c}_j(m|x) \\ i\mathbf{s}(m|x) \end{pmatrix} = 0 \text{ for } x^2 < 0.$$

The two distributions are combinable as a second "exponential" in the Dirac algebra

$$\begin{aligned}
e^{imt} \hookrightarrow \mathbf{exp}(im|x) &= \gamma^j \mathbf{c}_j(m|x) + i\mathbf{1}_4 \mathbf{s}(m|x) \\
&= \int \frac{d^4q}{(2\pi)^3} \epsilon(q_0) \delta(\gamma q - m) e^{iqx} = \epsilon(x_0) \int \frac{d^4q}{i\pi(2\pi)^3} \frac{1}{\gamma q_{\mathrm{P}} - m} e^{iqx} \\
\text{with} \quad & \frac{1}{\gamma q_{\mathrm{P}} - m} = \frac{\gamma q + m}{q_{\mathrm{P}}^2 - m^2}.
\end{aligned}$$

The crossover sums are a Lorentz scalar and vector with definite energy

$$e^{\pm i|m|t} \hookrightarrow \begin{pmatrix} \mathbf{C}(m|x) \pm \epsilon(m) i\mathbf{s}(m|x) \\ \mathbf{c}_j(m|x) \pm \epsilon(m) i\mathbf{S}_j(m|x) \end{pmatrix} = \int \frac{d^4q}{(2\pi)^3} \vartheta(\pm q_0) 2 \begin{pmatrix} |m| \\ \pm q_j \end{pmatrix} \delta(m^2 - q^2) e^{iqx}.$$

All these Lorentz compatible distributions of the time representation matrix elements fulfill a *homogeneous Klein-Gordon equation*

$$(d_t^2 + m^2) \begin{pmatrix} \cos mt \\ i \sin mt \end{pmatrix} = 0 \hookrightarrow \left\{ \begin{array}{ll} (\partial^2 + m^2) \begin{pmatrix} \mathbf{c}_j(m|x) \\ i\mathbf{s}(m|x) \end{pmatrix} & = 0, \\ (\partial^2 + m^2) \begin{pmatrix} \mathbf{C}(m|x) \\ i\mathbf{S}_j(m|x) \end{pmatrix} & = 0. \end{array} \right.$$

The crossover sums with an additional causal order occur as Feynman propagators for relativistic quantum particle fields and distribute the causal time representations

$$\begin{aligned} \begin{pmatrix} 1 \\ \epsilon(t) \end{pmatrix} e^{\pm i|mt|} \hookrightarrow \begin{pmatrix} \mathbf{E}(\pm i|m||x) \\ \mathbf{E}_j(\pm i|m||x) \end{pmatrix} &= \begin{pmatrix} \mathbf{C}(m|x) \pm \epsilon(mx_0) i\mathbf{s}(m|x) \\ \epsilon(x_0)\mathbf{c}_j(m|x) \pm \epsilon(m) i\mathbf{S}_j(m|x) \end{pmatrix} \\ &= \int \frac{d^4q}{(2\pi)^3} \vartheta(\pm q_0 x_0) 2 \begin{pmatrix} |m| \\ \pm q_j \end{pmatrix} \delta(m^2 - q^2) e^{iqx} \\ &= \int \frac{d^4q}{i\pi(2\pi)^3} \begin{pmatrix} \pm |m| \\ q_j \end{pmatrix} \frac{1}{q^2 \mp io - m^2} e^{iqx}. \end{aligned}$$

Causal support and spacelike contributions go with the real and imaginary part. With the causal order $\epsilon(t) \hookrightarrow \epsilon(x_0)\vartheta(x^2)$, the Feynman propagators obey *inhomogeneous Klein-Gordon equations*

$$\begin{aligned} \left. \begin{array}{ll} d_t e^{\pm i|mt|} & = \pm i|m|\epsilon(t) e^{\pm i|mt|}, \\ (d_t^2 + m^2) e^{\pm i|mt|} & = \pm 2i|m|\delta(t) \end{array} \right\} \\ \hookrightarrow \left\{ \begin{array}{ll} \partial_j \mathbf{E}(\pm i|m||x) & = \pm i|m| \mathbf{E}_j(\pm i|m||x), \\ (\partial^2 + m^2) \mathbf{E}(\pm i|m||x) & = \pm 2i|m|\delta(x). \end{array} \right. \end{aligned}$$

In the following the distributed time representation matrix elements are considered in more detail, especially with respect to the accompanying position representation properties, which come in spherical form $\mathbb{R} \ni z \longmapsto e^{imz} \in \mathbf{U}(1)$ and $\cos mz, \sin mz \in \mathbf{SO}(2)$, analoguous to the time representations above, and with the hyperbolic representation matrix elements

$$\mathbb{R} \ni z \longmapsto e^{-|mz|} = \int \frac{dq}{\pi} \frac{|m|}{q^2 + m^2} e^{-iqz}.$$

3.3 Fourier Transforms of Energy-Momentum Distributions

Spacetime translations $\mathbb{R}^d$, $d \geq 1$, analogous energy-momenta as dual vector space, come with Lebesgue measure, d^dx and d^dq respectively, invariant under the action of $\mathbf{SL}(\mathbb{R}^d) \vec{\times} \mathbb{R}^d$ with the Poincaré group $\mathbf{SO}_0(1,s) \vec{\times} \mathbb{R}^{1+s}$, $d = 1+s$, $s \geq 0$, as subgroup. The measure normalization is not fixed. In the following, integrations over the full space have the shorthand notation $\int_{\mathbb{R}^d} = \int$. A decomposition $\mathbb{R}^{1+s} \cong \mathbb{R} \oplus \mathbb{R}^s$ into time and position translations (analogue into energy and momenta) is induced by a rest system.

Energy-momenta measures, expressible by generalized mappings and a Lebesgue measure $d^{1+s}q$ give rise, by Fourier transforms via the translation representations $\mathbb{R}^{1+s} \ni x \longmapsto e^{iqx} \in \mathbf{U}(1)$, to distributions on spacetime, valued in a complex vector space U with Lorentz group action

$$\begin{array}{l} \mathbb{R}^{1+s} \ni x \longmapsto \mu(x) = \int d^{1+s}q \; e^{iqx} \; \tilde{\mu}(q) \in U, \\ \Lambda \in \mathbf{SO}_0(1,s): \quad \mu_\Lambda(x) = D(\Lambda).\mu(\Lambda^{-1}.x). \end{array}$$

A Dirac energy-momentum integration for one mass gives the $\mathbf{SO}_0(1,s)$-*invariant measures of the energy-momentum hyperboloid* $\mathcal{Y}^s$, $s = 1, 2, \dots$, *and of the kinetic energies over the mass threshold*:

$$\mathcal{Y}^s \cong \mathbf{SO}_0(1,s)/\mathbf{SO}(s), \quad \Omega^{s-1} \cong \mathbf{SO}(s)/\mathbf{SO}(s-1),$$

$$\int d^{1+s}q \; \vartheta(q_0)\delta(q^2 - m^2) = \left\{ \begin{array}{l} \int \frac{d^s q}{2q_0} = \int_0^\infty \frac{q^{s-1}dq}{2q_0} \int d^{s-1}\omega \\ \qquad\qquad \text{with } q_0 = \sqrt{\vec{q}^2 + m^2}, \\ \int_{|m|}^\infty dq_0 \; |\vec{q}|^{s-2} \int d^{s-1}\omega \\ \qquad\qquad \text{with } |\vec{q}| = \sqrt{q_0^2 - m^2}. \end{array} \right.$$

For energies below the threshold there is the integration measure with *imaginary "momentum"*

$$\int_{-|m|}^{|m|} dq_0 |Q|^{s-2} \int d^{s-1}\omega \text{ with } |Q| = \sqrt{m^2 - q_0^2}.$$

Distributions of Minkowski spacetime will be called relativistic distributions of time and position representations if they are Lorentz invariant integrations over corresponding representation coefficients:

$$\mu(x) = \left\{ \begin{array}{l} \int \frac{d^s q}{q_0} \; e^{-i\vec{q}\vec{x}} \; f(q_0, x_0), \\ \int dq_0 \; e^{iq_0 x_0} \int d^{s-1}\omega \begin{pmatrix} \vartheta(q_0^2 - m^2)|\vec{q}|^{s-2} & g(\vec{q}, \vec{x}) \\ \vartheta(m^2 - q_0^2)|Q|^{s-2} g(\vec{\omega}|Q|, \vec{x}) \end{pmatrix}, \end{array} \right.$$

$$\text{with } \left\{ \begin{array}{lll} \text{time} & \mathbb{R} \ni x_0 \longmapsto f(q_0, x_0), \\ \text{position} & \mathbb{R}^s \ni \vec{x} \longmapsto \begin{pmatrix} g(\vec{q}, \vec{x}) \\ g(\vec{\omega}|Q|, \vec{x}) \end{pmatrix}, \; \vec{\omega} = \frac{\vec{q}}{|\vec{q}|} \in \Omega^{s-1}. \end{array} \right.$$

The *projection on time and position representation coefficients* is defined by integration over position and time respectively

$$\begin{array}{lll} \text{time projection:} & \int \frac{d^s x}{(2\pi)^s} \mu(x) & = \frac{1}{m} f(m, x_0), \\ \text{position projection:} & \int \frac{dx_0}{2\pi} \; \mu(x) & = \int d^{s-1}\omega \begin{pmatrix} 0 \\ m^{s-2} g(\vec{\omega}m, \vec{x}) \end{pmatrix}. \end{array}$$

In an integral for 4-dimensional Minkowski energy-momentum

$$\mathbb{R}^4 \cong \mathbb{R} \oplus (\mathbb{R}_+ \times \Omega^2), \quad d^4 q = dq_0 \vec{q}^2 d|\vec{q}| \; d^2\omega,$$

the integration $d^2\omega$ over the 2-sphere momentum directions $\Omega^2 \cong \mathbf{SO}(3)/\mathbf{SO}(2)$ leaves a Cartan coordinate $(q_0, |\vec{q}|)$-dependence. Hence the Poincaré group loses its rotations and is reduced to a self-dual dilation group $\mathbf{O}(1,1)$ acting on energy and one momentum dimension:

$$\mathbf{O}(1,3)\ \vec{\times}\ \mathbb{R}^4 \hookrightarrow \mathbf{O}(1,1)\ \vec{\times}\ \mathbb{R}^2.$$

For a 4-dimensional Lorentz scalar integral the integration over the 2-sphere yields the characteristic derivative $-\frac{\partial}{\partial \frac{r^2}{4\pi}}$, which can be used as Lorentz invariant derivative $\frac{\partial}{\partial \frac{x^2}{4\pi}}$ of the corresponding *integral for 2-dimensional spacetime.* This gives for the two kinds of Fourier transforms of generalized functions in ordered energy-momentum space the *not-ordered and the ordered* one:

$$\begin{aligned}\int d^4q \begin{pmatrix} 1 \\ \epsilon(q_0)\vartheta(q^2)\end{pmatrix} \mu(q^2) e^{iqx} &= -\frac{\partial}{\partial \frac{r^2}{4\pi}} \int dq_0 dq_3 \begin{pmatrix} 1 \\ \epsilon(q_0)\vartheta(q_0^2 - q_3^2)\end{pmatrix} \mu(q_0^2 - q_3^2) e^{iq_0x_0 - iq_3 r} \\ &= \frac{\partial}{\partial \frac{x^2}{4\pi}} \int d^2q \begin{pmatrix} 1 \\ \epsilon(q_0)\vartheta(q^2)\end{pmatrix} \mu(q^2) e^{iqx}\Big|_{x=(x_0,r)}.\end{aligned}$$

The integrals both over the energy q_0 and the hemisphere-directed momentum modulus $q_3 = \epsilon(q_3)|\vec{q}|$ are over all reals $\int_\infty^\infty$. The Kepler factor $\frac{1}{r}$-proportional contributions are characteristic for position distributions

$$\int d^3q\ e^{-i\vec{q}\vec{x}} \mu(\vec{q}^2) = -\frac{\partial}{\partial \frac{r^2}{4\pi}} \int dq_3\ e^{-iq_3 r} \mu(q_3^2) = -\frac{2\pi}{r}\frac{\partial}{\partial r} \int dq_3\ e^{-iq_3 r} \mu(q_3^2)$$

with the projection on one axis

$$\int \frac{dxdy}{2\pi}\ \frac{e^{\pm i|q_3|r}}{r} = \frac{e^{\pm i|q_3 z|}}{\mp i|q_3|}.$$

Nonscalar integrals arise with derivations $\frac{\partial}{\partial x} = 2x\frac{\partial}{\partial x^2}$.

The time antisymmetric integral with $d^4q\ \epsilon(q_0)\vartheta(q^2)$ is trivial for spacelike x as seen for $x_0 = 0$, i.e., Fourier transforms of ordered energy-momentum measures have *causal support*:

$$\begin{aligned}\int d^2q\ \epsilon(q_0)\vartheta(q^2)\mu(q^2)e^{iqx} &= \int d^2q\ \epsilon(q_0)\vartheta(q^2)\mu(q^2)e^{-iq_3 r} \\ &= 0 \text{ for } x^2 = -r^2 < 0.\end{aligned}$$

It is useful to tabulate some Fourier transforms for energy-momentum distributions (P is a polynomial):

$\mu(q) = \int d^4x \tilde{\mu}(x) e^{-iqx}$	$\tilde{\mu}(x) = \int \frac{d^4q}{(2\pi)^4} \mu(q) e^{iqx}$
$\mu(-q),\ \overline{\mu(q)}$	$\tilde{\mu}(-x),\ \overline{\tilde{\mu}(-x)}$
$\mu(\alpha q),\ \alpha > 0;\ \mu(q+p),\ e^{iqy}\mu(q)$	$\frac{1}{\alpha^4}\tilde{\mu}(\frac{x}{\alpha});\ e^{-ipx}\tilde{\mu}(x),\ \tilde{\mu}(x+y)$
$P(iq)\mu(q),\ P(i\frac{\partial}{\partial q})\mu(q)$	$P(\frac{\partial}{\partial x})\tilde{\mu}(x),\ P(x)\tilde{\mu}(x)$
1	$\delta(x)$
$\vartheta(q^2)$	$\frac{1}{\pi^3}\frac{1}{(x_{\rm P}^2)^2}$
$\epsilon(q_0)\vartheta(q^2)$	$\frac{i}{\pi^2}\epsilon(x_0)\delta'(x^2)$
$\delta(q^2)$	$-\frac{1}{4\pi^3}\frac{1}{x_{\rm P}^2}$
$\epsilon(q_0)\delta(q^2)$	$-\frac{i}{4\pi^2}\epsilon(x_0)\delta(x^2)$
$\frac{\Gamma(1+\nu)}{(q^2+io)^{1+\nu}}$ $\nu \in \mathbb{R},\quad \nu \neq \pm 1, \pm 2, \ldots$	$\frac{i}{(4\pi)^2}\frac{\Gamma(1-\nu)}{(\frac{x^2}{4}-io)^{1-\nu}}$
$\frac{i}{\pi}\frac{1}{q^2+io} = \delta(q^2) + \frac{i}{\pi}\frac{1}{q_{\rm P}^2}$	$-\frac{1}{4\pi^3}\frac{1}{x^2-io} = \frac{i}{4\pi^2}[\frac{i}{\pi}\frac{1}{x_{\rm P}^2} - \delta(x^2)]$
$\frac{1}{i}\frac{q\ \Gamma(1+\nu)}{(q^2+io)^{1+\nu}}$ $\nu \in \mathbb{R},\quad \nu \neq -1, \pm 2, \pm 3 \ldots$	$\frac{i}{(4\pi)^2}\frac{\frac{x}{2}\ \Gamma(2-\nu)}{(\frac{x^2}{4}-io)^{2-\nu}}$
$\frac{1}{i\pi}\frac{q}{q^2+io} = q[-\delta(q^2) + \frac{1}{i\pi}\frac{1}{q_{\rm P}^2}]$	$\frac{i}{2\pi^3}\frac{x}{(x^2-io)^2} = \frac{1}{2\pi^2}x[\frac{i}{\pi}\frac{1}{(x_{\rm P}^2)^2} + \delta'(x^2)]$
$\frac{1}{i\pi}\frac{q}{(q^2+io)^2} = q[\delta'(q^2) + \frac{1}{i\pi}\frac{1}{(q_{\rm P}^2)^2}]$	$\frac{i}{8\pi^3}\frac{x}{x^2-io} = \frac{1}{8\pi^2}x[\frac{i}{\pi}\frac{1}{x_{\rm P}^2} - \delta(x^2)]$

The Fourier transformation exchanges with each other Dirac and principal value contribution. In contrast to a positive definite product, e.g., for energies to $(E^2 - io)^\nu = E^{2\nu}$ with trivial Dirac distribution $\delta(E^2) = 0$, the imaginary parts of $(q^2 - io)^\nu$ with indefinite energy-momenta $q^2 = q_0^2 - \vec{q}^2$ are nontrivial, e.g., the Dirac distribution $\delta(q^2)$. Hence there can be the nontrivial Fourier transformation on the light cone, e.g., from $\delta(q^2)$ to $\frac{1}{x_\mathrm{P}^2}$.

One obtains for simultaneous spacetime and energy-momentum order with $2\vartheta(\pm q_0 x_0) = 1 \pm \epsilon(q_0)\epsilon(x_0) = 1 \pm \epsilon(q_0 x_0)$,

$$\begin{array}{rcccl} & & \pm i\pi\vartheta(x^2) + 1 & = & 2 \int \frac{d^4q}{\pi}\, \vartheta(\pm q_0 x_0)\delta'(q^2) e^{iqx}, \\ \frac{1}{x^2 \mp io} & = & \pm i\pi\delta(x^2) + \frac{1}{x_\mathrm{P}^2} & = & - \int \frac{d^4q}{2\pi} \vartheta(\pm q_0 x_0)\delta(q^2) e^{iqx}, \\ \frac{1}{(x^2 \mp io)^2} & = & \mp i\pi\delta'(x^2) + \frac{1}{(x_\mathrm{P}^2)^2} & = & \int \frac{d^4q}{16\pi} \vartheta(\pm q_0 x_0)\vartheta(q^2) e^{iqx}. \end{array}$$

3.4 Scattering Waves (on Shell)

The $\mathbf{O}(1,s)$-invariant Dirac measure, supported by the energy-momenta hyperboloid $\mathcal{Y}^s \subset \check{\mathbb{R}}^{1+s}$ for $m^2 > 0$, gives the *scalar cosine* of spacetime translations. It is a representation coefficient of the Poincaré group $\mathbf{SO}_0(1,s) \vec{\times} \mathbb{R}^{1+s}$ (chapter "Harmonic Analysis"):

$$\begin{aligned} \mathbf{C}(m|x) &= \int \frac{d^{1+s}q}{(2\pi)^s} |m| \delta(m^2 - q^2) e^{iqx} = \int \frac{d^{1+s}q}{(2\pi)^s} |m| \delta(m^2 - q^2) \cos qx \\ &= |m| \int \frac{d^s q}{q_0 (2\pi)^s} \cos qx \Big|_{q_0 = \sqrt{\vec{q}^2 + m^2}} = |m| \int \frac{d^s q}{q_0 (2\pi)^s} e^{-i\vec{q}\vec{x}} \cos q_0 x_0 \Big|_{q_0 = \sqrt{\vec{q}^2 + m^2}}, \\ s = 0: \quad & \mathbf{C}(m|x) = \cos m x_0. \end{aligned}$$

$\frac{|m|}{(2\pi)^s}$ is a convenient normalization factor. One has the cosine properties and the invariance

$$\begin{aligned} &\mathbf{C}(m|x) = \mathbf{C}(m|-x) = \mathbf{C}(-m|x) = \overline{\mathbf{C}(m|x)}, \\ &\mathbf{C}(m|\Lambda.x) = \mathbf{C}(m|x), \ \Lambda \in \mathbf{O}(1,s). \end{aligned}$$

For time translations with $s = 0$, there is no momentum integration left, and one has to put $\vec{q} = 0$.

In the scalar cosine the energy q_0 has to surpass the mass threshold m, i.e., there is a positive kinetic energy $\vec{q}^2 = q_0^2 - m^2$. The time representations in $\mathbf{U}(1) \cong \mathbf{SO}(2)$ are paired with standing spherical waves from compact position translation representations in $\mathbf{SO}(2)$, forming together coefficients of Hilbert space representations of the Poincaré group:

$$\begin{aligned} \frac{\mathbf{C}(m|x)}{|m|} &= \int \frac{d^4q}{(2\pi)^3} \delta(m^2 - q^2) e^{iqx} = 2 \frac{\partial}{\partial x^2} \int \frac{d^2q}{(2\pi)^2} \delta(q^2 - m^2) e^{iqx} \Big|_{x = (x_0, r)} \\ &= -2 \frac{\partial}{\partial r^2} \int \frac{dq_0}{(2\pi)^2} e^{iq_0 x_0}\, \vartheta(q_0^2 - m^2) \frac{\cos |\vec{q}| r}{|\vec{q}|} \Big|_{|\vec{q}| = \sqrt{q_0^2 - m^2}} \\ &= \int \frac{dq_0}{(2\pi)^2} e^{iq_0 x_0}\, \vartheta(q_0^2 - m^2) \frac{\sin |\vec{q}| r}{r} \Big|_{|\vec{q}| = \sqrt{q_0^2 - m^2}}. \end{aligned}$$

The full energy-momentum integration gives a function, defined Lebesgue-almost everywhere in spacetime:

$$\begin{array}{lll} s=1: & 2\pi\frac{\mathbf{C}(m|x)}{|m|} & = \vartheta(-x^2)2\mathcal{K}_0(m|x|) - \vartheta(x^2)\pi\mathcal{N}_0(m|x|), \quad |x| = \sqrt{|x^2|}, \\ s=3: & 2\pi^2\frac{\mathbf{C}(m|x)}{|m|} & = \frac{\partial}{\partial x^2}[\vartheta(-x^2)2\mathcal{K}_0(m|x|) - \vartheta(x^2)\pi\mathcal{N}_0(m|x|)]. \end{array}$$

The functions involved are given explicitly in the next section.

By derivations $\partial_k = 2x_k\frac{\partial}{\partial x^2}$ of the scalar cosine one obtains functions with nontrivial $\mathbf{O}(1,s)$-behavior, e.g., the *vector sine* of spacetime translations (for $s=0$ without momentum integration one has to take the component $j=0$ with $\vec{q}=0$),

$$\begin{aligned} i\mathbf{S}_j(m|x) &= \epsilon(m)\int\frac{d^{1+s}q}{(2\pi)^s}q_j\delta(m^2-q^2)e^{iqx} = \epsilon(m)\int\frac{d^{1+s}q}{(2\pi)^s}q_j\delta(m^2-q^2)i\sin qx \\ &= \epsilon(m)\int\frac{d^sq}{q_0(2\pi)^s}q_j\ i\sin qx\Big|_{q_0=\sqrt{\vec{q}^2+m^2}} = \epsilon(m)\int\frac{d^sq}{q_0(2\pi)^s}\ e^{-i\vec{q}\vec{x}}\begin{pmatrix} q_0 i\sin q_0x_0 \\ \vec{q}\cos q_0x_0\end{pmatrix}\Big|_{q_0=\sqrt{m^2+\vec{q}^2}} \\ &= \epsilon(m)\int\frac{dq_0}{(2\pi)^2}e^{iq_0x_0}\ \vartheta(q_0^2-m^2)\begin{pmatrix} q_0\frac{\sin|\vec{q}|r}{r} \\ i\frac{\vec{x}}{r}\ \frac{\sin|\vec{q}|r-|\vec{q}|r\cos|\vec{q}|r}{r^2}\end{pmatrix}\Big|_{|\vec{q}|=\sqrt{q_0^2-m^2}} \text{ for } s=3, \end{aligned}$$

with the transformation properties

$$\begin{array}{ll} \mathbf{S}_j(m|x) & = -\mathbf{S}_j(m|-x) = -\mathbf{S}_j(-m|x) = \overline{\mathbf{S}_j(m|x)}, \\ \mathbf{S}_j(m|\Lambda.x) & = \Lambda_j^k\mathbf{S}_k(m|x), \quad \Lambda\in\mathbf{O}(1,s). \end{array}$$

$\mathbf{C}$ and $\mathbf{S}_j$ involve spherical Bessel functions, multiplied by the matching spherical harmonics $\mathrm{Y}^L_m(\varphi,\theta)j_L(|\vec{q}|r)$ (chapter "The Kepler Factor").

3.5 Macdonald, Neumann, and Bessel Functions

Exponential, cosine, and sine in $\mathbb{R}$-representations are embedded in Macdonald, Neumann, and Bessel functions as $\mathbb{R}^2$-representations with an acting orthogonal group (chapter "Residual Spacetime Representations").

The 2-dimensional Fourier transforms of the energy-momenta hyperboloids as $\mathbf{SO}_0(1,1)$-orbits can be computed in hyperbolic coordinates $(\epsilon(q_0)q_0, q_3) = (\cosh\psi, \sinh\psi)$,

$$\begin{aligned} \int d^2q\begin{pmatrix}1\\ \epsilon(q_0)\end{pmatrix}\delta(q^2-1)e^{iqx} &= \int d\psi\begin{pmatrix}\cos(x_0\cosh\psi)\\ i\sin(x_0\cosh\psi)\end{pmatrix}e^{-ix_3\sinh\psi} \\ &= \begin{pmatrix}\vartheta(-x^2)2\mathcal{K}_0(|x|)-\vartheta(x^2)\pi\mathcal{N}_0(|x|)\\ i\epsilon(x_0)\vartheta(x^2)\pi\mathcal{J}_0(|x|)\end{pmatrix}, \end{aligned}$$

with the real *Macdonald, Neumann, and Bessel functions* for index 0 and real argument $0\neq\xi\in\mathbb{R}$:

$$\begin{array}{lll} 2\mathcal{K}_0(\xi) & = \int d\psi\ e^{-|\xi|\cosh\psi} & = \int d\psi\ \cos(\xi\sinh\psi), \\ -\pi\mathcal{N}_0(\xi) & = \int d\psi\ \cos(\xi\cosh\psi) & = 2\int_0^\infty d\psi\ e^{-|\xi|\sinh\psi} - \int_0^\pi d\chi\ \sin(|\xi|\sin\chi), \\ \pi\mathcal{J}_0(\xi) & = \int d\psi\ \sin(|\xi|\cosh\psi). & \end{array}$$

The functions arise also for $\mathbf{SO}(2)$-orbits:

$$\begin{aligned}
\int \frac{d^2q}{\pi}\, \delta(\vec{q}^2 - 1)e^{-i\vec{q}\vec{x}} &= \mathcal{J}_0(|\vec{x}|) &= \int_0^\pi \frac{d\chi}{\pi}\, \cos(|\vec{x}|\cos\chi) &= \int_0^{2\pi} \frac{d\chi}{2\pi}\, e^{i|\vec{x}|\cos\chi},\\
\int \frac{d^2q}{\pi} \frac{1}{\vec{q}^2+1} e^{-i\vec{q}\vec{x}} &= 2\mathcal{K}_0(|\vec{x}|) &= \int d\psi\, \cos(|\vec{x}|\sinh\psi) &= \int d\psi\, e^{-|\vec{x}|\cosh\psi}.
\end{aligned}$$

The Macdonald function with squared dependence of complex argument $\zeta \in \mathbb{C}$ can be decomposed together with the embedded $\mathbb{C}$ and $\mathbb{R}$-representations $e^{-\zeta}$ and $e^{i\xi} = \cos\xi + i\sin\xi$:

$$\begin{aligned}
|\arg\zeta| < \tfrac{\pi}{2}:\ 2\mathcal{K}_0(\zeta) &= \int d\psi\, e^{-\zeta\cosh\psi} = -\sum_{n=0}^{\infty} \frac{(\frac{\zeta^2}{4})^n}{(n!)^2}[\log\tfrac{\zeta^2}{4} - 2\Gamma'(1) - 2\varphi(n)],\\
\xi > 0:\ 2\mathcal{K}_0(-i\xi) &= -\pi\mathcal{N}_0(\xi) + i\pi\mathcal{J}_0(\xi)\\
= \int d\psi\, e^{i|\xi|\cosh\psi} &= \int d\psi\, [\cos(\xi\cosh\psi + i\sin(|\xi|\cosh\psi)].
\end{aligned}$$

Its expansion involves Euler's constant $-\Gamma'(1)$:

$$\begin{aligned}
\varphi(0) &= 0,\ \ \varphi(n) = 1 + \tfrac{1}{2} + \cdots + \tfrac{1}{n},\ \ n = 1, 2, \ldots,\\
-\Gamma'(1) &= \lim_{n\to\infty}[\varphi(n) - \log n] = 0.5772\ldots.
\end{aligned}$$

$\mathcal{K}_0$ and $\mathcal{N}_0$ have a logarithmic singularity for $\zeta^2 = 0$:

$$\xi \in \mathbb{R}:\quad -\log(-\tfrac{\xi^2}{4}) = -\log\tfrac{\xi^2}{4} + i\pi.$$

The regular Bessel function $\mathcal{J}_0$ is defined also for complex argument

$$\mathcal{J}_0(\zeta) = \sum_{n=0}^{\infty} \frac{(-\frac{\zeta^2}{4})^n}{(n!)^2}.$$

The functions with *integer index* $L = 0, 1, 2\ldots$ arise by derivation:

$$\begin{aligned}
\frac{\mathcal{J}_L(\zeta)}{(\frac{\zeta}{2})^L} &= (-\frac{\partial}{\partial\frac{\zeta^2}{4}})^L \mathcal{J}_0(\zeta) = \sum_{n=0}^{\infty} \frac{(-\frac{\zeta^2}{4})^n}{(L+n)!n!} = \frac{\sqrt{\pi}}{\Gamma(\frac{1}{2}+L)} \int_{-1}^{1} \frac{d\mu}{\pi} \frac{\cos\mu\zeta}{(1-\mu^2)^{\frac{1}{2}-L}},\\
\mathcal{J}_L(\zeta) &= \int_{-1}^{1} \frac{d\mu}{\pi} \frac{(-4i\mu)^L}{\sqrt{1-\mu^2}} e^{i\mu\zeta},\\
\mathcal{J}_{-L}(\zeta) &= (-1)^L \mathcal{J}_L(\zeta),\ \ \lim_{\zeta\to 0} \frac{\mathcal{J}_L(\zeta)}{(\frac{\zeta}{2})^L} = \frac{1}{\Gamma(1+L)},\\
e^{\frac{\zeta^2-1}{2}} &= \sum_{z=-\infty}^{\infty} \zeta^z \mathcal{J}_z(\zeta),\ \ e^{i\xi\sin\chi} = \sum_{z=-\infty}^{\infty} e^{iz\chi}\mathcal{J}_z(\xi),\ \ \mathcal{J}_L(\xi) = \int_{-\pi}^{\pi} \frac{d\chi}{2\pi} e^{i(\xi\sin\chi - L\chi)},\\
\frac{\mathcal{J}_L(\zeta)}{(\frac{\zeta}{2})^L} &= \mathcal{E}_L(\tfrac{\zeta^2}{4}) = (1+L)\mathcal{E}_{1+L}(\tfrac{\zeta^2}{4}) - \tfrac{\zeta^2}{4}\mathcal{E}_{2+L}(\tfrac{\zeta^2}{4}).
\end{aligned}$$

The Neumann and Macdonald functions with integer index $L = 1, 2, \ldots$ have order-L singularities for $\zeta \to 0$, from derivations of the logarithm $\pi\mathcal{N}_0(\zeta) = -\log\zeta^2 + \cdots$:

$$\begin{aligned}
\frac{\pi\mathcal{N}_L(\zeta)}{(\frac{\zeta}{2})^L} &= (-\frac{\partial}{\partial\frac{\zeta^2}{4}})^L \pi\mathcal{N}_0(\zeta)\\
&= -\sum_{n=1}^{L} \frac{(n-1)!(\frac{\zeta^2}{4})^{-n}}{(L-n)!} + \sum_{n=0}^{\infty} \frac{(-\frac{\zeta^2}{4})^n}{(L+n)!n!}[\log\tfrac{\zeta^2}{4} - 2\Gamma'(1) - \varphi(L+n) - \varphi(n)],\\
\mathcal{N}_{-L}(\zeta) &= (-1)^L \mathcal{N}_L(\zeta),\ \ \lim_{\zeta\to 0}(\tfrac{\zeta}{2})^L \pi\mathcal{N}_L(\zeta) = -\Gamma(L).
\end{aligned}$$

The Bessel functions with real index

$$\lambda \in \mathbb{R}: \ \frac{\mathcal{J}_\lambda(\zeta)}{(\frac{\zeta}{2})^\lambda} = \sum_{n=0}^\infty \frac{(-\frac{\zeta^2}{4})^n}{\Gamma(1+\lambda+n)n!}, \begin{cases} \zeta d_\zeta \mathcal{J}_\lambda(\zeta) &= \lambda\mathcal{J}_\lambda(\zeta) - \zeta\mathcal{J}_{\lambda+1}(\zeta) \\ &= -\lambda\mathcal{J}_\lambda(\zeta) + \zeta\mathcal{J}_{\lambda-1}(\zeta), \\ \frac{2\lambda}{\zeta}\mathcal{J}_\lambda(\zeta) &= \mathcal{J}_{\lambda-1}(\zeta) + \mathcal{J}_{\lambda+1}(\zeta), \end{cases}$$

are cylinder functions, i.e., solutions of the Bessel differential equation

$$[(\zeta\tfrac{d}{d\zeta})^2 + \zeta^2 - \lambda^2]\mathcal{Z}_\lambda(\zeta) = 0, \ \ [\tfrac{\zeta^2}{4}(\tfrac{d}{d\frac{\zeta^2}{4}})^2 + \lambda^2\tfrac{d}{d\frac{\zeta^2}{4}} + 1]\tfrac{\mathcal{Z}_\lambda(\zeta)}{\zeta^\lambda} = 0.$$

The Bessel function partners are

$$\begin{array}{rrl} \textit{Macdonald functions:} & 2\mathcal{K}_\lambda(\zeta) & = \frac{e^{i\lambda\frac{\pi}{2}}\pi\mathcal{J}_{-\lambda}(i\zeta) - e^{-i\lambda\frac{\pi}{2}}\pi\mathcal{J}_\lambda(i\zeta)}{\sin\lambda\pi} = 2\mathcal{K}_{-\lambda}(\zeta) \\ & & = \int d\psi\ \cosh\lambda\psi\ e^{-\zeta\cosh\psi}, \ |\arg\ \zeta| < \frac{\pi}{2}, \\ \textit{Neumann functions:} & \mathcal{N}_\lambda(\zeta) & = \frac{\cos\lambda\pi\mathcal{J}_\lambda(\zeta) - \mathcal{J}_{-\lambda}(\zeta)}{\sin\lambda\pi} \\ \textit{Hankel functions:} & \mathcal{H}_\lambda^{1,2}(\zeta) & = \mathcal{J}_\lambda(\zeta) \pm i\mathcal{N}_\lambda(\zeta) = e^{\mp i\lambda\pi}\mathcal{H}_{-\lambda}^{1,2}(\zeta) = -\mathcal{H}_{-\lambda}^{2,1}(-\zeta), \\ \zeta \leftrightarrow i\zeta: & 2\mathcal{K}_\lambda(\zeta) & = e^{i\lambda\frac{\pi}{2}}i\pi\mathcal{H}_\lambda^1(i\zeta) = e^{i\lambda\frac{\pi}{2}}\pi[i\mathcal{J}_\lambda(i\zeta) - \mathcal{N}_\lambda(i\zeta)]. \end{array}$$

The integer index functions arise as limits. All solutions of Bessel's differential equation are spanned by $\{\mathcal{J}_\lambda, \mathcal{N}_\lambda\}$, for noninteger λ also by $\{\mathcal{J}_{\pm\lambda}\}$.

The hyperbolic Macdonald and the spherical Bessel and Neumann functions are the *half-integer index* functions. They arise by derivation from the irreducible $\mathbb{C}$-representation matrix elements:

$$\left.\begin{array}{lll} \sqrt{\pi}\frac{\frac{2}{\pi}\mathcal{K}_{L-\frac{1}{2}}(\zeta)}{(\frac{\zeta}{2})^{L-\frac{1}{2}}} & = (-\frac{d}{d\frac{\zeta^2}{4}})^L e^{-\zeta} & = 2\frac{k_{L-1}(\zeta)}{(\frac{\zeta}{2})^{L-1}}, \\ \sqrt{\pi}\frac{\mathcal{N}_{L-\frac{1}{2}}(\zeta)}{(\frac{\zeta}{2})^{L-\frac{1}{2}}} & = (-\frac{d}{d\frac{\zeta^2}{4}})^L \sin\zeta & = -2\frac{n_{L-1}(\zeta)}{(\frac{\zeta}{2})^{L-1}}, \\ \sqrt{\pi}\frac{\mathcal{J}_{L-\frac{1}{2}}(\zeta)}{(\frac{\zeta}{2})^{L-\frac{1}{2}}} & = (-\frac{d}{d\frac{\zeta^2}{4}})^L \cos\zeta & = 2\frac{j_{L-1}(\zeta)}{(\frac{\zeta}{2})^{L-1}}, \end{array}\right\} (k_0, n_0, j_0)(\zeta) = \frac{(e^{-\zeta}, \cos\zeta, \sin\zeta)}{\zeta},$$

$$\begin{array}{l} \mathcal{J}_{\pm(L-\frac{1}{2})}(\zeta) = \mp(-1)^L\mathcal{N}_{\mp(L-\frac{1}{2})}(\zeta), \ \mathcal{K}_{L-\frac{1}{2}}(\zeta) = \mathcal{K}_{-L+\frac{1}{2}}(\zeta), \\ \lim_{\zeta\to 0}\frac{j_L(\zeta)}{(2\zeta)^L} = \frac{L!}{(1+2L)!}, \ \ \lim_{\zeta\to 0}(2\zeta)^{1+L}n_L(\zeta) = 2\frac{(2L)!}{L!}; \end{array}$$

j_L, n_L, and k_L are used as 3-position representation coefficients (chapter "The Kepler Factor"), related to spherical waves and to interactions.

3.6 Yukawa Potential and Force (off Shell)

In ordered spacetime $\mathbb{R}^{1+s}$ with the orthochronous group $\mathbf{SO}_0(1,s)$ there exists the *scalar sine* of spacetime translations for $m \in \mathbb{R}$,

$$\begin{array}{rl} i\mathbf{s}(m|x) & = \int \frac{d^{1+s}q}{(2\pi)^s}\epsilon(q_0)m\delta(m^2 - q^2)e^{iqx} = \int \frac{d^{1+s}q}{(2\pi)^s}\epsilon(q_0)m\delta(m^2 - q^2)i\sin qx \\ & = m\int \frac{d^s q}{q_0(2\pi)^s}i\sin qx\Big|_{q_0=\sqrt{m^2+\vec{q}^2}} = m\int \frac{d^s q}{q_0(2\pi)^s}e^{-i\vec{q}\vec{x}}\ i\sin q_0x_0\Big|_{q_0=\sqrt{m^2+\vec{q}^2}}, \end{array}$$

with the transformation properties

$$\begin{array}{l} \mathbf{s}(m|x) = -\mathbf{s}(m|-x) = -\mathbf{s}(-m|x) = \overline{\mathbf{s}(m|x)}, \\ \mathbf{s}(m|\Lambda.x) = \mathbf{s}(m|x), \ \Lambda \in \mathbf{SO}_0(1,s). \end{array}$$

It has *causal support* and involves the lightcone Dirac distribution for 4-dimensional Minkowski spacetime:

$$\begin{array}{lll} s=0: & \mathbf{s}(m|x) & = \sin mx_0, \\ s=1: & \frac{\mathbf{s}(m|x)}{m} & = \frac{\epsilon(x_0)}{2}\vartheta(x^2)\mathcal{E}_0(\frac{m^2x^2}{4}), \\ s=3: & \frac{\mathbf{s}(m|x)}{m} & = \frac{\epsilon(x_0)}{2\pi}\frac{\partial}{\partial x^2}\vartheta(x^2)\mathcal{E}_0(\frac{m^2x^2}{4}) \\ & & = \frac{\epsilon(x_0)}{2\pi}\frac{m^2}{4}[\delta(\frac{m^2x^2}{4}) - \vartheta(x^2)\mathcal{E}_1(\frac{m^2x^2}{4})]. \end{array}$$

An $\epsilon(x_0)$-multiplied ordered integration $d^{1+s}q\epsilon(q_0)$ with an energy-momentum Dirac measure coincides with an integration with an energy-momentum principal value P pole function as shown by the identities

$$\epsilon(x_0)\int d^{1+s}q\ \epsilon(q_0)\delta^{(N)}(m^2-q^2)e^{iqx} = \int \frac{d^{1+s}q}{i\pi}\frac{\Gamma(1+N)}{(q_{\rm P}^2-m^2)^{1+N}}e^{iqx}, \quad N=0,1,\ldots.$$

The *causal sine*

$$\epsilon(x_0)\frac{\mathbf{s}(m|x)}{m} = \epsilon(x_0)\int \frac{d^{1+s}q}{i(2\pi)^s}\epsilon(q_0)\delta(m^2-q^2)e^{iqx} = \int \frac{d^{1+s}q}{\pi(2\pi)^s}\frac{1}{-q_{\rm P}^2+m^2}e^{iqx}$$

is a relativistic distribution of a causal time representation.

The energy-momentum support of the quantization distribution $\mathbf{s}(m|x)$ with the Dirac measure is the mass shell $\{q \,|\, q^2 = m^2\}$ whereas $\epsilon(x_0)\mathbf{s}(m|x)$ with the principal value has support for all energy-momenta:

$$\begin{array}{llll} \int \frac{d^{1+s}x}{2\pi} & i\frac{\mathbf{s}(m|x)}{m} & = \epsilon(q_0)\delta(q^2-m^2) & \textit{on shell, i.e., } q^2=m^2, \\ \int \frac{d^{1+s}x}{2\pi} & \epsilon(x_0)i\frac{\mathbf{s}(m|x)}{m} & = \frac{i}{\pi}\frac{1}{-q_{\rm P}^2+m^2} & \textit{off shell for } q^2\neq m^2. \end{array}$$

The harmonic analysis of both the sine and the ordered sine displays time representations in $\mathbf{U}(1)\cong\mathbf{SO}(2)$ paired with 2-sphere distributions of compact position translation representations in $\mathbf{SO}(2)$ for energies q_0^2 over the threshold m^2 (in quantum mechanics scattering waves $E_{\rm kin}=E-V_0>0$). For energies with q_0^2 below the threshold m^2 (in quantum mechanics bound waves $E-V_0<0$) the ordered sine involves Yukawa potentials as 2-sphere distributions of noncompact position translation representations in $\mathbf{SO}_0(1,1)$ with the position eigenvalue from an imaginary "momentum"

$$\begin{aligned} \begin{pmatrix} 1 \\ \epsilon(x_0) \end{pmatrix}\frac{i\mathbf{s}(m|x)}{m} &= \int \frac{d^4q}{(2\pi)^3}\begin{pmatrix} \epsilon(q_0)\delta(q^2-m^2) \\ \frac{i}{\pi}\frac{1}{-q_{\rm P}^2+m^2} \end{pmatrix}e^{iqx} \\ &= 2\frac{\partial}{\partial x^2}\int \frac{d^2q}{(2\pi)^2}\begin{pmatrix} \epsilon(q_0)\delta(q^2-m^2) \\ \frac{i}{\pi}\frac{1}{-q_{\rm P}^2+m^2} \end{pmatrix}e^{iqx}|_{x=(x_0,r)} \\ = -2\frac{\partial}{\partial r^2}\int \frac{dq_0}{(2\pi)^2}e^{iq_0x_0} &\left[\vartheta(q_0^2-m^2)\begin{pmatrix} \epsilon(q_0)\frac{\cos|\vec q|r}{|\vec q|} \\ -i\frac{\sin|\vec q|r}{|\vec q|} \end{pmatrix} + \vartheta(m^2-q_0^2)\begin{pmatrix} 0 \\ i\frac{e^{-|Q|r}}{|Q|} \end{pmatrix}\right] \\ = \int \frac{dq_0}{(2\pi)^2}e^{iq_0x_0} &\left[\vartheta(q_0^2-m^2)\begin{pmatrix} \epsilon(q_0)\frac{\sin|\vec q|r}{r} \\ i\frac{\cos|\vec q|r}{r} \end{pmatrix} + \vartheta(m^2-q_0^2)\begin{pmatrix} 0 \\ i\frac{e^{-|Q|r}}{r} \end{pmatrix}\right] \\ &\text{with } |\vec q| = \sqrt{q_0^2-m^2} \text{ and } |Q| = \sqrt{m^2-q_0^2}. \end{aligned}$$

The sine gives $r=0$-regular spherical Bessel functions, whereas the ordered sine contains $r=0$-singular spherical Neumann and hyperbolic Macdonald functions.

The causal sine obeys an inhomogeneous Klein-Gordon equation. Causal time representations have the Yukawa potential $\frac{e^{-|m|r}}{r} \leftrightarrow \sin|mt|$ as relativistic supplement for position translations

$$\left.\begin{array}{ll} (\partial_0^2 + m^2)\epsilon(t)\frac{\sin mt}{m} & = 2\delta(t) \\ (-\vec{\partial}^2 + m^2)\frac{e^{-|m|r}}{2\pi r} & = 2\delta(\vec{x}) \end{array}\right\} \hookrightarrow (\partial^2 + m^2)\epsilon(x_0)\frac{\mathbf{s}(m|x)}{m} = 2\delta(x).$$

The causal *vector cosine* arises by derivation,

$$\begin{aligned} \mathbf{c}_j(m|x) &= \int \frac{d^{1+s}q}{(2\pi)^{1+s}}\epsilon(q_0)q_j\delta(m^2 - q^2)e^{iqx} = \int \frac{d^{1+s}q}{(2\pi)^{1+s}}\epsilon(q_0)q_j\delta(m^2 - q^2)\cos qx \\ &= \int \frac{d^sq}{q_0(2\pi)^s} q_j \cos qx \Big|_{q_0=\sqrt{m^2+\vec{q}^2}} = \int \frac{d^sq}{q_0(2\pi)^s} e^{-i\vec{q}\vec{x}} \begin{pmatrix} q_0 \cos q_0x_0 \\ \vec{q}\, i\sin q_0x_0 \end{pmatrix}\Big|_{q_0=\sqrt{m^2+\vec{q}^2}} \\ &= \int \frac{dq_0}{(2\pi)^2} e^{iq_0x_0}\ \epsilon(q_0)\vartheta(q_0^2 - m^2) \begin{pmatrix} q_0 \frac{\sin|\vec{q}|r}{r} \\ i\frac{\vec{x}}{r}\ \frac{\sin|\vec{q}|r - |\vec{q}|r\cos|\vec{q}|r}{r^2} \end{pmatrix}\Big|_{|\vec{q}|=\sqrt{q_0^2-m^2}} \text{ for } s = 3. \end{aligned}$$

It has the transformation properties

$$\begin{aligned} &\mathbf{c}_j(m|x) = \mathbf{c}_j(m|-x) = \mathbf{c}_j(-m|x) = \overline{\mathbf{c}_j(m|x)}, \\ &\mathbf{c}_j(m|\Lambda.x) = \Lambda_j^k\mathbf{c}_k(m|x),\ \Lambda \in \mathbf{SO}_0(1,s). \end{aligned}$$

For time $x_0 = 0$ there arises a Dirac measure supported by the position space origin

$$\mathbf{c}_j(m|\vec{x}) = \delta_j^0\delta(\vec{x}).$$

The explicit spacetime expressions read

$$\begin{aligned} s = 0:\ \ \mathbf{c}_j(m|x) &= \cos mx_0, \\ s = 1:\ \ \mathbf{c}_j(m|x) &= \epsilon(x_0)x_j\frac{\partial}{\partial x^2}\vartheta(x^2)\mathcal{E}_0(\tfrac{m^2x^2}{4}) \\ &= \epsilon(x_0)x_j\frac{m^2}{4}[\delta(\tfrac{m^2x^2}{4}) - \vartheta(x^2)\mathcal{E}_1(\tfrac{m^2x^2}{4})], \\ s = 3:\ \ \mathbf{c}_j(m|x) &= \frac{\epsilon(x_0)}{\pi}x_j(\frac{\partial}{\partial x^2})^2\vartheta(x^2)\mathcal{E}_0(\tfrac{m^2x^2}{4}) \\ &= \frac{\epsilon(x_0)}{\pi}x_j\frac{m^4}{16}[\delta'(\tfrac{m^2x^2}{4}) - \delta(\tfrac{m^2x^2}{4}) + \vartheta(x^2)\mathcal{E}_2(\tfrac{m^2x^2}{4})]. \end{aligned}$$

The harmonic analysis of the $\epsilon(x_0)$-multiplied vector cosine for $1 + s = 4$ involves *Yukawa forces* (hyperbolic Macdonald function) with the Coulomb force for $m = 0$:

$$\begin{aligned} \epsilon(x_0)\mathbf{c}_j(m|x) &= \frac{1}{i\pi}\int \frac{d^4q}{(2\pi)^s}\frac{q_j}{q_{\text{P}}^2 - m^2}e^{iqx} \\ &= \int \frac{dq_0}{(2\pi)^2}e^{iq_0x_0}\left[\vartheta(q_0^2 - m^2)\begin{pmatrix} iq_0\frac{\cos|\vec{q}|r}{r} \\ -\frac{\vec{x}}{r}\ \frac{\cos|\vec{q}|r + |\vec{q}|r\sin|\vec{q}|r}{r^2} \end{pmatrix} + \vartheta(m^2 - q_0^2)\begin{pmatrix} iq_0 \\ -\frac{\vec{x}}{r}\ \frac{1+r|Q|}{r} \end{pmatrix}\frac{e^{-|Q|r}}{r}\right]. \end{aligned}$$

3.7 Feynman Propagators

The *scalar Feynman propagators* have causally supported imaginary part:

$$\begin{aligned} \mathbf{E}(\pm i|m||x) &= \mathbf{C}(m|x) \pm \epsilon(mx_0)i\mathbf{s}(m|x) = \overline{\mathbf{E}(\mp i|m||x)} \\ &= \int \frac{d^4q}{(2\pi)^3}\vartheta(\pm q_0x_0)2|m|\delta(m^2 - q^2)e^{iqx} \\ &= \pm\int \frac{d^4q}{i\pi(2\pi)^3}\frac{|m|}{q^2 \mp io - m^2}e^{iqx} = |m|\int \frac{d^3q}{q_0(2\pi)^3}e^{-i\vec{q}\vec{x}}\ e^{\pm iq_0|x_0|}\Big|_{q_0=\sqrt{m^2+\vec{q}^2}}. \end{aligned}$$

They arise as Fourier transforms of an on shell particle Dirac distribution and a principal value function that contains both on shell particle and off shell interaction contributions in the decomposition

$$\pm\frac{1}{i\pi}\frac{1}{q^2\mp io-m^2} = \delta(q^2-m^2) \pm \frac{1}{i\pi}\frac{1}{q_{\rm P}^2-m^2}.$$

The harmonic analysis

$$\begin{aligned}\frac{\mathbf{E}(\pm i|m||x)}{|m|} &= \pm\int \frac{d^4q}{i\pi(2\pi)^3}\frac{1}{q^2\mp io-m^2}e^{iqx} = \pm 2\frac{\partial}{\partial x^2}\int \frac{d^2q}{i(2\pi)^2}\frac{1}{q^2\mp io-m^2}e^{iqx}\Big|_{x=(x_0,r)}\\ &= -2\frac{\partial}{\partial r^2}\int \frac{dq_0}{(2\pi)^2}e^{iq_0x_0}\left[\vartheta(q_0^2-m^2)\frac{e^{\mp i|\vec q|r}}{|\vec q|} \pm i\vartheta(m^2-q_0^2)\frac{e^{-|Q|r}}{|Q|}\right]\\ &= \pm i\int \frac{dq_0}{(2\pi)^2}e^{iq_0x_0}\Big[\underbrace{\vartheta(q_0^2-m^2)\frac{e^{\mp i|\vec q|r}}{r}}_{\text{on shell } |\vec q|=\sqrt{q_0^2-m^2}}\quad \underbrace{+\vartheta(m^2-q_0^2)\frac{e^{-|Q|r}}{r}}_{\text{off shell } |Q|=i|\vec q|=\sqrt{m^2-q_0^2}}\Big]\end{aligned}$$

displays a Yukawa potential (hyperbolic Macdonald functions) from the causally supported part. The choice of $\pm io$ (*Sommerfeld condition*) in $\frac{1}{q^2\pm io-m^2}$ connects the causal time structure with the preorder of position, i.e., with the out- or ingoing particle waves.

Derivation gives the *vector Feynman propagators*

$$\begin{aligned}\mathbf{E}_j(\pm i|m||x) &= \epsilon(x_0)\mathbf{c}_j(m|x) \pm \epsilon(m)i\mathbf{S}_j(m|x) = \overline{\mathbf{E}_j(\mp i|m||x)}\\ &= \pm\int \frac{d^4q}{(2\pi)^3}\vartheta(\pm q_0x_0)2q_j\delta(m^2-q^2)e^{iqx}\\ &= \int \frac{d^4q}{i\pi(2\pi)^3}\frac{q_j}{q^2\mp io-m^2}e^{iqx} = \int \frac{d^3q}{q_0(2\pi)^3}e^{-i\vec q\vec x}\begin{pmatrix}\epsilon(x_0)q_0\\ \pm\vec q\end{pmatrix}e^{\pm iq_0|x_0|}\Big|_{q_0=\sqrt{m^2+\vec q^2}}\end{aligned}$$

involving causal time representations and Yukawa forces.

The two poles of Feynman propagators lie centrally reflected in the complex energy plane

$$\begin{aligned}&\mathbf{E}(+i|m||x),\ \mathbf{E}_j(+i|m||x): &&\text{poles at } q_0 = \pm(\sqrt{m^2+\vec q^2}+io),\\ &\mathbf{E}(-i|m||x),\ \mathbf{E}_j(-i|m||x): &&\text{poles at } q_0 = \pm(\sqrt{m^2+\vec q^2}-io).\end{aligned}$$

The crossover on shell representations use one pole only:

$$\begin{aligned}\frac{\mathbf{C}(m|x)\pm\epsilon(m)i\mathbf{s}(m|x)}{2|m|} &= \int \frac{d^4q}{(2\pi)^3}\vartheta(\pm q_0)\delta(m^2-q^2)e^{iqx} = \int \frac{d^3q}{2q_0(2\pi)^3}e^{\pm iqx}\Big|_{q_0=\sqrt{m^2+\vec q^2}}\\ &= \int \frac{dq_0}{(2\pi)^2}e^{iq_0x_0}\ \vartheta(\pm q_0)\vartheta(q_0^2-m^2)\frac{\sin|\vec q|r}{r}\Big|_{|\vec q|=\sqrt{q_0^2-m^2}}.\end{aligned}$$

3.8 Summary

Spacetime translation distributions as Fourier transformed energy-momentum distributions embed – Lorentz compatibly – time representation matrix elements. Causally supported distributions in Minkowski spacetime are obtained from Fourier transformed principal value energy poles. They involve Yukawa potentials and forces that originate from off-shell imaginary "momenta."

<table>
<tr><td rowspan="2">$\begin{pmatrix} \cos mt \\ i \sin mt \end{pmatrix} \hookrightarrow$</td><td>$\begin{pmatrix} \mathbf{C}(m|x) \\ i\mathbf{S}_j(m|x) \end{pmatrix}$</td><td>$\int d^3x\ \mathbf{C}(m|x) = \cos mx_0$</td></tr>
<tr><td>$\begin{pmatrix} \mathbf{c}_j(m|x) \\ i\mathbf{s}(m|x) \end{pmatrix} = 0 \text{ for } x^2 < 0$</td><td>$\begin{aligned} \int d^3x\ \mathbf{s}(m|x) &= \sin mx_0 \\ \int dx_0 \epsilon(x_0) \frac{\mathbf{s}(m|x)}{m} &= -\frac{\partial}{\partial r^2} \frac{e^{-|m|r}}{\pi|m|} = \frac{e^{-|m|r}}{2\pi r} \end{aligned}$</td></tr>
</table>

distribution of time representations

$\begin{pmatrix} \mathbf{C}(m\|x) \\ i\mathbf{S}_j(m\|x) \end{pmatrix} = \epsilon(m) \int \frac{d^4q}{(2\pi)^3} \begin{pmatrix} m \\ q_j \end{pmatrix} \delta(m^2 - q^2) e^{iqx}$
$\frac{\mathbf{C}(m\|x)}{\|m\|} = \int \frac{dq_0}{(2\pi)^2} e^{iq_0 x_0}\ \vartheta(q_0^2 - m^2) \frac{\sin \|\vec{q}\| r}{r}$

Fock form functions (on shell $|\vec{q}| = \sqrt{q_0^2 - m^2}$)

$\begin{aligned} \begin{pmatrix} \mathbf{c}_j(m\|x) \\ i\mathbf{s}(m\|x) \end{pmatrix} &= \int \frac{d^4q}{(2\pi)^3} \epsilon(q_0) \begin{pmatrix} q_j \\ m \end{pmatrix} \delta(m^2 - q^2) e^{iqx} \\ &= \epsilon(x_0) \int \frac{d^4q}{i\pi(2\pi)^3} \begin{pmatrix} q_j \\ m \end{pmatrix} \frac{1}{q_{\mathrm{P}}^2 - m^2} e^{iqx} \end{aligned}$
$i\frac{\mathbf{s}(m\|x)}{m} = \int \frac{dq_0}{(2\pi)^2} e^{iq_0 x_0}\ \epsilon(q_0) \vartheta(q_0^2 - m^2) \frac{\sin \|\vec{q}\| r}{r}$

quantization distributions (on shell)

$\begin{aligned} \begin{pmatrix} e^{\pm i\|mt\|} \\ \epsilon(t) e^{\pm i\|mt\|} \end{pmatrix} &\hookrightarrow \begin{pmatrix} \mathbf{E}(\pm i\|m\|\|x) \\ \mathbf{E}_j(\pm i\|m\|\|x) \end{pmatrix} = \begin{pmatrix} \mathbf{C}(m\|x) \pm \epsilon(mx_0) i\mathbf{s}(m\|x) \\ \epsilon(x_0)\mathbf{c}_j(m\|x) \pm \epsilon(m) i\mathbf{S}_j(m\|x) \end{pmatrix} \\ &= \pm \int \frac{d^4q}{i\pi(2\pi)^3} \begin{pmatrix} \|m\| \\ \pm q_j \end{pmatrix} \frac{1}{q^2 \mp io - m^2} e^{iqx} = \int \frac{d^4q}{(2\pi)^3} 2\vartheta(\pm q_0 x_0) \begin{pmatrix} \|m\| \\ \pm q_j \end{pmatrix} \delta(m^2 - q^2) e^{iqx} \end{aligned}$
$\begin{aligned} \frac{\epsilon(x_0)\mathbf{s}(m\|x)}{m} &= \int \frac{dq_0}{(2\pi)^2} e^{iq_0 x_0}\ \frac{\vartheta(q_0^2 - m^2)\cos\|\vec{q}\|r - \vartheta(m^2 - q_0^2) e^{-\|Q\|r}}{r} \\ &= 2\frac{\partial}{\partial r^2} \int \frac{dq_0}{(2\pi)^2} e^{iq_0 x_0} \left[\vartheta(q_0^2 - m^2) \frac{\sin\|\vec{q}\|r}{\|\vec{q}\|} - \vartheta(m^2 - q_0^2) \frac{e^{-\|Q\|r}}{\|Q\|} \right] \\ \frac{\mathbf{E}(\pm i\|m\|\|x)}{\|m\|} &= \pm i \int \frac{dq_0}{(2\pi)^2} e^{iq_0 x_0}\ \frac{\vartheta(q_0^2 - m^2) e^{\mp i\|\vec{q}\|r} + \vartheta(m^2 - q_0^2) e^{-\|Q\|r}}{r} \end{aligned}$

Feynman propagators (on shell and off shell $|Q| = \sqrt{m^2 - q_0^2}$)

MATHEMATICAL TOOLS

3.9 Distributions

Topological vector spaces $\mathcal{F} \in \underline{\mathbf{tvec}}_{\mathbb{C}}$ have as topological duals the vector subspaces $\mathcal{F}' \subseteq \mathcal{F}^T$ (algebraic dual) with the continuous linear forms.

If the space is given by a *test function* subspace of the continuous complex-valued functions of an open real set $T \subseteq \mathbb{R}^d$,

$$\mathcal{F}(T) \subseteq \mathcal{C}(T) = \{f : T \longrightarrow \mathbb{C} \text{ continuous}\} \in \star\underline{\mathbf{vec}}_{\mathbb{C}},$$

the *distributions* in the topologial dual $\mathcal{F}'(T)$ are expressed by integration of *generalized functions* with Lebesgue measure

$$\mathcal{F}'(T) \times \mathcal{F}(T) \longrightarrow \mathbb{C}, \quad \langle \mu, f\rangle = \int \mu(x)\ d^dx\ f(x).$$

Distributions are the adequate formulation for disjoint additive mappings on measurable sets (chapter "The Kepler Factor"). The test function spaces stand for the measure subrings used there. Distributions have no values for points, but - in some sense - are characterized by values on open sets.

Properties of and operations with test functions can be rolled over, via the dual product, to the distributions, e.g.,

$$\begin{array}{rll} \text{conjugation:} & \langle \overline{\mu}, f\rangle & = \overline{\langle \mu, \overline{f}\rangle}, \\ \mu \text{ positive:} & \langle \mu, f\rangle \geq 0 & \text{for all } f \text{ with } f(T) \subseteq \mathbb{R}_+. \end{array}$$

In this way the derivations of distributions are also defined: $\langle \partial\mu, f\rangle = \langle \mu, -\partial f\rangle$.

The *support of a function* $f \in \mathcal{F}(T)$ is the closure of the set of points with nontrivial value:

$$\operatorname{supp} f = \overline{\{x \in T \mid f(x) \neq 0\}}.$$

A distribution μ vanishes on an open subset $S \subseteq T$ if $\langle \mu, f\rangle = 0$ for all functions with $\operatorname{supp} f \subseteq S$. The complement $C_{\mathbb{R}^d}T$ of the largest open S where μ vanishes is called the *support of the distribution* μ.

For a continuous linear mapping the transposition, restricted to the topological dual, is a linear mapping

$$F \in \mathbf{tvec}_{\mathbb{C}}(\mathcal{F}_1(T), \mathcal{F}_2(T)) \Rightarrow F^T \in \mathbf{vec}_{\mathbb{C}}(\mathcal{F}_2'(T), \mathcal{F}_1'(T)).$$

The dual $\mathcal{D}'(T) = \mathcal{C}_c^\infty(T)'$ of the complex-valued *infinitely continuously differentiable and compactly supported* test functions $\mathcal{C}_c^\infty(T)$, equipped with the limiting Fréchet topology, defines the *T-distributions.* In this "very large" space there are distribution subspaces, defined as duals of less-restricted test functions having the "very small" $\mathcal{C}_c^\infty(T)$ as subspace (more test functions have fewer distributions and vice versa). For example, the distributions with *compact support* $\mathcal{D}_c'(T)$ arise for the *infinitely continuously differentiable* test functions (without the support condition) or the *Radon measures* $\mathcal{M}(T) = \mathcal{C}_c(T)'$ for the *compactly supported continuous* functions

$$\begin{array}{llll} \mathcal{C}_c^\infty(T) & \subset \mathcal{C}^\infty(T) & \Rightarrow \mathcal{D}'(T) & \supset \mathcal{D}_c'(T) \quad = \mathcal{C}^\infty(T)', \\ \mathcal{C}_c^\infty(T) & \subset \mathcal{C}_c(T) & \Rightarrow \mathcal{D}'(T) & \supset \mathcal{M}(T). \end{array}$$

The test functions on $\mathbb{R}^d$, *rapidly decreasing* as expressed with polynomials P in the variables and derivatives and the Euclidean scalar product $r^2 = \vec{x}^2$,

$$\mathcal{S}(\mathbb{R}^d) = \{f \in \mathcal{C}^\infty(\mathbb{R}^d) \mid \sup_{x\in\mathbb{R}^n} |P_1(x)P_2(\tfrac{\partial}{\partial x})f(\vec{x})| \text{ finite for all polynomials}\},$$

e.g., for the harmonic Bose oscillators starting from the ground state wave function $f_0(\vec{x}) \sim e^{-\frac{r^2}{2}}$ (chapter "Quantum Probability"), equipped with a topology defined by the corresponding supremum seminorms, give rise to the *tempered (slowly increasing) distributions* $\mathcal{S}'(\mathbb{R}^d)$. Since the Hausdorff topology of a vector space $\mathbb{R}^d$ is unique, the tempered distributions can be defined also for $\mathbf{O}(1,3)$-Minkowski space $\mathbb{R}^4$, where the norm $x_0^2 + \vec{x}^2$, incompatible with the Lorentz transformations, is used for the topological properties only. Distributions like $\delta(q^2 - m^2)$ with $q^2 = q_0^2 - \vec{q}^2$ are tempered, however not with compact support. All functions, e.g., the Fock form functions, and distributions, e.g., the quantization and Feynman propagator distributions used above, are tempered distributions $\mathcal{S}'(\mathbb{R}^4)$.

The *structure of distributions* is locally characterizable by functions and their derivatives, therefore "generalized functions." Since in the integral form $\int \mu(x)d^dx f(x)$ derivatives can be rolled over, it is understandable that derivatives play a decisive role in characterizing distributions by functions: Every distribution $\mu \in \mathcal{D}'(T)$ is locally equal to a finite sum of derivatives of locally integrable functions. Every Radon measure $\mathcal{M}(\mathbb{R}^d)$ with Lebesgue basis d^dx is a finite sum of derivatives (maximally ∂^{2d}) of continuous functions, e.g., the $\mathbb{R}$-Dirac measure $\delta(x) = \frac{d^2}{dx^2}\frac{|x|}{2}$. The distributions with support at the origin 0 are finite linear combinations of the derivatives of the $\mathbb{R}$-Dirac measure $d^dx\ \delta(x) \in \mathcal{D}'_c(\mathbb{R}^d)$. The tempered distributions $\mu \in \mathcal{S}'(\mathbb{R}^d)$ are locally equal to finite sums of derivatives of continuous functions with the absolute value growing at infinity more slowly than some polynomial, e.g.,

$$\begin{aligned}
\frac{\Gamma(1+N)}{(x-io)^{1+N}} &= \frac{\Gamma(1+N)}{x_{\mathrm{P}}^{1+N}} + i\pi\delta^{(N)}(-x) &&= (-\tfrac{d}{dx})^{1+N}\log(-x+io),\ N = 0,1,\dots,\\
\log(-x+io) &= \log|x| + i\pi\vartheta(x) &&= \tfrac{d}{dx}[x(\log|x|-1) + i\pi x\vartheta(x)].
\end{aligned}$$

Embedding the functions into distributions (generalized functions), the following inclusions hold with the topological duals arising by "central" reflection at $\bowtie$,

$$\begin{array}{ccccc}
\mathcal{C}_c(T) \supset \mathcal{C}_c^\infty(T) & \subset & \mathcal{C}^\infty(T) \\
\cap & \bowtie & \cap \\
\mathcal{D}'_c(T) & \subset & \mathcal{D}'(T) \supset \mathcal{M}(T)
\end{array}$$

and including the topologically self-dual Hilbert spaces $L^2_{dx}(\mathbb{R}^d)$ (all functions and distributions are defined up to sets with trivial Lebesgue measure)

$$\begin{array}{ccccc}
 & \mathcal{C}_c^\infty \ \subset & \mathcal{S} & \subset\ \mathcal{C}^\infty \\
 & & \cap & \\
\text{on } \mathbb{R}^d: & \cap & L^2_{dx} \cong [L^2_{dx}]' & \cap \\
 & & \cap & \\
 & \mathcal{D}'_c \ \subset & \mathcal{S}' & \subset\ \mathcal{D}'
\end{array}$$

3.10 Fourier Transformation

For test functions and distributions on the real vector space $x \in \mathbb{R}^d$ (translations) there exist the corresponding structures on the isomorphic dual space $q \in \check{\mathbb{R}}^d$ (energy-momenta), related to each other via the dual product $\langle q, x\rangle = qx$. Functions and distributions on dual spaces can be connected by the $\mathbf{U}(1)$-representations ($\mathcal{C}^\infty$-functions) of the translations or the energy-momenta:

$$\begin{array}{ll} D^x : \check{\mathbb{R}}^d \ni q \longmapsto e^{2i\pi xq} \in \mathbf{U}(1), & \langle D^x|D^{x'}\rangle = \int d^d q \overline{D^x(q)} D^{x'}(q) = \delta(x-x'), \\ D^q : \mathbb{R}^d \ni x \longmapsto e^{2i\pi xq} \in \mathbf{U}(1), & \langle D^q|D^{q'}\rangle = \int d^d x \overline{D^q(x)} D^{q'}(x) = \delta(q-q'). \end{array}$$

The Fourier transform connects with each other the rapidly decreasing test functions on the translations and the energy-momenta by a topological vector space *isomorphism*. The inverse Fourier transform is related to the conjugation (antilinear):

$$\mathcal{S}(\check{\mathbb{R}}^d) \overset{\mathbf{F}}{\cong} \mathcal{S}(\mathbb{R}^d), \begin{cases} f \longmapsto \mathbf{F}.f = \tilde{f} : & \begin{cases} \tilde{f}(x) &= \int d^d q\ f(q) e^{-2i\pi qx}, \\ \tilde{f}(x) &= \int \frac{d^d q}{(2\pi)^d}\ f(\frac{q}{2\pi}) e^{-iqx}, \end{cases} \\ \mathbf{F}^{-1} \cong \overline{\mathbf{F}} & \\ \tilde{f} \longmapsto \mathbf{F}^{-1}.\tilde{f} = f : & \begin{cases} f(q) &= \int d^d x\ \tilde{f}(x) e^{2i\pi qx}, \\ f(\frac{q}{2\pi}) &= \int d^d x\ \tilde{f}(x) e^{iqx}. \end{cases} \end{cases}$$

One has the *Plancherel theorem* for the $\mathbf{U}(1)$-induced scalar product

$$\begin{array}{l} \mathcal{S}(\check{\mathbb{R}}^d) \times \mathcal{S}(\check{\mathbb{R}}^d) \longrightarrow \mathbb{C}, \quad \mathcal{S}(\mathbb{R}^d) \times \mathcal{S}(\mathbb{R}^d) \longrightarrow \mathbb{C}, \\ \langle g|f\rangle = \int d^d q\ \overline{g(q)} f(q) = \int d^d x\ \overline{\tilde{g}(x)} \tilde{f}(x) = \langle \mathbf{F}.g|\mathbf{F}.f\rangle = \langle \tilde{g}|\tilde{f}\rangle; \end{array}$$

$\mathcal{S}(\check{\mathbb{R}}^d)$ is dense in the topologically self-dual Hilbert space $L^2_{dq}(\check{\mathbb{R}}^d)$:

$$\text{on } \check{\mathbb{R}}^d : \quad \begin{array}{lll} \mathcal{S} & \subset L^2_{dq} \subset & \mathcal{S}', \\ \overline{\mathcal{S}} & = L^2_{dq} \cong & [L^2_{dq}]', \end{array}$$

on which the Fourier transform can be extended as an isometry:

$$L^2_{dq}(\check{\mathbb{R}}^d) \overset{\mathbf{F}}{\cong} L^2_{dx}(\mathbb{R}^d).$$

The Fourier transform can be rolled over from functions $\mathcal{F}_2(\mathbb{R}^d)$ to distributions $\mathcal{F}'_2(\mathbb{R}^d)$ if there exists a test function space $\mathcal{F}_1(\check{\mathbb{R}}^d)$ whose Fourier transform is valued in the test function space $\mathcal{F}_2(\mathbb{R}^d)$:

$$\left.\begin{array}{ll} \mathbf{F} : \mathcal{F}_1(\check{\mathbb{R}}^d) & \longrightarrow \mathcal{F}_2(\mathbb{R}^d) \\ \mathbf{F}^T : \mathcal{F}'_2(\mathbb{R}^d) & \longrightarrow \mathcal{F}'_1(\check{\mathbb{R}}^d) \end{array}\right\} \text{ with } \begin{array}{ll} \langle \mathbf{F}^T.\mu_2, f_1\rangle & = \langle \mu_2, \mathbf{F}.f_1\rangle, \\ \langle \tilde{\mu}_2, f_1\rangle & = \langle \mu_2, \tilde{f}_1\rangle. \end{array}$$

The transposed Fourier transform is a vector space isomorphism for the tempered distributions:

$$\mathcal{S}'(\mathbb{R}^d) \overset{\mathbf{F}^T}{\cong} \mathcal{S}'(\check{\mathbb{R}}^d), \quad \tilde{\mu} \longmapsto \mathbf{F}^T.\tilde{\mu} = \mu, \begin{cases} \mu(q) &= \int d^d x\ \tilde{\mu}(x) e^{-2i\pi qx}, \\ \langle \mu, f\rangle &= \langle \mathbf{F}^T.\tilde{\mu}, f\rangle \\ &= \langle \tilde{\mu}, \mathbf{F}.f\rangle = \langle \tilde{\mu}, \tilde{f}\rangle. \end{cases}$$

The Fourier transform of compactly supported distributions gives $\mathcal{C}^\infty$-functions in the self-dual relation $\mathcal{D}'_c = [\mathcal{C}^\infty]'$:

$$\mathbf{F} : \mathcal{D}'_c(\mathbb{R}^d) \longrightarrow \mathcal{C}^\infty(\check{\mathbb{R}}^d).$$

3.11 Measures of Symmetric Spaces

The orbit $G \bullet q_0$ of a real Lie group G acting on a vector $q_0 \in V \cong \mathbb{R}^n$, $n \geq 1$, parametrizes the equivalence classes of the corresponding fixgroup (closed subgroup)

$$G \bullet q_0 \cong G/H, \quad H = G_{q_0}.$$

Lebesgue integration, restricted to the orbit, gives a positive *G-invariant measure of the cosets G/H*

$$\text{for } G/H : \quad \int d^n q \; \delta(q \in G \bullet q_0).$$

For a bilinear or sesquilinear form with invariance group G (chapter "Spacetime Translations"), the orbit is parametrizable by the vectors with equal form value,

$$\begin{array}{l} \zeta : V \times V \longrightarrow \mathbb{K}, \quad \zeta(q,q) = \overline{q}q, \quad G = \mathbf{UL}(V,\zeta), \\ G \bullet q_0 = \{q \in V \mid \overline{q}q = \overline{q_0}q_0\}, \\ \text{for } G/H : \quad \int d^n q \; \delta(\overline{q}q - \overline{q_0}q_0). \end{array}$$

For unitary invariance group $\mathbf{SU}(r,s)$, acting on $\mathbb{C}^n \cong \mathbb{R}^{2n}$ with Lebesque measure $d^n z \; d^n \overline{z}$, $dz \; d\overline{z} = dx \; dy$, the Dirac measure for a fixed square yields an invariant measure of the real $(2n-1)$-dimensional coset spaces:

$$\begin{array}{rl} \text{for } \mathbf{SU}(r,s)/\mathbf{SU}(r-1,s) : & \int d^n z d^n \overline{z} \; \delta(\overline{z}z - 1), \\ \text{for } \mathbf{SU}(2) : & \int dz_1 d\overline{z_1} dz_2 d\overline{z_2} \; \delta(|z_1|^2 + |z_2|^2 - 1) \\ & \sim \int_{-2\pi}^{2\pi} d\chi \int_0^{2\pi} d\varphi \int_0^\pi \sin\theta d\theta \\ & \text{with } \begin{pmatrix} z_1 \\ z_2 \end{pmatrix} = e^{i\frac{\chi}{2}} \begin{pmatrix} e^{i\frac{\varphi}{2}} \cos\frac{\theta}{2} \\ e^{-i\frac{\varphi}{2}} \sin\frac{\theta}{2} \end{pmatrix}. \end{array}$$

For orthogonal group $\mathbf{SO}_0(r,s)$, acting on $\mathbb{R}^n$, there are three proper fixgroup types with the invariant measures of the real $(n-1)$-dimensional symmetric spaces:

$$\begin{array}{ll} \int d^n q \; \delta(\vec{q}_r^{\,2} - \vec{q}_s^{\,2} - 1) & \text{for } \mathbf{SO}_0(r,s)/\mathbf{SO}_0(r-1,s), \\ \int d^n q \; \delta(\vec{q}_r^{\,2} - \vec{q}_s^{\,2} + 1) & \text{for } \mathbf{SO}_0(r,s)/\mathbf{SO}_0(r,s-1), \\ \int d^n q \; \delta(\vec{q}_r^{\,2} - \vec{q}_s^{\,2}) & \text{for } \mathbf{SO}_0(r,s)/\mathbf{SO}_0(r-1,s-1) \;\vec{\times}\; \mathbb{R}^{n-2}, \\ d^n q = d^r q \; d^s q, \; \int d^r q & = \int_0^\infty |\vec{q}|^{r-1} d|\vec{q}| \int d^{r-1}\omega. \end{array}$$

3.11.1 Spherical and Hyperbolic Measures

Measures for spheres $\Omega^s \cong \mathbf{SO}(1+s)/\mathbf{SO}(s)$, $s = 1, 2\ldots$, and hyperboloids $\mathcal{Y}^s \cong \mathbf{SO}_0(1,s)/\mathbf{SO}(s)$ can be constructed with the defining representations of $\mathbf{SO}(1+s)$ and $\mathbf{SO}_0(1,s)$ acting on energy-momenta (imaginary "momenta" for spheres):

$$\begin{array}{rl}\log \mathbf{SO}(1+s) \ \vec{\oplus} \ \mathbb{R}^{1+s}: & \begin{pmatrix} 0 & i\vec{\chi} \\ \hline i\vec{\chi} & \log \mathbf{SO}(s)\end{pmatrix}\begin{pmatrix} q_0 \\ \hline i\vec{q}\end{pmatrix}, \\ & q^2 = q_0^2 - (i\vec{q})^2 = q_0^2 + \vec{q}^2, \\ \log \mathbf{SO}(1,s) \ \vec{\oplus} \ \mathbb{R}^{1+s}: & \begin{pmatrix} 0 & \vec{\psi} \\ \hline \vec{\psi} & \log \mathbf{SO}(s)\end{pmatrix}\begin{pmatrix} q_0 \\ \hline \vec{q}\end{pmatrix}, \\ & q^2 = q_0^2 - \vec{q}^2.\end{array}$$

Positive vectors $q^2 > 0$ have the fixgroup $\mathbf{SO}(s)$. The representations of the fixgroup classes are parametrizable by unit vectors. For the hyperboloid with $q_0 = \vartheta(q_0)q_0 > 0$, one obtains

$$q^2 = 1: \quad \begin{cases} \begin{pmatrix} q_0 \\ \hline i\vec{q}\end{pmatrix} = \begin{pmatrix} \cos\chi \\ \frac{\vec{q}}{|\vec{q}|} i \sin\chi\end{pmatrix}, & \begin{pmatrix} q_0 & iq_a \\ \hline iq_b & \delta_{ab} - \frac{q_a q_b}{1+q_0}\end{pmatrix} \in \mathbf{SO}(1+s)/\mathbf{SO}(s), \\ \begin{pmatrix} q_0 \\ \hline \vec{q}\end{pmatrix} = \begin{pmatrix} \cosh\psi \\ \frac{\vec{q}}{|\vec{q}|} \sinh\psi\end{pmatrix}, & \begin{pmatrix} q_0 & q_a \\ \hline q_b & \delta_{ab} + \frac{q_a q_b}{1+q_0}\end{pmatrix} \in \mathbf{SO}_0(1,s)/\mathbf{SO}(s).\end{cases}$$

Unit vectors $q^2 = 1$ can be used to parametrize the positive $\mathbf{SO}(1+s)$ and $\mathbf{SO}_0(1,s)$-invariant measures, unique up to a factor. In addition, there are other parametrizations, both with a finite and infinite range, with a trigonometric or hyperbolic "angle" (χ, ψ), with imaginary "momenta", and real momenta (ip, p) and with imaginary and real Poincaré parameters (iv, v) for spheres and hyperboloids respectively.

The parametrizations for semi-1-sphere and semi-1-hyperboloid are

$$\begin{array}{llll}\Omega^1 \ni \begin{pmatrix} q_0 \\ iq\end{pmatrix} & \sim \begin{pmatrix} \cos\chi \\ i\sin\chi\end{pmatrix}_{-\frac{\pi}{2}}^{\frac{\pi}{2}} & = \frac{1}{\sqrt{1+p^2}}\begin{pmatrix} 1 \\ ip\end{pmatrix}_{-\infty}^{\infty} & = \frac{1}{1+v^2}\begin{pmatrix} 1-v^2 \\ 2iv\end{pmatrix}_{-1}^{1}, \\ \mathcal{Y}^1 \ni \begin{pmatrix} q_0 \\ q\end{pmatrix} & \sim \begin{pmatrix} \cosh\psi \\ \sinh\psi\end{pmatrix}_{0}^{\infty} & = \frac{1}{\sqrt{1-p^2}}\begin{pmatrix} 1 \\ p\end{pmatrix}_{0}^{1} & = \frac{1}{1-v^2}\begin{pmatrix} 1+v^2 \\ 2v\end{pmatrix}_{0}^{1}.\end{array}$$

From the 1-forms

$$\begin{array}{llll}\begin{pmatrix} -\sin\chi d\chi \\ i\cos\chi d\chi\end{pmatrix} & = \begin{pmatrix} dq_0 \\ idq\end{pmatrix} & = \frac{1}{\sqrt{1+p^2}^3}\begin{pmatrix} -pdp \\ idp\end{pmatrix} & = \frac{1}{(1+v^2)^2}\begin{pmatrix} -4vdv \\ i2(1-v^2)dv\end{pmatrix}, \\ \begin{pmatrix} \sinh\psi d\psi \\ \cosh\psi d\psi\end{pmatrix} & = \begin{pmatrix} dq_0 \\ dq\end{pmatrix} & = \frac{1}{\sqrt{1-p^2}^3}\begin{pmatrix} pdp \\ dp\end{pmatrix} & = \frac{1}{(1-v^2)^2}\begin{pmatrix} 4vdv \\ 2(1+v^2)dv\end{pmatrix},\end{array}$$

one obtains the measures

$$\begin{array}{llll}\int d^1\omega & = \int_{-\pi}^{\pi} d\chi = 2\int_{-\frac{\pi}{2}}^{\frac{\pi}{2}} d\chi & = 2\int_0^1 \frac{dq_0}{\sqrt{1-q_0^2}} & = 2\int_0^1 \frac{dq}{\sqrt{1-q^2}} \\ & & = 2\int_0^{\infty} \frac{2dp}{1+p^2} & = 2\int_0^1 \frac{4dv}{1+v^2} = \oint_{p=i} \frac{2dp}{1+p^2} = 2\pi, \\ \int d^1\mathrm{y} & = \int_{-\infty}^{\infty} d\psi = 2\int_0^{\infty} d\psi & = 2\int_1^{\infty} \frac{dq_0}{\sqrt{q_0^2-1}} & = 2\int_0^{\infty} \frac{dq}{\sqrt{1+q^2}} \\ & & = 2\int_0^1 \frac{dp}{1-p^2} & = 2\int_0^1 \frac{2dv}{1-v^2}.\end{array}$$

The abelian case $s = 1$ is embedded in the general case with the $(s-1)$-sphere and factors q^{s-1} with $q = |\vec{q}|$, where $\mathbf{SO}_0(1,s)/\mathbf{SO}_0(1,s-1) \cong \mathcal{Y}^{(1,s-1)} \cong \mathcal{Y}^1 \times \Omega^{s-1}$:

$$\text{for } \begin{pmatrix} \Omega^s \\ \mathcal{Y}^s \\ \mathcal{Y}^{(1,s-1)} \end{pmatrix}: \quad \int \begin{pmatrix} d^s\omega \\ d^s\mathrm{y} \\ d^s\mathrm{s} \end{pmatrix} = \int 2d^{1+s}q \begin{pmatrix} \delta(q_0^2 + \vec{q}^2 - 1) \\ \vartheta(q_0)\delta(q_0^2 - \vec{q}^2 - 1) \\ \delta(q_0^2 - \vec{q}^2 + 1) \end{pmatrix}$$

$$= \int d^{s-1}\omega \begin{pmatrix} \int_0^\pi \sin^{s-1}\chi \ d\chi \\ \int_0^\infty \sinh^{s-1}\psi \ d\psi \\ \int_{-\infty}^\infty \cosh^{s-1}\psi \ d\psi \end{pmatrix}.$$

The 0-sphere consists of two points:

$$\begin{aligned} &\int d^s\omega = |\Omega^s| = \tfrac{2\pi^{\frac{1+s}{2}}}{\Gamma(\frac{1+s}{2})} = 2, 2\pi, 4\pi, 2\pi^2, \tfrac{8\pi^2}{3}, \ldots, \\ &\int d^0\omega = |\Omega^0| = \text{card}\,\{1,-1\} = 2, \ |\Omega^1| = 2\pi, \quad \tfrac{|\Omega^s|}{|\Omega^{s-2}|} = \tfrac{2\pi}{s-1}. \end{aligned}$$

The hyperboloid $\mathcal{Y}^2$ parametrization with $\vec{v} \in \mathbb{R}^2$ yields F. Klein's Euclidean model of Lobachevsky's non-Euclidean plane.

The parametrizations above are generalized with $(q,p,v) \to (\vec{q},\vec{p},\vec{v}) \in \mathbb{R}^s$ for Ω^s and $\mathcal{Y}^s$. The momentum p-parametrization is square-root-free for odd position dimension $s = 1, 3, \ldots$, the v-parametrization for all $s = 1, 2, \ldots$, for the *spherical measures*

$$\begin{aligned} \int d^s\omega &= \int d^{s-1}\omega \int_0^1 \frac{dq_0}{\sqrt{1-q_0^2}^{\,2-s}} &&= \int d^{s-1}\omega \int_0^1 \frac{q^{s-1}dq}{\sqrt{1-q^2}} \\ &= \int d^{s-1}\omega \int_0^\infty \frac{2p^{s-1}dp}{\sqrt{1+p^2}^{\,1+s}} &&= \int d^{s-1}\omega \int_{\vec{v}^2 \le 1} 2v^{s-1}dv \ (\tfrac{2}{1+v^2})^s, \end{aligned}$$

and the *hyperbolic measures*

$$\begin{aligned} \int d^s\mathrm{y} &= \int d^{s-1}\omega \int_1^\infty \frac{dq_0}{\sqrt{q_0^2-1}^{\,2-s}} &&= \int d^{s-1}\omega \int_0^\infty \frac{q^{s-1}dq}{\sqrt{1+q^2}} \\ &= \int d^{s-1}\omega \int_0^1 \frac{2p^{s-1}dp}{\sqrt{1-p^2}^{\,1+s}} &&= \int d^{s-1}\omega \int_0^1 v^{s-1}dv \ (\tfrac{2}{1-v^2})^s, \\ \int d^s\mathrm{s} &= 2\int d^{s-1}\omega \int_1^\infty \frac{q_0^{s-1}dq_0}{\sqrt{q_0^2-1}} &&= 2\int d^{s-1}\omega \int_0^\infty \frac{dq}{\sqrt{1+q^2}^{\,2-s}} \\ &= 2\int d^{s-1}\omega \int_0^1 \frac{2dp}{\sqrt{1-p^2}^{\,1+s}} &&= 2\int d^{s-1}\omega \int_0^1 \frac{2dv}{1+v^2}\left(\frac{1+v^2}{1-v^2}\right)^s, \end{aligned}$$

and, if possible, with a volume integration

$$\begin{aligned} \int d^s\omega &= \int_{\vec{q}^2 \le 1} \frac{d^sq}{\sqrt{1-\vec{q}^2}} &&= \int \frac{2d^sp}{\sqrt{1+\vec{p}^2}^{\,1+s}} &&= \int_{\vec{v}^2 \le 1} 2d^sv \ (\tfrac{2}{1+\vec{v}^2})^s, \\ \int d^s\mathrm{y} &= \int \frac{d^sq}{\sqrt{1+\vec{q}^2}} &&= \int_{\vec{p}^2 \le 1} \frac{2d^sp}{\sqrt{1-\vec{p}^2}^{\,1+s}} &&= \int_{\vec{v}^2 \le 1} d^sv \ (\tfrac{2}{1-\vec{v}^2})^s, \\ \int d^s\mathrm{s} &= 2\int_{\vec{q}^2 \ge 1} \frac{d^sq}{\sqrt{\vec{q}^2-1}}. \end{aligned}$$

Bibliography

[1] N. Bourbaki, *Intégration, Chapitres 1-8* (1952-1959), Hermann, Paris.

[2] I.M. Gel'fand, G.E. Shilov, *Generalized Functions I (Properties and Operations) and II (Spaces of Fundamental and Generalized Functions)* (1958, English translation 1963), Academic Press, New York and London.

[3] F. Treves, *Topological Vector Spaces, Distributions and Kernels*, Academic Press, New York, London (1967).

[4] N.Ja. Vilenkin, A.U. Klimyk, *Representations of Lie Groups and Special Functions* (1991), Kluwer Academic Publishers, Dordrecht, Boston, London.

4

MASSIVE PARTICLE QUANTUM FIELDS

With the experience of quantum mechanics on the one hand, where time comes as a real parameter and position as an operator in a Bose quantum algebra $\mathbf{Q}_-(\mathbb{C}^2)$, and special relativity on the other hand, where time and position constitute Minkowski spacetime, one may expect a relativistic quantum structure with both time and position as operators. However both time and position $x = (t, \vec{x})$ are used not as operators, but "only" as real parameters for the translation behavior of relativistic fields. The value space of the fields carries the quantum degrees of freedom, with the example of a free scalar quantum field $\mathbf{\Phi}$ for massive neutral particles:

time translations $t \in \mathbb{R}$	spacetime translations $x \in \mathbb{R}^4$
$\downarrow$	$\downarrow$
$\mathbf{x}(t)$	$\mathbf{\Phi}(x)$

$$\text{quantum mechanics} \quad \begin{array}{ccc} \mathbb{R} & \hookrightarrow & \mathbb{R}^4 \\ \mathbf{Q}_-(\mathbf{W}) & \hookrightarrow & \mathbf{Q}_-(\mathbf{W}^{\mathcal{Y}^3}) \end{array} \quad \text{quantum field theory}$$

The quantum algebras come as value spaces for mappings $\{\mathcal{Y}^3 \longrightarrow W\}$ of the energy-momentum hyperboloids $q^2 = m^2 > 0$ for free particles. For each momentum $\vec{q}$ and spin J, there is a quantum algebra over a representation space $W \supseteq \mathbb{C}^{1+2J}$ for spin and chargelike operations. In retrospect, quantum mechanics is characterizable by quantum orbits of time with the quantum structure implemented by position. The quantum structure of the orbits of relativistic spacetime is implemented by field degrees of freedom, e.g., by spin, electromagnetic charge, isospin, etc.

Free particles are characterized by irreducible Hilbert representations of the Poincaré group $\mathbf{SL}(\mathbb{C}^2) \vec{\times} \mathbb{R}^4$. The infinite-dimensional representations for particles are induced by and embed Hilbert representations of spacetime translations $\mathbb{R}^4$ and of position rotations, i.e., of spin $\mathbf{SU}(2)$ for massive particles $m^2 > 0$ (for convenience $m = |m|$ throughout this chapter) and of circularity (helicity, polarization) $\mathbf{SO}(2)$ for massless particles (chapter "Massless

Quantum Fields"). More mathematical details are discussed in the chapter "Harmonic Analysis."

Particles are embedded into quantum fields. The Feynman propagator for a particle field describes its spacetime behavior. It has two parts which are analogous to the two wave function types in quantum mechanics (chapter "The Kepler Factor"): For kinetic energy $E - V_0 = \frac{\vec{q}^2}{2} > 0$, there arise scattering waves $\frac{\sin|\vec{q}|r}{r}$, whereas bound waves $e^{-|Q|r}$ come with binding energy $E - V_0 = -\frac{Q^2}{2} < 0$. In a Feynman propagator, the scattering part is embedded into the Fock form function of the quantization opposite commutators. It involves matrix elements of Hilbert representations of the translations $\mathbb{R}^4$. This part describes free particles: on-shell with kinetic energy $q_0^2 - m^2 = \vec{q}^2 > 0$. The relativistic correspondence to the nonrelativistic bound waves is the $\epsilon(x_0)$-multiplied quantization (anti-) commutator distribution, which contains off-shell contributions ("virtual particles"), $q^2 \neq m^2$. The embedded Yukawa interactions $\frac{e^{-|Q|r}}{r}$ and forces are distributions (2-sphere spreads) of representation coefficients of position with imaginary "momentum" $q_0^2 - m^2 = -Q^2 < 0$ as eigenvalues, the analogue to the nonrelativistic binding energy. These "virtual particle" contributions have small-distance $r = 0$ singularities; they are not representation coefficients of the spacetime translations.

In addition to the translation properties, i.e., the invariant mass, and the energy-momenta eigenvalues, particle fields have homogeneous Lorentz transformation properties. Massive particle fields come with decompositions of Minkowski translations into time and position translations, induced by a rest system of the field embedded particle and determined up to rotations $\mathbf{SO}(3) \cong \mathbf{SU}(2)/\mathbb{I}(2)$ The $\mathbf{SU}(2)$-representations determine the spin of the particle.

Relativistic quantum fields for massive particles have particle degrees of freedom only. Their Hilbert representations allow a complete probability interpretation. This is in contrast to massless fields (chapter "Massless Quantum Fields").

The complex representations of the real-spacetime-related groups are in unitary groups that contain $\mathbf{U}(1)$-phase groups. An "internal" $\mathbf{U}(1)$-group in addition to the translations representing $\mathbf{U}(1)$ describes particles and antiparticles. For example, the Dirac representation of the Lorentz group in the anticonjugation group $\mathbf{SL}(\mathbb{C}^2) \longrightarrow \mathbf{U}(2,2)$ uses the probability-inducing $\mathbf{U}(\mathbf{1}_4)$ for the translations and the additional relative phase in $\mathbf{SU}(2,2) \supset \mathbf{U}(\mathbf{1}_2)_3 \times \mathbf{SL}(\mathbb{C}^2)$ for an internal "chargelike" $\mathbf{U}(1)$. With the exception of Majorana structures, all half-integer spin particles have an additional internal $\mathbf{U}(1)$-charge, arising, e.g., for neutrino-antineutrino as fermion number and for electron-positron as electromagnetic charge.

For relativistic quantum fields, the Lie algebras of the external Poincaré group and of the internal operations are implemented by position integrals of generator distributions, their currents, which are written with the quantization opposite commutators.

After a review of the $\mathbf{U}(1)$-representations for the time group $\mathbf{D}(1) \cong \mathbb{R}$ in quantum algebras, i.e., of the harmonic Fermi and Bose quantum oscillator, and of the quantum representation of an additional particle-antiparticle $\mathbf{U}(1)$-transformation group, the relativistic distribution of these compact time translation representations (chapter "Propagators") is used to define quantum particle fields with external spacetimelike and internal chargelike degrees of freedom.

4.1 Quantum Bose and Fermi Oscillators

The irreducible Hilbert representations of time and their self-dual combinations are described by the harmonic oscillators.

4.1.1 Time Translations and Particle-Antiparticle Transformations

A $\mathbb{C}$-quartet of complex 1-dimensional vector spaces $(W, \overline{W}, W^T, \overline{W}^T)$ (chapter "Antistructures; The Real in the Complex") with dual and antibases is related to each other by two conjugations: the definite dual space conjugation $\star$ for creation-annihilation and the indefinite antidual conjugation $\times$ for particle-antiparticle:

$$\begin{array}{ccc} \mathrm{u} \in W & \overset{\times}{\leftrightarrow} & \mathrm{a}^\star = \mathrm{u}^\times \in \overline{W} \\ \star \updownarrow & & \updownarrow \star \\ \mathrm{u}^\star = \mathrm{a}^\times \in W^T & \overset{\times}{\leftrightarrow} & \mathrm{a} \in \overline{W}^T \end{array} .$$

Compact time representations with energy (frequency) $m \in \mathbb{R}$ act on the antidoubled vector space $W_{\text{doub}} = W \oplus \overline{W}^T \cong \mathbb{C}^2$:

$$\text{time } \mathbb{R} \ni t \longmapsto e^{imt}\begin{pmatrix} 1 & 0 \\ 0 & 1 \end{pmatrix} \in \mathbf{U}(\mathbf{1}_2) \subset \mathbf{U}(2).$$

The antidoubling allows the representation of a particle-antiparticle $\mathbf{U}(1)$ with winding number $z \in \mathbb{Z}$:

$$\text{internal } \mathbf{U}(1) \ni e^{i\alpha} \longmapsto \begin{pmatrix} e^{iz\alpha} & 0 \\ 0 & e^{-iz\alpha} \end{pmatrix} \in \mathbf{U}(1)_3 \subset \mathbf{U}(1,1).$$

The Fermi and Bose quantum algebras $\epsilon = \pm 1$ for the complex quartet are characterized by the (anti-) commutators

$$\text{in } \mathbf{Q}_\epsilon(\mathbf{W}_{\text{doub}}): \quad [\mathrm{u}^\star, \mathrm{u}]_\epsilon = 1 = [\mathrm{a}^\star, \mathrm{a}]_\epsilon.$$

The adjoint action of Hamiltonian and internal charge operator implement the Lie algebra of $\mathbb{R} \times \mathbf{U}(1)$:

$$\mathbb{R} \oplus \log \mathbf{U}(1) \longrightarrow \mathbf{Q}_\epsilon(\mathbf{W}_{\text{doub}}), \begin{cases} H_0 &= H_0^\star &= m\frac{[\mathrm{u},\mathrm{u}^\star]_{-\epsilon}+[\mathrm{a},\mathrm{a}^\star]_{-\epsilon}}{2}, \\ Q &= Q^\star &= z\frac{[\mathrm{u},\mathrm{u}^\star]_{-\epsilon}-[\mathrm{a},\mathrm{a}^\star]_{-\epsilon}}{2}. \end{cases}$$

It determines with $d_t = \frac{d}{dt} = i \operatorname{ad} H_0$ the equations of motion and the time orbits,

$$\begin{aligned} d_t\begin{pmatrix}\mathrm{u}\\ \mathrm{a}\end{pmatrix} &= im\begin{pmatrix}\mathrm{u}\\ \mathrm{a}\end{pmatrix}, & d_t(\mathrm{u}^\star,\mathrm{a}^\star) &= -im(\mathrm{u}^\star,\mathrm{a}^\star),\\ \begin{pmatrix}\mathrm{u}\\ \mathrm{a}\end{pmatrix}(t) &= e^{imt}\begin{pmatrix}\mathrm{u}\\ \mathrm{a}\end{pmatrix}, & (\mathrm{u}^\star,\mathrm{a}^\star)(t) &= (\mathrm{u}^\star,\mathrm{a}^\star)e^{-imt}, \end{aligned}$$

and displays with $d_{i\alpha} = -i\frac{d}{d\alpha} = \operatorname{ad} Q$ the internal charge number:

$$d_{i\alpha}\begin{pmatrix}\mathrm{u}\\ \mathrm{a}\end{pmatrix} = \begin{pmatrix} z & 0\\ 0 & -z\end{pmatrix}\begin{pmatrix}\mathrm{u}\\ \mathrm{a}\end{pmatrix},\quad d_{i\alpha}(\mathrm{u}^\star,\mathrm{a}^\star) = (\mathrm{u}^\star,\mathrm{a}^\star)\begin{pmatrix} -z & 0\\ 0 & z\end{pmatrix}.$$

A charge Q-compatible basis allows *dual normalization factors* $\ell, \rho > 0$ for real and imaginary part,

$$\epsilon = +1: \left\{\begin{array}{ll} \mathbf{U}_+ = \frac{1}{\ell}\frac{\mathrm{u}+\mathrm{a}^\star}{\sqrt{2}}, & i\mathbf{U}_- = \frac{1}{\rho}\frac{\mathrm{u}-\mathrm{a}^\star}{\sqrt{2}},\\ \mathbf{A}_+ = \ell\frac{\mathrm{a}+\mathrm{u}^\star}{\sqrt{2}}, & i\mathbf{A}_- = \rho\frac{\mathrm{a}-\mathrm{u}^\star}{\sqrt{2}} \end{array}\right\} \Rightarrow \{\mathbf{A}_+,\mathbf{U}_+\} = 1 = \{\mathbf{A}_-,\mathbf{U}_-\},$$

$$\epsilon = -1: \left\{\begin{array}{ll} \mathbf{U}_+ = \frac{1}{\rho}\frac{\mathrm{u}+\mathrm{a}^\star}{\sqrt{2}}, & i\mathbf{U}_- = \frac{1}{\ell}\frac{\mathrm{u}-\mathrm{a}^\star}{\sqrt{2}},\\ \mathbf{A}_+ = \ell\frac{\mathrm{a}+\mathrm{u}^\star}{\sqrt{2}}, & i\mathbf{A}_- = \rho\frac{\mathrm{a}-\mathrm{u}^\star}{\sqrt{2}} \end{array}\right\} \Rightarrow [-i\mathbf{A}_-,\mathbf{U}_+] = 1 = [-i\mathbf{U}_-,\mathbf{A}_+],$$

which for $\star$-compatible combinations have to be 1 in the Fermi case and inverse to each other in the Bose case:

$$(\mathbf{A}_\pm)^\star = \mathbf{U}_\pm \iff \left\{\begin{array}{ll} \ell^2 = \rho^2 = 1, & \epsilon = +1,\\ \ell^2 = \frac{1}{\rho^2}, & \epsilon = -1. \end{array}\right.$$

For the combinations in the Fermi case the notations $(\mathbf{r},\mathbf{l})$ for "real-imaginary", or "right-left" with a Lorentz group action (below), will be used and $(\mathbf{x},\mathbf{p})$ for "position-momentum" in the Bose case:

$$\epsilon = +1: \left\{\begin{array}{llll} \mathbf{r} &= \frac{\mathrm{u}+\mathrm{a}^\star}{\sqrt{2}}, & \mathbf{l} &= \frac{\mathrm{u}-\mathrm{a}^\star}{\sqrt{2}},\\ \mathbf{r}^\star &= \frac{\mathrm{a}+\mathrm{u}^\star}{\sqrt{2}}, & -\mathbf{l}^\star &= \frac{\mathrm{a}-\mathrm{u}^\star}{\sqrt{2}} \end{array}\right\} \quad \Rightarrow \{\mathbf{r}^\star,\mathbf{r}\} = 1 = \{\mathbf{l}^\star,\mathbf{l}\},$$

$$\epsilon = -1: \left\{\begin{array}{llll} \mathbf{x} &= \ell\frac{\mathrm{u}+\mathrm{a}^\star}{\sqrt{2}}, & -i\mathbf{p} &= \frac{1}{\ell}\frac{\mathrm{u}-\mathrm{a}^\star}{\sqrt{2}},\\ \mathbf{x}^\star &= \ell\frac{\mathrm{a}+\mathrm{u}^\star}{\sqrt{2}}, & -i\mathbf{p}^\star &= \frac{1}{\ell}\frac{\mathrm{a}-\mathrm{u}^\star}{\sqrt{2}} \end{array}\right\} \quad \Rightarrow [i\mathbf{p}^\star,\mathbf{x}] = 1 = [i\mathbf{p},\mathbf{x}^\star],$$

$$[Q, \begin{pmatrix}\mathbf{r}\\ \mathbf{l}\\ \mathbf{x}\\ \mathbf{p}\end{pmatrix}] = z\begin{pmatrix}\mathbf{r}\\ \mathbf{l}\\ \mathbf{x}\\ \mathbf{p}\end{pmatrix},\quad [Q, \begin{pmatrix}\mathbf{r}^\star\\ \mathbf{l}^\star\\ \mathbf{x}^\star\\ \mathbf{p}^\star\end{pmatrix}] = -z\begin{pmatrix}\mathbf{r}^\star\\ \mathbf{l}^\star\\ \mathbf{x}^\star\\ \mathbf{p}^\star\end{pmatrix}.$$

In the reduced case without antidoubling, i.e., with particle-antiparticle identity, $\mathrm{u} = \mathrm{a}$, and trivial internal charge $Q = 0$, one has to take only "one half" of the structure. With Lorentz group representations, the reduced case is possible only for trivial two-ality, i.e., for integer spin particles.

In the charge-compatible combinations, the Hamiltonian and internal $\mathbf{U}(1)$-operator look as follows:

$$H_0 = \left\{\begin{array}{l} m\frac{[\mathbf{l},\mathbf{r}^\star]+[\mathbf{r},\mathbf{l}^\star]}{2},\\ m\frac{\ell^2\{\mathbf{p},\mathbf{p}^\star\}+\frac{1}{\ell^2}\{\mathbf{x},\mathbf{x}^\star\}}{2} \end{array}\right. \quad Q = \left\{\begin{array}{ll} z\frac{[\mathbf{l},\mathbf{l}^\star]+[\mathbf{r},\mathbf{r}^\star]}{2}, & \epsilon = +1,\\ z\frac{\{\mathbf{x},i\mathbf{p}^\star\}+\{-i\mathbf{p},\mathbf{x}^\star\}}{2}, & \epsilon = -1, \end{array}\right.$$

with the equations of motion

$$\begin{array}{ll}\epsilon=+1: \left\{\begin{array}{llll} d_t\mathbf{r} &= im\mathbf{l}, & d_t\mathbf{l} &= im\mathbf{r},\\ d_t\mathbf{r}^\star &= -im\mathbf{l}^\star, & d_t\mathbf{l}^\star &= -im\mathbf{r}^\star \end{array}\right. & \Rightarrow (d_t^2+m^2)(\mathbf{r},\mathbf{r}^\star,\mathbf{l},\mathbf{l}^\star)=0,\\ \epsilon=-1: \left\{\begin{array}{llll} d_t\mathbf{x} &= \ell^2 m\mathbf{p}, & d_t\mathbf{p} &= -\frac{m}{\ell^2}\mathbf{x},\\ d_t\mathbf{x}^\star &= \ell^2 m\mathbf{p}^\star, & d_t\mathbf{p}^\star &= -\frac{m}{\ell^2}\mathbf{x}^\star \end{array}\right. & \Rightarrow (d_t^2+m^2)(\mathbf{x},\mathbf{x}^\star,\mathbf{p},\mathbf{p}^\star)=0.\end{array}$$

They are derivable by variation from a classical Lagrangian, for Fermi with anticommuting Grassmann vectors:

$$\begin{array}{lll}\epsilon=+1: & L_F(\mathbf{r},\mathbf{r}^\star,\mathbf{l},\mathbf{l}^\star) & = i\mathbf{r}d_t\mathbf{r}^\star + i\mathbf{l}d_t\mathbf{l}^\star - m(\mathbf{r}\mathbf{l}^\star+\mathbf{l}\mathbf{r}^\star),\\ \epsilon=-1: & L_B(\mathbf{x},\mathbf{x}^\star,\mathbf{p},\mathbf{p}^\star) & = \mathbf{p}d_t\mathbf{x}^\star - \mathbf{x}d_t\mathbf{p}^\star - m(\ell^2\mathbf{p}\mathbf{p}^\star + \frac{1}{\ell^2}\mathbf{x}\mathbf{x}^\star).\end{array}$$

The kinetic terms in a classical Lagrangian couples and defines dual pairs.

In general, the *quantization opposite commutators* implement the basic space endomorphism Lie algebra in the quantum algebra (chapter "Quantum Algebras"):

$$[\mathrm{u},\mathrm{u}^\star]_{-\epsilon},\ [\mathrm{a},\mathrm{a}^\star]_{-\epsilon} \text{ and } \left\{\begin{array}{ll}\begin{pmatrix}[\mathbf{l}^\star,\mathbf{l}] & [\mathbf{r}^\star,\mathbf{l}]\\ [\mathbf{l}^\star,\mathbf{r}] & [\mathbf{r}^\star,\mathbf{r}]\end{pmatrix}, & \epsilon=+1,\\ \begin{pmatrix}\{i\mathbf{p}^\star,\mathbf{x}\} & \{\mathbf{x}^\star,\mathbf{x}\}\\ \{\mathbf{p}^\star,\mathbf{p}\} & \{\mathbf{x}^\star,-i\mathbf{p}\}\end{pmatrix}, & \epsilon=-1.\end{array}\right.$$

Examples of generators with internal degrees of freedom in addition to the internal $\mathbf{U}(1)$ are given below.

4.1.2 Fock Forms and Causal Ordering

Harmonic Fermi and Bose oscillators have time orbits. The following shorthand notation is used for the time dependence of (anti-) commutators and their Fock forms:

$$\begin{array}{lll}[a,b]_\epsilon(t-s) & = [a(s),b(t)]_\epsilon & = \epsilon[b,a]_\epsilon(s-t),\\ \langle[a,b]_\epsilon\rangle_{\mathrm{F}}(t-s) & = \langle[a(s),b(t)]_\epsilon\rangle_{\mathrm{F}} & = \epsilon\langle[b,a]_\epsilon\rangle_{\mathrm{F}}(s-t).\end{array}$$

The time-dependent quantization is a $\mathbf{U}(1)$-representation

$$D(t)=\begin{pmatrix}[\mathrm{a}^\star,\mathrm{a}]_\epsilon & [\mathrm{u}^\star,\mathrm{a}]_\epsilon\\ [\mathrm{a}^\star,\mathrm{u}]_\epsilon & [\mathrm{u}^\star,\mathrm{u}]_\epsilon\end{pmatrix}(t) = e^{imt}\begin{pmatrix}1 & 0\\ 0 & 1\end{pmatrix},$$

or, with the Q-compatible combinations, an $\mathbf{SO}(2)$-representation, for the Fermi case

$$\left(\begin{array}{cc|cc}\{\mathbf{l},\mathbf{l}^\star\} & \{\mathbf{r},\mathbf{l}^\star\} & \{\mathbf{l}^\star,\mathbf{l}^\star\} & \{\mathbf{r}^\star,\mathbf{l}^\star\}\\ \{\mathbf{l},\mathbf{r}^\star\} & \{\mathbf{r},\mathbf{r}^\star\} & \{\mathbf{l}^\star,\mathbf{r}^\star\} & \{\mathbf{r}^\star,\mathbf{r}^\star\}\\ \hline \{\mathbf{l},\mathbf{l}\} & \{\mathbf{r},\mathbf{l}\} & \{\mathbf{l}^\star,\mathbf{l}\} & \{\mathbf{r}^\star,\mathbf{l}\}\\ \{\mathbf{l},\mathbf{r}\} & \{\mathbf{r},\mathbf{r}\} & \{\mathbf{l}^\star,\mathbf{r}\} & \{\mathbf{r}^\star,\mathbf{r}\}\end{array}\right)(t) = \left(\begin{array}{cc|cc}\cos mt & -i\sin mt & 0 & 0\\ -i\sin mt & \cos mt & 0 & 0\\ \hline 0 & 0 & \cos mt & i\sin mt\\ 0 & 0 & i\sin mt & \cos mt\end{array}\right),$$

and for the Bose case

$$\left(\begin{array}{cc|cc} [i\mathbf{p},\mathbf{x}^\star] & [\mathbf{x},\mathbf{x}^\star] & [i\mathbf{p}^\star,\mathbf{x}^\star] & [\mathbf{x}^\star,\mathbf{x}^\star] \\ [\mathbf{p},\mathbf{p}^\star] & [\mathbf{x},-i\mathbf{p}^\star] & [\mathbf{p}^\star,\mathbf{p}^\star] & [\mathbf{x}^\star,-i\mathbf{p}^\star] \\ \hline [i\mathbf{p},\mathbf{x}] & [\mathbf{x},\mathbf{x}] & [i\mathbf{p}^\star,\mathbf{x}] & [\mathbf{x}^\star,\mathbf{x}] \\ [\mathbf{p},\mathbf{p}] & [\mathbf{x},-i\mathbf{p}] & [\mathbf{p}^\star,\mathbf{p}] & [\mathbf{x}^\star,-i\mathbf{p}] \end{array}\right)(t) = \left(\begin{array}{cc|cc} \cos mt & \ell^2 i\sin mt & 0 & 0 \\ \frac{i}{\ell^2}\sin mt & \cos mt & 0 & 0 \\ \hline 0 & 0 & \cos mt & \ell^2 i\sin mt \\ 0 & 0 & \frac{i}{\ell^2}\sin mt & \cos mt \end{array}\right).$$

The nontrivial (anti-) commutators can be read off from one quarter of the matrices, e.g., from $\left(\begin{array}{c|c} & \\ \hline & \times \end{array}\right)$.

The $\mathbf{U}(1)$-scalar-product-induced *Fock state* of a quantum algebra (chapter "Quantum Probability") defines a scalar product $\langle a|b\rangle_{\mathrm{F}} = \langle a^\star b\rangle_{\mathrm{F}}$:

$$\begin{pmatrix} \langle \mathrm{a}|\mathrm{a}\rangle_{\mathrm{F}} & \langle \mathrm{u}|\mathrm{a}\rangle_{\mathrm{F}} \\ \langle \mathrm{a}|\mathrm{u}\rangle_{\mathrm{F}} & \langle \mathrm{u}|\mathrm{u}\rangle_{\mathrm{F}} \end{pmatrix} = \begin{pmatrix} 1 & 0 \\ 0 & 1 \end{pmatrix} = d_{\mathrm{F}}(0).$$

It is the $t = 0$ value for the time-dependent Fock form of the quantization opposite (anti-) commutator

$$d_{\mathrm{F}}(t) = \begin{pmatrix} \langle [\mathrm{a}^\star,\mathrm{a}]_{-\epsilon}\rangle_{\mathrm{F}} & \langle [\mathrm{u}^\star,\mathrm{a}]_{-\epsilon}\rangle_{\mathrm{F}} \\ \langle [\mathrm{a}^\star,\mathrm{u}]_{-\epsilon}\rangle_{\mathrm{F}} & \langle [\mathrm{u}^\star,\mathrm{u}]_{-\epsilon}\rangle_{\mathrm{F}} \end{pmatrix}(t) = e^{imt}\begin{pmatrix} 1 & 0 \\ 0 & 1 \end{pmatrix}.$$

The Fock forms of the Q-compatible combinations are given by the time representation matrix elements:

$$\epsilon = +1: \left\{ \begin{array}{rl} \begin{pmatrix} \langle [\mathbf{l}^\star,\mathbf{l}]\rangle_{\mathrm{F}} & \langle [\mathbf{r}^\star,\mathbf{l}]\rangle_{\mathrm{F}} \\ \langle [\mathbf{l}^\star,\mathbf{r}]\rangle_{\mathrm{F}} & \langle [\mathbf{r}^\star,\mathbf{r}]\rangle_{\mathrm{F}} \end{pmatrix}(t) = d_{\mathrm{F}}(t) & = \begin{pmatrix} i\sin mt & \cos mt \\ \cos mt & i\sin mt \end{pmatrix}, \\ d_{\mathrm{F}}(0) = \begin{pmatrix} 0 & 1 \\ 1 & 0 \end{pmatrix}, \ D(t) & = \begin{pmatrix} \cos mt & i\sin mt \\ i\sin mt & \cos mt \end{pmatrix}, \end{array}\right.$$

$$\epsilon = -1: \left\{ \begin{array}{rl} \begin{pmatrix} \langle \{i\mathbf{p}^\star,\mathbf{x}\}\rangle_{\mathrm{F}} & \langle \{\mathbf{x}^\star,\mathbf{x}\}\rangle_{\mathrm{F}} \\ \langle \{\mathbf{p}^\star,\mathbf{p}\}\rangle_{\mathrm{F}} & \langle \{\mathbf{x}^\star,-i\mathbf{p}\}\rangle_{\mathrm{F}} \end{pmatrix}(t) = d_{\mathrm{F}}(t) & = \begin{pmatrix} i\sin mt & \ell^2\cos mt \\ \frac{1}{\ell^2}\cos mt & i\sin mt \end{pmatrix}, \\ d_{\mathrm{F}}(0) = \begin{pmatrix} 0 & \ell^2 \\ \frac{1}{\ell^2} & 0 \end{pmatrix}, \ D(t) & = \begin{pmatrix} \cos mt & \ell^2 i\sin mt \\ \frac{i}{\ell^2}\sin mt & \cos mt \end{pmatrix}. \end{array}\right.$$

Time representations, time-dependent Fock forms, and the Fock value of the Hamiltonian matrices $md_{\mathrm{F}}(0)$ are related to each other as follows:

$$d_{\mathrm{F}}(t) = \tfrac{1}{im}\tfrac{d}{dt}D(t) = D(t)\circ d_{\mathrm{F}}(0), \quad imd_{\mathrm{F}}(0) = D^{-1}(t)\circ \tfrac{d}{dt}D(t).$$

The Fock value of a *time-ordered product* connects the time order with the creation-annihilation order, first creation and only afterward annihilation:

$$\vartheta(t_1-t_2)\langle \mathrm{u}^\star(t_1)\mathrm{u}(t_2)\rangle_{\mathrm{F}} \pm \vartheta(t_2-t_1)\langle \mathrm{u}^\star(t_2)\mathrm{u}(t_1)\rangle_{\mathrm{F}}$$

$$= \vartheta(t)e^{-imt} \pm \vartheta(-t)e^{imt} = \left\{ \begin{array}{ll} \cos mt - \epsilon(t)i\sin mt & = e^{-im|t|}, \\ -i\sin mt + \epsilon(t)\cos mt & = \epsilon(t)e^{-im|t|}, \end{array}\right.$$

with $t = t_1 - t_2$.

This distinguishes one sign in the two possibilities $e^{\pm im|t|}$, here the minus sign. The time-ordered product can be written as the sum of the Fock form of the quantization opposite commutator and the ordered quantization (anti-) commutator,

$$e^{-im|t|} = \left\{ \begin{array}{ll} \langle [\mathbf{r}^\star, \mathbf{l}](t) - \epsilon(t)\{\mathbf{r}^\star, \mathbf{l}\}(t)\rangle_{\mathrm{F}}, & \epsilon = +1, \\ \frac{1}{\ell^2}\langle \{\mathbf{x}^\star, \mathbf{x}\}(t) - \epsilon(t)[\mathbf{x}^\star, \mathbf{x}](t)\rangle_{\mathrm{F}}, & \epsilon = -1, \end{array} \right.$$

$$\epsilon(t)e^{-im|t|} = \left\{ \begin{array}{ll} -\langle [\mathbf{r}^\star, \mathbf{r}](t) - \epsilon(t)\{\mathbf{r}^\star, \mathbf{r}\}(t)\rangle_{\mathrm{F}}, & \epsilon = +1, \\ \frac{1}{\ell^2}\langle \{\mathbf{x}^\star, i\mathbf{p}\}(t) - \epsilon(t)[\mathbf{x}^\star, i\mathbf{p}](t)\rangle_{\mathrm{F}}, & \epsilon = -1, \end{array} \right.$$

and for the time representations in $\mathbf{SO}(2)$ ($\ell^2 = 1$ for Fermi),

$$d_{\mathrm{F}}(t) - \epsilon(t)D(t) = \begin{pmatrix} \frac{d_t}{im} & \ell^2 \\ \frac{1}{\ell^2} & \frac{d_t}{im} \end{pmatrix} e^{-im|t|} = \begin{pmatrix} -\epsilon(t) & \ell^2 \\ \frac{1}{\ell^2} & -\epsilon(t) \end{pmatrix} e^{-im|t|}.$$

In this section and in the following one has to be aware that the (anti) commutators are valued in the quantum algebra, e.g., $[\mathrm{u}^\star, \mathrm{u}]_\epsilon = 1 \in \mathbf{Q}_\epsilon(\mathbb{C}^2)$, in contrast to the number-valued Fock state matrix elements, e.g., $\langle [\mathrm{u}^\star, \mathrm{u}]_{-\epsilon}\rangle_{\mathrm{F}} = 1 \in \mathbb{C}$.

4.2 Relativistic Distribution of Time Representations

For relativistic particle fields the Hilbert representations of time, as given for the harmonic oscillators, are embedded into Lorentz compatible functions and distributions of spacetime translations (chapter "Propagators").

The time representation

$$\mathbb{R} \ni t \longmapsto \begin{pmatrix} \cos mt & i \sin mt \\ i \sin mt & \cos mt \end{pmatrix} = \begin{pmatrix} \frac{d_t}{im} & 1 \\ 1 & \frac{d_t}{im} \end{pmatrix} i \sin mt = \begin{pmatrix} 1 & \frac{d_t}{im} \\ \frac{d_t}{im} & 1 \end{pmatrix} \cos mt$$

leads to the *quantization distributions* with causal support

$$\begin{aligned} \mathbb{R}^4_+ \ni x\vartheta(x^2) \longmapsto \mathbf{D}(m|x) &= \begin{pmatrix} \mathbf{c}_k & i\mathbf{s} \\ i\mathbf{s} & \mathbf{c}_k \end{pmatrix}(m|x) = \begin{pmatrix} \frac{\partial_k}{im} & 1 \\ 1 & \frac{\partial_k}{im} \end{pmatrix} i\mathbf{s}(m|x) \\ &= \int \frac{d^4q}{(2\pi)^3} \epsilon(q_0) \begin{pmatrix} \frac{q_k}{m} & 1 \\ 1 & \frac{q_k}{m} \end{pmatrix} m\delta(q^2 - m^2) e^{iqx}. \end{aligned}$$

As seen below, the minimal embedding (2×2) matrix $\begin{pmatrix} q_k & m \\ m & q_k \end{pmatrix}$ with diagonal Lorentz vectors and skew-diagonal Lorentz scalars has to be modified for more complicated Lorentz group representations.

The *Fock form functions* are defined for all translations:

$$\begin{aligned} \mathbb{R}^4 \ni x \longmapsto \mathbf{d}_{\mathrm{F}}(m|x) &= \begin{pmatrix} i\mathbf{s}_k & \mathbf{C} \\ \mathbf{C} & i\mathbf{s}_k \end{pmatrix}(m|x) = \begin{pmatrix} \frac{\partial_k}{im} & 1 \\ 1 & \frac{\partial_k}{im} \end{pmatrix} \mathbf{C}(m|x) \\ &= \int \frac{d^4q}{(2\pi)^3} \begin{pmatrix} \frac{q_k}{m} & 1 \\ 1 & \frac{q_k}{m} \end{pmatrix} m\delta(q^2 - m^2) e^{iqx}. \end{aligned}$$

They cannot be written as time and position derivatives of the quantization, in analogy to the time derivative $d_{\mathrm{F}}(t) = \frac{1}{im}\frac{d}{dt}D(t)$ of the last section. This relation arises for the time projections which is given by position Fourier transformation

$$\begin{aligned}\int d^3x\ \mathbf{D}(m|x) &= \begin{pmatrix}\delta_k^0 \cos mx_0 & i\sin mx_0\\ i\sin mx_0 & \delta_k^0\cos mx_0\end{pmatrix} &= D(x_0),\\ \int d^3x\ \mathbf{d}_{\rm F}(m|x) &= \begin{pmatrix}\delta_k^0 i\sin mx_0 & \cos mx_0\\ \cos mx_0 & \delta_k^0 i\sin mx_0\end{pmatrix} &= d_{\rm F}(x_0).\end{aligned}$$

With the time order related to the creation-annihilation order the relativistic time-ordered products (Feynman propagators) are

$$\begin{aligned}\mathbf{d}_{\rm F}(m\ x) - \epsilon(x_0)\mathbf{D}(m|x) &= \begin{pmatrix}-\mathbf{E}_k & \mathbf{E}\\ \mathbf{E} & -\mathbf{E}_k\end{pmatrix}(-im|x)\\ &= \begin{pmatrix}\frac{\partial_k}{im} & 1\\ 1 & \frac{\partial_k}{im}\end{pmatrix}\mathbf{E}(-im|x),\\ \mathbf{E}(-im|x) &= \mathbf{C}(m|x) - \epsilon(x_0)i\mathbf{s}(m|x),\\ \mathbf{E}_k(-im|x) &= \epsilon(x_0)\mathbf{c}_k(m|x) - i\mathbf{S}_k(m|x),\\ \int d^3x[\mathbf{d}_{\rm F}(m|x) - \epsilon(x_0)\mathbf{D}(m|x)] &= \begin{pmatrix}-\epsilon(x_0)\delta_k^0 & 1\\ 1 & -\epsilon(x_0)\delta_k^0\end{pmatrix}e^{-im|x_0|}.\end{aligned}$$

The position projection, given by time integration of the $\epsilon(x_0)$-multiplied quantization distribution gives Yukawa potential and force:

$$2\pi\int dx_0\ \epsilon(x_0)\mathbf{D}(m|x) = \begin{pmatrix}\delta_k^a\partial_a & im\\ im & \delta_k^a\partial_a\end{pmatrix}\frac{e^{-mr}}{r} = \begin{pmatrix}-\delta_k^a\frac{x_a}{r}\frac{1+mr}{r} & im\\ im & -\delta_k^a\frac{x_a}{r}\frac{1+mr}{r}\end{pmatrix}\frac{e^{-mr}}{r}.$$

4.3 Quantum Fields for Massive Particles

Particles are acted on with infinite-dimensional irreducible Hilbert representations of the Poincaré group $\mathbf{SL}(\mathbb{C}^2)\ \vec{\times}\ \mathbb{R}^4$ (Lorentz transformations and translations), induced by finite-dimensional irreducible Hilbert representations of the direct product $\mathbf{SU}(2)\times\mathbb{R}^4$ of rotation group and translations (chapter "Harmonic Analysis"). The representations of the massive particle group $\mathbf{SU}(2)\times\mathbb{R}^4$ for each energy-momentum are Lorentz compatibly integrated over the energy-momentum hyperboloid for one fixed mass. All particle quantum fields are Lorentz compatibly distributed time orbits and can be considered to be relativistically embedded Fermi or Bose oscillators, possibly with additional degrees of freedom (spin, circularity, charge, isospin, etc.).

The definition of relativistic particle fields for the noncompact Poincaré group involves two complications, the infinite dimensionality of the representations used (this section) and the embedding of the rotation group in the Lorentz group (next section). For easier access and an illustration, it is useful to consult the simplest examples with a massive scalar and vector field below.

4.3.1 The Hilbert Spaces for Massive Particles

Particles are described by *momentum operators* in complex quartets, called

$\mathrm{u}(m^2, J; \vec{q})^a$ *particle* *creator*	$\overset{\times}{\leftrightarrow}$	$\mathrm{a}^\star(m^2, J; \vec{q})^a$ *antiparticle* *annihilator*
$\star \updownarrow$		$\updownarrow \star$
$\mathrm{u}^\star(m^2, J; \vec{q})_a$ *particle* *annihilator*	$\overset{\times}{\leftrightarrow}$	$\mathrm{a}(m^2, J; \vec{q})_a$ *antiparticle* *creator*

for the invariants *mass* m and *spin* J
with the eigenvalues *momentum* $\vec{q}$ and *spin component* a.

A notation with $(m^2, J; \vec{q}, a)$ involves the two Poincaré group invariants m^2 for translations $\mathbf{p}^2$ and J for rotations $\mathbf{S}^2$ (Pauli-Lubanski vector $\mathbf{S}$), and the corresponding eigenvalues, momenta $\vec{q}$ for position translations $\vec{\mathbf{p}}$ and eigenvalue a for rotations around the momentum $\frac{\vec{\mathbf{p}}}{|\vec{\mathbf{p}}|}$. If obvious from the context, the Poincaré group invariants are omitted and only the eigenvalues in an (m^2, J)-"multiplet" are explicitly given, e.g., $\mathrm{u}(m^2, J; \vec{q})^a = \mathrm{u}(\vec{q})^a \in W(\vec{q}) \cong \mathbb{C}^{1+2J}$. The momentum operators are acted on irreducibly by the spin-translation group

$$(x, u) \in \mathbb{R}^4 \times \mathbf{SU}(2): \quad \mathrm{u}(\vec{q})^a \longmapsto e^{iqx} D^{2J}(u)^a_b \mathrm{u}(\vec{q})^b, \quad q^2 = m^2, \quad a, b = -J, \ldots, J.$$

The momenta parametrize the boosts for the rotation group orientations $\mathcal{Y}^3 \cong \mathbf{SO}_0(1,3)/\mathbf{SO}(3) \cong \mathbf{SL}(\mathbb{C}^2)/\mathbf{SU}(2) \cong \mathbb{R}^3$ with Lorentz invariant measure of the 3-dimensional energy-momentum hyperboloid with positive energy:

$$\text{for } \mathcal{Y}^3(m^2): \quad d^3\mathrm{y}(\tfrac{\vec{q}}{m}) = \tfrac{d^3 q}{2q_0(2\pi)^3}, \quad q_0 = \sqrt{m^2 + \vec{q}^2}.$$

The "huge" infinite-dimensional vector space $\prod_{\vec{q} \in \mathbb{R}^3} W(\vec{q})$ is a direct integral (chapter "The Kepler Factor"), where the integral sums up the "little" vector spaces for each momentum, e.g., for particle creators:

$$w : \mathcal{Y}^3 \longrightarrow W, \quad w = {}^\oplus\!\!\int d^3\mathrm{y}(\tfrac{\vec{q}}{m})\, \mathrm{u}(\vec{q})^a w(\vec{q})_a \in W^{\mathcal{Y}^3} = {}^\oplus\!\!\int d^3\mathrm{y}(\tfrac{\vec{q}}{m}) W(\vec{q}), \quad \mathrm{u}(\vec{q})^a \in W(\vec{q}) = W \times \{\vec{q}\},$$

where w is a W-valued spin $\mathbf{SU}(2)$-intertwiner on the Lorentz group $\mathbf{SL}(\mathbb{C}^2)$ or, equivalently, a W-valued mapping of the energy-momentum hyperboloid $\mathcal{Y}^3$. The direct integral is the distributive generalization for the basis expansion of a finite-dimensional vector $w = \sum_{q=1}^{n} \mathrm{u}^q w_q$.

The measure of the energy-momentum hyperboloid has an associated Dirac distribution

$$\frac{d^4q}{(2\pi)^3}\,\vartheta(q_0)\delta(q^2-m^2) = \frac{1}{2q_0}\,d^3\frac{q}{2\pi} \leftrightarrow 2q_0\,\delta(\frac{\vec q}{2\pi}).$$

A possible constant normalization factor is chosen as 1.

The quantization is induced by the duality distribution of the momentum operators, given by the asscociated Dirac distribution:

$$[\mathrm{u}^\star(\vec p)_a, \mathrm{u}(\vec q)^b]_\epsilon = [\mathrm{a}^\star(\vec p)^b, \mathrm{a}(\vec q)_a]_\epsilon = \delta_a^b 2q_0\delta(\tfrac{\vec q-\vec p}{2\pi}).$$

The $\mathbf{U}(1)$-scalar product induces the Fock form with the scalar product distribution

$$\begin{aligned} &\langle \mathrm{u}^\star(\vec p)_a\mathrm{u}(\vec q)^b\rangle_{\mathrm F} &&= \langle \mathrm{a}^\star(\vec p)^b\mathrm{a}(\vec q)_a\rangle_{\mathrm F} \\ =\ &\langle [\mathrm{u}^\star(\vec p)_a, \mathrm{u}(\vec q)^b]_{-\epsilon}\rangle_{\mathrm F} &&= \langle [\mathrm{a}^\star(\vec p)^b, \mathrm{a}(\vec q)_a]_{-\epsilon}\rangle_{\mathrm F} \ = \delta_a^b 2q_0\delta(\tfrac{\vec q-\vec p}{2\pi}). \end{aligned}$$

The creation operators define a measure-related distributive basis (not Hilbert space vectors) for a Hilbert space:

$$\begin{aligned} |m^2, J;\vec q, a\rangle = |\vec q, a\rangle &= \mathrm{u}(m^2, J;\vec q)^a|0\rangle, \\ \langle m^2, J_2;\vec q_2, a_2|m^2, J_1;\vec q_1, a_1\rangle &= \delta_{J_2}^{J_1}\delta_{a_2}^{a_1}2q_0\delta(\tfrac{\vec q_1-\vec q_2}{2\pi}). \end{aligned}$$

Stable particles for different masses $m_1^2 \neq m_2^2$ have Schur-orthogonal spaces. The direct integral with wave packets, square integrable on the energy-momentum hyperboloid $L^2_{d^3\mathrm{y}(\frac{\vec q}{m})}(\mathcal{Y}^3) = L^2(\mathcal{Y}^3)$ gives vectors in the Fock-Hilbert space for the particle with invariants (m^2, J):

$$\begin{aligned} &|m^2, J; w\rangle = {}^\oplus\!\!\int d^3\mathrm{y}(\tfrac{\vec q}{m})\ w(\vec q)_a|m^2, J;\vec q, a\rangle \in H(m^2, J) = L^2(\mathcal{Y}^3)\otimes\mathbb{C}^{1+2J} \\ \Rightarrow\ &\langle m^2, J_2; w_2|m^2, J_1; w_1\rangle = \delta_{J_2}^{J_1}\int d^3\mathrm{y}(\tfrac{\vec q}{m})\ \overline{w_2(\vec q)_a}w_1(\vec q)_a \\ &\qquad = \delta_{J_2}^{J_1}\int \tfrac{d^4q}{(2\pi)^3}\ \overline{w_2(\vec q)_a}\vartheta(q_0)\delta(q^2-m^2)w_1(\vec q)_a. \end{aligned}$$

The distributive completeness allows the sesquilinear decomposition of the unit operator in (projector on) the particle Hilbert space $H(m^2, J)$:

$$\begin{aligned} \mathcal{P}(m^2, J) \cong |m^2, J\rangle\langle m^2, J| &= \bigoplus_{a=-J}^{J}{}^\oplus\!\!\int \tfrac{d^4q}{(2\pi)^3}\vartheta(q_0)\delta(q^2-m^2)|q, a\rangle\langle q, a| \\ &= \bigoplus_{a=-J}^{J}{}^\oplus\!\!\int d^3\mathrm{y}(\tfrac{\vec q}{m})|\vec q, a\rangle\langle\vec q, a|, \\ \mathcal{P}(m^2, J)\circ\mathcal{P}(m^2, J) &= \mathcal{P}(m^2, J). \end{aligned}$$

4.3.2 Relativistic Quantum Fields

The embedding of the particle momentum operators with $\mathbf{SU}(2)$-action into fields with Lorentz group representations on finite-dimensional spaces $V \subset$

$W^{\mathcal{Y}^3}$ with basis $\{(\mathbf{D}^\iota)^j\}$ is effected by momentum-parametrized *transmutators* $D^\iota(q)$ for fixed mass $q^2 = m^2$, which are discussed in the next section:

$$\begin{array}{l}\mathbf{D}^\iota(m)^j = {}^\oplus\!\!\int d^3\mathrm{y}(\frac{\vec{q}}{m})\mathrm{u}(\vec{q})^a D^\iota(q)^j_a \in V \subset W^{\mathcal{Y}^3}, \quad q_0 = \sqrt{m^2+\vec{q}^2},\\ \mathbf{SL}(\mathbb{C}^2) \ni \lambda \longmapsto D(\lambda) \in \mathbf{GL}(V), \quad \mathbf{D}^\iota(m)^j \longmapsto D(\lambda)^j_k \mathbf{D}^\iota(m)^k.\end{array}$$

With the infinite volume of the energy-momentum hyperboloid the transmutators are not square integrable, $D^\iota(q)^k_a \notin L^2(\mathcal{Y}^3)$.

The spacetime translation orbits $\mathbb{R}^4 \ni x \longmapsto e^{iqx} \in \mathbf{U}(1)$, $q^2 = m^2$, of the momentum operators give the embedding particle fields. For a comparison with the harmonic oscillators as used above in the quantum algebras, they are given for Fermi and Bose$\binom{\epsilon=+1}{\epsilon=-1}$ with a "right-left" and a "position-momentum"-notation:

$$\begin{array}{rll}\binom{\mathbf{r}}{\mathbf{x}}(m|x) &= \mathbf{U}_+(m|x)^j &= Z_m {}^\oplus\!\!\int d^3\mathrm{y}(\frac{\vec{q}}{m})\ell\ \frac{e^{iqx}\mathrm{u}(\vec{q})^a + e^{-iqx}\mathrm{a}^\star(\vec{q})^a}{\sqrt{2}}\ D^+(q)^j_a,\\ \binom{\mathbf{l}}{-i\mathbf{p}}(m|x) &= i\mathbf{U}_-(m|x)^j &= Z_m {}^\oplus\!\!\int d^3\mathrm{y}(\frac{\vec{q}}{m})\frac{1}{\ell}\ \frac{e^{iqx}\mathrm{u}(\vec{q})^a - e^{-iqx}\mathrm{a}^\star(\vec{q})^a}{\sqrt{2}}\ D^-(q)^j_a,\\ \binom{\mathbf{r}^\star}{\mathbf{x}^\star}(m|x) &= \mathbf{A}_+(m|x)_j &= Z_m {}^\oplus\!\!\int d^3\mathrm{y}(\frac{\vec{q}}{m})\ell\ \frac{e^{iqx}\mathrm{a}(\vec{q})_a + e^{-iqx}\mathrm{u}^\star(\vec{q})_a}{\sqrt{2}}\ \check{D}^+(q)^a_j,\\ \binom{-\mathbf{l}^\star}{-i\mathbf{p}^\star}(m|x) &= i\mathbf{A}_-(m|x)_j &= Z_m {}^\oplus\!\!\int d^3\mathrm{y}(\frac{\vec{q}}{m})\frac{1}{\ell}\ \frac{e^{iqx}\mathrm{a}(\vec{q})_a - e^{-iqx}\mathrm{u}^\star(\vec{q})_a}{\sqrt{2}}\ \check{D}^-(q)^a_j,\\ \text{with } q_0 &= \sqrt{m^2+\vec{q}^2}, & \mathbf{A}^\star_\pm(m|x)_j = \mathbf{U}_\pm(m|x)^j.\end{array}$$

The Lorentz transformation properties for relativistic fields may be different for the transmutators D^+ and D^- in the $\mathbf{U}(1,1)$-symmetric and antisymmetric combinations respectively. In the Fermi case, the dual normalization factor is fixed, $\ell^2 = 1$. For massive particles, the free normalization factor will be chosen as $Z_m^2 = m > 0$. The Lorentz transformation behavior of particle fields combines the Lorentz action on the translations and on the value space

$$\mathbb{M} \ni x \longmapsto \mathbf{D}^\iota(m|x) \in V, \quad \mathbf{D}^\iota(m|0) = \mathbf{D}^\iota(m),$$

$$\begin{array}{ccc}\mathbb{M} & \xrightarrow{\Lambda(\lambda)} & \mathbb{M}\\ {\scriptstyle \mathbf{D}^\iota(m|\ )}\downarrow & & \downarrow{\scriptstyle \mathbf{D}^\iota_\lambda(m|\ )}\\ V & \xrightarrow[D(\lambda)]{} & V\end{array}, \qquad \begin{array}{l}\mathbf{SL}(\mathbb{C}^2) \ni \lambda \longmapsto \Lambda(\lambda) \in \mathbf{SO}_0(1,3),\\ \mathbf{D}^\iota_\lambda(m|x) = D(\lambda).\mathbf{D}^\iota(m|\Lambda^{-1}.x).\end{array}$$

In the following, shorthand notation is used for the translation dependence of (anti-)commutators and their Fock forms:

$$\begin{array}{lll}[A(y), B(x)]_\epsilon &= [A, B]_\epsilon(x-y) &= \epsilon[B, A]_\epsilon(y-x),\\ \langle [A(y), B(x)]_\epsilon\rangle_{\mathrm{F}} &= \langle [A, B]_\epsilon\rangle_{\mathrm{F}}(x-y) &= \epsilon\langle [B, A]_\epsilon\rangle_{\mathrm{F}}(y-x).\end{array}$$

Without specification of the Lorentz properties, Fermi fields have the quantization anticommutators and commutator Fock forms;

$$\epsilon = +1: \left\{\begin{array}{lll}\begin{pmatrix}\{\mathbf{l}^\star,\mathbf{l}\} & \{\mathbf{r}^\star,\mathbf{l}\}\\ \{\mathbf{l}^\star,\mathbf{r}\} & \{\mathbf{r}^\star,\mathbf{r}\}\end{pmatrix}(x) &= \begin{pmatrix}\mathbf{c}_k & i\mathbf{s}\\ i\mathbf{s} & \mathbf{c}_k\end{pmatrix}(m|x) &= \mathbf{D}(m|x),\\ \\ \begin{pmatrix}\langle[\mathbf{l}^\star,\mathbf{l}]\rangle_{\mathrm{F}} & \langle[\mathbf{r}^\star,\mathbf{l}]\rangle_{\mathrm{F}}\\ \langle[\mathbf{l}^\star,\mathbf{r}]\rangle_{\mathrm{F}} & \langle[\mathbf{r}^\star,\mathbf{r}]\rangle_{\mathrm{F}}\end{pmatrix}(x) &= \begin{pmatrix}i\mathbf{s}_k & \mathbf{C}\\ \mathbf{C} & i\mathbf{s}_k\end{pmatrix}(m|x) &= \mathbf{d}_{\mathrm{F}}(m|x),\end{array}\right.$$

and Bose fields the quantization commutators and anticommutator Fock forms,

$$\epsilon = -1 : \begin{cases} \begin{pmatrix} [i\mathbf{p}^\star, \mathbf{x}] & [\mathbf{x}^\star, \mathbf{x}] \\ [\mathbf{p}^\star, \mathbf{p}] & [\mathbf{x}^\star, -i\mathbf{p}] \end{pmatrix}(x) & = \begin{pmatrix} \mathbf{c}_k & \ell^2 i\mathbf{s} \\ \frac{i}{\ell^2}\mathbf{s} & \mathbf{c}_k \end{pmatrix}(m|x) \quad \sim \mathbf{D}(m|x), \\ \begin{pmatrix} \langle\{i\mathbf{p}, \mathbf{x}^\star\}\rangle_{\mathrm{F}} & \langle\{\mathbf{x}, \mathbf{x}^\star\}\rangle_{\mathrm{F}} \\ \langle\{\mathbf{p}, \mathbf{p}^\star\}\rangle_{\mathrm{F}} & \langle\{\mathbf{x}, -i\mathbf{p}^\star\}\rangle_{\mathrm{F}} \end{pmatrix}(x) & = \begin{pmatrix} i\mathbf{S}_k & \ell^2\mathbf{C} \\ \frac{1}{\ell^2}\mathbf{C} & i\mathbf{S}_k \end{pmatrix}(m|x) \quad \sim \mathbf{d}_{\mathrm{F}}(m|x). \end{cases}$$

The time derivative of mechanics is embedded into the spacetime derivative $d_t \hookrightarrow \partial_j$. Therefore, the dual pairing for the quantization in mechanics $\mathbf{p} = d_t\mathbf{x}$ (chapter "Quantum Algebras") arises as a timelike component of a Lorentz vector. With the Lorentz vector property of the quantization in the diagonal of $\mathbf{D}(m|x)$ the products of the fields have to contain the following Lorentz representations:

$$\begin{array}{ll} \epsilon = +1 : & \begin{cases} \text{Lorentz vectors } [1|1] \text{ in } & \mathbf{l} \otimes \mathbf{l}^\star, \quad \mathbf{r} \otimes \mathbf{r}^\star, \\ \text{Lorentz scalars } [0|0] \text{ in } & \mathbf{r} \otimes \mathbf{l}^\star, \quad \mathbf{l} \otimes \mathbf{r}^\star, \end{cases} \\ \epsilon = -1 : & \begin{cases} \text{Lorentz vectors } [1|1] \text{ in } & \mathbf{p} \otimes \mathbf{x}^\star, \quad \mathbf{x} \otimes \mathbf{p}^\star, \\ \text{Lorentz scalars } [0|0] \text{ in } & \mathbf{p} \otimes \mathbf{p}^\star, \quad \mathbf{x} \otimes \mathbf{x}^\star. \end{cases} \end{array}$$

4.4 Lorentz Group Embedding of Spin

Massive particles keep as homogeneous action group the position rotation fixgroup in the Lorentz transformations:

$$\text{Sylvester: } q^2 = m^2 > 0 \Rightarrow \begin{cases} \mathbb{R}^4 \cong \mathbb{T} \perp \mathbb{S}^3, \\ \text{fixgroup } \mathbf{SO}(3) \cong \mathbf{SU}(2)/\mathbb{I}(2). \end{cases}$$

The fixgroup action on particles induces the Lorentz group action on their relativistic fields.

The transition from the "little" spin group $\mathbf{SU}(2)$ to the "large" group $\mathbf{SL}(\mathbb{C}^2)$ uses rotation to Lorentz transmutators (chapters "Spacetime as Unitary Operation Classes" and "Harmonic Analysis"). They are energy-momentum parametrized representations of the mass hyperboloids $\mathbf{SL}(\mathbb{C}^2)/\mathbf{SU}(2) \cong \mathcal{Y}^3$ which is the fixgroup orientation manifold. The boost representations connect an irreducible spin representation $d : \mathbf{SU}(2) \longrightarrow \mathbf{SU}(W)$ with a Lorentz group representation on a finite-dimensional vector space $V \cong \mathbb{C}^n$:

$$\lambda \in \mathbf{SL}(\mathbb{C}^2), \quad D(\lambda) : V \longrightarrow V, \quad D(\lambda)_{k=1,\dots,n}^{j=1,\dots,n}.$$

In the decomposition of the Lorentz group representation with respect to irreducible spin $\mathbf{SU}(2)$-representations on subspaces $W^\iota \cong \mathbb{C}^{m_\iota}$ with square matrices $d^\iota(u)_{a=1,\dots,m_\iota}^{b=1,\dots,m_\iota}$,

$$\begin{array}{l} V \overset{\mathbf{SU}(2)}{\cong} \bigoplus\limits_\iota W^\iota, \quad D|_{\mathbf{SU}(2)} = \bigoplus\limits_\iota d^\iota, \\ u \in \mathbf{SU}(2) : \; d^\iota(u) : W^\iota \longrightarrow W^\iota, \end{array}$$

there has to occur the inducing spin representation d. The $\mathbf{SU}(2)$-decomposition of a finite-dimensional irreducible $\mathbf{SL}(\mathbb{C}^2)$-representation reads

$$\mathbf{SL}(\mathbb{C}^2) \cong \bigoplus \mathbf{SU}(2): \ [2L|2R] \cong \bigoplus_{J=|L-R|}^{L+R} [2J].$$

The corresponding decomposition of the boost representation, for all $(q_0, \vec{q})$ on a fixed energy-momentum hyperboloid $q^2 = m^2$,

$$\mathbf{SO}_0(1,3)/\mathbf{SO}(3) \cong \mathcal{Y}^3 \ni \ \tfrac{q}{m} \longmapsto D(\tfrac{q}{m}) = \bigoplus_\iota D^\iota(\tfrac{q}{m}),$$

defines the transmutators with rectangular matrices $D^\iota(\frac{q}{m})^{j=1,\dots n}_{a=1,\dots,m_\iota}$ as transformation from the "little" spin representation spaces W^ι to the "large" Lorentz group representation space V. The action of the Lorentz group on the transmutator gives the transmutator for the Lorentz transformed energy-momenta and the represented associated *Wigner rotation* $u = u(\lambda, \frac{q}{m}) \in \mathbf{SU}(2)$:

$$\lambda \in \mathbf{SL}(\mathbb{C}^2): \ \ D(\lambda)^k_j D^\iota(\tfrac{q}{m})^j_a = D^\iota(\lambda \circ \tfrac{q}{m} \circ \lambda^\star)^k_b d^\iota(u)^b_a.$$

The Wigner rotation defines - as $\mathbf{SU}(2)$-transformation - the Lorentz group action on the spaces $W^\iota(q) = W^\iota \times \{q\}$, e.g., for a basis $\{\mathrm{e}^\iota(q)^a\}_{a=1,\dots,m_\iota}$ - for the fields given by the creation and annihilation momentum operators,

$$\mathrm{e}^\iota(\lambda \circ q \circ \lambda^\star) = d^\iota(u).\mathrm{e}^\iota(q), \quad u = u(\lambda, \tfrac{q}{m}).$$

With the Lorentz invariant integration there arise vectors from a finite-dimensional space with the Lorentz group action as given on the vector space V:

$$\begin{aligned} \mathbf{D}^\iota(m)^k &= {}^\oplus\!\!\int d^3\mathrm{y}(\tfrac{\vec{q}}{m}) \mathrm{e}^\iota(q)^a D^\iota(\tfrac{q}{m})^k_a \in V \subset W^{\iota \mathcal{Y}^3}, \\ \mathbf{D}^\iota(m)^k &\longmapsto D(\lambda)^k_j \mathbf{D}^\iota(m)^j. \end{aligned}$$

Now the simplest examples: The fundamental transmutators from Lorentz group to rotation groups relate to each other the Pauli $\mathbf{SU}(2)$-representation and the two Weyl $\mathbf{SL}(\mathbb{C}^2)$-representations; they are the fundamental boost representations

$$\begin{aligned} V^{[1|0]} &\cong W^{[1]} \text{ with respect to } \mathbf{SU}(2), \\ [1|0](\tfrac{q}{m}) &= s(\tfrac{q}{m}): \ \ \mathrm{e}(\vec{q})^C \hookrightarrow s(\tfrac{q}{m})^A_C \mathrm{e}(\vec{q})^C, \\ & \qquad\qquad\qquad\qquad A, C = 1, 2, \\ V^{[0|1]} &\cong \hat{W}^{[1]} \text{ with respect to } \mathbf{SU}(2), \\ [0|1](\tfrac{q}{m}) &= \hat{s}(\tfrac{q}{m}): \ \ \hat{\mathrm{e}}(\vec{q})^C \hookrightarrow \hat{s}(\tfrac{q}{m})^{\dot{A}}_C \hat{\mathrm{e}}(\vec{q})^C, \\ & \qquad\qquad\qquad\qquad \dot{A}, C = 1, 2. \end{aligned}$$

They are nondecomposable since both the Weyl $\mathbf{SL}(\mathbb{C}^2)$-representation and the Pauli $\mathbf{SU}(2)$-representation are irreducible complex 2-dimensional.

Lorentz vector representation gives rise to two rectangular transmutators from Lorentz group to rotation groups. The 4-dimensional spaces are decomposable with respect to rotations $\mathbf{SO}(3)$ into a 1-dimensional and a 3-dimensional space for spin 0 and spin 1 respectively:

$$\begin{aligned} V^{[1|1]} &\cong W^{[0]} \oplus W^{[2]} \text{ with respect to } \mathbf{SO}(3),\\ [1|1](\tfrac{q}{m}) &= \Lambda(\tfrac{q}{m}) = s(\tfrac{q}{m}) \otimes \hat{s}^{-1}(\tfrac{q}{m}) : \quad \mathrm{e}(\vec{q})^{0,a} \hookrightarrow \Lambda(\tfrac{q}{m})^j_{0,a}\mathrm{e}(\vec{q})^{0,a},\\ & \qquad\qquad j = 0,1,2,3,\ a = 1,2,3. \end{aligned}$$

The energy-momenta of a massive particle give projector decompositions of the identity into $\mathbf{SO}(3)$-nondecomposable projectors

$$\mathbf{1}_4 = \mathcal{P}_{[0]}(q) + \mathcal{P}_{[2]}(q), \quad \delta^j_k = \tfrac{q^j q_k}{m^2} + (\delta^j_k - \tfrac{q^j q_k}{m^2}), \quad q^2 = m^2 > 0.$$

In general, the embedding of spin J particles into a relativistic field is not unique since spin J-representations come in all induced irreducible finite-dimensional Lorentz group representations. The "minimal" Lorentz group representation for a given spin J has "left and right spin" L and R as "close as possible" to each other:

$$J = L + R \text{ and } |L - R| = \begin{cases} 0 & \text{for spin } J = 0, 1, \dots,\\ \frac{1}{2} & \text{for spin } J = \frac{1}{2}, \frac{3}{2}, \dots, \end{cases}$$

This gives equal left-right spin for the embedding of integer spin J particles and the difference $\frac{1}{2}$ for half-integer spin

$$\begin{aligned} &\text{minimal: } \mathbf{irrep\,SU}(2) \hookrightarrow \mathbf{irrep}_{\,\mathrm{fin}}\mathbf{SL}(\mathbb{C}^2),\\ &[2J] \hookrightarrow \begin{cases} [J|J] & \text{for } J = 0, 1, \dots,\\ [\underset{2L}{J + \frac{1}{2}}|J - \frac{1}{2}] \oplus [J - \frac{1}{2}|\underset{2R}{J + \frac{1}{2}}] & \text{for } J = \frac{1}{2}, \frac{3}{2} \dots, \end{cases}\\ &\text{with } [2L|2R](\tfrac{q}{m}) = \bigvee s(\tfrac{q}{m}) \otimes \bigvee \hat{s}(\tfrac{q}{m}). \end{aligned}$$

The examples $J = \frac{1}{2}, 1$ are given above. For half-integer spin the $\times$-symmetric sum of two conjugated representations has to be used (chapter "Lorentz Symmetry").

In the quantizations and Fock forms of the particle embedding fields the unit $\mathbf{1}_{2J+1}$ of the particle $\mathbf{SU}(2)$-representation space is embedded as a scalar unit for the Lorentz group or as the timelike part of a Lorentz vector. For example, for a spin $\frac{1}{2}$ particle, relevant for a Dirac field, the unit $\mathbf{1}_2$ from the rest system is embedded into Lorentz scalars $\delta^B_A, \delta^{\dot{A}}_{\dot{B}}$ and into Lorentz vectors $\frac{(\sigma^j)^{\dot{A}}_B q_j}{m}$ and $\frac{(\check{\sigma}^j)^A_{\dot{B}} q_j}{m}$:

$$\begin{aligned} [1] \hookrightarrow [1|0] \oplus [0|1] \text{ with } & s(\tfrac{q}{m}), \hat{s}(\tfrac{q}{m}) = s^{-1\star}(\tfrac{q}{m})\\ \Rightarrow & \begin{pmatrix} ss^{-1} & ss^\star \\ \hat{s}s^{-1} & \hat{s}\hat{s}^{-1} \end{pmatrix}(\tfrac{q}{m}) = \begin{pmatrix} \mathbf{1}_2 & \frac{\sigma^j q_j}{m} \\ \frac{\check{\sigma}^j q_j}{m} & \mathbf{1}_2 \end{pmatrix}. \end{aligned}$$

For a spin-1 particle the metrical tensor $\mathbf{1}_3 \cong \delta^{ab}$ is embedded into a Lorentz tensor:

$$[2] \hookrightarrow [1|1] \text{ with } s(\tfrac{q}{m}) \otimes \hat{s}(\tfrac{q}{m}) = \Lambda(\tfrac{q}{m}) \ \Rightarrow \Lambda(\tfrac{q}{m})^i_a \delta^{ab} \Lambda(\tfrac{q}{m})^j_b = -\eta^{ij} + \tfrac{q^i q^j}{m^2}.$$

4.5 Massive Spin-0 Particle Fields

Scalar fields for massive spin-0 particles embed compact translation representations with trivial spin. An example is the neutral π^0-meson, considered as stable.

For Hermitian scalar Bose fields, particles and antiparticles coincide:

$$\begin{matrix} m > 0 \\ J = 0 \\ z = 0 \\ \epsilon = -1 \end{matrix} : \left\{ \begin{array}{rcl} \mathbf{\Phi}(x) &=& {}^{\oplus}\!\int \frac{d^3q}{2q_0(2\pi)^3} \; \gamma \; [e^{iqx}\mathrm{u}(\vec q) + e^{-iqx}\mathrm{u}^\star(\vec q)], \\ -i\mathbf{\Phi}(x)_k &=& {}^{\oplus}\!\int \frac{d^3q}{2q_0(2\pi)^3} \; \frac{m}{\gamma} \; [e^{iqx}\mathrm{u}(\vec q) - e^{-iqx}\mathrm{u}^\star(\vec q)] \quad \Lambda(\frac{q}{m})^0_k, \\ \text{with } q_0 &=& \sqrt{m^2 + \vec q^{\,2}}. \end{array} \right.$$

$\Lambda(\frac{q}{m})^0_k = \frac{q_k}{m}$ is a transmutator embedding an $\mathbf{SO}(3)$ scalar representation $[0]$ into a Lorentz vector representation $[1|1]$. Scalar fields have the Lorentz behavior

$$\Lambda \in \mathbf{SO}_0(1,3), \quad \mathbf{\Phi}_\Lambda(x) = \mathbf{\Phi}(\Lambda^{-1}.x).$$

The dual normalization has been chosen in such a way that the limit $m \to 0$ is finite $m\Lambda(\frac{q}{m})^0_k \to q_k$ (chapter "Massless Quantum Fields").

Scalar particle fields embed $\vec q$-indexed harmonic Bose oscillators: The Hermitian scalar field is the simplest relativistic distribution of an irreducible time representation orbit with Euclidean conjugation $\star$

$$\mathbb{R} \ni t \longmapsto e^{imt} \in \mathbf{U}(1) \text{ with } \mathrm{u} \overset{\star}{\leftrightarrow} \mathrm{u}^\star$$

as given by the quantum Bose oscillator with position-momentum $(\mathbf{x}\ \mathbf{p})$, inertial mass $M > 0$, and spring constant $k > 0$, i.e., frequency m with $m^2 = \frac{k}{M}$ and intrinsic length ℓ with $\ell^4 = \frac{1}{kM} = \frac{1}{m^2M^2}$:

$$\begin{array}{rlrl} \int d^3x \; \mathbf{\Phi}(x) &= \frac{\gamma}{m} \frac{e^{imx_0}\mathrm{u}(0) + e^{-imx_0}\mathrm{u}^\star(0)}{2}, & \mathbf{x}(t) &= \ell \frac{e^{imt}\mathrm{u} + e^{-imt}\mathrm{u}^\star}{\sqrt 2}, \\ \int d^3x \; \mathbf{\Phi}(x)_k &= \delta^0_k \frac{i}{\gamma} \frac{e^{imx_0}\mathrm{u}(0) - e^{-imx_0}\mathrm{u}^\star(0)}{2}, & \mathbf{p}(t) &= \frac{i}{\ell} \frac{e^{imt}\mathrm{u} - e^{-imt}\mathrm{u}^\star}{\sqrt 2}. \end{array}$$

The basic operators are the starting points of the translation orbits

$$\begin{array}{rcll} \mathbf{\Phi}(0) &=& {}^{\oplus}\!\int \frac{d^3q}{2q_0(2\pi)^3} \gamma \; [\mathrm{u}(\vec q) + \mathrm{u}^\star(\vec q)], & \mathbf{x}(0) = \ell \frac{\mathrm{u} + \mathrm{u}^\star}{\sqrt 2}, \\ \mathbf{\Phi}(0)_k &=& {}^{\oplus}\!\int \frac{d^3q}{2q_0(2\pi)^3} \frac{iq_k}{\gamma} \; [\mathrm{u}(\vec q) - \mathrm{u}^\star(\vec q)], & \mathbf{p}(0) = \frac{i}{\ell} \frac{\mathrm{u} - \mathrm{u}^\star}{\sqrt 2}. \end{array}$$

The field quantization commutator is the vector cosine with the origin-supported distribution for the equal time quantization:

$$\text{oscillator:} \left\{ \begin{array}{rl} [\mathrm{u}^\star, \mathrm{u}] &= 1, \\ [i\mathbf{p}, \mathbf{x}](t) &= \cos mt = \frac{d_t}{i\ell^2 m}[\mathbf{x}, \mathbf{x}](t) \\ &= 1 \text{ for } t = 0, \\ [\mathbf{x}, \mathbf{x}](t) &= \ell^2 i \sin mt, \end{array} \right.$$

$$\text{scalar field:} \left\{ \begin{array}{rl} [\mathrm{u}^\star(\vec p), \mathrm{u}(\vec q)] &= 2q_0\delta(\frac{\vec q - \vec p}{2\pi}), \\ [i\mathbf{\Phi}_k, \mathbf{\Phi}](x) &= \mathbf{c}_k(m|x) = \frac{\partial_k}{i\gamma^2}[\mathbf{\Phi}, \mathbf{\Phi}](x) \\ &= \delta^0_k \delta(\vec x) \text{ for } x_0 = 0, \\ [\mathbf{\Phi}, \mathbf{\Phi}](x) &= \gamma^2 \frac{i\mathbf{s}(m|x)}{m} = \gamma^2 \int \frac{d^4q}{(2\pi)^3} \epsilon(q_0)\delta(q^2 - m^2) e^{iqx} \\ &= \gamma^2 \int \frac{d^3q}{q_0(2\pi)^3} \; e^{-i\vec q\vec x} i \sin q_0 x_0 \Big|_{q_0 = \sqrt{m^2 + \vec q^{\,2}}} \\ &= 0 \text{ for spacelike } x^2 < 0. \end{array} \right.$$

The quantization commutator has causal support and is not a function $2\pi\frac{\mathbf{s}(m|x)}{m} = \epsilon(x_0)\delta(x^2) + \cdots$.

The anticommutator Fock form contains the Hilbert space scalar product. It is a function of the spacetime translations (coefficient of a Poincaré group representation)

$$\text{oscillator: } \begin{cases} \langle\{\mathrm{u}^\star, \mathrm{u}\}\rangle_{\mathrm{F}} &= \langle \mathrm{u}^\star \mathrm{u}\rangle_{\mathrm{F}} = 1, \\ \langle\{\mathbf{x}, \mathbf{x}\}\rangle_{\mathrm{F}}(t) &= \ell^2 \cos mt, \end{cases}$$

$$\text{scalar field: } \begin{cases} \langle\{\mathrm{u}^\star(\vec{p}), \mathrm{u}(\vec{q})\}\rangle_{\mathrm{F}} &= \langle \mathrm{u}^\star(\vec{p})\mathrm{u}(\vec{q})\rangle_{\mathrm{F}} = \langle\vec{p}|\vec{q}\rangle = 2q_0\delta(\frac{\vec{q}-\vec{p}}{2\pi}), \\ \langle\{\mathbf{\Phi}, \mathbf{\Phi}\}\rangle_{\mathrm{F}}(x) &= \gamma^2\frac{\mathbf{C}(m|x)}{m} = \gamma^2\int\frac{d^4q}{(2\pi)^3}\delta(q^2-m^2)e^{iqx} \\ &= \gamma^2\int\frac{d^3q}{q_0(2\pi)^3}\, e^{-i\vec{q}\vec{x}}\cos q_0x_0\Big|_{q_0=\sqrt{m^2+\vec{q}^2}}. \end{cases}$$

The intrinsic length unit for the oscillator is given by the Fock scalar product $2\langle\mathbf{x}|\mathbf{x}\rangle_{\mathrm{F}} = \ell^2$.

The field commutator and the Fock form of the anticommutator involve on-shell contributions with $\delta(q^2-m^2)$ ("real particles") with $r = 0$-regular distributions of compact representation coefficients of position translations

$$\begin{aligned} \begin{pmatrix} \langle\{\mathbf{\Phi}(y), \mathbf{\Phi}(x)\}\rangle_{\mathrm{F}} \\ [\mathbf{\Phi}(y), \mathbf{\Phi}(x)] \end{pmatrix} &= \gamma^2 \oplus\!\!\int \frac{d^3q d^3p}{4q_0p_0(2\pi)^6}\Big[\begin{pmatrix} \langle\vec{p}|\vec{q}\rangle \\ [\mathrm{u}^\star(\vec{p}), \mathrm{u}(\vec{q})] \end{pmatrix} e^{i(qx-py)} + \begin{pmatrix} \langle\vec{q}|\vec{p}\rangle \\ -[\mathrm{u}^\star(\vec{q}), \mathrm{u}(\vec{p})] \end{pmatrix} e^{-i(qx-py)}\Big] \\ &= \gamma^2\int\frac{dq_0}{(2\pi)^2}\begin{pmatrix} 1 \\ \epsilon(q_0) \end{pmatrix}\vartheta(q_0^2-m^2)e^{iq_0(x_0-y_0)}\,\frac{\sin|\vec{q}||\vec{x}-\vec{y}|}{|\vec{x}-\vec{y}|}\Big|_{|\vec{q}|=\sqrt{q_0^2-m^2}}. \end{aligned}$$

The coefficients for the Poincaré group $\mathbf{SO}_0(1,3)\ \vec{\times}\ \mathbb{R}^4$ embed the coefficients for the Euclidean group $\mathbf{SO}(3)\ \vec{\times}\ \mathbb{R}^3$ for scattering waves (chapter "The Kepler Factor").

The relativistic distribution of the causal time representations gives the Feynman propagator

$$\begin{aligned} \text{oscillator: } \langle\{\mathbf{x}, \mathbf{x}\}(t) - \epsilon(t)[\mathbf{x}, \mathbf{x}](t)\rangle_{\mathrm{F}} &= \ell^2[\cos mt - \epsilon(t)i\sin mt] \\ &= \ell^2 e^{-im|t|}, \\ \text{scalar field: } \langle\{\mathbf{\Phi}, \mathbf{\Phi}\}(x) - \epsilon(x_0)[\mathbf{\Phi}, \mathbf{\Phi}](x)\rangle_{\mathrm{F}} &= \gamma^2\frac{\mathbf{C}(m|x) - \epsilon(x_0)i\mathbf{s}(m|x)}{m} \\ = \frac{i}{\pi}\int\frac{d^4q}{(2\pi)^3}\frac{\gamma^2}{q^2+io-m^2}e^{iqx} &= \gamma^2\frac{\mathbf{E}(-im|x)}{m} \\ = -i\gamma^2\int\frac{dq_0}{(2\pi)^2}e^{iq_0x_0}\frac{\vartheta(q_0^2-m^2)\ e^{i|\vec{q}|r} + \vartheta(m^2-q_0^2)\ e^{-|Q|r}}{r} &= \gamma^2\int\frac{d^3q}{q_0(2\pi)^3}\, e^{-iq_0|x_0| - i\vec{q}\vec{x}}, \\ \text{with } |\vec{q}| = \sqrt{q_0^2-m^2} &\quad \text{and } |Q| = \sqrt{m^2-q_0^2}. \end{aligned}$$

Here on- and off-shell contributions are added:

$$\frac{i}{\pi}\frac{\gamma^2}{q^2+io-m^2} = \gamma^2\delta(q^2-m^2) + \frac{i}{\pi}\frac{\gamma^2}{q_{\mathrm{P}}^2-m^2}.$$

The principal value distribution with the off-shell contributions does not lead to Poincaré group representation coefficients. The Yukawa potential and forces involved are attributed to "virtual particles" with imaginary "momenta" (binding energy) and $r = 0$-singular distributions of noncompact position representation coefficients

$$\begin{aligned} \epsilon(x_0)[\mathbf{\Phi}, \mathbf{\Phi}](x) &= \int\frac{d^4q}{(2\pi)^3}\frac{i}{\pi}\frac{\gamma^2}{-q_{\mathrm{P}}^2+m^2}e^{iqx} \\ &= i\gamma^2\int\frac{dq_0}{(2\pi)^2}e^{iq_0x_0}\frac{\vartheta(q_0^2-m^2)\ \cos|\vec{q}|r + \vartheta(m^2-q_0^2)e^{-|Q|r}}{r}. \end{aligned}$$

The Yukawa potentials $\frac{e^{-|Q|r}}{r}$ differ by the Kepler factor, with $r = 0$-singularity, from the bound waves in quantum mechanics. The Feynman propagator has an explicit translation dependence via the order factor $\epsilon(x_0)$, which is not implemented by the field.

The free dual normalization factor γ^2 is introduced in such a way that it gives the normalization of the pole at the particle mass. Free massive Bose fields can be renormalized, e.g., with $\gamma^2 = 1$. For interacting theories γ^2 can be related to the coupling constant of the interaction via $\mathbf{\Phi}$.

The property to be coefficients of compact translation representations, time $\mathbb{R}$ or spacetime $\mathbb{R}^4$, is expressed by homogeneous free equations of motion. They are derivable from a classical Lagrangian function and Lagrangian density:

$$\text{oscillator:} \left\{ \begin{array}{l} L(\mathbf{x},\mathbf{p}) = \mathbf{p}d_t\mathbf{x} - \frac{m}{2}(\ell^2\mathbf{p}^2 + \frac{\mathbf{x}^2}{\ell^2}) \\ \Rightarrow \left\{ \begin{array}{ll} (d_t\frac{\partial}{\partial d_t\mathbf{x}} - \frac{\partial}{\partial \mathbf{x}})L(\mathbf{x},\mathbf{p}) = 0 & \Rightarrow d_t\mathbf{x} = \ell^2 m\mathbf{p}, \\ (d_t\frac{\partial}{\partial d_t\mathbf{p}} - \frac{\partial}{\partial \mathbf{p}})L(\mathbf{x},\mathbf{p}) = 0 & \Rightarrow d_t\mathbf{p} = -\frac{m}{\ell^2}\mathbf{x}, \end{array}\right. \end{array}\right.$$

$$\text{scalar field:} \left\{ \begin{array}{l} \mathbf{L}(\mathbf{\Phi},\mathbf{\Phi}_k) = \mathbf{\Phi}_k\partial^k\mathbf{\Phi} - \frac{1}{2}(\gamma^2\mathbf{\Phi}_k\mathbf{\Phi}^k + m^2\frac{\mathbf{\Phi}^2}{\gamma^2}) \\ \Rightarrow \left\{ \begin{array}{ll} (\partial_k\frac{\partial}{\partial\partial_k\mathbf{\Phi}} - \frac{\partial}{\partial\mathbf{\Phi}})\mathbf{L}(\mathbf{\Phi},\mathbf{\Phi}_k) = 0 & \Rightarrow \partial^k\mathbf{\Phi} = \gamma^2\mathbf{\Phi}^k, \\ (\partial_k\frac{\partial}{\partial\partial_k\mathbf{\Phi}_j} - \frac{\partial}{\partial\mathbf{\Phi}_j})\mathbf{L}(\mathbf{\Phi},\mathbf{\Phi}_k) = 0 & \Rightarrow \partial^k\mathbf{\Phi}_k = -\frac{m^2}{\gamma^2}\mathbf{\Phi}. \end{array}\right. \end{array}\right.$$

A second order derivative formalism uses

$$\begin{array}{ll} L(\mathbf{x}) = (d_t\mathbf{x})^2 - m^2\frac{\mathbf{x}^2}{2} & \Rightarrow (d_t^2 + m^2)\mathbf{x} = 0, \\ \mathbf{L}(\mathbf{\Phi}) = (\partial_k\mathbf{\Phi})^2 - m^2\frac{\mathbf{\Phi}^2}{2} & \Rightarrow (\partial^2 + m^2)\mathbf{\Phi} = 0. \end{array}$$

The on-shell quantization commutator and Fock value anticommutator obey homogeneous Klein-Gordon equations (free field equations), the causally ordered contribution in the Feyman propagator an inhomogenous one. The "virtual particles" (off-shell) that induce the Yukawa interactions are not free:

$$(\partial^2 + m^2)\begin{pmatrix} [\mathbf{\Phi},\mathbf{\Phi}](x) \\ \langle\{\mathbf{\Phi},\mathbf{\Phi}\}(x)\rangle_{\mathrm{F}} \end{pmatrix} = 0, \quad (\partial^2 + m^2)\epsilon(x_0)[\mathbf{\Phi},\mathbf{\Phi}](x) = 2i\gamma^2\delta(x).$$

For different particles and antiparticles with translation $\mathbb{R}^4$ and internal charge group $\mathbf{U}(1)$ representations

$$\mathbb{R}^4 \times \mathbf{U}(1) \ni (x, e^{i\alpha}) \longmapsto e^{iqx}\begin{pmatrix} e^{iz\alpha} & 0 \\ 0 & e^{-iz\alpha} \end{pmatrix} \in \mathbf{U}(1) \times \mathbf{U}(1)_3,$$

the full complex quartet is necessary. The spinless case gives a *non-Hermitian scalar Bose field*

$$\begin{array}{l} m > 0 \\ J = 0 \\ z = \pm 1 \\ \epsilon = -1 \end{array} : \left\{ \begin{array}{rcll} \mathbf{\Phi}(x) & = & {}^{\oplus}\!\!\int \frac{d^3q}{2q_0(2\pi)^3} & \gamma\ [e^{iqx}\mathrm{u}(\vec{q}) + e^{-iqx}\mathrm{a}^\star(\vec{q})], \\ -i\mathbf{\Phi}(x)_k & = & {}^{\oplus}\!\!\int \frac{d^3q}{2q_0(2\pi)^3} & \frac{q_k}{\gamma}\ [e^{iqx}\mathrm{u}(\vec{q}) - e^{-iqx}\mathrm{a}^\star(\vec{q})], \\ \mathbf{\Phi}^\star(x) & = & {}^{\oplus}\!\!\int \frac{d^3q}{2q_0(2\pi)^3} & \gamma\ [e^{-iqx}\mathrm{u}^\star(\vec{q}) + e^{iqx}\mathrm{a}(\vec{q})], \\ i\mathbf{\Phi}^\star(x)_k & = & {}^{\oplus}\!\!\int \frac{d^3q}{2q_0(2\pi)^3} & \frac{q_k}{\gamma}\ [e^{-iqx}\mathrm{u}^\star(\vec{q}) - e^{iqx}\mathrm{a}(\vec{q})], \\ \text{with } q_0 & = & \sqrt{m^2 + \vec{q}^2}. & \end{array}\right.$$

Examples are the charged $\pi^\pm$-mesons, considered as stable. Non-Hermitian scalar particle fields embed $\vec{q}$-indexed harmonic Bose oscillators with a charge.

They have as quantization commutators and anticommutator Fock forms

$$\begin{pmatrix} [i\mathbf{\Phi}_k^\star,\mathbf{\Phi}] & [\mathbf{\Phi}^\star,\mathbf{\Phi}] \\ [\mathbf{\Phi}_k^\star,\mathbf{\Phi}_j] & [\mathbf{\Phi}^\star,-i\mathbf{\Phi}_j]\end{pmatrix}(x) = (-i\partial)\,\frac{i\mathbf{s}(m|x)}{m} = \int \frac{d^4q}{(2\pi)^3}(\mathbf{q})\epsilon(q_0)\delta(q^2-m^2)e^{iqx},$$

$$\begin{pmatrix} \langle\{i\mathbf{\Phi}_k^\star,\mathbf{\Phi}\}\rangle_{\mathrm{F}} & \langle\{\mathbf{\Phi}^\star,\mathbf{\Phi}\}\rangle_{\mathrm{F}} \\ \langle\{\mathbf{\Phi}_k^\star,\mathbf{\Phi}_j\}\rangle_{\mathrm{F}} & \langle\{\mathbf{\Phi}^\star,-i\mathbf{\Phi}_j\}\rangle_{\mathrm{F}}\end{pmatrix}(x) = (-i\partial)\frac{\mathbf{C}(m|x)}{m} = \int \frac{d^4q}{(2\pi)^3}(\mathbf{q})\delta(q^2-m^2)e^{iqx},$$

$$\text{with } (-i\partial) = \begin{pmatrix} -i\partial_k & \gamma^2 \\ -\frac{\partial_k\partial_j}{\gamma^2} & -i\partial_j \end{pmatrix},\quad (\mathbf{q}) = \begin{pmatrix} q_k & \gamma^2 \\ \frac{q_kq_j}{\gamma^2} & q_j \end{pmatrix},$$

and Feynman propagators

$$\langle \begin{pmatrix} \{i\mathbf{\Phi}_k^\star,\mathbf{\Phi}\} & \{\mathbf{\Phi}^\star,\mathbf{\Phi}\} \\ \{\mathbf{\Phi}_k^\star,\mathbf{\Phi}_j\} & \{\mathbf{\Phi}^\star,-i\mathbf{\Phi}_j\}\end{pmatrix}(x) - \epsilon(x_0)\begin{pmatrix} [i\mathbf{\Phi}_k^\star,\mathbf{\Phi}] & [\mathbf{\Phi}^\star,\mathbf{\Phi}] \\ [\mathbf{\Phi}_k^\star,\mathbf{\Phi}_j] & [\mathbf{\Phi}^\star,-i\mathbf{\Phi}_j]\end{pmatrix}(x)\rangle_{\mathrm{F}}$$

$$= \frac{i}{\pi}\int \frac{d^4q}{(2\pi)^3}\frac{(\mathbf{q})}{q^2+io-m^2}e^{iqx}.$$

A Lagrangian density for classical fields reads

$$\mathbf{L}(\mathbf{\Phi},\mathbf{\Phi}^\star,\mathbf{\Phi}_k,\mathbf{\Phi}_k^\star) = \mathbf{\Phi}_k^\star\partial^k\mathbf{\Phi} + \mathbf{\Phi}_k\partial^k\mathbf{\Phi}^\star - (\gamma^2\mathbf{\Phi}_k^\star\mathbf{\Phi}^k + m^2\frac{\mathbf{\Phi}^\star\mathbf{\Phi}}{\gamma^2}).$$

For scalar particles $\{\mathrm{u}^\alpha,\mathrm{a}^{\star\alpha},\mathrm{u}_\alpha^\star,\mathrm{a}_\alpha\}_{\alpha=1,2,\ldots}$ with internal degrees of freedom the relativistic embedding is unchanged, one has to insert only additional indices, e.g., $\mathbf{\Phi}^\alpha,\mathbf{\Phi}_\alpha^\star$. If there exists an isomorphism ζ between dual and antispace, the identification of particles and antiparticles is possible:

$$\epsilon = -1:\quad \mathrm{a}(\vec{q})_\alpha = \zeta_{\alpha\beta}\mathrm{u}(\vec{q})^\beta,\quad \mathbf{\Phi}^\star(x)_\alpha = \zeta_{\alpha\beta}\mathbf{\Phi}(x)^\beta.$$

Two examples: The π-meson triplet $(\pi^0,\pi^\pm)$, taken above as example for Hermitian and non-Hermitian fields, can also be formulated with a triplet representation [2] of internal $\mathbf{SU}(2)$-isospin $\{\mathbf{\Phi}^\alpha\}_{\alpha=1,2,3}$. Here $\mathbf{\Phi}^3$ is a Hermitian scalar field and $\mathbf{\Phi}^{1,2}$ are given by the Hermitian combinations $(\frac{\mathbf{\Phi}+\mathbf{\Phi}^\star}{2},\frac{\mathbf{\Phi}-\mathbf{\Phi}^\star}{2i})$ of a non-Hermitian scalar field. And the quartet of K-mesons (K^+,K^0) and $(\overline{K}^0,K^-)$ with conjugated Pauli doublet representations $[\pm\frac{1}{2}||1]$ of internal $\mathbf{U}(2)$-hyperisospin is embedded into two non-Hermitian relativistic scalar fields $\{\mathbf{\Phi}^\alpha,\mathbf{\Phi}_\alpha^\star\}_{\alpha=1,2}$. Here the Cartan subgroup $\mathbf{U}(1)_+\subset\mathbf{U}(2)$ describes the electromagnetic action, it takes into account the central correlation $\mathbb{I}(2)$ of hypercharge $\mathbf{U}(\mathbf{1}_2)$ and isospin $\mathbf{SU}(2)$ (chapter "Rational Quantum Numbers" and "Gauge Interactions").

4.6 Massive Spin-1 Particle Fields

For massive spin 1 particles with triplet $\mathbf{SU}(2)$-representation [2], i.e., for an $\mathbf{SO}(3)$-vector, e.g., the neutral weak Z-boson or the charged weak $W^\pm$-bosons, considered as stable particles, the fixgroup $\mathbf{SO}(3)$-representations are embedded by rectangular (3×4) transmutators from Lorentz group to rotation groups. One takes the last three columns from the 4×4 matrix $\Lambda(\frac{q}{m})$ with the decomposition $[1|1] \cong [0] \oplus [2]$:

$$\Lambda(\tfrac{q}{m}) \cong \tfrac{1}{m}\left(\begin{array}{c|c} q_0 & q_a \\ q_b & \delta_{ab}m + \frac{q_a q_b}{m+q_0} \end{array}\right) \in \mathbf{SO}_0(1,3)/\mathbf{SO}(3)$$
$$\text{with } q_0 = \sqrt{m^2 + \vec{q}^2}.$$

Also, the antisymmetric product representation $[1|1] \wedge [1|1] = [2|0] \oplus [0|2]$ contains an $\mathbf{SO}(3)$-vector, $[2|0] \cong [2]$:

$$\text{spin } 1: \begin{cases} \Lambda(\frac{q}{m})^j_a \subset [1|1](\frac{q}{m}), \quad a = 1,2,3; \; j = 0,1,2,3, \\ \Lambda(\frac{q}{m})^l_0 \epsilon^{kj}_{lr} \Lambda(\frac{q}{m})^r_a \subset [2|0](\frac{q}{m}) \oplus [0|2](\frac{q}{m}), \\ \epsilon^{kj}_{lr} = (\delta \wedge \delta)^{kj}_{lr} = \delta^k_l \delta^j_r - \delta^j_l \delta^k_r \text{ (Clebsch-Gordan coefficients)}. \end{cases}$$

The $\mathbf{SO}(3)$-units are transmuted into the Lorentz compatible projectors for spin 1 and spin 0, for the metrical tensors:

$$\Lambda(\tfrac{q}{m})^k_a \delta^{ab} \Lambda(\tfrac{q}{m})^j_b = -\eta^{kj} + \tfrac{q^k q^j}{m^2}, \quad \Lambda(\tfrac{q}{m})^k_0 \Lambda(\tfrac{q}{m})^j_0 = \tfrac{q^k q^j}{m^2}.$$

For neutral coinciding particles and antiparticles, one has for Bose fields

$$\begin{matrix} m > 0 \\ J = 1 \\ z = 0 \\ \epsilon = -1 \end{matrix} : \begin{cases} \mathbf{Z}(x)^j &= {}^{\oplus}\!\int \frac{d^3q}{2q_0(2\pi)^3} & \gamma \, [e^{iqx}\mathrm{u}(\vec{q})^a + e^{-iqx}\mathrm{u}^\star(\vec{q})^a] & \Lambda(\frac{q}{m})^j_a, \\ i\mathbf{G}(x)^{kj} &= {}^{\oplus}\!\int \frac{d^3q}{2q_0(2\pi)^3} & \frac{m}{\gamma} \, [e^{iqx}\mathrm{u}(\vec{q})^a - e^{-iqx}\mathrm{u}^\star(\vec{q})^a] & \Lambda(\frac{q}{m})^l_0 \epsilon^{kj}_{lr} \Lambda(\frac{q}{m})^r_a, \\ \text{with } q_0 &= \sqrt{m^2 + \vec{q}^2}, \end{cases}$$

with the translation representations

$$\mathbb{R}^4 \ni x \longmapsto e^{iqx}\mathbf{1}_3 \in \mathbf{U}(\mathbf{1}_3) \text{ with } \mathrm{u}^a \overset{\star}{\leftrightarrow} \mathrm{u}^{\star a}.$$

The direct integrals $\mathbf{Z}(0)^j$ and $\mathbf{G}(0)^{kj}$ are $\mathbf{SO}(3)$-intertwiners on $\mathbf{SO}_0(1,3)$. Hermitian vector particle fields embed a spin triplet of $\vec{q}$-indexed harmonic Bose oscillators:

$$\begin{aligned} \int d^3x \; \mathbf{Z}(x)^j &= \delta^j_a \tfrac{\gamma}{m} \tfrac{e^{imx_0}\mathrm{u}(0)^a + e^{-imx_0}\mathrm{u}^\star(0)^a}{2}, \\ \int d^3x \; \mathbf{G}(x)^{kj} &= \epsilon^{kj}_{a0} \tfrac{i}{\gamma} \tfrac{e^{imx_0}\mathrm{u}(0)^a - e^{-imx_0}\mathrm{u}^\star(0)^a}{2}. \end{aligned}$$

Like for the quantum-mechanical 3-dimensional isotropic harmonic oscillator, the translation and spin group for the momentum operators are represented in $\mathbf{U}(3)$ as a subgroup of $\mathbf{U}(1,3)$, which arises by a complex embedding of the Lorentz group

$$\mathbb{R}^4 \times \mathbf{SO}(3) \longrightarrow \mathbf{U}(3), \quad \mathbf{U}(3) \subset \mathbf{U}(1,3) \supset \mathbf{SO}_0(1,3).$$

The equations of motion for the translation orbits and the Lorentz properties are as follows:

$$\begin{aligned} &\epsilon^{jk}_{lr} \partial^l \mathbf{Z}^r = \partial^j \mathbf{Z}^k - \partial^k \mathbf{Z}^j = \gamma^2 \mathbf{G}^{kj}, \\ &\partial_k \mathbf{G}^{jk} = -\tfrac{m^2}{\gamma^2} \mathbf{Z}^j, \\ &\Lambda \in \mathbf{SO}_0(1,3): \begin{cases} \mathbf{Z}_\Lambda(x)^j &= \Lambda^j_k \mathbf{Z}(\Lambda^{-1}.x)^k, \\ \mathbf{G}_\Lambda(x)^{kj} &= \Lambda^k_l \Lambda^j_r \mathbf{G}(\Lambda^{-1}.x)^{lr}. \end{cases} \end{aligned}$$

Since the four columns of $\Lambda(\frac{q}{m})^j_k$ are a Sylvester basis, the divergence of the field vanishes:

$$\Lambda(\tfrac{q}{m})^k_0 \eta_{kj} \Lambda(\tfrac{q}{m})^j_a = 0 \iff q_j \Lambda(\tfrac{q}{m})^j_a = 0 \Rightarrow \partial_j \mathbf{Z}^j = 0.$$

A classical Lagrangian density reads

$$\begin{array}{ll} \mathbf{L}(\mathbf{Z}^j, \mathbf{G}^{jk}) & = \frac{1}{2}\mathbf{G}^{jk}\epsilon^{lm}_{jk}\partial_l \mathbf{Z}_m + (\gamma^2 \frac{\mathbf{G}^{jk}\mathbf{G}_{jk}}{4} + m^2 \frac{\mathbf{Z}^j \mathbf{Z}_j}{2\gamma^2}), \\ \mathbf{L}(\mathbf{Z}^j) & = \frac{1}{2}\epsilon^{lm}_{jk}(\partial^j \mathbf{Z}^k)(\partial_l \mathbf{Z}_m) + m^2 \frac{\mathbf{Z}^j \mathbf{Z}_j}{2}. \end{array}$$

The quantization commutators and in the anticommutator Fock forms

$$\begin{array}{rl} \begin{pmatrix} [i\mathbf{G}^{kl}, \mathbf{Z}^j] & [\mathbf{Z}^k, \mathbf{Z}^j], \\ [\mathbf{G}^{kl}, \mathbf{G}^{jm}] & [\mathbf{Z}^k, -i\mathbf{G}^{jm}] \end{pmatrix}(x) & = (-i\partial)\frac{[\mathbf{Z}^t, \mathbf{Z}^s](x)}{\gamma^2}, \\ \begin{pmatrix} \langle\{i\mathbf{G}^{kl}, \mathbf{Z}^j\}\rangle_{\mathrm{F}} & \langle\{\mathbf{Z}^k, \mathbf{Z}^j\}\rangle_{\mathrm{F}} \\ \langle\{\mathbf{G}^{kl}, \mathbf{G}^{jm}\}\rangle_{\mathrm{F}} & \langle\{\mathbf{Z}^k, -i\mathbf{G}^{jm}\}\rangle_{\mathrm{F}} \end{pmatrix}(x) & = (-i\partial)\frac{\langle\{\mathbf{Z}^t, \mathbf{Z}^s\}\rangle_{\mathrm{F}}(x)}{\gamma^2}, \\ (-i\partial) & = \begin{pmatrix} -i\epsilon^{kl}_{tu}\delta^j_s\partial^u & \gamma^2\delta^k_t\delta^j_s \\ -\epsilon^{kl}_{tu}\epsilon^{jm}_{sr}\frac{\partial^r\partial^u}{\gamma^2} & -i\delta^k_t\epsilon^{jm}_{sr}\partial^r \end{pmatrix}. \end{array}$$

can be computed from

$$\begin{array}{rl} \begin{pmatrix} [\mathbf{Z}^k, \mathbf{Z}^j] \\ \langle\{\mathbf{Z}^k, \mathbf{Z}^j\}\rangle_{\mathrm{F}} \end{pmatrix}(x) & = -\gamma^2(\eta^{kj} + \frac{\partial^k\partial^j}{m^2})\begin{pmatrix} \frac{i\mathbf{s}(m|x)}{m} \\ \frac{\mathbf{C}(m|x)}{m} \end{pmatrix} \\ & = \gamma^2 \int \frac{d^4q}{(2\pi)^3}\begin{pmatrix} \epsilon(q_0) \\ 1 \end{pmatrix}(-\eta^{kj} + \frac{q^k q^j}{m^2})\delta(q^2 - m^2)e^{iqx} \\ & = \gamma^2 \int \frac{d^3q}{q_0(2\pi)^3}\, e^{-i\vec{x}\vec{q}}\, \Lambda(\frac{q}{m})^k_a\, \delta^{ab}\begin{pmatrix} i\sin x_0 q_0 \\ \cos q_0 x_0 \end{pmatrix}\Lambda(\frac{q}{m})^j_b, \end{array}$$

and the Feynman propagators from

$$\langle\{\mathbf{Z}^k, \mathbf{Z}^j\}(x) - \epsilon(x_0)[\mathbf{Z}^k, \mathbf{Z}^j](x)\rangle_{\mathrm{F}} = \frac{i}{\pi}\int \frac{d^4q}{(2\pi)^3}\frac{\gamma^2(-\eta^{kj} + \frac{q^k q^j}{m^2})}{q^2 + io - m^2}e^{iqx}.$$

4.7 Massive Spin-$\frac{1}{2}$ Dirac Particle Fields

To embed Pauli $\mathbf{SU}(2)$-spinors for massive spin-$\frac{1}{2}$ particles into $\mathbf{SL}(\mathbb{C}^2)$-representations with the left and right Weyl matrices $\sigma^j \cong (\mathbf{1}_2, \vec{\sigma}) \cong \check{\sigma}_j$, one uses the Weyl transmutators for $q^2 = m^2$:

$$\begin{array}{l} s(\frac{q}{m}) = \sqrt{\frac{m+q_0}{2m}}[\mathbf{1}_2 + \frac{\vec{q}}{m+q_0}], \quad \hat{s}(\frac{q}{m}) = s^{\star-1}(\frac{q}{m}) = \sqrt{\frac{m+q_0}{2m}}[\mathbf{1}_2 - \frac{\vec{q}}{m+q_0}], \\ s(\frac{q}{m})\hat{s}^{-1}(\frac{q}{m}) = \frac{q_k\sigma^k}{m} = \frac{q_0 - \vec{q}}{m}, \quad \hat{s}(\frac{q}{m})s^{-1}(\frac{q}{m}) = \frac{q_k\check{\sigma}^k}{m} = \frac{q_0 + \vec{q}}{m}. \end{array}$$

Hence two dual pairs with left- and right- handed *Weyl spinor particle fields*, in chapter "Lorentz Symmetry" with the $\mathbf{U}(2,2)$-anticonjugation notation $(\mathbf{r}, \mathbf{l}, \mathbf{r}^\star, \mathbf{l}^\star) = (\mathrm{r}, \mathrm{l}, \mathrm{r}^\times, \mathrm{l}^\times)$, of Fermi type are defined:

$$\begin{array}{l} m > 0 \\ J = \frac{1}{2} \\ z = \pm 1 \\ \epsilon = +1 \end{array} : \left\{ \begin{array}{lll} \mathbf{r}(x)^A & = \sqrt{m} \oplus\!\int \frac{d^3q}{2q_0(2\pi)^3} & [e^{iqx}\mathrm{u}(\vec{q})^C + e^{-iqx}\mathrm{a}^\star(\vec{q})^C] \quad \hat{s}^{-1}(\frac{q}{m})^A_C, \\ \mathbf{l}(x)^{\dot{A}} & = \sqrt{m} \oplus\!\int \frac{d^3q}{2q_0(2\pi)^3} & [e^{iqx}\mathrm{u}(\vec{q})^C - e^{-iqx}\mathrm{a}^\star(\vec{q})^C] \quad s^{-1}(\frac{q}{m})^{\dot{A}}_C, \\ \mathbf{r}^\star(x)_{\dot{A}} & = \sqrt{m} \oplus\!\int \frac{d^3q}{2q_0(2\pi)^3} & [e^{-iqx}\mathrm{u}^\star(\vec{q})_C + e^{iqx}\mathrm{a}(\vec{q})_C] \quad s(\frac{q}{m})^C_{\dot{A}}, \\ \mathbf{l}^\star(x)_A & = \sqrt{m} \oplus\!\int \frac{d^3q}{2q_0(2\pi)^3} & [e^{-iqx}\mathrm{u}^\star(\vec{q})_C - e^{iqx}\mathrm{a}(\vec{q})_C] \quad \hat{s}(\frac{q}{m})^C_A, \\ \text{with } q_0 & = \sqrt{m^2 + \vec{q}^2}. & \end{array} \right.$$

The product $\sqrt{m}s(\frac{q}{m})$ with Weyl transmutators is finite in the massless limit $m \to 0$ (chapter "Massless Quantum Fields").

In the representations of spacetime translations and charge group

$$\mathbb{R}^4 \times \mathbf{U}(1) \ni (x, e^{i\alpha}) \longmapsto e^{iqx}\begin{pmatrix} e^{iz\alpha}\mathbf{1}_2 & 0 \\ 0 & e^{-iz\alpha}\mathbf{1}_2 \end{pmatrix} \in \mathbf{U}(\mathbf{1}_4) \circ \mathbf{U}(\mathbf{1}_2)_3 \subset \mathbf{U}(2,2),$$

the internal charge group $\mathbf{U}(1)$ for particle-antiparticles comes as the subgroup $\mathbf{U}(\mathbf{1}_2)_3$. The spacetime translations, charge, and spin group are represented as subgroups of $\mathbf{U}(2,2)$. Dirac spinor particle fields embed spin doublets of $\vec{q}$-indexed harmonic Fermi oscillators, e.g., for the electron-positron

$$\int d^3x\ \mathbf{r}(x)^A = \frac{e^{imx_0}\mathrm{u}(0)^A + e^{-imx_0}\mathrm{a}^\star(0)^A}{2\sqrt{m}} \text{ etc.}$$

Both dual pairs $(\mathbf{r}, \mathbf{r}^\star)$ and $(\mathbf{l}, \mathbf{l}^\star)$ have one left-handed field $(\mathbf{l}, \mathbf{r}^\star)$ and one right-handed field $(\mathbf{r}, \mathbf{l}^\star)$.

The Weyl represented boosts $\check{\sigma}^k q_k = s(\frac{q}{m}) m \hat{s}(\frac{q}{m})$ lead to the *Weyl equations*, characterizing the translation orbits

$$\begin{array}{rclrcl} (\check{\sigma}^k \partial_k)^{\dot{A}}_B \mathbf{r}^B &=& im\mathbf{l}^{\dot{A}}, & (\sigma^k \partial_k)^A_{\dot{B}} \mathbf{l}^{\dot{B}} &=& im\mathbf{r}^A, \\ (\check{\sigma}^k \partial_k)^{\dot{B}}_A \mathbf{r}^\star_{\dot{B}} &=& -im\mathbf{l}^\star_A, & (\sigma^k \partial_k)^B_{\dot{A}} \mathbf{l}^\star_B &=& -im\mathbf{r}^\star_{\dot{A}}. \end{array}$$

The Lorentz transformations are

$$\lambda \in \mathbf{SL}(\mathbb{C}^2): \left\{ \begin{array}{ll} \mathbf{r}_\lambda(x)^A &= \lambda^A_B\ \mathbf{r}(\Lambda^{-1}.x)^B, \\ \mathbf{l}_\lambda(x)^{\dot{A}} &= \hat{\lambda}^{\dot{A}}_{\dot{B}}\ \mathbf{l}(\Lambda^{-1}.x)^{\dot{B}}. \end{array}\right.$$

The quantization anticommutators and the commutator Fock forms are

$$\begin{pmatrix} \{\mathbf{r}^\star_{\dot{B}}, \mathbf{r}^A\} & \{\mathbf{r}^\star_{\dot{B}}, \mathbf{l}^{\dot{A}}\} \\ \{\mathbf{l}^\star_B, \mathbf{r}^A\} & \{\mathbf{l}^\star_B, \mathbf{l}^{\dot{A}}\} \end{pmatrix}(x) = (-i\partial)\ \frac{\mathbf{is}(m|x)}{m} = \int \frac{d^4q}{(2\pi)^3}\ (\mathbf{q})\ \epsilon(q_0)\delta(q^2 - m^2)e^{iqx},$$

$$\begin{pmatrix} \langle[\mathbf{r}^\star_{\dot{B}}, \mathbf{r}^A]\rangle_\mathrm{F} & \langle[\mathbf{r}^\star_{\dot{B}}, \mathbf{l}^{\dot{A}}]\rangle_\mathrm{F} \\ \langle[\mathbf{l}^\star_B, \mathbf{r}^A]\rangle_\mathrm{F} & \langle[\mathbf{l}^\star_B, \mathbf{l}^{\dot{A}}]\rangle_\mathrm{F} \end{pmatrix}(x) = (-i\partial)\ \frac{\mathbf{C}(m|x)}{m} = \int \frac{d^4q}{(2\pi)^3}\ (\mathbf{q})\ \delta(q^2 - m^2)e^{iqx},$$

$$\text{with } (-i\partial) = \begin{pmatrix} -i\sigma^k{}^A_{\dot{B}}\partial_k & m\delta^{\dot{A}}_{\dot{B}} \\ m\delta^A_B & -i\check{\sigma}^k{}^{\dot{A}}_B\partial_k \end{pmatrix}, \quad (\mathbf{q}) = \begin{pmatrix} \sigma^k{}^A_{\dot{B}} q_k & m\delta^{\dot{A}}_{\dot{B}} \\ m\delta^A_B & \check{\sigma}^k{}^{\dot{A}}_B q_k \end{pmatrix}.$$

Four-component *Dirac spinor fields* with chiral Dirac matrices

$$\begin{array}{rl} \mathbf{\Psi} &= (\mathbf{r}^A, \mathbf{l}^{\dot{A}}), \quad \overline{\mathbf{\Psi}} = \gamma^0 \mathbf{\Psi}^\star = \begin{pmatrix} \mathbf{l}^\star_A \\ \mathbf{r}^\star_{\dot{A}} \end{pmatrix}, \quad \gamma^k = \begin{pmatrix} 0 & \sigma^k \\ \check{\sigma}^k & 0 \end{pmatrix}, \\ \mathbf{\Psi}(x) &= \sqrt{m}\ {}^\oplus\!\!\int \frac{d^3q}{2q_0(2\pi)^3} \left(e^{iqx}\mathrm{u}(\vec{q}), -e^{-iqx}\mathrm{a}^\star(\vec{q})\right)\ \sqrt{2}w_U\ s^{-1}_{\mathrm{doub}}(\frac{q}{m}), \\ \overline{\mathbf{\Psi}}(x) &= \sqrt{m}\ {}^\oplus\!\!\int \frac{d^3q}{2q_0(2\pi)^3}\ s_{\mathrm{doub}}(\frac{q}{m})\ \sqrt{2}w_U^{-1}\ \begin{pmatrix} e^{-iqx}\mathrm{u}^\star(\vec{q}) \\ e^{iqx}\mathrm{a}(\vec{q}) \end{pmatrix}, \end{array}$$

involve the Dirac boost representations $s_{\mathrm{doub}}(\frac{q}{m}) \in \mathbf{SU}(2,2)$ and the transformations $w_U \in \mathbf{GL}(\mathbb{C}^4)$ to the time diagonal basis with the $\mathbf{U}(\mathbf{1}_4)$-scalar product (chapter "Lorentz Symmetry")

$$s_{\mathrm{doub}}(\frac{q}{m}) = \begin{pmatrix} s(\frac{q}{m}) & 0 \\ 0 & \hat{s}(\frac{q}{m}) \end{pmatrix}, \quad w_U = \frac{1}{\sqrt{2}}\begin{pmatrix} \mathbf{1}_2 & \mathbf{1}_2 \\ -\mathbf{1}_2 & \mathbf{1}_2 \end{pmatrix}.$$

The boost representation embeds the harmonic Fermi oscillators with spin $\frac{1}{2}$ into Dirac fields. Its property $\frac{\gamma^k q_k}{m} s_{\text{doub}}(\frac{q}{m}) = s_{\text{doub}}(\frac{q}{m})$ is expressed by the *Dirac equation*

$$\mathbf{\Psi}(m + i\gamma^k \partial_k) = 0, \quad (m - i\gamma^k \partial_k)\overline{\mathbf{\Psi}} = 0.$$

Dirac particle fields have as quantization anticommutator

$$\begin{aligned}\{\overline{\mathbf{\Psi}}, \mathbf{\Psi}\}(x) &= (\mathbf{1}_4 - \tfrac{\gamma^k \partial_k}{im})i\mathbf{s}(m|x) = (\gamma^k \mathbf{c}_k + i\mathbf{s})(m|x) = \mathbf{exp}(im|x) \\ &= \int \tfrac{d^4q}{(2\pi)^3}\epsilon(q_0)(\gamma^k q_k + m)\delta(q^2 - m^2)e^{iqx} \\ &= \gamma_0 m \int \tfrac{d^3q}{q_0(2\pi)^3} e^{-i\vec{q}\vec{x}} \begin{pmatrix} \frac{q_0}{m}\cos q_0x_0 + \frac{\vec{q}}{m}\, i\sin q_0x_0 & \mathbf{1}_2\, i\sin q_0x_0 \\ \mathbf{1}_2\, i\sin q_0x_0 & \frac{q_0}{m}\cos q_0x_0 - \frac{\vec{q}}{m}\, i\sin q_0x_0 \end{pmatrix},\end{aligned}$$

and as commutator Fock form

$$\begin{aligned}\langle[\overline{\mathbf{\Psi}}, \mathbf{\Psi}]\rangle_{\text{F}}(x) &= (\mathbf{1}_4 + \tfrac{\gamma^k \partial_k}{im})\mathbf{C}(m|x) = (\mathbf{C} + i\gamma^k \mathbf{S}_k)(m|x) = \mathbf{EXP}(im|x) \\ &= \int \tfrac{d^4q}{(2\pi)^3}(\gamma^k q_k + m)\delta(q^2 - m^2)e^{iqx} \\ &= \gamma_0 m \int \tfrac{d^3q}{q_0(2\pi)^3} e^{-i\vec{q}\vec{x}} \begin{pmatrix} \frac{q_0}{m} i\sin q_0x_0 + \frac{\vec{q}}{m}\cos q_0x_0 & \mathbf{1}_2 \cos q_0x_0 \\ \mathbf{1}_2 \cos q_0x_0 & \frac{q_0}{m} i\sin q_0x_0 - \frac{\vec{q}}{m}\cos q_0x_0 \end{pmatrix}.\end{aligned}$$

In contrast to the on-shell compact representation matrix elements of the embedded time translations and Euclidean group:

$$\begin{aligned}\begin{pmatrix} \{\overline{\mathbf{\Psi}}, \mathbf{\Psi}\} \\ \langle[\overline{\mathbf{\Psi}}, \mathbf{\Psi}]\rangle_{\text{F}} \end{pmatrix}(x) = \gamma_0 \int \tfrac{dq_0}{(2\pi)^2} \begin{pmatrix} \epsilon(q_0) \\ 1 \end{pmatrix} & \vartheta(q_0^2 - m^2)e^{iq_0x_0} \\ & \begin{pmatrix} -i\frac{\sin|\vec{q}|r - |\vec{q}|r\cos|\vec{q}|r}{r^2}\frac{\vec{x}}{r} & (m+q_0)\frac{\sin|\vec{q}|r}{r}\mathbf{1}_2 \\ (m+q_0)\frac{\sin|\vec{q}|r}{r}\mathbf{1}_2 & i\frac{\sin|\vec{q}|r - |\vec{q}|r\cos|\vec{q}|r}{r^2}\frac{\vec{x}}{r} \end{pmatrix},\end{aligned}$$

with the time projections

$$\begin{aligned}\int d^3x \quad \gamma_0\{\overline{\mathbf{\Psi}}, \mathbf{\Psi}\}(x) &= \mathbf{1}_2 \otimes \begin{pmatrix} \cos mx_0 & i\sin mx_0 \\ i\sin mx_0 & \cos mx_0 \end{pmatrix}, \\ \int d^3x \gamma_0\langle[\overline{\mathbf{\Psi}}, \mathbf{\Psi}]\rangle_{\text{F}}(x) &= \mathbf{1}_2 \otimes \begin{pmatrix} i\sin mx_0 & \cos mx_0 \\ \cos mx_0 & i\sin mx_0 \end{pmatrix},\end{aligned}$$

there are off-shell contributions in the time-ordered anticommutator with Yukawa potential and force

$$\begin{aligned}\epsilon(x_0)\{\overline{\mathbf{\Psi}}, \mathbf{\Psi}\}(x) &= \tfrac{i}{\pi}\int \tfrac{d^4q}{(2\pi)^3}\tfrac{\gamma^k q_k + m}{-q_{\text{P}}^2 + m^2}e^{iqx} \\ = \gamma_0 \int \tfrac{dq_0}{(2\pi)^2}e^{iq_0x_0}\Big[& \vartheta(q_0^2 - m^2)\begin{pmatrix} \frac{\cos|\vec{q}|r + |\vec{q}| - \sin|\vec{q}|r}{r^2}\frac{\vec{x}}{r} & i(m+q_0)\frac{\cos|\vec{q}|r}{r}\mathbf{1}_2 \\ i(m+q_0)\frac{\cos|\vec{q}|r}{r}\mathbf{1}_2 & -\frac{\cos|\vec{q}|r + r|\vec{q}|\sin|\vec{q}|r}{r^2}\frac{\vec{x}}{r} \end{pmatrix} \\ & + \vartheta(m^2 - q_0^2)\begin{pmatrix} \frac{1+|Q|r}{r}\frac{\vec{x}}{r} & i(m+q_0)\mathbf{1}_2 \\ i(m+q_0)\mathbf{1}_2 & -\frac{1+|Q|r}{r}\frac{\vec{x}}{r} \end{pmatrix}\frac{e^{-|Q|r}}{r}\Big],\end{aligned}$$

with the position projection

$$\int dx_0 \epsilon(x_0)\langle\{\overline{\mathbf{\Psi}}, \mathbf{\Psi}\}(x)\rangle_{\text{F}} = \begin{pmatrix} im\mathbf{1}_2 & \frac{1+mr}{r}\frac{\vec{x}}{r} \\ -\frac{1+mr}{r}\frac{\vec{x}}{r} & im\mathbf{1}_2 \end{pmatrix}\frac{e^{-mr}}{2\pi r}.$$

The Feynman propagator reads

$$\begin{aligned}
\langle[\overline{\mathbf{\Psi}},\mathbf{\Psi}](x)-\epsilon(x_0)\{\overline{\mathbf{\Psi}},\mathbf{\Psi}\}(x)\rangle_{\mathrm{F}} &= \mathbf{EXP}(im|x)-\epsilon(x_0)\mathbf{exp}(im|x)\\
&= \mathbf{1}_4[\mathbf{C}(m|x)-\epsilon(x_0)i\mathbf{s}(m|x)]+\gamma^k[i\mathbf{S}_k(m|x)-\epsilon(x_0)\mathbf{c}_k(m|x)]\\
&= (\mathbf{1}_4+\tfrac{\gamma^k\partial_k}{im})\mathbf{E}(-im|x)=\tfrac{i}{\pi}\int\tfrac{d^4q}{(2\pi)^3}\tfrac{\gamma^k q_k+m}{q^2+io-m^2}e^{iqx}\\
&= -\gamma_0\int\tfrac{dq_0}{(2\pi)^2}e^{iq_0x_0}\Big[\vartheta(q_0^2-m^2)\begin{pmatrix}\frac{1-i|\vec{q}|r}{r}\frac{\vec{x}}{r} & i(m+q_0)\mathbf{1}_2\\ i(m+q_0)\mathbf{1}_2 & -\frac{1-i|\vec{q}|r}{r}\frac{\vec{x}}{r}\end{pmatrix}\frac{e^{i|\vec{q}|r}}{r}\\
&\qquad +\vartheta(m^2-q_0^2)\begin{pmatrix}\frac{1+|Q|r}{r}\frac{\vec{x}}{r} & i(m+q_0)\mathbf{1}_2\\ i(m+q_0)\mathbf{1}_2 & -\frac{1+|Q|r}{r}\frac{\vec{x}}{r}\end{pmatrix}\frac{e^{-|Q|r}}{r}\Big].
\end{aligned}$$

Its time projection comes with the projectors $\frac{\mathbf{1}_4\pm\gamma^0}{2}$,

$$\int\tfrac{d^3x}{2}\langle[\overline{\mathbf{\Psi}},\mathbf{\Psi}](x)-\epsilon(x_0)\{\overline{\mathbf{\Psi}},\mathbf{\Psi}\}(x)\rangle_{\mathrm{F}}=[\tfrac{\mathbf{1}_4-\gamma^0}{2}\vartheta(x_0)+\tfrac{\mathbf{1}_4+\gamma^0}{2}\vartheta(-x_0)]e^{-im|x_0|}.$$

Classical Lagrangian densities for anticommuting Grassmann vectors are

$$\begin{aligned}
\mathbf{L}(\mathbf{r},\mathbf{l}) &= i\mathbf{r}\check{\sigma}_k\partial^k\mathbf{r}^\star+i\mathbf{l}\sigma_k\partial^k\mathbf{l}^\star-m(\mathbf{r}\mathbf{l}^\star+\mathbf{l}\mathbf{r}^\star),\\
\mathbf{L}(\mathbf{\Psi}) &= i\mathbf{\Psi}\gamma_k\partial^k\overline{\mathbf{\Psi}}-m\mathbf{\Psi}\overline{\mathbf{\Psi}}.
\end{aligned}$$

With their decomposability into Weyl representations, Dirac particle fields have an additional $\mathbf{U}(1)$ degree of freedom, which can be used for internal operations, e.g., for electromagnetic $\mathbf{U}(1)$-transformations. For more internal degrees of freedom, $\alpha=1,2\ldots$, one has only to write more indices, $\mathbf{l}^{\dot{A}\alpha},\mathbf{r}^{A\alpha},\mathbf{l}^\star_{A\alpha},\mathbf{r}^\star_{\dot{A}\alpha}$ and $\mathbf{\Psi}^\alpha,\overline{\mathbf{\Psi}}_\alpha$.

4.8 Massive Spin-$\frac{1}{2}$ Majorana Particle Fields

Majorana particles are massive and have fixgroup $\mathbf{SU}(2)$ for spin. So far no Majorana particles have been found. Neutrinos, if massive, are discussed as possible candidates.

Since the dual Lorentz group representations acting on the irreducible Weyl spinors $\mathbf{r}^A$ and $\mathbf{l}^\star_A$ in a Dirac field are equivalent via the bilinear spinor metric

$$\lambda\in\mathbf{SL}(\mathbb{C}^2):\quad \epsilon_{AB}\lambda^B_C\epsilon^{CD}=(\lambda^{-1})^D_A,\quad \epsilon_{AB}=-\epsilon_{BA},$$

one can consider the case in which the four Weyl fields $(\mathbf{r},\mathbf{l}^\star;\mathbf{l},\mathbf{r}^\star)$ involve only one irreducible right- and left-handed Weyl representation: $(\mathbf{r},\mathbf{l})$

$$\mathbf{r}(x)^A=i\epsilon^{AB}\mathbf{l}^\star(x)_B,\ \mathbf{r}^\star(x)_{\dot{A}}=-i\mathbf{l}(x)^{\dot{B}}\epsilon_{\dot{B}\dot{A}}.$$

Hence one parametrizes massive *Majorana fields*:

$$\begin{matrix} m>0\\ J=\frac{1}{2}\\ z=0\\ \epsilon=+1\end{matrix}:\left\{\begin{aligned}
\mathbf{r}(x)^A &= \sqrt{m}\,{}^{\oplus}\!\!\int\tfrac{d^3q}{2q_0(2\pi)^3}\quad [e^{iqx}\mathrm{u}(\vec{q})^C+e^{-iqx}i\epsilon^{CB}\check{\mathrm{u}}(\vec{q})_B]\ \ \hat{s}^{-1}(\tfrac{q}{m})^A_C,\\
\mathbf{r}^\star(x)_{\dot{A}} &= \sqrt{m}\,{}^{\oplus}\!\!\int\tfrac{d^3q}{2q_0(2\pi)^3}\ [-e^{iqx}\mathrm{u}(\vec{q})^B i\epsilon_{BC}+e^{-iqx}\check{\mathrm{u}}(\vec{q})_C]\ \ s(\tfrac{q}{m})^C_{\dot{A}},\\
\text{with } q_0 &= \sqrt{m^2+\vec{q}^2}.
\end{aligned}\right.$$

Spacetime translations $\mathbb{R}^4$ and spin $\mathbf{SU}(2)$ are represented in $\mathbf{U}(2)$ for the momentum operators and as a subgroup of $\mathbf{U}(\mathbf{1}_2) \circ \mathbf{SL}(\mathbb{C}^2)$ for the relativistic fields

$$\mathbb{R}^4 \times \mathbf{SU}(2) \longrightarrow \mathbf{U}(2) \subset \mathbf{U}(\mathbf{1}_2) \circ \mathbf{SL}(\mathbb{C}^2).$$

Majorana particles involve the spinor metric as seen in the time projection

$$\sqrt{m} \int d^3x \; \mathbf{r}(x)^A = \frac{e^{imx_0}\mathrm{u}(0)^C + e^{-imx_0} i\epsilon^{CB}\check{\mathrm{u}}(0)_B}{2} \text{ etc.}$$

Particles and antiparticles from the Dirac fields now come in the two spin components

$$\mathrm{a}^A = -i\epsilon^{AB}\delta_{BC}\mathrm{u}^C = (\sigma^2)^A_C \mathrm{u}^C.$$

Therefore the $\mathbf{U}(1)$-degree of freedom $\mathbf{U}(1)_3 = \mathbf{SO}(2) \subset \mathbf{SU}(2)$ cannot be used as an additional charge group since it operates as spin subgroup. Particles coincide with antiparticles. With the left-right field identification the particle-antiparticle group $\mathbf{U}(\mathbf{1}_2)_3 \subset \mathbf{SU}(2,2)$, nontrivial for Dirac fields, is trivially represented for Majorana fields.

If one keeps the Dirac field $\boldsymbol{\Psi}$ language for Majorana fields, the identities above between left- and right-handed fields lead to

$$\boldsymbol{\Psi} = (\mathbf{r}, \mathbf{l}) \cong (\mathbf{r}^A, i\mathbf{r}^\star_{\dot{B}}\epsilon^{\dot{B}\dot{A}}), \quad \overline{\boldsymbol{\Psi}} = \begin{pmatrix} \mathbf{l}^\star \\ \mathbf{r}^\star \end{pmatrix} \cong \begin{pmatrix} i\epsilon_{AB}\mathbf{r}^B \\ \mathbf{r}^\star_{\dot{A}} \end{pmatrix} \Rightarrow \overline{\boldsymbol{\Psi}}\gamma_j\boldsymbol{\Psi} = 0.$$

The only $\mathbf{SL}(\mathbb{C}^2)$-invariant Lagrangian density reads

$$\mathbf{L}(\mathbf{r}) = i\mathbf{r}\check{\sigma}_k\partial^k\mathbf{r}^\star - im(\epsilon_{BA}\mathbf{r}^A\mathbf{r}^B - \mathbf{r}^\star_{\dot{A}}\mathbf{r}^\star_{\dot{B}}\epsilon^{\dot{B}\dot{A}}).$$

4.9 Spacetime Reflections of Spinor Fields

The spacetime reflection of a field Ψ combines the reflection $\mathtt{R}$ on the spacetime translations with the reflection $\mathtt{R}_V$ represented on the complex value space V:

$$\begin{array}{ccc} \mathbb{R}^4 & \xrightarrow{\mathtt{R}} & \mathbb{R}^4 \\ {\scriptstyle\Psi}\downarrow & & \downarrow{\scriptstyle \mathtt{R}\Psi}, \\ V & \xrightarrow[\mathtt{R}_V]{} & V \end{array} \qquad \Psi(x) \overset{\mathtt{R}}{\leftrightarrow} \mathtt{R}_V.\Psi(\mathtt{R}.x).$$

Therefore the $\mathtt{T}, \mathtt{P}, \mathtt{C}$-reflections of time and position translations and particle-antiparticle (chapter "Lorentz Symmetry") are given for Weyl fields as follows:

$$(\mathbf{l}, \mathbf{r})(x_0, \vec{x}) \left\{ \begin{array}{llll} \overset{\mathtt{T}}{\longleftrightarrow} & (\mathbf{r}^\star, \mathbf{l}^\star) & (-x_0, \vec{x}) & \text{(antilinear)}, \\ \overset{\mathtt{P}}{\longleftrightarrow} & (\mathbf{r}, \mathbf{l}) & (x_0, -\vec{x}) & \text{(linear)}, \\ \overset{\mathtt{T}\circ\mathtt{P}}{\longleftrightarrow} & (\mathbf{r}^\star, \mathbf{l}^\star) & (-x_0, -\vec{x}) & \text{(antilinear)}, \\ \overset{\mathtt{C}}{\longleftrightarrow} & (\epsilon\mathbf{r}^\star, \epsilon\mathbf{l}^\star) & (x_0, \vec{x}) & \text{(linear)}. \end{array} \right.$$

From the reflections of the fundamental Weyl spinor fields one can derive the reflections for fields with product Lorents group representations.

For left- and right-handed massive particle fermion fields

$$\begin{aligned}
\mathbf{r}(x)^A &= \sqrt{m}\ {}^{\oplus}\!\!\int \frac{d^3q}{2q_0(2\pi)^3}\ [e^{iqx}\mathrm{u}(\vec q)^C + e^{-iqx}\mathrm{a}^\star(\vec q)^C]\ \ e^{-\vec\beta(\vec q)}{}^A_C,\\
\mathbf{l}(x)^{\dot A} &= \sqrt{m}\ {}^{\oplus}\!\!\int \frac{d^3q}{2q_0(2\pi)^3}\ [e^{iqx}\mathrm{u}(\vec q)^C - e^{-iqx}\mathrm{a}^\star(\vec q)^C]\ \ e^{\vec\beta(\vec q)}{}^{\dot A}_C,\\
\mathbf{r}^\star(x)_{\dot A} &= \sqrt{m}\ {}^{\oplus}\!\!\int \frac{d^3q}{2q_0(2\pi)^3}\ [e^{-iqx}\mathrm{u}^\star(\vec q)_C + e^{iqx}\mathrm{a}(\vec q)_C]\ \ e^{-\vec\beta(\vec q)}{}^C_{\dot A},\\
\mathbf{l}^\star(x)_A &= \sqrt{m}\ {}^{\oplus}\!\!\int \frac{d^3q}{2q_0(2\pi)^3}\ [e^{-iqx}\mathrm{u}^\star(\vec q)_C - e^{iqx}\mathrm{a}(\vec q)_C]\ \ e^{\vec\beta(\vec q)}{}^C_A,
\end{aligned}$$

the reflections are effected with the boost reflections $\vec\beta(-\vec q) = -\vec\beta(\vec q)$ by the reflections of the momentum- and spin-direction-dependent (anti)particle creation and annihilation operators

$$(\mathrm{u}^C, \mathrm{a}^{\star C})(\vec q) \begin{cases} \overset{\mathsf{T}}{\longleftrightarrow} & \delta^{CA}(\mathrm{u}^\star_A, -\mathrm{a}_A) \quad (-\vec q),\\ \overset{\mathsf{P}}{\longleftrightarrow} & (\mathrm{u}^C, -\mathrm{a}^{\star C}) \quad (-\vec q),\\ \overset{\mathsf{T\circ P}}{\longleftrightarrow} & \delta^{CA}(\mathrm{u}^\star_A, \ \ \mathrm{a}_A) \quad (\vec q),\\ \overset{\mathsf{C}}{\longleftrightarrow} & \epsilon^{CA}(\mathrm{a}_A, -\mathrm{u}^\star_A) \quad (\vec q). \end{cases}$$

State space vectors $|z; \vec q, A\rangle$, e.g., for electron-positron with charge numbers $z = \mp 1$ and third spin component $A = \pm\frac{1}{2}$, have the reflection behavior

$$\begin{aligned} \text{electron: } \mathrm{u}(\vec q)^{1,2}|0\rangle &\cong |-1; \vec q, \pm\tfrac{1}{2}\rangle,\\ \text{positron: } \mathrm{a}(\vec q)_{1,2}|0\rangle &\cong |+1; \vec q, \mp\tfrac{1}{2}\rangle, \end{aligned} \quad \begin{cases} |\mp 1; \vec q, \pm\frac{1}{2}\rangle & \overset{\mathsf{T}}{\leftrightarrow} \quad \pm\langle\pm 1; \vec q, \pm\frac{1}{2}|,\\ |\mp 1; \vec q, \pm\frac{1}{2}\rangle & \overset{\mathsf{P}}{\leftrightarrow} \quad \pm|\mp 1; -\vec q, \pm\frac{1}{2}\rangle,\\ |-1; \vec q, \pm\frac{1}{2}\rangle & \overset{\mathsf{C}}{\leftrightarrow} \quad \pm|+1; \vec q, \pm\frac{1}{2}\rangle. \end{cases}$$

4.10 Representation Currents

With the embedding of time into spacetime translations, Lie algebras, acting on relativistic particle fields, come as position integrals over currents. Relativistic currents involve dual pairs of quantum fields.

4.10.1 Internal Lie Algebra

The representations of a real internal Lie algebra, e.g., $L = \log \mathbf{U}(N)$,

$$[\mathcal{D}(l^a), \mathcal{D}(l^b)] = \epsilon^{ab}_c \mathcal{D}(l^c), \quad l^a \in L \in \underline{\mathbf{lag}}_{\mathbb{R}}, \quad a, b, c = 1, \ldots, d,$$

come in a quantum algebra for the complex $\mathbb{C}^{1+2J} \otimes \mathbb{C}^n$-quartets with internal degrees of freedom $\{\mathrm{u}^\beta, \mathrm{u}^\star_\beta, \mathrm{a}_\beta, \mathrm{a}^{\star\beta}\}_{\beta=1}^n$ as twofold decomposable representations

$$\mathcal{D}(l^a) = iQ^a = \mathcal{D}^{a\beta}_{\ \gamma} \frac{[\mathrm{u}^\gamma, \mathrm{u}^\star_\beta]_{-\epsilon} - [\mathrm{a}_\beta, \mathrm{a}^{\star\gamma}]_{-\epsilon}}{2} = \mathcal{D}^{a\beta}_{\ \gamma} \begin{cases} \frac{[\mathbf{l}^\gamma, \mathbf{l}^\star_\beta] + [\mathbf{r}^\gamma, \mathbf{r}^\star_\beta]}{2}, & \epsilon = +1,\\ \frac{\{\mathbf{x}^\gamma, i\mathbf{p}^\star_\beta\} + \{-i\mathbf{p}^\gamma, \mathbf{x}^\star_\beta\}}{2}, & \epsilon = -1. \end{cases}$$

They involve the *quantization opposite commutators* $[\ ,\ ]_{-\epsilon}$.

In the case of relativistic particle quantum fields the quantum representations with the corresponding expressions in the momentum operators are integrated directly over the energy-momentum hyperboloid:

$$\mathcal{D}(l^a) = {}^{\oplus}\!\!\int \frac{d^3q}{2q_0(2\pi)^3} iQ(\vec{q})^a = {}^{\oplus}\!\!\int \frac{d^3q}{2q_0(2\pi)^3} \mathcal{D}^{a\beta}_{\ \gamma} \frac{[\mathrm{u}(\vec{q})^\gamma, \mathrm{u}^\star(\vec{q})_\beta]_{-\epsilon} - [\mathrm{a}(\vec{q})_\beta, \mathrm{a}^\star(\vec{q})^\gamma]_{-\epsilon}}{2}.$$

The adjoint action reads

$$\begin{array}{rclrcl} [\mathcal{D}(l^a), \mathrm{u}(\vec{q})^\beta] &=& \mathcal{D}^{a\beta}_{\ \gamma}\mathrm{u}(\vec{q})^\gamma, & [\mathcal{D}(l^a), \mathrm{u}^\star(\vec{q})_\gamma] &=& -\mathcal{D}^{a\beta}_{\ \gamma}\mathrm{u}^\star(\vec{q})_\beta, \\ [\mathcal{D}(l^a), \mathrm{a}(\vec{q})_\gamma] &=& -\mathcal{D}^{a\beta}_{\ \gamma}\mathrm{a}(\vec{q})_\beta, & [\mathcal{D}(l^a), \mathrm{a}^\star(\vec{q})^\beta] &=& \mathcal{D}^{a\beta}_{\ \gamma}\mathrm{a}^\star(\vec{q})^\gamma. \end{array}$$

Duality for relativistic fields is expressed by the Lorentz vector quantization distribution for equal time $\mathbf{c}_k(\vec{x}|m) = \delta^0_k\delta(\vec{x})$ (for Weyl and Dirac fields without external indices):

$$\begin{array}{lll} \text{scalar:} & [i\mathbf{\Phi}^\star_{k\gamma}, \mathbf{\Phi}^\beta](\vec{x}) & = [i\mathbf{\Phi}^\beta_k, \mathbf{\Phi}^\star](\vec{x}) = \delta^\beta_\gamma\delta^0_k\delta(\vec{x}), \\ \text{vector:} & [i\mathbf{G}_{kj\gamma}, \mathbf{Z}^{l\beta}](\vec{x}) & = \delta^\beta_\gamma\delta^a_j\delta^l_b\delta^b_a\delta^0_k\delta(\vec{x}), \\ \text{Weyl, right:} & \{\mathbf{l}^\star_\gamma, \mathbf{l}^\beta\}(\vec{x}) & = \delta^\beta_\gamma\sigma^0\delta(\vec{x}), \\ \text{Weyl, left} & \{\mathbf{r}^\star_\gamma, \mathbf{r}^\beta\}(\vec{x}) & = \delta^\beta_\gamma\check{\sigma}^0\delta(\vec{x}), \\ \text{Dirac:} & \{\overline{\mathbf{\Psi}}_\gamma, \mathbf{\Psi}^\beta\}(\vec{x}) & = \delta^\gamma_\beta\gamma^0\delta(\vec{x}). \end{array}$$

Therefore a represented nonabelian Lie algebra $\mathcal{D}(l^a)$ in terms of relativistic particle fields is embedded via Lorentz vector field *currents* $\{\mathbf{J}^a_k\}^{a=1,\dots,d}_{k=0,1,2,3}$. The representation consists of position integrals of the time component (time projection):

$$\mathcal{D}(l^a) = \int d^3x \; i\mathbf{J}(x)^a_0.$$

The currents for an internal Lie algebra representations for particle fields arise with the Lorentz vector modification from the quantum-mechanical Lie algebra representations above

$$i\mathbf{J}^a_k = \left\{ \begin{array}{ll} \mathcal{D}^{c\beta}_{\ \gamma}\frac{\{\mathbf{\Phi}^\gamma, i\mathbf{\Phi}^\star_{k\beta}\} + \{-i\mathbf{\Phi}^\gamma_k, \mathbf{\Phi}^\star_\beta\}}{2}, & \text{scalar}, \\ \mathcal{D}^{c\beta}_{\ \gamma}\frac{\{\mathbf{Z}^{j\gamma}, i\mathbf{G}_{kj\beta}\}}{2}, & \text{vector}, \\ \mathcal{D}^{c\beta}_{\ \gamma}\frac{[\mathbf{l}^\gamma\sigma_k, \mathbf{l}^\star_\beta] + [\mathbf{r}^\gamma\check{\sigma}_k, \mathbf{r}^\star_\beta]}{2} = \mathcal{D}^{a\beta}_{\ \gamma}\frac{[\mathbf{\Psi}^\gamma\gamma_k, \overline{\mathbf{\Psi}}_\beta]}{2}, & \text{Weyl, Dirac}. \end{array} \right.$$

The currents for Dirac fields are decomposable into currents for right- and left-handed Weyl fields. More explicit examples are given in the chapter "Gauge Interactions."

Via the equations of motion the fields $\mathbf{\Phi}_k, \mathbf{\Phi}^\star_k$, and $\mathbf{G}_{kj}$ may be replaced by first order derivatives of their dual partners $\mathbf{\Phi}, \mathbf{\Phi}^\star$, and $\mathbf{Z}_j$ respectively.

The currents have the adjoint action (analogue for the vector fields)

$$\begin{array}{ll} [i\mathbf{J}(\vec{x})^a_k, \mathbf{\Phi}(\vec{y})^\beta] & = \delta^0_k\delta(\vec{y}-\vec{x})\mathcal{D}^{a\beta}_{\ \gamma}\mathbf{\Phi}(\vec{y})^\gamma, \\ [i\mathbf{J}(\vec{x})^a_k, \mathbf{\Phi}^\star(\vec{y})_\gamma] & = -\delta^0_k\delta(\vec{y}-\vec{x})\mathcal{D}^{a\beta}_{\ \gamma}\mathbf{\Phi}^\star(\vec{y})_\beta, \\ [i\mathbf{J}(\vec{x})^a_0, \mathbf{\Psi}(\vec{y})^\beta] & = \delta(\vec{y}-\vec{x})\mathcal{D}^{a\beta}_{\ \gamma}\mathbf{\Psi}(\vec{y})^\gamma, \\ [i\mathbf{J}(\vec{x})^a_0, \overline{\mathbf{\Psi}}(\vec{y})_\gamma] & = -\delta(\vec{y}-\vec{x})\mathcal{D}^{a\beta}_{\ \gamma}\overline{\mathbf{\Psi}}(\vec{y})_\beta. \end{array}$$

The Lie bracket on the current level reads

$$[i\mathbf{J}(\vec{x})^a_0, i\mathbf{J}(\vec{y})^b_0] = \delta(\vec{x}-\vec{y})\epsilon^{ab}_c i\mathbf{J}(\vec{x})^c_0.$$

4.10.2 External Lie Algebra

For particle fields the quantization opposite commutators of the momentum operators implement the Lie algebra of the direct group $\mathbf{U} \times \mathbb{R}^4$ with the fixgroup $\mathbf{U} \in \{\mathbf{SU}(2), \mathbf{SO}(2)\}$ in the Lorentz group for massive and massless particles. Thus the representations of the Poincaré Lie algebra $\log \mathbf{SL}(\mathbb{C}^2) \vec{\times} \mathbb{R}^4$ are obtained by the corresponding transmutators.

First the currents for the homogeneous operations: The Lie algebra of the rotation fixgroup $\mathbf{U}$ is represented in the quantum algebras for the complex quartets,

$$\log \mathbf{U} \ni \mathbf{l}^a \longmapsto \mathcal{D}^{aB}_{C} \frac{[\mathrm{u}^C, \mathrm{u}^\star_B]_{-\epsilon} - [\mathrm{a}_B, \mathrm{a}^{\star C}]_{-\epsilon}}{2} = \mathcal{D}^{aB}_{C} \begin{cases} \frac{[\mathbf{l}^C, \mathbf{l}^\star_B] + [\mathbf{r}^C, \mathbf{r}^\star_B]}{2}, & \epsilon = +1, \\ \frac{\{\mathbf{x}^C, i\mathbf{p}^\star_B\} + \{-i\mathbf{p}^C, \mathbf{x}^\star_B\}}{2}, & \epsilon = -1, \end{cases}$$

with representations of the Lie algebra of $\mathbf{SU}(2)$ or $\mathbf{SO}(2)$,

$$\begin{array}{lll} m > 0,\ a = 1,2,3: & \mathcal{D}(l^a) & \cong 0,\ \frac{i}{2}\sigma^{aB}_{C},\ \epsilon^{abc}, \dots, \\ m = 0,\ a = 3: & \mathcal{D}(l^3) & \cong 0,\ \frac{i}{2}\sigma^{3B}_{C},\ \epsilon^{3bc},\ \dots. \end{array}$$

For quantum particle fields it is implemented with the direct integral for the corresponding momentum operator expressions

$$\log \mathbf{U} \ni \mathbf{l}^a \longmapsto {}^{\oplus}\!\!\int \frac{d^3q}{2q_0(2\pi)^3} \mathbf{l}(\vec{q})^a = {}^{\oplus}\!\!\int \frac{d^3q}{2q_0(2\pi)^3} \mathcal{D}^{aB}_{C} \frac{[\mathrm{u}(\vec{q})^C, \mathrm{u}^\star(\vec{q})_B]_{-\epsilon} - [\mathrm{a}(\vec{q})_B, \mathrm{a}^\star(\vec{q})^C]_{-\epsilon}}{2}$$

with obvious adjoint actions on the momentum operators. The induced Lorentz Lie algebra representations have currents $\{\mathbf{J}^{mn}_k\}_{m,n,k=0,1,2,3}$:

$$\log \mathbf{SL}(\mathbb{C}^2) \ni \mathbf{l}^{mn} \longmapsto \mathcal{D}(\mathbf{l}^{mn}) = \int d^3x\ i\mathbf{J}(x)^{mn}_0.$$

The antisymmetric Lorentz Lie algebra currents in terms of the fields

$$i\mathbf{J}^{mn}_k = \begin{cases} \frac{\{\mathbf{Z}^j (\mathcal{L}^{mn})^l_j, i\mathbf{G}_{kl}\}}{2}, & \text{vector}, \\ \frac{[\mathbf{l}\sigma_k \hat{\sigma}^{mn}, \mathbf{l}^\star] + [\mathbf{r}\check{\sigma}_k \sigma^{mn}, \mathbf{r}^\star]}{2} = \frac{[\mathbf{\Psi}\gamma_k \gamma^{mn}, \overline{\mathbf{\Psi}}]}{2}, & \text{Weyl, Dirac}, \end{cases}$$

contain the Lorentz Lie algebra representation $\gamma^{mn} = -\frac{1}{4}[\gamma^m, \gamma^n] = \begin{pmatrix} \sigma^{mn} & 0 \\ 0 & \hat{\sigma}^{mn} \end{pmatrix}$ for Dirac fields with the chiral projections for the Weyl fields and the Minkowski representation $\mathcal{L}^{mn}$ for vector fields.

Now the currents for the spacetime translations: The time translations are represented by integrating the Hamiltonians for the momentum operators over the energy-momentum hyperboloid:

$$\begin{aligned} \mathbb{R} \ni \mathbf{p}^0 \longmapsto\ & iH_0 = im \frac{[\mathrm{u}, \mathrm{u}^\star]_{-\epsilon} + [\mathrm{a}, \mathrm{a}^\star]_{-\epsilon}}{2} = im \begin{cases} \mathbf{l}\mathbf{r}^\star + \mathbf{r}\mathbf{l}^\star, & \epsilon = +1, \\ \ell^2 \mathbf{p}\mathbf{p}^\star + \frac{1}{\ell^2}\mathbf{x}\mathbf{x}^\star, & \epsilon = -1, \end{cases} \\ \longmapsto\ & {}^{\oplus}\!\!\int \frac{d^3q}{2q_0(2\pi)^3} iH_0(\vec{q}) = {}^{\oplus}\!\!\int \frac{d^3q}{2q_0(2\pi)^3} iq_0 \frac{[\mathrm{u}(\vec{q}), \mathrm{u}^\star(\vec{q})]_{-\epsilon} + [\mathrm{a}(\vec{q}), \mathrm{a}^\star(\vec{q})]_{-\epsilon}}{2}. \end{aligned}$$

The embedding of all four spacetime translations $\mathbb{R}^4$ acting on relativistic particle fields requires a Lorentz tensor. Therefore the translation representations have four currents that constitute the *energy-momentum tensor* $\{\mathbf{T}^j_k\}^3_{k,j=0}$:

$$\begin{aligned} \mathbb{R}^4 \ni \mathbf{p}^j \longmapsto \mathcal{D}(\mathbf{p}^j) &= {}^{\oplus}\!\!\int \frac{d^3q}{2q_0(2\pi)^3} iP(\vec{q})^j = \int d^3x\ i\mathbf{T}(x)^j_0, \\ P(\vec{q})^j &= q^j \frac{[\mathrm{u}(\vec{q}), \mathrm{u}^\star(\vec{q})]_{-\epsilon} + [\mathrm{a}(\vec{q}), \mathrm{a}^\star(\vec{q})]_{-\epsilon}}{2}. \end{aligned}$$

The adjoint action gives the energy-momenta as eigenvalues:

$$\begin{array}{rclrcl} [\mathcal{D}(\mathbf{p}^j), \mathrm{u}(\vec{q})] &=& iq^j\mathrm{u}(\vec{q}), & [\mathcal{D}(\mathbf{p}^j), \mathrm{u}^\star(\vec{q})] &=& -iq^j\mathrm{u}^\star(\vec{q}), \\ [\mathcal{D}(\mathbf{p}^j), \mathrm{a}(\vec{q})] &=& iq^j\mathrm{a}(\vec{q}), & [\mathcal{D}(\mathbf{p}^j), \mathrm{a}^\star(\vec{q})] &=& -iq^j\mathrm{a}^\star(\vec{q}). \end{array}$$

4.10.3 Gauge Construction of Currents

Instead of by applying transmutators, the explicit expression of energy-momentum tensors for classical relativistic fields will be obtained by a variational method that is applicable for any Lie algebra, internal or external.

The Lie algebra $\log G$ elements, implemented for fields, are position integrals (time projections) of spacetime-dependent currents. A direct construction of the currents in a classical framework replaces the one global group transformations by a spacetime-dependent one, i.e., a transformation for each translation. For example, a unitary internal transformation group for the momentum operators acts spacetime dependently on the fields:

$$\begin{array}{l} U = e^{i\gamma} \in \mathbf{U}(N), \ \gamma = \gamma_a \mathcal{D}(l^a), \\ \mathbb{R}^4 \ni x \longmapsto U(x) \in \mathbf{U}(N) \ \Rightarrow \left\{ \begin{array}{llll} \mathbf{\Phi} & \longmapsto \mathbf{\Phi} U, & \partial^k \mathbf{\Phi} & \longmapsto [\partial^k + l^k(U)]\mathbf{\Phi} U, \\ \mathbf{\Phi}^\star & \longmapsto U^\star \mathbf{\Phi}^\star, & \partial^k \mathbf{\Phi}^\star & \longmapsto U^\star[\partial^k - l^k(U)]\mathbf{\Phi}^\star. \end{array} \right. \end{array}$$

Because of the translation-dependence the field derivatives are changed by a *pure gauge* (chapters "Spin, Rotations. and Position" and "Gauge Interactions"):

$$\begin{array}{l} \partial^k \longmapsto \partial^k \pm l^k(U), \ \ l^k(U) = (\partial^k U)U^\star = -U\partial^k U^\star = i\partial^k\gamma + \cdots, \\ \qquad\qquad\qquad\qquad [\partial^k - l^k(U)]U = 0. \end{array}$$

This defines a translation dependent transformation $x \longmapsto U_*(x)$ of the Lie algebra, i.e., one obtains a Lie algebra $\{l^k(U(x))\}$ for each translation

$$\begin{array}{l} l^a \longmapsto l^k(U) = (U_*)^k_a l^a, \ \ (U_*)^k_a = i\partial^k\gamma_a + \cdots, \\ [l^k(U), l^j(U)] = \partial^k l^j(U) - \partial^j l^k(U), \ \ l^k(\mathbf{1}_N) = \delta^k_a l^a. \end{array}$$

To obtain the currents, one starts from a group G-invariant action $\int d^4x\, \mathbf{L}(x)$ with the Lagrangian in terms of relativistic fields, kinetic term and a potential assumed without spacetime derivatives, e.g.,

$$\mathbf{L} = \left\{ \begin{array}{ll} (\partial_k \mathbf{\Phi})(\partial^k \mathbf{\Phi}^\star) - m^2 \mathbf{\Phi}\mathbf{\Phi}^\star & -\mathcal{V}(\mathbf{\Phi}), \\ \frac{1}{2}\epsilon^{lm}_{jk}(\partial^j \mathbf{Z}^k)(\partial_l \mathbf{Z}_m) + m^2 \frac{\mathbf{Z}^j \mathbf{Z}_j}{2} & -\mathcal{V}(\mathbf{Z}), \\ i\mathbf{r}\check{\sigma}_k \partial^k \mathbf{r}^\star + i\mathbf{l}\sigma_k \partial^k \mathbf{l}^\star - m(\mathbf{r}\mathbf{l}^\star + \mathbf{l}\mathbf{r}^\star) & -\mathcal{V}(\mathbf{r}, \mathbf{l}), \\ i\mathbf{\Psi}\gamma_k \partial^k \overline{\mathbf{\Psi}} - m\mathbf{\Psi}\overline{\mathbf{\Psi}} & -\mathcal{V}(\mathbf{\Psi}). \end{array} \right.$$

By translation-dependent transformations, the kinetic terms of the $\mathbf{U}(N)$-invariant Lagrangians are changed, for three examples from above

$$\mathbf{L} \longmapsto \mathbf{L}(l(U)) = \left\{ \begin{array}{l} [\partial_k \mathbf{\Phi} + \mathbf{\Phi} l_k(U)][\partial^k - l^k(U)]\mathbf{\Phi}^\star - m^2 \mathbf{\Phi}\mathbf{\Phi}^\star - \mathcal{V}(\mathbf{\Phi}), \\ i\mathbf{r}\check{\sigma}_k[\partial^k - l^k(U)]\mathbf{r}^\star + i\mathbf{l}\sigma_k[\partial^k - l^k(U)]\mathbf{l}^\star \\ \qquad\qquad\qquad\qquad -m(\mathbf{r}\mathbf{l}^\star + \mathbf{l}\mathbf{r}^\star) - \mathcal{V}(\mathbf{r}, \mathbf{l}), \\ i\mathbf{\Psi}\gamma_k[\partial^k - l^k(U)]\overline{\mathbf{\Psi}} - m\mathbf{\Psi}\overline{\mathbf{\Psi}} - \mathcal{V}(\mathbf{\Psi}). \end{array} \right.$$

The classical current is the variation of the Lagrangian with respect to the spacetime orientation of the Lie algebra at $U_* = 0$:

$$i\mathbf{J}^a_k = i\frac{\partial \ \mathbf{L}(l(U))}{\partial \ (U_*)^k_a}\Big|_{U_*=0} = \begin{cases} i\mathbf{\Phi}\mathcal{D}(l^a)\partial_k\mathbf{\Phi}^\star - i(\partial_k\mathbf{\Phi})\mathcal{D}(l^a)\mathbf{\Phi}^\star, \\ \mathbf{r}\check{\sigma}_k\mathcal{D}(l^a)\mathbf{r}^\star + \mathbf{l}\sigma_k\mathcal{D}(l^a)\mathbf{l}^\star, \\ \mathbf{\Psi}\gamma_k\mathcal{D}(l^a)\overline{\mathbf{\Psi}}. \end{cases}$$

The currents for the Lorentz Lie algebra can be derived analogously.

To obtain, with this method, the energy-momentum tensor for the translation generators $\mathbf{p}^j \longmapsto {}^\oplus\!\int \frac{d^3q}{2q_0(2\pi)^3} iP(\vec{q})^j = \int d^3x \ i\mathbf{T}(x)^j_0$, one starts from the transformation of the spacetime fields which is effected by translation-dependent spacetime derivatives and 1-forms, expressed by a tetrad $h \cong h^k_\mu$ and its inverse $h^{-1} \cong h^\mu_k$:

$$\check{\mathbb{R}}^4 \ni \partial^k \longmapsto h^k_\mu(x)\partial^\mu, \quad \mathbb{R}^4 \ni dx_k \longmapsto h^\mu_k(x)dx_\mu.$$

Since derivatives are affected, a first order formalism is used:

$$\mathbf{L}(\partial^k) = \begin{cases} \mathbf{\Phi}_k\partial^k\mathbf{\Phi} & -\frac{1}{2}\mathbf{\Phi}_k\mathbf{\Phi}^k - \frac{m^2}{2}\mathbf{\Phi}^2 & -\mathcal{V}(\mathbf{\Phi}), \\ \frac{1}{2}\mathbf{G}_{jl}\epsilon^{jl}_{km}\partial^k\mathbf{Z}^m & +\frac{\mathbf{G}_{jl}\mathbf{G}^{jl}}{4} + m^2\frac{\mathbf{Z}^m\mathbf{Z}_m}{2} & -\mathcal{V}(\mathbf{Z}), \\ i\mathbf{r}\check{\sigma}_k\partial^k\mathbf{r}^\star + i\mathbf{l}\sigma_k\partial^k\mathbf{l}^\star & -m(\mathbf{r}\mathbf{l}^\star + \mathbf{l}\mathbf{r}^\star) & -\mathcal{V}(\mathbf{r},\mathbf{l}), \\ i\mathbf{\Psi}\gamma_k\partial^k\overline{\mathbf{\Psi}} & -m\mathbf{\Psi}\overline{\mathbf{\Psi}} & -\mathcal{V}(\mathbf{\Psi}). \end{cases}$$

The local translation transformation is nontrivial not only for the Lagrangian density but also for the integration measure

$$d^4x \ \mathbf{L}(\partial^k) \longmapsto \det h^{-1} d^4x \ \mathbf{L}(h^k_\mu\partial^\mu).$$

The energy-momentum tensor is the variation with respect to the tetrad at the trivial transformation $h^k_\mu = \delta^k_\mu$:

$$\mathbf{T}^\mu_k = \frac{\partial \ \det h^{-1}\mathbf{L}(h^l_\nu\partial^\nu)}{\partial \ h^k_\mu}\Big|_{h=\mathbf{1}_4} = -\delta^k_\mu\mathbf{L}(\partial) + \begin{cases} \mathbf{\Phi}_k\partial^\mu\mathbf{\Phi}, \\ \frac{1}{2}\mathbf{G}_{jl}\epsilon^{jl}_{km}\partial^\mu\mathbf{Z}^m, \\ i\mathbf{r}\check{\sigma}_k\partial^\mu\mathbf{r}^\star + i\mathbf{l}\sigma_k\partial^\mu\mathbf{l}^\star, \\ i\mathbf{\Psi}\gamma_k\partial^\mu\overline{\mathbf{\Psi}}. \end{cases}$$

The measure transformation gives the scalar term $\frac{\partial \ \det h^{-1}}{\partial \ h^k_\mu}\Big|_{h=\mathbf{1}_4} = -\delta^k_\mu$.

The Lagrangian density is determined up to spacetime derivatives. The translation behavior can be generated by different energy-momentum tensors. The construction uses only the linear dependence in the tetrad.

Analogous to such a construction of the relativistic energy-momentum tensor is the following construction of the Hamiltonian in mechanics (without explicit time-dependence):

$$\begin{array}{l} \mathbf{L}(d_t) = \mathbf{p}d_t\mathbf{x} - \mathbf{H}(\mathbf{x},\mathbf{p}), \text{ e.g., } \mathbf{H}(\mathbf{x},\mathbf{p}) = \frac{\mathbf{p}^2}{2M} + \mathcal{V}(\mathbf{x}), \\ \left.\begin{array}{l} \frac{d}{dt} \longmapsto h(t)\frac{d}{dt} \\ dt \longmapsto h^{-1}(t)dt \end{array}\right\} \Rightarrow \begin{array}{rl} dt \ \mathbf{L}(d_t) \longmapsto & h^{-1}dt \ \mathbf{L}(hd_t) \\ = & dt \ [\mathbf{p}d_t\mathbf{x} - h^{-1}\mathbf{H}(\mathbf{x},\mathbf{p})]. \end{array} \\ \frac{\partial h^{-1}\mathbf{L}(hd_t)}{\partial h}\Big|_{h=1} = -\mathbf{L}(d_t) + \mathbf{p}d_t\mathbf{x} = \mathbf{H}(\mathbf{x},\mathbf{p}). \end{array}$$

4.11 Relativistic Scattering

For a relativistic description of scattering phenomena (chapter "The Kepler Factor") the scalar products in the scattering matrix and the projectors for particles have to be considered with their momentum-dependence. This kinematic part for the Hilbert space of free particles, considered in the following, has to be used together with matrix elements of interactions, usually taken from Lagrangians, e.g., for gauge interactions. The perturbative Dyson expansion for the scattering matrix is treated in many textbooks.

4.11.1 Stable Particle Hilbert Spaces

The momentum operators $\mathrm{u}(m^2, \vec{q}) = \mathrm{u}(\vec{q})$, etc. for a stable particle build a direct integral Hilbert space $L^2(\mathcal{Y}^3) \otimes \mathbb{C}^{1+2J}$ with the Lorentz-invariant non-renormalizable positive measure of the energy-momentum hyperboloid $\mathcal{Y}^3(m^2)$:

$$\begin{aligned} \int \frac{d^4q}{(2\pi)^3}\vartheta(q^0)\delta(q^2 - m^2) &= \int d^3\mathrm{y}(\tfrac{\vec{q}}{m}), \\ d^3\mathrm{y}(\tfrac{\vec{q}}{m}) &= \frac{d^3q}{(2\pi)^3 2E} = \frac{q^2dqd^2\omega}{(2\pi)^3 2E}, \quad E = \sqrt{m^2 + \vec{q}^2}. \end{aligned}$$

One has the distributive orthogonality and completeness with the corresponding projector (all direct sums are orthogonal direct):

$$\begin{aligned} |m_a^2, \vec{q}_a\rangle &= |\vec{q}_a\rangle = \mathrm{u}(\vec{q}_a)|0\rangle, \quad \mathrm{u}^\star(\vec{q}_a)|0\rangle = 0, \\ \langle \mathrm{u}^\star(\vec{p}_b)\mathrm{u}(\vec{q}_a)\rangle_{\mathrm{F}} &= \langle \vec{p}_b|\vec{q}_a\rangle = \delta^{m_b^2}_{m_a^2} 2E_a\delta(\tfrac{\vec{q}_a - \vec{p}_b}{2\pi}), \\ {}^{\oplus}\!\int d^3\mathrm{y}(\tfrac{\vec{q}}{m})|\vec{q}\rangle\langle\vec{q}| &\cong \mathcal{P}(m^2) = \mathcal{P}(m^2) \circ \mathcal{P}(m^2) \text{ on } L^2(\mathcal{Y}^3) = H(m^2); \end{aligned}$$

$\{|\vec{q}\rangle \bigm| \vec{q} \in \mathbb{R}^3\}$ are not Hilbert space vectors, they constitute a measure-related distributive basis. The Hilbert space consists of $d^3\mathrm{y}(\frac{\vec{q}}{m})$ square integrable momentum functions (wave packets). Additional spin and chargelike quantum numbers for compact groups with finite-dimensional irreducible representation are left out for simplicity. They can be easily included.

From the quantum algebras for Fermi and Bose there arise the distributive basis for multiparticles, possibly also different particles:

$$\begin{aligned} |\vec{q}_1, \dots, \vec{q}_n\rangle &= |\vec{q}_1\rangle \cdots |\vec{q}_n\rangle = \mathrm{u}(\vec{q}_1) \otimes \cdots \otimes \mathrm{u}(\vec{q}_n)|0\rangle, \\ \langle \vec{p}_m, \dots, \vec{p}_1|\vec{q}_1, \dots, \vec{q}_n\rangle &= \delta_{nm} \sum_{\text{permutations}} (-1)^{\mathrm{sign}\pi} 2E_1\delta(\tfrac{\vec{q}_1 - \vec{p}_{\pi(1)}}{2\pi}) \cdots 2E_n\delta(\tfrac{\vec{q}_n - \vec{p}_{\pi(n)}}{2\pi}). \end{aligned}$$

For Fermi, sign π is the signature of the permutation; for Bose, one has to take $(-1)^{\mathrm{sign}\ \pi} = +1$.

The product space gives rise to relativistic phase space integrals as convolution *products of the particle measures* (chapter "Spectrum of Spacetime")

$$\int \frac{d^4q_a}{(2\pi)^3}\vartheta(q_1^0)\delta(q_1^2 - m_1^2) \cdots \int \frac{d^4q_c}{(2\pi)^3}\vartheta(q_n^0)\delta(q_n^2 - m_n^2)\delta(\tfrac{q_1 + \cdots + q_n - Q}{2\pi}),$$

e.g., for two particles with the threshold factor Δ,

$$\begin{aligned}
(\mu_1 * \mu_2)(Q) &= \int \frac{d^4q_1 d^4q_2}{(2\pi)^6} \vartheta(q_1^0)\delta(q_1^2 - m_1^2)\delta(\tfrac{q_1+q_2-Q}{2\pi})\vartheta(q_2^0)\delta(q_2^2 - m_2^2) \\
&= \tfrac{1}{4\pi}\vartheta(Q^0)\vartheta(Q^2 - m_+^2)\frac{\sqrt{\Delta(Q^2, m_1^2, m_2^2)}}{2Q^2}, \\
\Delta(s_{12}, m_1^2, m_2^2) &= (s_{12} - m_+^2)(s_{12} - m_-^2), \quad \left\{ \begin{array}{ll} s_{12} &= (q_1 + q_2)^2, \\ m_\pm^2 &= (m_1 \pm m_2)^2, \end{array} \right. \\
\Delta(a, b, c) &= a^2 + b^2 + c^2 - 2(ab + ac + bc),
\end{aligned}$$

computable in a rest system $Q = (M, 0)$:

$$\begin{aligned}
(\mu_1 * \mu_2)(Q) &= \int_0^\infty \frac{q^2 dq}{4\pi E_1 E_2}\delta(E_1 + E_2 - M) = \frac{q^2}{4\pi E_1 E_2} \frac{1}{\frac{d(E_1+E_2)}{dq}}\Big|_{q=q_\delta} = \frac{q_\delta}{4\pi M} \\
\text{with} \quad & \left. \begin{array}{r} \delta(E_1 + E_2 - M) = 0 \\ E_{1,2} = \sqrt{q^2 + m_{1,2}^2} \end{array} \right\} \Rightarrow q^2 = \frac{\Delta(M^2, m_1^2, m_2^2)}{4M^2} = q_\delta^2, \\
& E_1 E_2 \frac{d(E_1+E_2)}{dq}\Big|_{q=q_\delta} = q(E_1 + E_2)\Big|_{q=q_\delta} = q_\delta M.
\end{aligned}$$

4.11.2 Scattering Scalar Products

The scattering operator $S = \mathbf{1} - 2\pi i T$ (chapter "The Kepler Factor") with the transition operator T is the double limit of the evolution operator involving a free and an interaction Lagrangian (Hamiltonian) $L = L^0 + L^{\text{int}}$:

$$S(t_f, t_i) = e^{iL^0 t_f} e^{-iL(t_f - t_i)} e^{iL^0 t_i}.$$

A perturbative Dyson expansion of the interaction with the free time development uses the time-ordered exponential of the interaction

$$\begin{aligned}
S = \lim_{t_i, t_f \to \mp\infty} \mathbf{T} e^{-i\int_{t_i}^{t_f} dt\, L^{\text{int}}(t)} &= \mathbf{1} - 2\pi i T = \mathbf{1} + 2\pi i \int dt\, L^{\text{int}}(t) + \cdots \\
L^{\text{int}}(t) &= e^{iL^0 t} L^{\text{int}} e^{-iL^0 t}.
\end{aligned}$$

For nontrivial position, Lagrangian densities are used with limits for interaction time $\mathcal{T}$ and interaction volume $\mathcal{V} = \mathcal{R}^3$:

$$S = \lim_{\mathcal{T}, \mathcal{R} \to \infty} \mathbf{T} e^{-i\int_{\mathcal{T} \times \mathcal{V}} d^4x\, \mathbf{L}^{\text{int}}(x)} = \mathbf{1} - 2\pi i T = 1 + 2\pi i \int d^4x\, \mathbf{L}^{\text{int}}(x) + \cdots.$$

Scattering matrix elements are transition amplitudes from inital to final state vectors. Their absolute squares for normalized vectors $\langle i|i\rangle = 1 = \langle f|f\rangle$ give transition probabilities

$$\begin{aligned}
p_{|i\rangle \to |f\rangle} &= \frac{\langle i|S^\star|f\rangle\langle f|S|i\rangle}{\langle i|i\rangle\langle f|f\rangle} = \operatorname{tr} \mathcal{P}_{|i\rangle} \circ S^\star \circ \mathcal{P}_{|f\rangle} \circ S, \quad \mathcal{P}_{|i\rangle} = \frac{|i\rangle\langle i|}{\langle i|i\rangle}, \\
\langle f|S|i\rangle &= \langle f|i\rangle - 2i\pi\langle f|T|i\rangle, \\
|\langle f|S|i\rangle|^2 &= |\langle f|i\rangle|^2 - 2\pi \operatorname{Re}[\langle i|f\rangle\langle f|T|i\rangle] + (2\pi)^2|\langle f|T|i\rangle|^2..
\end{aligned}$$

For orthogonal vectors $\langle f|i\rangle = 0$ there remain only the transition matrix elements. With several final vectors, one obtains the probability with the projector to the corresponding Hilbert subspace:

$$\begin{aligned}
\text{final subspace } \mathcal{P}_{|F\rangle} &= |F\rangle\langle F| = \sum_{a,b=1}^{n} |f_a\rangle(\zeta^{-1})^{ab}\langle f_b| \text{ if } \langle f_a|f_b\rangle = \zeta_{ab},\\
\langle i|S^\star \circ \mathcal{P}_{|F\rangle} \circ S|i\rangle &= |\langle F|S|i\rangle|^2 = \sum_{a,b=1}^{n} \langle i|S^\star|f_a\rangle(\zeta^{-1})^{ab}\langle f_b|S|i\rangle,
\end{aligned}$$

especially simple for orthogonal vectors $\langle f_a|f_b\rangle = \delta_{ab}$, e.g., for stable particles.

4.11.3 Momentum Scalar Products

The momentum wave packet Hilbert spaces $H(m^2)$ are based on the momentum "eigenvectors" (not Hilbert vectors)

$$|\vec{Q}\rangle = |\vec{q}_1,\dots,\vec{q}_n\rangle,\quad d\mu(\vec{Q}) = d^3\mathrm{y}(\tfrac{\vec{q}_1}{m_1})\cdots d^3\mathrm{y}(\tfrac{\vec{q}_n}{m_n}).$$

The S-operator has a sesquilinear decomposition with its continuously indexed momentum scalar products (matrix elements)

$$S \cong {}^{\oplus}\!\!\int d\mu(\vec{P})d\mu(\vec{Q})\ \ |\vec{Q}\rangle\langle\vec{Q}|S|\vec{P}\rangle\langle\vec{P}|$$

with the product projectors

$${}^{\oplus}\!\!\int d\mu(\vec{Q})|\vec{Q}\rangle\langle\vec{Q}| \cong \bigotimes_{a=1}^{n}\mathcal{P}(m_a^2) = \mathcal{P}(m_1^2,\dots,m_n^2).$$

For identical particles, Bose or Fermi structures have to be taken into account. Nontrivial spin has to be summed over as well.

The analogous subspace projectors have to be used for probabilities, e.g., for a transition from an initial Hilbert space to a final Hilbert space, both with all possible momenta:

$$\begin{aligned}
|\langle M_1^2,\dots,M_m^2|S|m_1^2,\dots,m_n^2\rangle|^2 &= \int d\mu(\vec{P})d\mu(\vec{Q})\langle\vec{P}|S|\vec{Q}\rangle\langle\vec{Q}|S^\star|\vec{P}\rangle\\
&= \int \frac{d^3p_1\cdots d^3p_m}{(2\pi)^{3m}2E(\vec{p}_1)\cdots 2E(\vec{p}_m)} \int \frac{d^3q_1\cdots d^3q_m}{(2\pi)^{3n}2E(\vec{q}_1)\cdots 2E(\vec{q}_n)}|\langle\vec{p}_1,\dots,\vec{p}_m|S|\vec{q}_1,\dots,\vec{q}_n\rangle|^2.
\end{aligned}$$

Thus one can pick probabilities for different experimental setups, e.g., for an initial "vector" with definite momenta $|\vec{P}^0\rangle = |\vec{p}_1^{\,0},\dots,\vec{p}_m^{\,0}\rangle$ (examples below):

$$\begin{aligned}
|\langle\vec{p}_1^{\,0},\dots,\vec{p}_m^{\,0}|S|m_a^2,\dots,m_c^2\rangle|^2 &= \int d\mu(\vec{P})\ \delta(\vec{P},\vec{P}^0)\ d\mu(\vec{Q})\langle\vec{P}|S|\vec{Q}\rangle\langle\vec{Q}|S^\star|\vec{P}\rangle\\
&= \int \frac{d^3p_1\cdots d^3p_m}{(2\pi)^{3m}2E(\vec{p}_1)\cdots 2E(\vec{p}_m)}\delta(\vec{p}_1-\vec{p}_1^{\,0})\cdots\delta(\vec{p}_m-\vec{p}_m^{\,0})\int d\mu(\vec{Q})|\langle\vec{P}|S|\vec{Q}\rangle|^2\\
&= \frac{1}{(2\pi)^{3m}2E(\vec{p}_1^{\,0})\cdots 2E(\vec{p}_m^{\,0})}\int \frac{d^3q_1\cdots d^3q_n}{(2\pi)^{3n}2E(\vec{q}_1)\cdots 2E(\vec{q}_n)}|\langle\vec{p}_1^{\,0},\dots,\vec{p}_m^{\,0}|S|\vec{q}_1,\dots,\vec{q}_n\rangle|^2.
\end{aligned}$$

Transition probabilities, computed naively with momentum "eigenvectors" (not Hilbert vectors), lead to meaningless products of Dirac distributions as seen already in the trivial term, the scalar product square

$$\langle\vec{p}|\vec{q}\rangle = 2E\delta(\tfrac{\vec{q}-\vec{p}}{2\pi}),\quad |\langle\vec{p}|\vec{q}\rangle|^2 = 4E^2\delta(\tfrac{\vec{q}-\vec{p}}{2\pi})\ \delta(\tfrac{\vec{q}-\vec{p}}{2\pi}).$$

This problem does not arise for a correct treatment using Hilbert space vectors with momentum wave packets. However, it is desirable to have a *quantum-compatible classical point particle language* where, e.g., an initial "vector" $|\vec{p}\rangle$ with one particle of definite momentum can be used for probabilities. This is possible by remembering the limits for interaction time and interaction volume:

$$\begin{array}{llllll} \int_{t_i}^{t_f} dt = \mathcal{T} & \to \int dt, & \int_{\mathbb{R}^3} d^3x = \mathcal{V} & \to \int d^3x, & \mathcal{T}\mathcal{V} & \to \int d^4x, \\ \frac{1}{\mathcal{T}} & \to \int \frac{dE}{2\pi}, & \frac{1}{\mathcal{V}} & \to \int \frac{d^3q}{(2\pi)^3}, & \frac{1}{\mathcal{T}\mathcal{V}} & \to \int \frac{d^4q}{(2\pi)^4}. \end{array}$$

Finite volume normalizations can be introduced to avoid the undefined products of distributions:

$$\begin{array}{rl} \delta(\frac{E}{2\pi}) = \int dt\ e^{iEt}, & \delta(\frac{\vec{q}}{2\pi}) = \int d^3x\ e^{-i\vec{q}\vec{x}}, \\ \delta(\frac{q}{2\pi})\delta(\frac{q}{2\pi}) = \delta(\frac{q}{2\pi}) \int d^4x\ e^{iqx} & = \delta(\frac{q}{2\pi}) \int d^4x = \delta(\frac{q}{2\pi})\mathcal{T}\mathcal{V}, \\ \text{e.g., } |\langle\vec{p}|\vec{q}\rangle|^2 & = \mathcal{V}4E^2\delta(\frac{\vec{q}-\vec{p}}{2\pi}). \end{array}$$

An energy-momenta diagonality of the S-operator can be made explicit by extracting a corresponding Dirac distribution

$$\langle\vec{Q}|S|\vec{P}\rangle = \langle\vec{Q}|\vec{P}\rangle - 2i\pi\langle\vec{Q}|T|\vec{P}\rangle = \langle\vec{Q}|\vec{P}\rangle + \delta(\tfrac{P-Q}{2\pi})iT^0_{if}(\vec{Q}),$$

e.g., for a constant transition element

$$2\pi\langle\vec{Q}|T|\vec{P}\rangle = T^0 \int d^4x\ e^{i(P-Q)x} = \delta(\tfrac{P-Q}{2\pi})T^0.$$

The related probability involves the limit of the interaction range, e.g., for orthogonal initial and final vectors:

$$\langle f|i\rangle = 0: \quad |\langle\vec{Q}|S|\vec{P}\rangle|^2 = [\delta(\tfrac{P-Q}{2\pi})]^2|T^0_{if}(\vec{Q})|^2 = \mathcal{T}\mathcal{V}\ \delta(\tfrac{P-Q}{2\pi})|T^0_{if}(\vec{Q})|^2.$$

4.11.4 Mean Lifetimes and Cross Sections

In the following examples, the transition probability from an initial "vector" $|\vec{p}_a,\ldots,\vec{p}_b\rangle$ with particles of definite momenta to a final "vector" $|m_1^2,\ldots,m_n^2\rangle$ with particles of any momenta is needed:

$$|\langle\vec{p}_a,\ldots,\vec{p}_b|T|m_1^2,\ldots,m_n^2\rangle|^2 = \mathcal{T}\mathcal{V}\ \int d\mu(\vec{Q})\ \delta(\tfrac{p_a+\cdots+p_b-Q}{2\pi})|T^0_{a\ldots b,f}(\vec{Q})|^2.$$

It is computed with the classical normalization, where a finite interaction volume $\mathcal{V}$ is used for the initial "vector"

$$\begin{array}{rl} {}^{\oplus}\!\int \frac{d^3q}{(2\pi)^3 2\sqrt{\vec{q}^2+m^2}}|\vec{q}\rangle\langle\vec{q}| \to & \frac{1}{\mathcal{V}2\sqrt{\vec{p}^2+m^2}} = |\vec{p}\rangle\langle\vec{p}|, \\ |\vec{p}_a,\ldots,\vec{p}_b\rangle\langle\vec{p}_a,\ldots,\vec{p}_b| = & \frac{1}{\mathcal{V}^n 2E_c\cdots 2E_b}. \end{array}$$

Such a transition probability occurs in the *partial width* for the decay of a particle at rest $p = (M, 0)$ into n particles. Interaction time and volume drop out in the combination $\frac{\mathcal{T}\mathcal{V}}{\mathcal{T}\mathcal{V}2E} = \frac{1}{2M}$:

$$\Gamma_{|i\rangle\to|f\rangle} = \frac{|\langle\vec{p}|T|m_1^2,\ldots,m_n^2\rangle|^2}{\mathcal{T}} = \frac{1}{2M}\int d\mu(\vec{Q})\ \delta(\tfrac{p-Q}{2\pi})|T^0_{i,f}(0)|^2, \quad \vec{p} = 0.$$

The two-particle decay $|m_1^2, m_2^2\rangle$ involves the projector $\mathcal{P}_{|f\rangle} = \mathcal{P}(m_1^2, m_2^2)$ and the product measure with the threshold (relativistic phase space) factor, computed above,

$$\Gamma_{|M\rangle\to|m_1+m_2\rangle} = \vartheta(M^2 - m_+^2)\frac{\sqrt{\Delta(M^2,m_1^2,m_2^2)}}{8\pi M^3}|T^0_{M,m_1+m_2}(0)|^2.$$

The total width is the sum over all decay channels $\mathcal{P}_{|F\rangle}$.

The flux $\Phi_{ab} = \frac{v_{ab}}{\mathcal{V}}$ for two incoming classical particles $|\vec{p}_1, \vec{p}_2\rangle$ contains their relative velocity v_{ab}. The invariant product of the flux with the energies and the interaction volume is the square root of the two-particle threshold factor Δ:

$$\Phi_{ab}2E_aE_b\mathcal{V} = 2E_aE_bv_{ab} \overset{\vec{p}_b=0}{=} 2E_am_bv_a = 2m_b|\vec{p}_a| = \sqrt{\Delta(s_{ab}, m_a^2, m_b^2)}.$$

The transition probability from two initial particles of definite momenta to final particles of any momenta, divided by interaction time $\mathcal{T}$ and flux is the *total cross section*, with orthogonal vectors $\langle i|f\rangle = 0$ and classical two-particle normalization $\frac{1}{\mathcal{V}^2 4E_aE_b}$,

$$\begin{aligned}
\sigma^{tot}_{|a+b\rangle\to|f\rangle}(Q) &= \frac{|\langle\vec{p}_a,\vec{p}_b|T|m_1^2,\dots,m_n^2\rangle|^2}{\mathcal{T}\Phi_{ab}} \\
&= \frac{1}{2\sqrt{\Delta(Q^2,m_a^2,m_b^2)}}\int d\mu(\vec{Q})\delta(\tfrac{p_a+p_b-Q}{2\pi})|T^0_{ab,f}(Q)|^2 \\
\text{with } \frac{\mathcal{T}\mathcal{V}}{\mathcal{T}\Phi_{ab}\mathcal{V}^2 4E_aE_b} &= \frac{1}{2\sqrt{\Delta(Q^2,m_a^2,m_b^2)}}.
\end{aligned}$$

The *scattering angle* θ_{cd} for two final particles arises in the relativistic invariant of the squared energy-momentum difference

$$\begin{aligned}
t_{cd} &= (q_c - q_d)^2 = m_c^2 + m_d^2 + 2(|\vec{q}_c||\vec{q}_d|\cos\theta_{cd} - E_cE_d), \\
\frac{d}{dt_{cd}} &= \frac{1}{|\vec{q}_c||\vec{q}_d|}\frac{d}{d\cos\theta_{cd}}.
\end{aligned}$$

Hence the *differential cross section* can be picked from the total one

$$\begin{aligned}
\frac{d\sigma_{|a+b\rangle\to|c+d\rangle}(Q,t_{cd})}{dt_{cd}} &= \frac{1}{2\sqrt{\Delta(Q^2,m_a^2,m_b^2)}}\int d\mu(\vec{Q})\delta((q_c - q_d)^2 - t_{cd}) \\
&\quad\times \delta(\tfrac{p_a+p_b-q_c-q_d}{2\pi})|T^0_{ab,cd}(Q)|^2.
\end{aligned}$$

4.12 Summary

Relativistic particle quantum fields (canonically quantized fields) are built with particle and antiparticle creation and annihilation momentum operators as distributive bases of a direct integral Hilbert space with Lorentz invariant boost (momentum) measure. The infinite-dimensional Hilbert representations of the Poincaré group $\mathbf{SL}(\mathbb{C}^2) \vec{\times} \mathbb{R}^4$ are induced by compact representations of the particle group (direct product of spacetime translations and spin) $\mathbb{R}^4 \times \mathbf{SU}(2) \to \mathbf{U}(1+2J)$ for mass $m > 0$. Transmutators relate the fields, acted on by finite-dimensional representations of the Lorentz group, to the creation and annihilation momentum operators, acted on by the spin group. The quantization (duality) comes in a Lorentz vector.

The Feynman particle propagator connects, via $+io$ in $\frac{1}{q^2+io-m^2}$, the causal order with the creation-annihilation order. The propagator contains the Fock form on-shell functions from $\delta(q^2 - m^2)$ (positive kinetic energy real particles with induced representations of the Poincaré group) and the $\epsilon(x_0)$-multiplied quantization distributions from $\frac{1}{q_{\rm P}^2-m^2}$ (also off-shell, "virtual particles," not Hilbert representations of the Poincaré group), which embeds Yukawa interactions.

Representations of Lie algebras for internal groups and for the external Poincaré group, induced by representations of the "little" Lie algebras acting on the momentum operators, are position integrals (time projection) of currents that are constructed with the quantization opposite commutators.

The momentum operators for particles and antiparticles define a momentum dependent algebra via tensor products like $\mathrm{u}(\vec{q})\otimes\mathrm{a}(\vec{p})$ with the momentum distributed structures of a quantum algebra.

quantization	$[\mathrm{u}^\star(\vec{p})_a, \mathrm{u}(\vec{q})^b]_\epsilon = \delta^b_a 2q_0\delta(\frac{\vec{q}-\vec{p}}{2\pi}) = [\mathrm{a}^\star(\vec{p})^b, \mathrm{a}(\vec{q})_a]_\epsilon$
Fock form	$\langle[\mathrm{u}^\star(\vec{p})_a, \mathrm{u}(\vec{q})^b]_{-\epsilon}\rangle_{\rm F} = \delta^b_a 2q_0\delta(\frac{\vec{q}-\vec{p}}{2\pi}) = \langle[\mathrm{a}^\star(\vec{p})^b, \mathrm{a}(\vec{q})_a]_{-\epsilon}\rangle_{\rm F}$

momentum operators

mass $m > 0$, spin J or circularity $\pm 2j_3$, charge z	name, Bose or Fermi	field with $q_0 = \sqrt{m^2+\vec{q}^2}$ and Feynman propagator
$J = 0$ – $z = 0$	scalar $\epsilon = -1$	$\mathbf{\Phi}(x) = \oplus\!\!\int \frac{d^3q}{2q_0(2\pi)^3}\, \gamma\, [e^{iqx}\mathrm{u}(\vec{q}) + e^{-iqx}\mathrm{u}^\star(\vec{q})]$ $\langle\{\mathbf{\Phi},\mathbf{\Phi}\} - \epsilon(x_0)[\mathbf{\Phi},\mathbf{\Phi}]\rangle_{\rm F}(x) = \frac{i}{\pi}\int \frac{d^4q}{(2\pi)^3}\frac{\gamma^2}{q^2+io-m^2}e^{iqx}$
$J = 1$ – $z = 0$	vector $\epsilon = -1$	$\mathbf{Z}(x)^j = \oplus\!\!\int \frac{d^3q}{2q_0(2\pi)^3}\, \gamma\Lambda(\frac{q}{m})^j_a\, [e^{iqx}\mathrm{u}(\vec{q})^a + e^{-iqx}\mathrm{u}^\star_a(\vec{q})]$ $\langle\{\mathbf{Z}^j,\mathbf{Z}^k\} - \epsilon(x_0)[\mathbf{Z}^j,\mathbf{Z}^k]\rangle_{\rm F}(x) = \frac{i}{\pi}\int \frac{d^4q}{(2\pi)^3}\frac{\gamma^2(-\eta^{jk}+\frac{q^jq^k}{m^2})}{q^2+io-m^2}e^{iqx}$
$J = \frac{1}{2}$ – $z = \pm 1$	Dirac $\epsilon = +1$	$\mathbf{\Psi}(x) = \begin{pmatrix}\mathbf{r}(x)^A\\ \mathbf{l}(x)^{\dot{A}}\end{pmatrix} = \oplus\!\!\int \frac{d^3q}{2q_0(2\pi)^3}\begin{pmatrix}\sqrt{m}\hat{s}^{-1}(\frac{q}{m})^A_C\, [e^{iqx}\mathrm{u}(\vec{q})^C + e^{-iqx}\mathrm{a}^\star(\vec{q})^C]\\ \sqrt{m}s^{-1}(\frac{q}{m})^{\dot{A}}_C\, [e^{iqx}\mathrm{u}(\vec{q})^C - e^{-iqx}\mathrm{a}^\star(\vec{q})^C]\end{pmatrix}$ $\langle[\overline{\mathbf{\Psi}},\mathbf{\Psi}] - \epsilon(x_0)\{\overline{\mathbf{\Psi}},\mathbf{\Psi}\}\rangle_{\rm F}(x) = \frac{i}{\pi}\int \frac{d^4q}{(2\pi)^3}\frac{\gamma_jq^j+m}{q^2+io-m^2}e^{iqx}$
$J = \frac{1}{2}$ – $z = 0$	Majorana $\epsilon = +1$	$\mathbf{r}(x)^A = \oplus\!\!\int \frac{d^3q}{2q_0(2\pi)^3}\, \sqrt{m}\hat{s}^{-1}(\frac{q}{m})^A_C\, [e^{iqx}\mathrm{u}(\vec{q})^C + e^{-iqx}i\epsilon^{CB}\check{\mathrm{u}}(\vec{q})_B]$

massive particle quantum fields

Bibliography

[1] J.D. Bjorken, S. Drell, *Relativistische Quantenmechanik* (1966), BI Hochschultaschenbücher, Mannheim.

[2] J.D. Bjorken, S. Drell, *Relativistische Quantenfelder* (1966), BI Hochschultaschenbücher, Mannheim.

[3] N.N. Bogliubov, D.V. Shirkov, *Introduction to the Theory of Quantized Fields* (1959), Interscience, New York.

[4] F. Scheck, *Theoretische Physik 4 (Quantisierte Felder)* (2001), Springer.

[5] S. Weinberg, *The Quantum Theory of Fields I + II* (1995, 1996), Cambridge University Press, Cambridge.

5

MASSLESS QUANTUM FIELDS

Massless particle use, for their irreducible Hilbert representation of the Poincaré group, spacetime decompositions into time and position translations with one distinguished rotational axis. The position axis is fixed by the momentum direction of the never-resting particle and determined up to axial rotations $\mathbf{SO}(2)$ whose representations determine the circularity (helicity, polarization) of the particle. Axial rotations act on particle pairs with opposite circularity $\pm 2J \in \mathbb{Z}$. Strictly speaking, massless particles have no $\mathbf{SU}(2)$-spin; they have $\mathbf{SO}(2)$-polarization or helicity.

An axial rotation fixgroup $\mathbf{SO}(2)$ in a noncompact Euclidean fixgroup with two boosts $\mathbf{SO}(2) \vec{\times} \mathbb{R}^2 \subset \mathbf{SO}_0(1,3)$ gives additional structures compared with the massive case and a rotation fixgroup in the Lorentz group $\mathbf{SO}(3) \subset \mathbf{SO}_0(1,3)$: With the embedding of particles with axial rotation $\mathbf{SO}(2)$ properties into quantum fields with finite-dimensional Lorentz group representations, there can arise translation representations not only in the probability group $\mathbf{U}(1)$, but also in the noncompact group $\mathbf{U}(1,1)$ (indefinite metric). Massless quantum particle fields can have degrees of freedom without probabilistic particle interpretation, i.e., without state vectors in a Hilbert space. Nonparticle degrees of freedom in relativistic fields describe genuine interactions, e.g., the Coulomb interaction, which comes in addition to the two photons in the four components of an electromagnetic vector field.

The spacetime translation development of a mass-zero vector field involves eigenvectors (particles) and lightcone-related nilvectors. The eigenvector property is expressible by a trivial action of the nil-Hamiltonian, which in a quantum theory is equivalent to a trivial action of the nilquadratic Becchi-Rouet-Stora charge, constructed with the probability interpretation securing Fermi Fadeev-Popov scalar fields. The classical limit of the BRS-transformation gives Lie algebra transformations with spacetime-dependent parameters, which replace the Fadeev-Popov fields and are familiar as "gauge transformations." The translation eigenvectors with trivial BRS-charge are "gauge invariant."

After a review of indefinite unitary time translations as implemented in quantum algebras and the definition of a Hilbert space for translation eigenvectors, the relativistic embedding of massless particles is given with definite and indefinite metric degrees of freedom in their quantum fields.

5.1 Noncompact Time Representations in Quantum Algebras

The nondecomposable complex 2-dimensional time representations with invariant energy (frequency) m and basis-dependent nilconstant ν on a complex 2-dimensional vector space are in the noncompact group $\mathbf{U}(1,1)$ (chapter "Time Representations"). They are faithful:

$$\mathbb{R} \ni t \longmapsto e^{imt}\begin{pmatrix} 1 & i\nu t \\ 0 & 1 \end{pmatrix} = \begin{pmatrix} 1 & \nu \frac{d}{dm} \\ 0 & 1 \end{pmatrix} e^{imt} \in \mathbf{U}(1,1) \subset \mathbf{GL}(\mathbb{C}^2), \quad m, \nu \in \mathbb{R}, \quad \nu \neq 0.$$

Dual bases of the representation spaces

$$\mathrm{b}, \mathrm{g} \in V \cong \mathbb{C}^2 \cong V^T \ni \mathrm{g}^\times, \mathrm{b}^\times$$

have the equations of motion, with $d_t = \frac{d}{dt}$, and time orbits

$$d_t\begin{pmatrix} \mathrm{b} \\ \mathrm{g} \end{pmatrix} = i\begin{pmatrix} m & \nu \\ 0 & m \end{pmatrix}\begin{pmatrix} \mathrm{b} \\ \mathrm{g} \end{pmatrix}, \quad d_t(\mathrm{g}^\times, \mathrm{b}^\times) = -i(\mathrm{g}^\times, \mathrm{b}^\times)\begin{pmatrix} m & \nu \\ 0 & m \end{pmatrix},$$

$$\begin{pmatrix} \mathrm{b} \\ \mathrm{g} \end{pmatrix}(t) = e^{imt}\begin{pmatrix} 1 & i\nu t \\ 0 & 1 \end{pmatrix}\begin{pmatrix} \mathrm{b} \\ \mathrm{g} \end{pmatrix}, \quad (\mathrm{g}^\times, \mathrm{b}^\times)(t) = (\mathrm{g}^\times, \mathrm{b}^\times)e^{-imt}\begin{pmatrix} 1 & -i\nu t \\ 0 & 1 \end{pmatrix}.$$

Only g and $\mathrm{g}^\times$ are time development eigenvectors, the letters $\{\mathrm{g}, \mathbf{g}, \mathbf{G}, \gamma\}$ stand for "good" (eigenvectors) and $\{\mathrm{b}, \mathbf{b}, \mathbf{B}, \beta\}$ for "bad" (nilvectors).

For trivial nilconstant $\nu = 0$ there remain two $\mathbf{U}(1)$-representations, which are compatible with $\mathbf{U}(1,1)$-conjugation $\times$ and $\mathbf{U}(1)$-conjugation $\star$:

$$\nu = 0 : \begin{cases} (\mathrm{b}, \mathrm{g}) & = (\mathrm{a}, \mathrm{u}), \\ (\mathrm{b}^\times, \mathrm{g}^\times) & = (\mathrm{a}^\times, \mathrm{u}^\times) = (\mathrm{u}^\star, \mathrm{a}^\star). \end{cases}$$

The notation $\{\mathrm{b}, \mathrm{g}\}$ will also be used for $\nu = 0$ with $\mathbf{U}(1,1)$-conjugation.

A $\mathbf{U}(1,1)$-symmetric basis of $\mathbf{V} = V \oplus V^T \cong \mathbb{C}^4$

$$\mathbf{b}_+ = \frac{\mathrm{b}+\mathrm{b}^\times}{\sqrt{2}}, \quad \mathbf{b}_- = \frac{\mathrm{b}-\mathrm{b}^\times}{i\sqrt{2}}, \quad \mathbf{g}_+ = \frac{\mathrm{g}+\mathrm{g}^\times}{\sqrt{2}}, \quad \mathbf{g}_- = \frac{\mathrm{g}-\mathrm{g}^\times}{i\sqrt{2}},$$

is acted on by a real time representation

$$\begin{aligned} d_t(\mathbf{g}_+, \mathbf{g}_-, \mathbf{b}_+, \mathbf{b}_-) &= (\mathbf{g}_+, \mathbf{g}_-, \mathbf{b}_+, \mathbf{b}_-)h_1(m, \nu), \\ h_1(m, \nu) &= \left(\begin{array}{cc|cc} 0 & m & 0 & \nu \\ -m & 0 & -\nu & 0 \\ \hline 0 & 0 & 0 & m \\ 0 & 0 & -m & 0 \end{array}\right) = \begin{pmatrix} m & \nu \\ 0 & m \end{pmatrix} \otimes \begin{pmatrix} 0 & 1 \\ -1 & 0 \end{pmatrix} \\ &= \begin{pmatrix} 1 & \nu \frac{d}{dm} \\ 0 & 1 \end{pmatrix} \otimes h_0(m), \quad h_0(m) = m\begin{pmatrix} 0 & 1 \\ -1 & 0 \end{pmatrix}, \end{aligned}$$

with noncompact spiraling orbits in $\mathbf{SO}_0(2,2)$:

$$(\mathbf{g}_+,\mathbf{g}_-,\mathbf{b}_+,\mathbf{b}_-)(t) = (\mathbf{g}_+,\mathbf{g}_-,\mathbf{b}_+,\mathbf{b}_-)e^{h_1(m,\nu)t},$$

$$e^{h_1(m,\nu)t} = \left(\begin{array}{cc|cc} \cos mt & \sin mt & -\nu t\sin mt & \nu t\cos mt \\ -\sin mt & \cos mt & -\nu t\cos mt & -\nu t\sin mt \\ \hline 0 & 0 & \cos mt & \sin mt \\ 0 & 0 & -\sin mt & \cos mt \end{array}\right) = \begin{pmatrix} 1 & \nu\frac{d}{dm} \\ 0 & 1 \end{pmatrix} \otimes e^{h_0(m)t},$$

$$\begin{aligned} \text{e.g., } \mathbf{b}_+(t) &= \frac{\mathrm{b}(t)+\mathrm{b}^\times(t)}{\sqrt{2}} = \frac{(\mathrm{b}+i\nu t\mathrm{g})e^{imt}+(\mathrm{b}^\times - i\nu t\mathrm{g}^\times)e^{-imt}}{\sqrt{2}} \\ &= \mathbf{b}_+\cos mt - \mathbf{b}_-\sin mt - \nu(\mathbf{g}_- t\sin mt + \mathbf{g}_- t\cos mt). \end{aligned}$$

Only for $m=0$ is the real time representation $e^{h_1(0,\nu)t}$ decomposable:

$$m = 0 \Rightarrow \begin{cases} d_t(-\mathbf{g}_-,\mathbf{b}_+) &= (-\mathbf{g}_-,\mathbf{b}_+)\begin{pmatrix} 0 & \nu \\ 0 & 0 \end{pmatrix}, \\ d_t(\mathbf{g}_+,\mathbf{b}_-) &= (\mathbf{g}_+,\mathbf{b}_-)\begin{pmatrix} 0 & \nu \\ 0 & 0 \end{pmatrix}. \end{cases}$$

The characteristic nontrivial (anti-) commutators in the quantum algebra for Fermi $\epsilon = +1$ and Bose $\epsilon = -1$ are

$$\text{in } \mathbf{Q}_\epsilon(\mathbb{C}^4): \quad [\mathrm{b}^\times,\mathrm{g}]_\epsilon = 1 = [\mathrm{g}^\times,\mathrm{b}]_\epsilon.$$

The noncompact Hamiltonian

$$H_1 = H_1^\times = m\frac{[\mathrm{g},\mathrm{b}^\times]_{-\epsilon}+[\mathrm{b},\mathrm{g}^\times]_{-\epsilon}}{2} + \nu\frac{[\mathrm{g},\mathrm{g}^\times]_{-\epsilon}}{2}, \quad \frac{d}{dt} = i\,\mathrm{ad}\,H_1$$

leads to the time representation as time-dependent quantization

$$\begin{pmatrix} [\mathrm{g}^\times,\mathrm{b}]_\epsilon & [\mathrm{b}^\times,\mathrm{b}]_\epsilon \\ [\mathrm{g}^\times,\mathrm{g}]_\epsilon & [\mathrm{b}^\times,\mathrm{g}]_\epsilon \end{pmatrix}(t) = \begin{pmatrix} 1 & \nu\frac{d}{dm} \\ 0 & 1 \end{pmatrix} e^{imt} = \begin{pmatrix} 1 & i\nu t \\ 0 & 1 \end{pmatrix} e^{imt}.$$

Both for the Fermi and the Bose cases *dual normalization factors* $\sigma,\rho > 0$ are possible in a $\mathbf{U}(1,1)$-symmetric formulation:

$$\text{Fermi}: \begin{Bmatrix} \mathbf{B}_+ = \sigma\frac{\mathrm{b}+\mathrm{b}^\times}{\sqrt{2}}, & \mathbf{B}_- = \rho\frac{\mathrm{b}-\mathrm{b}^\times}{i\sqrt{2}}, \\ \mathbf{G}_+ = \frac{1}{\sigma}\frac{\mathrm{g}+\mathrm{g}^\times}{\sqrt{2}}, & \mathbf{G}_- = \frac{1}{\rho}\frac{\mathrm{g}-\mathrm{g}^\times}{i\sqrt{2}}, \end{Bmatrix} \Rightarrow \{\mathbf{B}_+,\mathbf{G}_+\} = 1 = \{\mathbf{B}_-,\mathbf{G}_-\},$$

$$\text{Bose}: \begin{Bmatrix} \mathbf{B}_+ = \sigma\frac{\mathrm{b}+\mathrm{b}^\times}{\sqrt{2}}, & \mathbf{B}_- = \rho\frac{\mathrm{b}-\mathrm{b}^\times}{i\sqrt{2}}, \\ \mathbf{G}_+ = \frac{1}{\rho}\frac{\mathrm{g}+\mathrm{g}^\times}{\sqrt{2}}, & \mathbf{G}_- = \frac{1}{\sigma}\frac{\mathrm{g}-\mathrm{g}^\times}{i\sqrt{2}}, \end{Bmatrix} \Rightarrow [-i\mathbf{B}_-,\mathbf{G}_+] = 1 = [-i\mathbf{G}_-,\mathbf{B}_+].$$

To obtain the time-dependent quantization as derivation of an irreducible representation also in the symmetric formulation, the dual normalization factors have to be trivial for the Fermi case:

$$\epsilon = +1: \begin{cases} \left(\begin{array}{cc|cc} \{\mathbf{G}_-,\mathbf{B}_-\} & \{\mathbf{G}_+,\mathbf{B}_-\} & \{\mathbf{B}_-,\mathbf{B}_-\} & \{\mathbf{B}_+,\mathbf{B}_-\} \\ \{\mathbf{G}_-,\mathbf{B}_+\} & \{\mathbf{G}_+,\mathbf{B}_+\} & \{\mathbf{B}_-,\mathbf{B}_+\} & \{\mathbf{B}_+,\mathbf{B}_+\} \\ \hline \{\mathbf{G}_-,\mathbf{G}_-\} & \{\mathbf{G}_+,\mathbf{G}_-\} & \{\mathbf{B}_-,\mathbf{G}_-\} & \{\mathbf{B}_+,\mathbf{G}_-\} \\ \{\mathbf{G}_-,\mathbf{G}_+\} & \{\mathbf{G}_+,\mathbf{G}_+\} & \{\mathbf{B}_-,\mathbf{G}_+\} & \{\mathbf{B}_+,\mathbf{G}_+\} \end{array}\right)(t) \\ = \left(\begin{array}{cc|cc} \cos mt & \frac{\rho}{\sigma}\sin mt & -\nu\rho^2 t\sin mt & \nu\sigma\rho t\cos mt \\ -\frac{\sigma}{\rho}\sin mt & \cos mt & -\nu\sigma\rho t\cos mt & -\nu\sigma^2 t\sin mt \\ \hline 0 & 0 & \cos mt & \frac{\sigma}{\rho}\sin mt \\ 0 & 0 & -\frac{\rho}{\sigma}\sin mt & \cos mt \end{array}\right) \\ = \begin{pmatrix} 1 & \nu\frac{d}{dm} \\ 0 & 1 \end{pmatrix}\begin{pmatrix} \cos mt & \sin mt \\ -\sin mt & \cos mt \end{pmatrix} \text{ for } \sigma = \frac{1}{\rho} = 1. \end{cases}$$

For the Bose parametrization one nontrivial normalization factor is possible:

$$\epsilon = -1: \left\{ \begin{array}{l} \left(\begin{array}{cc|cc} [-i\mathbf{G}_-,\mathbf{B}_+] & [\mathbf{G}_+,\mathbf{B}_+] & [-i\mathbf{B}_-,\mathbf{B}_+] & [\mathbf{B}_+,\mathbf{B}_+] \\ [\mathbf{G}_-,\mathbf{B}_-] & [\mathbf{G}_+,i\mathbf{B}_-] & [\mathbf{B}_-,\mathbf{B}_-] & [\mathbf{B}_+,i\mathbf{B}_-] \\ \hline [-i\mathbf{G}_-,\mathbf{G}_+] & [\mathbf{G}_+,\mathbf{G}_+] & [-i\mathbf{B}_-,\mathbf{G}_+] & [\mathbf{B}_+,\mathbf{G}_+] \\ [\mathbf{G}_-,\mathbf{G}_-] & [\mathbf{G}_+,i\mathbf{G}_-] & [\mathbf{B}_-,\mathbf{G}_-] & [\mathbf{B}_+,i\mathbf{G}_-] \end{array}\right)(t) \\ = \left(\begin{array}{cc|cc} \cos mt & i\frac{\sigma}{\rho}\sin mt & -\nu\sigma\rho t\sin mt & i\nu\sigma^2 t\cos mt \\ i\frac{\rho}{\sigma}\sin mt & \cos mt & i\nu\rho^2 t\cos mt & -\nu\sigma\rho t\sin mt \\ \hline 0 & 0 & \cos mt & i\frac{\sigma}{\rho}\sin mt \\ 0 & 0 & i\frac{\rho}{\sigma}\sin mt & \cos mt \end{array}\right) \\ = \begin{pmatrix} 1 & \nu\frac{d}{dm} \\ 0 & 1 \end{pmatrix}\begin{pmatrix} \cos mt & i\ell^2\sin mt \\ \frac{i}{\ell^2}\sin mt & \cos mt \end{pmatrix} \text{ for } \sigma = \frac{1}{\rho} = \ell > 0. \end{array}\right.$$

In the following, the different dual structures for Bose and Fermi are looked at in more detail.

5.1.1 The Bose Case

A doubled position-momentum notation for the $\times$-symmetric vectors in the Bose case,

$$\epsilon = -1: \left\{ \begin{array}{ll} \check{\mathrm{x}} = \ell\frac{\mathrm{b}+\mathrm{b}^\times}{\sqrt{2}}, & \mathrm{x} = \frac{1}{\ell}\frac{\mathrm{b}-\mathrm{b}^\times}{i\sqrt{2}} \\ \mathrm{p} = \ell\frac{\mathrm{g}+\mathrm{g}^\times}{\sqrt{2}}, & \check{\mathrm{p}} = \frac{1}{\ell}\frac{\mathrm{g}-\mathrm{g}^\times}{i\sqrt{2}} \end{array}\right\} \Rightarrow [i\mathrm{p},\mathrm{x}] = 1 = -[i\check{\mathrm{p}},\check{\mathrm{x}}],$$

gives the Hamiltonian

$$H_1 = m\frac{\frac{1}{\ell^2}\{\mathrm{p},\check{\mathrm{x}}\}+\ell^2\{\check{\mathrm{p}},\mathrm{x}\}}{2} + \nu\frac{\frac{1}{\ell^2}\{\mathrm{p},\mathrm{p}\}+\ell^2\{\check{\mathrm{p}},\check{\mathrm{p}}\}}{4}$$

and the equations of motion

$$\begin{array}{ll} & \left\{ \begin{array}{ll} d_t\mathrm{x} = \frac{1}{\ell^2}(m\check{\mathrm{x}}+\nu\mathrm{p}), & d_t\mathrm{p} = -m\ell^2\check{\mathrm{p}}, \\ d_t\check{\mathrm{x}} = -\ell^2(m\mathrm{x}+\nu\check{\mathrm{p}}), & d_t\check{\mathrm{p}} = \frac{m}{\ell^2}\mathrm{p} \end{array}\right. \\ \Rightarrow & \left\{ \begin{array}{ll} (d_t^2+m^2)\mathrm{x} = -2m\nu\check{\mathrm{p}}, & (d_t^2+m^2)\check{\mathrm{p}} = 0, \\ (d_t^2+m^2)\check{\mathrm{x}} = -2m\nu\mathrm{p}, & (d_t^2+m^2)\mathrm{p} = 0 \end{array}\right. \\ \Rightarrow & (d_t^2+m^2)^2(\mathrm{x},\mathrm{p},\check{\mathrm{x}},\check{\mathrm{p}}) = 0. \end{array}$$

They can be derived from a classical Lagrangian

$$L_B(\mathrm{x},\check{\mathrm{x}},\mathrm{p},\check{\mathrm{p}}) = \mathrm{p}d_t\mathrm{x} - \check{\mathrm{p}}d_t\check{\mathrm{x}} - m(\tfrac{1}{\ell^2}\mathrm{p}\check{\mathrm{x}} - \ell^2\check{\mathrm{p}}\mathrm{x}) - \tfrac{\nu}{2}(\tfrac{1}{\ell^2}\mathrm{p}^2 + \ell^2\check{\mathrm{p}}^2).$$

The time-dependent quantization reads

$$\left(\begin{array}{cc|cc} [-i\check{\mathrm{p}},\check{\mathrm{x}}] & [\mathrm{p},\check{\mathrm{x}}] & [-i\mathrm{x},\check{\mathrm{x}}] & [\mathrm{x},\check{\mathrm{x}}] \\ [\check{\mathrm{p}},\mathrm{x}] & [i\mathrm{p},\mathrm{x}] & [\mathrm{x},\mathrm{x}] & [i\mathrm{x},\mathrm{x}] \\ \hline [-i\check{\mathrm{p}},\mathrm{p}] & [\mathrm{p},\mathrm{p}] & [-i\mathrm{x},\mathrm{p}] & [\mathrm{x},\mathrm{p}] \\ [\check{\mathrm{p}},\check{\mathrm{p}}] & [i\mathrm{p},\check{\mathrm{p}}] & [\mathrm{x},\check{\mathrm{p}}] & [i\mathrm{x},\check{\mathrm{p}}] \end{array}\right)(t) = \left(\begin{array}{cc|cc} \cos mt & \ell^2 i\sin mt & -\nu t\sin mt & \nu\ell^2 it\cos mt \\ \frac{i}{\ell^2}\sin mt & \cos mt & \frac{i\nu}{\ell^2}t\cos mt & -\nu t\sin mt \\ \hline 0 & 0 & \cos mt & \ell^2 i\sin mt \\ 0 & 0 & \frac{i}{\ell^2}\sin mt & \cos mt \end{array}\right).$$

The *Bose parametrization for $m = 0$ is decomposable* into two real 2-dimensional nondecomposable representations with the dual pairs (x,p) and $(\check{\mathrm{x}},-\check{\mathrm{p}})$, reflecting the dynamics of two free mass points:

$$L_B(\mathrm{x}, \check{\mathrm{x}}, \mathrm{p}, \check{\mathrm{p}}) = \mathrm{p} d_t \mathrm{x} - \check{\mathrm{p}} d_t \check{\mathrm{x}} - (\tfrac{\mathrm{p}^2}{2M} + \tfrac{\check{\mathrm{p}}^2}{2M'}), \quad \begin{cases} d_t \mathrm{x} = \frac{\mathrm{p}}{M}, & d_t \mathrm{p} = 0, \\ d_t \check{\mathrm{x}} = -\frac{\check{\mathrm{p}}}{M'}, & d_t \check{\mathrm{p}} = 0, \end{cases}$$

with $\frac{1}{M} = \frac{\nu}{\ell^2}$, $\frac{1}{M'} = \nu \ell^2$,

$$\left(\begin{array}{cc|cc} [-i\check{\mathrm{p}}, \check{\mathrm{x}}] & [\mathrm{p}, \check{\mathrm{x}}] & [-i\mathrm{x}, \check{\mathrm{x}}] & [\check{\mathrm{x}}, \check{\mathrm{x}}] \\ \hline [\check{\mathrm{p}}, \mathrm{x}] & [i\mathrm{p}, \mathrm{x}] & [\mathrm{x}, \mathrm{x}] & [i\check{\mathrm{x}}, \mathrm{x}] \\ [-i\check{\mathrm{p}}, \mathrm{p}] & [\mathrm{p}, \mathrm{p}] & [-i\mathrm{x}, \mathrm{p}] & [\check{\mathrm{x}}, \mathrm{p}] \\ \hline [\check{\mathrm{p}}, \check{\mathrm{p}}] & [i\mathrm{p}, \check{\mathrm{p}}] & [\mathrm{x}, \check{\mathrm{p}}] & [i\check{\mathrm{x}}, \check{\mathrm{p}}] \end{array}\right)(t) = \left(\begin{array}{c|cc|c} 1 & 0 & 0 & \frac{it}{M'} \\ \hline 0 & 1 & \frac{it}{M} & 0 \\ 0 & 0 & 1 & 0 \\ \hline 0 & 0 & 0 & 1 \end{array}\right).$$

5.1.2 The Fermi Case

For the Fermi case the $\times$-symmetric vectors

$$\epsilon = +1 : \begin{Bmatrix} \check{\beta} = \frac{\mathrm{b}+\mathrm{b}^\times}{\sqrt{2}}, & \beta = \frac{\mathrm{b}-\mathrm{b}^\times}{i\sqrt{2}}, \\ \gamma = \frac{\mathrm{g}+\mathrm{g}^\times}{\sqrt{2}}, & \check{\gamma} = \frac{\mathrm{g}-\mathrm{g}^\times}{i\sqrt{2}} \end{Bmatrix} \Rightarrow \{\check{\beta}, \gamma\} = 1 = \{\beta, \check{\gamma}\},$$

are acted on by the Hamiltonian

$$H_1 = im\frac{[\check{\gamma}, \check{\beta}] + [\beta, \gamma]}{2} + i\nu\frac{[\check{\gamma}, \gamma]}{2}.$$

The equations of motion

$$\begin{aligned} &\begin{cases} d_t\beta = m\check{\beta} + \nu\gamma, & d_t\gamma = -m\check{\gamma}, \\ d_t\check{\beta} = -(m\beta + \nu\check{\gamma}), & d_t\check{\gamma} = m\gamma \end{cases} \\ \Rightarrow &\begin{cases} (d_t^2 + m^2)\beta = -2m\nu\check{\gamma}, & (d_t^2 + m^2)\check{\gamma} = 0, \\ (d_t^2 + m^2)\check{\beta} = -2m\nu\gamma, & (d_t^2 + m^2)\gamma = 0 \end{cases} \\ \Rightarrow &\ (d_t^2 + m^2)^2(\beta, \check{\beta}, \gamma, \check{\gamma}) = 0, \end{aligned}$$

can be derived from a classical Lagrangian with anticommuting Grassmann vectors

$$L_F(\beta, \check{\beta}, \gamma, \check{\gamma}) = i\gamma d_t\check{\beta} + i\check{\gamma} d_t\beta - im(\check{\gamma}\check{\beta} + \beta\gamma) - i\nu\check{\gamma}\gamma.$$

The time-dependent anticommutators are

$$\left(\begin{array}{cc|cc} \{\check{\gamma}, \beta\} & \{\gamma, \beta\} & \{\beta, \beta\} & \{\check{\beta}, \beta\} \\ \{\check{\gamma}, \check{\beta}\} & \{\gamma, \check{\beta}\} & \{\beta, \check{\beta}\} & \{\check{\beta}, \check{\beta}\} \\ \hline \{\check{\gamma}, \check{\gamma}\} & \{\gamma, \check{\gamma}\} & \{\beta, \check{\gamma}\} & \{\check{\beta}, \check{\gamma}\} \\ \{\check{\gamma}, \gamma\} & \{\gamma, \gamma\} & \{\beta, \gamma\} & \{\check{\beta}, \gamma\} \end{array}\right)(t) = \left(\begin{array}{cc|cc} \cos mt & \sin mt & -\nu t \sin mt & \nu t \cos mt \\ -\sin mt & \cos mt & -\nu t \cos mt & -\nu t \sin mt \\ \hline 0 & 0 & \cos mt & \sin mt \\ 0 & 0 & -\sin mt & \cos mt \end{array}\right).$$

The Fermi parametrization with dual pairs $(\gamma, \check{\beta})$ and $(\check{\gamma}, \beta)$ is never decomposable; it needs always four symmetric operators, also for the massless case $m = 0$:

$$L_F(\beta, \check{\beta}, \gamma, \check{\gamma}) = i\gamma d_t\check{\beta} + i\check{\gamma} d_t\beta - i\nu\check{\gamma}\gamma, \quad \begin{cases} d_t\beta = \nu\gamma, & d_t\gamma = 0, \\ d_t\check{\beta} = -\nu\check{\gamma}, & d_t\check{\gamma} = 0, \end{cases}$$

$$\left(\begin{array}{cc|cc} \{\check{\gamma}, \beta\} & \{\gamma, \beta\} & \{\beta, \beta\} & \{\check{\beta}, \beta\} \\ \{\check{\gamma}, \check{\beta}\} & \{\gamma, \check{\beta}\} & \{\beta, \check{\beta}\} & \{\check{\beta}, \check{\beta}\} \\ \hline \{\check{\gamma}, \check{\gamma}\} & \{\gamma, \check{\gamma}\} & \{\beta, \check{\gamma}\} & \{\check{\beta}, \check{\gamma}\} \\ \{\check{\gamma}, \gamma\} & \{\gamma, \gamma\} & \{\beta, \gamma\} & \{\check{\beta}, \gamma\} \end{array}\right)(t) = \left(\begin{array}{cc|cc} 1 & 0 & 0 & \nu t \\ 0 & 1 & -\nu t & 0 \\ \hline 0 & 0 & 1 & 0 \\ 0 & 0 & 0 & 1 \end{array}\right).$$

5.2 Indefinite Metric in Quantum Algebras

The construction of Hilbert spaces from quantum algebras $\mathbf{Q}_\epsilon(\mathbb{C}^4)$ with indefinite unitary time representations $\mathbb{R} \longrightarrow \mathbf{U}(1,1)$ is not so obvious. A product Fock space for $\mathbf{Q}_\epsilon(\mathbb{C}^4)$ with $\mathbf{U}(1,1)$-conjugation

$$\mathrm{FOCK}_\epsilon(\mathbb{C}^2) = \mathrm{FOCK}_\epsilon(\mathbb{C}) \otimes \mathrm{FOCK}_\epsilon(\mathbb{C}) = \mathbf{Q}_\epsilon(\mathbb{C}^4)/\mathbf{Q}_\epsilon(\mathbb{C}^4)(\mathrm{b}^\times + \mathrm{g}^\times)$$

with the left ideal $\mathbf{Q}_\epsilon(\mathbb{C}^4)(\mathrm{b}^\times + \mathrm{g}^\times) = \mathbf{Q}_\epsilon(\mathbb{C}^4)\mathrm{b}^\times + \mathbf{Q}_\epsilon(\mathbb{C}^4)\mathrm{g}^\times$ carries, induced by the $\mathbf{U}(1,1)$-conjugation, an indefinite sesquilinear form

$$\left.\begin{array}{ll} \langle \mathrm{b}|\mathrm{g}\rangle_{\mathrm{F}} = \langle \mathrm{b}^\times \mathrm{g}\rangle_{\mathrm{F}} = 1, & \langle \mathrm{b}|\mathrm{b}\rangle_{\mathrm{F}} = 0 \\ \langle \mathrm{g}|\mathrm{b}\rangle_{\mathrm{F}} = \langle \mathrm{g}^\times \mathrm{b}\rangle_{\mathrm{F}} = 1, & \langle \mathrm{g}|\mathrm{g}\rangle_{\mathrm{F}} = 0 \end{array}\right\} \Rightarrow \langle \mathrm{g} \pm \mathrm{b}|\mathrm{g} \pm \mathrm{b}\rangle_{\mathrm{F}} = \pm 2.$$

This prevents a probability interpretation for the full quantum algebra.

To find a probability interpretation also for a noncompact time development with the time translation representation

$$H_1 = m\frac{[\mathrm{b},\mathrm{g}^\times]_{-\epsilon} + [\mathrm{g},\mathrm{b}^\times]_{-\epsilon}}{2} + \nu \mathrm{g}\mathrm{g}^\times = mI + \nu N,$$

a crucial difference between the Bose and the Fermi quantum algebras has to be taken into account. The quantum nil-Hamiltonian N whose trivial adjoint action characterizes time translation eigenvectors

$$N = \mathrm{g}\mathrm{g}^\times : \quad [N, \mathrm{g}] = 0, \quad [N, \mathrm{b}] \neq 0, \quad [N, H_1] = 0$$

is nilquadratic for the Fermi case as well with the quantum algebra product

$$\text{in Fermi } \mathbf{Q}_+(\mathbb{C}^4) : \quad N^2 = 0 \text{ since } \{\mathrm{g}, \mathrm{g}\} = 2\mathrm{g}^2 = 0.$$

This is not the case in the Bose quantum algebra, which has no nontrivial zero divisors, i.e., $ab = 0 \Rightarrow a = 0$ or $b = 0$,

$$\text{in Bose } \mathbf{Q}_-(\mathbb{C}^4) : \quad N^2 \neq 0.$$

5.2.1 The Bose Case

To obtain a nilquadratic operator, which defines eigenvectors also in the Bose quantum algebra, the Bose structures are doubled by Fermi structures as done for quantum gauge fields, which are paired with Fadeev-Popov fields (below). There arises a twin structure in a $\mathbb{Z}_2$-graded quantum algebra with both a Bose sector (upper case letters $\{\mathrm{G}, \mathrm{B}\}$) and a Fermi sector (lower case letters $\{\mathrm{g}, \mathrm{b}\}$). There are eight basic degrees of freedom:

$$\mathbf{Q}_\pm(\mathbb{C}^8) = \mathbf{Q}_-(\mathbb{C}^4) \otimes \mathbf{Q}_+(\mathbb{C}^4) \text{ with } \left\{\begin{array}{ll} [\mathrm{G}^\times, \mathrm{B}] = 1, & [\mathrm{B}^\times, \mathrm{G}] = 1, \\ \{\mathrm{g}^\times, \mathrm{b}\} = 1, & \{\mathrm{b}^\times, \mathrm{g}\} = 1, \end{array}\right.$$

$$H_{B+F} = H_B + H_F, \quad \left\{\begin{array}{ll} H_B & = m\frac{\{\mathrm{B},\mathrm{G}^\times\} + \{\mathrm{G},\mathrm{B}^\times\}}{2} + \nu \mathrm{G}\mathrm{G}^\times, \\ H_F & = m\frac{[\mathrm{b},\mathrm{g}^\times] + [\mathrm{g},\mathrm{b}^\times]}{2} + \nu \mathrm{g}\mathrm{g}^\times. \end{array}\right.$$

By mixing basic Bose and Fermi degrees of freedom, a nilquadratic operator of Fermi type can be constructed, called *Becchi-Rouet-Stora operator*:

$$N_{BF} = \mathrm{gG}^\times + \mathrm{Gg}^\times \Rightarrow [H_{B+F}, N_{BF}] = 0, \quad N_{BF}^2 = 0.$$

Its graded adjoint action

$$\begin{array}{l} \mathrm{ad}\, N_{BF}(a) = \left\{ \begin{array}{ll} [N_{BF}, a] & \text{for } a \text{ Bose}, \\ \{N_{BF}, a\} & \text{for } a \text{ Fermi}, \end{array} \right. \\ [N_{BF}, \mathrm{G}] = 0, \quad \{N_{BF}, \mathrm{g}\} = 0, \quad [N_{BF}, H_{B+F}] = 0, \end{array}$$

defines, by a trivial eigenvalue, the unital subalgebra spanned by the time translation eigenvectors (time eigenalgebra)

$$\mathrm{INV}_{N_{BF}} \mathbf{Q}_\pm(\mathbb{C}^8) = \{p \in \mathbf{Q}_\pm(\mathbb{C}^8) \mid \mathrm{ad}\, N_{BF}(p) = 0\}.$$

With the doubling and a 4-dimensional basic vector space $\begin{pmatrix} \mathrm{b} \\ \mathrm{g} \\ \mathrm{B} \\ \mathrm{G} \end{pmatrix}$ one has the block-diagonal Hamiltonian and the block-skew-diagonal BRS-matrix

$$\begin{aligned} h_{B+F} &= h_B \oplus h_F = \begin{pmatrix} h & 0 \\ 0 & h \end{pmatrix} = \left(\begin{array}{cc|cc} m & \nu & 0 & 0 \\ 0 & m & 0 & 0 \\ \hline 0 & 0 & m & \nu \\ 0 & 0 & 0 & m \end{array}\right), \\ n_{BF} &= \begin{pmatrix} 0 & n \\ n & 0 \end{pmatrix} = \left(\begin{array}{cc|cc} 0 & 0 & 0 & 1 \\ 0 & 0 & 0 & 0 \\ \hline 0 & 1 & 0 & 0 \\ 0 & 0 & 0 & 0 \end{array}\right). \end{aligned}$$

The product Fock space has an indefinite metric both for Fermi and Bose:

$$\begin{array}{l} \mathrm{FOCK}_\pm(\mathbb{C}^4) = \mathrm{FOCK}_-(\mathbb{C}^2) \otimes \mathrm{FOCK}_+(\mathbb{C}^2), \\ \text{with } \langle \mathrm{G} \pm \mathrm{B} | \mathrm{G} \pm \mathrm{B} \rangle_{\mathrm{F}} = \pm 2, \quad \langle \mathrm{g} \pm \mathrm{b} | \mathrm{g} \pm \mathrm{b} \rangle_{\mathrm{F}} = \pm 2. \end{array}$$

The subspace with the time translation eigenvectors

$$\{|p\rangle_{\mathrm{F}} \in \mathrm{FOCK}_\pm(\mathbb{C}^4) \mid N_{BF}|p\rangle_{\mathrm{F}} = 0\}$$

contains, up to $|1\rangle_{\mathrm{F}}$ (the class of the algebra unit 1) with $\langle 1|1\rangle_{\mathrm{F}} = 1$, only normless vectors (ghosts), e.g., $\langle \mathrm{g}|\mathrm{g}\rangle_{\mathrm{F}} = 0 = \langle \mathrm{G}|\mathrm{G}\rangle_{\mathrm{F}}$, i.e., its metric is semidefinite. The associated Hilbert space $\mathbb{C}|1\rangle_{\mathrm{F}}$ contains only the classes of the scalars.

5.2.2 The Fermi Case

For the Fermi quantum algebra $\mathbf{Q}_+(\mathbb{C}^4) \cong \mathbb{C}^{16}$ the trace-induced irreducible nonabelian form, compatible with the $\mathbf{U}(1,1)$-conjugation and the quantization, is given by (chapter "Quantum Probability"):

$$\text{on } \mathbf{Q}_+(\mathbb{C}^4): \left\{ \begin{array}{l} \langle 1 \rangle_{\mathrm{H}} = 1, \\ \langle \mathrm{bg}^\times \rangle_{\mathrm{H}} = \frac{1}{2} \mathrm{tr} \begin{pmatrix} 0 & 0 \\ 0 & 1 \end{pmatrix} = \frac{1}{2}, \quad \langle \mathrm{gb}^\times \rangle_{\mathrm{H}} = \frac{1}{2} \mathrm{tr} \begin{pmatrix} 1 & 0 \\ 0 & 0 \end{pmatrix} = \frac{1}{2}, \\ \langle \mathrm{gg}^\times \rangle_{\mathrm{H}} = \frac{1}{2} \mathrm{tr} \begin{pmatrix} 0 & 1 \\ 0 & 0 \end{pmatrix} = 0, \quad \langle \mathrm{bb}^\times \rangle_{\mathrm{H}} = \frac{1}{2} \mathrm{tr} \begin{pmatrix} 0 & 0 \\ 1 & 0 \end{pmatrix} = 0, \\ \langle \mathrm{gg}^\times \mathrm{bb}^\times \rangle_{\mathrm{H}} = \frac{1}{2} \mathrm{tr} \begin{pmatrix} 0 & 1 \\ 0 & 0 \end{pmatrix} \begin{pmatrix} 0 & 0 \\ 1 & 0 \end{pmatrix} = \frac{1}{2}, \\ \Rightarrow \langle \{\mathrm{g}^\times, \mathrm{b}\} \rangle_{\mathrm{H}} = \langle \{\mathrm{b}^\times, \mathrm{g}\} \rangle_{\mathrm{H}} = 1, \quad \langle [\mathrm{g}^\times, \mathrm{b}] \rangle_{\mathrm{H}} = \langle [\mathrm{b}^\times, \mathrm{g}] \rangle_{\mathrm{H}} = 0. \end{array} \right.$$

This form is in analogy to the Killing form for semisimple Lie algebras, which is trivial for the traceless Lie algebra elements, e.g., $\operatorname{tr}\vec{\sigma} = 0$, but may be nontrivial for the "double" trace, e.g., $\frac{1}{2}\operatorname{tr}\sigma^a\sigma^t = \delta^{ab}$. The induced sesquilinear quantum algebra form is indefinite:

$$\begin{array}{l} \mathbf{Q}_+(\mathbb{C}^4) \times \mathbf{Q}_+(\mathbb{C}^4) \longrightarrow \mathbb{C}, \quad \langle a|b\rangle_{\rm H} = \langle a^\times b\rangle_{\rm H}, \\ \left. \begin{array}{l} \langle \mathrm{b}|\mathrm{g}\rangle_{\rm H} = \langle \mathrm{b}^\times \mathrm{g}\rangle_{\rm H} = \frac{1}{2}, \quad \langle \mathrm{b}|\mathrm{b}\rangle_{\rm H} = 0, \\ \langle \mathrm{g}|\mathrm{b}\rangle_{\rm H} = \langle \mathrm{g}^\times \mathrm{b}\rangle_{\rm H} = \frac{1}{2}, \quad \langle \mathrm{g}|\mathrm{g}\rangle_{\rm H} = 0 \end{array} \right\} \Rightarrow \langle \mathrm{g} \pm \mathrm{b}|\mathrm{g} \pm \mathrm{b}\rangle_{\rm H} = \pm 1. \end{array}$$

The trivial adjoint action of the nil-Hamiltonian defines the unital time eigenalgebra spanned by the time translation eigenvectors:

$$\begin{array}{rl} N = \mathrm{gg}^\times, & \mathrm{INV}_N\mathbf{Q}_+(\mathbb{C}^4) = \{a \in \mathbf{Q}_+(\mathbb{C}^4) \mid [N, a] = 0\} \cong \mathbb{C}^{10}, \\ \text{basis of} & \mathrm{INV}_N\mathbf{Q}_+(\mathbb{C}^4) : \{1, \mathrm{g}, \mathrm{g}^\times, I, \mathrm{gg}^\times, \mathrm{bg}, \mathrm{g}^\times\mathrm{b}^\times, \mathrm{g}I, I\mathrm{g}^\times, I^2\}. \end{array}$$

The sesquilinear form restricted to the time eigenalgebra $\langle p_1|p_2\rangle_{\rm H}$ can be nontrivial only on the unital subalgebra with the grade-0 elements,

$$\begin{array}{rl} & \mathrm{INV}_{I,N}\mathbf{Q}_+(\mathbb{C}^4) = \{a \in \mathbf{Q}_+(\mathbb{C}^4) \mid [I, a] = 0 = [N, a]\} \cong \mathbb{C}^4, \\ \text{basis of} & \mathrm{INV}_{I,N}\mathbf{Q}_+(\mathbb{C}^4) : \{1, I, \mathrm{gg}^\times, I^2\}. \end{array}$$

It is semidefinite and nontrivial only on the nontrivial scalars:

$$\begin{array}{lll} p_1^\times p_2 \notin \mathrm{INV}_{I,N}\mathbf{Q}_+(\mathbb{C}^4) & \Rightarrow \langle p_1^\times p_2\rangle_{\rm H} = 0, \\ p_1^\times p_2 \in \mathrm{INV}_{I,N}\mathbf{Q}_+(\mathbb{C}^4) & \iff p_1^\times p_2 = \alpha_0 \cdot 1 + \alpha_1 I + \alpha_2 \mathrm{gg}^\times + \alpha_3 I^2, \; \alpha_i \in \mathbb{C} \\ & \Rightarrow \langle p_1^\times p_2\rangle_{\rm H} = \alpha_0 \langle 1\rangle_{\rm H}. \end{array}$$

The associated Hilbert space is given by the left ideal classes:

$$\begin{array}{l} \mathrm{INV}_N\mathbf{Q}_+(\mathbb{C}^4)/\,\mathrm{INV}_N\mathbf{Q}_+(\mathbb{C}^4)(\mathrm{g} + \mathrm{g}^\times) \\ \cong \{|p\rangle_{\rm H} \mid N|p\rangle_{\rm H} = 0, \; \langle p|p\rangle_{\rm H} \ge 0\} \cong \{\alpha_0|1\rangle_{\rm H} \mid \alpha_0 \in \mathbb{C}\} \cong \mathbb{C}. \end{array}$$

From the basis of $\mathrm{INV}_N\mathbf{Q}_+(\mathbb{C}^4)$ only the algebra unit 1 gives a nontrivial norm vector $|1\rangle_{\rm H}$. All the other nine nontrivial time translation eigenvectors in the given basis are ghosts, e.g., $\langle \mathrm{g}^\times\mathrm{g}\rangle_{\rm H} = \langle \mathrm{gg}^\times\rangle_{\rm H} = 0$, and orthogonal to 1. As for the doubled Bose case, the associated Hilbert space $\mathbb{C}|1\rangle_{\rm H}$ contains only the classes of the scalars

5.3 Relativistic Distributions of Noncompact Time Representations

For relativistic fields, the matrix elements of the noncompact time representations $\mathbb{R} \ni t \longmapsto e^{imt}\begin{pmatrix} 1 & i\nu t \\ 0 & 1 \end{pmatrix} \in \mathbf{U}(1,1)$ are embedded into Lorentz compatible distributions of spacetime translations.

As to be expected from the residual representation of the characteristic noncompact coefficients by energy dipoles,

$$ite^{imt} = \oint \frac{dE}{2i\pi} \frac{1}{(E-m)^2} e^{iEt}, \quad -t\sin mt = \oint \frac{dE}{i\pi} \frac{2mE}{(E^2-m^2)^2} e^{iEt},$$

also the embedded causally supported spacetime distributions with the derivatives of the distributions $(\mathbf{c}_j, \mathbf{s})$ for massive particles $m \geq 0$ (chapter "Massive Quantum Particle Fields")

$$\mathbb{R}^4 \ni x \longmapsto \mathbf{D}_{\rm doub}(m|x) = \begin{pmatrix} \mathbf{D} & \nu\frac{d}{dm}\mathbf{D} \\ \clubsuit & \mathbf{D} \end{pmatrix}(m|x), \quad \mathbf{D}(m|x) = \begin{pmatrix} \mathbf{c}_j & i\mathbf{s} \\ i\mathbf{s} & \mathbf{c}_j \end{pmatrix}(m|x),$$

involve dipoles, in the form of derived Dirac distributions

$$\begin{aligned} \frac{d}{dm}\begin{pmatrix} i\mathbf{s} \\ \mathbf{c}_j \end{pmatrix}(m|x) &= \frac{d}{dm}\int \frac{d^4q}{(2\pi)^3}\begin{pmatrix} m \\ q_j \end{pmatrix}\epsilon(q_0)\delta(m^2-q^2)e^{iqx} \\ &= \int \frac{d^4q}{(2\pi)^3}\epsilon(q_0)\begin{pmatrix} \delta(m^2-q^2) - 2m^2\delta'(q^2-m^2) \\ -2mq_j\delta'(q^2-m^2) \end{pmatrix}e^{iqx}. \end{aligned}$$

They embed the characteristic time representation matrix elements $x_0 \sin q_0 x_0$ and $x_0 \cos q_0 x_0$ with $q_0 = \sqrt{m^2 + \vec{q}^2}$ ("on shell"):

$$\frac{d}{dm}\begin{pmatrix} i\mathbf{s} \\ \mathbf{c}_j \end{pmatrix}(m|x) = \int \frac{d^3q}{q_0(2\pi)^3}e^{-i\vec{q}\vec{x}}\begin{pmatrix} i\sin q_0x_0 - im^2\frac{\sin q_0x_0 - q_0x_0\cos q_0x_0}{q_0^2} \\ \begin{pmatrix} -mx_0\sin q_0x_0 \\ -im\vec{q}\,\frac{\sin q_0x_0 - q_0x_0\cos q_0x_0}{q_0^2} \end{pmatrix} \end{pmatrix}.$$

The additional position-related 2×2 contribution $\clubsuit(m|x)$ will be discussed in more detail below.

The time projection by position integration displays the noncompact time representations

$$\begin{aligned} \int d^3x \frac{d}{dm}\mathbf{D}(m|x) &= \frac{d}{dm}\begin{pmatrix} \delta_j^0\cos mx_0 & i\sin mx_0 \\ i\sin mx_0 & \delta_j^0\cos mx_0 \end{pmatrix} = \begin{pmatrix} -\delta_j^0 x_0\sin mx_0 & ix_0\cos mx_0 \\ ix_0\cos mx_0 & -\delta_j^0 x_0\sin mx_0 \end{pmatrix} \\ &\qquad \longrightarrow \begin{pmatrix} 0 & i \\ i & 0 \end{pmatrix}x_0 \text{ for } m \to 0, \\ \int d^3x\, \clubsuit(m|x) &= 0. \end{aligned}$$

The position projection of the $\epsilon(x_0)$-multiplied coeficients ("off shell") by time integration leads to Yukawa and exponential potential and force:

$$\begin{aligned} 2\pi\int dx_0\epsilon(x_0)\frac{d}{dm}\mathbf{D}(m|x) &= \frac{d}{dm}\begin{pmatrix} -\delta_j^a\frac{x_a}{r}\frac{1+mr}{r} & im \\ im & -\delta_j^a\frac{x_a}{r}\frac{1+mr}{r} \end{pmatrix}\frac{e^{-mr}}{r} \\ &= \begin{pmatrix} \delta_j^a m x_a & i(1-mr) \\ i(1-mr) & \delta_j^a m x_a \end{pmatrix}\frac{e^{-mr}}{r} \\ &\longrightarrow \begin{pmatrix} 0 & i \\ i & 0 \end{pmatrix}\frac{1}{r} \text{ for } m \to 0. \end{aligned}$$

The generalized functions for the massless case

$$\begin{aligned} \mathbf{D}(0|x) &= \begin{pmatrix} \mathbf{c}_j(0|x) & 0 \\ 0 & \mathbf{c}_j(0|x) \end{pmatrix}, \quad \lim_{m\to 0}\frac{d}{dm}\mathbf{D}(m|x) = \begin{pmatrix} 0 & i\mathbf{s}(x) \\ i\mathbf{s}(x) & 0 \end{pmatrix}, \\ \mathbf{D}_{\rm doub}(0|x) &= \left(\begin{array}{cc|cc} \mathbf{c}_j(0|x) & 0 & 0 & \nu i\mathbf{s}(x) \\ 0 & \mathbf{c}_j(0|x) & \nu i\mathbf{s}(x) & 0 \\ \hline 0 & \clubsuit & \mathbf{c}_j(0|x) & 0 \\ \clubsuit & 0 & 0 & \mathbf{c}_j(0|x) \end{array}\right), \end{aligned}$$

have no dipoles and no derived Dirac distributions $\delta'(q^2)$,

$$\begin{aligned} i\mathbf{s}(x) = \lim_{m\to 0}\frac{i\mathbf{s}(m|x)}{m} &= \int \frac{d^4q}{(2\pi)^3}\epsilon(q_0)\delta(q^2)e^{iqx} = \frac{i\epsilon(x_0)}{2\pi}\delta(x^2) \\ &= \int \frac{d^3q}{|\vec{q}|(2\pi)^3}e^{-i\vec{q}\vec{x}}i\sin|\vec{q}|x_0, \\ \mathbf{c}_j(0|x) = \partial_j\mathbf{s}(x) &= \int \frac{d^4q}{(2\pi)^3}q_j\epsilon(q_0)\delta(q^2)e^{iqx} = \frac{\epsilon(x_0)}{\pi}x_j\delta'(x^2) \\ &= \int \frac{d^3q}{|\vec{q}|(2\pi)^3}e^{-i\vec{q}\vec{x}}\begin{pmatrix} |\vec{q}|\cos|\vec{q}|x_0 \\ \vec{q}\, i\sin|\vec{q}|x_0 \end{pmatrix}. \end{aligned}$$

$\mathbf{s}(x)$ is not a ∂^j derivative of $\mathbf{c}_j(0|x)$. The massless dipole is given by

$$\begin{aligned} i\mathbf{s}^{\text{dip}}(x) = \lim_{m\to 0} \frac{i\mathbf{s}^{\text{dip}}(m|x)}{m} &= 2\int \frac{d^4q}{(2\pi)^3}\epsilon(q_0)\delta'(q^2)e^{iqx} = i\frac{\epsilon(x_0)}{4\pi}\vartheta(x^2) \\ &= \int \frac{d^3q}{|\vec{q}|(2\pi)^3}e^{-i\vec{q}\vec{x}}\, i\frac{\sin|\vec{q}|x_0 - |\vec{q}|x_0\cos|\vec{q}|x_0}{|\vec{q}|^2}. \end{aligned}$$

The position projection of the $\epsilon(x_0)$-multiplied distributions gives Coulomb potential and force:

$$\begin{array}{llll} \int d^3x\ \mathbf{s}(x) & = x_0, & 2\pi\int dx_0\epsilon(x_0)\mathbf{s}(x) & = \frac{1}{r}, \\ \int d^3x\ \mathbf{c}_j(0|x) & = \delta_j^0, & 2\pi\int dx_0\epsilon(x_0)\mathbf{c}_j(0|x) & = \delta_j^a\partial_a\frac{1}{r} = -\delta_j^a\frac{x_a}{r^3}, \\ \int d^3x\ \mathbf{s}^{\text{dip}}(x) & = \frac{x_0^3}{3}. & & \end{array}$$

5.4 The Hilbert Spaces for Massless Particles

Hilbert representations of the Poincaré group $\mathbf{SL}(\mathbb{C}^2)\ \vec{\times}\ \mathbb{R}^4$ for trivial translation-invariant $m^2 = 0$ are induced from Hilbert representations of $\mathbf{SO}(2)\times\mathbb{R}^4$. The energy-momentum fixgroup for trivial mass is the Euclidean group in two dimensions,

$$q \in \mathbb{R}^4,\ q^2 = 0,\ q \neq 0 \Rightarrow \mathbf{SO}_0(1,3)_q = \mathbf{SO}(2)\ \vec{\times}\ \mathbb{R}^2,$$

where the "translations" $\mathbb{R}^2$ originate from the boosts that are orthogonal to the axial rotations, e.g., rotation $\mathbf{l}^3$ acting on boosts $\mathbf{b}^{1,2}$ (chapter "Spacetime as Unitary Operation Classes"). The fixgroup in the fixgroup for a trivial "homogeneous" translation (boost) is the axial group

$$0 \in \mathbb{R}^2 \Rightarrow \mathbf{SO}(2)_0 = \mathbf{SO}(2).$$

The Hilbert space structure of massless particles is similar to that of massive ones (chapter "Particle Quantum Fields"). The momenta in creation and annihilation operators for massless particles with quantization (anti-)commutator

$$[\mathrm{u}^\star(\vec{p})_a, \mathrm{u}(\vec{q})^b]_\epsilon = \delta_a^b 2|\vec{q}|\delta(\tfrac{\vec{q}-\vec{p}}{2\pi})$$

parametrize the forward lightcone V^3:

$$0 \neq q = (|\vec{q}|, \vec{q}) \in \mathbf{SO}_0(1,3)/\mathbf{SO}(2)\ \vec{\times}\ \mathbb{R}^2 \cong \mathrm{V}^3 \cong \mathbf{D}(1)\times\Omega^2.$$

The Lorentz invariant measure of the lightcone

$$\int \frac{d^4q}{(2\pi)^3}\ \vartheta(q_0)\delta(q^2) = \int d^3\mathrm{v}(\vec{q}) = \int \frac{d^3q}{2|\vec{q}|(2\pi)^3} = \frac{1}{2(2\pi)^3}\int d^2\omega \int_0^\infty q\ dq,$$

integrates the "little" vector spaces $W(\vec{q}) \cong \mathbb{C}^{2-\delta_{L0}} = (\mathbb{C}, \mathbb{C}^2)$ for each momentum either with trivial or nontrivial representation $a = 0, \pm L$ of the axial rotations $\mathbf{SO}(2)$ around the momentum direction with invariant $L = 0, 1, 2, \ldots$:

$$\begin{aligned} w : \mathrm{V}^3 \longrightarrow W \quad w = {}^{\oplus}\!\!\int d^3\mathrm{v}(\vec{q})\ \mathrm{u}(\vec{q})^a w(\vec{q})_a &\in {}^{\oplus}\!\!\int d^3\mathrm{v}(\vec{q})\ W(\vec{q}), \\ \mathrm{u}(\vec{q})^a &\in W(\vec{q}) = W\times\{\vec{q}\}; \end{aligned}$$

w is a W-valued spin $\mathbf{SO}(2)$-intertwiner on the Lorentz group $\mathbf{SL}(\mathbb{C}^2)$, or, equivalently, a W-valued mapping of the energy-momentum lightcone.

The $\mathbf{U}(1)$-scalar product induces the Fock form with the scalar product distribution on the lightcone

$$\langle \mathrm{u}^\star(\vec{p})_a \mathrm{u}(\vec{q})^b\rangle_{\mathrm{F}} = \langle [\mathrm{u}^\star(\vec{p})_a, \mathrm{u}(\vec{q})^b]_{-\epsilon}\rangle_{\mathrm{F}} = \delta_a^b 2|\vec{q}|\delta(\tfrac{\vec{q}-\vec{p}}{2\pi}).$$

The creation operators define a measure-related distributive basis (not Hilbert space vectors):

$$\begin{aligned} |0, L; \vec{q}, a\rangle = |\vec{q}, a\rangle &= \mathrm{u}(0, L; \vec{q})^a|0\rangle, \\ \langle 0, L_2; \vec{q}_2, a_2|0, L_1; \vec{q}_1, a_1\rangle &= \delta_{L_2}^{L_1}\delta_{a_1 a_2} 2|\vec{q}|\delta(\tfrac{\vec{q}_1-\vec{q}_2}{2\pi}). \end{aligned}$$

The direct integral gives vectors in the Fock-Hilbert space, which contains the momentum functions square integrable on the forward lightcone multiplied with $(\mathbb{C}^1, \mathbb{C}^2)$ for the helicity components:

$$\begin{aligned} &|0, L; w\rangle = {}^\oplus\!\!\int d^3\mathrm{v}(\vec{q})\ \mathrm{u}(0, L; \vec{q})^a w(\vec{q})_a|0\rangle \in H(0, L) = L^2_{d^3\mathrm{v}(\vec{q})}(\mathrm{V}^3) \otimes \mathbb{C}^{2-\delta_{L0}} \\ \Rightarrow\ &\langle 0, L_2; w_2|0, L_1; w_1\rangle = \delta_{L_2}^{L_1} \int d^3\mathrm{v}(\vec{q})\ \overline{w_2(\vec{q})_a} w_1(\vec{q})_a \\ &\qquad = \delta_{L_2}^{L_1} \int \tfrac{d^4q}{(2\pi)^3}\ \overline{w_2(\vec{q})_a}\vartheta(q_0)\delta(q^2) w_1(\vec{q})_a. \end{aligned}$$

The distributive completeness allows the sesquilinear decomposition of the unit operator in the particle Hilbert space $H(0, L)$:

$$\begin{aligned} \mathcal{P}(0, L) \cong |0, L\rangle\langle 0, L| &= \bigoplus_{a=\pm L} {}^\oplus\!\!\int \tfrac{d^4q}{(2\pi)^3}\vartheta(q_0)\delta(q^2)|q, a\rangle\langle q, a| \\ &= \bigoplus_{a=\pm L} {}^\oplus\!\!\int d^3\mathrm{v}(\vec{q})\ |\vec{q}, a\rangle\langle\vec{q}, a|, \\ \mathcal{P}(0, L) \circ \mathcal{P}(0, L) &= \mathcal{P}(0, L). \end{aligned}$$

5.5 Massless Scalar Bose Particle Fields

The simplest case of a massless scalar field has trivial representation for the homogeneous groups. The field has particle degrees of freedom only.

A Hermitian scalar particle quantum field with nontrivial mass $m > 0$,

$$\begin{aligned} \mathbf{\Phi}(x) &= {}^\oplus\!\!\int \tfrac{d^3q}{2q_0(2\pi)^3}\ \gamma[e^{iqx}\mathrm{u}(\vec{q}) + e^{-iqx}\mathrm{u}^\star(\vec{q})], \\ [\mathbf{\Phi}, \mathbf{\Phi}](x) &= \gamma^2 \int \tfrac{d^4q}{(2\pi)^3}\epsilon(q_0)\delta(q^2 - m^2)e^{iqx} \\ &= \gamma^2 \int \tfrac{d^3q}{q_0(2\pi)^3}e^{-i\vec{q}\vec{x}} i \sin q_0 x_0\Big|_{q_0=\sqrt{m^2+\vec{q}^2}}, \end{aligned}$$

leads in the limit

$$m \longrightarrow 0,\quad \gamma^2 \longrightarrow g^2,\quad \mathbf{\Phi} \longrightarrow \varphi,\quad \mathbf{\Phi}_k \longrightarrow \varphi_k$$

to a Hermitian *massless scalar field*,

$$\begin{array}{ll} m = 0 \\ j = 0 \\ \epsilon = -1 \end{array} : \left\{ \begin{aligned} \varphi(x) &= {}^\oplus\!\!\int \tfrac{d^3q}{2|\vec{q}|(2\pi)^3} & g\ [e^{iqx}\mathrm{u}(\vec{q}) + e^{-iqx}\mathrm{u}^\star(\vec{q})], \\ -i\varphi_k(x) &= {}^\oplus\!\!\int \tfrac{d^3q}{2|\vec{q}|(2\pi)^3} & \tfrac{q_k}{g}\ [e^{iqx}\mathrm{u}(\vec{q}) - e^{-iqx}\mathrm{u}^\star(\vec{q})], \\ \text{with}\quad q_0 &= |\vec{q}|. \end{aligned} \right.$$

The factor $\frac{q^k}{|\vec{q}|}$ is part of an axial rotation transmutator from a Lorentz vector to an $\mathbf{SO}(2)$-scalar. For free fields a normalization with $g^2 = 1$ can be used.

The equations of motion and a classical Lagrangian density are

$$\mathbf{L}(\varphi, \varphi_k) = \varphi_k \partial^k \varphi - g^2 \frac{\varphi_k \varphi^k}{2}, \quad \begin{cases} \partial_k \varphi &= g^2 \varphi_k, \\ \partial^k \varphi_k &= 0, \end{cases}$$
$$\mathbf{L}(\varphi) = \tfrac{1}{2}(\partial^k \varphi)^2, \quad \partial^2 \varphi = 0.$$

The massless Bose field constitutes the relativistic distribution of a free nonrelativistic mass point:

$$\begin{aligned} L(\mathrm{x}, \mathrm{p}) &= \mathrm{p} d_t \mathrm{x} - \frac{\mathrm{p}^2}{2M}, \quad \begin{cases} d_t \mathrm{x} &= \frac{\mathrm{p}}{M}, \\ d_t \mathrm{p} &= 0, \end{cases} \quad \begin{pmatrix} [i\mathrm{p}, \mathrm{x}] & [\mathrm{x}, \mathrm{x}] \\ [\mathrm{p}, \mathrm{p}] & [\mathrm{x}, -i\mathrm{p}] \end{pmatrix}(t) = \begin{pmatrix} 1 & \frac{it}{M} \\ 0 & 1 \end{pmatrix}, \\ L(\mathrm{x}) &= \tfrac{1}{2}(d_t \mathrm{x})^2, \quad d_t^2 \mathrm{x} = 0. \end{aligned}$$

The commutators of the momentum operators

$$[\mathrm{u}^\star(\vec{p}), \mathrm{u}(\vec{q})] = 2|\vec{q}|\delta(\tfrac{\vec{q}-\vec{p}}{2\pi})$$

give the field commutators

$$\begin{aligned} \begin{pmatrix} [i\varphi_k, \varphi] & [\varphi, \varphi] \\ [\varphi_k, \varphi_j] & [\varphi, -i\varphi_j] \end{pmatrix}(x) &= (-i\partial) i\mathbf{s}(x) = \begin{pmatrix} \mathbf{c}_k(0|x) & g^2 i\mathbf{s}(x) \\ -\frac{\partial_j \partial_k}{g^2} i\mathbf{s}(x) & \mathbf{c}_j(0|x) \end{pmatrix} \\ &= \int \frac{d^4 q}{(2\pi)^3} (\mathbf{q}) \epsilon(q_0) \delta(q^2) e^{iqx} \end{aligned}$$

$$\text{with } (\mathbf{q}) = \begin{pmatrix} q_k & g^2 \\ \frac{q_k q_j}{g^2} & q_j \end{pmatrix}, \quad (-i\partial) = \begin{pmatrix} -i\partial_k & g^2 \\ -\frac{\partial_k \partial_j}{g^2} & -i\partial_j \end{pmatrix}.$$

Their time projection is a noncompact time representation, the position projection leads to the Coulomb potential and force

$$\begin{aligned} \int d^3x \begin{pmatrix} \mathbf{c}_k(0|x) & g^2 i\mathbf{s}(x) \\ -\frac{\partial_j \partial_k}{g^2} i\mathbf{s}(x) & \mathbf{c}_j(0|x) \end{pmatrix} &= \begin{pmatrix} \delta_k^0 & g^2 i x_0 \\ 0 & \delta_j^0 \end{pmatrix}, \\ 2\pi \int dx_0 \epsilon(x_0) \begin{pmatrix} \mathbf{c}_k(0|x) & g^2 i\mathbf{s}(x) \\ -\frac{\partial_j \partial_k}{g^2} i\mathbf{s}(x) & \mathbf{c}_j(0|x) \end{pmatrix} &= \begin{pmatrix} -\delta_k^a \frac{x_a}{r^2} & i g^2 \\ \frac{i}{g^2} \delta_k^a \delta_j^b \frac{3x_a x_b - \delta_{ab} r^2}{r^4} & -\delta_j^b \frac{x_b}{r^2} \end{pmatrix} \frac{1}{r}. \end{aligned}$$

Massless scalar Bose fields have an interpretation with a $\mathbf{U}(1)$-time development of creation and annihilation operators with a Euclidean conjugation

$$\mathbb{R}^4 \longrightarrow \mathbf{U}(1), \quad \mathrm{u} \overset{\star}{\leftrightarrow} \mathrm{u}^\star.$$

However, the time projection $\int d^3x \; \varphi(x)$ of the relativistic massless field is not defined.

The $\mathbf{U}(1)$-induced Fock value of the anticommutators

$$\langle \mathrm{u}^\star(\vec{p}) \mathrm{u}(\vec{q}) \rangle_{\mathrm{F}} = \langle \{\mathrm{u}^\star(\vec{p}), \mathrm{u}(\vec{q})\} \rangle_{\mathrm{F}} = 2|\vec{q}|\delta(\tfrac{\vec{q}-\vec{p}}{2\pi})$$

leads for the fields to

$$\begin{aligned} \begin{pmatrix} \langle\{i\varphi_k, \varphi\}\rangle_{\mathrm{F}} & \langle\{\varphi, \varphi\}\rangle_{\mathrm{F}} \\ \langle\{\varphi_k, \varphi_j\}\rangle_{\mathrm{F}} & \langle\{\varphi, -i\varphi_j\}\rangle_{\mathrm{F}} \end{pmatrix}(x) &= (-i\partial)\mathbf{C}(x) = \begin{pmatrix} i\mathbf{S}_k(0|x) & g^2 \mathbf{C}(x) \\ -\frac{\partial_j \partial_k}{g^2} \mathbf{C}(x) & i\mathbf{S}_j(0|x) \end{pmatrix} \\ &= \int \frac{d^4 q}{(2\pi)^3} (\mathbf{q}) \delta(q^2) e^{iqx}, \end{aligned}$$

which contains the scalar cosine in the limit:

$$\mathbf{C}(x) = \lim_{m\to 0}\frac{\mathbf{C}(m|x)}{m} = \int \frac{d^4q}{(2\pi)^3}\delta(q^2)e^{iqx} = -\frac{1}{2\pi^2}\frac{1}{x_P^2}.$$

The time projection of the Fock values is infinite:

$$\int d^3x\mathbf{C}(x) = \lim_{m\to 0}\frac{\cos x_0 m}{m},\quad \int d^3x\begin{pmatrix} i\mathbf{S}_k(0|x) & g^2\mathbf{C}(x)\\ \frac{\partial_j\partial_k}{g^2}\mathbf{C}(x) & i\mathbf{S}_j(0|x)\end{pmatrix} = \begin{pmatrix} 0 & \infty\\ \infty & 0\end{pmatrix}.$$

The Feynman propagator reads

$$\langle\{\varphi,\varphi\}(x) - \epsilon(x_0)[\varphi,\varphi](x)\rangle_{\mathrm{F}} = \frac{i}{\pi}\int\frac{d^4q}{(2\pi)^3}\frac{g^2}{q^2+io}e^{iqx}.$$

5.6 Massless Scalar Fermi Fields (Fadeev-Popov Fields)

A simple case of a massless field without any particle degrees of freedom is given by *Fadeev-Popov fields.* Their "creation-annihilation" operators are not acted on by a Hilbert representation of the Poincaré group. Fadeev-Popov fields are massless scalar Fermi fields with Lagrangian and equations of motion

$$\mathbf{L}_F(\beta,\check\beta,\gamma^k,\check\gamma^k) = i\gamma^k\partial_k\check\beta + i\check\gamma^k\partial_k\beta - ig^2\check\gamma^k\gamma_k,\quad \begin{cases}\partial^k\beta &= g^2\gamma^k,\\ \partial_k\gamma^k &= 0,\\ \partial^k\check\beta &= -g^2\check\gamma^k,\\ \partial_k\check\gamma^k &= 0,\end{cases}$$

$$\mathbf{L}_F(\beta,\check\beta) = i\partial^k\beta\partial_k\check\beta,\quad \begin{cases}\partial^2\beta &= 0,\\ \partial^2\check\beta &= 0.\end{cases}$$

The anticommutators embed the $\mathbf{U}(1,1)$ representations

$$\left(\begin{array}{cc|cc}\{\check\gamma^k,\beta\} & \{\gamma^k,\beta\} & \{\beta,\beta\} & \{\check\beta,\beta\}\\ \{\check\gamma^k,\check\beta\} & \{\gamma^k,\check\beta\} & \{\beta,\check\beta\} & \{\check\beta,\check\beta\}\\ \hline \{\check\gamma^k,\check\gamma^j\} & \{\gamma^k,\check\gamma^j\} & \{\beta,\check\gamma^j\} & \{\check\beta,\check\gamma^j\}\\ \{\check\gamma^k,\gamma^j\} & \{\gamma^k,\gamma^j\} & \{\beta,\gamma^j\} & \{\check\beta,\gamma^j\}\end{array}\right)(x)$$
$$= \lim_{m\to 0}\begin{pmatrix}\mathbf{D} & g^2\frac{d}{dm}\mathbf{D}\\ \clubsuit & \mathbf{D}\end{pmatrix}(m|x) \text{ with } \mathbf{D}(m|x) = \begin{pmatrix}\mathbf{c}^k & \mathbf{s}\\ -\mathbf{s} & \mathbf{c}^k\end{pmatrix}(m|x)$$
$$= \left(\begin{array}{cc|cc}\mathbf{c}^k(0|x) & 0 & 0 & g^2\mathbf{s}(x)\\ 0 & \mathbf{c}^k(0|x) & -g^2\mathbf{s}(x) & 0\\ \hline 0 & -\frac{\partial^j\partial^k}{g^2}\mathbf{s}(x) & \mathbf{c}^j(0|x) & 0\\ \frac{\partial^j\partial^k}{g^2}\mathbf{s}(x) & 0 & 0 & \mathbf{c}^j(0|x)\end{array}\right).$$

The time projection gives a representation for trivial energy in $\mathbf{SO}_0(2,2)$.

The four momentum operators in

$$\begin{array}{ll} m &= 0\\ j &= 0\\ \epsilon &= +1\end{array} : \begin{cases} i\beta(x) &= \oplus\int\frac{d^3q}{2|\vec q|(2\pi)^3} & g\,[e^{iqx}\,\mathrm{g}(\vec q) - e^{-iqx}\,\mathrm{g}^\times(\vec q)],\\ \check\beta(x) &= \oplus\int\frac{d^3q}{2|\vec q|(2\pi)^3} & g\,[e^{iqx}\,\mathrm{b}(\vec q) + e^{-iqx}\,\mathrm{b}^\times(\vec q)],\\ i\check\gamma^k(x) &= \oplus\int\frac{d^3q}{2|\vec q|(2\pi)^3} & \frac{q^k}{g}\,[e^{iqx}\,\mathrm{b}(\vec q) - e^{-iqx}\,\mathrm{b}^\times(\vec q)],\\ \gamma^k(x) &= \oplus\int\frac{d^3q}{2|\vec q|(2\pi)^3} & \frac{q^k}{g}\,[e^{iqx}\,\mathrm{g}(\vec q) + e^{-iqx}\,\mathrm{g}^\times(\vec q)],\end{cases}$$
with $q_0 = |\vec q|$.

have the quantization

$$\{\mathrm{g}^\times(\vec p),\mathrm{b}(\vec q)\}=\{\mathrm{b}^\times(\vec p),\mathrm{g}(\vec q)\}=2|\vec q|\delta(\tfrac{\vec q-\vec p}{2\pi}).$$

The momentum operators are interpretable with a time development in $\mathbf{U}(1)$. However, they have to be used with a $\mathbf{U}(1,1)$-conjugation $\times$ in order to have symmetric fields $(\beta,\check\beta)=(\beta^\times,\check\beta^\times)$.

A Fock form with $\mathbf{U}(1,1)$-conjugation is indefinite:

$$\begin{aligned}\langle \mathrm{g}^\times(\vec p)\mathrm{b}(\vec q)\rangle_{\mathrm F} &= \langle[\mathrm{g}^\times(\vec p),\mathrm{b}(\vec q)]\rangle_{\mathrm F} &= 2|\vec q|\delta(\tfrac{\vec q-\vec p}{2\pi})\\ =\langle \mathrm{b}^\times(\vec p)\mathrm{g}(\vec q)\rangle_{\mathrm F} &= \langle[\mathrm{b}^\times(\vec p),\mathrm{g}(\vec q)]\rangle_{\mathrm F} &= 2|\vec q|\delta(\tfrac{\vec q-\vec p}{2\pi}).\end{aligned}$$

It leads to the Feynman propagator

$$\begin{aligned}&\{\check\beta,\beta\}(x)=-g^2 i\int\tfrac{d^4q}{(2\pi)^3}\epsilon(q_0)\delta(q^2)e^{iqx},\quad \langle[\check\beta,\beta](x)\rangle_{\mathrm F}=-g^2 i\int\tfrac{d^4q}{(2\pi)^3}\delta(q^2)e^{iqx},\\ &\langle[\check\beta,\beta](x)-\epsilon(x_0)\{\check\beta,\beta\}(x)\rangle_{\mathrm F}=\tfrac{1}{\pi}\int\tfrac{d^4q}{(2\pi)^3}\tfrac{g^2}{q^2+io}e^{iqx}.\end{aligned}$$

5.7 Polarization (Helicity) in Spacetime

Transmutators connect with each other finite-dimensional axial rotation $\mathbf{SO}(2)$-representations for massless particles and finite-dimensional Lorentz group $\mathbf{SL}(\mathbb{C}^2)$-representations for the embedding quantum fields.

In the decomposition of a finite-dimensional irreducible $\mathbf{SL}(\mathbb{C}^2)$-representation into irreducible $\mathbf{SO}(2)$-representations,

$$\mathbf{SL}(\mathbb{C}^2)\cong\bigoplus\mathbf{SU}(2)\cong\bigoplus\mathbf{SO}(2):\ [2L|2R]\cong\bigoplus_{J=|L-R|}^{L+R}[2J]\cong\bigoplus_{r,s=-L,-R}^{L,R}D_{r-s},$$

the $\mathbf{U}(1)$-representations come in dual pairs $D_{2J}(e^{i\alpha_3})=e^{2Ji\alpha_3}$, $\pm 2J\in\mathbb{Z}$. The $\mathbf{U}(1)$-winding numbers $\{\pm 2J\}$ (e.g., for Weyl $\{\pm 1\}$ and for Lorentz vector $\{\pm 2,0,0\}$) for the rotation around the momentum direction $\frac{\vec q}{|\vec q|}$ will be called *circularity (polarization, helicity)*. Massless particles have no $\mathbf{SU}(2)$-spin; they have $\mathbf{SO}(2)$-circularity.

The relativistic embedding of the axial rotations around the momentum direction goes in two steps, $\mathbf{SO}(2)\hookrightarrow\mathbf{SU}(2)\hookrightarrow\mathbf{SL}(\mathbb{C}^2)$ (inducing in stages, chapter "Harmonic Analysis").

5.7.1 Rotation Group Embedding of Axial Rotations

The fundamental Pauli representation of the 2-sphere with the momentum directions,

$$\begin{aligned}\Omega^2\ni\tfrac{\vec q}{|\vec q|}\longmapsto u(\tfrac{\vec q}{|\vec q|}) &= \tfrac{1}{\sqrt{2|\vec q|(|\vec q|+q^3)}}\begin{pmatrix}|\vec q|+q^3 & -q^1+iq^2\\ q^1+iq^2 & |\vec q|+q^3\end{pmatrix}\in\mathbf{SU}(2)/\mathbf{SO}(2),\\ u(0,0,1) &= \mathbf{1}_2,\end{aligned}$$

is decomposable into two $\mathbf{SU}(2)/\mathbf{U}(1)$ axial-to-rotation transmutators for the opposite winding number parts of the $\mathbf{SO}(2)$-representation in the $\mathbf{SU}(2)$-representation [1]:

$$\begin{aligned} V^{[1]} &\cong W_{+1} \oplus W_{-1} \text{ with respect to } \mathbf{SO}(2), \\ [1](\tfrac{\vec{q}}{|\vec{q}|}) &= u(\tfrac{\vec{q}}{|\vec{q}|})_+ \oplus u(\tfrac{\vec{q}}{|\vec{q}|})_- : \ \mathrm{e}(\vec{q})^\pm \longmapsto u(\tfrac{\vec{q}}{|\vec{q}|})^A_\pm \mathrm{e}(\vec{q})^\pm, \quad A = 1,2. \end{aligned}$$

The adjoint $\mathbf{SO}(3)/\mathbf{SO}(2)$-representation is the symmetric product of two Pauli representations

$$\begin{aligned} [2](\tfrac{\vec{q}}{|\vec{q}|}) &= O(\tfrac{\vec{q}}{|\vec{q}|}) \cong \tfrac{1}{2} \operatorname{tr} u(\tfrac{\vec{q}}{|\vec{q}|})\sigma^b u^\star(\tfrac{\vec{q}}{|\vec{q}|})\sigma^a \\ &= \tfrac{1}{|\vec{q}|}\left(\begin{array}{c|c} \delta^{\alpha\beta}|\vec{q}| - \frac{q^\alpha q^\beta}{|\vec{q}|+q^3} & q^\alpha \\ -q^\beta & q^3 \end{array} \right) \in \mathbf{SO}(3)/\mathbf{SO}(2) \text{ with } \left\{ \begin{array}{ll} a,b &= 1,2,3, \\ \alpha,\beta &= 1,2. \end{array} \right. \end{aligned}$$

It is used in an $\mathbf{SO}(2)$-eigenvector basis $\{\sigma^0 = \sigma^3, \sigma^\pm = \frac{\sigma^1 \pm i\sigma^2}{2}\}$ for the $\mathbf{SO}(3)/\mathbf{SO}(2)$-transmutation

$$\begin{aligned} V^{[2]} &\cong W_{+2} \oplus W_{-2} \oplus W_0 \text{ with respect to } \mathbf{SO}(2), \\ \mathrm{e}(\vec{q})^\pm &\longmapsto u(\tfrac{\vec{q}}{|\vec{q}|})^A \sigma^\pm u^\star(\tfrac{\vec{q}}{|\vec{q}|})_B \mathrm{e}(\vec{q})^\pm, \\ \mathrm{e}(\vec{q})^0 &\longmapsto u(\tfrac{\vec{q}}{|\vec{q}|})^A \sigma^0 u^\star(\tfrac{\vec{q}}{|\vec{q}|})_B \mathrm{e}(\vec{q})^0. \end{aligned}$$

The momenta of a massless particle give projector decompositions of the identity into $\mathbf{SO}(2)$-nondecomposable projectors

$$\mathbf{1}_3 = \mathcal{P}_0(\vec{q}) + \mathcal{P}_{\pm 2}(\vec{q}), \quad \delta^a_b = \tfrac{q^a q_b}{\vec{q}^2} + (\delta^c_b - \tfrac{q^a q_b}{\vec{q}^2}), \quad \vec{q}^2 > 0.$$

For $\mathbf{SO}(2)$-winding numbers $\pm 2J$, $J = 0, \frac{1}{2}, \dots$, the transmutator to an $\mathbf{SU}(2)$-representation with maximal winding number $\pm 2J$ is given by the corresponding totally symmetric power of the Pauli representation,

$$\begin{aligned} \text{minimal: } \mathbf{irrep}\,\mathbf{SO}(2) &\hookrightarrow \mathbf{irrep}\,\mathbf{SU}(2), \\ D_{\pm 2J} &\hookrightarrow [2J] \text{ with } [2J](\tfrac{\vec{q}}{|\vec{q}|}) = \bigvee^{2J} u(\tfrac{\vec{q}}{|\vec{q}|}). \end{aligned}$$

The embedding of $\mathbf{SO}(2)$-polarization $\{\pm 2J\}$ into an $\mathbf{SU}(2)$-representation is not unique since $\{\pm 2J\}$ comes in all induced irreducible $\mathbf{SU}(2)$-representations (chapter "Harmonic Analysis").

The transmutators between the compact axial rotations $\mathbf{SO}(2)$ and the spin group $\mathbf{SU}(2)$ are square integrable functions $L^2_{d^2\omega}(\Omega^2)$ of the momentum directions $\Omega^2 \cong \mathbf{SU}(2)/\mathbf{SO}(2)$.

5.7.2 Lorentz Group Embedding of Axial Rotations

For the massless case, the embedding of spin group representations into finite-dimensional Lorentz group representations is achieved by a reinterpretation of the $\mathbf{SU}(2)$-representations $[2J](\frac{\vec{q}}{|\vec{q}|})$ in terms of $\mathbf{SL}(\mathbb{C}^2)$-representations. $u(\frac{\vec{q}}{|\vec{q}|})$ is a faithful representation of the 2-sphere $\mathbf{SU}(2)/\mathbf{SO}(2)$. As a representation of the 5-dimensional orientation manifold $\mathbf{SL}(\mathbb{C}^2)/\mathbf{SO}(2)$ it is trivial for the 3-dimensional boost manifold $\mathbf{SL}(\mathbb{C}^2)/\mathbf{SU}(2)$,

$$\mathbf{SL}(\mathbb{C}^2)/\mathbf{SO}(2) \cong \mathbf{SU}(2)/\mathbf{SO}(2) \times \mathbf{SL}(\mathbb{C}^2)/\mathbf{SU}(2) \cong \Omega^2 \times \mathcal{Y}^3.$$

As for the massive case, a "minimal" embedding representation of the Lorentz group is determined as follows:

$$\begin{array}{l} \text{minimal: } \mathbf{irrep}\,\mathbf{SU}(2) \longrightarrow \mathbf{irrep}_{\text{fin}}\mathbf{SL}(\mathbb{C}^2),\ \ [2J] \longmapsto [2L|2R]_0 = [2R|2L]_0, \\ [2L|2R]_0(\frac{\vec q}{|\vec q|}) = \bigvee^{2L} u(\frac{\vec q}{|\vec q|}) \otimes \bigvee^{2R} u(\frac{\vec q}{|\vec q|}) \text{ with } \left\{ \begin{array}{ll} L+R & = J, \\ |L-R| & = \left\{ \begin{array}{ll} 0, & J = 0, 1, \ldots, \\ \frac{1}{2}, & J = \frac{1}{2}, \frac{3}{2}, \ldots. \end{array} \right. \end{array} \right. \end{array}$$

The two fundamental Weyl representations are used for the minimal embedding of an $\mathbf{SO}(2)$-pair with $J = \pm\frac{1}{2}$, e.g., for massless Weyl fields (below)

$$\begin{array}{rl} V^{[1|0]} & \cong W_{+1} \oplus W_{-1} \text{ with respect to } \mathbf{SO}(2), \\ [1|0](\frac{q}{M}) & = u(\frac{\vec q}{|\vec q|}):\ \ \mathrm{e}(\vec q)^{\pm} \longmapsto u(\frac{\vec q}{|\vec q|})^A_{\pm}\mathrm{e}(\vec q)^{\pm},\ \ A = 1, 2, \\ V^{[0|1]} & \cong \hat W_{+1} \oplus \hat W_{-1} \text{ with respect to } \mathbf{SO}(2), \\ [0|1](\frac{q}{M}) & = u(\frac{\vec q}{|\vec q|}):\ \ \mathrm{e}(\vec q)^{\pm} \longmapsto u(\frac{\vec q}{|\vec q|})^{\dot A}_{\pm}\mathrm{e}(\vec q)^{\pm},\ \ \dot A = 1, 2. \end{array}$$

The Lorentz vector representation embeds an $\mathbf{SO}(2)$-pair with $J = \pm 1$ in an $\mathbf{SO}(2) \times \mathbf{SO}_0(1,1)$-eigenvector basis (Cartan representation) $\{\pi^{\pm} = \frac{\mathbf{1}_2 \pm \sigma^3}{2}, \sigma^{\pm} = \frac{\sigma^1 \pm i\sigma^2}{2}\}$, e.g., for massless vector fields (below)

$$\begin{array}{rl} V^{[1|1]} & \cong W_0^+ \oplus W_2 \oplus W_{-2} \oplus W_0^- \text{ with respect to } \mathbf{SO}(2), \\ [1|1](\frac{\vec q}{|\vec q|}) & = O_4(\frac{\vec q}{|\vec q|}) = \left(\begin{array}{c|c} 1 & 0 \\ \hline 0 & O(\frac{\vec q}{|\vec q|}) \end{array} \right) \cong [0](\frac{\vec q}{|\vec q|}) \oplus [2](\frac{\vec q}{|\vec q|}) \\ & \hfill \text{with respect to } \mathbf{SO}(3), \\ (\mathrm{e}^0 \pm \mathrm{e}^3)(\vec q) & \longmapsto u(\frac{\vec q}{|\vec q|})^A \pi^{\pm} u^\star(\frac{\vec q}{|\vec q|})_{\dot A} (\mathrm{e}^0 \pm \mathrm{e}^3)(\vec q), \\ (\mathrm{e}^1 \pm i\mathrm{e}^2)(\vec q) & \longmapsto u(\frac{\vec q}{|\vec q|})^A \sigma^{\pm} u^\star(\frac{\vec q}{|\vec q|})_{\dot A} (\mathrm{e}^1 \pm i\mathrm{e}^2)(\vec q). \end{array}$$

A projector decomposition into nondecomposable $\mathbf{SO}(3)$-projectors as used for massive vector particles, e.g., weak bosons:

$$\mathbf{1}_4 = \mathcal{P}_{[0]}(q) + \mathcal{P}_{[2]}(q),\ \ \delta^j_k = \frac{q^j q_k}{q^2} + (\delta^j_k - \frac{q^j q_k}{q^2}),\ \ q^2 = m^2 > 0,$$

is impossible for the massless case. This is a general property: Always, if the $\mathbf{SO}(2)$-embedding finite-dimensional $\mathbf{SL}(\mathbb{C}^2)$-representation is decomposable as an $\mathbf{SU}(2)$-representation, the embedding field for a massless particle also contains nonparticle degrees of freedom. That is the case for all nontrivial $\mathbf{SO}_0(1,3)$-representations, i.e., Lorentz vector (below), tensor, etc.

5.8 Massless Weyl Particle Fields

For *massless Weyl fields* the limit of the Weyl transmutators, used for massive fields (chapter "Massive Particle Quantum Fields"), leads to *helicity projectors*

$$\lim_{m \to 0} \sqrt{m}\ (s, \hat s)(\tfrac{q}{m}) = \sqrt{2|\vec q|}\ p^{\pm}(q) \text{ with } q_0 = |\vec q|.$$

The helicity projectors are products of Pauli transmutators to the third momentum basic vector $\vec{q} = (0, 0, |\vec{q}|)$, which characterizes the $\mathbf{SO}(2)$-circularity,

$$p^{\pm}(q) = \frac{1_2 + \frac{\vec{q}\vec{\sigma}}{|\vec{q}|}}{2} = u(\tfrac{\vec{q}}{|\vec{q}|})\frac{1_2 \pm \sigma^3}{2}u^\star(\tfrac{\vec{q}}{|\vec{q}|}).$$

From the spin doublets in the massive case there remains one component for the massless particles, e.g.,

$$\begin{aligned} p^+(q)^A_C \mathrm{u}(\vec{q})^C &= u(\tfrac{\vec{q}}{|\vec{q}|})^A_+ \mathrm{u}(\vec{q}) \text{ with } \mathrm{u}(\vec{q}) = u^\star(\tfrac{\vec{q}}{|\vec{q}|})^+_C \mathrm{u}(\vec{q})^C, \\ \sqrt{2|\vec{q}|}\,u(\tfrac{\vec{q}}{|\vec{q}|}) &= \frac{1}{\sqrt{|\vec{q}|+q^3}}\begin{pmatrix} |\vec{q}|+q^3 & -q^1+iq^2 \\ q^1+iq^2 & |\vec{q}|+q^3 \end{pmatrix}. \end{aligned}$$

The $m \to 0$ transmutator limit for Dirac fields gives massless Weyl fields, either $(\mathbf{r}, \mathbf{r}^\star)$ or $(\mathbf{l}, \mathbf{l}^\star)$, with a momentum-dependent normalization factor $\sqrt{2|\vec{q}|}$:

$$\begin{matrix} m = 0 \\ 2J = \pm 1 \\ z = \pm 1 \\ \epsilon = +1 \end{matrix} : \quad \begin{aligned} &\begin{cases} \mathbf{r}(x)^A &= {}^{\oplus}\!\int \frac{d^3q}{2|\vec{q}|(2\pi)^3}\sqrt{2|\vec{q}|}\; u(\tfrac{\vec{q}}{|\vec{q}|})^A_+ \quad [e^{iqx}\mathrm{u}(\vec{q}) + e^{-iqx}\mathrm{a}^\star(\vec{q})], \\ \mathbf{r}^\star(x)_{\dot{A}} &= {}^{\oplus}\!\int \frac{d^3q}{2|\vec{q}|(2\pi)^3}\sqrt{2|\vec{q}|}\; u^\star(\tfrac{\vec{q}}{|\vec{q}|})^+_{\dot{A}} \quad [e^{iqx}\mathrm{a}(\vec{q}) + e^{-iqx}\mathrm{u}^\star(\vec{q})], \end{cases} \\ &\text{with } q_0 = |\vec{q}|, \\ &\begin{cases} \mathbf{l}(x)^{\dot{A}} &= {}^{\oplus}\!\int \frac{d^3q}{2|\vec{q}|(2\pi)^3}\sqrt{2|\vec{q}|}\; u(\tfrac{\vec{q}}{|\vec{q}|})^{\dot{A}}_- \quad [e^{iqx}\mathrm{u}(\vec{q}) - e^{-iqx}\mathrm{a}^\star(\vec{q})], \\ \mathbf{l}^\star(x)_A &= {}^{\oplus}\!\int \frac{d^3q}{2|\vec{q}|(2\pi)^3}\sqrt{2|\vec{q}|}\; u^\star(\tfrac{\vec{q}}{|\vec{q}|})^-_A \quad [-e^{iqx}\mathrm{a}(\vec{q}) + e^{-iqx}\mathrm{u}^\star(\vec{q})]. \end{cases} \end{aligned}$$

Massless Weyl fields have only particle degrees of freedom. Particles and antiparticles rotate in opposite directions around the flight direction:

$$\text{for } \mathbf{SO}(2): \quad 2J(\mathrm{u}, \mathrm{a}) = \begin{cases} (1, -1) \text{ in } \mathbf{r}, \\ (-1, 1) \text{ in } \mathbf{l}. \end{cases}$$

The equations of motion and the classical Lagrangian densities are

$$\begin{aligned} \check{\sigma}^k \partial_k \mathbf{r} = 0, \quad \mathbf{L}(\mathbf{r}) = i\mathbf{r}\check{\sigma}^k \partial_k \mathbf{r}^\star, \\ \sigma^k \partial_k \mathbf{l} = 0, \quad \mathbf{L}(\mathbf{l}) = i\mathbf{l}\sigma^k \partial_k \mathbf{l}^\star. \end{aligned}$$

The massless Weyl field pairs have quantization anticommutators and commutator Fock forms

$$\begin{aligned} \begin{pmatrix} \{\mathbf{r}^\star_{\dot{B}}, \mathbf{r}^A\} & \{\mathbf{r}^\star_{\dot{B}}, \mathbf{l}^{\dot{A}}\} \\ \{\mathbf{l}^\star_B, \mathbf{r}^A\} & \{\mathbf{l}^\star_B, \mathbf{l}^{\dot{A}}\} \end{pmatrix}(x) &= \begin{pmatrix} \sigma^{k\,A}_{\ \dot{B}} & 0 \\ 0 & \check{\sigma}^{k\,\dot{A}}_{\ B} \end{pmatrix} \mathbf{c}_k(0|x), \\ \begin{pmatrix} \langle[\mathbf{r}^\star_{\dot{B}}, \mathbf{r}^A]\rangle_{\mathrm{F}} & \langle[\mathbf{r}^\star_{\dot{B}}, \mathbf{l}^{\dot{A}}]\rangle_{\mathrm{F}} \\ \langle[\mathbf{l}^\star_B, \mathbf{r}^A]\rangle_{\mathrm{F}} & \langle[\mathbf{l}^\star_B, \mathbf{l}^{\dot{A}}]\rangle_{\mathrm{F}} \end{pmatrix}(x) &= \begin{pmatrix} \sigma^{k\,A}_{\ \dot{B}} & 0 \\ 0 & \check{\sigma}^{k\,\dot{A}}_{\ B} \end{pmatrix} i\mathbf{S}_k(0|x). \end{aligned}$$

The time projection gives two trivial time representations. The position projection leads to a Coulomb force.

5.9 Massless Vector Bose Fields (Gauge Fields)

Massless vector fields have, in contrast to massive ones, both particle and nonparticle degrees of freedom. Therefore, it is useful to describe the quantum structure of a massless vector field $\{\mathbf{A}^j\}_{j=0}^3$, $m = 0$, side by side with and in contrast to the quantum structure of a massive vector field $\{\mathbf{Z}^j\}_{j=0}^3$, $m > 0$. Massless particles have fixgroup $\mathbf{SO}(2)$ (polarization) with a decomposition of spacetime into time and position with one distinguished polarization axis (momentum direction), whereas massive particles use a rest-system-induced decomposition into time and position with fixgroup $\mathbf{SO}(3)$ (spin).

Fields for massive spin 1 Bose particles have the classical Lagrangian and the field equations

$$\begin{aligned} \mathbf{L}(\mathbf{Z}^j, \mathbf{G}^{jk}) &= \tfrac{1}{2}\mathbf{G}^{jk}\epsilon_{jk}^{lm}\partial_l \mathbf{Z}_m + (\gamma^2 \tfrac{\mathbf{G}^{jk}\mathbf{G}_{jk}}{4} + m^2 \tfrac{\mathbf{Z}^j\mathbf{Z}_j}{2\gamma^2}), \\ &\left\{ \begin{array}{ll} \epsilon_{lr}^{jk}\partial^l \mathbf{Z}^r &= \partial^j \mathbf{Z}^k - \partial^k \mathbf{Z}^j = \gamma^2 \mathbf{G}^{kj}, \\ \partial_k \mathbf{G}^{jk} &= -\tfrac{m^2}{\gamma^2}\mathbf{Z}^j. \end{array} \right. \end{aligned}$$

In the limit of vanishing mass

$$m \to 0, \ \gamma^2 \to g^2, \ \ \mathbf{Z} \to \mathbf{A}, \ \ \mathbf{G} \to \mathbf{F}$$

one obtains the classical Lagrangian for massless vector fields ("gauge fields")

$$\mathbf{L}(\mathbf{A}^j, \mathbf{F}^{jk}) = \tfrac{1}{2}\mathbf{F}^{jk}\epsilon_{jk}^{lm}\partial_l \mathbf{A}_m + g^2 \tfrac{\mathbf{F}^{jk}\mathbf{F}_{jk}}{4}, \quad \left\{ \begin{array}{ll} \epsilon_{lr}^{jk}\partial^l \mathbf{A}^r &= g^2 \mathbf{F}^{kj}, \\ \partial_k \mathbf{F}^{jk} &= 0. \end{array} \right.$$

For free fields a normalization with $g^2 = 1$ can be used. In interacting gauge theories g^2 is the gauge coupling constant (chapter "Gauge Interactions").

The classical Lagrangian has to be modified for a quantum theory with duality pairs. Since the Lagrangian for a massless vector field does not contain dual partners for all four components $\{\mathbf{A}^j\}_{j=0}^3$, i.e., since there is no equation of motion involving $\partial^0 \mathbf{A}^0$ or the Lorentz scalar $\partial_j \mathbf{A}^j$ in contrast to $\partial_j \mathbf{Z}^j = 0$ in the massive case, the theory of massless vector quantum fields requires a *Lorentz scalar Bose field* $\mathbf{S}$ *("gauge fixing" field)* in addition to the field strength $\mathbf{F}$

$$\begin{aligned} \mathbf{L}(\mathbf{A}^j, \mathbf{F}^{jk}, \mathbf{S}) &= \tfrac{1}{2}\mathbf{F}^{jk}\epsilon_{jk}^{lm}\partial_l \mathbf{A}_m + \mathbf{S}\partial_j \mathbf{A}^j + g^2 \tfrac{\mathbf{F}^{jk}\mathbf{F}_{jk}}{4} - g^2 \lambda \tfrac{\mathbf{S}^2}{2}, \\ &\left\{ \begin{array}{ll} \epsilon_{lr}^{jk}\partial^l \mathbf{A}^r &= g^2 \mathbf{F}^{kj}, \\ \partial_j \mathbf{A}^j &= g^2 \lambda \mathbf{S}, \end{array} \right. , \quad \partial_k \mathbf{F}^{jk} - \partial^j \mathbf{S} = 0 \end{aligned}$$

with a *"gauge fixing" parameter* $\lambda \neq 0$.

The Lorentz vector field (gauge potential) $\{\mathbf{A}^j\}_{j=0}^3 \sim \{\mathbf{A}^0, \mathbf{A}^a\}$ has a scalar and vector potential as rotation group decomposition, $[1|1] \cong [0] \oplus [2]$, the "gauge fixing" field $\mathbf{S}$ a trivial representation $[0|0] \cong [0]$. The antisymmetric tensor field (field strength) $\{\mathbf{F}^{kj}\}_{k,j=0}^3 \sim \{\mathbf{F}^{a0}, \mathbf{F}^{ab}\}$, acted on by the adjoint representation, is decomposable into electric and magnetic field, $[0|2] \oplus [2|0] \cong [2] \oplus [2]$.

In contrast to the commutator of the massive vector field

$$\begin{aligned} [\mathbf{Z}^k, \mathbf{Z}^j](x) &= \gamma^2 \int \tfrac{d^4q}{(2\pi)^3}\epsilon(q_0)(-\eta^{kj} + \tfrac{q^k q^j}{m^2})\delta(q^2 - m^2)e^{iqx} \\ &= -\gamma^2(\eta^{kj} + \tfrac{\partial^k \partial^j}{m^2})\tfrac{i\mathbf{s}(m|x)}{m}, \end{aligned}$$

the massless vector field has the quantization

$$\begin{aligned} [\mathbf{A}^k, \mathbf{A}^j](x) &= g^2 \int \tfrac{d^4q}{(2\pi)^3}\epsilon(q_0)(-\eta^{kj} - 2\nu q^k q^j \tfrac{\partial}{\partial q^2})\delta(q^2)e^{iqx} \\ &= -g^2[\eta^{kj} i\mathbf{s}(x) - \nu \partial^k \partial^j i\mathbf{s}^{\text{dip}}(x)] \\ &\text{with } 1 - \lambda = 2\nu. \end{aligned}$$

Its commutator involves no spin projectors, as does that of the massive field. In addition to the massless pole structure $\eta^{kj}\delta(q^2)$, relevant for the $\mathbf{U}(1)$-time representations, there occurs the characteristic *dipole structure* $q^k q^j \delta'(q^2)$, which embeds a $\mathbf{U}(1,1)$-time development. There are no dipoles for the particularly simple "gauge fixing" parameter

$$\lambda = 1, \quad \nu = 0 \Rightarrow [\mathbf{A}^k, \mathbf{A}^j] = -g^2 \eta^{kj} \int \frac{d^4 q}{(2\pi)^3} \epsilon(q_0)\delta(q^2) e^{iqx}.$$

All commutators can be computed by derivations of the gauge field commutator. The nontrivial ones are

$$\begin{pmatrix} [i\mathbf{F}^{kl}, \mathbf{A}^j] & [\mathbf{A}^k, \mathbf{A}^j] \\ [\mathbf{F}^{kl}, \mathbf{F}^{jn}] & [\mathbf{A}^k, -i\mathbf{S}] \end{pmatrix}(x) = \begin{pmatrix} -i\epsilon^{lk}_{ut}\delta^j_s \partial^u & g^2 \delta^k_t \delta^j_s \\ -\epsilon^{lk}_{ut}\epsilon^{nj}_{rs} \frac{\partial^r \partial^u}{g^2} & -\frac{i}{\lambda}\delta^k_t \partial_s \end{pmatrix} \frac{[\mathbf{A}^t, \mathbf{A}^s](x)}{g^2}$$

$$= \int \frac{d^4 q}{(2\pi)^3} \begin{pmatrix} \epsilon^{kl}_{ut}\eta^{tj} q^u & g^2[-\eta^{kj} - 2\nu q^k q^j \frac{\partial}{\partial q^2}] \\ \epsilon^{kl}_{ut}\epsilon^{nj}_{rs}\eta^{ts}\frac{q^r q^u}{g^2} & -\frac{q^k}{\lambda} \end{pmatrix} \epsilon(q_0)\delta(q^2) e^{iqx}.$$

The massless field theory embeds two nonrelativistic free mass points acted on by noncompact time representations for trivial energy:

$$L_B(\mathrm{x}, \check{\mathrm{x}}, \mathrm{p}, \check{\mathrm{p}}) = \mathrm{p} d_t \mathrm{x} - \check{\mathrm{p}} d_t \check{\mathrm{x}} - \frac{\mathrm{p}^2}{2M} - \frac{\check{\mathrm{p}}^2}{2M'}, \quad \begin{cases} d_t \mathrm{x} &= \frac{\mathrm{p}}{M}, \quad d_t \mathrm{p} = 0, \\ d_t \check{\mathrm{x}} &= -\frac{\check{\mathrm{p}}}{M'}, \quad d_t \check{\mathrm{p}} = 0. \end{cases}$$

The commutators of the quantum mechanical model are embedded into spacetime commutators

$$\left(\begin{array}{cc|cc} [-i\check{\mathrm{p}}, \check{\mathrm{x}}] & [\mathrm{p}, \check{\mathrm{x}}] & [-i\mathrm{x}, \check{\mathrm{x}}] & [\check{\mathrm{x}}, \check{\mathrm{x}}] \\ [\check{\mathrm{p}}, \mathrm{x}] & [i\mathrm{p}, \mathrm{x}] & [\mathrm{x}, \mathrm{x}] & [\check{\mathrm{x}}, i\mathrm{x}] \\ \hline [-i\check{\mathrm{p}}, \mathrm{p}] & [\mathrm{p}, \mathrm{p}] & [\mathrm{x}, -i\mathrm{p}] & [\check{\mathrm{x}}, \mathrm{p}] \\ [\check{\mathrm{p}}, \check{\mathrm{p}}] & [\mathrm{p}, i\check{\mathrm{p}}] & [\mathrm{x}, \check{\mathrm{p}}] & [\check{\mathrm{x}}, i\check{\mathrm{p}}] \end{array}\right)(t) = \left(\begin{array}{cc|cc} 1 & 0 & 0 & \frac{it}{M'} \\ 0 & 1 & \frac{it}{M} & 0 \\ \hline 0 & 0 & 1 & 0 \\ 0 & 0 & 0 & 1 \end{array}\right)$$

$$\hookrightarrow \lim_{m\to 0} \begin{pmatrix} \mathbf{D} & \nu \frac{d}{dm}\mathbf{D} \\ \clubsuit & \mathbf{D} \end{pmatrix}(m|x)$$

$$\text{with } \mathbf{D}(m|x) = \begin{pmatrix} \mathbf{c}^j & i\ell^2 \mathbf{s} \\ \frac{i}{\ell^2}\mathbf{s} & \mathbf{c}^j \end{pmatrix}(m|x) = \left(\begin{array}{cc|cc} \mathbf{c}^j(0|x) & 0 & 0 & \nu i\mathbf{s}(x) \\ 0 & \mathbf{c}^j(0|x) & \nu i\mathbf{s}(x) & 0 \\ \hline 0 & \clubsuit & \mathbf{c}^k(0|x) & 0 \\ \clubsuit & 0 & 0 & \mathbf{c}^k(0|x) \end{array}\right).$$

The positions $(\mathrm{x}, \check{\mathrm{x}})$ as time nilvectors are embedded into the vector field, the momenta $(\mathrm{p}, \check{\mathrm{p}})$ as time eigenvectors into the field strengths and the "gauge fixing" field:

$$\begin{aligned} \{\check{\mathrm{x}}, \mathrm{x}\} &\hookrightarrow \mathbf{A}^k = \{\mathbf{A}^0, \mathbf{A}^a\}, \\ \{\check{\mathrm{p}}, \mathrm{p}\} &\hookrightarrow \{\mathbf{S}, \mathbf{F}^{0a}\}, \quad a = 1, 2, 3. \end{aligned}$$

The two free mass points are used for the two lightlike degrees of freedom $\mathbf{A}^0 \pm \mathbf{A}^3$. The time projection of the spacetime commutators, rearranged with respect to time and position components, displays $(4 = 1 + 3)$ noncompact time representations of type $\begin{pmatrix} 1 & ix_0 \\ 0 & 1 \end{pmatrix}$, i.e., with trivial energy:

$$\int d^3x \left(\begin{array}{cc|cc} [i\mathbf{S}, \mathbf{A}^0] & [i\mathbf{F}^{0a}, \mathbf{A}^0] & [\mathbf{A}^a, \mathbf{A}^0] & [\mathbf{A}^0, \mathbf{A}^0] \\ [i\mathbf{S}, \mathbf{A}^b] & [i\mathbf{F}^{0a}, \mathbf{A}^b] & [\mathbf{A}^a, \mathbf{A}^b] & [\mathbf{A}^0, \mathbf{A}^b] \\ \hline [\mathbf{S}, \mathbf{F}^{b0}] & [\mathbf{F}^{0a}, \mathbf{F}^{b0}] & [\mathbf{A}^a, -i\mathbf{F}^{b0}] & [\mathbf{A}^0, -i\mathbf{F}^{b0}] \\ [\mathbf{S}, \mathbf{S}] & [\mathbf{F}^{0a}, \mathbf{S}] & [\mathbf{A}^a, -i\mathbf{S}] & [\mathbf{A}^0, -i\mathbf{S}] \end{array}\right) = \left(\begin{array}{c|cc|c} -1 & 0 & 0 & -g^2\lambda i x_0 \\ \hline 0 & \delta^{ab} & \delta^{ab} g^2 i x_0 & 0 \\ 0 & 0 & \delta^{ab} & 0 \\ \hline 0 & 0 & 0 & -1 \end{array}\right).$$

The position projections can be derived from the modified Coulomb potential in the projection of the $\epsilon(x_0)$-multiplied massless vector field commatator

$$2i\pi \int dx_0 \epsilon(x_0)[\mathbf{A}^k, \mathbf{A}^j](x) = g^2[\eta^{kj} + 2\nu\delta_a^k\delta_b^j(\delta^{ab} - \tfrac{x^a x^b}{r^2})]\tfrac{1}{r}.$$

Now the momentum operators will be given for the massless vector field: With energy-momenta decomposition $q = (|\vec{q}|, \vec{q})$ one obtains with the matrix

$$\partial^k\partial^j i\mathbf{s}^{\mathrm{dip}}(x) = \int \frac{d^3q}{|\vec{q}|(2\pi)^3} e^{-i\vec{q}\vec{x}} i \left(\begin{array}{c|c} \sin|\vec{q}|x_0 + |\vec{q}|x_0\cos|\vec{q}|x_0 & -iq^a x_0 \sin|\vec{q}|x_0 \\ \hline -iq^b x_0 \sin|\vec{q}|x_0 & -\frac{q^a q^b}{\vec{q}^2}(\sin|\vec{q}|x_0 - |\vec{q}|x_0\cos|\vec{q}|x_0) \end{array}\right)$$

the time representation matrix elements in the vector field commutator

$$[\mathbf{A}^k, \mathbf{A}^j](x) = \int \frac{d^3q}{|\vec{q}|(2\pi)^3} e^{-i\vec{q}\vec{x}}\, g^2 i \left(\begin{array}{c|c} \begin{array}{c} -(1-\nu)\sin|\vec{q}|x_0 \\ +\nu|\vec{q}|x_0\cos|\vec{q}|x_0 \end{array} & -i\nu q^a x_0 \sin|\vec{q}|x_0 \\ \hline -i\nu q^b x_0 \sin|\vec{q}|x_0 & \begin{array}{c} \delta^{ab}\sin|\vec{q}|x_0 \\ -\nu\frac{q^a q^b}{\vec{q}^2}(\sin|\vec{q}|x_0 - |\vec{q}|x_0\cos|\vec{q}|x_0) \end{array} \end{array}\right).$$

With the transmutator from the rotation group $\mathbf{SO}(3)$ to the axial rotations $\mathbf{SO}(2)$ the third momentum axis is used as polarization axis $q = (|\vec{q}|, 0, 0, |\vec{q}|)$:

$$\begin{aligned}
O_4(\tfrac{\vec{q}}{|\vec{q}|}) &= \left(\begin{array}{c|c} 1 & 0 \\ \hline 0 & O(\frac{\vec{q}}{|\vec{q}|}) \end{array}\right) \in \mathbf{SO}(3)/\mathbf{SO}(2), \\
[\mathbf{A}^k, \mathbf{A}^j](x) &\cong \int \frac{d^3q}{|\vec{q}|(2\pi)^3} e^{-i\vec{q}\vec{x}}\, O_4(\tfrac{\vec{q}}{|\vec{q}|}) \circ [\mathbf{AA}](|\vec{q}|x_0) \circ O_4(\tfrac{\vec{q}}{|\vec{q}|}), \\
[\mathbf{AA}](|\vec{q}|x_0) &= g^2 i \left(\begin{array}{c|c|c} \begin{array}{c} -(1-\nu)\sin|\vec{q}|x_0 \\ +\nu|\vec{q}|x_0\cos|\vec{q}|x_0 \end{array} & 0 & -i\nu|\vec{q}|x_0\sin|\vec{q}|x_0 \\ \hline 0 & \mathbf{1}_2\sin|\vec{q}|x_0 & 0 \\ \hline -i\nu|\vec{q}|x_0\sin|\vec{q}|x_0 & 0 & \begin{array}{c} (1-\nu)\sin|\vec{q}|x_0 \\ +\nu|\vec{q}|x_0\cos|\vec{q}|x_0 \end{array} \end{array}\right).
\end{aligned}$$

This is compared with the commutator of a massive vector field with the transmutator from Lorentz group to rotation groups $\Lambda(\frac{q}{m}) \in \mathbf{SO}_0(1,3)/\mathbf{SO}(3)$ for the three spin components

$$\begin{aligned}
[\mathbf{Z}^k, \mathbf{Z}^j](x) &\cong \int \frac{d^3q}{q_0(2\pi)^3} e^{-i\vec{q}\vec{x}}\, \Lambda(\tfrac{q}{m}) \circ [\mathbf{ZZ}](q_0x_0) \circ \Lambda(\tfrac{q}{m}), \\
[\mathbf{ZZ}](q_0x_0) &= \gamma^2 i \left(\begin{array}{c|c} 0 & 0 \\ \hline 0 & \mathbf{1}_3 \sin q_0x_0 \end{array}\right), \quad q_0 = \sqrt{m^2 + \vec{q}^2}, \\
\frac{\partial}{\partial i q_0 x_0}\Big|_{x_0=0} [\mathbf{ZZ}](q_0x_0) &= \gamma^2 \left(\begin{array}{c|c} 0 & 0 \\ \hline 0 & \mathbf{1}_3 \end{array}\right).
\end{aligned}$$

For the massless field the axial rotation trivial contributions (0th and 3rd components) are transformed into a nonorthogonal lightlike basis with $\mathbf{SO}_0(1,1)$-eigenvectors

$$w \circ \begin{pmatrix} -1 & 0 \\ 0 & 1 \end{pmatrix} \circ w^T = \begin{pmatrix} 0 & 1 \\ 1 & 0 \end{pmatrix}, \quad w = \tfrac{1}{\sqrt{2}}\begin{pmatrix} 1 & 1 \\ -1 & 1 \end{pmatrix}.$$

One obtains with the transmutator $O_w(\frac{\vec{q}}{|\vec{q}|}) = O_4(\frac{\vec{q}}{|\vec{q}|}) \circ w$ in the new basis

$$\begin{aligned}
[\mathbf{A}^k, \mathbf{A}^j](x) &= \int \frac{d^3q}{|\vec{q}|(2\pi)^3} e^{-i\vec{q}\vec{x}}\, O_w(\tfrac{\vec{q}}{|\vec{q}|})^k\, [\mathbf{AA}](|\vec{q}|x_0)\, O_w(\tfrac{\vec{q}}{|\vec{q}|})^j, \\
[\mathbf{AA}](|\vec{q}|x_0) &= g^2 i \left(\begin{array}{c|c|c} \nu|\vec{q}|x_0 e^{-i|\vec{q}|x_0} & 0 & (1-\nu)\sin|\vec{q}|x_0 \\ \hline 0 & \mathbf{1}_2\sin|\vec{q}|x_0 & 0 \\ \hline (1-\nu)\sin|\vec{q}|x_0 & 0 & \nu|\vec{q}|x_0 e^{-i|\vec{q}|x_0} \end{array}\right), \quad q_0 = |\vec{q}|, \\
\frac{\partial}{\partial i|\vec{q}|x_0}\Big|_{x_0=0} [\mathbf{AA}](|\vec{q}|x_0) &= g^2 \left(\begin{array}{c|c|c} \nu & 0 & 1-\nu \\ \hline 0 & \mathbf{1}_2 & 0 \\ \hline 1-\nu & 0 & \nu \end{array}\right).
\end{aligned}$$

The 1. and 2. components of the massless field with nontrivial polarization around the momentum $\vec{q}$ carry two $\mathbf{U}(1)$ time representations with energy $q_0 = |\vec{q}|$. They constitute a harmonic $\mathbf{U}(2)$-oscillator

$$\alpha, \beta \in \{1,2\} : \left\{ \begin{array}{rl} \mathbb{R}^4 & \longrightarrow \mathbf{U}(\mathbf{1}_2) \ni e^{i|\vec{q}|x_0 - i\vec{q}\vec{x}} \begin{pmatrix} 1 & 0 \\ 0 & 1 \end{pmatrix}, \\ \mathrm{u}(\vec{q}, x_0)^\alpha & = e^{i|\vec{q}|x_0} \mathrm{u}(\vec{q})^\alpha, \\ \mathbb{R}^4 \times \mathbf{SO}(2) & \longrightarrow \mathbf{U}(\mathbf{1}_2) \circ \mathbf{SU}(2) = \mathbf{U}(2), \\ [\mathrm{u}^\star(\vec{p})_\alpha, \mathrm{u}(\vec{q})^\beta] & = \delta_\alpha^\beta 2|\vec{q}|\delta(\frac{\vec{q}-\vec{p}}{2\pi}), \end{array} \right.$$

The 0th and 3rd components have trivial polarization $\mathbf{SO}(2)$ and are connected in a $\mathbf{U}(1,1)$ time representations with energy $q_0 = |\vec{q}|$ and nilconstant $\nu|\vec{q}|$ involving the "gauge fixing" constant $2\nu = 1 - \lambda$:

$$j, k \in \{0,3\} : \left\{ \begin{array}{rl} \mathbb{R}^4 & \longrightarrow \mathbf{U}(1,1) \ni e^{i|\vec{q}|x_0 - i\vec{q}\vec{x}} \begin{pmatrix} 1 & i\nu|\vec{q}|x_0 \\ 0 & 1 \end{pmatrix}, \\ \mathrm{B}(\vec{q}, x_0) & = e^{i|\vec{q}|x_0} [\mathrm{B}(\vec{q}) + i\nu|\vec{q}|x_0 \mathrm{G}(\vec{q})], \\ \mathrm{G}(\vec{q}, x_0) & = e^{i|\vec{q}|x_0} \mathrm{G}(\vec{q}), \\ [\mathrm{B}^\times(\vec{p}), \mathrm{G}(\vec{q})] & = [\mathrm{G}^\times(\vec{p}), \mathrm{B}(\vec{q})] = 2|\vec{q}|\delta(\frac{\vec{q}-\vec{p}}{2\pi}). \end{array} \right.$$

The massive vector field has three momentum operators for the spin components with $\mathbf{U}(1)$-representations of the spacetime translations

$$\begin{array}{rl} m & > 0 \\ J & = 1 \\ \epsilon & = -1 \end{array} : \left\{ \begin{array}{rl} \mathbf{Z}(x)^j & = {}^\oplus\!\int \frac{d^3q}{2q_0(2\pi)^3} \Lambda(\frac{q}{m})^j_a \gamma \, [e^{iqx} \mathrm{u}(\vec{q})^a + e^{-iqx} \mathrm{u}^\star(\vec{q})_a] \\ \text{with } q_0 & = \sqrt{m^2 + \vec{q}^2}, \quad a = 1,2,3, \end{array} \right.$$

whereas the harmonic analysis of a massless vector field contains four momentum operators, two with a time representation in $\mathbf{U}(\mathbf{1}_2)$ (1st and 2nd components) and two with a time representations in $\mathbf{U}(1,1)$ (0th and 3rd components)

$$\begin{array}{rl} m & = 0 \\ J & = \pm 1 \\ \epsilon & = -1 \end{array} : \left\{ \begin{array}{l} \mathbf{A}(x)^j = {}^\oplus\!\int \frac{d^3q}{2|\vec{q}|(2\pi)^3} O_w(\frac{\vec{q}}{|\vec{q}|})^j g \begin{pmatrix} e^{iqx}[\mathrm{B}(\vec{q}) + i\nu|\vec{q}|x_0 \mathrm{G}(\vec{q})] + (1-\nu)e^{-iqx}\mathrm{G}^\times(\vec{q}) \\ e^{iqx}\mathrm{u}(\vec{q})^1 + e^{-iqx}\mathrm{u}^\star(\vec{q})_1 \\ e^{iqx}\mathrm{u}(\vec{q})^2 + e^{-iqx}\mathrm{u}^\star(\vec{q})_2 \\ (1-\nu)e^{iqx}\mathrm{G}(\vec{q}) + e^{-iqx}[\mathrm{B}^\times(\vec{q}) - i\nu|\vec{q}|x_0 \mathrm{G}^\times(\vec{q})] \end{pmatrix} \\ \text{with } q_0 = |\vec{q}|. \end{array} \right.$$

For a trivial nilconstant $\nu = 0$ one has four $\mathbf{U}(1)$-representations, where two of them have a $\mathbf{U}(1,1)$-conjugation.

The "gauge fixing" field involves only the eigenvectors of the $\mathbf{U}(1,1)$-time representation

$$i\mathbf{S}(x) = \sqrt{2}\, {}^\oplus\!\int \frac{d^3q}{2|\vec{q}|(2\pi)^3} \frac{|\vec{q}|}{g} \, [e^{iqx}\mathrm{G}(\vec{q}) - e^{-iqx}\mathrm{G}^\times(\vec{q})], \quad q_0 = |\vec{q}|, \quad \partial^2 \mathbf{S} = 0.$$

The basic vector spaces $W_A(\vec{q}), W_A^T(\vec{q}) \cong \mathbb{C}^4$ at each momentum of the massless field are spanned by four conjugated pairs of momentum operators. With the complex spacetime representation, the Lorentz group $\mathbf{SO}_0(1,3)$

comes in the indefinite unitary group $\mathbf{U}(1,3)$. The massless vector fields involve an $\mathbf{SO}(2)$-polarized particle pair (left and right polarized photons) with Hilbert representations (harmonic $\mathbf{U}(2)$-oscillator) and an $\mathbf{SO}(2)$-trivial pair with translation representations in an indefinite unitary subgroup and without particle interpretation,

$$\begin{array}{l} \text{for } \mathbf{A}^j: \ \mathbb{R}^4 \times \mathbf{SO}(2) \longrightarrow \mathbf{U}(1,1) \times \mathbf{U}(2) \\ \qquad \mathbf{U}(1,1) \times \mathbf{U}(2) \subset \mathbf{U}(1,3) \supset \mathbf{SO}_0(1,3). \end{array}$$

This is in contrast to the massive vector field that comes with three creation-annihilation pairs $W_Z(\vec{q}), W_Z^T(\vec{q}) \cong \mathbb{C}^3$ of particle operators. The massive $\mathbf{SO}(3)$-spin triplet particles have Hilbert representations (harmonic $\mathbf{U}(3)$-oscillator)

$$\begin{array}{l} \text{for } \mathbf{Z}^j: \ \mathbb{R}^4 \times \mathbf{SO}(3) \longrightarrow \mathbf{U}(3), \\ \qquad \mathbf{U}(3) \subset \mathbf{U}(1,3) \supset \mathbf{SO}_0(1,3). \end{array}$$

A $\mathbf{U}(1)$-time development with particle interpretation has a Fock state

$$\langle \mathrm{u}^\star(\vec{p})_\alpha \mathrm{u}(\vec{q})^\beta \rangle_{\mathrm{F}} = \langle \{\mathrm{u}^\star(\vec{p})_\alpha, \mathrm{u}(\vec{q})^\beta\} \rangle_{\mathrm{F}} = \delta_\alpha^\beta 2|\vec{q}|\delta(\tfrac{\vec{q}-\vec{p}}{2\pi}).$$

A Fock form also for the $\mathbf{U}(1,1)$-time representations leads to an indefinite metric

$$\begin{array}{rll} \langle \mathrm{B}^\times(\vec{p})\mathrm{G}(\vec{q})\rangle_{\mathrm{F}} & = \langle \{\mathrm{B}^\times(\vec{p}), \mathrm{G}(\vec{q})\}\rangle_{\mathrm{F}} & = 2|\vec{q}|\delta(\tfrac{\vec{q}-\vec{p}}{2\pi}) \\ = \langle \mathrm{G}^\times(\vec{p})\mathrm{B}(\vec{q})\rangle_{\mathrm{F}} & = \langle \{\mathrm{G}^\times(\vec{p}), \mathrm{B}(\vec{q})\}\rangle_{\mathrm{F}} & = 2|\vec{q}|\delta(\tfrac{\vec{q}-\vec{p}}{2\pi}). \end{array}$$

It gives a Fock value for the anticommutator of the massless vector field

$$\begin{array}{ll} \langle \{\mathbf{A}^k, \mathbf{A}^j\}\rangle_{\mathrm{F}}(x) & = g^2 \int \frac{d^4q}{(2\pi)^3}(-\eta^{kj} - 2\nu q^k q^j \frac{\partial}{\partial q^2})\delta(q^2) e^{iqx}, \\ \langle \{\mathbf{Z}^k, \mathbf{Z}^j\}\rangle_{\mathrm{F}}(x) & = \gamma^2 \int \frac{d^4q}{(2\pi)^3}(-\eta^{kj} + \frac{q^k q^j}{m^2})\delta(q^2 - m^2) e^{iqx}, \end{array}$$

and hence the Feynman propagator, always to be compared with the massive vector field structures

$$\begin{array}{ll} \langle \{\mathbf{A}^k, \mathbf{A}^j\}(x) - \epsilon(x_0)[\mathbf{A}^k, \mathbf{A}^j](x)\rangle_{\mathrm{F}} & = g^2 \frac{i}{\pi} \int \frac{d^4q}{(2\pi)^3}\left[\frac{-\eta^{kj}}{q^2+io} + 2\nu\frac{q^k q^j}{(q^2+io)^2}\right] e^{iqx}, \\ \langle \{\mathbf{Z}^k, \mathbf{Z}^j\}(x) - \epsilon(x_0)[\mathbf{Z}^k, \mathbf{Z}^j](x)\rangle_{\mathrm{F}} & = \gamma^2 \frac{i}{\pi} \int \frac{d^4q}{(2\pi)^3} \frac{(-\eta^{kj} + \frac{q^k q^j}{m^2})}{q^2+io-m^2} e^{iqx}. \end{array}$$

5.10 Eigenvectors and Nilvectors in a Gauge Dynamics

The spacetime dependence of a classical gauge transformation with parameter β as invariance of the Lagrangian

$$\begin{array}{l} \mathbf{L}(\mathbf{A}^j, \mathbf{F}^{jk}) = \frac{1}{2}\mathbf{F}^{jk}\epsilon_{jk}^{lm}\partial_l \mathbf{A}_m + g^2 \frac{\mathbf{F}^{jk}\mathbf{F}_{jk}}{4}, \\ \mathbf{A}^k \longmapsto \mathbf{A}^k + \partial^k \beta, \quad \mathbf{F}_{kj} \longmapsto \mathbf{F}_{kj}, \end{array}$$

is drastically reduced for a quantum gauge theory with a duality-completing scalar field ("gauge fixing" field) $\mathbf{S}$. There remains a transformation with a "massless" Lie parameter field β:

$$\begin{array}{l} \mathbf{L}(\mathbf{A}^j, \mathbf{F}^{jk}, \mathbf{S}) = \frac{1}{2}\mathbf{F}^{jk}\epsilon_{jk}^{lm}\partial_l \mathbf{A}_m + \mathbf{S}\partial_k \mathbf{A}^k + g^2 \frac{\mathbf{F}^{jk}\mathbf{F}_{jk}}{4} - g^2 \lambda \frac{\mathbf{S}^2}{2}, \\ \mathbf{A}^k \longmapsto \mathbf{A}^k + \partial^k \beta, \quad \mathbf{F}_{kj} \longmapsto \mathbf{F}_{kj}, \quad \mathbf{S} \longmapsto \mathbf{S} \text{ with } \partial^2 \beta = 0. \end{array}$$

5.10.1 Fadeev-Popov Ghosts in Quantum Mechanics

The "gauge fixing" part of the dynamics with the gauge transformations for a free gauge field theory

$$\mathbf{L}(\mathbf{A},\mathbf{S}) = \mathbf{S}\partial_k\mathbf{A}^k - g^2\lambda\tfrac{\mathbf{S}^2}{2}, \quad \begin{cases} \partial_k\mathbf{A}^k = g^2\lambda\mathbf{S}, & \partial_k\mathbf{S} = 0, \\ \mathbf{A}^k \longmapsto \mathbf{A}^k + \gamma^k, & \mathbf{S} \longmapsto \mathbf{S}, \\ \partial^k\beta = \gamma^k, & \partial_k\gamma^k = 0, \end{cases}$$

is the Lorentz compatible spacetime distribution of the noncompact time development for a free mass point:

$$L(\mathbf{x},\mathbf{p}) = \mathbf{p}d_t\mathbf{x} - \tfrac{\mathbf{p}^2}{2}, \quad \begin{cases} d_t\mathbf{x} = \mathbf{p}, & d_t\mathbf{p} = 0, \\ \mathbf{x} \longmapsto \mathbf{x} + \gamma, & \mathbf{p} \longmapsto \mathbf{p}, \\ d_t\beta = \gamma, & d_t\gamma = 0. \end{cases}$$

The gauge transformation is the relativistic distribution of a position translation transformation for the free mass point position.

A noncompact time development has eigenvectors and nilvectors. The subspace built by the eigenvectors has a trivial eigenvalue (nildimension) for the action of the nilpotent part of the Hamiltonian. In the self-dual space, spanned by position and momentum, the Hamiltonian matrix of the free mass point, a linear 2×2 transformation, is nilquadratic:

$$\begin{aligned} &\mathbf{x} \cong x = \begin{pmatrix} 0 \\ 1 \end{pmatrix}, \quad \mathbf{p} \cong p = \begin{pmatrix} 1 \\ 0 \end{pmatrix}, \quad \langle p, x\rangle = 1. \\ &H_B = \tfrac{\mathbf{p}^2}{2} \cong h = \begin{pmatrix} 0 & 1 \\ 0 & 0 \end{pmatrix} \Rightarrow h(x) = p, \quad h(p) = 0, \quad h \circ h = 0. \end{aligned}$$

In a Bose quantum algebra, the Hamiltonian is not nilquadratic with respect to the quantum product

$$[i\mathbf{p},\mathbf{x}] = 1, \quad H_B = \tfrac{\mathbf{p}^2}{2} \Rightarrow [iH_B,\mathbf{x}] = \mathbf{p}, \quad [iH_B,\mathbf{p}] = 0 \quad \text{but } H^2 \neq 0.$$

By introducing additional Fermi degrees of freedom as partners for the Bose position-momentum pair it is possible to construct nontrivial nilquadratic quantum operators. To formulate the distinction between eigenvectors and nilvectors (particle and not particle interpretable) a quantum gauge theory has a *Bose-Fermi twin structure* as discussed above. The spinless part of the gauge Bose field and its "gauge fixing" dual partner are accompanied by *Fadeev-Popov fields* as their Fermi counterparts, whose classical limits are the spacetime-dependent Lie parameters of the gauge group.

The Bose-Fermi twin structure is discussed first in the nonrelativistic quantum-mechanical model: A noncompact time development for the additional Fermi degrees of freedom needs two dual pairs:

$$\begin{aligned} &\text{Bose: } [i\mathbf{p},\mathbf{x}] = 1, \quad \text{Fermi: } \{\beta, \check{\gamma}\} = 1 = \{\gamma, \check{\beta}\}, \\ &\text{Hamiltonian: } H_{B+F} = H_B + H_F = \tfrac{\mathbf{p}^2}{2} + i\check{\gamma}\gamma. \end{aligned}$$

The equations of motion are

$$\text{Bose: } \begin{cases} d_t\mathbf{x} &= [iH_{B+F},\mathbf{x}] &= \mathbf{p}, \\ d_t\mathbf{p} &= [iH_{B+F},\mathbf{p}] &= 0, \end{cases} \quad \text{Fermi: } \begin{cases} d_t\beta &= [iH_{B+F},\beta] &= \gamma, \\ d_t\gamma &= [iH_{B+F},\gamma] &= 0, \\ d_t\check{\beta} &= [iH_{B+F},\check{\beta}] &= -\check{\gamma}, \\ d_t\check{\gamma} &= [iH_{B+F},\check{\gamma}] &= 0. \end{cases}$$

They can be derived from a classical Lagrangian (first or second order time derivatives)

$$\begin{aligned} L(\mathbf{x}, \mathbf{p}, \beta, \gamma) = &\ \mathbf{p} d_t \mathbf{x} - \frac{\mathbf{p}^2}{2} + i\gamma d_t \check{\beta} + i\check{\gamma} d_t \beta - i\check{\gamma}\gamma, \\ L(\mathbf{x}, \beta) = &\ \frac{1}{2}(d_t \mathbf{x})^2 + i(d_t \beta)(d_t \check{\beta}). \end{aligned}$$

The nilquadratic *Becchi-Rouet-Stora charge* N_{BF} implementing the gauge transformation $\delta \mathbf{x} = \gamma$ is given by the time development invariant

$$N_{BF} = \gamma \mathbf{p} \Rightarrow N_{BF}^2 = 0, \quad [H_{B+F}, N_{BF}] = 0.$$

Its linear hybrid adjoint action in a hybrid algebra generated by Bose and Fermi vectors,

$$[\![a, b]\!] = \begin{cases} [a, b] & \iff a \text{ or } b \text{ are Bose}, \\ \{a, b\} & \iff a \text{ and } b \text{ are Fermi}, \end{cases}$$

defines the BRS-transformations

$$\text{Bose: } \begin{cases} \delta \mathbf{x} & = [iN_{BF}, \mathbf{x}] & = \gamma, \\ \delta \mathbf{p} & = [iN_{BF}, \mathbf{p}] & = 0, \end{cases} \quad \text{Fermi: } \begin{cases} \delta \beta & = \{iN_{BF}, \beta\} & = 0, \\ \delta \gamma & = \{iN_{BF}, \gamma\} & = 0, \\ \delta \check{\beta} & = \{iN_{BF}, \check{\beta}\} & = i\mathbf{p}, \\ \delta \check{\gamma} & = \{iN_{BF}, \check{\gamma}\} & = 0. \end{cases}$$

With the *Fadeev-Popov number operator* for the Fermi degrees of freedom

$$\begin{aligned} &F = i(\check{\gamma}\beta + \check{\beta}\gamma) \Rightarrow \begin{cases} [iF, \beta] & = \beta, & [iN_{BF}, \check{\gamma}] & = -\check{\gamma}, \\ [iF, \check{\beta}] & = -\check{\beta}, & [iF, \gamma] & = \gamma, \end{cases} \\ &[F, H_{B+F}] = 0, \end{aligned}$$

the space $\mathbf{Q}(H_{B+F})$ spanned by the eigenvectors of the Hamiltonian is defined by trivial eigenvalues both for the BRS-charge N_{BF} and the Fadeev-Popov number F

$$\mathbf{Q}(H_{B+F}) = \{a \mid [F, a] = 0 \text{ and } [\![N_{BF}, a]\!] = 0\}.$$

5.10.2 Fadeev-Popov Ghosts for Quantum Gauge Fields

The Lorentz compatible distribution of the nonrelativistic model for the electromagnetic quantum gauge field,

$$\begin{aligned} \mathbf{L}(\mathbf{A}, \mathbf{F}, \mathbf{S}, \beta, \gamma) = &\ \mathbf{F}_{kj} \frac{\partial^k \mathbf{A}^j - \partial^j \mathbf{A}^k}{2} + \mathbf{S} \partial_k \mathbf{A}^k + g^2 \left(\frac{\mathbf{F}_{kj}\mathbf{F}^{kj}}{4} - \lambda \frac{\mathbf{S}^2}{2}\right) \\ &+ i\gamma^k \partial_k \check{\beta} + i\check{\gamma}^k \partial_k \beta - ig^2 \lambda \check{\gamma}^k \gamma_k, \end{aligned}$$

$$\text{Bose: } \begin{cases} \partial^k \mathbf{A}^j - \partial^j \mathbf{A}^k & = g^2 \mathbf{F}^{jk}, \\ \partial_k \mathbf{A}^k & = g^2 \lambda \mathbf{S}, \\ \partial^j \mathbf{F}_{kj} - \partial_k \mathbf{S} & = 0, \end{cases} \quad \text{Fermi: } \begin{cases} \partial^k \beta & = g^2 \lambda \gamma^k, \\ \partial_k \gamma^k & = 0, \\ \partial^k \check{\beta} & = -g^2 \lambda \check{\gamma}^k, \\ \partial_k \check{\gamma}^k & = 0, \end{cases}$$

uses *Fadeev-Popov fields* $(\beta, \check{\beta}, \gamma^k, \check{\gamma}^k)$ with the Fermi quantization

$$[i\mathbf{S}, \mathbf{A}^k](\vec{x}) = \{\beta, \check{\gamma}^k\}(\vec{x}) = \{\gamma^k, \check{\beta}\}(\vec{x}) = \delta_0^k \delta(\vec{x}).$$

A second order derivative Lagrangian reads

$$\mathbf{L}(\mathbf{A}, \beta) = -\tfrac{1}{4g^2}(\partial^j \mathbf{A}^k - \partial^k \mathbf{A}^j)(\partial_j \mathbf{A}_k - \partial_k \mathbf{A}_j) + \tfrac{1}{2g^2\lambda}(\partial_k \mathbf{A}^k)^2 + \tfrac{i}{g^2\lambda}(\partial^k \beta)(\partial_k \check{\beta}).$$

The hybrid adjoint action of the nilquadratic linear BRS-charge generates the linear BRS-transformations

$$\begin{matrix} N_{BF} = g^2\lambda \int d^3x \gamma_0(x)\mathbf{S}(x), \\ N_{BF}^2 = 0 \end{matrix} \Rightarrow \begin{matrix} \text{Bose:} \left\{ \begin{array}{lll} \delta \mathbf{A}^k &= [iN_{BF}, \mathbf{A}^k] &= g^2\lambda\delta_0^k\gamma_0, \\ \delta \mathbf{S} &= [iN_{BF}, \mathbf{S}] &= 0, \\ \delta \mathbf{F}_{kj} &= [iN_{BF}, \mathbf{F}_{kj}] &= 0, \end{array} \right. \\ \text{Fermi:} \left\{ \begin{array}{lll} \delta \beta &= \{iN_{BF}, \beta\} &= 0, \\ \delta \gamma^k &= \{iN_{BF}, \gamma^k\} &= 0, \\ \delta \check{\beta} &= \{iN_{BF}, \check{\beta}\} &= ig^2\lambda\mathbf{S}, \\ \delta \check{\gamma}^k &= \{iN_{BF}, \check{\gamma}^k\} &= 0. \end{array} \right. \end{matrix}$$

The subspace with the particle interpretable degrees of freedom, i.e., without nilvectors, is characterized by trivial BRS-charge and a trivial Faddev-Popov number

$$F = \int d^3x \mathbf{F}_0(x), \quad \mathbf{F}_k = i(\check{\gamma}_k \beta + \check{\beta}\gamma_k).$$

"Gauge invariant" fields are characterizable as translation eigenvectors.

The spinless and "gauge fixing" Bose degrees of freedom and the Fermi Fadeev-Popov ones display a twin structure. The BRS-current $\mathbf{N}_k(x)$ of Fermi type has its counterpart in the nonderivative part $\mathbf{H}(x)$ of the Lagrangian (Bose type)

$$\mathbf{N}_k = g^2\lambda\gamma_k\mathbf{S}, \quad \mathbf{H}_{B+F} = g^2\lambda[\tfrac{\mathbf{S}^2}{2} + i\check{\gamma}^k\gamma_k].$$

The dynamics H_{B+F} in the mass point model arises by BRS-transformation from an operator K connecting Bose and Fermi degrees of freedom

$$\begin{array}{l} N_{BF} = \gamma\mathbf{p}, \quad H_{B+F} = \tfrac{\mathbf{p}^2}{2} + i\check{\gamma}\gamma, \\ H_{B+F} = \{N_{BF}, K\}, \quad K = \tfrac{\check{\beta}\mathbf{p}}{2} + \check{\gamma}\mathbf{x}. \end{array}$$

Since $N_{BF}^2 = 0$ the BRS-invariance of the Hamiltonian is obvious:

$$[N_{BF}, H_{B+F}] = [N_{BF}, \{N_{BF}, K\}] = 0.$$

The corresponding relativistic field operators are the position distributions

$$\mathbf{K} = \tfrac{\check{\beta}\mathbf{S}}{2} + \check{\gamma}_k\mathbf{A}^k, \quad (H_{B+F}, N_{BF}, K) = \int d^3x(\mathbf{H}, \mathbf{N}_0, \mathbf{K})(x).$$

5.11 Summary

Massless spacetime vector fields ("gauge fields") $\{\mathbf{A}^k\}_{k=0}^3$ are acted on by the 4-dimensional Minkowski representation of $\mathbf{SO}_0(1,3)$, like the spacetime translations. They realize, together with the field strengths $\{\mathbf{F}_{jk}\}$ in the real 6-dimensional adjoint representation, the two fundamental representations of the Lorentz group. Duality pairing for a quantum theory requires a scalar field ("gauge fixing" field) $\mathbf{S}$ to complete four $(4 = 3 + 1)$ dual pairs $(\mathbf{F}_{a0}, \mathbf{A}^a)_{a=1}^3$ and $(\mathbf{S}, \mathbf{A}^0)$.

The translation representations acting on the four components of the gauge field are in the indefinite unitary group $\mathbf{U}(1,3) \supset \mathbf{U}(1,1) \times \mathbf{U}(2)$ as supgroup of the indefinite metric Lorentz group $\mathbf{SO}_0(1,3)$. The Minkowski metric shows up in the indefinite signature $(1,3)$ metric for the gauge field inner product space. A projection to a probability interpretable vector subspace with the two particle degrees of freedom for left and right circularly polarized photons requires the transition to translation eigenvectors that are determined by a trivial action of the nilpotent part of the dynamics. To define a nilquadratic projection (Becchi-Rouet-Stora transformation) in the quantum algebra, the Bose type gauge fields $(\mathbf{A}^0, \mathbf{S})$ have to be paired with Lorentz scalar fields (β, γ) of Fermi type (Fadeev-Popov fields). They have no particle degrees of freedom. Translation eigenvectors have trivial Becchi-Rouet-Stora charge; they are "gauge invariant."

Bibliography

[1] C. Becchi, A. Rouet, R. Stora, *Ann. of Phys.* 98 (1976), 287.

[2] T. Kugo, I. Ojima, *Progr. of Theor. Phys.* 60 (1978), 1869.

6

GAUGE INTERACTIONS

With the work of Weyl and London on gauge theories, Maxwell's equations for Faraday's electromagnetic field concepts proved to be a theory of phase $\mathbf{U}(1)$-operations that act, compatibly with spacetime translations, on complex representation spaces. In quantum electrodynamics the electromagnetic $\mathbf{U}(1)$ is implemented by the electromagnetic potential (field) interacting with Dirac fields for electrons and positrons. The standard model of the electroweak and strong interactions for lepton and quark quantum fields embeds quantum electrodynamics into a representation theory for the compact internal action groups $\mathbf{U}(1)$ (hypercharge), $\mathbf{SU}(2)$ (isospin) and $\mathbf{SU}(3)$ (color), implemented by twelve gauge fields acting on left- and right-handed Weyl fields and, for nonabelian groups, on themselves:

electrostatics $\mathbf{SO}(3) \vec{\times} \mathbb{R}^3$	$\hookrightarrow$	electrodynamics $\mathbf{SO}_0(1,3) \vec{\times} \mathbb{R}^4$

$\hookrightarrow$	quantum electrodynamics $\mathbf{U}(1) \times [\mathbf{SO}_0(1,3) \vec{\times} \mathbb{R}^4]$	$\hookrightarrow$	standard gauge interactions $\mathbf{U}(2 \times 3) \times [\mathbf{SO}_0(1,3) \vec{\times} \mathbb{R}^4]$

It is remarkable that each of the incomplete theories shows its own esthetics and beauty.

All spacetime translations have to take into account the orientation of the internally acting group, there is no spacetime translation without internal group action. In quantum field theory, Lorentz compatible distributions of Lie algebra representations define currents (chapter "Massive Particle Quantum Fields'). The representation of a Lie algebra on a vector space is a power-three tensor whose spacetime distribution comes as a product of the current with the gauge field. Such a power three tensor constitutes a gauge interaction vertex for a field theory. This is used for the real 12-parametric standard model Lie symmetry with its $(1+3+8)$ gauge fields.

The internal action groups for hypercharge, isospin, and color come in centrally correlated representations, the eigenvalue of the abelian hypercharge $\mathbf{U}(1)$-action is related to the center representation of the nonabelian isospin-color group $\mathbf{SU}(2) \times \mathbf{SU}(3)$. For example, isospin doublets and color triplets come with hypercharge factors $\frac{1}{2}$ and $\frac{1}{3}$, a doublet-triplet quark with hypercharge $\frac{1}{6}$.

From the 12-parametric internal symmetry operations for interactions there remains only an electromagnetic $\mathbf{U}(1)$-symmetry for particles. The isospin $\mathbf{SU}(2)$-symmetry is broken ("bleached"), it leaves its trace in particle multiplicities. This is in contrast to color $\mathbf{SU}(3)$, where experiments show only trivial color representations for particles which is interpreted as color confinement. A simultaneous diagonalization of the rank $1+1+2=4$ centrally correlated symmetry structure of the interaction is possible for a maximal abelian subgroup that is trivial either for isospin $\mathbf{SU}(2)$ or for color $\mathbf{SU}(3)$. Taking a color-trivial maximal diagonalization, the electroweak $\mathbf{U}(2)$-operations require a projection to an electromagnetic $\mathbf{U}(1)$ Cartan subgroup, correlating hypercharge and isospin, as remaining internal particle symmetry group. In the standard model, this projection (electroweak symmetry breakdown) is effected by a ground state, degenerate with the Goldstone manifold $\mathbf{U}(2)/\mathbf{U}(1)$ and implemented by a scalar field (Higgs field).

After a short review of classical and quantum electrodynamics, its embedding into the standard model of electroweak and strong interactions is discussed together with the ground state induced rearrangement of the interactions to the particle language.

6.1 Classical Maxwell Equations

Experiences with amber (electron) and stones from Magnesia (Greek town in Asia Minor) and experiments have shown the existence of "nonmechanical" interactions, especially nongravitational ones, which, in today's language, cannot be related to Poincaré group, i.e., external, operations. On the "spacetime screen," i.e., with each spacetime translation, there also act "internal" operations. The electric and magnetic interactions were taken as a first hint for "charge" related operation groups.

In the beginning, it was enough to characterize these properties (eigenvalues) by an *electric charge* Q, first measured[1] in an ad hoc unit, e.g., $[Q]=\text{C}$ (coulomb), introduced for a dimensional grading in addition to units for length, time and mass, which can be measured with, e.g., the ad hoc human order of magnitude units $[L]=\text{m}$ (meter), $[T]=\text{s}$ (second), and $[M]=\text{kg}$ (kilogram). In the course of this chapter an independent charge unit will be replaced by an $[L],[T],[M]$-derived unit and two of the remaining "human" units will be replaced by natural or structurally intrinsic units.

Charges change in time $t\longmapsto Q(t)$ (time orbits), which leads to the definition of an *electric current* $I(t)$:

$$I=-d_tQ \text{ with } [I]=\tfrac{[Q]}{\text{s}}.$$

Now the historical transition to a framework with position-dependent fields: It appeared possible to distribute charge in position with an $\mathbf{SO}(3)$-scalar

[1] For $[a]$ read "unit of a,"

volume density $\rho(t,\vec{x})$ and the current with an $\mathbf{SO}(3)$-vectorial area density $\vec{J}(t,\vec{x})$:

$$\begin{array}{ll} Q = \int_V d^3x\ \rho & \text{with } [\rho] = \frac{[Q]}{\text{m}^3}, \\ I = \int_{\partial V} d^2\vec{x} \vec{J} & \text{with } [\vec{J}] = \frac{[Q]}{\text{sm}^2}. \end{array}$$

From now on, in addition to the time translation group $\mathbb{R}$, the Euclidean position group $\mathbf{SO}(3) \vec{\times} \mathbb{R}^3$ is assumed as action group. All fields considered in the following depend on time and position translations $\mathbb{R}^4 \ni (t,\vec{x}) \longmapsto \Phi(t,\vec{x})$ (translation orbits) and are valued in, for classical electrodynamics, real vector spaces.

If the current definition makes sense for all volumes,

$$0 = d_t Q + I = \int_V d^3x\ \partial_t\rho + \int_{\partial V} d^2\vec{x}\ \vec{J} = \int_V d^3x\ (\partial_t\rho + \text{div}\ \vec{J}),$$

one obtains the continuity equation which characterizes a *conserved current*

$$\partial_t\rho + \text{div}\ \vec{J} = 0.$$

Following Faraday, the sources ρ define a vector field $\vec{D}$, called an *electric field* (electric displacement). Hence the conserved current leads to a source-free field $\partial_t\vec{D} + \vec{J}$, which can be written as the curl of another vector field $\vec{H}$, called a *magnetic field*:

$$\begin{array}{l} \text{div}\ \vec{D} = \rho \Rightarrow \text{div}\ (\partial_t\vec{D} + \vec{J}) = 0 \Rightarrow \partial_t\vec{D} + \vec{J} = \text{rot}\ \vec{H} \\ \text{with } [\vec{D}] = \frac{[Q]}{\text{m}^2},\quad [\vec{H}] = \frac{[Q]}{\text{s m}}. \end{array}$$

As seen from the units (no length unit), the vector fields are not valued in position space.

The definitions and assumptions used so far are summarized in the

$$\textit{inhomogeneous Maxwell equations:} \left\{ \begin{array}{rl} \text{div}\ \vec{D} & = \rho, \\ -\partial_t\vec{D} + \text{rot}\ \vec{H} & = \vec{J}, \end{array} \right.$$

with the displacement current $\partial_t\vec{D}$ introduced by Maxwell in the manner given above.

A dynamical spacetime theory has to give the action of time and position translations on all fields involved, i.e., on $\{\rho, \vec{J}, \vec{D}, \vec{H}\}$: This is classically expressed by differential equations with the translation-action-implementing derivatives $\frac{\partial}{\partial t}, \frac{\partial}{\partial \vec{x}}$.

One is working with $\mathbf{SO}(3)$-scalars and vectors, coming with both position parities, i.e., with eigenvalues $\mathsf{p} = \pm 1$ for the represented position reflection $\mathbf{O}(3)/\mathbf{SO}(3) \cong \mathbb{I}(2) \ni \mathsf{P} : \vec{x} \leftrightarrow -\vec{x}$. With parity conservation and the parities of a time and position translation basis $\{\mathbf{p}^0, \mathbf{p}^a\}$, the parities of the three position areas, the position volume, and the fields involved are given as follows:

	$\mathbf{SO}(3)$-scalar $L = 0$	$\mathbf{SO}(3)$-vector $L = 1$
$\mathsf{p} = +1$	scalar $\mathbf{p}^0 \cong \frac{\partial}{\partial t}$, ρ	axial vector $\mathbf{p}^a \wedge \mathbf{p}^b$, $\vec{H}$
$\mathsf{p} = -1$	pseudoscalar $\mathbf{p}^a \wedge \mathbf{p}^b \wedge \mathbf{p}^c$	polar vector $\mathbf{p}^a \cong \frac{\partial}{\partial x_a}$, $\vec{J}$, $\vec{D}$

Acted on by the $\mathbf{SO}(3)$-scalar time derivative ∂_t and the $\mathbf{SO}(3)$-vector position derivative $\frac{\partial}{\partial \vec{x}}$, a vector field $\vec{\Phi}$, here $\vec{D}$ and $\vec{H}$, gives rise to two vector fields and one scalar field $\{\partial_t \vec{\Phi}, \text{rot } \vec{\Phi}; \text{div } \vec{\Phi}\}$:

	$L = 0$	$L = 1$
$\mathsf{p} = +1$	div $\vec{D}$	$\partial_t \vec{H}$, rot $\vec{D}$
$\mathsf{p} = -1$	div $\vec{H}$	$\partial_t \vec{D}$, rot $\vec{H}$

Additional $L = 2$ fields have no invariant coupling to charge ρ and current $\vec{J}$.

So far, equations with space and time translation action for the axial vector and pseudoscalar fields (parity $\mathsf{p} = (-1)^{1+L}$) are missing, i.e., for one $\mathbf{SO}(3)$-scalar div $\vec{H}$ and for two $\mathbf{SO}(3)$-vectors $\partial_t \vec{H}$ and rot $\vec{D}$. If there do not exist further sources in addition to a charge Q, the simplest equations, compatible with $\mathbf{O}(3)$, i.e., with the rotations $\mathbf{SO}(3)$ and the position reflection P, require the still undetermined scalar derivative field to be trivial (no magnetic monopoles) and equate the two vector derivative fields (law of Faraday and Lenz)

$$\textit{homogeneous Maxwell equations:} \quad \left\{ \begin{array}{rl} \text{div } \vec{H} & = 0, \\ \frac{1}{c^2}\partial_t \vec{H} + \text{rot } \vec{D} & = 0. \end{array} \right.$$

The free constant c^2 has the units of velocity squared

$$\left. \begin{array}{rl} [\partial_t \vec{H}] & = \frac{[Q]}{\text{s}^2 \text{m}}, \\ [\text{rot } \vec{D}] & = \frac{[Q]}{\text{m}^3} \end{array} \right\} \Rightarrow [\frac{\partial_t \vec{H}}{\text{rot } \vec{D}}] = [c^2] = \frac{\text{m}^2}{\text{s}^2}.$$

One ad hoc unit, e.g., second for time, is traded for the intrinsic

$$\textit{fundamental unit: } c = 299\ 792\ 459\ \frac{\text{m}}{\text{s}}.$$

Those "simplest" equations with positive $c^2 > 0$ for the field theory of a conserved charge in spacetime have proved to be physically relevant as Maxwell's equations for the electromagnetic field strengths $\{\vec{D}, \vec{H}\}$ in the vacuum with c the speed of light (electromagnetic wave). To define the dynamics completely in the case of a nontrivial charge-current, there have to be added dynamical equations for the space-time behavior of the charge-current $\{\rho, \vec{J}\}$ in addition to the continuity equation above, i.e., for charged mass points in mechanics and for charged fields in field theories, which will be done below.

Historically, the velocity of light was found to be related to the product of the dielectricity and permeability constants of the vacuum, ϵ_0 and μ_0 respectively

$$c^2 = \frac{1}{\epsilon_0 \mu_0},$$

arising in the transformations from the electric displacement $\vec{D}$ and magnetic field $\vec{H}$ to the electric field $\vec{E}$ and magnetic induction $\vec{B}$:

$$\vec{E} = \frac{1}{\epsilon_0}\vec{D}, \quad \vec{B} = \mu_0 \vec{H},$$
$$\left.\begin{array}{rl} \text{div } \vec{D} & = \rho, \\ -\partial_t \vec{D} + \text{rot } \vec{H} & = \vec{J}, \end{array}\right\} \text{ and } \left\{ \begin{array}{rl} \text{div } \vec{B} & = 0, \\ \partial_t \vec{B} + \text{rot } \vec{E} & = 0. \end{array}\right.$$

For electromagnetism in materials the introduction of (3×3) transformations $\epsilon, \mu \in \mathbf{GL}(\mathbb{R}^3)$ with $\vec{E} = \frac{1}{\epsilon_0 \epsilon}\vec{D}$ and $\vec{B} = \mu_0 \mu \vec{H}$, in the simplest case with constants ϵ, μ, may be convenient for a phenomenological parametrization.

6.2 The Electromagnetic Gauge Field

The homogeneous Maxwell equations allow, in analogy to the dual position-momentum pairs $(\mathbf{x}, \mathbf{p})$ in mechanics, the definition of dual partners ("positions" for the field strengths $\vec{H}$ and $\vec{D}$ as "momenta"), called, with respect to position rotations $\mathbf{SO}(3)$, *scalar and vector potential* V and $\vec{A}$ with parity $\mathrm{p} = +1$ and $\mathrm{p} = -1$ respectively:

$$\text{homogeneous equations} \Rightarrow \left\{ \begin{array}{rl} \text{rot } \vec{A} & = \vec{H}, \\ \text{grad } V - \frac{1}{c^2}\partial_t \vec{A} & = \vec{D}. \end{array}\right.$$

Those equations are the analogue of the mechanical $M d_t \mathbf{x} = \mathbf{p}$. The position-momentum analogue dual pairs are given by the four potentials and the six field strengths $(\mathbf{x}; \mathbf{p}) \sim (V, \vec{A}; \vec{D}, \vec{H})$.

The potentials are determined up to the derivatives of an $\mathbf{SO}(3)$-scalar field $(t, \vec{x}) \longmapsto \gamma(t, \vec{x})$ (*gauge transformations*):

$$\begin{array}{ll} \vec{A} & \longmapsto \vec{A} + \text{grad } \gamma \text{ with } [\gamma] = \frac{[Q]\text{m}}{\text{s}}, \\ V & \longmapsto V + \frac{1}{c^2}\partial_t \gamma. \end{array}$$

The inhomogeneous Maxwell equations, the analogue of the mechanical $d_t \mathbf{p} = \mathbf{F}$ (force), are second order equations for the electromagnetic potential, in analogy to the mechanical $M d_t^2 \mathbf{x} = \mathbf{F}$:

$$\text{inhomogeneous equations} \Rightarrow \left\{ \begin{array}{rl} \vec{\partial}^2 V - \frac{1}{c^2}\partial_t \text{div } \vec{A} & = \rho, \\ (\frac{1}{c^2}\partial_t^2 - \vec{\partial}^2)\vec{A} - \text{grad } (\partial_t V - \text{div } \vec{A}) & = \vec{J}. \end{array}\right.$$

They can be rearranged in the form

$$\begin{array}{rlrl} - & (\partial_{ct}^2 - \vec{\partial}^2) cV & +\partial_{ct} & (\partial_t V - \text{div } \vec{A}) = c\rho, \\ & (\partial_{ct}^2 - \vec{\partial}^2)\vec{A} & -\text{grad} & (\partial_t V - \text{div } \vec{A}) = \vec{J}. \end{array}$$

These second order Maxwell equations with the characteristic derivative combination $\partial_{ct}^2 - \vec{\partial}^2$ display a representation structure for the Lorentz group

$\mathbf{SO}_0(1,3)$ with the Minkowski representation $[1|1]$ for the translations, for derivatives

$$\begin{array}{llll} (ct,\vec{x}) & = x_k, \quad (ct,-\vec{x}) & = x^k = \eta^{kj}x_j & \text{with } [x] = \text{m}, \\ (\frac{\partial}{\partial ct},\frac{\partial}{\partial \vec{x}}) & = \partial^k, \quad (\frac{\partial}{\partial ct},-\frac{\partial}{\partial \vec{x}}) & = \partial_k & \text{with } [\partial] = \frac{1}{\text{m}}, \\ \partial_{ct}^2 - \vec{\partial}^2 = \eta_{kj}\partial^k\partial^j, & k = 0,1,2,3 & & \end{array}$$

and for potentials and charge-current densities

$$\begin{array}{lll} (-cV,\vec{A}) & = A_k & \text{with } [A] = \frac{[Q]}{\text{s}}, \\ (c\rho,\vec{J}) & = J_k & \text{with } [J] = \frac{[Q]}{\text{sm}^2}. \end{array}$$

The equations of motion

$$\partial_j\partial^j A_k - \partial_k\partial^j A_j = \partial^j(\partial_j A_k - \partial_k A_j) = J_k$$

collect the field strengths as follows:

$$\begin{array}{l} cD_a = \partial_a A_0 - \partial_0 A_a, \quad a = 1,2,3, \\ H_a = \epsilon_{abc}\partial_b A_c \Rightarrow \epsilon_{abc}H_c = \partial_a A_b - \partial_b A_a. \end{array}$$

They constitute an antisymmetric 6-component Lorentz tensor acted on by an adjoint $\mathbf{SO}_0(1,3)$-representation $[2|0] \oplus [0|2] = [1|1] \wedge [1|1]$:

$$\partial_k A_j - \partial_j A_k = F_{jk} = -F_{kj}$$

$$F_{a0} = cD_a, \quad F_{ab} = -\epsilon_{abc}H_c, \quad F_{kj} = \begin{pmatrix} 0 & -cD_1 & -cD_2 & -cD_3 \\ cD_1 & 0 & -H_3 & H_2 \\ cD_2 & H_3 & 0 & -H_1 \\ cD_3 & -H_2 & H_1 & 0 \end{pmatrix} \text{ with } [F] = \frac{[Q]}{\text{sm}}.$$

Dynamical theories with all the concepts involved have to be characterized by representations of the corresponding group with their invariants. In Newtonian mechanics, the inhomogeneous Galileo group relates to each other equivalent reference frames for position and time. It is expanded to and simplified by the inhomogeneous Lorentz group (Poincaré group)

$$[\mathbf{SO}(3) \vec{\times} \mathbb{R}^3] \vec{\times} [\mathbb{R}^3 \oplus \mathbb{R}] \overset{\frac{1}{c^2}>0}{\longrightarrow} \mathbf{SO}_0(1,3) \vec{\times} \mathbb{R}^4,$$

which relates to each other equivalent spacetime reference frames in special relativity. Nonrelativistic theories can be recovered in the Inönü-Wigner contraction. The contracted boosts $\mathbf{SO}_0(1,3)/\mathbf{SO}(3) \to \mathbb{R}^3$ (chapter "Spacetime Translations"),

$$\mathbf{SO}_0(1,3) \overset{c\to\infty}{\longrightarrow} \mathbf{SO}(3) \vec{\times} \mathbb{R}^3:$$

$$\Lambda = \begin{pmatrix} C_\psi & C_\psi \frac{\vec{v}^T}{c} \\ C_\psi \frac{\vec{v}}{c} & \mathbf{1}_3 + \frac{C_\psi}{1+C_\psi}\frac{\vec{v}\otimes\vec{v}^T}{c^2} \end{pmatrix} \overset{c\to\infty}{\longrightarrow} \Lambda_\infty = \begin{pmatrix} 1 & 0 \\ \vec{v} & \mathbf{1}_3 \end{pmatrix} \text{ with } \cosh\psi = C_\psi = \frac{1}{\sqrt{1-\frac{\vec{v}^2}{c^2}}},$$

are derived with a renormalization with $\frac{1}{c}$ for the time translations

$$\begin{array}{ll} \mathbf{SO}_0(1,3): & \begin{pmatrix} ct \\ \vec{x} \end{pmatrix} \longmapsto \Lambda\begin{pmatrix} ct \\ \vec{x} \end{pmatrix} = \begin{pmatrix} C_\psi c(t + \frac{\vec{v}\vec{x}}{c^2}) \\ \vec{x} + C_\psi(\vec{v}t + \frac{C_\psi}{1+C_\psi}\frac{\vec{v}(\vec{v}\vec{x})}{c^2}) \end{pmatrix}, \\ \mathbf{SO}(3) \vec{\times} \mathbb{R}^3: & \begin{pmatrix} t \\ \vec{x} \end{pmatrix} \longmapsto \Lambda_\infty\begin{pmatrix} t \\ \vec{x} \end{pmatrix} = \begin{pmatrix} t \\ \vec{x} + \vec{v}t \end{pmatrix}. \end{array}$$

The Lorentz behavior of the electromagnetic field strengths reads (with the matrix $F_{ab} = -\epsilon_{abc}H_c = \mathbf{H}_3$ and $\mathbf{H}_3\vec{v} = \vec{H} \times \vec{v} = -\vec{v}^T\mathbf{H}_3$)

$$\mathbf{SO}_0(1,3): \begin{cases} F \cong \begin{pmatrix} 0 & -c\vec{D}^T \\ c\vec{D} & \mathbf{H}_3 \end{pmatrix} \longmapsto & \Lambda^T \circ F \circ \Lambda \\ \Rightarrow c\vec{D} \longmapsto & C_\psi c(\vec{D} - \frac{\vec{v}\times\vec{H}}{c^2} + \frac{C_\psi}{1+C_\psi}\frac{\vec{v}(\vec{v}\vec{D})}{c^2}), \\ \vec{H} \longmapsto & C_\psi(\vec{H} + \vec{v} \times \vec{D} + \frac{C_\psi}{1+C_\psi}\frac{\vec{v}(\vec{v}\vec{H})}{c^2}). \end{cases}$$

The Galileo contraction limit of electrodynamics, i.e., nonrelativistic electrodynamics, changes the second homogeneous Maxwell equation

$$\mathbf{SO}(3) \vec{\times} \mathbb{R}^3: \begin{cases} \text{div } \vec{D} = \rho, & -\partial_t\vec{D} + \text{rot } \vec{H} = \vec{J}, \\ \text{div } \vec{H} = 0, & \text{rot } \vec{D} = 0, \\ \vec{D} \longmapsto \vec{D}, & \vec{H} \longmapsto \vec{H} + \vec{v} \times \vec{D}, \\ \rho \longmapsto \rho, & \vec{J} \longmapsto \vec{J} + \vec{v}\rho. \end{cases}$$

For the Lorentz action group $\mathbf{SO}_0(1,3)$ with rank 2, there are the two signature $(3,3)$ invariants:

$$\eta F \cong \begin{pmatrix} 0 & c\vec{D}^T \\ c\vec{D} & \mathbf{H}_3 \end{pmatrix}, \quad \det(\eta F - \lambda \mathbf{1}_4) = \lambda^4 + (\vec{H}^2 - c^2\vec{D}^2)\lambda^2 - (c\vec{D}\vec{H})^2.$$

Both the positive parity Killing invariant

$$F_{kj}F^{kj} = 2(\vec{H}^2 - c^2\vec{D}^2), \quad \mathrm{p} = +1$$

and the negative parity chiral (volume) invariant with the dual field strength tensor $\epsilon^{kjlm}F_{lm}$ exchanging $c\vec{D} \leftrightarrow \vec{H}$,

$$\epsilon^{kjlm}F_{lm} = \begin{pmatrix} 0 & -H_1 & -H_2 & -H_3 \\ H_1 & 0 & -cD_3 & cD_2 \\ H_2 & cD_3 & 0 & -cD_1 \\ H_3 & -cD_2 & cD_1 & 0 \end{pmatrix} = \frac{1}{c\mu_0}\begin{pmatrix} 0 & -cB_1 & -cB_2 & -cB_3 \\ cB_1 & 0 & -E_3 & E_2 \\ cB_2 & E_3 & 0 & -E_1 \\ cB_3 & -E_2 & E_1 & 0 \end{pmatrix},$$
$$\epsilon^{kjlm}F_{lm}F_{kj} = -8c\vec{D}\vec{H}, \quad \mathrm{p} = -1,$$

involve the highest action velocity c as relative normalization of the positive and negative "metric" sector. With the field strengths F_{kj} being derivatives of the gauge potential, the homogeneous Maxwell equations are identities for the derivatives of the dual

$$\partial_l\epsilon^{kjlm}F_{kj} = 0.$$

The first order Maxwell equations

$$\epsilon^{kj}_{lr}\partial^l A^r = \partial^k A^j - \partial^j A^k = F^{jk}, \quad \partial^j F_{kj} = J_k \text{ with } \epsilon^{kj}_{lr} = \delta^k_l\delta^j_r - \delta^j_l\delta^k_r$$

are gauge invariant with a Lorentz scalar field γ (representation $[0|0]$)

$$A^k \longmapsto A^k + \partial^k\gamma, \quad F_{kj} \longmapsto F_{kj}, \quad J_k \longmapsto J_k.$$

They can be derived from the Lagrangian

$$\mathbf{L}(A,F,J) = F_{kj}\frac{\partial^k A^j - \partial^j A^k}{2} + \frac{F_{kj}F^{kj}}{4} - A^k J_k$$

with the gauge behavior

$$\mathbf{L}(A,F,J) \longmapsto \mathbf{L}(A,F,J) - \partial^k(\gamma J_k) + \gamma\partial^k J_k.$$

A Lagrangian has the unit of an action spacetime density measurable with Planck's unit for actions (duality normalization, chapter "Quantum Algebras")

$$\textit{fundamental unit: } \hbar = 1.0545\ldots \times 10^{-34}\frac{\text{kg m}^2}{\text{s}}.$$

One more ad hoc unit can be traded for the fundamental intrinsic unit $\hbar$, e.g., kilogram for mass, leaving two ad hoc units, e.g., meter for length and coulomb for charge. The unit of the Lagrangian can be expressed also with the charge unit $[Q]$:

$$[\mathbf{L}(A,F,J)] = \frac{[Q]^2}{\text{s}^2\text{m}^2} = \frac{[\hbar]}{\text{m}^4} = \frac{\text{kg}}{\text{s m}^2}.$$

Hence the charge unit is a derived unit

$$[Q]^2 = [\tfrac{\hbar}{c^2}] = \text{kg s} = [M][T].$$

The gauge function, the gauge potential, the field strengths, and the current have the units

$$\begin{aligned}
[\gamma] &= \sqrt{\tfrac{\text{kg m}^2}{\text{s}}} &&= [\sqrt{\hbar}],\\
[A] &= \sqrt{\tfrac{\text{kg}}{\text{s}}} &&= [\sqrt{\hbar}]\tfrac{1}{\text{m}},\\
[F] &= \sqrt{\tfrac{\text{kg}}{\text{s m}^2}} &&= [\sqrt{\hbar}]\tfrac{1}{\text{m}^2},\\
[J] &= \sqrt{\tfrac{\text{kg}}{\text{s m}^4}} &&= [\sqrt{\hbar}]\tfrac{1}{\text{m}^3}.
\end{aligned}$$

Summarizing the Lorentz invariant Maxwell dynamics: Imposing Lorentz symmetry $\mathbf{O}(1,3)$ and its representations, a conserved vector current J_k in the defining $[1|1]$-representation can be written - necessary smoothness assumed, as a derivative ∂_l (Lorentz representation $[1|1]$) of a nonscalar field only with an antisymmetric tensor field, i.e., in the adjoint representation $[2|0] \oplus [0|2]$,

$$\partial^k J_k = 0 \Rightarrow J_k = \partial^j F_{kj}.$$

The derivative of a scalar field $J_k = \partial_k\Phi$ does not have to be conserved. If the other $[1|1]$-transforming derivative, constructible from ∂_l and F_{jk}, i.e., the derivative of the dual tensor, is assumed to be trivial, the tensor field is defined, up to gauge invariance, by the antisymmetric derivative of a vector field in the defining $[1|1]$-representation

$$\epsilon^{lmkj}\partial_l F_{kj} = 0 \iff F^{jk} = \epsilon^{kj}_{lr}\partial^l A^r.$$

6.3 The Charged Relativistic Mass Point

To have the dynamics complete, the translation behavior of the electromagnetic fields has to be supplemented by that of the charge and current implementing matter, in the simplest idealization a charged mass point. The relativistic dynamics of a charged mass point in an electromagnetic field is a hybrid theory: It uses both mass points as eigentime orbits and fields as spacetime translation orbits.

The defining relations between the time-dependent position $\{X_k\}_{k=0}^3$ as $\mathbf{SO}_0(1,3)$-vector,

$$\begin{aligned} X_k &= (ct, \vec{X}(t)),\ \ dX_k = (cdt, d\vec{X}) = (1, \tfrac{\vec{V}}{c})cdt,\ \ \vec{V} = \tfrac{d\vec{X}}{dt},\\ d\tau^2 &= dX_k dX^k = c^2dt^2 - d\vec{X}^2 = c^2dt^2(1-\tfrac{\vec{V}^2}{c^2}),\ \ d_\tau = \tfrac{d}{d\tau} = \tfrac{1}{\sqrt{1-\frac{\vec{V}^2}{c^2}}}d_{ct},\\ \text{with}\ \ &[X_k] = [\tau] = \mathrm{m}, \end{aligned}$$

and, with the invariant *eigentime* τ, the momentum $\{P_k\}_{k=0}^3$ of a mass point with rest mass m,

$$\begin{aligned} P_k &= m\tfrac{dX_k}{d\tau} = \tfrac{m}{\sqrt{1-\frac{\vec{V}^2}{c^2}}}(1, \tfrac{\vec{V}}{c}),\ \ P_kP^k = m^2,\ \ d_\tau P_k = 0,\\ \text{with}\ \ &[P_k] = [d_\tau] = [m] = \tfrac{1}{\mathrm{m}}, \end{aligned}$$

are expressible by the Lagrangian of a free relativistic mass point

$$L(X,P) = -P^k d_\tau X_k + \tfrac{P_kP^k}{2m}\ \text{with}\ [L(X,P)] = \tfrac{1}{\mathrm{m}}.$$

With Planck's unit $\hbar$ and the highest velocity of action c, the mass is measured in inverse length units:

$$m = \tfrac{c}{\hbar}\underline{m},\ \ [\underline{m}] = \mathrm{kg},\ \ [m] = \tfrac{1}{\mathrm{m}}.$$

The current for a mass point is proportional to the momentum with *charge number* q

$$\begin{aligned} J_k(x) &= q\sqrt{\hbar}\int dX_k\delta(x-X) = q\sqrt{\hbar}\sqrt{1-\tfrac{\vec{V}^2}{c^2}}\tfrac{P_k}{m}\delta(\vec{x}-\vec{X}(t)),\\ &\int d^3x J_0 = q\sqrt{\hbar},\ \ q\in\mathbb{R},\ \ \partial^k J_k = 0. \end{aligned}$$

It involves a Dirac distribution for the pointlike position density $\mathbf{J}_0(x)$:

$$J_k(x) = q\sqrt{\hbar}\mathbf{J}_k(x),\ \ \mathbf{J}_k(x) = (1, \tfrac{\vec{V}}{c})\delta(\vec{x}-\vec{X}(t)),\ \ [\mathbf{J}] = \tfrac{1}{\mathrm{m}^3}.$$

The Lorentz invariant interaction is the line integral along the spacetime coordinates of the mass point:

$$q\sqrt{\hbar}\int d\tau(d_\tau X_k)A^k(X) = q\sqrt{\hbar}\int dX_kA^k(X) = \int d^4x A^k(x)J_k(x).$$

From an action as sum of the free actions and the interaction

$$\begin{array}{lll} \int d^4x\, \mathbf{L}(A,F) & +\hbar\int d\tau L(X,P) & -\int d^4x A^k J_k \\ = \int d^4x [F_{kj}\frac{\partial^k A^j - \partial^j A^k}{2} + \frac{F_{kj}F^{kj}}{4}] & +\hbar\int d\tau \Big[-P^k d_\tau X_k + \frac{P_k P^k}{2m} & -\frac{q}{\sqrt{\hbar}} d_\tau X_j A^k(X)\Big] \end{array}$$

one obtains the equations of motion:

$$\begin{array}{llll} \partial^k A^j - \partial^j A^k & = F^{jk}, & \partial^j F_{kj} & = J_k = q\sqrt{\hbar}\int dX_k \delta(x-X), \\ d_\tau X_k & = \frac{P_k}{m}, & d_\tau P^k & = \frac{q}{\sqrt{\hbar}} d_\tau X_j \Big(\frac{\partial A^j}{\partial X_k} - \frac{\partial A^k}{\partial X_j}\Big). \end{array}$$

The *Lorentz force* as effected by the fields strengths at the mass-point coordinates $X_k = (ct, \vec{X}(t))$ arises as a consequence of the Lorentz invariant minimal coupling $A_k J^k$,

$$d_\tau P^k = \frac{q}{\sqrt{\hbar}} d_\tau X_j F^{kj}(X) \Rightarrow \left\{ \begin{array}{ll} d_t \frac{m}{\sqrt{1-\frac{\vec{V}^2}{c^2}}} & = \frac{q}{\sqrt{\hbar}} \vec{V}\vec{D}, \\ d_t \frac{m\vec{V}}{\sqrt{1-\frac{\vec{V}}{c}^2}} & = \frac{q}{\sqrt{\hbar}} (c^2\vec{D} + \vec{V}\times\vec{H}). \end{array} \right.$$

Via the Lorentz force, the amplitudes of electric and magnetic fields (not valued in position space) can be observed as position valued-amplitudes of a charged mass point motion.

6.4 Electrodynamics as U(1)-Representation

The "arbitrariness" of the vector potential $A^k \longmapsto A^k + \partial^k\gamma$ can be interpreted as a transformation behavior with respect to a real 1-dimensional Lie group with Lie parameter $\gamma(x)$ for each spacetime translation. The symmetry connected with such a transformation group becomes the cornerstone for understanding the origin of the electromagnetic interactions which are necessary for the compatibility of the related "internal" operations with translations. Reversing the historical order of the arguments, the local internal transformations can be used to establish the corresponding gauge interactions. In addition to reference frames, equivalent with respect to external operations, there are also internally equivalent reference frames.

There are two locally isomorphic real 1-dimensional Lie groups, the noncompact simply connected dilation group $\mathbf{D}(1)$ and the compact phase group $\mathbf{U}(1)$. Weyl, after a wrong attempt with the dilation group $\mathbf{D}(1)$, whence the name "gauge", and London initiated the interpretation of electrodynamics as $\mathbf{U}(1)$-actions, operating Lorentz compatibly with the translations. The electromagnetic vector current J_k is, up to a normalization constant $g > 0$ and the square root of Planck's unit, the position distribution of the Lie algebra representation of an *electromagnetic group* $\mathbf{U}(1)$, coming as internal ("chargelike") group together with the external ("spacetimelike") Poincaré group, of which the gauge function $\gamma(x)$ is the translation-dependent Lie parameter

$$\begin{array}{llll} J_k = g\sqrt{\hbar}\mathbf{J}_k, & \log\mathbf{U}(1) & \ni i & \longmapsto iQ = i\int d^3x \mathbf{J}_0(x), \\ & \mathbf{U}(1) & \ni e^{i\gamma} & \longmapsto e^{iQ\gamma} \text{ (if defined)}. \end{array}$$

With **J** having a position density unit, the charge Q and also the Lie parameter γ has trivial unit 1, replacing the old $\frac{\gamma}{\sqrt{\hbar}}$.

The simplest example for a **U**(1)-representation is given by the action on a non-Hermitian scalar particle field $(\mathbf{\Phi}, \mathbf{\Phi}^\star)$ with integer **U**(1)-winding numbers $\pm z$:

$$\mathbf{\Phi} \longmapsto e^{iz\gamma}\mathbf{\Phi}, \quad \mathbf{\Phi}^\star \longmapsto e^{-iz\gamma}\mathbf{\Phi}^\star, \quad z \in \mathbb{Z}.$$

In a quantum field theory (chapter "Massive Particle Quantum Fields") $(\mathbf{\Phi}, \mathbf{\Phi}^\star)$ come with dual partners $(\mathbf{\Phi}_k^\star, \mathbf{\Phi}_k)$ and commutators in analogy to $[i\mathbf{p}, \mathbf{x}] = 1$,

$$[i\mathbf{\Phi}_k^\star, \mathbf{\Phi}](\vec{x}) = [i\mathbf{\Phi}_k, \mathbf{\Phi}^\star](\vec{x}) = \delta_k^0\delta(\vec{x}) \text{ with } [\mathbf{\Phi}] = \tfrac{1}{\mathrm{m}}, \quad [\mathbf{\Phi}_k] = \tfrac{1}{\mathrm{m}^2},$$

and the **U**(1)-current

$$\mathbf{J}_k = iz(\mathbf{\Phi}\mathbf{\Phi}_k^\star - \mathbf{\Phi}^\star\mathbf{\Phi}_k) \Rightarrow \begin{cases} [Q, \mathbf{\Phi}](\vec{x}) &= z\mathbf{\Phi}(\vec{x}), \\ [Q, \mathbf{\Phi}_k](\vec{x}) &= z\mathbf{\Phi}_k(\vec{x}), \\ [Q, \mathbf{\Phi}^\star](\vec{x}) &= -z\mathbf{\Phi}^\star(\vec{x}), \\ [Q, \mathbf{\Phi}_k^\star](\vec{x}) &= -z\mathbf{\Phi}_k^\star(\vec{x}). \end{cases}$$

The electromagnetic interaction is described by the Lagrangian

$$\mathbf{L}(A, F, \mathbf{\Phi}, \mathbf{\Phi}_k^\star) = F_{kj}\frac{\partial^k A^j - \partial^j A^k}{2} + \frac{F_{kj}F^{kj}}{4} - g\sqrt{\hbar}A^k\mathbf{J}_k + \hbar\mathbf{L}(\mathbf{\Phi}, \mathbf{\Phi}_k^\star).$$

The constant g and Planck's unit $\hbar$ will be used to renormalize the gauge fields, the field strengths, and the current with units of a position density for a length, an area, and a volume respectively

$$\begin{array}{rllll} & \mathbf{A} = \frac{g}{\sqrt{\hbar}}A, & \mathbf{F} = \frac{1}{g\sqrt{\hbar}}F, & \mathbf{J} = \frac{1}{g\sqrt{\hbar}}J, \\ \text{with } [\gamma] = 1, & [\mathbf{A}] = \frac{1}{\mathrm{m}}, & [\mathbf{F}] = \frac{1}{\mathrm{m}^2}, & [\mathbf{J}] = \frac{1}{\mathrm{m}^3}, \end{array}$$

with **U**(1)-gauge behavior

$$\begin{array}{ll} \mathbf{A}^k \longmapsto \mathbf{A}^k + \partial^k\gamma, & \mathbf{F}_{kj} \longmapsto \mathbf{F}_{kj}, \\ \mathbf{\Phi} \longmapsto e^{iz\gamma}\mathbf{\Phi}, & \mathbf{\Phi}_k \longmapsto e^{iz\gamma}\mathbf{\Phi}_k \Rightarrow \mathbf{J}_k \longmapsto \mathbf{J}_k. \end{array}$$

The **U**(1)-gauge invariant Lagrangian

$$\begin{aligned} \mathbf{L}(\mathbf{A}, \mathbf{F}, \mathbf{\Phi}, \mathbf{\Phi}_k^\star) &= \tfrac{1}{\hbar}\mathbf{L}(A, F, \mathbf{\Phi}, \mathbf{\Phi}_k^\star) = \mathbf{L}(\mathbf{A}, \mathbf{F}) - \mathbf{A}^k\mathbf{J}_k + \mathbf{L}(\mathbf{\Phi}, \mathbf{\Phi}_k^\star) \\ &\overset{\mathbf{U}(1)}{\longmapsto} \mathbf{L}(\mathbf{A}, \mathbf{F}, \mathbf{\Phi}, \mathbf{\Phi}_k^\star) \end{aligned}$$

comprises the individual free Lagrangians

$$\begin{aligned} \mathbf{L}(\mathbf{A}, \mathbf{F}) &= \mathbf{F}_{kj}\frac{\partial^k\mathbf{A}^j - \partial^j\mathbf{A}^k}{2} + g^2\frac{\mathbf{F}_{kj}\mathbf{F}^{kj}}{4}, \\ \mathbf{L}(\mathbf{\Phi}, \mathbf{\Phi}_k^\star) &= \mathbf{\Phi}_k^\star\partial^k\mathbf{\Phi} + \mathbf{\Phi}_k\partial^k\mathbf{\Phi}^* - \mathbf{\Phi}_k^\star\mathbf{\Phi}^k - m^2\mathbf{\Phi}\mathbf{\Phi}^*, \end{aligned}$$

and the interaction

$$\mathbf{A}^k\mathbf{J}_k = iz\mathbf{A}^k(\mathbf{\Phi}\mathbf{\Phi}_k^\star - \mathbf{\Phi}^*\mathbf{\Phi}_k).$$

The constant g^2 is the *electromagnetic coupling (Sommerfeld's fine structure) constant*, experimentally given by

$$\frac{g^2}{4\pi} \sim \frac{1}{137.036}, \quad g^2 \sim \frac{1}{10.9}.$$

It is the normalization of the gauge fields. All electromagnetic interactions are quantitatively determined by the value of the gauge coupling constant g^2 and the representation characteristic integer $\mathbf{U}(1)$-winding numbers involved. For example, the Coulomb potential between two mass points with integer charge numbers as $\mathbf{U}(1)$-winding numbers is given by

$$\frac{1}{\hbar c}\mathcal{V}(\vec{x}) = z_1 \frac{g^2}{4\pi|\vec{x}|} z_2, \quad z_{1,2} \in \mathbb{Z}.$$

The defining charge $\mathbf{U}(1)$-representation comes with charge number $z = -1$ attributed to the electron.

Historically and also conveniently for everyday use, the electron charge number is expressed in a charge unit, e.g., with the coulomb, with the value e. Thus the winding number z is replaced by a charge Q,

$$z = \frac{Q}{e} \in \mathbb{Z} \Rightarrow \mathcal{V}(\vec{x}) = \frac{\hbar c g^2}{4\pi e^2} \frac{Q_1 Q_2}{|\vec{x}|}.$$

The historical use of the dielectricity and permeability constant for the vacuum leads with the fundamental charge to the fine structure constant parametrization

$$\frac{\hbar c g^2}{e^2} = \frac{1}{\epsilon_0} = \frac{c^2}{\mu_0}, \quad \frac{g^2}{4\pi} = \frac{e^2}{4\pi\epsilon_0 \hbar c}.$$

The unit coulomb, historically motivated by an electrolysis involving silver atoms,[2] is defined by an experimentally convenient value for $\mu_0 = 4\pi \times 10^{-7}\frac{\text{mkg}}{\text{C}^2}$ leading to the $\mathbf{U}(1)$-number for one coulomb $-\frac{\text{C}}{e} \sim 6 \times 10^{18}$.

The field equations for the charged scalar particle fields,

$$\begin{aligned} \partial^k \mathbf{A}^j - \partial^j \mathbf{A}^k &= g^2 \mathbf{F}^{jk}, & \partial^j \mathbf{F}_{kj} &= \mathbf{J}_k, \\ (\partial^k - iz\mathbf{A}^k)\mathbf{\Phi} &= \mathbf{\Phi}^k, & (\partial^k - iz\mathbf{A}^k)\mathbf{\Phi}_k &= -m^2\mathbf{\Phi}, \end{aligned}$$

contain the *covariant derivative*, which implements simultaneously the action of spacetime translations and internal Lie algebra via the sum of spacetime derivative and $\mathbf{U}(1)$-gauge field, multiplied by the $\mathbf{U}(1)$-eigenvalue. A covariantly derived particle field keeps a homogeneous $\mathbf{U}(1)$-behavior with translation-dependent Lie parameters $\gamma(x)$, e.g.,

$$\mathbf{U}(1): \quad \left.\begin{array}{ll} \mathbf{A}^k & \longmapsto \mathbf{A}^k + \partial^k\gamma, \\ \mathbf{\Phi} & \longmapsto e^{iz\gamma}\mathbf{\Phi}, \end{array}\right\} \Rightarrow (\partial^k - iz\mathbf{A}^k)\mathbf{\Phi} \longmapsto e^{iz\gamma}(\partial^k - iz\mathbf{A}^k)\mathbf{\Phi}.$$

The electromagnetic interaction can be written with second order derivatives for the gauge and the scalar fields:

$$\begin{aligned} \mathbf{L}(\mathbf{A}) &= -\frac{1}{4g^2}(\partial^j \mathbf{A}^k - \partial^k \mathbf{A}^j)(\partial_j \mathbf{A}_k - \partial_k \mathbf{A}_j), \\ \mathbf{L}(\mathbf{\Phi}, \mathbf{A}) &= [(\partial_k - iz\mathbf{A}_k)\mathbf{\Phi})][(\partial^k + iz\mathbf{A}^k)\mathbf{\Phi}^\star] - m^2\mathbf{\Phi}\mathbf{\Phi}^*. \end{aligned}$$

[2] The Faraday charge is given with Avogadro's number by $Q^{\text{Faraday}} = N^{\text{Avogadro}}e \sim 96.5\frac{\text{C}}{\text{millimol}}$ and the atomic mass number for silver by $Z(\text{Ag}) \sim 108$.

The additional factor 2 in the current with first order derivatives

$$-\frac{\partial \mathbf{L}(\mathbf{\Phi},\mathbf{A})}{\partial \mathbf{A}^k} = \mathbf{J}_k = iz(\mathbf{\Phi}\partial_k\mathbf{\Phi}^\star - \mathbf{\Phi}^\star\partial_k\mathbf{\Phi}) - 2z^2\mathbf{A}_k\mathbf{\Phi}\mathbf{\Phi}^\star$$

does not arise in the Lagrangian. This statistical factor is a consequence of the power-2 gauge field product $\mathbf{A}^k\mathbf{A}^j$ in the Lagrangian.

By the current-gauge field coupling $\mathbf{A}^k\mathbf{J}_k$ the global internal $\mathbf{U}(1)$-invariance (first kind gauge transformation) $\mathbf{\Phi}(x) \longmapsto e^{iz\gamma}\mathbf{\Phi}(x)$ is embedded into a local invariance (second kind gauge transformation) $\mathbf{\Phi}(x) \longmapsto e^{iz\gamma(x)}\mathbf{\Phi}(x)$ with translation dependent Lie parameters.

Quantum electrodynamics describes the electromagnetic interaction of charged particles in quantum fields, e.g., of charged spinless pions in fields $(\mathbf{\Phi},\mathbf{\Phi}^\star)$ or of positrons and electrons with charge number $z = \pm 1$, spin $\frac{1}{2}$, and mass $m^2 > 0$ in a Dirac field $\mathbf{\Psi}$:

$$\begin{array}{l}
\mathbf{\Psi} \longmapsto e^{iz\gamma}\mathbf{\Psi}, \quad \overline{\mathbf{\Psi}} \longmapsto e^{-iz\gamma}\overline{\mathbf{\Psi}}, \quad z \in \mathbb{Z}, \\
\{\overline{\mathbf{\Psi}},\mathbf{\Psi}\}(\vec{x}) = \gamma^0\delta(\vec{x}) \text{ with } [\mathbf{\Psi}] = [\overline{\mathbf{\Psi}}] = \frac{1}{\sqrt{\mathrm{m}^3}}, \\
\mathbf{J}_k = z\frac{[\mathbf{\Psi}\gamma_k,\overline{\mathbf{\Psi}}]}{2}, \quad Q = \int d^3x \mathbf{J}_0(x) \Rightarrow \left\{ \begin{array}{ll} [Q,\mathbf{\Psi}](\vec{x}) & = z\mathbf{\Psi}(\vec{x}), \\ [Q,\overline{\mathbf{\Psi}}](\vec{x}) & = -z\overline{\mathbf{\Psi}}(\vec{x}). \end{array} \right.
\end{array}$$

Its Lagrangian includes the free Lagrangian and the electromagnetic interaction

$$\begin{array}{l}
\mathbf{L}(\mathbf{A},\mathbf{F}) - \mathbf{A}^k\mathbf{J}_k + \mathbf{L}(\mathbf{\Psi}), \\
\mathbf{L}(\mathbf{\Psi}) = i\overline{\mathbf{\Psi}}\partial^k\gamma_k\mathbf{\Psi} + m\overline{\mathbf{\Psi}}\mathbf{\Psi}, \quad \mathbf{A}^k\mathbf{J}_k = z\mathbf{A}^k\frac{[\mathbf{\Psi}\gamma_k,\overline{\mathbf{\Psi}}]}{2},
\end{array}$$

with the field equations

$$(\partial^k - iz\mathbf{A}^k)\mathbf{\Psi}\gamma_k = im\mathbf{\Psi}, \quad (\partial^k + iz\mathbf{A}^k)\gamma_k\overline{\mathbf{\Psi}} = -im\overline{\mathbf{\Psi}}.$$

A *pure gauge* as derivative of a "sufficiently smooth" Lorentz scalar field $ia(x) \in \log \mathbf{U}(1)$ or as internal derivative of a translation-dependent group element $e^{ia(x)} \in \mathbf{U}(1)$ has trivial field strenghts, i.e., commuting translation and gauge action

$$i\mathbf{A}^k = i\partial^k a = (\partial^k e^{ia})e^{-ia} \Rightarrow \mathbf{F}^{jk} = \partial^k\mathbf{A}^j - \partial^j\mathbf{A}^k = 0.$$

A pure gauge can be absorbed by a redefinition of the local $\mathbf{U}(1)$-phase properties of the fields

$$\begin{array}{l}
(\partial^k - iz\partial^k a)\mathbf{\Phi} = e^{iza}\partial^k e^{-iza}\mathbf{\Phi}, \\
\mathbf{\Phi} \longmapsto e^{iz\gamma}\mathbf{\Phi}, \quad a \longmapsto a + \gamma \Rightarrow e^{-iza}\mathbf{\Phi} \longmapsto e^{-iza}\mathbf{\Phi}.
\end{array}$$

A sufficiently smooth gauge field on time $t \longmapsto i\mathbf{A}(t)$ can always be written as a pure gauge

$$i\mathbf{A} = (d_t e^{ia})e^{-ia} \text{ with } a(t) = \int^t dT\mathbf{A}(T).$$

The analogous procedure for spacetime is different because of the nontrivial position degrees of freedom. There arises a path-dependent line integral, familiar from the classical charged mass point above:

$$i\mathbf{A}^k = (\partial^k e^{i\mathbf{a}})e^{-i\mathbf{a}} \text{ with } \mathbf{a}(x) = \int_\infty^x dX_j\mathbf{A}^j(X).$$

6.5 Quantum Gauge Fields

The canonical quantization for a $\mathbf{U}(1)$-gauge field with the field strengths as dual partner,

$$\begin{aligned} [i\mathbf{F}_{kj}, \mathbf{A}^l](\vec{x}) &= \delta^0_k \delta^a_j \delta^l_b \delta^b_a \delta(\vec{x}) \Rightarrow [i\mathbf{F}_{0a}, \mathbf{A}^b](\vec{x}) = \delta^b_a \delta(\vec{x}), \\ g^2 \mathbf{F}^{a0} &= \partial^0 \mathbf{A}^a - \partial^a \mathbf{A}^0 \end{aligned}$$

does not involve a dual partner for the field $\mathbf{A}^0(x)$. In contrast to a theory with massive vector fields $\mathbf{Z}^k$ with $\partial_k \mathbf{Z}^k = 0$, the dynamics yields no equation for the time translation action $\partial_0 \mathbf{A}^0$, i.e., it does not determine the Lorentz scalar $\partial_k \mathbf{A}^k$. The Lorentz scalar dual partner (chapter "Massless Quantum Fields") related to the time derivative is the *"gauge fixing" field* $\mathbf{S}$

$$[i\mathbf{S}, \mathbf{A}^l](\vec{x}) = \delta^l_0 \delta(\vec{x}) \Rightarrow [i\mathbf{S}, \mathbf{A}^0](\vec{x}) = \delta(\vec{x}).$$

A Lagrangian involves a dimensionless "gauge fixing" constant $g^2\lambda \in \mathbb{R}$:

$$\begin{aligned} \mathbf{L}(\mathbf{A}, \mathbf{F}, \mathbf{S}, \mathbf{J}) &= \mathbf{F}_{kj} \tfrac{\partial^k \mathbf{A}^j - \partial^j \mathbf{A}^k}{2} + \mathbf{S} \partial_k \mathbf{A}^k + g^2 \left(\tfrac{\mathbf{F}_{kj} \mathbf{F}^{kj}}{4} - \lambda \tfrac{\mathbf{S}^2}{2} \right) - \mathbf{A}^k \mathbf{J}_k, \\ \partial^k \mathbf{A}^j - \partial^j \mathbf{A}^k &= g^2 \mathbf{F}^{jk}, \quad \partial^j \mathbf{F}_{kj} - \partial_k \mathbf{S} = \mathbf{J}_k. \\ \partial_k \mathbf{A}^k &= g^2 \lambda \mathbf{S}, \end{aligned}$$

With the introduction of a duality-pairing-completing scalar field $\mathbf{S}$ the translation-dependence of a classical gauge transformation is drastically reduced for a quantum gauge theory to a "massless" Lie parameter field:

$$\begin{aligned} &\mathbf{A}^k \longmapsto \mathbf{A}^k + \partial^k \gamma, \quad \mathbf{F}_{kj} \longmapsto \mathbf{F}_{kj}, \quad \mathbf{S} \longmapsto \mathbf{S}, \quad \mathbf{J}_k \longmapsto \mathbf{J}_k \\ &\text{with } \partial^2 \gamma = 0. \end{aligned}$$

In a quantum gauge theory, the spacetime dependent group parameter is interpreted as Fadeev-Popov field to be added as Fermi twins to the Bose gauge field (appendix).

6.6 Representation Currents

Gauge theories connect spacetime translations and internal Lie algebra operations. With dual bases and Lie bracket,

$$\underline{\mathbf{lag}}_{\mathbb{R}} \ni L \cong \mathbb{R}^d : \quad \langle \check{l}_a, l^b \rangle = \delta^b_a, \quad [l^a, l^b] = \epsilon^{ab}_c l^c,$$

the Lie algebra is represented in endomorphisms of a vector space and its dual $W, W^T \cong \mathbb{C}^n$ with dual bases $\langle \check{e}_\beta, e^\gamma \rangle = \delta^\gamma_\beta = \epsilon \langle e^\gamma, \check{e}_\beta \rangle$, $\epsilon = \pm 1$ (Fermi and Bose):

$$\begin{aligned} &\mathcal{D} : L \longrightarrow \mathbf{AL}(W), && l^a \longmapsto \mathcal{D}(l^a) = \mathcal{D}^{a\beta}_\gamma e^\gamma \otimes \check{e}_\beta, \\ &L \times W \longrightarrow W, && l^a \bullet e^\beta = \mathcal{D}^{a\beta}_\gamma e^\gamma, \\ &\check{\mathcal{D}} : L \longrightarrow \mathbf{AL}(W^T), && l^a \longmapsto -\mathcal{D}(l^a)^T = -\mathcal{D}^{a\beta}_\gamma \epsilon \check{e}_\beta \otimes e^\gamma, \\ &L \times W^T \longrightarrow W^T, && l^a \bullet \check{e}_\gamma = -\mathcal{D}^{a\beta}_\gamma \check{e}_\beta, \end{aligned}$$

e.g., the Lie algebras of $\mathbf{U}(1)$ and $\mathbf{SU}(n)$ with the irreducible and defining representations in a Pauli basis respectively:

$$\begin{array}{lll} \log \mathbf{U}(1) & \longrightarrow \mathbf{AL}(\mathbb{C}), & \mathcal{D}(l^0) \cong iz, \quad z \in \mathbb{Z}, \\ \log \mathbf{SU}(n) & \longrightarrow \mathbf{AL}(\mathbb{C}^n), & \mathcal{D}(l^a) \cong \frac{i}{2}\tau(n)^{a\beta}_{\gamma}, \\ \text{Pauli matrices: } \{\tau(n)^a \mid a = 1, \ldots, n^2 - 1\}. & & \end{array}$$

For spacetime fields the Lie algebra representation is given by the charges

$$l^a \longmapsto iQ^a = i \int d^3x \mathbf{J}^a_0(x), \quad [iQ^a, iQ^b] = \epsilon^{ab}_c iQ^c.$$

They are position integrals over the currents, which are defined with the quantization opposite (anti)commutators - exemplified by

field	quantization	current
scalar (Hermitian)	$[i\mathbf{\Phi}_{k\beta}, \mathbf{\Phi}^\gamma](\vec{x}) = \delta^\gamma_\beta \delta^0_k \delta(\vec{x})$	$i\mathbf{J}^a_k = \mathcal{D}^{a\beta}_{\gamma} \frac{\{\mathbf{\Phi}^\gamma, i\mathbf{\Phi}_{k\beta}\}}{2}$
scalar (complex)	$[i\mathbf{\Phi}^\star_{k\beta}, \mathbf{\Phi}^\gamma](\vec{x}) = [i\mathbf{\Phi}^\gamma_k, \mathbf{\Phi}^\star_\beta](\vec{x}) = \delta^\gamma_\beta \delta^0_k \delta(\vec{x})$	$i\mathbf{J}^a_k = \mathcal{D}^{a\beta}_{\gamma} \frac{\{\mathbf{\Phi}^\gamma, i\mathbf{\Phi}^\star_{k\beta}\} + \{-i\mathbf{\Phi}^\gamma_k, \mathbf{\Phi}^\star_\beta\}}{2}$
vector (Hermitian)	$[i\mathbf{G}^\gamma_{kj}, \mathbf{A}^l_\beta](\vec{x}) = \delta^\gamma_\beta \delta^0_k \delta^a_j \delta^l_b \delta^b_a \delta(\vec{x})$	$i\mathbf{J}^a_k = \mathcal{D}^{a\beta}_{\gamma} \frac{\{\mathbf{A}^j_\beta, i\mathbf{G}^\gamma_{kj}\}}{2}$
Weyl (left)	$\{\mathbf{l}^\star_\beta, \mathbf{l}^\gamma\}(\vec{x}) = \delta^\gamma_\beta \sigma^0 \delta(\vec{x})$	$i\mathbf{J}^a_k = \mathcal{D}^{a\beta}_{\gamma} \frac{[\mathbf{l}^\gamma \check{\sigma}_k, \mathbf{l}^\star_\beta]}{2}$
Weyl (right)	$\{\mathbf{r}^\star_\beta, \mathbf{r}^\gamma\}(\vec{x}) = \delta^\gamma_\beta \check{\sigma}^0 \delta(\vec{x})$	$i\mathbf{J}^a_k = \mathcal{D}^{a\beta}_{\gamma} \frac{[\mathbf{r}^\gamma \sigma_k, \mathbf{r}^\star_\beta]}{2}$
Dirac	$\{\overline{\mathbf{\Psi}}_\beta, \mathbf{\Psi}^\gamma\}(\vec{x}) = \delta^\gamma_\beta \gamma^0 \delta(\vec{x})$	$i\mathbf{J}^a_k = \mathcal{D}^{a\beta}_{\gamma} \frac{[\mathbf{\Psi}^\gamma \gamma_k, \overline{\mathbf{\Psi}}_\beta]}{2}$

The adjoint action on the fields reads

$$[iQ^a, \mathbf{\Phi}^\beta] = \mathcal{D}^{a\beta}_{\gamma} \mathbf{\Phi}^\gamma, \quad [iQ^a, \mathbf{\Phi}^\star_\gamma] = -\mathcal{D}^{a\beta}_{\gamma} \mathbf{\Phi}^\star_\beta.$$

The simultaneous external-internal action (covariant derivatives)

$$(\partial^k \delta^\beta_\gamma - \mathcal{D}^{a\beta}_{\gamma} \mathbf{A}^k_a)\mathbf{\Phi}^\gamma, \quad (\partial^k \delta^\beta_\gamma + \mathcal{D}^{a\beta}_{\gamma} \mathbf{A}^k_a)\mathbf{\Phi}^\star_\beta$$

is implemented by *gauge vertices* (gauge interactions). They are Lorentz compatible spacetime distributions of the power-three Lie algebra representation tensor:

$$\begin{array}{rcccccc} \mathcal{D} = & \check{l}_a \otimes \mathcal{D}(l^a) & = & \check{l}_a & \otimes & \mathcal{D}^{a\beta}_{\gamma} & e^\gamma \otimes \check{e}_\beta, \\ \text{implemented by} & \mathbf{A}^k_a \ \mathbf{J}^a_k & \overset{\text{e.g.}}{=} & \mathbf{A}^k_a & & \mathcal{D}^{a\beta}_{\gamma} & \frac{[\mathbf{\Psi}^\gamma \gamma_k, \overline{\mathbf{\Psi}}_\beta]}{2}. \end{array}$$

Gauge vertices with their Fermi twins, the Becchi-Rouet-Stora vertices, are discussed in the appendix.

6.7 Lie-Algebra-Valued Gauge Fields

Gauge fields go with a Lie algebra and its currents: The number of gauge fields is given by the Lie algebra dimension $L \cong \mathbb{R}^d$, they transform under the adjoint Lie algebra representation (for field strength $\mathbf{F}^b$ and the currents $\mathbf{j}^b, \mathbf{J}^b$) and its dual coadjoint representation (for gauge fields $\mathbf{A}_c$)

$$\begin{array}{lll} \mathrm{ad}\, : L \longrightarrow \mathbf{AL}(L), & \mathrm{ad}\, l^a = \epsilon_c^{ab} l^c \otimes \check{l}_b, & l^b \longmapsto \epsilon_c^{ab} l^c, \\ \check{\mathrm{ad}}\, : L \longrightarrow \mathbf{AL}(L^T), & \check{\mathrm{ad}}\, l^a = -\epsilon_c^{ab} \check{l}_b \otimes l^c, & \check{l}_c \longmapsto \epsilon_c^{ab} \check{l}_b. \end{array}$$

The gauge field currents are products of the dual pairs $(\mathbf{A}_a, \mathbf{F}^a)$:

$$\begin{array}{ll} \mathbf{j}_k^a & = \mathbf{A}_b^j \epsilon_c^{ab} \mathbf{F}_{kj}^c, \\ Q^a & = \int d^3x\, (\mathbf{j}_0^a + \mathbf{J}_0^a) \end{array} \Rightarrow \left\{ \begin{array}{rl} [iQ^a, (\mathbf{F}_{kj}^b, \mathbf{j}^b, \mathbf{J}^b)] & = \epsilon_c^{ab} (\mathbf{F}_{kj}^c, \mathbf{j}^c, \mathbf{J}^c), \\ {[iQ^a, \mathbf{A}_c^k]} & = -\epsilon_c^{ab} \mathbf{A}_b^k. \end{array} \right.$$

With the adjoint Lie algebra representation the gauge field self-coupling is nontrivial only for a nonabelian Lie algebra.

In the Lagrangian for the gauge field sector

$$\mathbf{L}(\mathbf{A}, \mathbf{F}) = \mathbf{F}_{kj}^c \frac{\partial^k \mathbf{A}_c^j - \partial^j \mathbf{A}_c^k}{2} + g_{cb} \frac{\mathbf{F}_{kj}^c \mathbf{F}^{kjb}}{4} - \tfrac{1}{2} \mathbf{A}_a^k \mathbf{j}_k^a$$

the statistical factor $\frac{1}{2}$ in $\frac{1}{2}\mathbf{A}_a^k \mathbf{j}_k^a$ takes into account the tensor power 2 of the gauge field $\mathbf{A} \vee \mathbf{A}$ in the interaction. The current arises by gauge field derivation

$$\tfrac{1}{2} \mathbf{A}_a^k \mathbf{j}_k^a = \epsilon_c^{ab} \frac{\mathbf{A}_a^k \mathbf{A}_b^j}{2} \mathbf{F}_{kj}^c, \quad \frac{\partial \frac{1}{2} \mathbf{A}_a^k \mathbf{j}_k^a}{\partial \mathbf{A}_a^k} = \epsilon_c^{ab} \mathbf{A}_b^j \mathbf{F}_{kj}^c.$$

In a second order derivative formulation, there occur derivatives and cubic gauge field products in the current

$$\mathbf{j}_k^a = \mathbf{A}_b^j \epsilon_c^{ab} (\partial_k \mathbf{A}_j^c - \partial_j \mathbf{A}_k^c + \epsilon^{cde} \mathbf{A}_k^d \mathbf{A}_j^e + \delta_{kj} \partial^l \mathbf{A}_l^c).$$

For the *gauge field normalization* there has to exist an invariant nondegenerate symmetric bilinear form of the Lie algebra and its dual,

$$L^T \times L^T \longrightarrow \mathbb{R}, \quad \langle \check{l}_a | \check{l}_b \rangle = \kappa_{ab} = \kappa_{ba}, = \pm \tfrac{1}{\kappa^2} \delta_{ab} \text{ (Sylvester basis)}$$

e.g., the Killing form for a semisimple Lie algebra like $\mathbf{SU}(n)$, $n \geq 2$, or a squared linear form for an abelian Lie algebra like $\mathbf{U}(1)$. In the following with compact gauge group $\mathbf{U}$, the normalization $\kappa_{ab} = \frac{1}{\kappa_{\mathbf{U}}^2} \delta_{ab}$ is assumed and Lie algebra bases with totally antisymmetric structure constants $\epsilon_d^{ab} \delta^{dc} = -\epsilon^{abc}$. The gauge field coupling constant in the field strength square is the normalization ratio of the represented internal Lie algebra $L = \log \mathbf{U}$ and the external Lorentz Lie algebra $\log \mathbf{SO}_0(1,3)$ with its Killing form $\eta \wedge \eta$, normalized by $\kappa^2_{\mathbf{SO}_0(1,3)}$:

$$\begin{array}{ll} \langle \mathbf{F} | \mathbf{F} \rangle & = g^2\, \delta_{ab} \eta^{kl} \eta^{jm} \mathbf{F}_{kj}^a \mathbf{F}_{lm}^b = g^2 \mathbf{F}_{kj}^a \mathbf{F}_a^{kj}, \\ g^2 & = \frac{\kappa^2_{\mathbf{SO}_0(1,3)}}{\kappa^2_{\mathbf{U}}}. \end{array}$$

The Lagrangian for the gauge field sector

$$\mathbf{L}(\mathbf{A}, \mathbf{F}) = \mathbf{F}_{kj}^c \frac{\partial^k \mathbf{A}_c^j - \partial^j \mathbf{A}_c^k - \epsilon_c^{ab} \mathbf{A}_a^k \mathbf{A}_b^j}{2} + g^2 \frac{\mathbf{F}_{kj}^c \mathbf{F}_c^{kj}}{4}$$

gives the field equations

$$\partial^k \mathbf{A}_c^j - \partial^j \mathbf{A}_c^k - \epsilon_c^{ab} \mathbf{A}_a^k \mathbf{A}_b^j = g^2 \mathbf{F}_c^{jk} , \quad \partial^j \mathbf{F}_{kj}^b + \epsilon_c^{ab} \mathbf{A}_a^j \mathbf{F}_{kj}^c = 0.$$

Via the current of a field $\mathbf{E} : \mathbb{R}^4 \longrightarrow W$, e.g., of a lepton or a quark field, there occur, in a gauge vertex, finite-dimensional faithful "matrix" representations of the Lie algebra $L \subset W \otimes W^T$ (for short $\mathcal{D}(l^a) = l^a$),

$$\text{gauge vertex for } \mathbf{E}\text{: } \mathbf{A}_a^j(x)\ l^{a\beta}_{\ \ \gamma}\ \mathbf{E}^\gamma(x) \mathbf{E}^\star_{\beta,j}(x).$$

Thus the gauge field and the current can be used as valued in the Lie algebra and its dual $\check{l}_a \in L^T \subset W \otimes W^T$ (denoted by underlining):

$$\begin{array}{lll} \mathbb{R}^4 \longrightarrow \log \mathbf{U}, & x \longmapsto & \underline{\mathbf{A}}^j(x) = \mathbf{A}_a^j(x) l^a, \\ \mathbb{R}^4 \longrightarrow (\log \mathbf{U})^T, & x \longmapsto & (\underline{\mathbf{F}}_{jk}, \underline{\mathbf{J}}_k, \underline{\mathbf{j}}_k)(x) = (\mathbf{F}_{jk}^a, \mathbf{J}_k{}^a, \mathbf{j}_k^a)(x) \check{l}_a, \end{array}$$

e.g., for $\mathbf{U}(2)$ in the defining Pauli representation

$$\log \mathbf{U}(2) : \quad \underline{\mathbf{A}} = i \frac{\mathbf{A}_0 \mathbf{1}_2 + \vec{\mathbf{A}} \vec{\tau}}{2} = \frac{i}{2} \begin{pmatrix} \mathbf{A}_0 + \mathbf{A}_3 & \mathbf{A}_1 - i\mathbf{A}_2 \\ \mathbf{A}_1 + i\mathbf{A}_2 & \mathbf{A}_0 - \mathbf{A}_3 \end{pmatrix}.$$

For each translation $x \in \mathbb{R}^4$, there is a transformation of a Lie algebra basis with the gauge fields the $(4 \times d)$ matrix elements $(\mathbf{A}_a^j)_{a=1,\dots,d}^{j=0,1,2,3}$

$$L \ni l^a \longmapsto \mathbf{A}_a^j(x) l^a = \underline{\mathbf{A}}^j(x) \in L.$$

The transformation leads to Lie algebra valued vectors with Lorentz representation, not necessarily to a Lie algebra basis.

Written with Lie-algebra-valued fields, the Lagrangian and the field equations involve two products: the Lie algebra dual product $\langle \check{l}_a, l^b \rangle = \delta_a^b$ (via the trace in $W \otimes W^T$) and the invariant bilinear form of the Lie algebra $\langle \check{l}_a | \check{l}_b \rangle = \delta_{ab}$:

$$\begin{array}{l} \mathbf{L}(\mathbf{A}, \mathbf{F}) = \langle \underline{\mathbf{F}}_{kj}, \frac{\partial^k \underline{\mathbf{A}}^j - \partial^j \underline{\mathbf{A}}^k - [\underline{\mathbf{A}}^k, \underline{\mathbf{A}}^j]}{2} \rangle + \frac{g^2}{4} \langle \underline{\mathbf{F}}_{kj} | \underline{\mathbf{F}}^{kj} \rangle, \\ \partial^k \underline{\mathbf{A}}^j - \partial^j \underline{\mathbf{A}}^k - [\underline{\mathbf{A}}^k, \underline{\mathbf{A}}^j] = g^2 \underline{\mathbf{F}}^{jk} , \quad \partial^j \underline{\mathbf{F}}_{kj} + [\underline{\mathbf{A}}^j, \underline{\mathbf{F}}_{kj}] = 0. \end{array}$$

The action of the Lie group $u \in \mathbf{U} = \exp L \subset \mathbf{GL}(W)$ on a field $\mathbf{E} \in W$ and, as (co-)adjoint action, on its Lie algebra and its dual leaves the Lagrangian invariant:

$$\begin{array}{rlrl} u : & W \longrightarrow W, & \mathbf{E} & \longmapsto u\, \mathbf{E}, \\ \mathrm{Ad}\, u : & L \longrightarrow L, & l & \longmapsto u \circ l \circ u^{-1}, \\ & & \underline{\mathbf{A}}^j & \longmapsto u \circ \underline{\mathbf{A}}^j \circ u^{-1}, \\ \check{\mathrm{Ad}}\, u : & L^T \longrightarrow L^T, & \check{l} & \longmapsto \check{u} \circ \check{l} \circ \check{u}^{-1}, \quad \check{u} = u^{-1T}, \\ & & \underline{\mathbf{F}}_{jk} & \longmapsto \check{u} \circ \underline{\mathbf{F}}_{jk} \circ \check{u}^{-1}. \end{array}$$

6.8 Lie Algebras of Spacetime and Gauge Group

The Lie algebra of the group $\mathbb{R}^4 \times \mathbf{U}$ with spacetime translations and internal operations is constituted by spacetime derivations and group $\mathbf{U}$-derivations.

The directed Lie group logarithms, via Lie group parameter derivatives $\partial^a = \frac{\partial}{\partial \gamma_a}$ with a local (at the group unit) exponential Lie algebra parametrization of the group $\mathbf{U}$ (chapter "Spin, Rotations, and Position"),

$$\begin{aligned} \log^a: \mathbf{U} \longrightarrow \log \mathbf{U}, \quad u(l) = e^{\gamma_a l^a} \longmapsto l^a(u) &= (\partial^a e^l) \circ e^{-l} = \sum_{k \geq 0} \frac{(\operatorname{ad} l)^k}{(1+k)!}(l^a) \\ &= l^a + \frac{[l,l^a]}{2} + \frac{[l,[l,l^a]]}{3!} + \frac{[l,[l,[l,l^a]]]}{4!} + \cdots, \\ \text{at } \mathbf{1} \in \mathbf{U}: \quad l^a(\mathbf{1}) &= l^a, \end{aligned}$$

define, with the Lie-Jacobi isomorphism, a Lie algebra basis $\{l^a(u)\}$ at each group element:

$$u_*: \log \mathbf{U} \longrightarrow \log \mathbf{U}, \quad l^a \longmapsto l^a(u) = (u_*)^a_b l^b, \quad (u_*)^a_b = \langle \check{l}_b, (\partial^a u) \circ u^{-1} \rangle,$$

e.g., for hyperisospin $\mathbf{U}(2)$,

$$\begin{aligned} \log \mathbf{U}(2) \ni l(u) &= (\partial e^{i\frac{\gamma_0 \mathbf{1}_2 + \vec{\gamma}\vec{\tau}}{2}}) \circ e^{-i\frac{\gamma_0 \mathbf{1}_2 + \vec{\gamma}\vec{\tau}}{2}} \\ &= \begin{cases} \frac{i}{2}\mathbf{1}_2 \in \log \mathbf{U}(1) \text{ for } \partial = \frac{\partial}{\partial \gamma_0}, \\ [\delta_{ab}\frac{\sin\gamma}{\gamma} + \epsilon^{abc}\frac{\gamma_c}{\gamma}\frac{1-\cos\gamma}{\gamma} - \frac{\gamma_a\gamma_b}{\gamma^2}(1 - \frac{\sin\gamma}{\gamma})]\frac{i}{2}\tau^b \in \log \mathbf{SU}(2) \\ \qquad \text{for } \partial = \frac{\partial}{\partial \gamma_a}. \end{cases} \end{aligned}$$

The Lie bracket at this group element is the antisymmetric derivative

$$\partial^a l^b(u) - \partial^b l^a(u) = [l^a(u), l^b(u)] = (u_*)^a_c \; [l^c, l^d] \; (u_*)^b_d.$$

With the translation-dependence of the Lie-algebra-valued gauge fields, $\mathbb{R}^4 \ni x \longmapsto \underline{\mathbf{A}}^j(x) \in \log \mathbf{U}$, the Lie parameters for the gauge group $\mathbf{U}$ also have to be parametrized by translations ("there is a $\mathbf{U}$-transformation at each spacetime point"):

$$\mathbb{R}^4 \longrightarrow \mathbf{U}, \quad x \longmapsto U(x) = u(\gamma(x)) = e^{\gamma_a(x) l^a}, \quad U = u \circ \gamma.$$

The spacetime derivatives $\partial^j = \frac{\partial}{\partial x_j}$ of the group $\mathbf{U}$ define *pure gauges*,

$$\mathbb{R}^4 \longrightarrow \log \mathbf{U}, \quad x \longmapsto l^j(U(x)), \quad l^j(U) = (\partial^j U) \circ U^{-1}.$$

The transition from Lie parameter derivatives (Lie algebra bases) to spacetime derivatives (translation bases) is given by the Jacobi transformation $(\partial^j \gamma_a)^{j=0,1,2,3}_{a=1\dots,d}$:

$$\begin{aligned} \log \mathbf{U} \ni \; & l^j(U) = (\partial^j \gamma_a) l^a(U) = (\partial^j \gamma_a)(u_*)^a_b l^b = (U_*)^j_b l^b, \\ \text{e.g., } \log \mathbf{U}(2) \ni \; & (\partial^j \gamma_0)\frac{i}{2}\mathbf{1}_2 + (\partial^j \gamma_a)[\delta_{ab}\frac{\sin\gamma}{\gamma} + \epsilon^{abc}\frac{\gamma_c}{\gamma}\frac{1-\cos\gamma}{\gamma} + \frac{\gamma_a\gamma_b}{\gamma^2}(1 - \frac{\sin\gamma}{\gamma})]\frac{i}{2}\tau^b. \end{aligned}$$

As a $(4 \times d)$ matrix, it can be bijective only for Lie algebra dimension smaller than spacetime dimension $d \leq 4$, e.g., for hyperisospin $\dim_{\mathbb{R}} \mathbf{U}(2) = 4$, not, however, for color $\dim_{\mathbb{R}} \mathbf{SU}(3) = 8$,

$$\text{rank}_{\mathbb{R}} \partial^j \gamma_a(x) \leq \min\{d, 4\}.$$

Also for pure gauges, the Lie bracket is the antisymmetric derivative

$$\partial^k l^j(U) - \partial^j l^k(U) = [l^k(U), l^j(U)] = (U_*)^k_c [l^c, l^d](U_*)^j_d$$

or, in a differential geometric language, pure gauges have a trivial curvature.

The group $\mathbf{U}$-action on pure gauges is represented by an *affine group* $\mathbf{U} \vec{\times} \log \mathbf{U}$:

$$\begin{aligned} U, V \in \mathbf{U}: \quad l^j(V) \longmapsto l^j(U \circ V) &= \partial^j(U \circ V) \circ (U \circ V)^{-1} \\ &= (\partial^j U) \circ U^{-1} + U \circ l^j(V) \circ U^{-1} \\ &= \operatorname{Ad} U . l^j(V) + l^j(U). \end{aligned}$$

The Lie-algebra-valued gauge field have the same homogeneous and translative contributions

$$\underline{\mathbf{A}}^j \longmapsto \operatorname{Ad} U . \underline{\mathbf{A}}^j + l^j(U), \quad \mathbf{A}^j_a \longmapsto (\operatorname{Ad} U)^b_a \mathbf{A}^j_b + (U_*)^b_a \partial^j \gamma_b.$$

The gauge field transformation is also suggested by its expression, analogous to a pure gauge, with a group element in the exponential path-dependent form

$$\underline{\mathbf{a}}(x) = \int_\infty^x dX_k \underline{\mathbf{A}}^k(X): \quad \underline{\mathbf{A}}^j = (\partial^j e^{\underline{\mathbf{a}}}) \circ e^{-\underline{\mathbf{a}}} \longmapsto (\partial^j U \circ e^{\underline{\mathbf{a}}}) \circ (U \circ e^{\underline{\mathbf{a}}})^{-1}.$$

The field strengths transform homogeneously:

$$\partial^k \underline{\mathbf{A}}^j - \partial^j \underline{\mathbf{A}}^k - [\underline{\mathbf{A}}^k, \underline{\mathbf{A}}^j] = g^2 \underline{\mathbf{F}}^{jk} \longmapsto \check{U} \circ g^2 \underline{\mathbf{F}}^{jk} \circ \check{U}^{-1}.$$

With the gauge vertex $\mathbf{A}^k(\frac{1}{2}\mathbf{j}_k + \mathbf{J}_k)$ for the interaction there is an invariance under gauge transformations with translation-dependent group elements: The covariant derivative has a homogeneous transformation behavior

$$\begin{aligned} \mathbf{E} \longmapsto U\,\mathbf{E}, \qquad \partial^j \mathbf{E} &\longmapsto \partial^j U\,\mathbf{E} = U[\partial^j + l^j(U)]\mathbf{E}, \\ (\partial^j - \underline{\mathbf{A}}^j)\mathbf{E} &\longmapsto U(\partial^j - \underline{\mathbf{A}}^j)\mathbf{E}. \end{aligned}$$

For quantum gauge interactions the spacetime-dependent $\mathbf{U}$-transformations with their geometric interpretation in a classical field theory are reduced to "global" $\mathbf{U}$-transformations and only one nilquadratic BRS-transformation for which the Fadeev-Popov field are the "quantum-field-valued Lie parameters." This BRS-transformation has its origin in the nilpotent part of the reducible, but nondecomposable generator for the spacetime translations acting on relativistic massless fields (chapter "Massless Quantum Fields"). It is discussed in more detail below. In a quantum theory, the geometric interpretation of the BRS-transformation as "local" $\mathbf{U}$-transformations remains possible only for the particle interpretable degrees of freedom, not, however, for the nonphotonic degrees of freedom of the quantum gauge field $\mathbf{A}$ and not for the Fadeev-Popov degrees of freedom. The nonparticle degrees of freedom are not acted on with a covariant derivative.

6.9 Electroweak and Strong Gauge Interactions

The *standard model* of the elementary interactions in Minkowski spacetime is a theory of compatibly represented external and internal operations. It embeds the electromagnetic interaction for a Dirac electron field (quantum electrodynamics) into the electroweak and strong gauge interactions of quark and lepton Weyl fields. The fields involved are acted on by irreducible representations $[2L|2R]$ of the Lorentz group $\mathbf{SL}(\mathbb{C}^2)$ and irreducible representations of the hypercharge group $\mathbf{U}(1)$ (rational hypercharge number in $[y]$), of the isospin group $\mathbf{SU}(2)$ (integer or halfinteger isospin in $[2T]$), and the color group $\mathbf{SU}(3)$ respectively as given in the following table:

field	symbol	$\mathbf{SL}(\mathbb{C}^2)$ $[2L\|2R]$	$\mathbf{U}(1)$ $[y]$	$\mathbf{SU}(2)$ $[2T]$	$\mathbf{SU}(3)$ $[2C_1, 2C_2]$
left lepton	$\mathbf{l}$	$[1\|0]$	$-\frac{1}{2}$	$[1]$	$[0,0]$
right lepton	$\mathbf{e}$	$[0\|1]$	-1	$[0]$	$[0,0]$
left quark	$\mathbf{q}$	$[1\|0]$	$\frac{1}{6}$	$[1]$	$[1,0]$
right up quark	$\mathbf{u}$	$[0\|1]$	$\frac{2}{3}$	$[0]$	$[1,0]$
right down quark	$\mathbf{d}$	$[0\|1]$	$-\frac{1}{3}$	$[0]$	$[1,0]$
hypercharge gauge	$\mathbf{A}_0$	$[1\|1]$	0	$[0]$	$[0,0]$
isospin gauge	$\vec{\mathbf{A}}$	$[1\|1]$	0	$[2]$	$[0,0]$
color gauge	$\mathbf{G}$	$[1\|1]$	0	$[0]$	$[1,1]$
Higgs	$\mathbf{\Phi}$	$[0\|0]$	$\frac{1}{2}$	$[1]$	$[0,0]$

the fields of the minimal standard model

The electromagnetic $\mathbf{U}(1)$ is embedded into the product of the abelian hypercharge $\mathbf{U}(1)$ and the nonabelian isospin-color group $\mathbf{SU}(2) \times \mathbf{SU}(3)$,

$$\mathbf{U}(1) \hookrightarrow \mathbf{U}(2 \times 3) = \mathbf{U}(\mathbf{1}_6) \circ [\mathbf{SU}(2) \times \mathbf{SU}(3)] = \frac{\mathbf{U}(1) \times \mathbf{SU}(2) \times \mathbf{SU}(3)}{\mathbb{I}(2) \times \mathbb{I}(3)}.$$

The fields are acted on homogeneously by a direct product of Lorentz and internal transformations

$$\mathbf{U}(2 \times 3) \times \mathbf{SL}(\mathbb{C}^2) : V \longrightarrow V, \quad \begin{cases} \mathbf{l} & \longmapsto u_2 \otimes s.\mathbf{l}, & u_2 \in \mathbf{U}(2), s \in \mathbf{SL}(\mathbb{C}^2), \\ \mathbf{e} & \longmapsto u_1 \otimes \hat{s}.\mathbf{e}, & u_1 \in \mathbf{U}(1), \\ \mathbf{q} & \longmapsto u_6 \otimes s.\mathbf{q}, & u_6 \in \mathbf{U}(2 \times 3), \\ \mathbf{u} & \longmapsto u_3 \otimes \hat{s}.\mathbf{u}, & u_3 \in \mathbf{U}(3), \\ \mathbf{d} & \longmapsto u_3' \otimes \hat{s}.\mathbf{d}, & u_3' \in \mathbf{U}(3), \\ \mathbf{A}_0 & \longmapsto \Lambda.\mathbf{A}_0, & \Lambda \in \mathbf{SL}(\mathbb{C}^2)/\mathbb{I}(2), \\ \vec{\mathbf{A}} & \longmapsto O_3 \otimes \Lambda.\vec{\mathbf{A}}, & O_3 \in \mathbf{U}(2)/\mathbf{U}(1), \\ \mathbf{G} & \longmapsto O_8 \otimes \Lambda.\vec{\mathbf{G}}, & O_8 \in \mathbf{U}(3)/\mathbf{U}(1), \\ \mathbf{\Phi} & \longmapsto u_2^\star.\mathbf{\Phi}. & \end{cases}$$

With the exception of the Higgs field, the isospin $\mathbf{SU}(2)$-representation is a subrepresentation of the Lorentz group $\mathbf{SL}(\mathbb{C}^2)$-representation. This is a characteristic structure of induced representations that start with the two-sided regular representation of the doubled group (chapter "Harmonic Analysis").

Both factors in the internal group $\mathbf{U}(\mathbf{1}_6) \circ [\mathbf{SU}(2) \times \mathbf{SU}(3)]$ are centrally correlated, i.e., the representations of hypercharge $\mathbf{U}(1)$ are related to the representations of the $\mathbf{SU}(2)\times\mathbf{SU}(3)$-center, the cyclotomic group $\mathbb{I}(2)\times\mathbb{I}(3) = \mathbb{I}(6)$ (hexality = two-triality, "star of David"). All $\mathbf{U}(2\times 3)$-representations $[y||2T;2C_1,2C_2]$ carried by the standard model fields with the isospin and color multiplicities

$$d_{\mathbf{SU}(2)} = 1+2T, \quad d_{\mathbf{SU}(3)} = (1+2C_1)(1+2C_2)(1+C_1+C_2)$$

can be generated by the dual defining representations of $\mathbf{U}(2\times 3)$,

$$u = [\tfrac{1}{6}||1;1,0], \quad \check{u} = [-\tfrac{1}{6}||1;0,1],$$

as seen in the powers $\bigwedge^{n} u \otimes \bigwedge^{m} \check{u}$ (all fermion fields are taken as left-handed)

field	$\mathbf{U}(2\times 3)$ $[y\|\|2T;2C_1,2C_2]$	(n,m)	$n-m$ $=6y$	$6y$ mod 2	$6y$ mod 3
$\mathbf{l}$	$[-\frac{1}{2}\|\|1;0,0]$	$(0,3)$	-3	1	0
$\mathbf{e}^\star$	$[1\|\|0;0,0]$	$(6,0)$	6	0	0
$\mathbf{q}$	$[\frac{1}{6}\|\|1;1,0]$	$(1,0)$	1	1	1
$\mathbf{u}^\star$	$[-\frac{2}{3}\|\|0;0,1]$	$(0,4)$	-4	0	-1
$\mathbf{d}^\star$	$[\frac{1}{3}\|\|0;0,1]$	$(2,0)$	2	0	-1
$\mathbf{A}_0$	$[0\|\|0;0,0]$	$(0,0)$	0	0	0
$\vec{\mathbf{A}}$	$[0\|\|2;0,0]$	$(1,1)$	0	0	0
$\mathbf{G}$	$[0\|\|0;1,1]$	$(1,1)$	0	0	0
$\mathbf{\Phi}$	$[\frac{1}{2}\|\|1;0,0]$	$(3,0)$	3	1	0

The central correlations of the internal symmetries are expressed by the modulo relations

$$\begin{aligned} &6y \mod 2 = 2T \mod 2, \quad 6y \mod 3 = 2(C_1-C_2) \mod 3 \\ &y\cdot d_{\mathbf{SU}(2)}\cdot d_{\mathbf{SU}(3)} \in \mathbb{Z}. \end{aligned}$$

The nongauge fields with the free Lagrangians

$$\begin{aligned} \text{left fermions:}\quad \mathbf{L}(\mathbf{l}) &= i\mathbf{l}^\star_\alpha \partial^k \check{\sigma}_k \mathbf{l}^\alpha + i\mathbf{q}^\star_{\alpha c}\partial^k\check{\sigma}_k\mathbf{q}^{\alpha c}, \quad \alpha = 1,2;\ c=1,2,3, \\ \text{right fermions:}\quad \mathbf{L}(\mathbf{r}) &= i\mathbf{e}^\star\partial^k\sigma_k\mathbf{e} + i\mathbf{u}^\star_c\partial^k\sigma_k\mathbf{u}^c + i\mathbf{d}^\star_c\partial^k\sigma_k\mathbf{d}^c, \\ \text{Higgs:}\quad \mathbf{L}(\mathbf{\Phi}) &= \mathbf{\Phi}^\star_{k\alpha}\partial^k\mathbf{\Phi}^\alpha + \mathbf{\Phi}^\alpha_k\partial^k\mathbf{\Phi}^\star_\alpha - \mathbf{\Phi}^\alpha_k\mathbf{\Phi}^{k\star}_\alpha \cong (\partial^k\mathbf{\Phi}^\star_\alpha)(\partial_k\mathbf{\Phi}^\alpha), \end{aligned}$$

interact with the four gauge fields $\mathbf{A}_0$ and $\vec{\mathbf{A}}$ for the electroweak interactions and the eight gauge fields $\mathbf{G}$ for the strong interactions,

$$\begin{aligned} \mathbf{L}(\mathbf{A}_0) &= \mathbf{F}_{kj}\frac{\partial^k\mathbf{A}_0^j-\partial^j\mathbf{A}_0^k}{2} && +g_1^2\frac{\mathbf{F}_{kj}\mathbf{F}^{kj}}{4}, \\ \mathbf{L}(\vec{\mathbf{A}}) &= \mathbf{F}^c_{kj}\frac{\partial^k\mathbf{A}_c^j-\partial^j\mathbf{A}_c^k-\epsilon_c^{ab}\mathbf{A}_a^k\mathbf{A}_b^j}{2} && +g_2^2\frac{\mathbf{F}^b_{kj}\mathbf{F}_b^{kj}}{4}, \\ \mathbf{L}(\mathbf{G}) &= \mathbf{F}^C_{kj}\frac{\partial^k\mathbf{G}_C^j-\partial^j\mathbf{G}_C^k-\epsilon_C^{AB}\mathbf{G}_A^k\mathbf{G}_B^j}{2} && +g_3^2\frac{\mathbf{F}^B_{kj}\mathbf{F}_B^{kj}}{4}. \end{aligned}$$

The indices differentiate between the different Lie algebras in the case of the field strengths $\mathbf{F}$. The structure constants are taken in a Pauli and Gell-Mann basis

$$\begin{aligned} \log\mathbf{SU}(2)\colon\ &\{\tfrac{i}{2}\tau^{a\beta}_{\gamma}|a=1,2,3,\ \beta=1,2\}, && [\tfrac{i}{2}\tau^a,\tfrac{i}{2}\tau^b] &&= \epsilon^{ab}_c\tfrac{i}{2}\tau^c, \\ \log\mathbf{SU}(3)\colon\ &\{\tfrac{i}{2}\lambda^{Ab}_{c}|A=1,\ldots,8;\ b=1,2,3\}, && [\tfrac{i}{2}\lambda^A,\tfrac{i}{2}\lambda^B] &&= \epsilon^{AB}_C\tfrac{i}{2}\lambda^C. \end{aligned}$$

The gauge field coupling constants are the normalization ratios of the internal Lie algebras and the Lorentz Lie algebra

$$(\frac{1}{g_1^2}, \frac{1}{g_2^2}, \frac{1}{g_3^2}) = \frac{(\kappa^2_{\mathbf{U}(1)}, \kappa^2_{\mathbf{SU}(2)}, \kappa^2_{\mathbf{SU}(3)})}{\kappa^2_{\mathbf{SO}_0(1,3)}}.$$

The gauge interactions of the matter fields,

$$\mathbf{L}(\mathbf{A}_0) + \mathbf{L}(\vec{\mathbf{A}}) + \mathbf{L}(\mathbf{G}) + \mathbf{L}(\mathbf{l}) + \mathbf{L}(\mathbf{r}) + \mathbf{L}(\mathbf{\Phi}) - (\mathbf{A}_0^k \mathbf{J}_k + \mathbf{A}_a^k \mathbf{J}_k^a + \mathbf{G}_A^k \mathbf{J}_k^A),$$

involve the currents for the nongauge fields

$$\begin{array}{lrl} \log \mathbf{U}(1): & \mathbf{J}_k = & -\frac{1}{2}\mathbf{l}\check{\sigma}_k\mathbf{l}^\star + \frac{1}{6}\mathbf{q}\check{\sigma}_k\mathbf{q}^\star - \mathbf{e}\sigma_k\mathbf{e}^\star + \frac{2}{3}\mathbf{u}\sigma_k\mathbf{u}^\star - \frac{1}{3}\mathbf{d}\sigma_k\mathbf{d}^\star \\ & & -\frac{i}{2}(\mathbf{\Phi}^\star\mathbf{\Phi}_k - \mathbf{\Phi}\mathbf{\Phi}_k^\star), \\ \log \mathbf{SU}(2): & \mathbf{J}_k^a = & \mathbf{l}\check{\sigma}_k\frac{\tau^a}{2}\mathbf{l}^\star + \mathbf{q}\check{\sigma}_k\frac{\tau^a}{2}\mathbf{q}^\star \\ & & +(i\mathbf{\Phi}\frac{\tau^a}{2}\mathbf{\Phi}_k^\star - i\mathbf{\Phi}_k\frac{\tau^a}{2}\mathbf{\Phi}^\star), \\ \log \mathbf{SU}(3): & \mathbf{J}_k^A = & \mathbf{q}\check{\sigma}_k\frac{\lambda^A}{2}\mathbf{q}^\star + \mathbf{u}\sigma_k\frac{\lambda^A}{2}\mathbf{u}^\star + \mathbf{d}\sigma_k\frac{\lambda^A}{2}\mathbf{d}^\star. \end{array}$$

The gauge field strengths equations are given with the corresponding currents

$$\partial^j \mathbf{F}_{kj} + \mathbf{F}_{kj} \times \mathbf{A}^j = \mathbf{J}_k.$$

The spacetime translations of lepton and quark come with internal gauge field actions

$$\begin{array}{rlrl} (\partial^k + \frac{i}{2}\mathbf{A}_0^k - i\frac{\tau^a}{2}\mathbf{A}_a^k)\mathbf{l}\check{\sigma}_k & = 0, & (\partial^k + i\mathbf{A}_0^k)\mathbf{e}\sigma_k & = 0, \\ (\partial^k - \frac{i}{6}\mathbf{A}_0^k - i\frac{\tau^a}{2}\mathbf{A}_a^k - i\frac{\lambda^A}{2}\mathbf{G}_A^k)\mathbf{q}\check{\sigma}_k & = 0, & (\partial^k - \frac{2i}{3}\mathbf{A}_0^k - i\frac{\lambda^A}{2}\mathbf{G}_A^k)\mathbf{u}\sigma_k & = 0, \\ & & (\partial^k + \frac{i}{3}\mathbf{A}_0^k - i\frac{\lambda^A}{2}\mathbf{G}_A^k)\mathbf{d}\sigma_k & = 0, \end{array}$$

as well as the Higgs field

$$(\partial^k - i\frac{\mathbf{1}_2\mathbf{A}_0^k + \tau^a\mathbf{A}_a^k}{2})\mathbf{\Phi} = \mathbf{\Phi}^k, \quad (\partial^k - i\frac{\mathbf{1}_2\mathbf{A}_0^k + \tau^a\mathbf{A}_a^k}{2})\mathbf{\Phi}_k = 0.$$

For the scalar Higgs field the second order Lagrangian reads

$$\mathbf{L}(\mathbf{\Phi}, \mathbf{A}) = [(\partial^k - i\frac{\mathbf{1}_2\mathbf{A}_0^k + \tau^a\mathbf{A}_a^k}{2})\mathbf{\Phi}][(\partial_k + i\frac{\mathbf{1}_2\mathbf{A}_{0k} + \tau^a\mathbf{A}_{ak}}{2})\mathbf{\Phi}^\star].$$

6.10 Ground State Degeneracy

Obviously, for a dynamics acted on by and invariant under a group $\mathbf{U}$, the individual solutions, e.g., the classical time orbits in position or the time translation and rotation eigenvectors in quantum mechanics, do not have to be $\mathbf{U}$-invariant. The classical elliptic planetary orbits as individual solutions of the $\mathbf{SO}(4)$-invariant Kepler Hamiltonian (chapter "The Kepler Factor") are not $\mathbf{SO}(4)$-invariant, not even $\mathbf{SO}(3)$-invariant. However, the set of all solutions, characterizable in the classical example by the set of all initial or boundary conditions, can be decomposed into irreducible $\mathbf{U}$-orbits (equivalent solutions) with an orbit-characterizing fixgroup. For example, for the classical Kepler dynamics the fixgroup for ellipses is $\mathbf{SO}(2) \subset \mathbf{SO}(4)$ and for hyperbolas $\mathbf{SO}_0(1,1) \subset \mathbf{SO}_0(1,3)$.

In a quantum dynamics, a ground state vector $|\Omega\rangle$ is defined as a solution (eigenvector) with minimal energy M. The ground state vector orbit in all energy eigenstate vector of the dynamics (Hamiltonian),

$$\{\text{ground state vectors}\} = \mathbf{U}|\Omega\rangle \cong \mathbf{U}/\mathbf{U}_\Omega,$$

is called *degenerate* for a proper fixgroup $\mathbf{U}_\Omega \neq \mathbf{U}$, i.e., for a nontrivial orbit of equivalent ground states (*degeneracy manifold*) $\mathbf{U}/\mathbf{U}_\Omega \neq \{1\}$. Without loss of generality the actual ground state vector chosen $|\Omega\rangle$ can be taken at the unit of the acting group $\mathbf{U}$, i.e., as starting point of the ground state vector orbit $|\Omega\rangle \in \mathbf{U}|\Omega\rangle$. By the choice of one special ground state for $\mathbf{U}_\Omega \neq \mathbf{U}$ there arises a *symmetry breakdown (rearrangement)*: Energy eigenvectors, defined with respect to the chosen ground state, are acted on only by representations of the *ground state fixgroup or invariance group* $\mathbf{U}_\Omega$, i.e., $\mathbf{U}$-representations D of the dynamics (interactions) are decomposed into subgroup $\mathbf{U}_\Omega$-representations for state vectors (particles) $D \overset{\mathbf{U}_\Omega}{\cong} \bigoplus_\iota d^\iota$. A transmutation from the "large" interaction symmetry $\mathbf{U}$ to the "little" ground state and particle symmetry $\mathbf{U}_\Omega$ involves a harmonic analysis of the degeneracy manifold $\mathbf{U}/\mathbf{U}_\Omega$ (chapter "Harmonic Analysis"). The eigenvectors (particles) are "stripped" or "frozen" or "bleached" with respect to the degrees of freedom in $\mathbf{U}/\mathbf{U}_\Omega$.

There are totally symmetric ground states, $\mathbf{U}_\Omega = \mathbf{U}$, i.e., lowest-energy states that transform trivially under the interaction group, e.g., an $\mathbf{SO}(4)$-scalar for the ground state in the Kepler potential $\mathcal{V}(\vec{x}) = -\frac{1}{r}$ or an $\mathbf{SU}(3)$-scalar for the harmonic oscillator $\mathcal{V}(\vec{x}) = \frac{\vec{x}^2}{2}$.

The simplest example for a ground state degeneracy $\mathbf{U}_\Omega \neq \mathbf{U}$ is given by a Lagrangian for a mass point with space reflection symmetric potential, i.e., with the discrete interaction symmetry group $\mathbf{U} = \mathbb{I}(2) : \mathbf{x} \leftrightarrow -\mathbf{x}$,

$$L(\mathbf{x}) = \tfrac{m}{2}(d_t\mathbf{x})^2 - \mathcal{V}(\mathbf{x}), \quad \mathcal{V}(\mathbf{x}) = \tfrac{g_0}{8}(\mathbf{x}^2 - M^2)^2, \quad g_0, M > 0.$$

The ground state orbit is characterized by two reflection-related potential minima, time-independent:

$$\mathbb{I}(2) : \mathbf{x} \longmapsto -\mathbf{x}, \quad \begin{cases} \mathcal{V}(\mathbf{x}) = \min \Rightarrow \langle \mathbf{x}^2 \rangle = \langle \Omega|\mathbf{x}^2|\Omega\rangle = M^2, \\ \text{ground state vectors: } \{|\Omega\rangle \mid \langle \mathbf{x}\rangle = \langle\Omega|\mathbf{x}|\Omega\rangle = \pm M\} \cong \mathbb{I}(2). \end{cases}$$

The ground state fixgroup is the trivial group $\mathbf{U}_\Omega = \{1\}$. An expansion around one minimum, e.g., $\langle \mathbf{x}\rangle = +M$, rearranges the $\mathbb{I}(2)$-representations

$$\begin{aligned} \mathbf{x}(t) = M + \underline{\mathbf{x}}(t) \Rightarrow L(\mathbf{x}) &= \tfrac{m}{2}(d_t\underline{\mathbf{x}})^2 - \tfrac{g_0}{8}(\underline{\mathbf{x}}^2 + 2M\underline{\mathbf{x}})^2 \\ &= \tfrac{m}{2}(d_t\underline{\mathbf{x}})^2 - \tfrac{g_0}{2}M^2\underline{\mathbf{x}}^2 - \tfrac{g_0}{2}M\underline{\mathbf{x}}^3 - \tfrac{g_0}{8}\underline{\mathbf{x}}^4. \end{aligned}$$

There is no symmetry under $\underline{\mathbf{x}} \leftrightarrow -\underline{\mathbf{x}}$. After rearrangement with respect to the chosen ground state there arises, in lowest order, a harmonic oscillator

$$L^0(\underline{\mathbf{x}}) = \frac{m}{2}(d_t\underline{\mathbf{x}})^2 - \frac{g_0 M^2}{2}\underline{\mathbf{x}}^2 \Rightarrow \omega_{\underline{\mathbf{x}}}^2 = \frac{g_0 M^2}{m}.$$

Embedding the discrete $\mathbb{I}(2)$-symmetric example into a continuous symmetry with two mass points

$$\mathbf{x}^2 + \mathbf{y}^2 = \mathbf{z}^\star \mathbf{z}, \quad \mathbf{z} = \mathbf{x} + i\mathbf{y}, \quad \begin{cases} L(\mathbf{z}) &= \frac{m}{2}(d_t\mathbf{z})(d_t\mathbf{z}^\star) - \mathcal{V}(\mathbf{z}), \\ \mathcal{V}(\mathbf{z}) &= \frac{g_0}{8}(\mathbf{z}\mathbf{z}^\star - M^2)^2. \end{cases}$$

The minima of the potential are degenerate with the $\mathbf{U}(1)$-invariance group of the interaction

$$\mathbf{U}(1) : \mathbf{z} \longmapsto e^{i\gamma}\mathbf{z}, \quad \begin{cases} \mathcal{V}(\mathbf{z}) = \min \Rightarrow \langle \mathbf{z}\mathbf{z}^\star \rangle = M^2, \\ \text{ground state vectors: } \{|\Omega\rangle \mid \langle \mathbf{z} \rangle = e^{i\gamma}M\} \cong \mathbf{U}(1). \end{cases}$$

The ground state circular orbit parametrizes the degeneracy manifold by the $\mathbf{U}(1)$-degree of freedom in $\mathbf{z}$. The time-parametrized $\mathbf{U}(1)$-Lie parameter $t \longmapsto \gamma(t)$ is called the Goldstone degree of freedom. Thus the Lagrangian is rearrangeable:

$$\mathbf{z} = e^{i\gamma}R \Rightarrow L(\mathbf{z}) = \frac{m}{2}[(d_t R)^2 + R^2(d_t\gamma)^2] - \frac{g_0}{8}(R^2 - M^2)^2.$$

The choice of one ground state from the $\mathbf{U}(1)$-degenerate minima decomposes the $\mathbf{U}(1)$-orbit into points. The expansion around one minimum gives, in addition to the dilation degree of freedom with nontrivial frequency, the Goldstone degree of freedom, whose trivial frequency reflects the degenerate minima ("flat oscillations"):

$$\begin{aligned} R(t) = M + \underline{R}(t) \Rightarrow L(\mathbf{z}) &= \frac{m}{2}[(d_t\underline{R})^2 + (M + \underline{R})^2(d_t\gamma)^2] - \frac{g_0}{8}(\underline{R}^2 + 2M\underline{R})^2 \\ &= \frac{m}{2}(d_t\underline{R})^2 + \frac{mM^2}{2}(d_t\gamma)^2 - \frac{g_0}{2}M\underline{R}^2 + \cdots \\ &\Rightarrow \omega_{\underline{R}}^2 = \frac{g_0 M^2}{m}, \quad \omega_\gamma = 0. \end{aligned}$$

Via the Goldstone degree of freedom, the degeneracy manifold, here the group $\mathbf{U}(1)$ and its tangent space, here the Lie algebra $\log \mathbf{U}(1)$, have a time parametrization

$$\begin{aligned} \mathbb{R} &\longrightarrow \mathbf{U}(1), & t &\longmapsto e^{i\gamma(t)} = \frac{\mathbf{z}(t)}{\sqrt{\mathbf{z}\mathbf{z}^\star(t)}}, \\ \mathbb{R} &\longrightarrow \log \mathbf{U}(1), & t &\longmapsto (d_t e^{i\gamma})e^{-i\gamma} = d_t\gamma(d_\gamma e^{i\gamma})e^{-i\gamma} = i d_t\gamma. \end{aligned}$$

The field-theoretic distribution to spacetime translations $\mathbb{R}^4$ implements a $\mathbf{U}(1)$-degenerate ground state by an appropriate potential for a complex Lorentz scalar field $\varphi : \mathbb{R}^4 \longrightarrow \mathbb{C}$:

$$\mathbf{U}(1) : \varphi \longmapsto e^{i\gamma}\varphi, \quad \begin{cases} \mathbf{L}(\mathbf{\Phi}) &= \frac{1}{2}(\partial^j\mathbf{\Phi})(\partial_j\mathbf{\Phi}^\star) - \mathcal{V}(\mathbf{\Phi}), \\ \mathcal{V}(\mathbf{\Phi}) &= \frac{g_0}{8}(\mathbf{\Phi}\mathbf{\Phi}^\star - M^2)^2, \\ \mathcal{V}(\mathbf{\Phi}) &= \min \Rightarrow \langle \mathbf{\Phi}\mathbf{\Phi}^\star \rangle = M^2, \\ \text{ground} & \text{state vectors: } \{|\Omega\rangle \mid \langle \varphi \rangle = e^{i\gamma}M\} \cong \mathbf{U}(1). \end{cases}$$

It leads to the rearrangement of the $\mathbf{U}(1)$-degree of freedom with the Goldstone field $x \longmapsto \gamma(x)$:

$$\mathbf{\Phi} = e^{i\gamma}R \Rightarrow \mathbf{L}(\mathbf{\Phi}) = \tfrac{1}{2}[(\partial^j R)^2 + R^2(\partial^j\gamma)^2] - \tfrac{g_0}{8}(R^2 - M^2)^2.$$

Expansion around the chosen ground state gives a massive radial field $\underline{R}$ for the dilation degree of freedom and a massless Goldstone field γ:

$$R(x) = M + \underline{R}(x) \Rightarrow \begin{cases} m^2_{\underline{R}} &= g_0 M^2, \\ m_\gamma &= 0. \end{cases}$$

Via the Goldstone field the ground state manifold and its tangent space, i.e., the Lie parameters, have a spacetime translation parametrization

$$\begin{array}{lll} \mathbb{R}^4 & \longrightarrow \mathbf{U}(1), & x \longmapsto e^{i\gamma(x)} = \frac{\varphi(x)}{\sqrt{\varphi\varphi^\star(x)}}, \\ \mathbb{R}^4 & \longrightarrow \log \mathbf{U}(1), & x \longmapsto (\partial^j e^{i\gamma})e^{-i\gamma} = \partial^j\gamma(d_\gamma e^{i\gamma})e^{-i\gamma} = i\partial^j\gamma. \end{array}$$

If the degeneracy transformations, here $\mathbf{U}(1)$, are gauged, there are Lie-algebra-valued translation dependent fields:

$$\begin{array}{rl} \mathbf{L}(\mathbf{\Phi}, \mathbf{A}) = & \frac{1}{2}[(\partial^j - i\mathbf{A}^j)\mathbf{\Phi}][(\partial_j + i\mathbf{A}_j)\mathbf{\Phi}^\star] - \mathcal{V}(\mathbf{\Phi}) \\ & + \mathbf{F}_{kj}\frac{\partial^k\mathbf{A}^j - \partial^j\mathbf{A}^k}{2} + g^2\frac{\mathbf{F}_{kj}\mathbf{F}^{kj}}{4}, \\ \mathbb{R}^4 \longrightarrow & \Big(\log \mathbf{U}(1),\ \mathbf{U}(1),\ \log \mathbf{U}(1)\Big),\ \ x \longmapsto \Big(i\mathbf{A}^j(x),\ e^{i\gamma(x)},\ i\partial^j\gamma(x)\Big). \end{array}$$

The stripping of the local $\mathbf{U}(1)$-property by $e^{i\gamma(x)}$

$$\begin{array}{rl} \mathbf{L}(\mathbf{\Phi}, \mathbf{A}) = & \frac{1}{2}[(\partial^j R)^2 + R^2\mathbf{Z}^j\mathbf{Z}_j] - \frac{g_0}{8}(R^2 - M^2)^2 \\ & + \mathbf{F}_{kj}\frac{\partial^k\mathbf{Z}^j - \partial^j\mathbf{Z}^k}{2} + g^2\frac{\mathbf{F}_{kj}\mathbf{F}^{kj}}{4} \end{array}$$

leads to the gauge-invariant combination of gauge field and a pure gauge which is given by the derivative Goldstone degree of freedom

$$\log \mathbf{U}(1) \ni i\mathbf{Z}^j = i\mathbf{A}^j + (\partial^j e^{-i\gamma})e^{i\gamma} = i(\mathbf{A}^j - \partial^j\gamma) \longmapsto i\mathbf{Z}^j.$$

With respect to the rotation structure of the particle involved, the Goldstone mode provides the third $J_3 = 0$-component for a massive $\mathbf{SU}(2)$-spin $J = 1$ particle field in addition the the two $\mathbf{SO}(2)$-polarization components $J_3 = \pm 1$:

$$\begin{array}{rl} R(x) = M + \underline{R}(x) \Rightarrow \mathbf{L}(\mathbf{\Phi}, \mathbf{A}) = & \frac{1}{2}(\partial_j\underline{R})^2 + \frac{M^2}{2}\mathbf{Z}_j^2 - \frac{g_0}{2}M^2\underline{R}^2 + \cdots \\ & + \mathbf{F}_{kj}\frac{\partial^k\mathbf{Z}^j - \partial^j\mathbf{Z}^k}{2} + g^2\frac{\mathbf{F}_{kj}\mathbf{F}^{kj}}{4} \\ \Rightarrow & m^2_{\underline{R}} = g_0 M^2,\ \ m^2_{\mathbf{Z}} = m^2 g^2. \end{array}$$

In the electroweak standard model (more details below), a potential for a scalar Higgs field $\mathbf{\Phi}$ with a "large" internal interaction $\mathbf{U}(2)$-symmetry defines as minimum remaining "little" particle symmetry the electromagnetic fixgroup $\mathbf{U}(2)_\Omega = \mathbf{U}(1)_+$. The ground state degeneracy is given by the electromagnetic orientation manifold $\mathbf{U}(2)/\mathbf{U}(1)_+$ (chapter "Spacetime as Unitary Operation Classes").

6.11 From Interactions to Particles

The interactions in the standard model of elementary particles are invariant with respect to the external Poincaré group and the internal hypercharge, isospin and color group,

$$\textit{interaction symmetry:}\ \underbrace{\mathbf{U}(2\times 3)}_{\text{internal}} \times \underbrace{\mathbf{SL}(\mathbb{C}^2)\ \vec{\times}\,\mathbb{R}^4}_{\text{external}}.$$

The symmetries for the particles are

$$\textit{particle symmetry:}\ \begin{cases} \underbrace{\mathbf{U}(1)}_{\text{internal}} \times \underbrace{\mathbf{SU}(2)\times\mathbb{R}^4}_{\text{external}}, & \textit{massive},\\ \overbrace{\mathbf{U}(1)} \times \overbrace{\mathbf{SO}(2)\times\mathbb{R}^4}, & \textit{massless}, \end{cases}$$

where Wigner's definition for free particles as irreducible Hilbert representations of the Poincaré group is used (chapter "Harmonic Analysis"). As familiar from eigenvectors in quantum mechanics, particles are constructed as eigenvectors with respect to a maximally diagonalizable subgroup with the corresponding weights the eigenvalues for the operations involved. For example, eigenvectors for electromagnetic $\mathbf{U}(1)$-operations are characterized by charge numbers z, spin $\mathbf{SU}(2)$-eigenvectors with respect to an $\mathbf{SO}(2)$-subgroup (third spin direction) by eigenvalues $|J_3| \le J$ for a spin $(1+2J)$-plet and translation eigenvectors by momenta $\vec{q}$ on the hyperboloid for the invariant mass $q^2 = m^2$. The state vectors are $|z, J, m^2; J_3, \vec{q}\rangle$ for massive particles and $|z, \pm J_3; \vec{q}\rangle$ for massless ones.

With Wigner's particle definition, confined quarks are not particles; they do not have a mass as invariant for translation eigenvectors.

The word symmetry, in connection with multiplicity, is used in its strict sense: For example, as particles, proton and neutron may be called an isospin-induced or isospin-related doublet, but not an isospin-symmetric doublet; with their different masses there is no $\mathbf{SU}(2)$-symmetry connecting those two particle vectors. Or, more obviously, the three weak bosons $\{Z^0, W^\pm\}$ and the photon γ do not constitute an isospin-symmetric triplet-singlet; there is no isospin symmetry transformation left between them.

The transition from the "large" internal interaction symmetry to the "little" internal particle symmetry necessitates a discussion of the problem of maximal diagonalizable subgroups of the interaction group, which will be done in the appendix (Cartan tori). The internal symmetry reduction has two aspects: nontrivial color $\mathbf{SU}(3)$-representations are confined; no nontrivial color-induced particle multiplets have been seen in the particle regime. Nontrivial isospin-induced muliplicities remain visible in the case of the hypercharge-isospin breakdown, which is asymptotically reduced to an electromagnetic $\mathbf{U}(1)$-symmetry.

6.11.1 Electroweak Symmetry Reduction

With respect to the electroweak action group $\mathbf{U}(2)$, the definition of particles requires the transition from the hyperisospin interaction symmetry to an abelian electromagnetic particle subsymmetry. Taking into account the nontrivial central correlation $\mathbf{U}(\mathbf{1}_2)\cap\mathbf{SU}(2) = \{\pm\mathbf{1}_2\}$, a Cartan subgroup of $\mathbf{U}(2)$ is $\mathbf{U}(1)_+ \times \mathbf{U}(1)_-$ with $e^{\frac{\mathbf{1}_2\pm\tau^3}{2}\alpha} \in \mathbf{U}(1)_\pm$ (projective generators). For a ground state fixgroup $\mathbf{U}(1)_+$, as anticipated in the chosen hypercharge representation numbers, the electromagnetic charge number is the sum of hypercharge number and third isospin eigenvalue

$$z = y + T^3.$$

Asymmetric boundary condition are implemented by choosing from a ground state manifold $\mathbf{U}(2)/\mathbf{U}(1)_+$ (Goldstone degrees of freedom) one representative and hence violating (stripping, rearranging) the $\mathbf{U}(2)/\mathbf{U}(1)_+$-related transformations in hyperisospin $\mathbf{U}(2)$. The symmetry rearrangement is implemented by an ad hoc scalar field, the Higgs field $\mathbf{\Phi}$ in the defining $\mathbf{U}(2)$-representation, with a Higgs potential

$$\mathcal{V}(\mathbf{\Phi}) = \tfrac{g_0}{8}(\mathbf{\Phi}^\star\mathbf{\Phi} - M^2)^2, \quad g_0 > 0.$$

The minima of the potential

$$\mathcal{V}(\mathbf{\Phi}) = \min \Rightarrow \langle\mathbf{\Phi}^\star\mathbf{\Phi}(x)\rangle = \langle\Omega|\mathbf{\Phi}^\star\mathbf{\Phi}(x)|\Omega\rangle = M^2$$

give the breakdown characterizing mass unit and, taking the appropriate representative, leaves an electromagnetic $\mathbf{U}(1)_+$ (ground state fixgroup) as the remaining symmetry,

$$\langle\mathbf{\Phi}(x)\rangle = \langle\Omega|\mathbf{\Phi}(x)|\Omega\rangle = \begin{pmatrix} 0 \\ M \end{pmatrix} = \tfrac{\mathbf{1}_2-\tau^3}{2}M.$$

The $\mathbf{U}(2)$-asymmetric effects in Weinberg's original "Model of Leptons"

$$\begin{aligned}\mathbf{L}(\mathbf{\Phi}) = \; & [(\partial^k - i\tfrac{\mathbf{A}_0^k\mathbf{1}_2+\vec{\mathbf{A}}^k\vec{\tau}}{2})\mathbf{\Phi}][(\partial_k + i\tfrac{\mathbf{A}_{0k}\mathbf{1}_2+\vec{\mathbf{A}}_k\vec{\tau}}{2})\mathbf{\Phi}^\star] - \mathcal{V}(\mathbf{\Phi}) \\ & -g_e(\mathbf{e}\mathbf{\Phi}\mathbf{l}^\star + \mathbf{l}\mathbf{\Phi}^\star\mathbf{e}^\star)\end{aligned}$$

come in the particle structure of the $\mathbf{U}(2)$-gauge fields via the covariant derivative of the Higgs field and in the lepton particles via a Yukawa interaction. The ground state value of the Higgs field gives the mass contributions

$$\mathbf{L}(\mathbf{\Phi})\Big|_{\mathbf{\Phi}=\langle\mathbf{\Phi}\rangle} = M^2 \operatorname{tr} \tfrac{\mathbf{1}_2-\tau^3}{2}(\tfrac{\mathbf{A}_0\mathbf{1}_2+\vec{\mathbf{A}}\vec{\tau}}{2})^2 - Mg_e(\mathbf{e}\mathbf{l}_2^\star + \mathbf{l}^2\mathbf{e}^\star).$$

For the lepton fields the $\mathbf{U}(1)_+$-trivial component (up component in the isospin doublet, neutrino) remains massless. The electron mass m_e for the massive electron Dirac field $\mathbf{\Psi}_e$ with the down component in the isospin doublet as left-handed part can replace the Yukawa coupling constant g_e, both are theoretically undetermined parameters in the model

$$\begin{array}{rlll}\text{massless neutrino:} & \nu_e & = \mathbf{l}^1, & m_{\nu_e} = 0, \\ \text{massive electron:} & \mathbf{\Psi}_e & = (\mathbf{e}_L, \mathbf{e}_R) = (\mathbf{l}^2, \mathbf{e}), & m_e = Mg_e.\end{array}$$

In general, a Yukawa coupling for a left-handed isodoublet $\mathbf{Q} = \begin{pmatrix} \mathbf{U}_L \\ \mathbf{D}_L \end{pmatrix}$ with hypercharge number y and two corresponding right-handed isosinglets $(\mathbf{U}_R, \mathbf{D}_R)$ with adapted hypercharge numbers $y \pm \frac{1}{2}$ generates a mass term after the symmetry reduction:

$$\begin{aligned} \mathbf{L}^{\text{Yuk}}(\mathbf{\Phi})\Big|_{\mathbf{\Phi}=\langle\mathbf{\Phi}\rangle} &= -g_D(\mathbf{D}_R\mathbf{\Phi}\mathbf{Q}^\star + \mathbf{Q}\mathbf{\Phi}^\star\mathbf{D}_R^\star) - g_U(\mathbf{U}_R\mathbf{\Phi}^\star\mathbf{Q}^\star + \mathbf{Q}\mathbf{\Phi}\mathbf{U}_R^\star)\Big|_{\mathbf{\Phi}=\langle\mathbf{\Phi}\rangle} \\ &= -m_D(\mathbf{D}_R\mathbf{D}_L^\star + \mathbf{D}_L\mathbf{D}_R^\star) - m_U(\mathbf{U}_R\mathbf{U}_L^\star + \mathbf{U}_L\mathbf{U}_R^\star) \\ &= -m_D\mathbf{\Psi}_D\overline{\mathbf{\Psi}}_D - m_U\mathbf{\Psi}_U\overline{\mathbf{\Psi}}_U. \end{aligned}$$

Left and right components constitute two massive Dirac fields

$$\mathbf{\Psi}_U = (\mathbf{U}_L, \mathbf{U}_R), \quad \mathbf{\Psi}_D = (\mathbf{D}_L, \mathbf{D}_R), \quad m_D = Mg_D, \quad m_U = Mg_U.$$

An $\mathbf{SU}(2)$-index notation for the Yukawa couplings looks as follows:

$$\mathbf{D}_R\mathbf{\Phi}\mathbf{Q}^\star = \mathbf{D}_R\mathbf{\Phi}^\alpha\mathbf{Q}^\star_\alpha, \quad \mathbf{U}_R\mathbf{\Phi}^\star\mathbf{Q}^\star = \mathbf{U}_R\mathbf{\Phi}^\star_\beta\epsilon^{\beta\alpha}\mathbf{Q}^\star_\alpha.$$

Pairing the left-handed lepton fields $\mathbf{l} = \begin{pmatrix} \nu_L \\ \mathbf{e}_L \end{pmatrix}$ not only with a right-handed electron $\mathbf{e}_R$ but also with a right-handed neutrino partner ν_R, a nontrivial neutrino mass $m_\nu = Mg_\nu$ is possible. Such a right-handed isosinglet field ν_R comes with trivial hypercharge $y = 0$, i.e., without any internal gauge interaction ("sterile neutrino").

The vector-field-related terms in the Higgs field coupling,

$$\begin{aligned} M^2 \operatorname{tr} \frac{\mathbf{1}_2-\tau^3}{2}\frac{(\mathbf{A}_0\mathbf{1}_2+\vec{\mathbf{A}}\vec{\tau})^2}{4} &= M^2 \operatorname{tr} \frac{\mathbf{1}_2-\tau^3}{2}\frac{[(\mathbf{A}_0)^2+(\vec{\mathbf{A}})^2]\mathbf{1}_2+2\mathbf{A}_0\vec{\mathbf{A}}\vec{\tau}}{4} \\ &= M^2\frac{(\mathbf{A}_1)^2+(\mathbf{A}_2)^2+(\mathbf{A}_3-\mathbf{A}_0)^2}{4}, \end{aligned}$$

contributes to the free theory of two massive charged vector fields $\mathbf{W} \in \{\mathbf{A}_{1,2}\}$,

$$\mathbf{L}(\mathbf{W}) = \mathbf{F}_{kj}\frac{\partial^k\mathbf{W}^j-\partial^j\mathbf{W}^k}{2} + g_2^2\frac{\mathbf{F}_{kj}\mathbf{F}^{kj}}{4} + \frac{M^2}{2}\frac{\mathbf{W}^k\mathbf{W}_k}{2} \Rightarrow m_W^2 = \frac{g_2^2}{2}M^2.$$

The two neutral vector fields come with the free theory

$$\begin{aligned} \mathbf{L}(\mathbf{A}_0, \mathbf{A}_3) = \ & \mathbf{F}^0_{kj}\frac{\partial^k\mathbf{A}_0^j-\partial^j\mathbf{A}_0^k}{2} + \mathbf{F}^3_{kj}\frac{\partial^k\mathbf{A}_3^j-\partial^j\mathbf{A}_3^k}{2} \\ & + g_1^2\frac{\mathbf{F}^0_{kj}\mathbf{F}_0^{kj}}{4} + g_2^2\frac{\mathbf{F}^3_{kj}\mathbf{F}_3^{kj}}{4} + M^2\frac{(\mathbf{A}_3^k-\mathbf{A}_0^k)^2}{4}. \end{aligned}$$

The diagonalization from interaction to particle fields, required by the nondiagonal mass term, is performed with the Weinberg $\mathbf{SO}(2)$-rotation

$$\begin{aligned} g_1^2\frac{\mathbf{F}_0^2}{4} + g_2^2\frac{\mathbf{F}_3^2}{4} &= \gamma^2\frac{\mathbf{G}^2}{4} + g^2\frac{\mathbf{F}^2}{4} \quad \text{with} \quad \begin{pmatrix} \cos\theta & -\sin\theta \\ \sin\theta & \cos\theta \end{pmatrix}\begin{pmatrix} g_2\mathbf{F}^3 \\ g_1\mathbf{F}^0 \end{pmatrix} = \begin{pmatrix} \gamma\mathbf{G} \\ g\mathbf{F} \end{pmatrix}, \\ \mathbf{F}_0\partial\mathbf{A}_0 + \mathbf{F}_3\partial\mathbf{A}_3 &= \mathbf{G}\partial\mathbf{Z} + \mathbf{F}\partial\mathbf{A} \quad \text{with} \quad \begin{pmatrix} \cos\theta & -\sin\theta \\ \sin\theta & \cos\theta \end{pmatrix}\begin{pmatrix} \frac{1}{g_2}\mathbf{A}_3 \\ \frac{1}{g_1}\mathbf{A}_0 \end{pmatrix} = \begin{pmatrix} \frac{1}{\gamma}\mathbf{Z} \\ \frac{1}{g}\mathbf{A} \end{pmatrix}. \end{aligned}$$

It involves the Weinberg angle θ and dual normalizations $(\kappa, \frac{1}{\kappa})$ for the coupling constants $\kappa \in \{g_1, g_2, \gamma, g\}$. The combination $\mathbf{Z} = \mathbf{A}_3 - \mathbf{A}_0$ arises as massive vector field; the massless gauge field $\mathbf{A}$ carries the ground state fixgroup $\mathbf{U}(1)_+$

transformations; this defines the particle field normalizations $\{g, \gamma\}$ in terms of the gauge field normalizations and the Weinberg angle:

$$\begin{array}{lll} \mathbf{Z} & = \mathbf{A}_3 - \mathbf{A}_0 & \Rightarrow \frac{\cos\theta}{g_2} = \frac{\sin\theta}{g_1} = \frac{1}{\gamma}, \\ \mathbf{A} & = \sin^2\theta\ \mathbf{A}_3 + \cos^2\theta\ \mathbf{A}_0 & \Rightarrow \sin\theta = \frac{g}{g_2}. \end{array}$$

The Weinberg rotation has as analogue in mechanics the rearrangement of two individual mass point motions into center of mass and relative motion (chapter "The Kepler Factor") $\frac{p_1^2}{2m_1} + \frac{p_2^2}{2m_2} = \frac{p^2}{2m} + \frac{P^2}{2M}$ with mass sum M and reduced mass m. The electromagnetic gauge field $\mathbf{A}$ with $g^2\frac{\mathbf{F}_{kj}\mathbf{F}^{kj}}{4}$ corresponds to the center of mass coordinate X with free Hamiltonian $H_0 = \frac{P^2}{2M}$, i.e., the electromagnetic coupling g^2 is the normalization analogue to the inverse center mass $\frac{1}{M}$:

$$\begin{array}{l} \text{electroweak:} \\ (\frac{1}{g_1^2}, \frac{1}{g_2^2} | \frac{1}{\gamma^2}, \frac{1}{g^2}) \end{array} \left\{ \begin{array}{rl} (\mathbf{A}_0, \mathbf{A}_3) & \longmapsto (\mathbf{Z}, \mathbf{A}), \\ \mathbf{U}(\mathbf{1}_2) \circ \mathbf{U}(1)_3 & = \mathbf{U}(1)_+ \times \mathbf{U}(1)_-, \\ e^{i(\gamma_0 \mathbf{1}_2 + \gamma_3 \tau^3)} & = e^{i(\gamma_0+\gamma_3)\frac{\mathbf{1}_2+\tau^3}{2}} \times e^{i(\gamma_0-\gamma_3)\frac{\mathbf{1}_2-\tau^3}{2}}, \end{array} \right.$$

$$\begin{array}{l} \text{center of mass:} \\ (m_1, m_2 | m, M) \end{array} \left\{ \begin{array}{rl} (x_1, x_2) & \longmapsto (x, X), \\ \mathbf{SO}(2)_1 \times \mathbf{SO}(2)_2 & = \mathbf{SO}(2)_+ \times \mathbf{SO}(2)_-, \\ e^{i(x_1 p_1 + x_2 p_2)} & = e^{i(x_1+x_2)\frac{p_1+p_2}{2}} \times e^{i(x_1-x_2)\frac{p_1-p_2}{2}}, \end{array} \right.$$

where $\frac{1}{g_{1,2}}$ as orthogonal sides define the *electroweak orthogonal triangle*, in which the hypotenuse $\frac{1}{g}$ is related to Sommerfeld's fine structure constant $\frac{g^2}{4\pi} \sim \frac{1}{137}$ for the electromagnetic $\mathbf{U}(1)_+$-gauge field $\mathbf{A}$:

$$g_1 g_2 = \gamma g, \quad \frac{g_2}{g_1} = \cot\theta \quad \left\{ \begin{array}{ll} \frac{1}{g^2} & = \frac{1}{g_1^2} + \frac{1}{g_2^2}, \\ \gamma^2 & = g_1^2 + g_2^2, \end{array} \right.$$
$$(\tfrac{g}{g_2}, \tfrac{g}{g_1}, \tfrac{g}{\gamma}) = (\tfrac{g_1}{\gamma}, \tfrac{g_2}{\gamma}, \tfrac{g}{\gamma}) = (\sin\theta, \cos\theta, \cos\theta\sin\theta).$$

Multiplication by the area dilation factor $\gamma g = g_1 g_2$ gives the similar dual triangle with the squared lengths $(g_2^2, g_1^2 | g^2, \gamma^2)$.

The Weinberg rotation diagonalizes the free theory with two neutral vector particle fields:

$$\left. \begin{array}{rl} \mathbf{L}(\mathbf{A}_0, \mathbf{A}_3) & = \mathbf{F}_{kj}\frac{\partial^k \mathbf{A}^j - \partial^j \mathbf{A}^k}{2} + g^2\frac{\mathbf{F}_{kj}\mathbf{F}^{kj}}{4} \\ & +\mathbf{G}_{kj}\frac{\partial^k \mathbf{Z}^j - \partial^j \mathbf{Z}^k}{2} + G^2\frac{\mathbf{G}_{kj}\mathbf{G}^{kj}}{4} + M^2\frac{\mathbf{Z}^k\mathbf{Z}_k}{4} \end{array} \right\} \Rightarrow \left\{ \begin{array}{ll} m_A^2 & = 0, \\ m_Z^2 & = \frac{\gamma^2}{2}M^2. \end{array} \right.$$

The electroweak gauge field interactions are rearranged with the neutral and charged vector particle fields:

$$\left. \begin{array}{ll} \mathbf{A}_0 & = \mathbf{A} - \sin^2\theta\ \mathbf{Z} \\ \mathbf{A}_3 & = \mathbf{A} + \cos^2\theta\ \mathbf{Z} \end{array} \right\}, \quad \mathbf{A}_1 \mp i\mathbf{A}_2 = \mathbf{W}_\pm.$$

In general, one obtains for a left-handed isodoublet $\mathbf{Q} = \begin{pmatrix} \mathbf{U}_L \\ \mathbf{D}_L \end{pmatrix}$ with hypercharge number y and two corresponding right-handed isosinglets $(\mathbf{U}_R, \mathbf{D}_R)$ with hypercharge numbers $y \pm \frac{1}{2}$ as interaction with the vector particle fields,

$$\begin{aligned}
&[y\mathbf{Q}\check\sigma\mathbf{Q}^\star + (y+\tfrac{1}{2})\mathbf{U}_R\sigma\mathbf{U}_R^\star + (y-\tfrac{1}{2})\mathbf{D}_R\sigma\mathbf{D}_R^\star]\mathbf{A}_0 + \mathbf{Q}\check\sigma\tfrac{\vec\tau}{2}\mathbf{Q}^\star\ \vec{\mathbf{A}}\\
&\quad= \left[(y+\tfrac{1}{2})\mathbf{\Psi}_U\gamma\overline{\mathbf{\Psi}}_U + (y-\tfrac{1}{2})\mathbf{\Psi}_D\gamma\overline{\mathbf{\Psi}}_D\right]\mathbf{A}\\
&\quad+\left[\mathbf{\Psi}_U\gamma\tfrac{1-4(y+\frac{1}{2})\sin^2\theta+i\gamma_5}{4}\overline{\mathbf{\Psi}}_U - \mathbf{\Psi}_D\gamma\tfrac{1+4(y-\frac{1}{2})\sin^2\theta+i\gamma_5}{4}\overline{\mathbf{\Psi}}_D\right]\mathbf{Z}\\
&\quad+\mathbf{U}_L\check\sigma\mathbf{D}_L^\star\mathbf{W}_- + \mathbf{D}_L\check\sigma\mathbf{U}_L^\star\mathbf{W}_+,
\end{aligned}$$

where the parity combinations have been used in Dirac fields $\mathbf{\Psi}_{U,D}$, e.g., for $\mathbf{\Psi}_U$,

$$\begin{aligned}
\mathbf{\Psi}_U\gamma_k\overline{\mathbf{\Psi}}_U &= \mathbf{U}_L\check\sigma_k\mathbf{U}_L^\star + \mathbf{U}_R\sigma_k\mathbf{U}_R^\star,\\
i\mathbf{\Psi}_U\gamma_k\gamma_5\overline{\mathbf{\Psi}}_U &= \mathbf{U}_L\check\sigma_k\mathbf{U}_L^\star - \mathbf{U}_R\sigma_k\mathbf{U}_R^\star.
\end{aligned}$$

This leads for the leptons with $y = -\frac{1}{2}$ to

$$\begin{aligned}
&(-\tfrac{1}{2}\mathbf{l}\check\sigma\mathbf{l}^\star - \mathbf{e}\sigma\mathbf{e}^\star)\mathbf{A}_0 + \mathbf{l}\check\sigma\tfrac{\vec\tau}{2}\mathbf{l}^\star\vec{\mathbf{A}}\\
&\quad= -\mathbf{\Psi}_e\gamma\overline{\mathbf{\Psi}}_e\mathbf{A} + \left(\tfrac{1}{2}\nu_e\check\sigma\nu_e^\star - \mathbf{\Psi}_e\gamma\tfrac{1-4\sin^2\theta+i\gamma_5}{4}\overline{\mathbf{\Psi}}_e\right)\mathbf{Z}\\
&\quad+\nu_e\check\sigma\mathbf{e}_L^\star\mathbf{W}_- + \mathbf{e}_L\check\sigma\nu_e^\star\mathbf{W}_+.
\end{aligned}$$

A "sterile neutrino" remains "sterile," The quark fields with $y = \frac{1}{6}$ have the electroweak interactions

$$\begin{aligned}
&\left(\tfrac{1}{6}\mathbf{q}\check\sigma\mathbf{q}^\star + \tfrac{2}{3}\mathbf{u}\sigma\mathbf{u}^\star - \tfrac{1}{3}\mathbf{d}\sigma\mathbf{d}^\star\right)\mathbf{A}_0 + \mathbf{q}\check\sigma\tfrac{\vec\tau}{2}\mathbf{q}^\star\ \vec{\mathbf{A}}\\
&\quad= \left(\tfrac{2}{3}\mathbf{\Psi}_u\gamma\overline{\mathbf{\Psi}}_u - \tfrac{1}{3}\mathbf{\Psi}_d\gamma\overline{\mathbf{\Psi}}_d\right)\mathbf{A} + \left(\mathbf{\Psi}_u\gamma\tfrac{1-\frac{8}{3}\sin^2\theta+i\gamma_5}{4}\overline{\mathbf{\Psi}}_u - \mathbf{\Psi}_d\gamma\tfrac{1-\frac{4}{3}\sin^2\theta+i\gamma_5}{4}\overline{\mathbf{\Psi}}_d\right)\mathbf{Z}\\
&\quad+\mathbf{u}_L\check\sigma\mathbf{d}_L^\star\mathbf{W}_- + \mathbf{d}_L\check\sigma\mathbf{u}_L^\star\mathbf{W}_+.
\end{aligned}$$

The electroweak model contains many basically unknown parameters, especially the gauge field normalizations $g_{1,2}^2$ and the ground state or electroweak mass unit M^2. The weak breakdown mass can be replaced by the experimentally determined Fermi constant for the four fermion interactions as low energy limit of the charged weak interaction, i.e., for the propagator

$$\begin{aligned}
q^2 \to 0:&\quad -\frac{g_2^2}{q^2-m_W^2} \longrightarrow \frac{g_2^2}{m_W^2} = \frac{2}{M^2},\\
\text{experiment:}&\quad M \sim 169\frac{\text{GeV}}{c^2}.
\end{aligned}$$

To determine the electroweak orthogonal triangle, one needs one constant in addition to the experimentally determined fine structure constant, e.g., the experimentally determined Weinberg angle

$$\text{experiment:}\left\{\begin{array}{ll}\frac{g^2}{4\pi} & \sim\frac{1}{137}\\ \sin^2\theta & \sim 0.23\end{array}\right\} \Rightarrow \left\{\begin{array}{rl}(\frac{1}{g_1^2},\frac{1}{g_2^2}|\frac{1}{\gamma^2},\frac{1}{g^2}) & \sim(8.4, 2.5|1.9, 10.9),\\ \gamma g = g_1 g_2 = \frac{2g^2}{\sin 2\theta} & \sim\frac{1}{4.6},\\ \frac{g_2^2}{g_1^2} = \cot^2\theta & \sim 3.35,\\ g_2^2 = \frac{2m_W^2}{M^2} & \sim\frac{1}{2.5},\end{array}\right.$$

from which the dual electroweak mass triangle can be computed:

$$\begin{aligned}
(m_W^2, m_1^2|m_0^2, m_Z^2) &= (g_2^2, g_1^2|g^2,\gamma^2)\frac{M^2}{2} = (\frac{1}{\sin^2\theta},\frac{1}{\cos^2\theta}|1,\frac{4}{\sin^2 2\theta})\frac{g^2M^2}{2},\\
(m_W, m_1|m_0, m_Z) &\sim (2.1, 1.2|1, 2.4)\ 37\frac{\text{GeV}}{c^2}.
\end{aligned}$$

The weak boson masses are in good agreement with the experimental results:

$$m_W \sim 80\frac{\text{GeV}}{c^2}, \quad m_Z \sim 91\frac{\text{GeV}}{c^2}.$$

6.11.2 U(2)-Value of the Weinberg Angle

The Weinberg angle involves, via the coupling constants, the normalization ratio of the hypercharge and isospin operations

$$\tan^2\theta = \frac{g_1^2}{g_2^2} = \frac{\kappa^2_{\mathbf{SU}(2)}}{\kappa^2_{\mathbf{U}(1)}} \sim 0.3.$$

A connection of both groups, e.g., in a larger group, can determine this ratio.

Also, the central correlation of both groups $\mathbf{U}(2) \cong \frac{\mathbf{U}(1)\times\mathbf{SU}(2)}{\mathbb{I}(2)}$, as seen in the hypercharge and isospin invariants, relates to each other the Lie algebra normalizations; in lowest order: $\mathbf{U}(2)$ is the invariance group of a scalar product in two dimensions, e.g., for the Higgs representations before fixing a ground state

$$H \times H \longrightarrow \mathbb{C}, \quad \langle \Phi^\alpha | \Phi^\beta \rangle = M^2 \delta^{\alpha\beta}, \quad \alpha, \beta = 1, 2.$$

The scalar product for the defining $\mathbf{U}(2)$-representation induces scalar products on the tensor powers of H and, with the inverse product $\sim \delta_{\alpha\beta}$, of the dual space H^T and its tensor powers. The representation of the gauge fields is isomorphic to $\mathbf{U}(2)$-representations on the tensor product $H \otimes H^T \cong \mathbb{C}^4$ with the induced scalar product (the overall normalization κ^2 is not determined by $\mathbf{U}(2)$-symmetry)

$$[H \otimes H^T] \times [H \otimes H^T] \longrightarrow \mathbb{C}, \quad \langle \Phi^\alpha \otimes \Phi^\star_\gamma | \Phi^\beta \otimes \Phi^\star_\delta \rangle = \kappa^2 \delta^{\alpha\beta}\delta_{\gamma\delta}.$$

The decomposition of the product space $H \otimes H^T \cong \mathbb{C} \oplus \mathbb{C}^3$ into $\mathbf{SU}(2)$-isospin singlet and triplet as used for the hypercharge and isospin gauge fields leads to the rearrangement of the metric tensor

$$\text{for } \mathbf{U}(2): \quad \delta^{\alpha\beta}\delta_{\gamma\delta} = \tfrac{1}{2}\delta^\alpha_\gamma\ \delta^\beta_\delta + \tfrac{1}{6}\vec{\tau}^\alpha_\gamma\ \vec{\tau}^\beta_\delta$$

with the relative normalization for $\mathbf{U}(1)$ and $\mathbf{SU}(2)$

$$\frac{\kappa^2_{\mathbf{SU}(2)}}{\kappa^2_{\mathbf{U}(1}} = \frac{\frac{1}{6}}{\frac{1}{2}} = \frac{1}{3}.$$

The analogous rearrangement for the general case $\mathbf{U}(n) \cong \frac{\mathbf{U}(1)\times\mathbf{SU}(n)}{\mathbb{I}(n)}$, $n \geq 2$, with generalized Pauli matrices (chapter "Spacetime as Unitary Operation Classes") gives as normalization ratio

$$\text{for } \mathbf{U}(n): \quad \delta^{\alpha\beta}\delta_{\gamma\delta} = \tfrac{1}{n}\delta^\alpha_\gamma\ \delta^\beta_\delta + \tfrac{1}{2(1+n)}\vec{\tau}(n)^\alpha_\gamma\ \vec{\tau}(n)^\beta_\delta \Rightarrow \frac{\kappa^2_{\mathbf{SU}(n)}}{\kappa^2_{\mathbf{U}(1)}} = \frac{n}{2(1+n)}.$$

6.11.3 Transmutation from Hyperisospin to Electromagnetic Symmetry

The electroweak symmetry "breakdown" (rearrangement) is a transmutation from a hyperisopin $\mathbf{U}(2)$-compatible framework for the interaction to a formulation with remaining electromagnetic fixgroup $\mathbf{U}(1)_+$-symmetry for the particles (chapter "Spacetime as Unitary Operation Classes"):

$$\{t \in \mathbf{U}(2) \mid \langle \mathbf{\Phi}(x) \rangle = \begin{pmatrix} 0 \\ M \end{pmatrix} = t \begin{pmatrix} 0 \\ M \end{pmatrix}\} \cong \mathbf{U}(1)_+.$$

The $\mathbf{U}(2)/\mathbf{U}(1)_+$-isomorphic orbit of the Higgs field in the Hilbert space $\mathbb{C}^2$ provides translation-dependent Lie parameters for the fixgroup classes (symmetric space):

$$\begin{array}{ll} \mathbf{\Phi}(x) = e^{\frac{i}{2}\gamma_\perp(x)} \frac{\mathbf{1}_2 - \tau^3}{2} R(x), & \mathbb{R}^4 \ni x \longmapsto \frac{i}{2}\gamma_\perp(x) = i\frac{-\gamma_3(x)\mathbf{1}_2 + \vec{\gamma}(x)\vec{\tau}}{2} \in \log \mathbf{U}(2), \\ & \mathbb{R}^4 \ni x \longmapsto V(x) = v(\frac{\mathbf{\Phi}(x)}{R(x)}) = e^{\frac{i}{2}\gamma_\perp(x)} \in \mathbf{U}(2). \end{array}$$

The representations of the electromagnetic orientation manifold (Goldstone or ground state manifold) on a vector space $W \cong \mathbb{C}^{1+2T}$ with $\mathbf{U}(2)$-representation

$$\begin{array}{l} (\mathbf{U}(2)/\mathbf{U}(1)_+)_{\text{repr}} = \mathcal{G}^3 \longrightarrow \mathbf{U}(1+2T), \quad \frac{\mathbf{\Phi}}{R} \longmapsto D(v(\frac{\mathbf{\Phi}}{R})), \quad R = |\mathbf{\Phi}|, \\ \qquad v(\frac{\mathbf{\Phi}}{R}) = \frac{1}{R}\begin{pmatrix} \mathbf{\Phi}_2^\star & \mathbf{\Phi}^1 \\ -\mathbf{\Phi}_1^\star & \mathbf{\Phi}^2 \end{pmatrix} = e^{\frac{i}{2}\gamma_\perp} \in \mathbf{U}(2)/\mathbf{U}(1)_+, \end{array}$$

are products of the fundamental representation. A "left" hyperisospin $\mathbf{U}(2)$ action gives the representation with the $\mathbf{U}(2)$-transformed Higgs vector up to a "right" action with the electromagnetic fixgroup $\mathbf{U}(1)_+$:

$$\begin{array}{l} u = e^{i\frac{\gamma_0 \mathbf{1}_2 + i\vec{\gamma}\vec{\tau}}{2}} \in \mathbf{U}(2) \Rightarrow u \circ v(\frac{\mathbf{\Phi}}{R}) = v(\frac{u.\mathbf{\Phi}}{R}) \circ t(u) \\ \qquad \text{with } t(u) = e^{i2\gamma_0 \frac{\mathbf{1}_2 + \tau^3}{2}} \in \mathbf{U}(1)_+. \end{array}$$

The $\mathcal{G}^3$-representations are decomposable into transmutators from $\mathbf{U}(2)$-vectors (boldface) to $\mathbf{U}(1)_+$-vectors with $\mathcal{G}^3$-frozen components (underlined) $\underline{\alpha} = 1, 2$,

$$\begin{array}{l} \underline{\mathbf{vec}}_{\mathbf{U}(2)} \ni W \overset{\mathbf{U}(1)_+}{\cong} \bigoplus_\iota W^\iota, \quad W^\iota \in \underline{\mathbf{vec}}_{\mathbf{U}(1)_+}, \\ \left. \begin{array}{ll} W \ni \mathbf{E}^\alpha = D(V)^\alpha_{\underline{\alpha}} E^{\underline{\alpha}} & \in \bigoplus_\iota W^\iota, \\ W^T \ni \mathbf{E}^\star_\alpha = E^\star_{\underline{\alpha}} D(V^\star)^{\underline{\alpha}}_\alpha & \in \bigoplus_\iota W^{\iota T}, \end{array} \right\} \text{with } \mathbf{E}^\star_\alpha \mathbf{E}^\alpha = E^\star_{\underline{\alpha}} E^{\underline{\alpha}}, \end{array}$$

especially for the Higgs field, where the dilation degree of freedom R constitutes the frozen field:

$$\begin{array}{lll} (\epsilon^{\alpha\beta}\mathbf{\Phi}^\star_\beta, \mathbf{\Phi}^\alpha) & = (V_1^\alpha, V_2^\alpha) R & \cong V \circ (\frac{\mathbf{1}_2 + \tau^3}{2}, \frac{\mathbf{1}_2 - \tau^3}{2}) R, \\ (\epsilon_{\alpha\beta}\mathbf{\Phi}^\beta, \mathbf{\Phi}^\star_\alpha) & = R(V_\alpha^{\star 1}, V_\alpha^{\star 2}) & \cong R(\frac{\mathbf{1}_2 + \tau^3}{2}, \frac{\mathbf{1}_2 - \tau^3}{2}) \circ V^\star. \end{array}$$

Hence $\mathbf{U}(2)$-invariants have the same form in the frozen fields as in the boldface unfrozen ones. For example, the Yukawa coupling above with a $\mathbf{U}(2)$-doublet fermion field $\mathbf{Q}$, like the left-handed lepton fields, is written with the $\mathbf{U}(1)_+$-fields $Q^{1,2} = V_\alpha^{\star 1,2}\mathbf{Q}^\alpha = (U_L, D_L)$:

$$\begin{array}{ll} \mathbf{L}^{\rm Yuk}(\mathbf{\Phi}) & = -g_D(\mathbf{D}_R\mathbf{\Phi}\mathbf{Q}^\star + \mathbf{Q}\mathbf{\Phi}^\star\mathbf{D}_R^\star) \quad -g_U(\mathbf{U}_R\mathbf{\Phi}^\star\mathbf{Q}^\star + \mathbf{Q}\mathbf{\Phi}\mathbf{U}_R^\star) \\ & = -g_D R(\mathbf{D}_R D_L^\star + D_L\mathbf{D}_R^\star) \quad -g_U R(\mathbf{U}_R U_L^\star + U_L\mathbf{U}_R^\star). \end{array}$$

The group element $V(x) = v(\frac{\mathbf{\Phi}(x)}{R(x)}) \in \mathbf{U}(2)/\mathbf{U}(1)_+$ defines a gauge transformation with a translation parametrized Lie algebra element as representative of the fix-Lie-algebra classes $\log \mathbf{U}(2)/\log \mathbf{U}(1)_+$

$$\begin{array}{rl} \mathbb{R}^4 & \longrightarrow \log \mathbf{U}(2), \quad x \longmapsto l^j(V(x)), \; l^j(V) = (\partial^j V) \circ V^\star, \\ l^j(V) & = i\frac{-(\partial^j\gamma_3)\mathbf{1}_2 + (\partial^j\gamma_a)[\delta_{ab}\frac{\sin\gamma}{\gamma} + \epsilon_{abc}\frac{\gamma_c}{\gamma}\frac{1-\cos\gamma}{\gamma} + \frac{\gamma_a\gamma_b}{\gamma^2}(1-\frac{\sin\gamma}{\gamma})]\tau^b}{2} \\ & = i\frac{(V_*)_0^j(\frac{\mathbf{\Phi}}{R})\mathbf{1}_2 + (V_*)_a^j(\frac{\mathbf{\Phi}}{R})\tau^a}{2}. \end{array}$$

The Lie-algebra-valued gauge fields, in the (2×2) Pauli representation

$$\begin{array}{lll} \mathbb{R}^4 \longrightarrow \log \mathbf{U}(2), & x \longmapsto \underline{\mathbf{A}}^j(x) = i\frac{\mathbf{A}_0^j(x)\mathbf{1}_2 + \vec{\mathbf{A}}^j(x)\vec{\tau}}{2}, \\ \mathbb{R}^4 \longmapsto \mathbf{U}(2), & x \longmapsto U(x), \\ \mathbb{R}^4 \longrightarrow \log \mathbf{U}(2), & x \longmapsto l^j(U(x)), \quad l^j(U) = (\partial^j U) \circ U^\star, \end{array}$$

with the affine $\mathbf{U} \vec{\times} \log \mathbf{U}$ transformation behavior

$$\underline{\mathbf{A}}^j \longmapsto U \circ \underline{\mathbf{A}}^j \circ U^\star + l^j(U)$$

are stripped of the Higgs-field-provided $\mathbf{U}(2)/\mathbf{U}(1)_+$-degrees of freedom

$$\mathbb{R}^4 \longrightarrow \log \mathbf{U}(1)_+, \quad x \longmapsto \underline{A}^j(x), \quad \underline{A}^j = l^j(V^\star) + V^\star \circ \underline{\mathbf{A}}^j \circ V.$$

There remains only the electromagnetic $\mathbf{U}(1)_+$-gauge degree of freedom

$$\underline{A}^j \longmapsto V^\star \circ U \circ \underline{\mathbf{A}}^j \circ U^\star \circ V + l^j(V^\star \circ U)$$

parametrized with spacetime translations

$$\begin{array}{l} \mathbb{R}^4 \ni x \longmapsto V(x)^\star \circ U(x) \in \mathbf{U}(1)_+, \\ \mathbb{R}^4 \ni x \longmapsto l^j(V(x)^\star \circ U(x)) \in \log \mathbf{U}(1)_+. \end{array}$$

The Higgs field derivative

$$(\partial^j - \underline{\mathbf{A}}^j)\mathbf{\Phi} = V(\partial^j - \underline{A}^j)\frac{\mathbf{1}_2 - \tau^3}{2}R$$

leads with the ground-state-characterizing mass $\langle R \rangle = M$ to the mass terms for the spin-1 particles (weak bosons) in the vector fields related to the three ground state degrees of freedom $\mathbf{U}(2)/\mathbf{U}(1)_+$.

The transition from the quantum fields for the electroweak interactions to particles uses two transmutations, internal and external: The internal hyperisospin-to-electromagnetism transmutators parametrize the $\mathbf{U}(2)/\mathbf{U}(1)_+$-degrees of freedom with the translation-dependent Higgs vectors. The external Lorentz-to-rotation transmutators parametrize the $\mathbf{SL}(\mathbb{C}^2)/\mathbf{SU}(2)$ and

$\mathbf{SL}(\mathbb{C}^2)/\mathbf{SO}(2)$-degrees of freedom for massive and massless particles respectively with momenta. For example, the internal-external freezing of the left-handed lepton isodoublet field leads, in the up-component, to a massless (anti)neutrino $(\mathrm{u}(\vec{q}), \mathrm{a}(\vec{q}))$ and, in the down component, to a massive electron-positron $(\mathrm{u}^C(\vec{q}), \mathrm{a}^C(\vec{q}))$:

$$\mathbf{l}(x)^{\dot{A}\alpha} = \left(e^{\frac{i}{2}\gamma_\perp(\mathbf{\Phi}(x))}\right)^{\alpha}_{1,2} \oplus\!\!\int \frac{d^3q}{(2\pi)^3} \begin{cases} \frac{e^{iqx}\mathrm{u}(\vec{q}) - e^{-iqx}\mathrm{a}^\star(\vec{q})}{\sqrt{2}}\Big|_{q^2=0} & \frac{1}{\sqrt{2|\vec{q}|}} & (e^{i\vec{\alpha}_\perp(\vec{q}_\perp)\vec{\sigma}})^{\dot{A}}_{-}, \\ \frac{e^{iqx}\mathrm{u}^C(\vec{q}) - e^{-iqx}\mathrm{a}^{\star C}(\vec{q})}{\sqrt{2}}\Big|_{q^2=m_e^2} & \sqrt{\frac{m}{2q_0^2}} & (e^{\vec{\beta}(\vec{q})\vec{\sigma}})^{\dot{A}}_{C}. \end{cases}$$

If the Higgs field is more than an ad hoc-implementation of the ground state degeneracy and if there is a Higgs particle with creation operator $\mathrm{U}(\vec{q})$, it comes in the dilation degree of freedom of the Higgs field

$$R(x) = \left(e^{-\frac{i}{2}\gamma_\perp(\mathbf{\Phi}(x))}\right)^2_\alpha \mathbf{\Phi}^\alpha(x) = \oplus\!\!\int \frac{d^3q}{(2\pi)^3 2q_0} \frac{e^{iqx}\mathrm{U}(\vec{q}) + e^{-iqx}\mathrm{U}^\star(\vec{q})}{\sqrt{2}}\Big|_{q^2=m^2_{\mathrm{Higgs}}}.$$

6.12 Reflections in the Standard Model

A relativistic dynamics may be invariant with respect to the $\mathtt{C}, \mathtt{P}, \mathtt{T}$ reflections (chapter "Lorentz Operations"). A lack of a reflection symmetry $\mathbb{I}(2) = \{1, \mathtt{R}\}$ can have two different reasons: With the reflection $\mathtt{R}$ represented on the field value space V

$$\begin{array}{ccc} \mathbb{R}^4 & \stackrel{\mathtt{R}}{\longrightarrow} & \mathbb{R}^4 \\ \mathbf{\Phi}\downarrow & & \downarrow{\scriptstyle \mathtt{R}\mathbf{\Phi}} \\ V & \underset{\mathtt{R}\bullet}{\longrightarrow} & V \end{array}, \quad \mathbf{\Phi}(x) \stackrel{\mathtt{R}}{\leftrightarrow} \mathtt{R} \bullet \mathbf{\Phi}(\mathtt{R}.x)$$

the interaction may not be $\mathtt{R}$-invariant (breakdown), or there does not even exist an $\mathtt{R}$-representation on V (nonimplementation). Both cases occur in the standard model for quark and lepton fields.

6.12.1 Position Reflection Breakdown

The $\mathbf{U}(1)$-vertex for a Dirac electron-positron field $\mathbf{\Psi} = (\mathrm{l}, \mathrm{r})$ interacting with an electromagnetic gauge field $\mathbf{A}$,

$$-\overline{\mathbf{\Psi}}\gamma\mathbf{\Psi}\mathbf{A} = -(\mathbf{l}^\star\sigma\mathbf{l} + \mathbf{r}^\star\check{\sigma}\mathbf{r})\mathbf{A},$$

is invariant under reflection $\mathtt{P}$ of position translations:

$$\mathbf{\Psi}(x_0, \vec{x}) \stackrel{\mathtt{P}}{\leftrightarrow} \gamma^0\mathbf{\Psi}(x_0, -\vec{x}), \quad \mathbf{A}_j(x_0, \vec{x}) \stackrel{\mathtt{P}}{\leftrightarrow} \begin{pmatrix} \mathbf{A}_0 \\ -\mathbf{A}_a \end{pmatrix}(x_0, -\vec{x}).$$

In the standard model of leptons with a left-handed isospin doublet field $\mathbf{l}$ and a right-handed isospin singlet field $\mathbf{e}$ the hyperisospin $\mathbf{U}(2)$ vertex with gauge fields $\mathbf{A}^0$ and $\vec{\mathbf{A}}$ and internal Pauli matrices reads

$$-(\tfrac{1}{2}\mathbf{l}^\star\sigma\mathbf{l} + \mathbf{e}^\star\check{\sigma}\mathbf{e})\mathbf{A}^0 + \mathbf{l}^\star\sigma\tfrac{\vec{\tau}}{2}\mathbf{l}\vec{\mathbf{A}}.$$

All gauge fields are assumed with the polar vector reflection behavior as given above. The position reflection P-invariance is broken in the electroweak standard model both by nonimplementation and by noninvariance of the interaction: One component of the lepton isodoublet, e.g., $\mathbf{e}_L = \frac{1-\tau_3}{2}\mathbf{l} \in V_L \cong \mathbb{C}^2$, can be used together with the right-handed isosinglet $\mathbf{e} = \mathbf{e}_R$ as a basis of a Dirac space $\mathbf{\Psi}_e \in V_L \oplus V_R \cong \mathbb{C}^4$ with a representation of P. This is impossible for the unpaired left-handed field $\nu_e = \frac{1+\tau_3}{2}\mathbf{l} \in W_L \cong \mathbb{C}^2$, here P is not implementable. However, also for the electron left-right pair $(\mathbf{e}_L, \mathbf{e}_R)$ the resulting gauge vertex breaks position reflection invariance via the neutral weak interactions mediated by the massive vector field $\mathbf{Z}$ arising in addition to the $\mathbf{U}(1)$-electromagnetic gauge field $\mathbf{A}$:

$$-\mathbf{e}_L^\star\sigma\mathbf{e}_L\frac{\mathbf{A}^0+\mathbf{A}^3}{2} - \mathbf{e}_R^\star\check{\sigma}\mathbf{e}_R\mathbf{A}^0 \quad = -\overline{\mathbf{\Psi}}_e\gamma\mathbf{\Psi}_e\mathbf{A} - \overline{\mathbf{\Psi}}_e\gamma\frac{1-4\sin^2\theta+i\gamma_5}{4}\mathbf{\Psi}_e\mathbf{Z}.$$

There is no parameter involved whose vanishing could restore a P-invariant dynamics.

6.12.2 GP-Invariance for Lepton Fields

The linear CP-reflection, involving the spinor "metric" (volume form) for the particle-antiparticle reflection, relates to each other Weyl pairs

$$V_L \overset{\mathtt{CP}}{\longleftrightarrow} V_R^T,\ \ \mathbf{l}^A \leftrightarrow \delta^A_{\dot{A}}\epsilon^{\dot{A}\dot{B}}\mathbf{l}^\star_{\dot{B}},$$
$$V_R \overset{\mathtt{CP}}{\longleftrightarrow} V_L^T,\ \ \mathbf{r}^{\dot{A}} \leftrightarrow \delta^{\dot{A}}_A\epsilon^{AB}\mathbf{r}^\star_B.$$

A $\mathbf{U}(1)$-gauge interaction for Weyl fields with terms like

$$-(z_l\mathbf{l}^\star\sigma\mathbf{l} + z_r\mathbf{r}^\star\check{\sigma}\mathbf{r})\mathbf{A},\quad z_{l,r} \in \mathbb{Z},$$

is CP-invariant.

The particle-antiparticle reflection C has to be extended by a reflection of the internal operation space in the case of Weyl spinors with nonabelian internal degrees of freedom. For isospin $\mathbf{SU}(2)$-doublets, e.g., the lepton isodoublet $\mathbf{l}$, the internal reflection is given by the, up to a factor unique, $\mathbf{SU}(2)$-invariant Pauli isospinor reflection, denoted as internal reflection by $\mathtt{I} = \epsilon$:

$$\begin{array}{ccc} U & \xrightarrow{u} & U \\ {\scriptstyle \mathtt{I}}\downarrow & & \downarrow{\scriptstyle \mathtt{I}} \\ U^T & \xrightarrow[\check{u}]{} & U^T \end{array},\qquad \begin{array}{l} u \in \mathbf{SU}(2)\ \text{(isospin)}, \\ \mathrm{u}^\alpha \overset{\mathtt{I}}{\leftrightarrow} \epsilon^{\alpha\beta}\mathrm{u}^\star_\beta,\quad \alpha = 1,2, \\ -\vec{\tau} = \epsilon^{-1}\circ\vec{\tau}^T\circ\epsilon. \end{array}$$

The linear particle-antiparticle reflection is the product $\mathtt{G} = \mathtt{IC}$ of external and internal reflection:

$$V_L \otimes U \overset{\mathtt{G}}{\leftrightarrow} V_L^T \otimes U^T, \quad \mathbf{l}^{A\alpha} \leftrightarrow \epsilon^{\alpha\beta}\epsilon^{AB}\mathbf{r}^\star_{B\beta}.$$

Thus the linear particle-antiparticle reflection $\mathtt{GP}$ reads for left-handed Weyl spinor-isospinors

$$V_L \otimes U \overset{\mathtt{GP}}{\longleftrightarrow} V_R^T \otimes U^T, \quad \mathbf{l}^{A\alpha} \leftrightarrow \epsilon^{\alpha\beta}\delta^A_{\dot{A}}\epsilon^{\dot{A}\dot{B}}\mathbf{l}^\star_{\dot{B}\beta}, \quad \mathtt{G} \circ \mathtt{P} = \mathtt{I} \circ \mathtt{C} \circ \mathtt{P}.$$

The standard model for leptons, i.e., with internal hypercharge-isospin action, is acted on by $\mathtt{GP}$; the gauge interaction is $\mathtt{GP}$-invariant.

6.12.3 $\mathtt{CP}$-Problems for Quark Fields

If nontrivial $\mathbf{SU}(3)$-representations, e.g., quark triplets and antitriplets, are included in the standard model, an extended $\mathtt{CP}$-reflection requires a linear reflection γ between dual representation spaces of color $\mathbf{SU}(3)$, i.e., an $\mathbf{SU}(3)$-invariant bilinear form of the representation space

$$\begin{array}{ccc} U & \overset{D(u)}{\longrightarrow} & U \\ \gamma\downarrow & & \downarrow\gamma \\ U^T & \underset{\check{D}(u)}{\longrightarrow} & U^T \end{array}, \qquad \begin{array}{l} D : \mathbf{SU}(3) \longrightarrow \mathbf{GL}(U) \text{ (color representation)}, \\ \gamma^{-1} \circ D(u)^T \circ \gamma = D(u^{-1}) \text{ for all } u \in \mathbf{SU}(3). \end{array}$$

The situation for isospin $\mathbf{SU}(2)$ and $\mathbf{SU}(1+r)$, $r \geq 2$, especially for color $\mathbf{SU}(3)$, is completely different with respect to the existence of such a linear dual isomorphism γ: All irreducible $\mathbf{SU}(2)$-representations $[2T]$ with isospin $T = 0, \frac{1}{2}, 1, \ldots$ are self-dual, i.e., they have an, up to a scalar factor, unique invariant bilinear form $\bigvee^{2T}\epsilon$ as product of the spinor "metric" discussed above. That is not the case for color representations: Some representations are linearly self-dual; some are not. Dual representations have reflected integer invariants

$$\text{dual reflection for } \mathbf{SU}(1+r)\text{: } [2C_1, \ldots, 2C_r] \leftrightarrow [2C_r, \ldots, 2C_1].$$

Only those $\mathbf{SU}(1+r)$-representations whose weight diagram is central reflection symmetric in the real r-dimensional weight vector space have an, up to a scalar factor unique, $\mathbf{SU}(1+r)$-invariant bilinear form, i.e., they are linearly self-dual. Therefore the self-dual irreducible $\mathbf{SU}(3)$-representations are $[2C, 2C]$:

$$C_1 = C_2 = C \Rightarrow \dim_{\mathbb{C}} U = (1+2C)^3 = 1, 8, 27, \ldots.$$

The Killing form κ of the octet $[1, 1]$ defines, by totally symmetric products, the related invariant bilinear forms $\bigvee^{2C}\kappa$.

It is impossible to define a CP-extending duality-induced linear reflection for the triplet-antitriplet quark spaces U, U^T with representations $[1,0]$ and $[0,1]$ since there does not exist a color-invariant bilinear form for triplets $U \cong \mathbb{C}^3$. Or equivalently, there does not exist a (3×3) matrix γ for the reflection $-\vec{\lambda} = \gamma^{-1} \circ \vec{\lambda}^T \circ \gamma$ of all eight Gell-Mann matrices, in contrast to the spinor "metric" (volume form) for all three Pauli matrices $-\vec{\sigma} = \epsilon^{-1} \circ \vec{\sigma}^T \circ \epsilon$. The $\mathbf{U}(3)$-scalar product cannot be used for a reflection; it is sesquilinear.

6.13 Summary

The power three tensor of a Lie algebra representation for an internal symmetry, Lorentz compatibly embedded into Minkowski spacetime, constitutes a gauge vertex (gauge dynamics), electrodynamics for the Lie algebra of $\mathbf{U}(1)$, the electroweak-strong standard model for the Lie algebra of $\mathbf{U}(2 \times 3)$. The real 12-dimensional standard model interaction symmetry group $\mathbf{U}(2 \times 3)$ is constituted by the classes of the hypercharge-isospin-color group $\mathbf{U}(1) \times \mathbf{SU}(2) \times \mathbf{SU}(3)$ with respect to the discrete normal subgroup $\mathbb{I}(6) = \mathbb{I}(2) \times \mathbb{I}(3) = \{e^{2i\pi y} \mid y = \frac{1}{6}, \frac{1}{3}, \frac{1}{2}, \frac{2}{3}, \frac{5}{6}, 1\}$ describing the central correlations between the abelian hypercharge group $\mathbf{U}(1)$ and the nonabelian isospin-color group $\mathbf{SU}(2) \times \mathbf{SU}(3)$.

In a spacetime field theory, the gauge vertices for a d-dimensional Lie algebra come as products $\mathbf{A}_a^k \mathbf{J}_k^a$ of the gauge fields $\{\mathbf{A}_a^k\}_{a=1}^d$ with the currents $\{\mathbf{J}_k^a\}_{a=1}^d$, the latter being position distributions of the Lie algebra representation. One spacetime component of the gauge fields $\{\mathbf{A}_a^0\}_{a=1}^d$ represents the Lie algebra forms, the gauge-fixing fields $\{\mathbf{S}^a\}_{a=1}^d$ as dual partners the Lie algebra.

In a quantum gauge theory, the classical gauge transformations are replaced by of the nilquadratic Becchi-Rouet-Stora transformations, whose origin lies in the nonparticle interpretable nilpotent structures of reducible nondecomposable time representations. The quantum gauge fields $\{\mathbf{A}_a^k, \mathbf{S}^a, \mathbf{F}_{jk}^a\}_{a=1}^d$ of Bose type are supplemented by Fadeev-Popov fields $\{\alpha_a, \check{\alpha}_a, \xi_k^a, \check{\xi}_k^a\}_{a=1}^d$ of Fermi type. All those fields in the gauge sector are acted on by the adjoint and coadjoint internal Lie algebra representations. Only for the particle degrees of freedom with covariant derivative, the BSR-transformations can be interpreted as gauge transformations.

The transition from the interaction symmetry to the particle symmetry requires a diagonalization with respect to a maximal abelian subgroup (Cartan torus). The central correlations in the standard model can be taken into account by particles either with trivial isospin or with trivial color. As motivated by experiments, a color confinement is chosen.

APPENDICES

6.14 Fadeev-Popov Degrees of Freedom

6.14.1 Abelian Fadeev-Popov Fields

An electromagnetic quantum gauge field

$$\begin{aligned}\mathbf{L}(\mathbf{A},\mathbf{F},\mathbf{S},\alpha,\xi) = &\ \mathbf{F}_{kj}\tfrac{\partial^k\mathbf{A}^j-\partial^j\mathbf{A}^k}{2} + \mathbf{S}\partial_k\mathbf{A}^k + g^2(\tfrac{\mathbf{F}_{kj}\mathbf{F}^{kj}}{4} - \lambda\tfrac{\mathbf{S}^2}{2})\\ &+ i\xi^k\partial_k\check{\alpha} + i\check{\xi}^k\partial_k\alpha - ig^2\lambda\check{\xi}^k\xi_k,\end{aligned}$$

$$\text{Bose: } \begin{cases} \partial^k\mathbf{A}^j - \partial^j\mathbf{A}^k &= g^2\mathbf{F}^{jk},\\ \partial_k\mathbf{A}^k &= g^2\lambda\mathbf{S},\\ \partial^j\mathbf{F}_{kj} - \partial_k\mathbf{S} &= 0,\end{cases} \quad \text{Fermi: } \begin{cases} \partial^k\alpha &= g^2\lambda\xi^k,\\ \partial_k\xi^k &= 0,\\ \partial^k\check{\alpha} &= -g^2\lambda\check{\xi}^k,\\ \partial_k\check{\xi}^k &= 0\end{cases}$$

is accompanied by *Fadeev-Popov fields* $(\alpha,\check{\alpha},\xi^k,\check{\xi}^k)$ (in chapter "Massless Quantum Fields" denoted by $(\beta,\check{\beta},\gamma^k,\check{\gamma}^k)$) with the Fermi quantization

$$\{\alpha,\check{\xi}^k\}(\vec{x}) = \delta_0^k\delta(\vec{x}) = \{\xi^k,\check{\alpha}\}(\vec{x}).$$

In a classical theory the gauge-fixing field $\mathbf{S}$ and all the Fadeev-Popov fields are set to zero.

The second order derivative Lagrangian reads

$$\mathbf{L}(\mathbf{A},\alpha) = -\tfrac{1}{4g^2}(\partial^j\mathbf{A}^k - \partial^k\mathbf{A}^j)(\partial_j\mathbf{A}_k - \partial_k\mathbf{A}_j) + \tfrac{1}{2g^2\lambda}(\partial_k\mathbf{A}^k)^2 + \tfrac{i}{g^2\lambda}(\partial^k\alpha)(\partial_k\check{\alpha}).$$

The nilquadratic *linear BRS-charge* generates the linear BRS-transformations

$$N = g^2\lambda\int d^3x\xi_0(x)\mathbf{S}(x),\quad N^2 = 0,$$

$$\text{Bose: } \begin{cases} \delta\mathbf{A}^k &= [iN,\mathbf{A}^k] &= g^2\lambda\delta_0^k\xi_0,\\ \delta\mathbf{S} &= [iN,\mathbf{S}] &= 0,\\ \delta\mathbf{F}_{kj} &= [iN,\mathbf{F}_{kj}] &= 0,\end{cases} \quad \text{Fermi: } \begin{cases} \delta\alpha &= \{iN,\alpha\} &= 0,\\ \delta\xi^k &= \{iN,\xi^k\} &= 0,\\ \delta\check{\alpha} &= \{iN,\check{\alpha}\} &= ig^2\lambda\mathbf{S},\\ \delta\check{\xi}^k &= \{iN,\check{\xi}^k\} &= 0.\end{cases}$$

In contrast to the classical gauge transformation, the BRS-transformation implemented by N is nontrivial only for the spinless component of the gauge field:

$$\mathbf{A}^0 \longmapsto \mathbf{A}^0 + \partial^0\alpha,\quad \mathbf{A}^a \longmapsto \mathbf{A}^a,\quad a = 1,2,3.$$

The Fadeev-Popov Fermi quantum field $\alpha(x)$ takes the role of the classical gauge parameter. The group element is reduced to the absolute and linear term

$$e^{i\alpha(x)} = 1 + i\alpha(x) \text{ for } \{\alpha(x),\alpha(x)\} = 0.$$

For an electromagnetic interaction $\mathbf{A}^k\mathbf{J}_k$ with additional fields the BRS-transfomation is effected by a *nonlinear BRS-current*, involving the $\mathbf{U}(1)$-current. Also, the nonlinear BRS-charge is nilquadratic:

$$\begin{array}{l} N = \int d^3x \mathbf{N}_0(x), \quad \mathbf{N}_k = g^2\lambda\xi_k\mathbf{S} + \alpha\mathbf{J}_k, \quad N^2 = 0, \\ \text{e.g., } [N, \mathbf{\Phi}] = \alpha z\mathbf{\Phi}, \quad \{N, \mathbf{\Psi}\} = \alpha z\mathbf{\Psi}. \end{array}$$

In addition to this nonlinear transformation involving the translation-dependent Fadeev-Popov quantum fields and replacing the translation-dependent classical gauge transformations, there remains the "global" $\mathbf{U}(1)$-transformation

$$Q = \int d^3x \mathbf{J}_0(x), \text{ e.g., } [Q, \mathbf{\Phi}] = z\mathbf{\Phi}, \quad [Q, \mathbf{\Psi}] = z\mathbf{\Psi}.$$

The spinless and gauge-fixing Bose degrees of freedom and the Fermi Fadeev-Popov ones display a twin structure. The BRS-current $\mathbf{N}_k(x)$ of Fermi type has its counterpart in the nonderivative part $\mathbf{H}(x)$ of the Lagrangian (Bose type)

$$\begin{array}{l} \mathbf{N}_k = g^2\lambda\xi_k\mathbf{S} + \alpha\mathbf{J}_k, \quad \mathbf{H} = g^2\lambda[\frac{\mathbf{S}^2}{2} + i\check{\xi}^k\xi_k] + \mathbf{A}^k\mathbf{J}_k, \\ [i\mathbf{S}, \mathbf{A}^k](\vec{x}) = \{\alpha, \check{\xi}^k\}(\vec{x}) = \{\xi^k, \check{\alpha}\}(\vec{x}) = \delta_0^k\delta(\vec{x}). \end{array}$$

The analogue in the mass point model (chapter "Massless Quantum Fields") is given by

$$\begin{array}{l} N = \xi\mathbf{p} + \alpha Q, \quad H = \frac{\mathbf{p}^2}{2} + i\check{\xi}\xi + \mathbf{x}Q, \\ [i\mathbf{p}, \mathbf{x}] = \{\alpha, \check{\xi}\} = \{\xi, \check{\alpha}\} = 1. \end{array}$$

Here the analogue to the current $\mathbf{J}_k$ built by fields is the $\mathbf{U}(1)$-charge Q built by some time-dependent charged degrees of freedom. The Bose dynamics H arises by BRS-transformation from an operator K connecting Bose and Fermi degrees of freedom

$$H = \{N, K\}, \quad K = \frac{\check{\alpha}\mathbf{p}}{2} + \check{\xi}\mathbf{x}.$$

Since $N^2 = 0$, the BRS-invariance of the Hamiltonian is obvious

$$[N, H] = [N, \{N, K\}] = 0.$$

The corresponding field operators are

$$\mathbf{K} = \frac{\check{\alpha}\mathbf{S}}{2} + \check{\xi}_k\mathbf{A}^k, \quad (H, N, K) = \int d^3x(\mathbf{H}, \mathbf{N}_0, \mathbf{K})(x).$$

6.14.2 Nonabelian Fadeev-Popov Fields

For abelian gauge theories the Fadeev-Popov degrees of freedom are used only to define the asymptotic particle interpretable space. In this case they have no interaction. In the nonabelian case with the nontrivial current of the gauge fields the Fadeev-Popov fields also have a nontrivial gauge interaction.

Like the gauge fields, the Fadeev-Popov fields also have to come in the adjoint representations of the Lie algebra L for $(\xi, \check{\xi})$ and in the dual adjoint one for $(\alpha, \check{\alpha})$ with the current

$$\begin{aligned} i\mathcal{J}_k^a &= -\epsilon_c^{ab}(\alpha_b \check{\xi}_k^c + \check{\alpha}_b \xi_k^c), \\ i\mathcal{J}_k &= \alpha \times \check{\xi}_k + \check{\alpha} \times \xi_k \text{ (index slimmed-down notation)}, \end{aligned}$$

leading to the full current as sum of Fadeev-Popov field current, gauge field current, and "nongauge" sector current for the internal Lie algebra action:

$$\begin{aligned} \mathcal{J}_k + \mathbf{j}_k + \mathbf{J}_k &= -i(\alpha \times \check{\xi}_k + \check{\alpha} \times \xi_k) + \mathbf{A}_k \times \mathbf{S} + \mathbf{A}^j \times \mathbf{F}_{kj} + \mathbf{J}_k, \\ Q^a &= \int d^3x (\mathcal{J}_k^a + \mathbf{j}_k^a + \mathbf{J}_k^a)(x). \end{aligned}$$

In the gauge field current coupling the contribution $\mathbf{A}^k(\mathbf{A}_k \times \mathbf{S})$ with the gauge-fixing field vanishes because of the antisymmetry of the Lie bracket. To preserve the BRS-symmetry, the corresponding term $\mathbf{A}^k(\check{\alpha} \times \xi_k)$ has to be omitted on the Fadeev-Popov sector involving the linear BRS-partner $\check{\alpha}$ for $\mathbf{S}$, i.e., $\delta\check{\alpha} = ig^2\lambda\mathbf{S}$. Therefore, the dynamically relevant gauge field current coupling involves not the full current, but

$$\mathbf{A}^k(-i\alpha \times \check{\xi}_k + \tfrac{1}{2}\mathbf{A}^j \times \mathbf{F}_{kj} + \mathbf{J}_k)$$

with the statistical factor $\frac{1}{2}$.

Together with the linear terms, the gauge field interaction is given by

$$\begin{aligned} \mathbf{H} &= \lambda g_{ab}(\tfrac{\mathbf{S}^a\mathbf{S}^b}{2} + i\check{\xi}^{ka}\xi_k^b) &&+\mathbf{A}_a^k\Big[-\epsilon_c^{ab}\Big(-i\alpha_b\check{\xi}_k^c + \tfrac{1}{2}\mathbf{A}_b^j\mathbf{F}_{kj}^c\Big) + \mathbf{J}_k^a\Big] \\ &= \lambda(\tfrac{\mathbf{S}\bullet\mathbf{S}}{2} + i\check{\xi}^k \bullet \xi_k) &&+\mathbf{A}^k[-i\alpha \times \check{\xi}_k + \tfrac{1}{2}\mathbf{A}^j \times \mathbf{F}_{kj} + \mathbf{J}_k]. \end{aligned}$$

$\mathbf{H}$ can be constructed as the BRS-transform $\mathbf{H} = \{N, \mathbf{K}\}$ with a nilquadratic BRS-charge $N^2 = 0$, which leads to its BRS-invariance $[N, \mathbf{H}] = 0$. The corresponding expression in the mechanical mass point model

$$H = \tfrac{\mathbf{p}\bullet\mathbf{p}}{2} + i\check{\xi} \bullet \xi + \mathbf{x}(-i\alpha \times \check{\xi} + Q)$$

is a BRS-transform

$$H = \{N, K\} \text{ with } \begin{cases} H = & g_{ab}(\frac{\mathbf{p}^a\mathbf{p}^b}{2} + i\check{\xi}^a\xi^b) \quad +\mathbf{x}_a(i\epsilon_c^{ab}\alpha_b\check{\xi}^c + Q^a), \\ K = & \frac{\check{\alpha}_a\mathbf{p}^a}{2} + \check{\xi}^a\mathbf{x}_a, \\ N = & g_{ab}\xi^a\mathbf{p}^b \quad +\alpha_a(\frac{i}{2}\epsilon_c^{ab}\alpha_b\check{\xi}^c + Q^a), \end{cases}$$

which holds also in the spacetime theory with

$$\begin{aligned} \mathbf{H} &= \lambda(\tfrac{\mathbf{S}\bullet\mathbf{S}}{2} + i\check{\xi}^k \bullet \xi_k) + \mathbf{A}^k[-i\alpha \times \check{\xi}_k + \tfrac{1}{2}\mathbf{A}^j \times \mathbf{F}_{kj} + \mathbf{J}_k], \\ \mathbf{K} &= \tfrac{\check{\alpha}\mathbf{S}}{2} + \check{\xi}_k\mathbf{A}^k, \\ \mathbf{N}_k &= \lambda\xi_k \bullet \mathbf{S} + \alpha[-\tfrac{i}{2}\alpha \times \check{\xi}_k + \mathbf{A}^j \times \mathbf{F}_{kj} + \mathbf{J}_k]. \end{aligned}$$

The statistical factors $\frac{1}{2}$ take care of the power-2 expressions $\mathbf{A}_a^k\mathbf{A}_b^j$ and $\alpha_a\alpha_b$.

It will be shown in the next section that the BRS-charge is nilquadratic,

$$N = \int d^3x \mathbf{N}_0(x), \quad N^2 = 0,$$

which ensures the BRS-invariance of the dynamics with the Lagrangian

$$\begin{aligned}\mathbf{L}(\mathbf{A},\mathbf{F},\mathbf{S},\alpha,\xi,\mathbf{J}) = &\ \dots\partial\dots(\text{kinetic terms}) - \mathbf{H}\\ = &\ \mathbf{F}_{kj}\frac{\partial^k\mathbf{A}^j-\partial^j\mathbf{A}^k+\mathbf{A}^k\wedge\mathbf{A}^j}{2}+\mathbf{S}\partial_k\mathbf{A}^k+\frac{\mathbf{F}_{kj}\bullet\mathbf{F}^{kj}}{4}-\lambda\frac{\mathbf{S}\bullet\mathbf{S}}{2}\\ &+i\xi_k\partial^k\check{\alpha}+i\check{\xi}_k(\partial^k\alpha+\mathbf{A}^k\wedge\alpha)-i\lambda\check{\xi}^k\bullet\xi_k\\ &-\mathbf{A}^k\mathbf{J}_k.\end{aligned}$$

The BRS-charge is a scalar with respect to the internal Lie algebra action

$$[Q^a, N] = 0.$$

The classical gauge transformations with translation dependent parameters $\{\alpha^a(x) \mid a = 1,\dots,d\}$ arise in a quantum theory with d "global" charges $\{Q^a\}_{a=1}^d$ and only one BRS-charge N.

The field equations for gauge Bose and Fermi (Fadeev-Popov) sector are

$$\text{Bose:}\ \begin{cases}\partial^k\mathbf{A}^j-\partial^j\mathbf{A}^k &= \bullet\mathbf{F}^{jk}-\mathbf{A}^k\wedge\mathbf{A}^j,\\ \partial_k\mathbf{A}^k &= \lambda\bullet\mathbf{S},\\ \partial^j\mathbf{F}_{kj}-\partial_k\mathbf{S} &= \mathbf{A}^j\times\mathbf{F}_{kj}-i\alpha\times\check{\xi}_k+\mathbf{J}_k,\end{cases}$$

$$\text{Fermi:}\ \begin{cases}\partial^k\alpha &= \lambda\bullet\xi^k-\mathbf{A}^k\wedge\alpha,\\ \partial_k\xi^k &= 0,\\ \partial_k\check{\alpha} &= \lambda\bullet\check{\xi}_k,\\ \partial^k\check{\xi}_k &= \mathbf{A}^k\times\check{\xi}_k.\end{cases}$$

The fields with a particle asymptotics come with the covariant derivative in contrast to the spinless gauge field component, the gauge fixing field $(\mathbf{A}^0,\mathbf{S})$, and the Fadeev-Popov field pair $(\check{\alpha},\xi^k)$, which are all fields without particle asymptotics. These nonparticle fields do not come with a covariant derivative.

The dynamics has to be compared with the BRS-transformations (indices $a = 1,2,3$ for position translations)

$$\text{Bose:}\ \begin{cases}\delta\mathbf{A}^k &= [iN,\mathbf{A}^k] &= \lambda\bullet\delta_0^k\xi^0-\delta_a^k\alpha\wedge\mathbf{A}^a,\\ \delta\mathbf{S} &= [iN,\mathbf{S}] &= 0,\\ \delta\mathbf{F}_{kj} &= [iN,\mathbf{F}_{kj}] &= \delta_k^0\delta_j^a\alpha\times\mathbf{F}_{0a},\end{cases}$$

$$\text{Fermi:}\ \begin{cases}\delta\alpha &= \{iN,\alpha\} &= -\frac{\alpha\wedge\alpha}{2},\\ \delta\xi^k &= \{iN,\xi^k\} &= 0,\\ \delta\check{\alpha} &= \{iN,\check{\alpha}\} &= i\lambda\bullet\mathbf{S},\\ \delta\check{\xi}_k &= \{iN,\check{\xi}_k\} &= \delta_k^0 i(\mathbf{A}^j\times\mathbf{F}_{kj}-i\alpha\times\check{\xi}_k+\mathbf{J}_k).\end{cases}$$

The gauge field has the BRS-behavior

$$\mathbf{A}_c^k \longmapsto \mathbf{A}_c^k+\delta_0^k\partial^0\alpha_c-\epsilon_c^{ab}\alpha_a\mathbf{A}_b^k.$$

In the second order formulation for the Fadeev-Popov fields, one has the equations

$$\begin{aligned}&\mathbf{L}(\alpha,\check{\alpha},\mathbf{A}) = i\partial_k\check{\alpha}\bullet(\partial^k\alpha+\mathbf{A}^k\wedge\alpha),\\ &\partial_k(\partial^k\alpha+\mathbf{A}^k\wedge\alpha)=0,\quad (\partial^k-\mathbf{A}^k\times)\partial_k\check{\alpha}=0,\end{aligned}$$

and the dynamically relevant gauge field-current coupling

$$-i\mathbf{A}^k(\alpha \times \bullet\partial_k\check{\alpha}) = i\mathbf{A}^k_a\epsilon^{ab}_c g^{cd}\alpha_b\partial_k\check{\alpha}_d.$$

6.15 Gauge and BRS-Vertices

A Lie algebra representation in a relativistic theory is distributed by currents, which are products of dual field pairs, e.g., for quantum electrodynamics as the electromagnetic current $\mathbf{J}_j = z\frac{[\boldsymbol{\Psi}\gamma_j,\overline{\boldsymbol{\Psi}}]}{2}$:

$$\mathcal{D}^a = \int d^3x\ i\mathbf{J}^a_0(x).$$

A gauge field current coupling is the relativistically compatible distribution and embedding of a Lie algebra representation with the gauge field $\{\mathbf{A}^0_a\}^d_{a=1}$ implementing the dual Lie algebra

$$\check{l}_a \otimes \mathcal{D}^a = \int d^3x\ \mathbf{A}^j_a(x)\ i\mathbf{J}^a_j(x).$$

A representation of a real Lie algebra L on a vector space V,

$$\begin{array}{rl} \mathcal{D}: L \longrightarrow \mathbf{AL}(V), & l \longmapsto \mathcal{D}(l) = \mathcal{D}(l)^\beta_\gamma e^\gamma \otimes \check{e}_\beta, \\ & \mathcal{D}(l^a) = \mathcal{D}^a = \mathcal{D}^{a\beta}_\gamma e^\gamma \otimes \check{e}_\beta, \\ \text{dual bases of } V, V^T: & \langle \check{e}_\beta, e^\gamma\rangle = \delta^\gamma_\beta, \\ \text{dual bases of } L, L^T: & \langle \check{l}_b, l^c\rangle = \delta^c_b, \quad \text{structure constants: } [l^a, l^b] = \epsilon^{ab}_c l^c, \end{array}$$

defines a power-three tensor

$$\mathcal{D} = \check{l}_a \otimes \mathcal{D}^a = \mathcal{D}^{a\beta}_\gamma \check{l}_a \otimes e^\gamma \otimes \check{e}_\beta \in L^T \otimes V \otimes V^T.$$

The *dual adjoint representation tensor* is given by the $(1,2)$-tensor

$$\check{\mathrm{ad}} = \check{l}_a \otimes (-\,\mathrm{ad}\,l^a)^T = \check{l}_a \otimes \mathcal{L}^a = -\epsilon^{ab}_c \check{l}_a \otimes \check{l}_b \otimes l^c \in L^T \otimes L^T \otimes L.$$

In the quantum algebra $\mathbf{Q}_\epsilon(\mathbf{V})$, Fermi or Bose $\epsilon = \pm 1$, of the self-dual vector space $\mathbf{V} = V \oplus V^T$, the Lie brackets are implemented by commutators

$$\begin{array}{rl} \text{in } \mathbf{Q}_\epsilon(\mathbf{V}): & [\check{e}_\beta, e^\gamma]_\epsilon = \delta^\gamma_\beta, \\ & \mathcal{D}^a = \mathcal{D}^{a\beta}_\gamma \frac{[e^\gamma, \check{e}_\beta]_{-\epsilon}}{2}, \quad [\mathcal{D}^a, \mathcal{D}^b] = \epsilon^{ab}_c \mathcal{D}^c, \\ & [\mathcal{D}^a, e^\beta] = \mathcal{D}^{a\beta}_\gamma e^\gamma, \quad [\mathcal{D}^a, \check{e}_\gamma] = -\mathcal{D}^{a\beta}_\gamma \check{e}^\beta, \end{array}$$

e.g., the adjoint representation of L in the adjoint quantum algebra $\mathbf{Q}_\epsilon(\mathbf{L})$ with the self-dual Lie algebra space $\mathbf{L} = L \oplus L^T$,

$$\begin{array}{rl} \text{in } \mathbf{Q}_\epsilon(\mathbf{L}): & [l^c, \check{l}_b]_\epsilon = \delta^c_b, \\ & \mathcal{L}^a = -\epsilon^{ab}_c \check{l}_b l^c, \quad [\mathcal{L}^a, \mathcal{L}^b] = \epsilon^{ab}_c \mathcal{L}^c, \\ & [\mathcal{L}^a, l^b] = \epsilon^{ab}_c l^c, \quad [\mathcal{L}^a, \check{l}_c] = -\epsilon^{ab}_c \check{l}_b. \end{array}$$

One has to distinguish between the Lie bracket $[l^a, l^b]$ in L, the endomorphism commutator $[\mathcal{L}^a, \mathcal{L}^b]$ in $\mathbf{AL}(L)$, coinciding with the commutator in the quantum algebra $\mathbf{Q}_\epsilon(\mathbf{L})$, and the quantization (anti) commutator $[\check{l}_c, l^b]_\epsilon$ in the quantum algebra (chapter "Quantum Algebra").

The representation tensor $\mathcal{D}$ is an element of the product quantum algebra of Lie algebra and representation space; it is invariant under Lie algebra action

$$\begin{array}{llll} \mathcal{D} = \check{l}_a \mathcal{D}^a = \mathcal{D}^{a\beta}_\gamma \check{l}_a \frac{[e^\gamma, \check{e}_\beta]_{-\epsilon}}{2} & \in \mathbf{Q}_{\epsilon'}(\mathbf{L}) \otimes \mathbf{Q}_\epsilon(\mathbf{V}), & [\mathcal{D}^a, \mathcal{D}] = 0, \\ \mathrm{ad}_{\epsilon'} = \frac{1}{2}\check{l}_a \mathcal{L}^a = -\epsilon^{ab}_c \frac{\check{l}_a \check{l}_b}{2} l^c & \in \mathbf{Q}_{\epsilon'}(\mathbf{L}), & [\mathcal{L}^a, \mathrm{ad}] = 0. \end{array}$$

The adjoint tensor involves a statistical factor $\frac{1}{2}$.

The two cases with Bose or Fermi statistics $\epsilon' = \mp 1$ of the quantum algebra for the Lie algebra $L \neq V$ have different properties.

The full representation tensor in the case of a Bose quantum algebra for the Lie algebra,

$$\begin{array}{rl} \text{in } \mathbf{Q}_-(\mathbf{L}) \otimes \mathbf{Q}_\epsilon(\mathbf{V}): & [l^b, \check{l}_c] = \delta^b_c, \\ l^a \longmapsto & \mathcal{L}^a_- + \mathcal{D}^a = -\epsilon^{ab}_c \check{l}_b l^c + \mathcal{D}^{a\beta}_\gamma \frac{[e^\gamma, \check{e}_\beta]_{-\epsilon}}{2}, \\ \text{representation tensor:} & \check{l}_a(\frac{1}{2}\mathcal{L}^a_- + \mathcal{D}^a) = \check{l}_a \mathcal{D}^a, \end{array}$$

does not contain the adjoint representation contribution because of the antisymmetry of the Lie bracket:

$$\mathrm{ad}_- = \tfrac{1}{2}\check{l}_a \mathcal{L}^a_- = 0 \quad \text{since } \epsilon^{ab}_c = -\epsilon^{ba}_c \text{ and } [\check{l}_a, \check{l}_b] = 0.$$

For the Bose case the representation tensor $\check{l}_a \mathcal{D}^a$ is called *gauge interaction vertex.*

In the case of a Fermi quantum algebra for the Lie algebra with a basis notation $(\check{l}_a, l^a) \to (\check{\xi}_a, \alpha^a)$ the representation tensor contains a nontrivial adjoint representation contribution,

$$\begin{array}{rl} \text{in } \mathbf{Q}_+(\mathbf{L}) \otimes \mathbf{Q}_\epsilon(\mathbf{V}): & \{\alpha^b, \check{\xi}_c\} = \delta^b_c, \\ \alpha^a \longmapsto & \mathcal{L}^a_+ + \mathcal{D}^a = -\epsilon^{ab}_c \check{\xi}_b \alpha^c + \mathcal{D}^{a\beta}_\gamma \frac{[e^\gamma, \check{e}_\beta]_{-\epsilon}}{2}, \\ \text{representation tensor:} & iN = \check{\xi}_a(\frac{1}{2}\mathcal{L}^a_+ + \mathcal{D}^a). \end{array}$$

For the Fermi case the representation tensor N is called *gauge BRS-vertex.* Because of the Jacobi-Leibniz property of the Lie bracket it is nilquadratic:

$$\begin{array}{rl} \mathrm{ad}^2_+ & = (\check{\xi}_a \frac{1}{2}\mathcal{L}^a_+)^2 = 0, \quad N^2 = 0, \\ -\{N, N\} & = \{\check{\xi}_a(\frac{1}{2}\mathcal{L}^a_+ + \mathcal{D}^a), \check{\xi}_b(\frac{1}{2}\mathcal{L}^b_+ + \mathcal{D}^b)\} \\ & = \check{\xi}_a \check{\xi}_b [\mathcal{D}^a, \mathcal{D}^b] + \check{\xi}_a [\mathcal{L}^a_+, \check{\xi}_b] \mathcal{D}^b + \frac{1}{2}\check{\xi}_a [\mathcal{L}^a_+, \check{\xi}_b] \mathcal{L}^b_+ + \frac{1}{4}\check{\xi}_a \check{\xi}_b [\mathcal{L}^a_+, \mathcal{L}^b_+] \\ & = -\frac{1}{4}\epsilon^{ab}_c \check{\xi}_a \check{\xi}_b \mathcal{L}^c_+ = \frac{1}{4}\epsilon^{ab}_c \epsilon^{cd}_e \check{\xi}_a \check{\xi}_b \check{\xi}_d \alpha^e = -\frac{1}{4}(\epsilon^{bd}_c \epsilon^{ca}_e + \epsilon^{da}_c \epsilon^{cb}_e) \check{\xi}_a \check{\xi}_b \check{\xi}_d l^e = 0, \end{array}$$

where the Jacobi identity for the adjoint representation, the antisymmetry of the Lie bracket, and the Fermi property $\{\check{\xi}_b, \check{\xi}_a\} = 0$ have been used.

For the mass point model, momenta-positions and a Fadeev-Popov pair implement the adjoint Lie algebra representation both in a Bose and Fermi quantum algebra:

$$\begin{array}{lllll} \text{Bose:} & [i\mathbf{p}^b, \mathbf{x}_c] = \delta^b_c, & \mathcal{L}^a_- = -\epsilon^{ab}_c \mathbf{x}_b i\mathbf{p}^c, & \mathrm{ad}_- = -\epsilon^{ab}_c \frac{\mathbf{x}_a \mathbf{x}_b}{2} i\mathbf{p}^c, & \mathrm{ad}_- = 0, \\ \text{Fermi:} & \{\check{\xi}^b, \alpha_c\} = \delta^b_c, & \mathcal{L}^a_+ = -\epsilon^{ab}_c \alpha_b \check{\xi}^c, & \mathrm{ad}_+ = -\epsilon^{ab}_c \frac{\alpha_a \alpha_b}{2} \check{\xi}^c, & \mathrm{ad}^2_+ = 0. \end{array}$$

Together with matter fields one has the gauge interaction vertex and the nilquadratic gauge BRS-vertex:

$$\begin{aligned} iH &= \mathbf{x}_a(-\epsilon_c^{ab}\alpha_b\check{\xi}^c + \mathcal{D}^{a\beta}_{\gamma}\frac{[e^\gamma,\check{e}_\beta]_{-\epsilon}}{2}),\\ iN &= \alpha_a(-\frac{1}{2}\epsilon_c^{ab}\alpha_b\check{\xi}^c + \mathcal{D}^{a\beta}_{\gamma}\frac{[e^\gamma,\check{e}_\beta]_{-\epsilon}}{2}),\\ K &= \check{\xi}^a\mathbf{x}_a + \frac{\check{\alpha}_a\mathbf{p}^a}{2},\\ H &= \{N,K\},\quad [N,H]=0. \end{aligned}$$

In the Lorentz compatible spacetime distribution of the gauge interaction vertex and the BRS-vertex,

$$\begin{aligned} i\mathbf{H} &= \mathbf{A}_a^k(-\epsilon_c^{ab}\alpha_b\check{\xi}_k^c + \frac{i}{2}\mathbf{j}_k^a + i\mathbf{J}_k^a),\\ i\mathbf{N}_k &= \alpha_a(-\frac{1}{2}\epsilon_c^{ab}\alpha_b\check{\xi}_k^c + i\mathbf{j}_k^a + i\mathbf{J}_k^a), \end{aligned}$$

the gauge field $\mathbf{A}_a^0(x)$ is the spacetime distribution of the Bose implemented dual Lie algebra elements $\mathbf{x}_a(t) \cong \check{l}_a \in L^T$, the gauge fixing fields $\mathbf{S}^a(x)$ for the Lie algebra elements $i\mathbf{p}^a(t) \cong l^a \in L$.

	time	spacetime	spacetime
	Lie algebra	gauge interaction (Bose)	gauge BRS (Fermi)
general representation	$\mathcal{D} = \check{l}_a \otimes \mathcal{D}(l^a)$ $= \mathcal{D}^{a\beta}_{\gamma}\check{l}_a \otimes e^\gamma \otimes \check{e}_\beta$	$\mathbf{H} = \mathbf{A}_a^k\mathbf{J}_k^a$ $= \mathbf{A}_a^k\mathbf{l}\check{\sigma}_k\frac{\tau^a}{2}\mathbf{l}^\star$	$\mathbf{N}_k = \alpha_a\mathbf{J}_k^a$ $= \alpha_a\mathbf{l}\check{\sigma}_k\frac{\tau^a}{2}\mathbf{l}^\star$
adjoint representation	$\check{\mathrm{ad}} = \check{l}_a \otimes (-\mathrm{ad}\, l^a)^T$ $= -\epsilon_c^{ab}\check{l}_a \otimes \check{l}_b \otimes l^c$	$\mathbf{ad}_- = \frac{1}{2}\mathbf{A}_a^k\mathbf{j}_k^a$ $= -\epsilon_c^{ab}\frac{\mathbf{A}_a^k\mathbf{A}_b^j}{2}\mathbf{F}_{kj}^c$	$\mathbf{ad}_+ = \frac{1}{2}\alpha_a\mathcal{J}_k^a$ $= -\epsilon_c^{ab}\frac{\alpha_a\alpha_b}{2}\check{\xi}_k^c$

representation tensors and gauge vertices

6.16 Cartan Tori

A Lie algebra has Cartan subalgebras, for semisimple Lie algebras given by maximal abelian subalgebras. Going from a Lie algebra to its exponent, a simply connected Lie group, a Cartan subalgebra gives rise to a Cartan subgroup. A maximal abelian *direct product* subgroup of a compact group

$$\mathbf{U}(1)^n = \underbrace{\mathbf{U}(1) \times \cdots \times \mathbf{U}(1)}_{n\text{times}}$$

will be called an *n-dimensional Cartan torus*, which can be parametrized for each direct factor ("circle") by $\mathbf{U}(1) = \{e^{i\alpha} \bigm| \alpha \in [0, 2\pi[\}$. Like a Cartan subalgebra, a Cartan torus is not unique.

If the dimension of a Cartan torus coincides with the rank of the Lie algebra, the Cartan torus is called *complete* for the group. There are situations in which there does exist a Cartan subalgebra, but no complete Cartan torus.

6.16.1 A Complete Cartan Torus for $\mathbf{SU}(n)$

The Lie algebra $\log \mathbf{SU}(n)$, $n \geq 2$, in the defining complex n-dimensional representation has a basis for a Cartan subalgebra $\log \mathbf{U}(1)^{n-1}$ given by the diagonal matrices

$$\text{Cartan subalgebra basis: } \{i\tau(n)^{m^2-1} \mid m = 2, 3, \ldots, n\}.$$

It is defined inductively for $n \geq 2$ by

$$\begin{aligned}
\tau(2)^3 = \begin{pmatrix} 1 & 0 \\ 0 & -1 \end{pmatrix}, \quad \tau(n)^{m^2-1} &= \left(\begin{array}{c|c} \tau(n-1)^{m^2-1} & 0 \\ \hline 0 & 0 \end{array}\right), \\
& m = 2, \ldots, n-1, \\
\tau(n)^{n^2-1} &= \frac{1}{\sqrt{\binom{n}{2}}} \left(\begin{array}{c|c} \mathbf{1}_{n-1} & 0 \\ \hline 0 & -(n-1) \end{array}\right), \\
\text{e.g., } n = 3: \ \tau(3)^3 = \left(\begin{array}{cc|c} 1 & 0 & 0 \\ 0 & -1 & 0 \\ \hline 0 & 0 & 0 \end{array}\right), \quad \tau(3)^8 &= \frac{1}{\sqrt{3}} \left(\begin{array}{c|c} \mathbf{1}_2 & 0 \\ \hline 0 & -2 \end{array}\right).
\end{aligned}$$

The exponent gives a complete Cartan torus of dimension $n-1$ (rank of $\log \mathbf{SU}(n)$).

The "maximal" diagonal element, characterized by a nontrivial determinant, generates the center of $\mathbf{SU}(n)$ and is normalized to display integer $\mathbf{U}(1)$-winding numbers

$$\begin{aligned}
& \mathbf{w}_n = \sqrt{\tbinom{n}{2}}\,\tau(n)^{n^2-1} = \left(\begin{array}{c|c} \mathbf{1}_{n-1} & 0 \\ \hline 0 & -(n-1) \end{array}\right), \quad \det \mathbf{w}_n = -(n-1), \\
& \mathbf{U}(1)_{n^2-1} = \{e^{i\alpha \mathbf{w}_n} \mid \alpha \in [0, 2\pi[\}, \\
& e^{\frac{2i\pi}{n}\mathbf{w}_n} = e^{\frac{2\pi i}{n}} \mathbf{1}_n \in \mathbf{U}(\mathbf{1}_n) \cap \mathbf{U}(1)_{n^2-1} = \mathbf{U}(\mathbf{1}_n) \cap \mathbf{SU}(n) \cong \mathbb{I}(n), \\
\text{e.g.,} \quad & \mathbf{w}_2 = \tau^3, \quad \mathbf{w}_3 = \sqrt{3}\lambda^8 = \begin{pmatrix} \mathbf{1}_2 & 0 \\ 0 & -2 \end{pmatrix},
\end{aligned}$$

with the scalar phase group $\mathbf{U}(\mathbf{1}_n) = \mathbf{U}(1)\mathbf{1}_n$.

6.16.2 A Complete Cartan Torus for $\mathbf{U}(n)$

Hypercharge and isospin symmetry with central correlation (chapter "Rational Quantum Numbers"), called hyperisospin

$$\frac{\mathbf{U}(1) \times \mathbf{SU}(2)}{\mathbb{I}(2)} \cong \mathbf{U}(2),$$

has a Cartan subalgebra in the defining complex 2-dimensional representation

$$\{i\alpha_0 \mathbf{1}_2 + i\alpha_3 \tau^3 \mid \alpha_{0,3} \in [0, 2\pi[\} \cong \mathbb{R}^2.$$

Its exponent has as factors the scalar and the third component phase group, which, however, are not direct factors for a torus:

$$e^{i\alpha_0 \mathbf{1}_2 + i\alpha_3 \tau^3} \in \mathbf{U}(\mathbf{1}_2) \circ \mathbf{U}(1)_3.$$

The parametrization has the following ambiguity for the common center $\mathbf{U}(\mathbf{1}_2) \cap \mathbf{SU}(2) \cong \mathbb{I}(2)$:

$$(\alpha_0, \alpha_3) = (\pi, 0) \cong (0, \pi), \quad e^{i\pi \mathbf{1}_2} = e^{i\pi\tau^3} = -\mathbf{1}_2 \in \mathbb{I}(2) \cong \mathbf{U}(\mathbf{1}_2) \cap \mathbf{SU}(2).$$

A Cartan torus of $\mathbf{U}(2)$ arises with a projector basis containing two orthogonal elements

$$\begin{array}{l} e^{i\alpha_+\frac{\mathbf{1}_2+\tau^3}{2}}e^{i\alpha_-\frac{\mathbf{1}_2-\tau^3}{2}} \in \mathbf{U}(1)_+\times\mathbf{U}(1)_-, \quad \alpha_\pm=\alpha_0\pm\alpha_3,\\ \mathcal{P}_\pm(2)=\frac{\mathbf{1}_2\pm\tau^3}{2},\quad \mathcal{P}_+(2)\mathcal{P}_-(2)=0. \end{array}$$

For the general case

$$\frac{\mathbf{U}(1)\times\mathbf{SU}(n)}{\mathbb{I}(n)}\cong\mathbf{U}(n),\quad \mathbf{U}(\mathbf{1}_n)\cap\mathbf{SU}(n)\cong\mathbb{I}(n),$$

the exponent of a Cartan subalgebra in the defining complex n-dimensional representation

$$\{i\alpha_0\mathbf{1}_n+i\sum_{m=2}^{n}\alpha_{m^2-1}\tau(n)^{m^2-1}\mid \alpha_{0,m}\in[0,2\pi[\}\cong\mathbb{R}^n$$

gives an abelian group for which the scalar phase factor is correlated with the center-generating factor

$$\begin{array}{rll} & \mathbf{U}(\mathbf{1}_n)\circ\mathbf{U}(1)_{n^2-1}, & e^{i\alpha\mathbf{w}_n}\in\mathbf{U}(1)_{n^2-1},\\ \text{e.g., for } \mathbf{U}(3): & \mathbf{U}(\mathbf{1}_3)\circ\mathbf{U}(1)_8, & e^{i\alpha\mathbf{w}_3}\in\mathbf{U}(1)_8. \end{array}$$

A Cartan torus comes with the appropriate projectors $\mathcal{P}_\pm(n)$ and parameters $\alpha_\pm$:

$$\begin{array}{l} \mathbf{U}(\mathbf{1}_{n-1})_+\times\mathbf{U}(1)_-,\quad e^{i\alpha_+\mathcal{P}_+(n)}e^{i\alpha_-\mathcal{P}_-(n)}\in\mathbf{U}(\mathbf{1}_{n-1})_+\times\mathbf{U}(1)_-,\\ \text{with}\left\{\begin{array}{ll} \mathcal{P}_+(n)=\frac{(n-1)\mathbf{1}_n+\mathbf{w}_n}{n}, & \alpha_+=\alpha_0+\frac{\alpha_{n^2-1}}{\sqrt{\binom{n}{2}}},\\ \mathcal{P}_-(n)=\frac{\mathbf{1}_n-\mathbf{w}_n}{n}, & \alpha_-=\alpha_0-(n-1)\frac{\alpha_{n^2-1}}{\sqrt{\binom{n}{2}}}, \end{array}\right.\\ \mathcal{P}_+(n)\mathcal{P}_-(n)=0,\quad (\mathbf{w}_n)^2=(n-1)\mathbf{1}_n-(n-2)\mathbf{w}_n,\\ \mathcal{P}_+(n)=\mathcal{P}_{++}(n)+\mathcal{P}_{+-}(n), \end{array}$$

with the example for $n=3$, relevant for hypercolor $\mathbf{U}(3)$:

$$\begin{array}{rl} \mathcal{P}_+(3) &= \begin{pmatrix}\mathbf{1}_2 & 0\\ 0 & 0\end{pmatrix},\quad \mathcal{P}_-(3)=\begin{pmatrix}\mathbf{0}_2 & 0\\ 0 & 1\end{pmatrix},\\ \mathcal{P}_{+\pm}(3) &= \begin{pmatrix}\mathcal{P}_\pm(2) & 0\\ 0 & 0\end{pmatrix}. \end{array}$$

For the groups $\mathbf{U}(n)$ with rank-n Lie algebras there exist complete Cartan tori.

6.16.3 A Complete Cartan Torus for the Hydrogen Atom

For the nonrelativistic hydrogen bound-state vectors (chapter "The Kepler Factor") an exponentiated Cartan subalgebra of $\log[\mathbf{SU}(2)\times\mathbf{SU}(2)]$ with basis $\{i\vec{\sigma}\otimes\mathbf{1}_2,\mathbf{1}_2\otimes i\vec{\tau}\}$ in the defining quartet representation

$$\begin{array}{l} \text{Cartan algebra } \{i\alpha_3\sigma^3\otimes\mathbf{1}_2+\mathbf{1}_2\otimes i\beta_3\tau^3\}\cong\mathbb{R}^2,\\ e^{i\alpha_3\sigma^3}\otimes e^{i\beta_3\tau^3}=\begin{pmatrix} e^{i(\alpha_3+\beta_3)} & 0 & 0 & 0\\ 0 & e^{i(\alpha_3-\beta_3)} & 0 & 0\\ 0 & 0 & e^{-i(\alpha_3-\beta_3)} & 0\\ 0 & 0 & 0 & e^{-i(\alpha_3+\beta_3)} \end{pmatrix}\in\mathbf{U}(1)_3\circ\mathbf{U}(1)_3,\\ \text{parameters: } \{\alpha_3+\beta_3,\ \alpha_3-\beta_3\}, \end{array}$$

leads to a complete Cartan torus via a basis of orthogonal generators $\mathcal{L}_\pm$ for coordinates $\gamma_\pm$:

$$\begin{array}{l} e^{i\alpha_3\sigma^3}\otimes e^{i\beta_3\tau^3} = e^{i\gamma_+\mathcal{L}^3_+}e^{i\gamma_-\mathcal{L}^3_-} \in \mathbf{U}(1)_+\times\mathbf{U}(1)_-,\\ \mathcal{L}^3_\pm = \frac{\sigma^3\otimes\mathbf{1}_2\pm\mathbf{1}_2\otimes\tau^3}{2},\quad \mathcal{L}^3_+\mathcal{L}^3_- = 0,\quad \gamma_\pm = \alpha_3\pm\beta_3, \end{array}$$

$\mathcal{L}^3_+ = \mathcal{L}^3$ is the third component of the angular momenta $\log\mathbf{SO}(3)$.

In the general case, two special groups are nontrivially centrally correlatable (chapter "Rational Quantum Numbers") for dimensions with a common nontrivial factor (not relatively prime)

$$\frac{\mathbf{SU}(n)\times\mathbf{SU}(m)}{\mathbb{I}(k)},\quad \mathbb{I}(k)\subset\mathbb{I}(n)\cap\mathbb{I}(m),\quad n,m,k\geq 2.$$

The exponent of a Cartan Lie subalgebra is centrally correlated by the $\mathbf{U}(1)$'s generated by $\mathbf{w}_{n,m}$:

$$\begin{array}{l} \mathbf{U}(1)_{n^2-1}\circ\mathbf{U}(1)_{m^2-1}\text{-Lie algebra: } \{i\alpha\mathbf{w}_n\otimes\mathbf{1}_m + \mathbf{1}_n\otimes i\beta\mathbf{w}_m\}\cong\mathbb{R}^2,\\ e^{i\alpha\mathbf{w}_n}\otimes e^{i\beta\mathbf{w}_m}\in\mathbf{U}(1)_{n^2-1}\circ\mathbf{U}(1)_{m^2-1}. \end{array}$$

The four parameters

$$\begin{array}{l} e^{i\alpha\mathbf{w}_n}\otimes e^{i\beta\mathbf{w}_m}\cong\begin{pmatrix} e^{i[\alpha+\beta]} & 0 & 0 & 0\\ 0 & e^{i[\alpha-(m-1)\beta]} & 0 & 0\\ 0 & 0 & e^{-i[(n-1)\alpha-\beta]} & 0\\ 0 & 0 & 0 & e^{-i[(n-1)\alpha+(m-1)\beta]}\end{pmatrix},\\ \text{parameters: } \{\alpha+\beta,\ \alpha-(m-1)\beta,\ (n-1)\alpha-\beta,\ (n-1)\alpha+(m-1)\beta\}, \end{array}$$

are reducible to two parameters only for $n = m = 2$, e.g., for the hydrogen symmetry. In this case, an orthogonal Cartan subalgebra basis leads to a complete Cartan torus.

6.16.4 No Complete Cartan Torus for Hypercharge-Isospin-Color

The internal interaction symmetry $\mathbf{U}(2\times 3) = \frac{\mathbf{U}(1)\times\mathbf{SU}(2)\times\mathbf{SU}(3)}{\mathbb{I}(2)\times\mathbb{I}(3)}$ has as defining complex 6-dimensional representation for its Lie algebra with rank 4,

$$\begin{array}{l} \log[\mathbf{U}(1)\times\mathbf{SU}(2)\times\mathbf{SU}(3)] = \{i\alpha_0\mathbf{1}_2\otimes\mathbf{1}_3 + i\vec{\alpha}\vec{\tau}\otimes\mathbf{1}_3 + \mathbf{1}_2\otimes i\vec{\beta}\vec{\lambda}\}\cong\mathbb{R}^{12},\\ \text{Cartan subalgebra: } \{i\alpha_0\mathbf{1}_2\otimes\mathbf{1}_3 + i\alpha_3\tau^3\otimes\mathbf{1}_3 + \mathbf{1}_2\otimes i(\beta_3\lambda^3+\beta_8\lambda^8)\}\cong\mathbb{R}^4, \end{array}$$

with three Pauli matrices $\vec{\tau}$ (isospin) and eight Gell-Mann matrices $\vec{\lambda}$ (color) as used for the left-handed quark isodoublet field in the standard model.

The exponentiated diagonal Cartan algebra has three correlated factors generated by $\mathbf{w}_2 = \tau^3$ and $\mathbf{w}_3 = \sqrt{3}\lambda^8$:

$$e^{i\alpha_0\mathbf{1}_2\otimes\mathbf{1}_3 + i\alpha_3\tau^3\otimes\mathbf{1}_3 + \mathbf{1}_2\otimes i\beta_8\lambda^8}\in\mathbf{U}(\mathbf{1}_6)\circ\mathbf{U}(1)_3\circ\mathbf{U}(1)_8.$$

The relevant parameter combinations in the four phases that arise,

$$\text{parameters: } \{(\alpha_0 \pm \alpha_3) + \tfrac{\beta_8}{\sqrt{3}}, \quad (\alpha_0 \pm \alpha_3) - \tfrac{2\beta_8}{\sqrt{3}}\},$$

cannot be disentangled with an orthogonal basis for a representation of the direct product $\mathbf{U}(1) \times \mathbf{U}(1) \times \mathbf{U}(1)$.

There exists a complete Cartan torus $\mathbf{U}(1)_+ \times \mathbf{U}(1)_-$ parametrized by $\{\alpha_0 \pm \alpha_3\}$ for hyperisospin $\mathbf{U}(2)$, there exists also a complete Cartan torus $\mathbf{U}(1)_{++} \times \mathbf{U}(1)_{+-} \times \mathbf{U}(1)_-$ parametrized by $\{\alpha_0 + \frac{\beta_8}{\sqrt{3}} \pm \beta_3, \alpha_0 - \frac{2\beta_8}{\sqrt{3}}\}$ for hypercolor $\mathbf{U}(3)$; however, there does not exist a complete Cartan torus $\mathbf{U}(1)^4$ for faithful representations of the internal $\mathbf{U}(2 \times 3)$-interaction symmetry. The hypercharge group $\mathbf{U}(\mathbf{1}_6)$ has to go either as $\mathbf{U}(\mathbf{1}_2)$ with isospin $\mathbf{U}(1)_3$ or as $\mathbf{U}(\mathbf{1}_3)$ with color $\mathbf{U}(1)_8$.

6.16.5 Eigenvector Bases for Correlated Groups

A semisimple Lie algebra, and also $\log \mathbf{U}(n)$, allows, for any finite-dimensional representation vector space, a basis of eigenvectors for a Cartan subalgebra. Eigenvectors of a Cartan subalgebra do not have to remain eigenvectors for the exponentiated Cartan algebra.

Since a correlation of two Lie groups $G_1 \times G_2$ via a discrete center C does not change the Lie algebra

$$\log \tfrac{G_1 \times G_2}{C} = \log[G_1 \times G_2] = \log G_1 \oplus \log G_2,$$

there can arise the case in which there exists an eigenvector basis for the Lie algebra representation space that is not an eigenvector basis for the correlated group. This is the case for compact groups without a complete Cartan torus, especially for the internal interaction symmetry group $\mathbf{U}(2 \times 3)$.

It is impossible to give an eigenvector basis for the internal group $\mathbf{U}(2 \times 3)$ in faithful representations, e.g., for the isodoublet color triplet representation $[\frac{1}{6}||1;1,0]$. It is possible to give either eigenvector bases for the internal quotient groups $\mathbf{U}(2 \times 3)/\mathbf{SU}(3) \cong \mathbf{U}(2)$ (hyperisospin) or for $\mathbf{U}(2 \times 3)/\mathbf{SU}(2) \cong \mathbf{U}(3)$ (hypercolor), e.g., for the representations either with the left-handed isodoublet color singlet lepton or with the right-handed isosinglet color triplet quark. A quark confinement can be interpreted as the decision, with respect to a particle classification, for the complete hyperisospin Cartan torus $\mathbf{U}(1)_+ \times \mathbf{U}(1)_- \subset \mathbf{U}(2)$ and against the complete hypercolor Cartan torus $\mathbf{U}(1)_{++} \times \mathbf{U}(1)_{+-} \times \mathbf{U}(1)_- \subset \mathbf{U}(3)$.

Bibliography

[1] C. Becchi, A. Rouet, R. Stora, *Ann. of Phys.* 98 (1976), 287.

[2] J. Hucks, *Phys. Rev.* D 43 (1991), 2709.

[3] J.D. Jackson, *Classical Electrodynamics* (1975), Wiley, New York etc.

[4] T. Kugo, I. Ojima, *Progr. of Theor. Phys.* 60 (1978), 1869.

[5] A. Sommerfeld, *Elektrodynamik* (1964), Akad. Verlagsges. Leipzig.

[6] S. Weinberg, *Phys. Rev. Lett.* 18 (1967), 507.

[7] H. Weyl, *Raum, Zeit, Materie* (1923), Wissenschaftliche Buchgesellschaft, Darmstadt.

7

HARMONIC ANALYSIS

In addition to physical properties that are characterized by a *rational number*, such as an integer electromagnetic charge number or a (half)integer spin (chapter "Rational Quantum Numbers"), particles have properties that seem[1] to be taken from a *continuous spectrum*, especially their masses and the gauge coupling constants, e.g., the fine structure constant normalizing the electromagnetic $\mathbf{U}(1)$-Lie algebra in the spacetime-formulated interaction.

All quantum numbers (invariants and eigenvalues) for compact Lie groups, including finite groups, are rational and, as well as the finite-dimensionality of their irreducible representation spaces, ultimately related to integer winding numbers as powers $e^{i\alpha} \longmapsto (e^{i\alpha})^Z$ of the circle group (torus) $\mathbf{U}(1) = \exp i\mathbb{R}$ (chapter "Rational Quantum Numbers"). Lie operations for continuous quantum numbers have to come from a *noncompact Lie group*, as familiar from the eigenvalues for the representations of the causal group $\mathbf{D}(1) = \exp \mathbb{R} \ni e^t \longmapsto (e^t)^{im} \in \mathbf{U}(1)$, e.g., the energies $m \in \mathbb{R}$ for scattering waves. The eigenvalues for irreducible representations are characterized in the following table displaying the twofold dichotomy compact-noncompact and abelian-nonabelian:

	compact	noncompact
abelian	$\mathbf{U}(1) \longrightarrow \mathbf{U}(1)$ $e^{i\alpha} \longmapsto e^{Zi\alpha}$ $Z \in \mathbb{Z}$	$\left\{ \begin{array}{l} \mathbf{D}(1) \longrightarrow \mathbf{U}(1) \\ \mathbf{D}(1) \longrightarrow \mathbf{SU}(1,1) \end{array} \right.$ $e^{\beta} \longmapsto e^{(im+\gamma)\beta}$ $im + \gamma \in i\mathbb{R} + \mathbb{R}$
nonabelian $n \geq 2$	$\mathbf{SU}(2) \longrightarrow \mathbf{SU}(1+2J)$ $u \longmapsto \bigvee^{2J} u,\ 2J = \|Z\|$ $Z \in \mathbb{Z}$	$\left\{ \begin{array}{l} \mathbf{SL}(\mathbb{C}^2) \longrightarrow \text{compact} \\ \mathbf{SL}(\mathbb{C}^2) \longrightarrow \text{noncompact} \end{array} \right.$ $\mathbb{Z} \times [i\mathbb{R} + \mathbb{R}]$

irreducible group representations with | example weights |

Throughout this chapter complex vector spaces $V \in \underline{\mathbf{vec}}_{\mathbb{C}}$ for representations of locally compact groups, especially of real finite-dimensional Lie groups $G \in \underline{\mathbf{lgrp}}_{\mathbb{R}}$ with positive and G-invariant *Haar measure*, an indispensible important tool for continuous groups, are considered. The complex numbers are

[1]Since experimental numbers come with errors, one can never be sure whether they are from a rational or a continuous spectrum. Numerologists develop great skills for integer graduations of experimental numbers by a few units, e.g., for the fine structure constant $\alpha \sim 2^{-2}\pi^{-3}$. With a few exceptions, e.g., Balmer's formula, numerologists find dead-end quantitative coincidences without qualitative insights.

used with the canonical conjugation, i.e., as the doubled reals. Complex represented real Lie group operations have to come as unitary automorphisms, i.e., with definite or indefinite conjugations. Definite unitary representations are also called *Hilbert representations*. Definite and indefinite unitary groups have a compact and noncompact parameter space respectively. Therefore, spaces with faithful Hilbert representations of noncompact groups have to be infinite-dimensional.

rep G denotes the equivalence classes of G-representations with respect to the intertwining isomorphisms, the classes of the irreducible (Hilbert) representations are called the *(definite) group dual* (also dual group space):

$$\mathbf{irrep}\, G = \check{G}, \quad \mathbf{irrep}_+ G = \check{G}_+.$$

The irreducible representations are characterized by the invariants of the group and its Lie algebra.

To give a first survey: Finite groups, e.g., cyclic or permutation groups, with the discrete topology are special cases of compact groups, e.g., unitary groups, which are special cases of locally compact groups, e.g., complex linear groups

group G:	finite	$\subset$	compact	$\subset$	locally compact
e.g., abelian	$\mathbb{I}(n)$	$\subset$	$\mathbf{U}(1)$	$\subset$	$\mathbf{GL}(\mathbb{C})$
special	$\mathbf{G}(n)_+$	$\subset$	$\mathbf{SU}(n)$	$\subset$	$\mathbf{SL}(\mathbb{C}^n)$
general	$\mathbf{G}(n)$	$\subset$	$\mathbf{U}(n)$	$\subset$	$\mathbf{GL}(\mathbb{C}^n)$
group dual $\check{G}_+$:	finite	$\subset$	countable	$\subset$	continuous

Harmonic analysis is the classification of the irreducible representations of a group and the decomposition of the group mappings, e.g., physical fields as spacetime mappings, taken as a "huge" representation vector space, into irreducible group representation spaces. Especially for noncompact groups, it connects intimately algebraic with topological and measure structures. It will be based on the rather straightforward stuctures for finite groups. Finite-dimensional Hilbert representations build up the representation structure of finite and compact groups. This is briefly recapitulated in the first part of this chapter. In contrast to the completely understood and explicitly known representations of compact and abelian groups with a general theory, the analogous situation is much more difficult and complicated for noncompact nonabelian groups with their individual pecularities. In the following, only an orientation - sometimes very sloppy - is given without doing justice to the many topological and measure-related structures, which should be looked at in more detail in the mathematical literature.

The harmonic analysis of functions on a locally compact group involves, as dual partner for a Haar measure $d^G k$ of the group, a positive and G-invariant *Plancherel measure* of the group dual with the characterizing invariants

$$\text{Haar measure } d^G k \leftrightarrow d^{\check{G}_+} D \text{ Plancherel measure.}$$

This is exemplified for the rotation group $\mathbf{SO}(3)$ with Euler angle parametrized normalized Haar measure $\int_{\mathbf{SO}(3)} \frac{d\chi d\varphi d\cos\theta}{(4\pi)^2}$ by the associated Plancherel measure

$\sum_{L=0}^{\infty}(1+2L)$ for the irreducible representations with the rotation-invariant L (angular momentum). Another basically important example is the Plancherel momentum measure $\int_{\check{\mathbb{R}}^n} d^n \frac{p}{2\pi}$ of the irreducible nonfaithful Hilbert representations $x \longmapsto e^{ipx}$ of the noncompact abelian translations with Haar measure $\int_{\mathbb{R}^n} d^n x$.

Also, the functions on group classes, i.e., on homogeneous spaces G/H with a closed subgroup, can be harmonically analyzed with respect to the G-representations involved. Of physical importance are

$$\begin{array}{lll} \text{spheres:} & \Omega^s & \cong \mathbf{SO}(1+s)/\mathbf{SO}(s), \\ \text{hyperboloids:} & \mathcal{Y}^s & \cong \mathbf{SO}_0(1,s)/\mathbf{SO}(s), \\ \text{affine ``planes'':} & \mathbb{R}^d & \cong \mathbf{SO}_0(t,s) \ \vec{\times}\ \mathbb{R}^d/\mathbf{SO}_0(t,s). \end{array}$$

For physics, a "prejudice" with respect to finiteness and simplicity tries to avoid, on a basic level, infinite-dimensional representations and all the related mathematical complications. In this case, one has to start with a compact, possibly even finite, operation structure. A conceptual justification has to be given for such a starting point and a way from the basic compact operations to the apparently successful noncompact operations for the experimental description of spacetime. In this way one may use a continuum limit or a group contraction procedure or, what will be done here, a relativization of compact real Lie group operations U in general complex linear ones G, formalized with the symmetric spaces G/U, called *U-relativity in G*. For a spacetime parametrization of physical phenomena, one has to face the mathematically demanding infinite-dimensional Hilbert representations of noncompact nonabelian groups. Prominent and important examples, discussed below, are the faithful quantum-mechanical representations of the Heisenberg group $\mathbf{H}(s)$ with s position-momentum pairs, of the Euclidean group $\mathbf{SO}(3) \ \vec{\times}\ \mathbb{R}^3$ for nonrelativistic scattering, and of the Lorentz group $\mathbf{SL}(\mathbb{C}^2)$, its extension $\mathbf{GL}(\mathbb{C}^2)$ and its unitary classes $\mathbf{D}(2) \cong \mathbf{GL}(\mathbb{C}^2)/\mathbf{U}(2)$ as model of nonlinear spacetime. Vectors acted on by irreducible Hilbert representations of the Poincaré group $\mathbf{SL}(\mathbb{C}^2) \ \vec{\times}\ \mathbb{R}^4$ with causal mass squares $m^2 \geq 0$ as continuous invariant for the noncompact translations and rational quantum numbers J (spin) or Z (polarization) as invariants for compact rotations are, according to *Wigner's classification*, what one calls elementary particles. The particle Hilbert representations can be induced from familiar finite-dimensional ones of Poincaré subgroups $\mathbf{U} \times \mathbb{R}^4$. Here $\mathbf{U} \in \{\mathbf{SU}(2), \mathbf{SO}(2)\}$ denotes a position rotation group as the stability group of the decomposition of the spacetime translations $\mathbb{R}^4$ in a rest system (with one fixed position direction for massless particles). The $\mathbf{U}$-representations determine the spin (polarization). The irreducible Hilbert representations of the Poincaré group connect with each other, Lorentz compatibly, the (eigen)time $\mathbb{R}$-representations $t \longmapsto e^{imt} \in \mathbf{U}(1)$ and the Hilbert representations of the scattering group $\mathbf{SO}(3) \ \vec{\times}\ \mathbb{R}^3$, e.g., $\vec{x} \longmapsto \frac{\sin|\vec{q}|r}{|\vec{q}|r}$. Their classification gives the types of all possible representations with their invariants. However, it does not give any information about which of the possible representations are used for elementary particles, i.e., which masses really occur with which spin and charge numbers.

7.1 Representations on Group Functions

7.1.1 Group Function Vector Spaces

All group representations involve complex group functions (also equivalence classes of functions or generalized functions). They inherit from the complex numbers as value space the vector space structure as well as a conjugation that combines the canonical number conjugation with the group inversion

$$\begin{aligned}\mathbb{C}(G) = \mathbb{C}^G = \ & \{f : G \longrightarrow \mathbb{C}\} \in \star\underline{\mathbf{vec}}_{\mathbb{C}}, \\ & f \leftrightarrow \hat{f}, \ \ \hat{f}(k) = \overline{f(k^{-1})}.\end{aligned}$$

The canonical number conjugation is definite with $\mathbf{U}(1)$-invariant scalar product $\langle\alpha|\alpha\rangle = |\alpha|^2 \geq 0$. The additional G-inherited product structure (group algebra with convolution) will be discussed below. Functions on a group transport group-specific properties into the numbers as value space, e.g., cyclicity into function periodicity $\mathbf{U}(1) \ni e^{i\alpha} \longmapsto f(\alpha) = f(\alpha + 2\pi)$. For the noncompact group $\mathbf{D}(1)$ any function $e^{\alpha} \longmapsto f(\alpha)$ belongs to $\mathbb{C}^{\mathbf{D}(1)} \cong \mathbb{C}^{\mathbb{R}}$.

The finite support functions constitute a direct vector space sum (chapter "Time Representations") with the *group elements* $\{k \mid k \in G\}$ *as canonical basis*:

$$\mathbb{C}^G \supseteq \mathbb{C}^{(G)} = \bigoplus_{k\in G} k\mathbb{C} \ \ni f = \bigoplus_{k\in G} kf(k)\,(\textit{canonical expansion})\ , \quad k\mathbb{C} \cong \mathbb{C}.$$

For a finite group $\mathbb{C}^{(G)} = \mathbb{C}^G$. For infinite cardinality, $\mathbb{C}^{(G)}$ is a proper subspace: e.g., the representation $\mathbf{U}(1) \ni e^{i\alpha} \longmapsto e^{2i\alpha}$ is not an element of $\mathbb{C}^{(\mathbf{U}(1))}$.

For any locally compact group the direct sum is replaced by a *direct integral* (chapter "The Kepler Factor") with Haar measure (counting measure for a finite group with discrete topology)

$$\text{notation: } \int_G d^G k = \begin{cases} \int dk, \\ \sum\limits_{k\in G} & \text{for finite groups.}\end{cases}$$

The group functions are expanded as direct integral with the group elements $\{k \mid k \in G\}$ as *Haar-measure-related distributive basis* and the function values as coefficients,

$$\begin{aligned}\mathbb{C}(G) = \prod_{k\in G} k\mathbb{C} \ &= {}^{\oplus}\!\!\int dk\ k\mathbb{C} \ni f = {}^{\oplus}\!\!\int dk\ kf(k)\,(\textit{canonical expansion}), \\ \dim_{\mathbb{C}} \mathbb{C}^G \ &= \ \operatorname{card} G.\end{aligned}$$

There exist left- and right-invariant Haar measures, $d_L k = d_L gk$ and $d_R k = d_R kg^{-1}$, both unique up to a scalar factor. The involutive *measure conjugation* defines antimeasures via inverse conjugation:

$$\mu \leftrightarrow \hat{\mu}, \quad \int dk\ \hat{\mu}(k) = \int dk\ \Delta(k^{-1})\overline{\mu(k^{-1})}.$$

It involves the modular G-representation $G \ni k \longmapsto \Delta(k) \in \mathbf{D}(1)$ in the dilation group, relating to each other left and right Haar measures:

$$d_L k \leftrightarrow \Delta(k^{-1}) d_R k^{-1}.$$

Only with a normal subgroup N and classes $G/N \cong \mathbf{D}(1)$ can a group be not unimodular. For a connected Lie group the modular function is the determinant of the adjoint representation $\mathrm{Ad} : G \longrightarrow \mathbf{GL}(L)$ on its Lie algebra,

$$\text{for } k = e^l : \quad \Delta(k) = \det \mathrm{Ad}\, k = \det e^{\mathrm{ad}\, l} = e^{\mathrm{tr}\ \mathrm{ad}\, l}.$$

Compact and connected semisimple Lie groups are unimodular, $\Delta(k) = 1$ for all $k \in G$.

If properties are taken almost everywhere with respect to Haar measure, there are these important spaces, especially for infinite, but locally compact groups: The *Radon measure algebra* $\mathcal{M}(G) = \mathcal{C}_c(G)'$ with the measures as Haar-measure-based generalized functions, the *Lebesgue spaces* $L^p(G)$, $1 \le p \le \infty$, where the *Lebesgue function algebra* $L^1(G)$ with the absolute integrable, and the *Lebesgue function algebra dual* $L^\infty(G)$ with the essentially bounded and the self-dual *Hilbert space* $L^2(G)$ with the square integrable function classes are of basic importance (more below).

7.1.2 Regular Representations

The *right and left regular representations* of a group G and the *two-sided regular representation of the squared group* $G \times G$ (group and "isogroup") are given by actions on the vector space $\mathbb{C}(G)$ with the group functions, often restricted to $L^2(G)$. They are defined by the left and right inverse multiplication, $L_h(k) = hk$ and $R_g(k) = kg^{-1}$.

$$\begin{array}{ccc} G & \stackrel{L_h, R_g}{\longrightarrow} & G \\ f \downarrow & & \downarrow {}_hf,\ f_g, \\ \mathbb{C} & \underset{\mathrm{id}_{\mathbb{C}}}{\longrightarrow} & \mathbb{C} \end{array} \qquad \begin{array}{rl} {}_hf(k) & = f(h^{-1}k), \\ f_g(k) & = f(kg), \\ \text{for all} & h, g, k \in G, \end{array}$$

$$G \times G \longrightarrow \mathbf{GL}(\mathbb{C}^G), \quad (h, g) \longmapsto \mathcal{R}_g \circ \mathcal{L}_h \text{ with } \mathcal{R}_g \circ \mathcal{L}_h(f) = {}_hf_g.$$
$${}_hf_g = {}^{\oplus}\!\!\int dk\ kf(h^{-1}kg) = {}^{\oplus}\!\!\int dk\ hkg^{-1}f(k).$$

With the regular representations the whole graph of f is transformed, e.g., for $\mathbf{U}(1)$ or $\mathbf{D}(1)$ the translation $f_\gamma(\alpha) = f(\alpha + \gamma)$.

Left- and right-invariant Haar measures have to be taken for the left and right actions. The related modifications are unnecessary for unimodular groups.

The isomorphism between left and right, e.g., left and right regular representation, left and right cosets, and the related squared structure with the inner automorphisms $k \longmapsto gkg^{-1}$ for the diagonal group $G \cong \Delta(G) \subset G \times G \ni (g, g)$, will show up at many points in the following.

7.1.3 Representation Matrix Elements (Coefficients)

The matrix elements of any representation (representation coefficients) of a locally compact group $D : G \longrightarrow \mathbf{GL}(V)$ on a complex vector space, finite- or infinite-dimensional, are group functions

$$\begin{aligned} v \in V, \omega \in V^T: \quad & D^v_\omega : G \longrightarrow \mathbb{C}, \quad D^v_\omega = {}^{\oplus}\!\!\int dk\ kD(k)^v_\omega \in \mathbb{C}(G), \\ D(k)^v_\omega \quad & = \langle \omega, D(k).v\rangle = \langle \check{D}(k^{-1}).\omega, v\rangle \\ & = \langle \omega, k \bullet v\rangle = \langle k^{-1} \bullet \omega, v\rangle \\ \text{“two point function”} \quad & = \langle k_2 \bullet \omega, k_1 \bullet v\rangle \text{ for all } k_{1,2} \text{ with } k = k_2^{-1}k_1. \end{aligned}$$

They are complex group orbits and can be harmonically analyzed with respect to the irreducible Hilbert representations. Such a Hilbert representation analysis may be even possible for functions that arise as matrix elements of finite-dimensional indefinite unitary representations of a noncompact locally compact group.

An example of nonabelian group functions is given by the finite-dimensional irreducible (iso)spin $\mathbf{SU}(2)$-representations with Euler angles $u \cong (\chi, \varphi, \theta)$ (chapter “Spin, Rotations, and Position”),

$$\mathbf{SU}(2) \ni u \quad = \begin{pmatrix} e^{i\frac{\chi+\varphi}{2}} \cos\frac{\theta}{2} & ie^{-i\frac{\chi-\varphi}{2}} \sin\frac{\theta}{2} \\ ie^{i\frac{\chi-\varphi}{2}} \sin\frac{\theta}{2} & e^{-i\frac{\chi+\varphi}{2}} \cos\frac{\theta}{2} \end{pmatrix} \longmapsto 2J(u) = \bigvee^{2J} u \in \mathbf{SU}(1+2J), \quad 2J = 0, 1, \dots,$$

$$\text{e.g., } 2(u)^a_b \quad = \left(\begin{array}{c|c|c} e^{i(\chi+\varphi)} \cos^2\frac{\theta}{2} & ie^{i\varphi} \frac{\sin\theta}{\sqrt{2}} & -e^{-i(\chi-\varphi)} \sin^2\frac{\theta}{2} \\ \hline ie^{i\chi} \frac{\sin\theta}{\sqrt{2}} & \cos\theta & ie^{-i\chi} \frac{\sin\theta}{\sqrt{2}} \\ \hline -e^{i(\chi-\varphi)} \sin^2\frac{\theta}{2} & ie^{-i\varphi} \frac{\sin\theta}{\sqrt{2}} & e^{-i(\chi+\varphi)} \cos^2\frac{\theta}{2} \end{array}\right) \in \mathbf{SU}(3),$$

$$\begin{aligned} 2J(u)^a_b \quad & = i^{b-a} e^{i(a\chi+b\varphi)} 2J(z)^a_b, \quad z = \cos\theta, \quad a, b \in \{-J, \dots, J\}, \\ 2J(z)^a_b \quad & = \frac{(-1)^{J-b}}{2^J} \sqrt{\frac{(J+b)!}{(J-a)!(J+a)!(J-b)!}} \frac{(1-z)^{\frac{a-b}{2}}}{(1+z)^{\frac{a+b}{2}}} \left(\frac{d}{dz}\right)^{J-b} (1-z)^{J-a}(1+z)^{J+a}. \end{aligned}$$

All representation coefficients are in $L^2(\mathbf{SU}(2))$. There are coefficients that do not depend on all group parameters, e.g., the middle column and line in $2(u)$, which depend only on 2-sphere parameters (φ, θ) and (χ, θ), i.e., they contain functions $L^2(\Omega^2)$ on the axial rotation classes $\Omega^2 \cong \mathbf{SU}(2)/\mathbf{SO}(2)$. The central, only θ-dependents, elements are functions $L^2(\Omega^1)$ on the left-right classes $\mathbf{SO}(2) \setminus \mathbf{SO}(3)/\mathbf{SO}(2) \cong \Omega^1$ with the Legendre polynomials $2L(z)^0_0 = \mathrm{P}^L(z)$ for $L = 0, 1, \dots$.

All matrix element functions for a representation D constitute the *matrix function space* $\mathbb{C}_D(G)$, a vector subspace of all functions, stable under both right and left group actions:

$$G \times \mathbb{C}_D(G) \times G \longrightarrow \mathbb{C}_D(G), \quad \mathbb{C}_D(G) \in \underline{\mathbf{vec}}_{G\times G},$$

$$\begin{array}{ccc} G & \stackrel{L_h \times R_g}{\longrightarrow} & G \\ D^v_\omega \downarrow & & \downarrow D^{h\bullet v}_{g\bullet\omega}, \\ \mathbb{C} & \underset{\mathrm{id}_{\mathbb{C}}}{\longrightarrow} & \mathbb{C} \end{array} \quad D(g^{-1}kh)^v_\omega = D(k)^{h\bullet v}_{g\bullet\omega}.$$

Therefore, dual representations are embedded via the "column-functions" and the "line-functions" into the right and left regular representations respectively

$$\begin{aligned} \mathbb{C}_D(G) &= \{D^v_\omega \mid v \in V,\ \omega \in V^T\} \\ &\cong V \otimes V^T \in \underline{\mathbf{vec}}_{G\times G}, \qquad D \otimes \check{D} \hookrightarrow R \times L. \end{aligned}$$

The matrix function spaces coincide for equivalent representations

$$f: V' \longrightarrow V,\ D' = f^{-1} \circ D \circ f \Rightarrow \mathbb{C}_{D'}(G) \ni D'^u_\theta = D^{f(u)}_{\check{f}(\theta)} \in \mathbb{C}_D(G).$$

Representations have as conjugation the antidual involution

$$D \leftrightarrow \hat{D},\quad D(k)^v_\omega \leftrightarrow \hat{D}(k)^v_\omega = \overline{D(k^{-1})^v_\omega},$$

e.g., the left-right Weyl representations $\mathbf{SL}(\mathbb{C}^2) \ni s = e^{i\vec{\alpha}+\vec{\beta}} \leftrightarrow \hat{s} = e^{i\vec{\alpha}-\vec{\beta}}$. A representation is definite unitary for $D = \hat{D}$.

7.2 Harmonic Analysis of Finite Groups

A finite group with discrete topology is compact. It is considered with the invariant counting measure (cardinality). The group functions contain the group

$$\text{finite } N \in \underline{\mathbf{grp}}\text{: } N \subset \mathbb{C}^N = \bigoplus_{k\in N} |k\}\mathbb{C}.$$

A bra-ket notation with $|\ \}$ ("braced" ket) is used in this section for the group functions with the left-right regular $N \times N$-action. $\mathbb{C}^N$ inherits the group product as convolution (direct sum notation for vectors and usual sum for numbers):

$$\begin{aligned} &(\mathbb{C}^N, *) \in \star\underline{\mathbf{aag}}_{\mathbb{C}}, \quad \text{product: } |f_1\} * |f_2\} = \bigoplus_k |k\}(f_1 * f_2)(k), \\ &\text{where } (f_1 * f_2)(k) = \sum_{(k_1,k_2)\in N\times N} f_1(k_1) f_2(k_2) \delta^k_{k_1k_2} \text{ with } \delta^k_g = \begin{cases} 1 & \text{for } k = g, \\ 0 & \text{for } k \neq g. \end{cases} \end{aligned}$$

The group algebra $\mathbb{C}^N$, unital with conjugation, coincides with all Lebesgue function spaces and with the group C*-algebra

$$\mathbb{C}^N = L^p(N) = \mathbf{C}^*(N),\quad 1 \le p.$$

$\mathbb{C}^N$ inherits the scalar product of the numbers, the canonical basis is a Hilbert basis

$$\{f|g\} = \sum_k \overline{f(k)} g(k),\quad \{k|k'\} = \delta^{kk'}.$$

The group N is the disjoint union of its conjugacy classes (inner automorphism orbits) with fixgroups N^{Int}_k:

$$N = \biguplus_{Z=1}^n \operatorname{Int} N(k_Z),\quad \operatorname{Int} N(k) = \{gkg^{-1} \mid g \in G\} \cong N/N^{\text{Int}}_k.$$

The complex span of the conjugacy classes with $g \operatorname{Int} N(k) g^{-1} = \operatorname{Int} N(k)$ are full matrix subalgebras in the decomposition of the group algebra $\mathbb{C}^N$ (Maschke-Burnside-Wedderburn theorem, chapter "Time Representations"): They are the minimal two-sided ideals and characterize the irreducible representation spaces of the group-"isogroup" $N \times N$ or of the group algebra $\mathbb{C}^N$. For example, the permutation group $\mathbf{G}(4)$ has five irreducible representations (Young ideals),

$$\begin{array}{ll} \mathbb{C}^N & = \bigoplus_{Z=1}^{n} \mathbb{C}(d_Z \times d_Z) \overset{\text{e.g.}}{\cong} \left(\begin{array}{c|c|c|c|c} 1\times 1 & 0 & 0 & 0 & 0 \\ \hline 0 & 3\times 3 & 0 & 0 & 0 \\ \hline 0 & 0 & 2\times 2 & 0 & 0 \\ \hline 0 & 0 & 0 & 3\times 3 & 0 \\ \hline 0 & 0 & 0 & 0 & 1\times 1 \end{array} \right) \sim \begin{pmatrix} 1 \\ \vdots \\ 24 \end{pmatrix}, \\ \operatorname{card} N & = \sum_{Z=1}^{n} d_Z^2, \qquad \text{for } \mathbb{C}^{\mathbf{G}(4)}: \quad 1+9+4+9+1 = 4! = 24. \end{array}$$

The group is represented by finite unitary groups,

$$N \longrightarrow \bigoplus_{Z=1}^{n} \exp \mathbb{C}(d_Z \times d_Z) \subseteq \mathbf{U}(\operatorname{card} N), \quad \exp \mathbb{C}(d_Z \times d_Z) \subset \mathbf{U}(d_Z).$$

The columns $L_Z^{\min}$ in a simple ideal are minimal left ideals and characterize the irreducible group representations $Z : N \longrightarrow \mathbf{U}(d_Z)$ for left action:

$$\begin{array}{rl} \mathbb{C}(d_Z \times d_Z) & = V_Z \otimes V_Z^T = L_Z^{\min} \otimes R_Z^{\min}, \quad V_Z \cong \mathbb{C}^{d_Z}, \\ \text{group dual: } \check{N} & = \check{N}_+ \cong \{Z = 1, \dots, n\}. \end{array}$$

The rows $R_Z^{\min}$ are acted on by irreducible right multiplications. The equivalence reflects the equivalent left-right action $\operatorname{Int} g = L_g \times R_g$ in the construction of the conjugacy classes.

There are *two different kinds of Hilbert spaces with three different types of bases*: In addition to the Hilbert bases of the irreducible representation spaces of the group N ("angled" kets $|\ \rangle$ with "angled" scalar product),

$$\text{Hilbert basis of } V_Z \cong \mathbb{C}^{d_Z}: \ \{|a\rangle \bigm| a = 1, \dots, d_Z\}, \ \left\{ \begin{array}{rl} \langle a'|a\rangle & = \delta^{aa'}, \\ \mathrm{id}_{V_Z} & \cong |a\rangle\langle a|, \end{array} \right.$$

there are two kinds of Hilbert bases of the group algebra ("braced" kets $|\ \}$), the canonical basis with the group elements and *harmonic bases (representation bases)* for the orthogonal simple ideals $\mathbb{C}(d_Z \times d_Z)$:

$$\begin{array}{rl} \text{Hilbert bases of } \mathbb{C}^N: & \left\{ \begin{array}{l} \text{canonical: } \{|k\} \bigm| k \in N\}, \ \{k'|k\} = \delta^{kk'}, \\ \text{harmonic: } \{\sqrt{d_Z}|Z_b^a\} \bigm| Z; a, b\}, \end{array} \right. \\ \text{Hilbert basis of } V_Z \otimes V_Z^T: & \{\sqrt{d_Z}|Z_b^a\}\}, \ |Z_b^a\} = Z|b\rangle\langle a|, \ \{Z_b^a| = |a\rangle\langle b|\overline{Z}. \end{array}$$

With the scalar product all direct sums are orthogonal $\oplus = \perp$. *Schur's orthonormality relations for the harmonic bases*, i.e., for the coefficients of the irreducible group representation classes, *sum over the group*

$$\{Z'^{a'}_{b'}|Z_b^a\} = \sum_{k\in N} \overline{Z'(k)^{a'}_{b'}} Z(k)^a_b = \tfrac{1}{d_Z}\delta^{ZZ'}\delta^{aa'}\delta_{bb'} \text{ with } Z(k)^a_b = \langle b|Z(k).a\rangle.$$

This defines the *Schur scalar product on the group dual* ("braced" scalar product)

$$\hat{N} \times \hat{N} \longmapsto \mathbb{C}\mathbf{1}, \quad \{Z'|Z\} = \sum_{k \in N} \overline{Z'(k)} Z(k) = \frac{1}{d_Z} \delta^{ZZ'} \mathbf{1}_{d_Z \times d_Z}.$$

The products are *Plancherel-normalized* with multiplicity factors $\frac{1}{d_Z}$ (representation dimensions) (more below at "Compact Groups").

Bases with group elements and representation bases are related to each other by unitary *Fourier transformation* $\mathbf{F} \in \mathbf{U}(\operatorname{card} N)$: the group elements (canonical basis) have a harmonic expansion, a harmonic basis has a canonical expansion ($\cong$ since the identity is bilinear and the ket-bra notation sesquilinear),

$$\begin{array}{ll} \mathrm{id}_{\mathbb{C}^N} \cong \bigoplus_{k \in N} |k\}\{k| & = \bigoplus_{Z=1}^{n} d_Z |Z_a^b\}\{Z_a^b|, \\ \mathbf{F}: \ \mathbb{C}^N = \bigoplus_{k \in N} |k\}\mathbb{C} & = \bigoplus_{Z=1}^{n} \mathbb{C}(d_Z \times d_Z) \end{array} \Rightarrow \left\{ \begin{array}{ll} |k\} & = \bigoplus_{Z=1}^{n} d_Z \overline{Z(k)_a^b} |Z_a^b\}, \\ |Z_a^b\} & = \bigoplus_{k \in N} Z(k)_a^b |k\}, \\ \{k|Z_a^b\} & = Z(k)_a^b = \langle b|Z(k)a\rangle. \end{array} \right.$$

The decomposition of the identity relates to each other the Haar measure of the group and the Plancherel measure of the group dual, both counting measures: d_Z is the number of irreducible representation spaces $V_Z \cong V_Z^T$ (columns or rows, each with dimension d_Z) in one irreducible $(d_Z \times d_Z)$ matrix subalgebra,

$$\check{N} \longrightarrow \mathbb{N}, \quad Z \longmapsto d_Z.$$

Consequently, the two-sided regular representation and hence any group function has two decompositions - the canonical expansion into group elements and the *harmonic expansion (Fourier analysis)* into irreducible representations, called *harmonic matrices* with the *harmonic (Fourier) coefficients* as entries. The isometry is *Plancherel's theorem*:

$$|f\} = \bigoplus_{k \in N} |k\} f(k) = \bigoplus_{Z=1}^{n} d_Z |\tilde{f}(Z)\} = \bigoplus_{Z=1}^{n} d_Z |Z_b^a\} \tilde{f}(Z)_b^a.$$

$$\begin{array}{rlll} \text{function values:} & f(k) & = \{k|f\} & = \sum_{Z=1}^{n} d_Z Z_b^a(k) \tilde{f}(Z)_b^a, \\ \text{harmonic coefficients:} & \tilde{f}(Z)_b^a & = \{Z_b^a|f\} & = \sum_{k \in N} \overline{Z_b^a(k)} f(k), \\ \text{Plancherel unitarity:} & \{f|f\} & = \sum_{k \in N} \overline{f(k)} f(k) & = \sum_{Z=1}^{n} d_Z \overline{\tilde{f}(Z)_b^a} \tilde{f}(Z)_b^a. \end{array}$$

The tilde $f \leftrightarrow \tilde{f}$ for the Fourier transformation is customary, but not really necessary. There is only one function $|f\}$, with different expansions in the canonical basis $f(k) = \{k|f\}$ and harmonic bases $\tilde{f}(Z) = \{Z|f\}$.

The simplest nontrivial example is the transposition group $\mathbf{G}(2) = \{|e\}, |a\}\}$ with harmonic basis $\{\frac{|e\}+|a\}}{\sqrt{2}}, \frac{|e\}-|a\}}{\sqrt{2}}\}$ used for the symmetric-antisymmetric representation decomposition of functions with two variables:

$$\begin{array}{rl} \mathbb{C}^{\mathbf{G}(2)} \ni |f\} &= f(e)|e\} \oplus f(a)|a\} = |\tilde{f}(+)\} \oplus |\tilde{f}(-)\} \\ \text{with } |\tilde{f}(\pm)\} &= \frac{f(e)\pm f(a)}{\sqrt{2}} \frac{|e\}\pm|a\}}{\sqrt{2}}, \\ \begin{pmatrix} \frac{|e\}+|c\}}{\sqrt{2}} \\ \frac{|e\}-|c\}}{\sqrt{2}} \end{pmatrix} &= \mathbf{F}\begin{pmatrix} |e\} \\ |a\} \end{pmatrix}, \quad \mathbf{F} = \frac{1}{\sqrt{2}}\begin{pmatrix} 1 & 1 \\ 1 & -1 \end{pmatrix} \in \mathbf{U}(2). \end{array}$$

The group product defines the convolution product, which is in the canonical basis the matrix product for the harmonic matrices, with the algebra automorphism:

$$\begin{array}{c} \mathbb{C}^N \leftrightarrow \mathbb{C}^N, \ f \leftrightarrow \tilde{f}, \\ \widetilde{f_1 * f_2} = \tilde{f}_1 \circ \tilde{f}_2 \end{array} \text{ with } \left\{ \begin{array}{rl} \widetilde{f_1 * f_2}(Z) &= \tilde{f}_1(Z) \circ \tilde{f}_2(Z) \\ = \{Z|f_1 * f_2\} &= \{Z|f_1\} \circ \{Z|f_2\} \\ \text{for all } Z &= 1, \cdots, n. \end{array} \right.$$

7.3 Algebras and Vector Spaces for Locally Compact Groups

Two products are important for the (generalized) group functions $\mu = {}^{\oplus}\!\int dg\, g\mu(g)$ of a locally compact group. ¿From the group, they inherit the group multiplication $g_1 g_2 = g \in G$ as *convolution product* (where defined):

$$\begin{array}{rll} \mu_1 * \mu_2 &= [{}^{\oplus}\!\int dg\, g\mu_1(g)] \ * \ [{}^{\oplus}\!\int dg\, g\mu_2(g)] &= {}^{\oplus}\!\int dg\, g(\mu_1 * \mu_2)(g), \\ \mu_1 * \mu_2(g) &= \int dg_1 dg_2\, \mu_1(g_1)\, \delta(g_1 g_2 g^{-1}) \mu_2(g_2) &= \int dg_1\, \mu_1(g_1)\mu_2(g_1^{-1}g). \end{array}$$

The associative convolution is abelian if and only if the group multiplication is abelian, e.g., for energy-momenta $q \in \mathbb{R}^d$ with Lebesgue measure dq,

$$\begin{array}{rl} (\mu_1 * \mu_2)(q) &= \int dq_1\, dq_2 \mu_1(q_1)\, \delta(q_1 + q_2 - q)\, \mu_2(q_2) \\ &= \int dq_1\, \mu_1(q_1)\mu_2(q - q_1) = \int dq_2\, \mu_1(q - q_2)\mu_2(q_2). \end{array}$$

From the complex number multiplication, the group functions inherit the abelian *pointwise product* (where defined), important for product representations

$$\mu_1 \cdot \mu_2(g) = \mu_1(g)\mu_2(g).$$

With respect to the convolution and pointwise product the *Lebesgue Banach spaces* $L^p(G)$, $1 \le p \le \infty$, with classes of Haar-measure-almost-everywhere defined functions, are related to each other as follows

$$1 \le p, r, s \le \infty : \left\{ \begin{array}{lll} L^p(G) * L^r(G) & \subseteq L^s(G) & \text{with } \frac{1}{p} + \frac{1}{r} - \frac{1}{s} = 1, \\ L^p(G) \cdot L^r(G) & \subseteq L^s(G) & \text{with } \frac{1}{p} + \frac{1}{r} - \frac{1}{s} = 0. \end{array} \right.$$

They are left-right modules for the *Lebesgue function algebra* $L^1(G)$, which is a convolution algebra, and for the *essentially bounded group functions* $L^\infty(G)$, which constitute a unital algebra with the pointwise product. The convolutive $L^1(G)$-action is linear and continuous with a norm smaller than $\| f \|_1$. The action on $L^2(G)$ is unitary, i.e., a Hilbert representation.

The Lebesgue spaces are left-right convolution modules even for the *Radon measure algebra* $\mathcal{M}(G)$, a unital convolution Banach algebra,

$$\begin{array}{rl} & \mathcal{M}(G) * \mathcal{M}(G) = \mathcal{M}(G), \\ 1 \le p \le \infty: & \mathcal{M}(G) * L^p(G) * \mathcal{M}(G) = L^p(G). \end{array}$$

The Radon measures, Haar measure with Radon distributions $\omega(g)dg$, embed the group by Dirac measures. δ_e is the convolution unit:

$$\begin{array}{rl} G \ni k \longmapsto \delta_k \in \mathcal{M}(G) \text{ with } \delta_k(g) & = \delta(gk^{-1}), \\ \delta_k * \delta_l & = \delta_{kl}, \quad \delta_e = \delta, \\ \mu * \delta_e & = \delta_e * \mu = \mu. \end{array}$$

The normalizations of Dirac distributions and Haar measure go in parallel,

$$\langle \delta_k, f \rangle = \int \delta_k(g) dg \ f(g) = f(k).$$

In general, the group elements are not $L^1(G)$-elements. However, as a replacement, there exist approximations of group elements. Especially, a *group unit approximation* $\tilde{\delta}_e$ is given by a series of group functions with support shrinking to $e \in G$, e.g., by characteristic functions on compact e-neighborhoods $\{k \longmapsto \frac{\chi_C(k)}{\mu(C)} \mid e \in C \text{ compact}\}$.

$\mathcal{M}(G)$ contains the function algebra $L^1(G)$ as a two-sided ideal. It constitutes the dual of the compactly supported continuous functions $\mathcal{C}_c(G)$, which are dense in all $L^p(G)$, $1 \le p < \infty$. The involutive convolution algebra $\mathcal{C}_c(G)$ is a subspace of the bounded continuous functions $\mathcal{C}_b(G)$, which can be considered as a closed, in general proper, subspace of the essentially bounded functions

$$\begin{array}{lll} L^1(G) & \subseteq \mathcal{M}(G) & \supset G, \\ \mathcal{C}_c(G) & \subseteq \mathcal{C}_b(G) & \subseteq L^\infty(G). \end{array}$$

All the (generalized) function vector spaces and algebras considered have an involution

$$\begin{array}{l} \mu \leftrightarrow \hat{\mu}, \quad \hat{\mu}(g) = \Delta(g^{-1})\overline{\mu(g^{-1})}, \quad \text{for } \mathbb{R}^d: \quad \mu(q) \leftrightarrow \overline{\mu(-q)}, \\ (\mu_1 * \mu_2)\hat{} = \hat{\mu}_2 * \hat{\mu}_1. \end{array}$$

With a group representation $D: G \longrightarrow \mathbf{GL}(V)$ there is the representation of the group algebras in the endomorphism algebra $\mathbf{AL}(V)$. It defines the *harmonic D-components* $D(\mu) = \{D|\mu\}$ of a function or a Radon distribution

$$\begin{array}{l} A(G) = \mathcal{C}_c(G), \ \mathcal{M}(G), \\ D: A(G) \longrightarrow \mathbf{AL}(V), \ \mu \longmapsto D(\mu), \end{array} \left\{ \begin{array}{rl} D(\mu) = \tilde{\mu}(D) = & \int dg \ D(g)\mu(g), \\ D(\delta_e) = \mathrm{id}_V, & D(\delta_g) = D(g), \\ D(\mu_1 * \mu_2) = & D(\mu_1) \circ D(\mu_2) \\ = \widetilde{\mu_1 * \mu_2}(D) = & \tilde{\mu}_1(D) \circ \tilde{\mu}_2(D). \end{array} \right.$$

There is a unique correpondence between Hilbert representations of the group G and conjugation-compatible representations of the Lebesgue function algebra $L^1(G)$ that are nondegenerate, i.e., $L^1(G) \bullet v = 0 \iff v = 0$.

The essentially bounded functions $L^\infty(G)$ constitute the dual space for the Lebesgue function algebra $L^1(G)$:

$$\text{duality:} \quad \begin{cases} L^p(G)' &= L^r(G), \ \frac{1}{p} + \frac{1}{r} = 1, \ 1 < p, r < \infty, \\ L^1(G)' &= L^\infty(G), \\ \mathcal{C}_c(G)' &= \mathcal{M}(G). \end{cases}$$

In the following, the Radon measure convolution algebra $\mathcal{M}(G)$ with the Lebesgue function algebra $L^1(G)$ as ideal, and the essentially bounded function pointwise product algebra $L^\infty(G)$, the dual of $L^1(G)$, play the most important roles. These spaces with two-sided convolutive action of $\mathcal{M}(G)$ and $L^1(G)$ and pointwise action of $L^\infty(G)$ will be used for the group G and its dual $\check{G}$:

$*$	$L^1(G)$	$\mathcal{M}(G)$	$L^\infty(G)$
$L^1(G)$	$L^1(G)$	$L^1(G)$	$L^\infty(G)$
$\mathcal{M}(G)$	$L^1(G)$	$\mathcal{M}(G)$	$L^\infty(G)$
$L^\infty(G)$	$L^\infty(G)$	$L^\infty(G)$	–

convolution product
$\mu_1 * \mu_2(g)$
from group product

$\cdot$	$L^1(G)$	$\mathcal{M}(G)$	$L^\infty(G)$
$L^1(G)$	–	–	$L^1(G)$
$\mathcal{M}(G)$	–	–	$\mathcal{M}(G)$
$L^\infty(G)$	$L^1(G)$	$\mathcal{M}(G)$	$L^\infty(G)$

pointwise product
$\mu_1 \cdot \mu_2(g)$
for product representations

The Lebesgue function algebra $L^1(G)$ is injected into its universal enveloping stellar algebra (chapter "Quantum Probability"), the *stellar group algebra*: (group C*-algebra)

$$\iota : L^1(G) \longrightarrow \mathbf{C}^\star(G) \in \underline{\mathbf{sa\hat{a}g}}_{\mathbb{C}}.$$

It can be considered as a convolutive subalgebra, dense in $\mathbf{C}^\star(G)$. A representation of $L^1(G)$ in a stellar algebra S, i.e., $S \subseteq \mathbf{AL}(H)$ with a Hilbert space H, is extendable to $\mathbf{C}^\star(G)$.

Every Radon distribution of an open subset $T \subseteq \mathbb{R}^n$ is a finite sum of derivatives of order $N \leq n$ of locally essentially bounded functions [7] $\mathcal{M}(T) \subseteq \{\alpha_N \partial^N L^\infty(T)\}$, e.g., the Dirac distributions as derivation of the step and sign functions $\vartheta, \epsilon \in L^\infty(\mathbb{R})$:

$$\begin{array}{rrl} \mathbb{R}: & \frac{d}{dx}\vartheta(x) & = \frac{d}{dx}\frac{\epsilon(x)}{2} = \delta(x), \\ \mathbb{R}^{2R}: & (\frac{d}{dx^2})^N \vartheta(x^2) & = \delta^{(N-1)}(x^2), \quad N = 1, \dots, R. \end{array}$$

7.4 Harmonic Analysis of Compact Groups

Harmonic analysis with respect to irreducible representations of a compact group in general takes up the concepts of finite groups as a subclass of compact groups.

A compact group U has unique normalized Haar measure $\int du = 1$, e.g.,

$$\begin{array}{lll} \mathbf{U}(1) \ni & u(\alpha): & \int_{\mathbf{U}(1)} du = \int_{-\pi}^{\pi} \frac{d\alpha}{2\pi} = 1, \\ \mathbf{SU}(2) \ni & u(\chi, \varphi, \theta): & \int_{\mathbf{SU}(2)} du = \int d^3u = \int_{-2\pi}^{2\pi} \frac{d\chi}{4\pi} \int_0^{2\pi} \frac{d\varphi}{2\pi} \int_{-1}^{1} \frac{d\cos\theta}{2} = 1. \end{array}$$

Its representations are, up to equivalence, in definite unitary groups, $U \longrightarrow \mathbf{U}(V)$, i.e., on Hilbert spaces V. The representation coefficients are continuous functions with compact support. The Lebesgue function algebra for a compact group is maximal. All spaces $L^p(U)$ are not only convolution modules, but even subalgebras of $L^1(U)$

$$\left.\begin{array}{l} L^1(U) \supseteq L^p(U) \supseteq L^q(U) \supseteq L^\infty(U) \text{ for } 1 \le p \le q \le \infty \\ \Rightarrow L^p(U) * L^p(U) \longrightarrow L^p(U) \end{array}\right\}, \quad L^p(U) \in \star\underline{\mathbf{na\hat{a}g}}_{\mathbb{C}}.$$

According to Weyl, all U-representations are semisimple Hilbert representations with the simple ones finite-dimensional:

$$\begin{array}{rl} \text{simple for compact group:} & U \longrightarrow \mathbf{U}(d_Z) \subset \mathbb{C}(d_Z \times d_Z), \\ \text{group dual:} & \check{U} = \check{U}_+ \cong \{Z\} \subseteq \mathbb{Z}^r. \end{array}$$

Not only for finite, but for compact groups in general, the irreducible U-representation spaces are characterized by the minimal left and right ideals $\mathbb{C}^{d_Z}$, arising as columns and rows with multiplicity d_Z in full matrix algebras $\mathbb{C}(d_Z \times d_Z)$. The set of irreducible representation classes (group dual) is isomorphic to winding numbers, e.g., to $\mathbb{Z}$ for $\mathbf{U}(1)$ or to the cone $\mathbb{N}_0^r$ for a semisimple compact rank-r Lie group.

The *Peter-Weyl theorem* generalizes the Maschke-Burnside-Wedderburn theorem for finite groups: The direct sum of the finite-dimensional matrix algebras for all irreducible representation classes is dense in all function spaces $L^p(U)$ and in the stellar algebra $\mathbf{C}^\star(U)$, with respect to their possibly different topologies

$$\bigoplus_{Z \in \hat{U}} \mathbb{C}(d_Z \times d_Z) \text{ dense in } L^p(U) \text{ and } \mathbf{C}^\star(U).$$

It exhausts the Hilbert space of the square integrable group functions; with $\mathbf{SU}(2)$ as example,

$$L^2(U) \overset{\text{dense}}{\supseteq} \bigoplus_{Z \in \check{U}} \mathbb{C}(d_Z \times d_Z) \overset{\text{e.g.}}{\cong} \left(\begin{array}{c|c|c|c|c} 1 \times 1 & 0 & \dots & 0 & \dots \\ \hline 0 & 2 \times 2 & \dots & 0 & \dots \\ \hline \dots & \dots & \dots & \dots & \dots \\ \hline 0 & 0 & \dots & (1+2J) \times (1+2J) & \dots \\ \hline \dots & \dots & \dots & \dots & \dots \end{array}\right) \text{ for } \mathbf{SU}(2),$$

$$\dim_{\mathbb{C}} L^2(U) = \sum_{Z \in \check{U}} d_Z^2 \overset{\text{e.g.}}{=} \sum_{2J=0}^{\infty} (1+2J)^2 = \aleph^0.$$

In *Schur's orthonormality relations for the harmonic components* the "braced" scalar products come as *group integrals*, which are Plancherel-normalized with the representation dimensions

$$\begin{array}{rl} \text{Hilbert basis of } L^2(U): & \{\sqrt{d_Z}|Z^a_b\} \bigm| a, b = 1, \dots, d_Z,\ Z \in \check{U}\}, \\ \check{U} \times \check{U} \longrightarrow \mathbf{C1},\ \{Z'|Z\} & = \frac{1}{d_Z} \delta^{ZZ'} \mathbf{1}_{d_Z \times d_Z}, \\ Z(u)^w_v & = \langle w|Z(u).v\rangle, \quad |v\rangle, \dots \in V_Z, \\ \{Z'^{v'}_{w'}|Z^w_v\} & = \int du\ \overline{Z'(u)^{v'}_{w'}} Z(u)^w_v = \frac{1}{d_Z} \delta^{ZZ'} \langle v'|v\rangle \langle w|w'\rangle, \\ \{Z'^{a'}_{b'}|Z^a_b\} & = \int du\ \overline{Z'(u)^{a'}_{b'}} Z(u)^a_b = \frac{1}{d_Z} \delta^{ZZ'} \delta^{aa'} \delta_{bb'}. \end{array}$$

All direct sums are orthogonal $\oplus = \perp$. For example, for the irreducible $\mathbf{U}(1)$- and $\mathbf{SU}(2)$-representations,

$$\begin{array}{rl}\mathbf{irrep}\,\mathbf{U}(1)\cong\mathbb{Z}: & Z(\alpha)=e^{i\alpha Z},\ \int_{-\pi}^{\pi}\frac{d\alpha}{2\pi}\overline{Z'(\alpha)}Z(\alpha)=\{Z|Z'\}=\delta^{ZZ'},\\ \mathbf{irrep}\,\mathbf{SU}(2)\cong\mathbb{N}_0: & \int d^3u\ \overline{2J'(u)^{a'}_{b'}}2J(u)^a_b=\{2J'^{a'}_{b'}|2J^a_b\}=\frac{1}{1+2J}\delta^{JJ'}\delta^{aa'}\delta_{bb'},\end{array}$$

one has explicitely Schur's orthonormality, illustrated with the doublet and triplet $\mathbf{SU}(2)$-representation coefficients

$$\begin{array}{rl}\int d^3u\begin{pmatrix}|e^{i\frac{\chi+\varphi}{2}}\cos\frac{\theta}{2}|^2 & |ie^{-i\frac{\chi-\varphi}{2}}\sin\frac{\theta}{2}|^2\\ |ie^{i\frac{\chi-\varphi}{2}}\sin\frac{\theta}{2}|^2 & |e^{-i\frac{\chi+\varphi}{2}}\cos\frac{\theta}{2}|^2\end{pmatrix} & =\frac{1}{2}\begin{pmatrix}1&1\\1&1\end{pmatrix},\\ \int d^3u\begin{pmatrix}e^{-i\frac{\chi+\varphi}{2}}\cos\frac{\theta}{2}\\ -ie^{-i\frac{\chi-\varphi}{2}}\sin\frac{\theta}{2}\end{pmatrix}(ie^{-i\frac{\chi-\varphi}{2}}\sin\frac{\theta}{2},e^{-i\frac{\chi+\varphi}{2}}\cos\frac{\theta}{2}) & =\begin{pmatrix}0&0\\0&0\end{pmatrix},\\ \int d^3u\begin{pmatrix}|e^{i(\chi+\varphi)}\cos^2\frac{\theta}{2}|^2 & |ie^{i\varphi}\frac{\sin\theta}{\sqrt{2}}|^2 & |-e^{-i(\chi-\varphi)}\sin^2\frac{\theta}{2}|^2\\ |ie^{i\chi}\frac{\sin\theta}{\sqrt{2}}|^2 & |\cos\theta|^2 & |ie^{-i\chi}\frac{\sin\theta}{\sqrt{2}}|^2\\ |-e^{i(\chi-\varphi)}\sin^2\frac{\theta}{2}|^2 & |ie^{-i\varphi}\frac{\sin\theta}{\sqrt{2}}|^2 & |e^{-i(\chi+\varphi)}\cos^2\frac{\theta}{2}|^2\end{pmatrix} & =\frac{1}{3}\begin{pmatrix}1&1&1\\1&1&1\\1&1&1\end{pmatrix},\\ \int d^3u\begin{pmatrix}e^{-i(\chi+\varphi)}\cos^2\frac{\theta}{2}\\ -ie^{-i\chi}\frac{\sin\theta}{\sqrt{2}}\\ -e^{-i(\chi-\varphi)}\sin^2\frac{\theta}{2}\end{pmatrix}(ie^{-i\frac{\chi-\varphi}{2}}\sin\frac{\theta}{2},e^{-i\frac{\chi+\varphi}{2}}\cos\frac{\theta}{2}) & =\begin{pmatrix}0&0\\0&0\\0&0\end{pmatrix}.\end{array}$$

In addition to this group-integration-orthonormality of the representation coefficients in $L^2(U)$ ("braced" scalar product) there is the $\mathbf{U}(d_z)$ orthonormality ("angled" scalar product) for Hilbert bases of the finite-dimensional irreducible representation spaces, e.g., the two and three columns (rows), are orthonormal for each group element:

$$|a\rangle\in V_Z:\quad\langle a|b\rangle=\delta^{ab}\Rightarrow\overline{Z(u)^a_b}Z(u)^a_c=\delta_{bc}\text{ for all }u\in U.$$

For finite groups, both kinds of Hilbert bases of the group algebra $\mathbb{C}^U$, the canonical one and the harmonic ones, are finite. For compact groups in general, e.g., for $\mathbf{U}(n)$, the group elements are a Haar-measure-related distributive basis; the harmonic Hilbert bases remain discrete. The Fourier expansion (decomposition of the two-sided regular $U\times U$-representation, harmonic analysis) with a discrete harmonic basis $\{\sqrt{d_Z}|Z^a_b\}\ \big|\ a,b=1,\dots,d_Z,\ Z\in\check{U}\}$ is described by the orthogonal Peter-Weyl decomposition of a square integrable function in $L^2(U)={}^{\perp}\!\!\int du\ |u\}\ \mathbb{C}$:

$$\begin{array}{rll}|f\}={}^{\oplus}\!\!\int du\ |u\}f(u)=\bigoplus_{Z\in\check{U}}d_Z|\tilde{f}(Z)\} & =\bigoplus_{Z\in\check{U}}d_Z|Z^a_b\}\tilde{f}(Z)^b_a, & \\ \text{function values:}\ \ f(u) & =\{u|f\} & =\sum_{Z\in\check{U}}d_Z Z(u)^a_b\tilde{f}(Z)^b_a,\\ \text{harmonic coefficients:}\ \ \tilde{f}(Z)^b_a & =\{Z^b_a|f\} & =\int du\ \overline{Z(u)^a_b}f(u),\\ \text{Plancherel unitarity:}\ \ \{f|f\} & =\int_U du\ \overline{f(u)}f(u) & =\sum_{Z\in\check{U}_+}d_Z\overline{\tilde{f}(Z)^a_b}\tilde{f}(Z)^a_b.\end{array}$$

The canonical distributive basis $\{|u\}\ \big|\ u\in U\}$, which is a Hilbert basis only for a finite group, has a harmonic expansion and vice versa:

$$\mathrm{id}_{L^2(U)}\cong{}^{\oplus}\!\!\int du\ |u\}\{u=\bigoplus_{Z\in\check{U}}d_Z|Z^b_a\}\{Z^b_a|\Rightarrow\left\{\begin{array}{rl}|u\} & =\bigoplus_{Z\in\check{U}}d_Z|Z^b_a\}\overline{Z(u)^a_b},\\ |Z^a_b\} & ={}^{\oplus}\!\!\int du\ |u\}Z(u)^a_b,\\ Z(u)^a_b & =\{u|Z^b_a\}=\langle a|Z(u)b\rangle.\end{array}\right.$$

It involves the normalized Haar measure of the group and the Plancherel measure of the group dual (invariants of the irreducible representations), which counts the multiplicity of irreducible group U-representation vector spaces, with either left or right U-action, in an irreducible algebra $L^2(U)$-representation with left-right $U \times U$-action.

Examples are the Fourier series for square integrable $\mathbf{U}(1)$- and $\mathbf{SU}(2)$-functions:

$$L^2(\mathbf{U}(1)) : \begin{cases} |f\} &= {}^{\oplus}\!\int_{-\pi}^{\pi} \frac{d\alpha}{2\pi} \, |\alpha\} \, f(\alpha) = \bigoplus_{Z\in\mathbb{Z}} |Z\} \, \tilde f(Z), \\ e^{i\alpha Z} &= \{\alpha|Z\}, \;\; e^{-i\alpha Z} = \{Z|\alpha\}, \\ f(\alpha) &= \{\alpha|f\} = \sum_{Z\in\mathbb{Z}} e^{i\alpha Z} \tilde f(Z), \\ \tilde f(Z) &= \{Z|f\} = \int_{-\pi}^{\pi} \frac{d\alpha}{2\pi} \, e^{-i\alpha Z} f(\alpha), \end{cases}$$

$$L^2(\mathbf{SU}(2)) : \begin{cases} |f\} &= {}^{\oplus}\!\int d^3u \; |u\} \; f(u) = \bigoplus_{2J\in\mathbb{N}} (1+2J) |2J_b^a\} \tilde f(2J)_a^b, \\ f(u) &= \{u|f\} = \sum_{2J\in\mathbb{N}} (1+2J) 2J(u)_b^a \tilde f(2J)_a^b, \\ \tilde f(2J)_a^b &= \{2J_a^b|f\} = \int d^3u \; \overline{2J(u)_b^a} f(u). \end{cases}$$

The irreducible representations decompose the Haar measure associate Dirac measure of the group. It is the generalization of the orthogonality $\{k|l\} = \delta^{kl}$ for the canonical finite group basis to the canonical distributive basis of the continuous group:

$$\begin{aligned} u, v \in U : \;\; f(u) &= \int dv \; \{u|v\} f(v) = \sum_{Z\in\check U} d_Z \int dv \;\; \mathrm{tr}\, [Z(u) \circ Z(v^{-1})] f(v), \\ \{u|v\} = \delta(vu^{-1}) &= \sum_{Z\in\check U} d_Z \, \mathrm{tr}\, Z(v) \circ Z(u^{-1}), \\ \delta(u) &= \sum_{Z\in\check U} d_Z \, \mathrm{tr}\, Z(u), \;\; \delta(e) = \sum_{Z\in\check U} d_Z^2 \text{ (if defined)}, \end{aligned}$$

in the examples

$$\begin{aligned} \mathbf{U}(1) : \qquad \delta(\tfrac{\alpha-\beta}{2\pi}) &= \sum_{Z\in\mathbb{Z}} e^{i\alpha Z} e^{-i\beta Z}, \;\; \delta(\tfrac{\alpha}{2\pi}) = \sum_{Z\in\mathbb{Z}} e^{i\alpha Z}, \\ \mathbf{SU}(2) : \qquad \delta(vu^{-1}) &= \sum_{2J\in\mathbb{N}} (1+2J) \, \mathrm{tr}\, 2J(v) \circ 2J(u^{-1}), \\ \delta(\tfrac{\chi}{4\pi}) \delta(\tfrac{\varphi}{2\pi}) \delta(\tfrac{\cos\theta}{2}) &= \sum_{2J\in\mathbb{N}} (1+2J) \, \mathrm{tr}\, 2J(\chi, \varphi, \cos\theta). \end{aligned}$$

The convolution of square integrable functions is the multiplication of the harmonic matrices in the Fourier algebra automorphism:

$$\begin{aligned} L^2(U) \leftrightarrow L^2(U), \;\; f \leftrightarrow \tilde f \text{ with } \widetilde{f_1 * f_2} &= \tilde f_1 \circ \tilde f_2, \\ \text{for all } Z\text{: } \{Z|f_1 * f_2\} &= \{Z|f_1\} \circ \{Z|f_2\}, \\ \text{e.g. } \mathbf{U}(1) : \;\; (f_1 * f_2)(\alpha) &= \sum_{Z\in\mathbb{Z}} \tilde f_1(Z) \tilde f_2(Z) e^{i\alpha Z}. \end{aligned}$$

7.5 Hilbert Representations and Scalar-Product-Inducing Functions

7.5.1 Cyclic Hilbert Representations

A group G determines, via its (closed) subgroups, its action spaces. Without linear structure: An irreducible G-realization acts on classes G/H with a fixgroup $H \subseteq G$. Any set with G-realization is decomposable into irreducible orbits $G \bullet x \cong G/G_x$. Now with linearity: Any vector in a Hilbert space V with a real Lie group G-representation defines by the closed span of its orbit $G \bullet v \cong G/G_v$ a *cyclic G-Hilbert space* $\overline{\mathbb{C}^{(G \bullet v)}} \subseteq V$. With a cyclic vector for V, the representation is called cyclic too. A G-invariant subspace and its orthogonal decompose the Hilbert space $V = W \oplus W^\perp$. Therefore, any Hilbert representation is decomposable into *a direct sum of cyclic Hilbert representations.*

7.5.2 Discrete Hilbert Representations

A locally compact Lie group has at most a countable set of Hilbert representations that have square integrable coefficients. Because of their discrete invariants, such representations are called *discrete representations*, all for a compact group where all coefficients are square integrable, e.g. for $\mathbf{U}(1)$. $\mathbf{D}(1)$ has no discrete representations. A nontrivial noncompact group example is the countable set of discrete $\mathbf{SU}(1,1)$-representations that are induced by representations of a compact Cartan subgroup $\mathbf{SO}(2)$ with their discrete winding numbers, and the not discrete $\mathbf{SU}(1,1)$-representations from the Cartan subgroup $\mathbf{SO}_0(1,1)$.

All Hilbert spaces for representations of a locally compact group can be constructed as equivalence classes $|L^1(G)\rangle$ of the Lebesgue function algebra $L^1(G)$ (more below). For noncompact groups, they do not have to be constituted by square integrable group functions. However, discrete representations act on square integrable Hilbert spaces. Group representations coefficients for square integrable Hilbert spaces do not have to be square integrable:

$$D^v_w(g) = \langle \imath | D(g) | w \rangle : \quad \begin{array}{ccc} \boxed{\begin{array}{c} D^v_w \in L^2(G) \\ \text{discrete} \end{array}} & \rightarrow & \boxed{|v\rangle \in |L^2(G)\rangle} \\ & \nearrow & \\ \boxed{D^v_w \notin L^2(G)} & \searrow & \boxed{|v\rangle \notin |L^2(G)\rangle} \end{array}$$

7.5.3 Inner Products of Representation Spaces

The dual product combined with the $\mathbf{U}(1)$-conjugation defines a sesquilinear form on dual Lebesgue spaces. It is the $L^\infty(G)$-valued convolution at the neutral element $e \in G$.

$$\begin{aligned} &\tfrac{1}{p} + \tfrac{1}{s} = 1, \quad 1 \le p, s \le \infty : \quad L^p(G) \times L^s(G) \xrightarrow{*} L^\infty(G) \xrightarrow{e} \mathbb{C}, \\ &\langle f_p, f_s \rangle \;= \int dk\; f_p(k) f_s(k) \;= f_p^- * f_s(e) \;= \int dk_1 dk_2\; f_p(k_1^{-1})\, \delta(k_1 k_2)\, f_s(k_2), \\ &\langle f_p | f_s \rangle \;= \int dk\; \overline{f_p(k)} f_s(k) \;= \hat{f}_p * f_s(e) \;= \int dk_1 dk_2\; \overline{f_p(k_1^{-1})}\, \delta(k_1 k_2)\, f_s(k_2). \end{aligned}$$

A *Hilbert metric for the Lebesgue function algebra* $L^1(G)$ of a locally compact group is defined by chosing from its dual $L^\infty(G)$, the essentially bounded Lebesgue functions,

$$L^\infty(G) * L^1(G) \longrightarrow L^\infty(G) \longrightarrow \mathbb{C}, \begin{cases} d * f(k) = \int dk_1 dk_2 d(k_1)\delta(k_1 k_2 k^{-1}) f(k_2), \\ d * f(e) = \langle d, f\rangle = \int dk\ d(k) f(k) = \langle f\rangle_d, \end{cases}$$

a function from the convex cone $d \in L^\infty(G)_+$: The *functions of positive type (scalar-product-inducing functions)* are defined by the property to give a positive linear form of the Lebesgue function algebra (chapter "Quantum Probability")

$$L^1(G) \longrightarrow \mathbb{C}, \quad \langle \hat{f} * f\rangle_d = \int dk_1 dk_2\ \overline{f(k_1)}\ d(k_1^{-1}k_2)\ f(k_2) \geq 0.$$

Positive-type functions and measures (below) are a very useful tool to characterize and to work with Hilbert representations.

A diagonal matrix element of a Hilbert representation D with any vector $|v\rangle \in H$ gives a positive-type function $D^v_v \in L^\infty(G)_+$:

$$\begin{aligned} D^v_v(k) = \langle v|k \bullet v\rangle : \quad \langle \hat{f} * f\rangle_{D^v_v} &= \int dk_1 dk_2 \overline{f(k_1)} \langle v|k_1^{-1}k_2 \bullet v\rangle f(k_2) \\ &= \int dk_1 dk_2 \langle f(k_1) k_1 \bullet v | f(k_2) k_2 \bullet v\rangle \\ &= \| \int dk f(k) k \bullet v \|^2, \end{aligned}$$

e.g., $\mathbb{R} \ni t \longmapsto e^{imt} \in \mathbf{U}(1)$. Compact group examples are the three diagonal elements in the $\mathbf{SO}(3)$-matrix above

$$\mathbf{SU}(2) \ni u(\chi, \varphi, \theta) \longmapsto \begin{pmatrix} e^{i(\chi+\varphi)} \cos^2 \frac{\theta}{2} & ie^{i\varphi} \frac{\sin\theta}{\sqrt{2}} & -e^{-i(\chi-\varphi)} \sin^2 \frac{\theta}{2} \\ ie^{i\chi} \frac{\sin\theta}{\sqrt{2}} & \cos\theta & ie^{-i\chi} \frac{\sin\theta}{\sqrt{2}} \\ -e^{i(\chi-\varphi)} \sin^2 \frac{\theta}{2} & ie^{-i\varphi} \frac{\sin\theta}{\sqrt{2}} & e^{-i(\chi+\varphi)} \cos^2 \frac{\theta}{2} \end{pmatrix} \in \mathbf{SO}(3).$$

Representation properties and, for quantum theory, probability normalization are related to each other:

$$\text{unit } G \ni e \longmapsto d_v(e) = \langle v|D(e)|v\rangle = \langle v|v\rangle.$$

A positive-type function defines a semiscalar product via the d-convolution at the group unit

$$\begin{aligned} L^1(G) * L^\infty(G) * L^1(G) &\longrightarrow L^\infty(G), \\ L^1(G) \times L^1(G) &\longrightarrow \mathbb{C}, \quad \langle f|f'\rangle_d = \int dk_1 dk_2\ \overline{f(k_1)}\ d(k_1^{-1}k_2)\ f'(k_2) \\ &\qquad\qquad = \langle \hat{f} * f'\rangle_d = \hat{f} * d * f'(e), \\ |\langle f|f'\rangle_d| &\leq \| d \|_\infty \| f \|_1 \| f' \|_1 . \end{aligned}$$

Therefore, the Lebesgue function algebra is a pre-Hilbert space with the representation of the group and its algebra by left multiplication, also on the scalar product space $|L^1(G)\rangle_d$ with the d-nontrivial right classes:

$$\begin{aligned} L^1(G) &\longrightarrow |L^1(G)\rangle_d, \quad f \longmapsto |f\rangle_d, \\ h \in G: \quad D(h) : |L^1(G)\rangle_d &\longrightarrow |L^1(G)\rangle_d, \quad D(h)|f\rangle_d = |{}_hf\rangle_d, \\ f' \in L^1(G): \quad D(f') : |L^1(G)\rangle_d &\longrightarrow |L^1(G)\rangle_d, \quad D(f')|f\rangle_d = |f' * f\rangle_d. \end{aligned}$$

The representations are extendable on the Hilbert space $H = \overline{|L^1(G)\rangle_d}$.

For this representation D there can be shown to exist a *cyclic vector* $|c\rangle$ that gives back the positive-type function started with. Therefore, *all positive-type functions are cyclic matrix elements*, i.e., diagonal representation matrix elements (group orbits) of a cyclic vector:

$$|c\rangle \in \overline{|L^1(G)\rangle_d} : \quad \begin{cases} D(f)|c\rangle &= |f\rangle_d, \text{ cyclic}, \\ d(k) &= \langle c|k \bullet c\rangle, \text{ positive-type function}, \\ \langle f_1|f_2\rangle_d &= \langle \int dk_1 f(k_1) k_1 \bullet c | \int dk_2 f(k_2) k_2 \bullet c\rangle. \end{cases}$$

For unital $L^1(G) \supset G \ni e$, e.g., for a finite group, the class of the unit is a cyclic vector. In general, the class of a unit approximation $\tilde{\delta}_e$ leads to a cyclic vector. The norm of the positive-type function is the norm squared of the cyclic vector and the function value at the group unit:

$$|d(k)| \leq \| c \|^2 = d(e).$$

Functions of positive type are, almost everywhere, bounded group functions $L^\infty(G)_+ \overset{dg}{=} \mathcal{C}_b(G)_+$. They do not have to be positive functions. They are the continuous generalization of scalar products $d \succeq 0$ for finite-dimensional spaces (chapter "Spacetime Translations"), which can by characterized, equivalently, by positive matrices, i.e., hermiticity $d = d^\star$ and positive spectral values $\operatorname{spec} d \subset \mathbb{R}_+$, or by a factorization (unit diagonalization) $d = \xi^\star \circ \xi$. In the case of positive-type functions, positive finite matrices for sequilinear forms can be built with any number of group elements:

$$d \in L^\infty(G)_+ \iff d(k_i^{-1} k_j)_{i,j=1}^n \succeq 0 \text{ for } \{k_j \in G\}_{j=1}^n \text{ and } n = 1, 2, \ldots.$$

Using two elements, there follow the conjugation invariance $d = \hat{d}$ and the absolute value restriction by the value at the neutral element:

$$\begin{aligned} n = 1 : &\quad d(e) \geq 0, \\ n = 2 : &\quad \begin{pmatrix} d(e) & d(k) \\ d(k^{-1}) & d(e) \end{pmatrix} \succeq 0 \Rightarrow \begin{cases} d(k) &= \overline{d(k^{-1})}, \\ |d(k)| &\leq d(e). \end{cases} \end{aligned}$$

The spectrum positivity for matrices $\operatorname{spec} d \subset \mathbb{R}_+$ can be generalized to the positivity of the coeffients in a harmonic analysis of the positive-type function (more below).

The trivial function $G \ni g \longmapsto d_1(g) = 1$ characterizes the trivial group representation. Conjugate functions $d \leftrightarrow \overline{d}$ characterize dual representations, a real function $d = \overline{d}$ a self-dual representation.

With Gel'fand and Raikov [9], there is a surjection of the functions of positive type to the equivalence classes of cyclic Hilbert representations of a locally compact group:

$$L^\infty(G)_+ \longrightarrow \mathbf{rep}_+ G \text{ (cyclic) } \supseteq \check{G}_+.$$

A square integrable function leads to a positive-type function

$$L^2(G) \ni \xi \longmapsto \hat{\xi} * \xi \in L^\infty(G)_+.$$

In the Hilbert spaces $\{\overline{|L^1(G)\rangle_d} \mid d \in L^\infty(G)_+\}$ the spaces with square integrable group functions are distinguished by this sort of positive-type functions: If and only if a positive-type function is "diagonalizable," i.e., a convolution product $d = \hat{\xi} * \xi$ with a square integrable one $\xi \in L^2(G)$, can the Hilbert space for the d-associated representation D be constructed with square integrable function classes $\xi * L^1(G) \subseteq L^2(G)$:

$$d = \hat{\xi} * \xi \text{ with } \xi \in L^2(G) \Rightarrow \begin{cases} L^1(G) \ni f \longmapsto f_\xi = \xi * f \in L^2(G), \\ \langle f|f'\rangle_d = (\hat{f} * \hat{\xi} * \xi * f')(e) = (\hat{f}_\xi * f'_\xi)(e), \\ \langle f'|\hat{\xi} * D * \xi|f\rangle = \langle f_\xi|D|f'_\xi\rangle. \end{cases}$$

A Hilbert product for square integrable functions replaces in $d(k_1^{-1}k_2) \to \delta(k_1^{-1}k_2)$ the positive-type function for $L^1(G)$ by the Dirac distribution for $L^2(G)$. This can be taken as a generalization of the factorization of positive matrices (diagonalization of sesquilinear forms) $d = \xi^\star \circ \mathbf{1} \circ \xi$ with $d(v, w) = \langle \xi.v|\xi.w\rangle$.

The sesquilinear form with a positive-type function $d \in L^\infty(G)_+$ can be extended to the Radon distributions, $\mathcal{M}(G) * L^\infty(G) * \mathcal{M}(G) \subseteq L^\infty(G)$,

$$\mathcal{M}(G) \times \mathcal{M}(G) \longrightarrow \mathbb{C}, \quad \langle \omega|\omega'\rangle_d = \int dk_1 dk_2 \ \overline{\omega(k_1)} d(k_1^{-1}k_2)\omega'(k_2).$$

The Hilbert spaces $\overline{|\mathcal{C}_c(G)\rangle_\omega}$ contain classes of functions from the convolution algebra with compact-support functions. They are constructed analogously by a scalar product inducing Radon distribution of positive type:

$$\begin{aligned} \omega \in \mathcal{M}(G)_+ : \quad \mathcal{C}_c(G) \times \mathcal{C}_c(G) \longrightarrow \mathbb{C}, \quad \langle f|f'\rangle_\omega &= \langle \hat{f} * f\rangle_\omega = \hat{f} * \omega * f'(e) \\ &= \int dk_1 dk_2 \ \overline{f(k_1)}\omega(k_1^{-1}k_2) f'(k_2). \end{aligned}$$

Positive-type functions yield only cyclic representations; positive-type measures yield more general representations. For example, the Dirac distribution at the unit δ_e leads to $L^2(G) = \overline{|\mathcal{C}_c(G)\rangle_{\delta_e}}$.

7.5.4 Irreducible Hilbert Representations

An irreducible G-Hilbert space is cyclic, but not the other way around. For example, for the decomposable representations $\mathbf{U}(1) \ni e^{i\alpha} \longrightarrow e^{i\sigma_3\alpha}$ or $\mathbb{R} \ni t \longrightarrow e^{i\sigma_3 t}$ the vector $|c\rangle = \binom{1}{0} \in V \cong \mathbb{C}^2$ is cyclic. A cyclic vector $\overline{\mathbb{C}^{G \bullet |c\rangle}}$ has a characteristic fixgroup $H \bullet |c\rangle = \{|c\rangle\}$. For $G = \mathbf{D}(1)$ there is only the trivial fixgroup $\{1\}$ for nontrivial cyclic representations.

The continuous functions of positive type are a convex cone. The extremal continuous normalized functions of positive type $d(e) = 1$, called *pure states*, give the equivalence classes of the irreducible Hilbert representations. The extremal positive-type functions have no nontrivial decomposition with cone vectors, they are the "corners" in the cone $L^\infty(G)_+$. Any Hilbert representation is decomposable into an orthogonal *direct integral of irreducible Hilbert representations.* For a compact group, the direct integral is a direct sum. A positive-type function and the related cyclic Hilbert space may be decomposable with Plancherel measure and a positive distribution $\tilde{d}$ of the group dual as (distributive) normalization coefficients into a direct integral of extremal positive-type functions, pure states with pure cyclic vectors:

$$G \ni k \longmapsto d(k) = \int_{\check{G}_+} dD\ \tilde{d}(D)D(k), \quad d = {}^{\oplus}\!\!\int_{\check{G}_+} dD\ \tilde{d}(D)\ D.$$

There is a *surjection of the pure states to the irreducible Hilbert representation classes*, which is a bijection for abelian groups:

$$d(e) = 1, \quad L^\infty(G)_+ \ni d \text{ (extremal)} \longrightarrow D \in \mathbf{irrep}_+ G = \check{G}_+.$$

The *Gel'fand-Raikov theorem* guarantees "enough" irreducible Hilbert representations of G: For different group elements $k_{1,2} \in G$ there exists a separating irreducible representation, i.e., $D(k_1) \neq D(k_2)$.

7.6 Harmonic Analysis of Noncompact Groups

For a locally compact noncompact group the group dual also has continuous contributions:

group:	finite	$\subset$	compact	$\subset$	locally compact
e.g.,	$\mathbf{G}(n)$	$\subset$	$\mathbf{U}(n)$	$\subset$	$\mathbf{GL}(\mathbb{C}^n)$
canonical "bases" card G	$\{k \mid k \in N\}$ finite	$\subset$	$\{u \mid u \in U\}$ continuous	$\subset$	$\{k \mid k \in G\}$ continuous
harmonic "bases": card $\check{G}_+$	$\{Z_a^b \mid Z, a, b\}$ finite	$\subset$	$\{Z_a^b \mid Z, a, b\}$ countable	$\subset$	$\{D_q^p \mid D, q, p\}$ continuous

The harmonic analysis for locally compact noncompact nonabelian groups in general is difficult (*Harish-Chandra theory*): As suggested by the possible occurrence of not square integrable Hilbert spaces, the harmonic analysis has to be formulated with care. The square integrable functions from the Lebesgue function algebra $L^{1\cap 2}(G) = L^1(G) \cap L^2(G)$ generate the convolution algebra $L(G)$. The Fourier transform (harmonic analysis) of these functions can be decomposed with a direct Plancherel integral into components from $G \times G$-irreducible subalgebras:

$$\begin{aligned}
L(G) \ni |f\} &= {}^{\perp}\!\!\int_G dk\ |k\}f(k) = {}^{\perp}\!\!\int_{\check{G}_+} dD\ |D\}\tilde{f}(D),\\
f(k) &= \int_{\check{G}_+} dD\ \{k|D\}\tilde{f}(D), \quad \{k|D\} = D(k),\\
\tilde{f}(D) &= \int_G dk\ \{D|k\}f(k), \qquad \{D|k\} = \hat{D}(k) = D(k^{-1})^\star,\\
\{f|f\} &= \int dk |f(k)|^2 = \int dD\ \operatorname{tr} \tilde{f}(D)\overline{\tilde{f}(D)},\\
\mathrm{id}_{L^2(G)} &\cong {}^{\perp}\!\!\int_G dk\ |k\}\{k| = {}^{\perp}\!\!\int_{\check{G}_+} dD\ |D\}\{D|.
\end{aligned}$$

The support of the Plancherel measure may exclude irreducible Hilbert representations (nonamenable groups), that is the case for noncompact semisimple groups, as visible, e.g., in the supplementary representations of the Lorentz group with nontrivial positive-type function Hilbert space, i.e., with not square integrable functions (below).

7.6.1 Flat Spacetime and Interaction-Free States

A basically important and not so difficult example is that of the translation representations. They characterize interaction-free states with time translations for the harmonic oscillator, position translations for free nonrelativistic scattering, and relativistic spacetime translations for free particles.

For the translation group, the energy-momenta characterize the group dual with the equivalence classes of the irreducible Hilbert representations. The harmonic analysis (Fourier integrals) for the noncompact abelian group $\mathbf{D}(1)^n \cong \mathbb{R}^n$ (translations) with definite dual group $\mathbf{irrep}_+\mathbb{R}^n = \check{\mathbb{R}}^n$ ((energy-)momenta) for the 1-dimensional irreducible representations $|p\} = \{x \longmapsto e^{ipx}\}$ (for $p \neq 0$ faithful only for the $\mathbf{U}(1)$-classes) is formulated in the language above:

$$\begin{array}{rcccl} L^2(\mathbb{R}^n) \ni |f\} &=& {}^{\perp}\!\!\int d^n x\, |x\}f(x) &=& {}^{\perp}\!\!\int d^n \frac{p}{2\pi}\, |p\}\tilde{f}(p), \\ \mathrm{id}_{L^2(\mathbb{R}^n)} &\cong& {}^{\perp}\!\!\int d^n x\, |x\}\{x| &=& {}^{\perp}\!\!\int d^n \frac{p}{2\pi}\, |p\}\{p|, \\ && \text{canonical} && \text{harmonic} \end{array}$$

$$\begin{array}{rl} \text{canonical distributive basis:} & \{|x\} \bigm| x \in \mathbb{R}^n\},\ \ \{y|x\} = \delta(x-y), \\ \text{harmonic distributive basis:} & \left\{ \begin{array}{l} \{|p\} \bigm| p \in \check{\mathbb{R}}^n\},\ \ \{q|p\} = \delta(\frac{p-q}{2\pi}), \\ \text{distributive Schur orthogonality,} \end{array} \right. \\ \text{functions values:} & f(x) = \{x|f\} = \int d^n \frac{p}{2\pi}\, e^{ipx} \tilde{f}(p) \\ \text{with} & e^{ipx} = \{x|p\},\ \ e^{-ipx} = \{p|x\}, \\ \text{harmonic coefficients:} & \tilde{f}(p) = \{p|f\} = \int d^n x\, e^{-ipx} f(x), \\ \text{Plancherel unitarity:} & \{f|f\} = \int d^n x\, |f(x)|^2 = \int d^n \frac{p}{2\pi}\, |\tilde{f}(p)|^2. \end{array}$$

The $\int d^n x$-associate Plancherel measure $\int d^n \frac{p}{2\pi}$ is Lebesgue.

According to a theorem of Lebesgue, the Fourier transformation of the convolution algebra $L^1(\mathbb{R}^n)$ is an injective algebra morphism, with a dense range, but not surjective, into the continuous functions $\mathcal{C}_0(\check{\mathbb{R}}^n)$ vanishing at infinity. $\mathcal{C}_0$ is a Banach space with the norm $\| f \| = \sup_{q \in \check{\mathbb{R}}^n} |f(q)|$ and a subspace of L^∞. The Fourier transformation can be extended to the Radon measure algebra with values in the bounded continuous functions $\mathcal{C}_b(\check{\mathbb{R}}^n)$. Positive Radon measures and the continuous functions of positive type are bijective (*Bochner's theorem* [2]):

$$\text{Fourier:} \quad \left\{ \begin{array}{rl} \widetilde{L^p(\mathbb{R}^n)} & \subseteq L^r(\check{\mathbb{R}}^n),\ \ \frac{1}{p} + \frac{1}{r} = 1,\ 1 \le p \le 2,\ \ \infty \ge r \ge 2 \\ & \quad \text{with } \| \tilde{f} \|_r \le \| f \|_p \ \ (\textit{Hausdorff-Young inequality}), \\ \widetilde{L^1(\mathbb{R}^n)} & = \dot{\mathcal{C}}_0(\check{\mathbb{R}}^n) \text{ dense in } \mathcal{C}_0(\check{\mathbb{R}}^n), \\ \widetilde{\mathcal{M}(\mathbb{R}^n)_+} & = \mathcal{C}_b(\check{\mathbb{R}}^n)_+ \overset{d^n q}{=} L^\infty(\check{\mathbb{R}}^n)_+, \\ \widetilde{\mathcal{M}(\mathbb{R}^n)} & = \dot{\mathcal{C}}_b(\check{\mathbb{R}}^n) = \text{complex span of } \mathcal{C}_b(\check{\mathbb{R}}^n)_+. \end{array} \right.$$

The convolution and pointwise product for the three representation-relevant spaces is exchanged in the Fourier spaces with the harmonic components

$G = \mathbb{R}^n$

$*$	$L^1(\mathbb{R}^n)$	$\mathcal{M}(\mathbb{R}^n)$	$\dot{\mathcal{C}}_b(\mathbb{R}^n)$
$L^1(\mathbb{R}^n)$	$L^1(\mathbb{R}^n)$	$L^1(\mathbb{R}^n)$	$\dot{\mathcal{C}}_b(\mathbb{R}^n)$
$\mathcal{M}(\mathbb{R}^n)$	$L^1(\mathbb{R}^n)$	$\mathcal{M}(\mathbb{R}^n)$	$\dot{\mathcal{C}}_b(\mathbb{R}^n)$
$\dot{\mathcal{C}}_b(\mathbb{R}^n)$	$\dot{\mathcal{C}}_b(\mathbb{R}^n)$	$\dot{\mathcal{C}}_b(\mathbb{R}^n)$	–

$\cdot$	$L^1(\mathbb{R}^n)$	$\mathcal{M}(\mathbb{R}^n)$	$\dot{\mathcal{C}}_b(\mathbb{R}^n)$
$L^1(\mathbb{R}^n)$	–	–	$L^1(\mathbb{R}^n)$
$\mathcal{M}(\mathbb{R}^n)$	–	–	$\mathcal{M}(\mathbb{R}^n)$
$\dot{\mathcal{C}}_b(\mathbb{R}^n)$	$L^1(\mathbb{R}^n)$	$\mathcal{M}(\mathbb{R}^n)$	$\dot{\mathcal{C}}_b(\mathbb{R}^n)$

$\mu_1 * \mu_2(x)$ $\qquad$ $\mu_1 \cdot \mu_2(x)$

$\updownarrow$ Fourier $\updownarrow$

$\check{G}_+ = \check{\mathbb{R}}^n$

$\cdot$	$\dot{\mathcal{C}}_0(\check{\mathbb{R}}^n)$	$\dot{\mathcal{C}}_b(\check{\mathbb{R}}^n)$	$\mathcal{M}(\check{\mathbb{R}}^n)$
$\dot{\mathcal{C}}_0(\check{\mathbb{R}}^n)$	$\dot{\mathcal{C}}_0(\check{\mathbb{R}}^n)$	$\dot{\mathcal{C}}_0(\check{\mathbb{R}}^n)$	$\mathcal{M}(\check{\mathbb{R}}^n)$
$\dot{\mathcal{C}}_b(\check{\mathbb{R}}^n)$	$\dot{\mathcal{C}}_0(\check{\mathbb{R}}^n)$	$\dot{\mathcal{C}}_b(\check{\mathbb{R}}^n)$	$\mathcal{M}(\check{\mathbb{R}}^n)$
$\mathcal{M}(\check{\mathbb{R}}^n)$	$\mathcal{M}(\check{\mathbb{R}}^n)$	$\mathcal{M}(\check{\mathbb{R}}^n)$	–

$*$	$\dot{\mathcal{C}}_0(\check{\mathbb{R}}^n)$	$\dot{\mathcal{C}}_b(\check{\mathbb{R}}^n)$	$\mathcal{M}(\check{\mathbb{R}}^n)$
$\dot{\mathcal{C}}_0(\check{\mathbb{R}}^n)$	–	–	$\dot{\mathcal{C}}_0(\check{\mathbb{R}}^n)$
$\dot{\mathcal{C}}_b(\check{\mathbb{R}}^n)$	–	–	$\dot{\mathcal{C}}_b(\check{\mathbb{R}}^n)$
$\mathcal{M}(\check{\mathbb{R}}^n)$	$\dot{\mathcal{C}}_0(\check{\mathbb{R}}^n)$	$\dot{\mathcal{C}}_b(\check{\mathbb{R}}^n)$	$\mathcal{M}(\check{\mathbb{R}}^n)$

$\tilde{\mu}_1 \cdot \tilde{\mu}_2(p)$ from group product $\qquad$ $\tilde{\mu}_1 * \tilde{\mu}_2(p)$ for product representations

The continuous translation functions of positive type are surjective to the positive energy-momentum Radon measures, which give all cyclic translation representations

$$\mathcal{C}_b(\mathbb{R}^n)_+ \cong \mathcal{M}(\check{\mathbb{R}}^n)_+ \longrightarrow \mathbf{rep}\,\mathbb{R}^n \text{ (cyclic)}.$$

The Hilbert-product-inducing functions

$$\mathcal{C}_b(\mathbb{R}^n)_+ \ni d \leftrightarrow \tilde{d} \in \mathcal{M}(\check{\mathbb{R}}^n)_+, \quad d(x) = \int \tilde{d}(p) \frac{d^n p}{(2\pi)^n}\, e^{ipx}$$

can be transformed into an integration of the pointwise product of the harmonic components with a representation-characteristic positive energy-momentum measure $\tilde{d}(p)\frac{d^n p}{(2\pi)^n}$:

$$\begin{aligned}
L^1(\mathbb{R}^n) \overset{d}{*} L^1(\mathbb{R}^n) &\longrightarrow \mathbb{C} \overset{\text{Fourier}}{\leftrightarrow} \dot{\mathcal{C}}_0(\check{\mathbb{R}}^n) \overset{\tilde{d}}{\cdot} \dot{\mathcal{C}}_0(\check{\mathbb{R}}^n) \longrightarrow \mathbb{C}, \\
\langle f|f'\rangle_d &= \int d^n x_1 d^n x_2\, \overline{f(x_1)} d(x_2 - x_1) f'(x_2) = \hat{f} * d * f'(0) \\
&= \int \tilde{d}(p) \frac{d^n p}{(2\pi)^n}\, \overline{\tilde{f}}(p)\, \tilde{f}'(p) = \frac{1}{(2\pi)^n}(\overline{\tilde{f}} \cdot \tilde{d}) * \tilde{f}'_-(0).
\end{aligned}$$

The extremal states for the irreducible translation representations on 1-dimensional Hilbert spaces are the Dirac measures $\delta_p \in \mathcal{M}(\check{\mathbb{R}}^n)_+$ supported by the invariant eigenvalue

$$\mathbb{R}^n \ni x \longmapsto d_p(x) = e^{ipx} = \int d^n q\ \delta(q-p) e^{iqx}, \quad d_p \in \mathcal{C}_b(\mathbb{R}^n)_+.$$

A cyclic Hilbert representation of translations $\mathbb{R}^n$ may be decomposable into a direct integral of irreducible $\mathbb{R}^n$-representations. A simple example is given by the cyclic self-dual translation representation, decomposable into dual pure states as used in the harmonic oscillator:

$$\begin{aligned}
\mathbb{R} \ni t \longmapsto d_{m^2}(t) &= 2\cos mt = \int dp\; 2|p|\delta(p^2 - m^2)e^{ipt}, \quad d_{m^2} \in \mathcal{C}_b(\mathbb{R})_+,\\
d(t) &= d_+(t) + d_-(t), \quad \begin{cases} |d(t)| \le d(0) = 2,\\ d(-t) = \overline{d(t)},\end{cases}\\
d_\pm(t) = e^{\pm imt} &= \int dp\; \delta(p \mp m)e^{ipt}, \; d_+ = \overline{d_-},\\
|c\rangle = \begin{pmatrix}1\\0\end{pmatrix} &= \frac{|c_+\rangle + |c_-\rangle}{\sqrt{2}}, \quad |c_\pm\rangle = \frac{1}{\sqrt{2}}\begin{pmatrix}1\\ \pm i\end{pmatrix} \cong \langle c_\mp|,\\
\{f|f\}_d &= \int dp\; 2|p|\delta(p^2 - m^2)|\tilde f(p)|^2\\
&= |\tilde f(m)|^2 + |\tilde f(-m)|^2 = \{f|f\}_+ + \{f|f\}_-.
\end{aligned}$$

Other examples are irreducible Hilbert representations for a tangent group $G \,\vec{\times}\, \mathbb{R}^n$, e.g., $\mathbf{SO}_0(t,s) \,\vec{\times}\, \mathbb{R}^{t+s}$. With the adjoint and codajoint group actions $G \,\vec{\times}\, \mathbb{R}^n$ of a group on its (dual) Lie algebra $\log G \cong \mathbb{R}^n \cong (\log G)^T$, affine groups are of more general interest (below and chapter "Residual Spacetime Representations").

7.7 Induced Group Representations

Harmonic analysis for a locally compact group G can be generalized to and rearranged for the harmonic analysis of its homogeneous (symmetric) spaces, i.e., of cosets $H \backslash G$ with closed subgroups H: Harmonic analysis of complex valued G-functions is the decomposition of the two-sided regular squared group $G \times G$-representation with respect to $G \times G$-irreducible subalgebras, e.g., the decomposition into full matrix algebras for a compact group. Harmonic analysis of $H \backslash G$-mappings into a vector space W with a given H-representation is the decomposition with respect to G-irreducible representation vector subspaces, the remaining action from right $H \backslash G \,\overleftarrow{\times}\, G$, all with the same given H-representation.

Explicit group G-representations, especially for noncompact G, can be induced from subgroup H-representations. Without linear structure: The theory of group G-realizations, all up to equivalence, relies on the fact that a group has both left and right multiplication. The *irreducible realizations are the right multiplications on the left*[2] *quotients $H \backslash G$ (H-orbits, H-classes)* with the subgroups $\{H \subseteq G\}$ as fixgroups. The theory of induced group representations formulates the realization analogous theorem with linearity as additional structure. Now all possible representations of the subgroup H have to be taken into account: *Each representation of each subgroup induces a G-representation*

$$\mathrm{ind}_H^G: \quad d\Big\downarrow^{H}_{\mathbf{GL}(W)} \longmapsto \Big\downarrow^{G}_{\mathbf{GL}(\mathrm{ind}_H^G(W))} \mathrm{ind}_H^G(d).$$

The G-representation $\mathrm{ind}_H^G(d)$ with its vector space $\mathrm{ind}_H^G(W)$ will be constructed in the following.

[2] Obviously, everything in the following can be done by interchanging left and right.

For induced group representations, left and right action do not have to be equivalent. The "quadratic" matrix algebras $V \otimes V^T$ with two-sided $G \times G$-action, e.g., $\mathbb{C}(d_Z \times d_Z)$ for compact groups are decomposed into "rectangular" vector spaces $W \otimes V^T$ (transmutators) with left-right $H \times G$-action.

7.7.1 Subgroup Intertwiners on a Group

With a subgroup H-representation $d : H \longrightarrow \mathbf{GL}(W)$ the group mappings $W(G) = W^G$ come with an H-action:

$$h \in H : \quad \begin{array}{ccc} G & \xrightarrow{L_h} & G \\ w \downarrow & & \downarrow {}_hw, \\ W & \xrightarrow[d(h)]{} & W \end{array} \quad \begin{array}{l} H \times W(G) \longrightarrow W(G), \quad w \longmapsto {}_hw \\ \text{with } {}_hw(k) = d(h).w(h^{-1}k). \end{array}$$

Like the group functions $\mathbb{C}(G)$, the mappings $W(G)$ also take into account the G-structures, e.g., periodicity $w(\alpha + 2\pi) = w(\alpha)$ for the compact degrees of freedom.

Those mappings that are compatible with the H-orbit structure constitute the vector subspace with the *H-intertwiners on G for the representation d*; they are the *invariants* for the H-action

$$\begin{aligned} w = {}_hw \in \mathbf{set}_H(G, W) &= \mathrm{INV}_H W(G) = W_H(G) = W^{H\backslash G}, \\ w(hk) &= d(h).w(k) \text{ for all } h \in H, k \in G. \end{aligned}$$

An H-intertwiner w on G maps the H-equivalence classes in the group, i.e., the left orbits $Hk \in H\backslash G$, into the H-orbits $H \bullet w(k)$ in the vector space, i.e., into the equivalence classes $W/d[H]$: Intertwiners give W-valued mappings on symmetric spaces $H\backslash G$.

In analogy to the canonical projection of the group to the H-classes $G \ni k \longmapsto Hk \ni H\backslash G$, all H-interwiners can be obtained by the projection of $W(G)$ to the H-orbits, effected by integration with Haar measure of H:

$$\begin{aligned} &W^G \ni f \longmapsto H \bullet f \ni W^{H\backslash G}, \quad (H \bullet f)(k) = \int_H dh \; {}_hf(k) = \int_H dh \; d(h)f(h^{-1}k), \\ &{}_{h'}(H \bullet f)(k) = d(h').(H \bullet f)(h'^{-1}k) = \int_H dh \; d(h'h)f((h'h)^{-1}k) = (H \bullet f)(k). \end{aligned}$$

Trivial H-representations lead to the class functions $\mathbb{C}^{H\backslash G}$, involving the Lebesgue functions $L^p(H\backslash G)$ on the homogeneous space with the subgroup classes

$$\mathbb{C}^G \supseteq \mathbb{C}^{H\backslash G}, \quad W^G \supseteq W^{H\backslash G}.$$

In the case of the trivial group $H = \{e\}$ with the H-orbits consisting of one element, one has all mappings $\mathbf{set}_{\{e\}}(G, W) = W(G)$.

For the H-intertwiners on G the group function algebra, the "huge" vector space $\mathbb{C}(G)$ with the left-right regular $G \times G$-action, is rearranged by collecting the 1-dimensional vector subspaces $k\mathbb{C}$ into "larger" W-isomorphic subspaces

$W(Hk)$ for each class, which are directly integrated with a measure of the H-orbits:

$$\begin{array}{l} W_H(G) = {}^{\oplus}\!\int_{H\backslash G} d\mu(Hk)\ W(Hk),\quad W(Hk) = W \times \{Hk\} \cong W \cong \mathbb{C}^m, \\ \dim_{\mathbb{C}} W^{H\backslash G} = \dim_{\mathbb{C}} W \cdot \operatorname{card} H\backslash G. \end{array}$$

If there exists a G-invariant positive $H\backslash G$-measure, it is unique up to a factor. For a compact group H there exists always one, for any closed H a *quasi-invariant measure*, which transforms with a positive continuous function:

$$\begin{array}{rll} G\text{-invariant:} & d\mu(Hkg) & = d\mu(Hk), \\ G \text{ quasi-invariant:} & d\mu(Hkg) & = f(Hk,g)d\mu(Hk) \text{ with } f(Hk,g) > 0, \\ \text{notation:} & {}^{\oplus}\!\int_{H\backslash G} d\mu(Hk) & = {}^{\oplus}\!\int dHk. \end{array}$$

Unless both the dimension of W and the index of H in G are finite, e.g., for the full group $W_G(G) \cong W$, the intertwiner space $W_H(G)$ is infinite-dimensional. With a basis $\{|a\rangle\}_{a=1}^m$ for W and $\{|Hk,a\rangle\}_{a=1}^m$ for $W(Hk)$ an H-intertwiner on G can be expanded with a measure-related distributive basis and the function values as coefficients (bra-ket notation),

$$\begin{array}{rll} W^{H\backslash G} & \ni |w\rangle & = {}^{\oplus}\!\int dHk\ |Hk,a\rangle w(Hk)_a \ \textit{(canonical expansion)}, \\ \text{generalizing } \mathbb{C}^G & \ni |f\rangle & = {}^{\oplus}\!\int dk\ \ |k\rangle f(k), \\ W^n & \ni |w\rangle & = \bigoplus_{\iota=1}^{n} |\iota;a\rangle\ w_a^\iota \text{ for } \operatorname{card} H\backslash G = n, \end{array}$$

or as a $(\dim_{\mathbb{C}} W \times \operatorname{card} H\backslash G)$-"matrix" with $H\backslash G$-indexed, possibly over-countably infinitely many columns of length m:

$$W^{H\backslash G} \ni |w\rangle \cong \underbrace{\left.\begin{pmatrix} w(Hk_1)_1 & w(Hk_2)_1 & \dots & w(Hk_r)_1 & \dots \\ w(Hk_1)_2 & w(Hk_2)_2 & \dots & w(Hk_r)_2 & \dots \\ \dots & & \dots & \dots & \dots \\ w(Hk_1)_m & w(Hk_2)_m & \dots & w(Hk_r)_m & \dots \end{pmatrix}\right\} m{=}\dim_{\mathbb{C}} W}_{\operatorname{card} H\backslash G \text{ times}}$$

$$\text{e.g., } \mathbb{C}^G \ni |f\rangle \cong \underbrace{\begin{pmatrix} f(k_1) & f(k_2) & \dots & f(k_r) & \dots \end{pmatrix}}_{\operatorname{card} G \text{ times},\ k_r \in G}$$

The column $w(Hk)$ arises with a "width" (multiplicity, measure) dHk.

7.7.2 Examples for Symmetric Space Mappings

In the following examples for H-intertwiners on a group G the classes $Hk \in H\backslash G$ are parametrized by vectors from G-spaces with fixgroup H.

An example, relevant for scattering structures, are the intertwiners $\mathbb{C}^2(\Omega^2, L)$, which map the orbits of an axial rotation subgroup $\mathbf{SO}(2)$ in all rotations $\mathbf{SO}(3)$, i.e., the 2-sphere Ω^2, into $\mathbf{SO}(2)$-orbits in $W \cong \mathbb{C}^2$, acted on by an $\mathbf{SO}(2)$-representation with winding numbers (polarization) $L = 0, 1, \dots$

and left or right polarization $h = \pm 1$:

$$\begin{array}{ccc} \mathbf{SO}(3) & \stackrel{O_3(\chi)}{\longrightarrow} & \mathbf{SO}(3) \\ |w\rangle \downarrow & & \downarrow |w\rangle, \\ \mathbb{C}^2 & \underset{O^L(\chi)}{\longrightarrow} & \mathbb{C}^2 \end{array} \qquad \begin{array}{rl} \mathbf{SO}(2)\backslash\mathbf{SO}(3) \cong \Omega^2 & \longrightarrow \mathbb{C}^2, \\ \vec{\omega} = \frac{\vec{q}}{|\vec{q}|} & \longmapsto w(\vec{\omega}), \end{array}$$

$$|w\rangle = {}^{\oplus}\!\!\int d^2\omega\ |\vec{\omega}, h\rangle w(\vec{\omega})_h \in {}^{\oplus}\!\!\int d^2\omega\ |\vec{\omega}\rangle\mathbb{C}^2.$$

The 2-sphere is parametrizable by the momenta directions in $\mathbb{R}^3 \cong \mathbb{R}_+ \times \Omega^2$.

Relevant for internal transformations are the intertwiners $\mathbb{C}(\mathcal{G}^3, Z)$, which map electromagnetic $\mathbf{U}(1)$-Cartan group orbits in hyperisospin $\mathbf{U}(2)$, i.e., the Goldstone manifold $\mathcal{G}^3$, into $\mathbf{U}(1)$-orbits in $W \cong \mathbb{C}$, acted on by an irreducible representation with winding (electromagnetic charge) numbers $Z \in \mathbb{Z}$:

$$\begin{array}{ccc} \mathbf{U}(2) & \stackrel{L_t}{\longrightarrow} & \mathbf{U}(2) \\ |w\rangle \downarrow & & \downarrow |w\rangle, \\ \mathbb{C} & \underset{d^Z(t)}{\longrightarrow} & \mathbb{C} \end{array} \qquad \begin{array}{rl} \mathbf{U}(1)_+\backslash\mathbf{U}(2) \cong \mathcal{G}^3 & \longrightarrow \mathbb{C}, \\ e^{i(-\gamma_3\mathbf{1}_2 + \vec{\gamma}\vec{\tau})} & \longmapsto w(\vec{\gamma}), \end{array}$$

$$|w\rangle = {}^{\oplus}\!\!\int d^3\gamma\ \ |\vec{\gamma}\rangle w(\vec{\gamma}) \in {}^{\oplus}\!\!\int d^3\gamma\ |\vec{\gamma}\rangle\mathbb{C}.$$

The Goldstone manifold is parametrizable by three Goldstone degrees of freedom in the Higgs vectors $\frac{\Phi}{|\Phi|}$ in $\mathbb{C}^2 \cong \mathbb{R}_+ \times \mathcal{G}^3$.

Relevant for massive particles are the intertwiners $\mathbb{C}^{1+2J}(\mathcal{Y}^3)$, which map the orbits of a rest-system-defined spin subgroup $\mathbf{SU}(2)$ in the Lorentz group $\mathbf{SL}(\mathbb{C}^2)$, i.e., the energy-momentum hyperboloid $\mathcal{Y}^3$ into spin orbits in a representation space $W \cong \mathbb{C}^{1+2J}$, acted on by an irreducible $\mathbf{SU}(2)$-representation for spin $2J = 0, 1, \ldots$:

$$\begin{array}{ccc} \mathbf{SL}(\mathbb{C}^2) & \stackrel{L_u}{\longrightarrow} & \mathbf{SL}(\mathbb{C}^2) \\ |w\rangle \downarrow & & \downarrow |w\rangle, \\ \mathbb{C}^{1+2J} & \underset{2J(u)}{\longrightarrow} & \mathbb{C}^{1+2J} \end{array} \qquad \begin{array}{rl} \mathbf{SU}(2)\backslash\mathbf{SL}(\mathbb{C}^2) \cong \mathcal{Y}^3 & \longrightarrow \mathbb{C}^{1+2J}, \\ \frac{q}{m} = \mathrm{y} = \begin{pmatrix} \cosh\psi \\ \frac{\vec{q}}{|\vec{q}|}\sinh\psi \end{pmatrix} & \longmapsto w(\frac{q}{m}) = w(\mathrm{y}), \end{array}$$

$$\begin{aligned} |w\rangle = {}^{\oplus}\!\!\int \frac{d^3q}{q_0}\ |\vec{q}, A\rangle w(\vec{q})_A \in {}^{\oplus}\!\!\int \frac{d^3q}{q_0}\ |\vec{q}\rangle\mathbb{C}^{1+2J} = {}^{\oplus}\!\!\int d^3\mathrm{y}\ |\mathrm{y}\rangle\mathbb{C}^{1+2J}, \\ d^3\mathrm{y} = \frac{d^3q}{q_0}, \quad \frac{q_0}{m} = \sqrt{1 + \frac{\vec{q}^2}{m^2}}. \end{aligned}$$

The boost classes $\mathcal{Y}^3$ can be parametrized by momenta or with hyberbolic coordinates.

The compact-noncompact polar decomposition $\mathbf{GL}(\mathbb{C}^n) = \mathbf{U}(n) \circ \mathbf{D}(n)$ for $n = 1, 2$ gives rise to intertwiners $\mathbb{C}(\mathbb{R}_+, Z)$ and $\mathbb{C}^{1+2T}(\mathbb{R}^4_+, y)$, which map the orbits of the compact subgroups $\mathbf{U}(1) \subset \mathbf{GL}(\mathbb{C})$ and hyperisospin $\mathbf{U}(2)$ in the extended Lorentz group $\mathbf{GL}(\mathbb{C}^2)$ into $\mathbf{U}(1)$- and $\mathbf{U}(2)$-orbits in representation spaces $\mathbb{C}$ and $\mathbb{C}^{1+2T}$ respectively:

$$\begin{array}{ccc} \mathbf{GL}(\mathbb{C}), \mathbf{GL}(\mathbb{C}^2) & \xrightarrow{L_u} & \mathbf{GL}(\mathbb{C}), \mathbf{GL}(\mathbb{C}^2) \\ |w\rangle \downarrow & & \downarrow |w\rangle, \\ \mathbb{C},\ \mathbb{C}^{1+2T} & \underset{d^Z, d^{[y|2T]}(u)}{\longrightarrow} & \mathbb{C},\ \mathbb{C}^{1+2T} \end{array}$$

$$k = u(k)|k| \in \mathbf{GL}(\mathbb{C}^n),$$
$$\left.\begin{array}{ll} \mathbf{U}(1)\backslash\mathbf{GL}(\mathbb{C}) \cong \mathbb{R}_+ & \longrightarrow \mathbb{C}, \\ \mathbf{U}(2)\backslash\mathbf{GL}(\mathbb{C}^2)) \cong \mathbb{R}^4_+ & \longrightarrow \mathbb{C}^{1+2T}, \end{array}\right\} \quad |k| \longmapsto w(|k|).$$

The classes $\mathbf{D}(1)$ and $\mathbf{D}(2)$ can be parametrized by the (space)time translations of the future cone or with Lie parameters:

$$\begin{array}{llll} \mathbb{R}_+ \ni t & = e^{\psi}, & \int_0^\infty \frac{dt}{t} & = \int d\psi, \\ \mathbb{R}^4_+ \ni x & = e^{\psi_0}(\cosh\psi + \frac{\vec{x}}{r}\sinh\psi), & \int_{\mathbb{R}^4_+} \frac{d^4x}{(x^2)^2} & = \int d\psi_0\ d^3\mathbf{y}. \end{array}$$

With distributive bases $\{|t\rangle\}$ for $W \cong \mathbb{C}$ and $\{|x,\alpha\rangle\}$ for $W \cong \mathbb{C}^{1+2T}$ the $\mathbf{U}(n)$-intertwiners on $\mathbf{D}(n)$ have the canonical expansion with a $\mathbf{GL}(\mathbb{C}^n)$-invariant positive measure of the future cone:

$$|w\rangle = \left\{ \begin{array}{lll} {}^\oplus\!\int_0^\infty \frac{dt}{t}\ |t\rangle w(t) & \in {}^\oplus\!\int d\psi\ |\psi\rangle\mathbb{C}, & n = 1, \\ {}^\oplus\!\int_{\mathbb{R}^4_+} \frac{d^4x}{(x^2)^2}\ |x,\alpha\rangle w(x)_\alpha & \in {}^\oplus\!\int d\psi_0\ d^3\mathbf{y}\ |\psi_0, \psi, \vec{\omega}\rangle\mathbb{C}^{1+2T}, & n = 2. \end{array}\right.$$

7.7.3 Inducing and Reducing Representations

The subgroup H-intertwiners $W^{H\backslash G}$ (orbit mappings) constitute the vector space acted on by the *G-representation, induced by the H-representation* $d : H \longrightarrow \mathbf{GL}(W)$ and defined by right multiplication, left over from the two-sided regular $G \times G$-representation. Induced representations are right regular representations on vector space H-orbits:

$$g \in G: \quad \begin{array}{ccc} G & \xrightarrow{R_g} & G \\ |w\rangle \downarrow & & \downarrow |w_g\rangle, \\ W & \underset{\mathrm{id}_W}{\longrightarrow} & W \end{array} \qquad \begin{array}{l} \mathrm{ind}_H^G(W) = W^{H\backslash G}, \quad \mathrm{ind}_H^G(d) = \mathcal{R}^a, \\ \mathcal{R}^d_g : W^{H\backslash G} \longmapsto W^{H\backslash G}, \quad |w\rangle \longmapsto |w_g\rangle, \end{array}$$

$$\begin{array}{rl} |w\rangle & = {}^\oplus\!\int dHk\ |Hk, a\rangle w(Hk)_a, \\ |w_g\rangle = |w\rangle \bullet g & = {}^\oplus\!\int dHk\ |Hk, a\rangle w(Hkg)_a. \end{array}$$

The induced G-representation structures come as direct integrals with the invariant $H\backslash G$-measure over the W-isomorphic spaces with the inducing H-representations.

The right regular G-representation is induced from the trivial $\{e\}$-representation $\mathrm{ind}_{\{e\}}^G(\mathrm{id}_\mathbb{C}) = \mathcal{R}$ and $\mathrm{ind}_{\{e\}}^G(\mathbb{C}) = \mathbb{C}^G$. For the full group $W_G(G) \cong W$ with $\mathrm{card}\, G\backslash G = 1$ the induced G-representation is equivalent to the inducing G-representation, $\mathrm{ind}_G^G(D) = D$ and $\mathrm{ind}_G^G(W) = W$.

Equivalent H-representations induce equivalent G-representations. The covariant *representation-inducing functor* relates to each other the representation classes

$$H \subseteq G: \quad \mathrm{ind}_H^G : \mathbf{rep}\, H = [\underline{\mathbf{vec}}_H] \longrightarrow [\underline{\mathbf{vec}}_G] = \mathbf{rep}\, G,$$
$$\begin{array}{ccc} H & & G \\ d \downarrow & \longmapsto & \downarrow \mathcal{R}^d . \\ \mathbf{GL}(W) & & \mathbf{GL}(W^{H\backslash G}) \end{array}$$

The functor is universal; the symmetric space mappings $W^{H\backslash G}$ are defined up to G-isomorphy. Any H-intertwiner $f : W \longrightarrow V$ into a G-representation vector space can be factorized with the embedding $W \ni |w\rangle \longmapsto |\iota(w)\rangle \in W^{H\backslash G}$ by the mappings with constant coefficients for each class,

$$H\backslash G \longrightarrow W, \quad Hk \longmapsto |w\rangle = |a\rangle w_a, \quad |\iota(w)\rangle = {}^\oplus\!\!\int dHk\; |Hk, a\rangle w_a,$$

and a unique G-intertwiner $\tilde{f}$,

$$\begin{array}{ccc} W & \xrightarrow{\iota} & W^{H\backslash G} \\ f \downarrow & & \downarrow \tilde{f} \\ V & \xrightarrow[\mathrm{id}_V]{} & V \end{array}, \qquad \begin{array}{rl} W, \iota, f & \in \underline{\mathbf{vec}}_H, \\ W^{H\backslash G}, V, \tilde{f} & \in \underline{\mathbf{vec}}_G. \end{array}$$

With a decomposable H-representation, the induced G-representation is also decomposable, i.e., the functor is additive:

$$(W_1 \oplus W_2)^{H\backslash G} = W_1^{H\backslash G} \oplus W_2^{H\backslash G}.$$

Inducing is compatible with direct products:

$$\mathrm{ind}_{H_1 \times H_2}^{G_1 \times G_2} \cong \mathrm{ind}_{H_1}^{G_1} \times \mathrm{ind}_{H_2}^{G_2}.$$

A subgroup representation is *reduced* from a group representation, denoted by $H \bullet W \subseteq G \bullet V$, if $H \subseteq G$ and $W \subseteq V$ (always up to isomorphism, e.g., a W-isomorphic subspace of V), and conversely the group representation $G \bullet V \supseteq H \bullet W$ is *induced* from the subgroup representation. The expression "induced" is justified: For a proper Lie subgroup $H \subset G$, the infinite-dimensional G-representation, induced from a finite-dimensional H-representation, contains all $H \bullet W$ extending finite-dimensional representations $G \bullet V$.

Reduction is "inverse" to induction with the additive covariant *representation-reducing functor*

$$G \supseteq H: \quad \mathrm{red}_H^G : \mathbf{rep}\, G = [\underline{\mathbf{vec}}_G] \longrightarrow [\underline{\mathbf{vec}}_H] = \mathbf{rep}\, H,$$
$$\begin{array}{ccc} G & & H \\ D \downarrow & \longmapsto & \downarrow \bigoplus_\iota d^\iota . \\ \mathbf{GL}(V) & & \mathbf{GL}(\bigoplus_\iota W^\iota) \end{array}$$

In general, the reduction of an induced representation $\mathcal{R}^d|_H$ leads to more H-representations than the original one d (Frobenius's multiplicity; more below):

$$\mathrm{red}_H^G \circ \mathrm{ind}_H^G \supseteq \mathrm{id}_{\mathbf{rep}\, H}.$$

Each G-subrepresentation in $\mathcal{R}^d$ contains an H-representation equivalent to the inducing representation d:

$$H \subset G: \quad R \subseteq \mathcal{R}^d \iff R[G].V \supseteq d[H].W.$$

7.7.4 Projection to Subgroup Representations

For a closed Lie subgroup $H \subseteq G$ and Haar measures, the (generalized) G-functions can be projected [5] to (generalized) G/H-functions by integration over the subgroup (where defined):

$$\begin{array}{rl} \mu(k) & \longmapsto \mu(Hk) = \int_H dh\, \mu(hk), \\ \text{e.g., } \delta_g(k) & \longmapsto \delta_g(Hk) = \int_H dh\, \delta(hkg^{-1}). \end{array}$$

Under certain conditions, related to unimodularity [5], there is the integral decomposition with respect to the subgroup with suitably normalized invariant measures,

$$\int_G dk\, \mu(k) = \int_{H\backslash G} dHk\, \mu(Hk) = \int_{H\backslash G} dHk \int_H dh\, \mu(hk).$$

With a group factorization (Cartan or Iwasawa) there exist corresponding measure decompositions, where in general the modularities of group and subgroup have to be taken into account:

$$\int_G dk = \int_H dh \int_{H\backslash G} dHk \;\; \text{e.g.,} \left\{ \begin{array}{rl} \int_{\mathbf{GL}(\mathbb{C}^2)} d^8 g & = \int_{\mathbf{U}(2)} d^4 u \int_{\mathbf{D}(2)} d^4 d, \\ \int_{\mathbf{U}(2)} d^4 u & = \int_{-\pi}^{\pi} d\alpha_0 \int_{\mathbf{SU}(2)} d^3 u, \\ \int_{\mathbf{SU}(2)} d^3 u & = \int_{-2\pi}^{2\pi} d\chi \int_{\Omega^2} d^2\omega, \\ \int_{\mathbf{D}(2)} d^4 d & = \int_{\mathbb{R}} d\psi_0 \int_{\mathcal{Y}^3} d^3\mathbf{y}, \\ \int_{\mathcal{Y}^3} d^3\mathbf{y} & = \int_{\mathbb{R}} \sinh^2 \psi\, d\psi \int_{\Omega^2} d^2\omega. \end{array} \right.$$

Other examples are (semi)direct product groups $G = L \vec{\times} H$ with normal subgroups H and subgroups $L \cong H\backslash G$: For example, translation representations are projected to representations of translation subgroups by integrations over the translation group $\mathbb{R}$, which gives the Dirac measure δ_0 of the trivial additive group $\{0\}$,

$$\begin{array}{lll} \mathbb{R} & \longrightarrow \{0\}: & \int \frac{dx}{2\pi} e^{iqx} = \delta(q) \sim \delta_0 \cong 1, \\ \mathbb{R}^n & \longrightarrow \mathbb{R}^{n-k}: & \int \frac{d^kx}{(2\pi)^k} e^{ipy+iqx} = [\delta(q)]^k e^{ipy} \cong e^{ipy}. \end{array}$$

Representations of Euclidean groups can be projected to lower-dimensional ones, e.g., for 3-dimensional position translations with the chain of positive-type functions for the subgroups $j_0 \longmapsto \mathcal{J}_0 \longmapsto \cos$,

$$\begin{array}{rl} \mathbf{SO}(3) \vec{\times} \mathbb{R}^3 \longrightarrow \mathbf{SO}(2) \vec{\times} \mathbb{R}^2: & \int \frac{dx_3}{2\pi} j_0(Pr) = \int \frac{dx_3}{2\pi} \int \frac{d^3q}{2\pi P} \delta(\vec{q}^2 - P^2) e^{i\vec{q}\vec{x}} \\ & = \int \frac{d^2q}{2\pi P} \delta(\vec{q}^2 - P^2) e^{i\vec{q}\vec{x}} = \frac{1}{2P} \mathcal{J}_0(|P\vec{x}|), \\ \longrightarrow \mathbb{R}: & \int \frac{dx_2 dx_3}{(2\pi)^2} j_0(Pr) = \frac{1}{2P} \int \frac{dx_2}{2\pi} \mathcal{J}_0(|P\vec{x}|) \\ & = \frac{1}{2P^2} \cos Px_1. \end{array}$$

Particle representations of the Poincaré group have nontrivial projections for time translations and trivial ones for the Euclidean group with momentum $\vec{q} = 0$,

$$\begin{array}{lll} \mathbf{SO}_0(1,3) \vec{\times} \mathbb{R}^4 & \longrightarrow \mathbb{R}: & \int \frac{d^3x}{(2\pi)^3} \int d^4q\, \delta(q^2 - m^2) e^{iqx} = \frac{1}{|m|} \cos mx_0, \\ \mathbf{SO}_0(1,3) \vec{\times} \mathbb{R}^4 & \longrightarrow \mathbf{SO}(3) \vec{\times} \mathbb{R}^3: & \int dx_0 \int d^4q\, \delta(q^2 - m^2) e^{iqx} = 0. \end{array}$$

The decomposition with respect to time representations shows the positive-type functions (spherical Bessel function) of irreducible representations of the Euclidean group for nontrivial momenta $\vec{q}^2 = q_0^2 - m^2 > 0$:

$$\begin{array}{rl} \mathbf{SO}_0(1,3) \vec{\times} \mathbb{R}^4 & \supset [\mathbf{SO}(3) \vec{\times} \mathbb{R}^3] \times \mathbb{R}, \\ \int \frac{d^4q}{2\pi} \delta(q^2 - m^2) e^{iqx} & = \int_{|m|}^\infty dq_0 \, \frac{\sin \sqrt{q_0^2 - m^2}\, r}{r} \cos q_0 x_0. \end{array}$$

7.7.5 Induced Positive-Type Measures

The embedding of a positive-type Radon distribution of a closed subgroup of a locally compact group $G \supseteq H$ defines a positive-type Radon G-distribution [3]:

$$\begin{array}{l} \mathcal{M}(H)_+ \ni \omega_H \longmapsto \omega_G \in \mathcal{M}(G)_+ \\ \text{with } \langle \omega_G, f \rangle = \langle \omega_H, f|_H \rangle = \int_H \omega_H(h) dh f(h) \text{ for } f \in \mathcal{C}_c(G). \end{array}$$

For non-unimodular groups the embedded measure has to be multiplied by $\sqrt{\frac{\Delta_G(h)}{\Delta_H(h)}}$ with the modular functions. If ω_H characterizes the class $[d]$ of the regular Hilbert representation of the subgroup, then ω_G characterizes the class $[\mathcal{R}^d]$ of the induced Hilbert representation The induced scalar product comes from:

$$\begin{array}{rl} f, f' \in \mathcal{C}_c(G): \quad \langle f | f' \rangle_{\omega_G} & = \langle \hat{f} * f' |_H \rangle_{\omega_H} = \int_G dk \int_H dh \, \overline{f(k)} \omega_H(h) f'(hk) \\ & = \int_{H\backslash G} dHk \int_{H \times H} dh_1 dh_2 \, \overline{f(h_1 k)} \omega_H(h_1^{-1} h_2) f'(h_2 k). \end{array}$$

Examples are given by the functions of a (semi)direct poduct group:

$$G = L \vec{\times} H : \ \langle f|f'\rangle_{\omega_G} = \int_L dl \int_{H\times H} dh_1 dh_2 \ \overline{f(l,h_1)} \ \omega_H(h_1^{-1}h_2) \ f'(l,h_2),$$

in the simplest case for the abelian product group $\mathbb{R}^{k+s}$ and their Fourier transforms, e.g., for time and position translations with $(X,x) \to (x_0,\vec{x})$ and $(Q,q) = (q_0,\vec{q})$:

$$\begin{aligned} G = \mathbb{R}^k \oplus \mathbb{R}^s : \ \langle f|f'\rangle_{\omega_{k+s}} &= \int d^kX \int d^s x_1 d^s x_2 \ \overline{f(X,x_1)} \ \omega_s(x_2 - x_1) \ f'(X,x_2) \\ &= \int \frac{d^kQ}{(2\pi)^t} \int \frac{d^sq}{(2\pi)^s} \ \overline{\tilde{f}(Q,q)} \ \tilde{\omega}_s(q) \ \tilde{f}'(Q,q). \end{aligned}$$

The regular representation of a subgroup with $\omega_H = \delta_e$ on $L^2(H)$ induces the regular representation of the full group on $L^2(G)$. The G-intertwiners $|Uf\rangle$ with $f \in \mathcal{C}_c(G)$ are valued in the H-space $\mathcal{C}_c(H)$

$$f \in \mathcal{C}_c(G), \quad \begin{array}{ccc} G & \xrightarrow{L_h} & G \\ {\scriptstyle |Uf\rangle}\downarrow & & \downarrow{\scriptstyle |Uf\rangle} \\ \mathcal{C}_c(G) & \xrightarrow[\mathcal{L}_h]{} & \mathcal{C}_c(G) \end{array}, \quad \begin{array}{l} |Uf\rangle(k) = f_k \in \mathcal{C}_c(H), \\ f_k(h) = f(hk), \\ \mathcal{R}_g.f_k = f_{kg} \end{array}$$

$$\Rightarrow \ \mathrm{ind}_H^G(\mathcal{R}_H) \cong \mathcal{R}_G.$$

Groups can be mapped to their representation classes as $\oplus$ -additive semigroups with the covariant functor

$$\begin{array}{ccc} H & & \mathbf{rep}\, H \\ {\scriptstyle \subseteq}\downarrow & \longmapsto & \downarrow{\scriptstyle \mathrm{ind}_H^G}. \\ G & & \mathbf{rep}\, G \end{array}$$

The morphism set for the groups used here involves only, if existent, the inclusion $\{H \hookrightarrow G\}$; otherwise it is empty $\emptyset$. The functor properties: Representation inducing is compatible with subgroup order; inducing can be performed in stages (transitivity):

$$\begin{aligned} & \mathrm{ind}_H^H = \mathrm{id}_{\mathbf{rep}\, H}, \\ \text{subgroups } K \subseteq H \subseteq G \Rightarrow \ & \mathrm{ind}_H^G \circ \mathrm{ind}_K^H \cong \mathrm{ind}_K^G. \end{aligned}$$

For example, a polarization representation can induce a spin representation can induce a Lorentz group representation:

$$\begin{array}{ccccc} \mathbf{SO}(2) & \subset & \mathbf{SU}(2) & \subset & \mathbf{SL}(\mathbb{C}^2), \\ \text{polarization} & & \text{spin} & & \text{“left-right spin”} \\ e^{\pm i\chi Z} & \hookrightarrow & \bigvee^{2J} u & \hookrightarrow & \bigvee^{2L} s \otimes \bigvee^{2R} \hat{s}, \\ |Z| & & J \geq \frac{|Z|}{2} & & L + R \geq J. \end{array}$$

7.8 Harmonic Analysis of Symmetric Spaces

In general, the G-representation on $W^{H\backslash G}$ is reducible, i.e., there are subspaces of H-intertwiners on G that are stable under G-action. It will be assumed that a decomposition of an induced G-representation, i.e., a *harmonic analysis of symmetric space mappings* $W^{H\backslash G}$, is possible with finite-dimensional transmutators $H \bullet W \leftrightarrow G \bullet V$ (next subsection) from the H-space W with bases $\{|Hk, a\rangle \mid a = 1, \ldots, m\}$ for $W(Hk)$ to irreducible G-spaces V_D with bases $\{|D; j\rangle \mid j = 1, \ldots, n\}$. Those subspaces arise with multiplicities n_D.

Such a decomposition of $W^{H\backslash G}$ into irreducible Hilbert spaces V_D is possible for compact groups G with Frobenius's theorem (below). For noncompact groups, finite-dimensional irreducible spaces V_D do not have to be Hilbert spaces, i.e., the representations on them can be indefinite unitary.

The intertwiners have the expansions

$$\begin{array}{rlll} & W^{H\backslash G} \ni w & \leftrightarrow & \tilde{w} \in W^{H\backslash G}, \\ & \text{canonical expansion} & & \text{harmonic expansion} \\ W^{H\backslash G} \ni |w\rangle & = {}^{\oplus}\!\int_{H\backslash G} dHk\ |Hk, a\rangle w(Hk)_a & = & \bigoplus_{d \subseteq D \in \check{G}} n_D |D; j\rangle\ \tilde{w}(D)_j, \\ W\text{-mapping values:}\ \ w(Hk)_a & = \langle Hk, a|w\rangle & = & \sum_{d \subseteq D \in \check{G}} n_D D(Hk)^j_a\ \tilde{w}(D)_j, \\ \text{harmonic coefficients:}\ \ \tilde{w}(D)_j & = \langle D; j|w\rangle & = & \int dHk\ \check{D}(Hk)^a_j\ w(Hk)_a. \end{array}$$

The decomposition of the identity with dual distributive bases looks as follows:

$$\begin{array}{rll} \mathrm{id}_{W^{H\backslash G}} & = {}^{\oplus}\!\int dHk\ \mathrm{id}_{W(Hk)} & = \bigoplus_{d \subseteq D \in \check{G}} n_D\, \mathrm{id}_{V_D} \\ & \cong {}^{\oplus}\!\int dHk\ |Hk, a\rangle\langle Hk, a| & = \bigoplus_{d \subseteq D \in \check{G}} n_D |D; j\rangle\langle D; j|. \end{array}$$

The sum goes over all G-representations D on spaces V_D that are suprepresentations of d:

$$W^{H\backslash G} \supseteq V_D \ni |D; j\rangle = {}^{\oplus}\!\int dHk\ |Hk, a\rangle D(Hk)^j_a \quad \text{for all } G \bullet V_D \supseteq H \bullet W.$$

The transmutators $D(Hk)^{j=1,\ldots,n}_{a=1,\ldots,m}$ mediate between the inducing H-spaces $W \cong W(Hk)$ and the G-spaces V_D:

$$\begin{array}{rl} \mathrm{id}_{W^{H\backslash G}} & \cong \bigoplus_{d \subseteq D \in \check{G}} n_D\ {}^{\oplus}\!\int dHk\ |Hk, a\rangle\ D(Hk)^j_a\ \langle D; j|, \\ D(Hk)^j_a & = \langle Hk, a|D; j\rangle, \quad \check{D}(Hk)^a_j = \langle D; j|Hk, a\rangle. \end{array}$$

In the harmonic analysis of group functions $\mathbb{C}^G$, both the canonical basis $|k\rangle \in G$ and the harmonic bases $|D\rangle \in \mathbf{rep}\, G$ have a two-sided (from left and right) G-action $G \bullet (V_D \otimes V_D^T) \bullet G$, whence for $W^{H\backslash G}$ in general, there are different one-sided actions: $H \bullet W$ (from left) and $V_D^T \bullet G$ (from right).

7.8.1 Transmutators for Induced Representations

An explicit description of a finite-dimensional extension-reduction structure $H \bullet W \subseteq G \bullet V$ for induced G-representations is given by transmutators (chapter "Spacetime as Unitary Operation Classes"). They were used in the chapters "Massive Particle Quantum Fields" and "Massless Quantum Fields" to induce a Lorentz group action $G = \mathbf{SL}(\mathbb{C}^2)$ on fields from a stabilgroup action $H \in \{\mathbf{SO}(2), \mathbf{SU}(2)\}$ on particles and in the chapter "Gauge Interactions" to reduce the hyperisospin $G = \mathbf{U}(2)$ action on fields to an electromagnetic $H = \mathbf{U}(1)$ action on particles.

For the classes $Hk \in H \backslash G$, *orbit representatives* can be chosen, e.g., a polar decomposition $k = u(k)|k| \in \mathbf{U}(n) \circ \mathbf{D}(n) = \mathbf{GL}(\mathbb{C}^n)$; in general there is no natural choice:

$$\begin{aligned} H\backslash G &\longrightarrow (H\backslash G)_{\text{repr}} \subseteq G, \quad Hk \longmapsto k_r \text{ with } Hk = Hk_r, \\ (H\backslash G)_{\text{repr}} &= \{k_r \in G \mid \text{coset representatives}\}, \\ G &= \biguplus_{\text{repr } k_r} Hk_r = H \circ (H\backslash G)_{\text{repr}}. \end{aligned}$$

The right G-realization on the left H-orbits

$$R_{g^{-1}} : H\backslash G \longrightarrow H\backslash G, \; Hk \longmapsto Hkg$$

has H-isomorphic fixgroups $\{g \in G \mid Hk = Hkg\} = kHk^{-1}$. The G-action may look complicated for the chosen representatives: It hits the chosen representative $(kg)_r$ up to a transformation with a *Wigner element* $h(k_r, g) \in H$ *from the subgroup* which depends on the acting element and the representatives chosen:

$$\begin{aligned} R_{g^{-1}} : (H\backslash G)_{\text{repr}} \longrightarrow (H\backslash G)_{\text{repr}}, \; k_r &\longmapsto k_r g = h(kg)_r \\ \text{with} \quad h &= h(k_r, g) = k_r g (kg)_r^{-1} \in H. \end{aligned}$$

A G-representation

$$G \ni k \longmapsto D(k) \overset{\text{e.g.}}{\cong} \begin{pmatrix} \bullet & \bullet & \bullet & \bullet & \bullet & \bullet & \bullet & \bullet \\ \bullet & \bullet & \bullet & \bullet & \bullet & \bullet & \bullet & \bullet \\ \bullet & \bullet & \bullet & \bullet & \bullet & \bullet & \bullet & \bullet \\ \bullet & \bullet & \bullet & \bullet & \bullet & \bullet & \bullet & \bullet \\ \bullet & \bullet & \bullet & \bullet & \bullet & \bullet & \bullet & \bullet \\ \bullet & \bullet & \bullet & \bullet & \bullet & \bullet & \bullet & \bullet \\ \bullet & \bullet & \bullet & \bullet & \bullet & \bullet & \bullet & \bullet \\ \bullet & \bullet & \bullet & \bullet & \bullet & \bullet & \bullet & \bullet \end{pmatrix} \in \mathbf{GL}(V), \quad V \cong \mathbb{C}^n \overset{\text{e.g.}}{=} \mathbb{C}^8,$$

is decomposable into subgroup H-representations with square $(m_\iota \times m_\iota)$ matrices, e.g., as $8 = 2 + 1 + 3 + 2$:

$$V \overset{H}{\cong} \bigoplus_{\iota=1}^N W^\iota, \quad H \bullet W^\iota \subseteq W^\iota, \quad W^\iota \cong \mathbb{C}^{m_\iota}, \quad \sum_{\iota=1}^N m_\iota = n,$$

$$H \ni h \longmapsto D(h) = \bigoplus_{\iota=1}^N d^\iota(h) \overset{\text{e.g.}}{\cong} \left(\begin{array}{cc|c|ccc|cc} \bullet & \bullet & 0 & 0 & 0 & 0 & 0 & 0 \\ \bullet & \bullet & 0 & 0 & 0 & 0 & 0 & 0 \\ \hline 0 & 0 & \bullet & 0 & 0 & 0 & 0 & 0 \\ \hline 0 & 0 & 0 & \bullet & \bullet & \bullet & 0 & 0 \\ 0 & 0 & 0 & \bullet & \bullet & \bullet & 0 & 0 \\ 0 & 0 & 0 & \bullet & \bullet & \bullet & 0 & 0 \\ \hline 0 & 0 & 0 & 0 & 0 & 0 & \bullet & \bullet \\ 0 & 0 & 0 & 0 & 0 & 0 & \bullet & \bullet \end{array}\right) \in \bigoplus_{\iota=1}^N \mathbf{GL}(W^\iota).$$

The *representation of the symmetric space* $H\backslash G$,

$$(H\backslash G)_{\rm repr} \ni k_r \longmapsto D(k_r) \in \mathbf{GL}(V) \subset V \otimes V^T,$$

is correspondingly decomposable into *transmutators from* $H \bullet W^\iota$ *to* $G \bullet V$ with rectangular $(m_\iota \times n)$ matrices from $W^\iota \otimes V^T$:

$$D(k_r) = \bigoplus_{\iota=1}^{N} D^\iota(k_r) = \overbrace{\begin{pmatrix} D^1(k_r) & m_1 \text{ lines} \\ \hline D^2(k_r) & m_2 \text{ lines} \\ \hline \dots & \dots \quad \dots \\ \hline D^N(k_r) & m_N \text{ lines} \end{pmatrix}}^{n \text{ columns}} \overset{\text{e.g.}}{\cong} \left(\begin{array}{cccccccc} \bullet&\bullet&\bullet&\bullet&\bullet&\bullet&\bullet&\bullet \\ \bullet&\bullet&\bullet&\bullet&\bullet&\bullet&\bullet&\bullet \\ \hline \bullet&\bullet&\bullet&\bullet&\bullet&\bullet&\bullet&\bullet \\ \hline \bullet&\bullet&\bullet&\bullet&\bullet&\bullet&\bullet&\bullet \\ \bullet&\bullet&\bullet&\bullet&\bullet&\bullet&\bullet&\bullet \\ \bullet&\bullet&\bullet&\bullet&\bullet&\bullet&\bullet&\bullet \\ \hline \bullet&\bullet&\bullet&\bullet&\bullet&\bullet&\bullet&\bullet \\ \bullet&\bullet&\bullet&\bullet&\bullet&\bullet&\bullet&\bullet \end{array}\right),$$

$$(H\backslash G)_{\rm repr} \ni k_r \longmapsto D^\iota(k_r) = |\iota; a\rangle D^\iota(k_r)^j_a \langle D; j| \in W^\iota \otimes V^T.$$

The transmutators are $(m_\iota \times n)$- dimensional vector spaces with $H \times G$-representations. They have a G-action from the right and an H-action from the left:

$$\begin{array}{ccc} (H\backslash G)_{\rm repr} & \overset{R_{g^{-1}}}{\longrightarrow} & (H\backslash G)_{\rm repr} \\ {\scriptstyle D^\iota}\downarrow & & \downarrow{\scriptstyle D^\iota_g}, \\ W^\iota \otimes V^T & \longrightarrow & W^\iota \otimes V^T \end{array} \qquad \begin{aligned} D^\iota_g(k_r) &= D^\iota(k_r g) = D^\iota(k_r) \circ D(g) \\ &= d^\iota(h(g,k_r)) \circ D((kg)_r),\ g \in G, \\ D^\iota(hk_r) &= d^\iota(h) \circ D^\iota(k_r),\ h \in H. \end{aligned}$$

For the examples above, the spin $\mathbf{SU}(2)$-representations on $\mathbf{SO}(2)$-intertwiners $\{w : \mathbf{SU}(2) \longrightarrow \mathbb{C}^2\}$, induced from an $\mathbf{SO}(2)$-representation with polarization $\pm Z \in \mathbb{Z}$, $Z \neq 0$, have to contain an $\mathbf{SO}(2)$-representation with polarization Z. Therefore, the minimal induced spin J is half the polarization $|Z|$:

$$\Omega^2 \longrightarrow \mathbf{SU}(1+2J),\ \tfrac{\vec q}{|\vec q|} \longmapsto \bigvee^{2J} u(\tfrac{\vec q}{|\vec q|}),\ \ J \geq \tfrac{|Z|}{2}.$$

For the hyperisospin $\mathbf{U}(2)$-representations on $\mathbf{U}(1)$-intertwiners $\{w : \mathbf{U}(2) \longrightarrow \mathbb{C}\}$, induced from a $\mathbf{U}(1)$-representation with charge number $Z \in \mathbb{Z}$, the minimal induced isospin T is half the charge $|Z|$:

$$\mathcal{G}^3 \longrightarrow \mathbf{U}(1+2T),\ \tfrac{\Phi}{|\Phi|} \longmapsto \bigvee^{2T} v(\tfrac{\Phi}{|\Phi|}),\ \ T \geq \tfrac{|Z|}{2}.$$

The Lorentz $\mathbf{SL}(\mathbb{C}^2)$-representations on $\mathbf{SU}(2)$-intertwiners $\{w : \mathbf{SL}(\mathbb{C}^2) \longrightarrow \mathbb{C}^{1+2J}\}$, induced from an $\mathbf{SU}(2)$-representation with spin J, have to contain a spin representation with J:

$$\mathcal{Y}^3 \longrightarrow \mathbf{SL}(\mathbb{C}^{(1+2L)(1+2R)}),\ \tfrac{q}{m} \longmapsto \bigvee^{2L} s(\tfrac{q}{m}) \otimes \bigvee^{2R} \hat s(\tfrac{q}{m}),\ \ L+R \geq J.$$

7.8.2 Lebesgue Spaces for Induced Representations

The Lebesgue spaces $L^p(G)$ for the complex group function classes are generalizable to the H-intertwiners, i.e., to the classes of mappings from the symmetric space $H\backslash G$ with G-invariant positive measure into a finite-dimensional Hilbert space $W \cong \mathbb{C}^n$ with scalar product $\langle a|b\rangle = \delta^{ab}$:

$$W^{H\backslash G} \ni |w\rangle = {}^{\oplus}\!\!\int dHk\ |Hk,a\rangle w(Hk)_a,$$

$$|w\rangle \in \mathcal{L}^p(H\backslash G, W) \iff \begin{cases} \| w \|_p = [\int_{H\backslash G} dHk\ |w(Hk)|^p]^{\frac{1}{p}} < \infty \\ \text{with } |w(Hk)|^2 = \overline{w(Hk)_a} w(Hk)_a. \end{cases}$$

Again, there are Hilbert spaces with square integrable mappings L^2. The convolution Lebesgue group algebra $L^1(G, A)$ of mappings, valued in a C*-algebra A, has Hilbert products with a positive-type mapping $L^\infty(G, A)_+$.

For square integrability with the G-invariant scalar product for the intertwiners

$$W^{H\backslash G} \times W^{H\backslash G} \longrightarrow \mathbb{C},\ \ \langle w_2|w_1\rangle = \int dHk\ \overline{w_2(Hk)_a} w_1(Hk)_a,$$

the bases for each coset element are, in general, not a basis with Hilbert vectors, only a *distributive basis* with the *measure-related Dirac distribution*, orthogonal and positive (chapter "The Kepler Factor"), as *scalar product distribution* and the corresponding completeness,

$$\{|Hk,a\rangle \mid Hk \in H\backslash G,\ a = 1,\dots,m\},\ \ \begin{cases} \langle Hk',b|Hk,a\rangle &= \delta^{ab}\delta(Hk,Hk'), \\ |Hk,a\rangle\langle Hk,a| &\cong \mathrm{id}_{L^2(H\backslash G,W)}, \end{cases}$$
$$\text{where } \int_{H\backslash G} dHk\ \delta(Hk,Hk')\langle Hk|w\rangle = \langle Hk'|w\rangle.$$

This leads to generalizations of Schur's orthonormality for representation coefficients.

G-subrepresentations of the intertwiners involve transmutators as W-valued functions:

$$W^{H\backslash G} \supseteq V_D \ni |D;j\rangle = {}^{\oplus}\!\!\int dHk\ |Hk,a\rangle D(Hk)^j_a.$$

The finite-dimensional transmutators are Hilbert representation spaces only for compact groups. There, they are complete for the harmonic analysis of the Hilbert spaces with square integrable functions $L^2(G/H)$. For noncompact groups G, the irreducible finite-dimensional spaces V_D do not have to be Hilbert spaces, and the transmutators do not have to be square integrable:

$$\langle D(Hk)^j|D(Hk)^l\rangle = \int dHk\ \overline{D(Hk)^j_a} D(Hk)^l_a.$$

7.9 Induced Representations of Compact Groups

The compact group U-representation, induced from an irreducible, i.e., finite-dimensional subgroup K-representation, is decomposable into irreducible ones.

The matrix elelements in finite-dimensional rectangular transmutators are square integrable. They are complete for the harmonic analysis of the group U and its homogeneous spaces U/K, i.e., they exhaust, in orthogonal direct Peter-Weyl decompositions with Schur orthogonality, all square integrable induced representations:

$$\text{compact} \;:\; L^2(U/K, W_d) \overset{\text{dense}}{\supseteq} \bigoplus_{D \supseteq d} n_D \, W_d \otimes V_D^T.$$

There occur all G-representations D (countably many) which contain the inducing K-representation. There is Frobenius's reciprocity theorem for the number n_D of d-induced U-representations D (below). With a basis $|a\rangle \in W_d$ and $|D;j\rangle \in V_D$ one obtains the harmonic D-components $\tilde{w}(D)_j$:

$$|w\rangle : (U/K)_r \longrightarrow W_d \quad |w\rangle = {}^{\oplus}\!\!\int_{(U/K)_r} du_r \; |u_r, a\rangle w(u_r)_a = \bigoplus_{D \supseteq d} n_D |D;j\rangle \tilde{w}(D)_j,$$

$$\text{with} \begin{cases} w(u_r)_a &= \langle u_r, a|w\rangle, \quad \tilde{w}(D)_j = \langle D;j|w\rangle, \\ |D;j\rangle &= {}^{\oplus}\!\!\int_{(U/K)_r} du_r \; |u_r, a\rangle D(u_r)^j_a, \\ w(gu_r)_a &= \bigoplus_{D \supseteq d} n_D \tilde{w}(D)_j D(gu_r)^j_a, \; g \in U, \end{cases}$$

e.g., the harmonic analysis of functions

$$L^2(U/K) \ni |f\rangle = \bigoplus_{D \supseteq d_0} n_D |D_0^j\rangle \tilde{f}(D)_j \text{ with } \begin{cases} |D_0^j\rangle &= {}^{\oplus}\!\!\int_{(U/K)_r} du_r \; |u_r\rangle D(u_r)^j_0, \\ \tilde{f}(D)_j &= \langle D;j|f\rangle. \end{cases}$$

7.9.1 Frobenius Reciprocity

Frobenius's reciprocity theorem for the K-intertwiners $L^2(U/K, W_d)$ generalizes the theorems related to the left-right action-induced square structure of the group algebras, i.e., the Maschke-Burnside-Wedderburn theorem for the group algebra $\mathbb{C}^U$ in the case of finite groups and the Peter-Weyl theorem for the square integrable functions $L^2(U)$ in the case of compact groups, now including the K-representation space $W_d \cong \mathbb{C}^m$: It states that the number n_D of equivalent irreducible U-representations of class $[D]$ induced on $L^2(U/K, W_d)$ equals the number n_d of equivalent K-representations of class $[d]$ in this $[D]$: Therefore a K-representation class in the induced U-representations comes always as a square matrix of K-representations: irreducible K-representations induce $K \times K$-representations

$$\check{U}: \quad L^2(U/K, W_d) \cong \underbrace{V_D \oplus \ldots \oplus V_D}_{n_D \text{ times}} \oplus \cdots,$$

$$\check{K}: \qquad V_D \cong \left.\begin{pmatrix} W_d \\ \oplus \\ \cdots \\ \oplus \\ W_d \end{pmatrix}\right\} n_d \text{ times} \oplus \ldots,$$

$$U \text{ compact} \Rightarrow \begin{cases} n_D = n_d, \\ L^2(U/K, W_d) \supseteq \begin{pmatrix} W_d & \ldots & W_d \\ \cdots & \cdots & \cdots \\ \cdots & \cdots & \cdots \\ W_d & \ldots & W_d \end{pmatrix} \cong W_d^{n_d} \otimes W_d^{n_d}. \end{cases}$$

With the universality of the inducing functor, one has for irreducible representations of compact groups the multiplicities given by the intertwining dimensions

$$n_d = \dim_{\mathbb{C}} \mathbf{vec}_K(W_d, V_D) = \dim_{\mathbb{C}} \mathbf{vec}_U(L^2(U/K, W_d), V_D) = n_D.$$

The Maschke-Burnside-Wedderburn-Peter-Weyl square structure is a special case: For the induced regular representation the space $L^2(U)$, i.e., $W_d = \mathbb{C}$, contains all irreducible U-representation spaces with the multiplicity equal to its dimension, $n_D = \dim_{\mathbb{C}} V_D$. The left-right isomorphism leads to a direct sum of full matrix algebras. Frobenius reciprocity gives the generalization of the harmonic analysis of the square integrable group U functions $L^2(U)$, now for the W_d-valued symmetric space mappings $L^2(U/K, W_d)$. The direct sum of matrix algebras is rearranged: In each full matrix subalgebra $V_L \otimes V_L^T \subset L^2(U)$ the irreducible U-representation space $V_L \cong \mathbb{C}^{d_L}$ contains n_d times the irreducible K-space W_d, whence also V_L^T, which leads to the square substructure

$$L^2(U) \supset V_L \otimes V_L^T \supseteq W_d^{n_d} \otimes W_d^{n_d} \subseteq L^2(U/K, W_d) \subseteq L^2(U).$$

7.9.2 Examples of Frobenius Reciprocity

The Peter-Weyl decomposition of the $\mathbf{SO}(3)$-functions

$$L^2(\mathbf{SO}(3)) \overset{\mathbf{SO}(3)}{\cong} \bigoplus_{L=0}^{\infty} \mathbb{C}^{1+2L} \otimes \mathbb{C}^{1+2L} : \left\{ \begin{array}{rl} f(u) & = \sum_{L=0}^{\infty} (1+2L) 2L(u)_a^b \tilde{f}(2L)_b^a, \\ f(u) & = \{u|f\},\ 2L(u)_a^b = \{u|2L_a^b\}, \\ |2L_a^b\} & = {}^{\oplus}\!\!\int d^3u\ |u\}\ 2L(u)_a^b \in V_L \otimes V_L^T, \end{array} \right.$$

is the reservoir to be used for the $\mathbf{SO}(3)$-representations, induced from $\mathbf{SO}(2)$-representations. Those $\mathbf{SO}(3)$-representations act on mappings of the 2-sphere into an $\mathbf{SO}(2)$-representation space W.

The defining $\mathbf{SO}(3)$-representation $L = 1$ is decomposable for trivial $\mathbf{SO}(2)$-action, i.e., for $\chi = 0$, into three axial-to-rotation transmutators:

$$\begin{array}{rl} \mathbf{SO}(3) \ni 2(u)_a^b & = (e^{i\chi} 2_+^b(\vec{\omega}), 2_0^b(\vec{\omega}), e^{-i\chi} 2_-^b(\vec{\omega})) \cong \mathbf{SO}(2) \circ \mathbf{SO}(3)/\mathbf{SO}(2), \\ (2_+^b, 2_0^b, 2_-^b)(\vec{\omega}) & = \left(\begin{array}{c|c|c} e^{i\varphi} \cos^2 \frac{\theta}{2} & ie^{i\varphi} \frac{\sin\theta}{\sqrt{2}} & -e^{i\varphi} \sin^2 \frac{\theta}{2} \\ i\frac{\sin\theta}{\sqrt{2}} & \cos\theta & i\frac{\sin\theta}{\sqrt{2}} \\ -e^{-i\varphi} \sin^2 \frac{\theta}{2} & ie^{-i\varphi} \frac{\sin\theta}{\sqrt{2}} & e^{-i\varphi} \cos^2 \frac{\theta}{2} \end{array} \right) \in \mathbf{SO}(3)/\mathbf{SO}(2). \end{array}$$

The functions $L^2(\Omega^2)$ have trivial $\mathbf{SO}(2)$-representations on the scalars $\mathbb{C}$. All irreducible complex $\mathbf{SO}(3)$-representations have exactly one trivial $\mathbf{SO}(2)$-representation: the "middle" column in the $(1+2L) \times (1+2L)$ algebra. Therefore (Frobenius) they arise exactly once in the harmonic analysis of the 2-sphere functions:

$$L^2(\Omega^2) \overset{\mathbf{SO}(3)}{\cong} \bigoplus_{L=0}^{\infty} \mathbb{C}^{1+2L} : \left\{ \begin{array}{ll} \mathrm{id}_{L^2(\Omega^2)} & \cong {}^{\oplus}\!\!\int d^2\omega |\vec{\omega}\rangle\langle\vec{\omega}| = \bigoplus_{L=0}^{\infty} |2L_0^b\rangle\langle 2L_0^b|, \\ w(\vec{\omega}) & = \sum_{L=0}^{\infty} 2L(\vec{\omega})_0^b \, \tilde{w}(2L)_b, \\ |2L_0^b\rangle & = {}^{\oplus}\!\!\int d^2\omega \, |\vec{\omega}\rangle \, 2L(\vec{\omega})_0^b \in V_L, \end{array} \right.$$

with the spherical harmonics as transmutators, the middle column $|2L_0^b\rangle = \sqrt{\frac{4\pi}{1+2L}}\mathrm{Y}_b^L$ starting with $\sqrt{\frac{4\pi}{3}}\mathrm{Y}^1$. They are the minimal left ideal types in the Peter-Weyl decomposition.

For the mappings $\{|w\rangle : \Omega^2 \longrightarrow \mathbb{C}_l^2\}$ with the nontrivial $\mathbf{SO}(2)$-representation on $\mathbb{C}_l^2 \cong \mathbb{C}^2$ by $e^{i\sigma^3\varphi} \longmapsto e^{il\sigma^3\varphi}$, $l = 1, 2, \dots$ the value space is decomposable with two irreducibles (polarization directions), $e^{\pm il\varphi}$ on $\mathbb{C}^\pm \cong \mathbb{C}$. Therefore, each nontrivial $\mathbf{SO}(3)$-representation comes for $\mathbb{C}^+$ and $\mathbb{C}^-$ with transmutators $2L_{hl}^b$ for $L \geq l$ and $h = \pm 1$:

$$L^2(\Omega^2, \mathbb{C}_l^2) \overset{\mathbf{SO}(3)}{\cong} 2 \times \bigoplus_{L=l}^{\infty} \mathbb{C}^{1+2L} : \left\{ \begin{array}{ll} \mathrm{id}_{L^2(\Omega^2, \mathbb{C}_l^2)} & \cong {}^{\oplus}\!\!\int d^2\omega |\vec{\omega}, hl\rangle\langle\vec{\omega}, hl| \\ & = \bigoplus_{L=l}^{\infty} |2L_{hl}^b\rangle\langle 2L_{hl}^b|, \\ w(\vec{\omega})_h & = \sum_{L=l}^{\infty} 2L(\vec{\omega})_{hl}^b \, \tilde{w}(2L)_b, \\ |2L_{hl}^b\rangle & = {}^{\oplus}\!\!\int d^2\omega \, |\vec{\omega}, hl\rangle \, 2L(\vec{\omega})_{hl}^b \in V_L. \end{array} \right.$$

For $l = 1$ it starts with the left and right columns in the defining representation for $\chi = 0$.

The Peter-Weyl decomposition of $\mathbf{SU}(2)$-functions

$$L^2(\mathbf{SU}(2)) \overset{\mathbf{SU}(2)}{\cong} \bigoplus_{2J=0,1,\dots} \mathbb{C}^{1+2J} \otimes \mathbb{C}^{1+2J}$$

is the reservoir for $\mathbf{SU}(2)$-representations induced from $\mathbf{SO}(2)$-representations. For the nontrivial irreducible representations $\mathbf{SO}(2) \ni e^{i\sigma^3\frac{\chi}{2}} \longmapsto e^{iZ\frac{\chi}{2}} \in \mathbf{U}(1)$ on $\mathbb{C}$ and for the decomposable ones $e^{i\sigma^3\frac{\chi}{2}} \longmapsto e^{iN\sigma^3\frac{\chi}{2}}$ on $\mathbb{C}^2$, one obtains the decomposition into all extending irreducible $\mathbf{SU}(2)$-representations with half-integer spin, once and twice respectively:

$$\begin{array}{lll} |Z| = 1, 2, \dots : & L^2(\Omega^2, \mathbb{C}_Z) \overset{\mathbf{SU}(2)}{\cong} & \bigoplus_{J=\frac{|Z|}{2}, \frac{|Z|}{2}+1, \dots} \mathbb{C}^{1+2J}, \\ N = 1, 2, \dots : & L^2(\Omega^2, \mathbb{C}_N^2) \overset{\mathbf{SU}(2)}{\cong} & 2 \times \bigoplus_{J=\frac{N}{2}, \frac{N}{2}+1, \dots} \mathbb{C}^{1+2J}. \end{array}$$

The Peter-Weyl decomposition of functions $L^2(\mathbf{SO}(4))$ gives all irreducible representations $[2J_1|2J_2]$ with integer $J_1 + J_2$. The harmonic analysis of functions $L^2(\Omega^3)$ on the 3-sphere $\Omega^3 \cong \mathbf{SO}(4)/\mathbf{SO}(3)$ ($\mathbf{SO}(3)$-intertwiners on

$\mathbf{SO}(4)$) involves - in the decomposition - only those irreducible $\mathbf{SO}(4)$-representations that contain a trivial $\mathbf{SO}(3)$-representation, i.e., $J_1 = J_2$. They occur as often as they contain a trivial $\mathbf{SO}(3)$-representation, i.e., once:

$$[2J_1|2J_2] \overset{\mathbf{SO}(3)}{\cong} \bigoplus_{L=|J_1-J_2|}^{J_1+J_2} [2L], \quad L^2(\Omega^3) \overset{\mathbf{SO}(4)}{\cong} \bigoplus_{2J=0}^{\infty} \mathbb{C}^{(1+2J)(1+2J)}.$$

The Peter-Weyl decomposition for $\mathbf{SU}(3)$-functions

$$\begin{array}{rcl} L^2(\mathbf{SU}(3)) & \overset{\mathbf{SU}(3)}{\cong} & \mathbb{C} \oplus \mathbb{C}(3\times 3) \oplus \mathbb{C}(\overline{3}\times\overline{3}) \oplus \mathbb{C}(6\times 6) \oplus \mathbb{C}(\overline{6}\times\overline{6}) \\ & & \oplus\ \mathbb{C}(8\times 8) \oplus \cdots \end{array}$$

is the reservoir for the $\mathbf{SU}(3)$-representation, e.g., induced by a doublet $\mathbf{SU}(2)$-Pauli representation $[1]$ on $W \cong \mathbb{C}^2$. The induced $\mathbf{SU}(3)$-representation on the $\mathbf{SU}(2)$-intertwiners is decomposed into irreducible $\mathbf{SU}(3)$-representations $[N_1, N_2]$ as follows: All $\mathbf{SU}(3)$-representations with $\mathbf{SU}(2)$-doublets arise and only those, i.e., $N_1 + N_2 \geq 1$. *No* singlet $[0,0]$ (it has *no* $\mathbf{SU}(2)$-doublet), *one* triplet $[1,0]$, and *one* antitriplet $[0,1]$ (since both triplet and antitriplet have *one* $\mathbf{SU}(2)$-doublet), equally one sextet $[2,0]$ and one antisextet $[0,2]$, *two* octets $[1,1]$ (since an octet has *two* $\mathbf{SU}(2)$-doublets), etc.:

$$\begin{array}{rcl} L^2(\mathbf{SU}(3)/\mathbf{SU}(2), \mathbb{C}^2) & \overset{\mathbf{SU}(3)}{\cong} & \mathbb{C}^3 \oplus \mathbb{C}^{\overline{3}} \oplus \mathbb{C}^6 \oplus \mathbb{C}^{\overline{6}} \\ & \oplus & [\mathbb{C}^8 \oplus \mathbb{C}^8] \oplus \cdots \\ & & \\ (\mathbf{SU}(2)\text{-doublets}) & \supset & \mathbb{C}^2 \oplus \mathbb{C}^2 \oplus \mathbb{C}^2 \oplus \mathbb{C}^2 \\ & \oplus & \begin{pmatrix} \mathbb{C}^2 & \mathbb{C}^2 \\ \mathbb{C}^2 & \mathbb{C}^2 \end{pmatrix} \oplus \cdots. \end{array}$$

With $[1] = [1]^\star$ for the $\mathbf{SU}(2)$-representation conjugated $\mathbf{SU}(3)$-representations, $[N_1, N_2]^\star = [N_2, N_1]$ have to arise with equal multiplicity.

7.10 Representations of Affine Groups

As pioneered by Wigner for the Poincaré group, representations of affine subgroups $G \vec{\times} \mathbb{R}^n$, like Euclidean or Poincaré groups

$$\mathbf{SO}(1+s) \vec{\times} \mathbb{R}^{1+s}, \quad \mathbf{SO}_0(1,s) \vec{\times} \mathbb{R}^{1+s}, \quad (g_1, x_1)(g_2, x_2) = (g_1 g_2, x_1 + g_1.x_2),$$

are inducible from those of direct product subgroups. An inducing procedure ind_H^G for the homogeneous group uses a parametrization of $H\backslash G$ by eigenvalues (characters, energy-momenta) of the translations.

The irreducible 1-dimensional unitary translation representations constitute the group dual

$$\chi^{iq} : \mathbb{R}^n \longrightarrow \mathbf{U}(1), \quad \chi^{iq}(x) = e^{i\langle q, x\rangle}.$$

The translation eigenvalues (energies, momenta) carry the dual action of the homogeneous group $\langle \check{g}.q, x\rangle = \langle q, g.x\rangle$, which defines the semidirect group $G \vec{\times} \check{\mathbb{R}}^n$ with fixgroups of the translation eigenvalues:

$$G_q = \{h \in G \mid \check{h}.q = q\}.$$

A translation representation χ^{iq} can be used in the phase group for the corresponding fixgroup representations $d : G_q \longrightarrow \mathbf{U}(W)$,

$$\begin{aligned} G_q \vec{\times} \mathbb{R}^n \longrightarrow \mathbf{U}(W), \quad d^{iq}(h,x) &= d(h)e^{i\langle q,x\rangle}, \\ d^{iq}((h_1,x_1)\circ(h_2,x_2)) &= d(h_1h_2)e^{i\langle q,x_1+h_1.x_2\rangle} \\ &= d(h_1h_2)e^{i\langle q,x_1\rangle}e^{i\langle \check{h}_1.q,x_2\rangle} \\ &= d^{iq}(h_1,x_1)\circ d^{iq}(h_2,x_2), \end{aligned}$$

since the two subgroup factors are "decoupled" in *a direct product* acting on the translation eigenvalues

$$d^{iq}[G_q \vec{\times} \mathbb{R}^n] = d[G_q] \times \chi^{iq}[\mathbb{R}^n],$$

e.g., for the Poincaré group $\mathbf{SO}_0(1,3) \vec{\times} \mathbb{R}^4$ with nontrivial character $q = (m, 0)$ having the rotation fixgroup $\mathbf{SO}(3)$,

$$\mathbf{SO}(3) \times \mathbb{R} \hookrightarrow \mathbf{SO}(3) \times \mathbb{R}^4 \ni (O,x) \longmapsto 2J(O)e^{iqx}.$$

A G-action on a translation eigenvalue gives the orbit with isomorphic fixgroup H for all elements $G.q \cong H\backslash G$. The intertwiners are acted on by the homogeneous group G:

$$G \times W^{H\backslash G} \longrightarrow W^{H\backslash G}, \quad |w\rangle \longmapsto |w\rangle \bullet g = {}^{\oplus}\!\!\int_{H\backslash G} dHq\ |q,a\rangle w(g.q)_a.$$

The representation Hilbert space is constituted by the (energy-)momentum functions $L^2(H\backslash G)$ (wave packets), supported by the orbit that is characterized by the translation invariant.

It has been shown that the $G \vec{\times} \mathbb{R}^n$-Hilbert representation induced from an *irreducible Hilbert* representation of the subgroup $H \times \check{\mathbb{R}}^n \longrightarrow \mathbf{U}(W)$ is irreducible too. The inequivalent irreducible Hilbert representations $\mathbf{irrep}_+ \ G \vec{\times} \mathbb{R}^n$, induced by translation eigenvalues, are given with their G-orbit decomposition

$$\check{\mathbb{R}}^n \cong \biguplus_{\text{repr } q_r} G_{q_r}\backslash G,$$

i.e., with one representative of each orbit, and all irreducible Hilbert representations of the fixgroup for this representative,

$$\biguplus_{\text{repr } q_r} \{d \times \chi^{iq_r} \mid d \in \mathbf{irrep}_+ \ G_{q_r}, \ \ q_r \in \check{\mathbb{R}}^n\} \overset{\text{ind}}{\longmapsto} \mathbf{irrep}_+ \ [G \vec{\times} \mathbb{R}^n].$$

It has also been shown that all irreducible Hilbert representations are inducable if there exist representatives $\{q_r\}$ that constitute a Borel set in all translation eigenvalues $\check{\mathbb{R}}^n$.

Irreducible $G \vec{\times} \mathbb{R}^n$ Hilbert representations are cyclic translation representations with positive Radon measures of the (energy-)momenta. The coefficients depend only on the translation parametrizable classes $x \in G \vec{\times} \mathbb{R}^n/G$, in general not a group. Matrix elements with nontrivial G-behavior arise by translation derivations $\frac{\partial}{\partial x}$.

7.10.1 Hilbert Spaces for Heisenberg Groups

The Heisenberg groups have classical and quantum representations. In the Heisenberg group $\mathbf{H}(1)$ as semidirect product, in a indefinite unitary faithful representation

$$\mathbf{H}(1) = \mathbb{R} \vec{\times} \mathbb{R}^2 \longrightarrow \mathbf{SL}(\mathbb{R}^3) \ni \left(\begin{array}{cc|c} 1 & p & t \\ 0 & 1 & x \\ \hline 0 & 0 & 1 \end{array}\right),$$

the homogeneous group $e^{p\mathbf{X}} \in \mathbb{R}$ with the position $\mathbf{X} \cong \begin{pmatrix} 0 & 1 \\ 0 & 0 \end{pmatrix}$ acts on the abelian normal subgroup $\mathbb{R}^2$ with momentum $\mathbf{P} \cong \begin{pmatrix} 0 \\ 1 \end{pmatrix}$ and the central action operator $[\mathbf{X}, \mathbf{P}] = \mathbf{I} \cong \begin{pmatrix} 1 \\ 0 \end{pmatrix}$,

$$\mathbb{R} \vec{\times} \mathbb{R}^2 \longrightarrow \mathbb{R}^2 : \quad \begin{pmatrix} 1 & p \\ 0 & 1 \end{pmatrix} \begin{pmatrix} t \\ x \end{pmatrix} = \begin{pmatrix} t + px \\ x \end{pmatrix};$$

$\mathbf{I}$ generates the invariants.

The position action on the dual space $(\hbar, q) \in \mathbb{R}^2$

$$\langle (\hbar, q), \begin{pmatrix} t \\ x \end{pmatrix} \rangle = \hbar t + qx, \quad (\hbar, x) \begin{pmatrix} 1 & -p \\ 0 & 1 \end{pmatrix} = (\hbar, x - \hbar p)$$

has two types of fixgroups with corresponding orbits: The fixgroups are characterized either by trivial or by nontrivial eigenvalue $\hbar \in \mathbb{R}$:

$$\begin{array}{lllllll} \hbar = 0: & (0, q) & \text{has full fixgroup} & \mathbb{R} & \text{and point orbit} & \{(0, q)\} & \cong \{1\}, \\ \hbar \neq 0: & (\hbar, q) & \text{has trivial fixgroup} & \{0\} & \text{and line orbit} & (\hbar, \mathbb{R}) & \cong \mathbb{R}. \end{array}$$

Correspondingly, the representations

$$\mathbf{irrep}_+ \mathbf{H}(1) = \{\hbar \mid \hbar \in \mathbb{R}\}$$

come in two types (*Stone-von Neumann theorem*): The first type with trivial representations of the central action operator $\mathbf{I} \in \operatorname{centr} \mathbf{H}(1)$, i.e., with invariant $\hbar = 0$, leads to classical unfaithful representations of the Heisenberg group with commuting position and momentum, i.e., of the abelian adjoint Heisenberg group $\operatorname{Int} \mathbf{H}(1) = \mathbf{H}(1)/\operatorname{centr} \mathbf{H}(1) \cong \mathbb{R}^2$. The Hilbert representations of $\mathbb{R}^2$ are in $\mathbf{U}(1)$.

The second type with trivial fixgroup and a nontrivial $\mathbf{I}$-eigenvalue $i\hbar$ induces the quantum representations of the Heisenberg group. There is a continuum of invariants of $0 \neq \hbar \in \mathbb{R}$. Different action quanta $\hbar \neq \hbar'$ define inequivalent representations with $\mathbf{I} \longmapsto i\hbar\mathbf{1}$. These irreducible faithful $\mathbf{H}(1)$-representations integrate over irreducible representations $\mathbb{R} \ni x \longmapsto e^{iqx} \in \mathbf{U}(1)$

for all momentum eigenvalues on the orbit line $q \in \mathbb{R}$ with orthogonal and positive scalar product distribution:

$$\hbar \neq 0: \quad \{|\hbar; q\rangle \mid q \in \mathbb{R}\} \text{ with } \begin{cases} {}^{\oplus}\!\!\int \frac{dq}{2\pi} |\hbar; q\rangle\langle\hbar; q| & \cong \mathrm{id}_{L^2(\mathbb{R})}, \\ \langle\hbar; q'|\hbar; q\rangle & = \delta(\frac{q-q'}{2\pi}), \\ |\hbar; q\rangle & \stackrel{\mathbb{R}}{\longmapsto} e^{iqx}|\hbar; q\rangle. \end{cases}$$

The Hilbert spaces consist of the square integrable momentum functions $f \in L^2(\mathbb{R})$ (wave packets). They are isomorphic to the square integrable position functions $\tilde{f}(x) = \int dq\ f(q)e^{iqx}$:

$$\begin{aligned} |\hbar; f\rangle &= {}^{\oplus}\!\!\int \tfrac{dq}{2\pi}\ f(q)|\hbar; q\rangle \\ \Rightarrow \langle\hbar; f'|\hbar; f\rangle &= \int \tfrac{dq}{2\pi}\ \overline{f'(q)}f(q) \quad = \int dx\ \overline{\tilde{f}'(x)}\tilde{f}(x). \end{aligned}$$

The action of the Lie algebra position or momentum operator is given by the derivatives $\mathbf{X} \longmapsto i\hbar\frac{d}{dq}$ and $\mathbf{P} \longmapsto -i\hbar\frac{d}{dx}$ respectively.

A harmonic analysis of functions on the Heisenberg group $\mathbf{H}(s)$ uses the classical Fourier components $|0; f\rangle$ with trivial Plancherel measure and the quantum components $|\hbar; f\rangle$ with Plancherel measure [3] $|\hbar|^s d\hbar$ for the invariant values of $\mathbf{I}$ which characterize the irreducible quantum representations.

7.10.2 Scattering Representations of Euclidean Groups

The irreducible Hilbert representations of the Euclidean group $\mathbf{SO}(s) \vec{\times} \mathbb{R}^s$ for $s \geq 2$, in a real $(s+1)$-dimensional indefinite unitary faithful representation

$$\mathbf{SO}(s) \vec{\times} \check{\mathbb{R}}^s = \{\begin{pmatrix} \mathbf{SO}(s) & \vec{q} \\ 0 & 1 \end{pmatrix} \mid \vec{q} \in \check{\mathbb{R}}^s\},$$

are induced from representations of the direct product subgroups that are related to the two translation character types (trivial and nontrivial)

$$\underset{\vec{q}=0}{\begin{pmatrix} \mathbf{SO}(s) & 0 \\ 0 & 1 \end{pmatrix}}, \qquad \underset{\vec{q}^2 = Q^2 > 0}{\begin{pmatrix} \mathbf{SO}(s-1) & 0 & 0 \\ 0 & 1 & Q \\ 0 & 0 & 1 \end{pmatrix}},$$

$$\mathbf{irrep}\,\mathbf{SO}(s) \uplus \mathbf{irrep}_+ \,[\mathbf{SO}(s-1) \times \mathbb{R}^s] \stackrel{\text{ind}}{\longmapsto} \mathbf{irrep}_+ \,[\mathbf{SO}(s) \vec{\times} \mathbb{R}^s].$$

The coefficients for Hilbert representations of general $\mathbf{SO}(s) \vec{\times} \mathbb{R}^s$ are given in the chapter "Residual Spacetime Representations."

The framework of nonrelativistic scattering is the representation theory of the Euclidean group $\mathbf{SU}(2) \vec{\times} \mathbb{R}^3$ (chapter "The Kepler Factor"). The irreducible Hilbert spaces induced by a trivial or faithful representation $\mathbf{SO}(2)$ on $W \cong \mathbb{C}^n$, $n = 2 - \delta_{J0} = 1, 2$, have a measure-related distributive basis:

$$\begin{aligned} \text{for } J = 0: &\quad \{|Q^2, 0; \vec{\omega}, h\rangle \mid \vec{\omega} \in \Omega^2,\ h = 0\}, \\ \text{for } J = \tfrac{1}{2}, 1, \cdots: &\quad \{|Q^2, J; \vec{\omega}, h\rangle \mid \vec{\omega} \in \Omega^2, h = \pm 1\}. \end{aligned}$$

By abuse of language, since not a Hilbert vector, an element $|Q^2, J; \vec{\omega}, h\rangle$ of the distributive basis is called a scattering "eigenstate" with momentum $\vec{q}$

(translation eigenvalues) of square $Q^2 > 0$ (translation invariant) and direction $\frac{\vec{q}}{|\vec{q}|} = \vec{\omega}$ and rotation invariant J with $\mathbf{SO}(2)$-eigenvalue h. The distributive basis is acted on by the inducing $\mathbf{SO}(2) \times \mathbb{R}^3$-representation

$$(e^{i\chi}, \vec{x}) \bullet |Q^2, J; \vec{\omega}, h\rangle \;= e^{hJi\chi} e^{-iQ\vec{\omega}\vec{x}} |Q^2, J; \vec{\omega}, h\rangle.$$

The momentum direction on the sphere $\vec{\omega} \in \Omega^2 \cong \mathbf{SU}(2)/\mathbf{SO}(2)$ is the axis for the fixgroup $\mathbf{SO}(2)$ rotations. The scalar product distribution involves the positive and orthogonal Dirac distribution on the 2-sphere,

$$\begin{aligned} \langle Q^2, J; \vec{\omega}', h' | Q^2, J; \vec{\omega}, h\rangle &= \delta_{hh'} 4\pi\delta(\vec{\omega} - \vec{\omega}') \text{ with } \delta(\vec{\omega}) = \tfrac{1}{\sin\theta}\delta(\theta)\delta(\varphi), \\ {}^{\oplus}\!\!\int \tfrac{d^2\omega}{4\pi} |Q^2, J; \vec{\omega}, h\rangle\langle Q^2, J; \vec{\omega}, h| &\cong \mathbf{1}_n \, \mathrm{id}_{L^2(\Omega^2)} = \mathrm{id}_{L^2(\Omega^2, \mathbb{C}^n)}, \quad n = 1, 2. \end{aligned}$$

The Hilbert space consists of 2-sphere square integrable momentum wave packets $L^2(\Omega^2, \mathbb{C}^n)$,

$$w \in L^2(\Omega^2, \mathbb{C}^n): \quad \left\{ \begin{aligned} |Q^2, J; w\rangle &= {}^{\oplus}\!\!\int \tfrac{d^2\omega}{4\pi}\, w(\vec{\omega})_h |Q^2, J; \vec{\omega}, h\rangle, \\ \langle Q^2, J; w_2 | Q^2, J; w_1\rangle &= \int \tfrac{d^2\omega}{4\pi}\, \overline{w_2(\vec{\omega})_h} w_1(\vec{\omega})_h. \end{aligned} \right.$$

The transformation behavior of the Hilbert vectors is built by that of the distributive basis.

All this can be seen as a distributive generalization of the finite-dimensional case, e.g., from one basic vector with irreducible translation dependence $|Q\rangle \longmapsto e^{iQz}|Q\rangle$ for the Hilbert space $\mathbb{C}|Q\rangle$ to a distributive basis $\{|Q^2, J; \vec{\omega}, h\rangle\}$ for the infinite-dimensional representation space $L^2(\Omega^2, \mathbb{C}^n)$. There is a pure vector for the irreducible representation, integrating the distributive basis:

$$\begin{aligned} |Q^2, J; 1, h\rangle &= {}^{\oplus}\!\!\int \tfrac{d^2\omega}{4\pi} |Q^2, J; \vec{\omega}, h\rangle \in L^2(\Omega^2, \mathbb{C}^n), \\ \langle Q^2, J; 1, h' | Q^2, J; 1, h\rangle &= \delta_{hh'}. \end{aligned}$$

The coefficient for translation representations $\mathbb{R} \ni z \longmapsto \langle Q|e^{iQz}|Q\rangle = e^{iQz}$ (function of positive type) has the analogous coefficient for the $\mathbf{SU}(2) \vec{\times} \mathbb{R}^3$-representations, which is a function of positive type for an irreducible $\mathbf{SU}(2) \vec{\times} \mathbb{R}^3$ and for a cyclic $\mathbb{R}^3$-representation

$$\begin{aligned} \vec{x} \longmapsto \int \tfrac{d^2\omega_1 d^2\omega_2}{(4\pi)^2} \langle Q^2, J; \vec{\omega}_2, h_2 | \cos\vec{q}\vec{x} | Q^2, J; \vec{\omega}_1, h_1\rangle &= \delta_{h_1h_2} \int \tfrac{d^2\omega}{4\pi} \cos Q\vec{\omega}\vec{x} \\ = \delta_{h_1h_2} \int \tfrac{d^3q}{2\pi Q} \delta(\vec{q}^2 - Q^2) e^{-i\vec{q}\vec{x}} &= \delta_{h_1h_2} \tfrac{\sin Qr}{Qr}. \end{aligned}$$

For nontrivial rotation properties, there are two fundamental Pauli transmutators from the fixgroup with the axial rotations around $\vec{\omega}$ to the full rotation group:

$$\begin{aligned} u(\tfrac{\vec{q}}{Q}) = \sqrt{\tfrac{Q+q_3}{2Q}} \begin{pmatrix} 1 & -\frac{q_1 - iq_2}{Q+q_3} \\ \frac{q_1 + iq_2}{Q+q_3} & 1 \end{pmatrix} &= \begin{pmatrix} \cos\frac{\theta}{2} & -e^{-i\varphi}\sin\frac{\theta}{2} \\ e^{i\varphi}\sin\frac{\theta}{2} & \cos\frac{\theta}{2} \end{pmatrix} \in \mathbf{SU}(2)/\mathbf{SO}(2), \\ u(\vec{\omega}) \circ \sigma^3 \circ u^\star(\vec{\omega}) &= \vec{\omega}. \end{aligned}$$

The induced $\mathbf{SU}(2) \vec{\times} \mathbb{R}^3$-representations involve the Wigner axial rotations

$$\begin{aligned} &\mathbf{SU}(2) \vec{\times} \mathbb{R}^3 \ni (r, \vec{x}) \longmapsto (o(r, \vec{\omega}), \vec{x}) \in \mathbf{SO}(2) \times \mathbb{R}^3, \\ &\text{Wigner element } o(r, \vec{\omega}) = u(r \bullet \vec{\omega})^\star \circ r \circ u(\vec{\omega}) \in \mathbf{SO}(2), \end{aligned}$$

acting on the distributive basis which is transmuted by $2J(\vec\omega)=\bigvee^{2J} u(\vec\omega)$ into an $\mathbf{SU}(2)$-basis:

$$2J(\vec\omega)^a_h|Q^2,J;\vec\omega,h\rangle,\ a=-J,-J+1,\dots,J.$$

The square integrable transmutators $2J_h\in L^2(\Omega^2,\mathbb{C}^n)$ lead to position fields $\mathbf{\Phi}$ with $\mathbf{SU}(2)\ \vec\times\,\mathbb{R}^3$-action, e.g., scalar and Pauli spinor fields:

$$\begin{array}{ll}\mathbf{\Phi}(Q^2,0|\vec x) &= {}^{\oplus}\!\int \frac{d^2\omega}{4\pi}\ \frac{e^{-iQ\vec\omega\vec x}\ |Q^2;\vec\omega\rangle+e^{iQ\vec\omega\vec x}\ \langle Q^2;\vec\omega|}{2},\\ \mathbf{\Phi}(Q^2,\frac12|\vec x)^a &= {}^{\oplus}\!\int \frac{d^2\omega}{4\pi}\ u(\vec\omega)^a_h\frac{e^{-iQ\vec\omega\vec x}\ |Q^2,\frac12;\vec\omega,h\rangle+e^{iQ\vec\omega\vec x}\ \langle Q^2,\frac12;\vec\omega,-h|}{2},\\ \mathbf{\Phi}^\star(Q^2,\frac12|\vec x)_a &= {}^{\oplus}\!\int \frac{d^2\omega}{4\pi}\ \frac{e^{iQ\vec\omega\vec x}\ \langle Q^2,\frac12;\vec\omega,h|+e^{-iQ\vec\omega\vec x}\ |Q^2,\frac12;\vec\omega,-h\rangle}{2}u^\star(\vec\omega)^h_a.\end{array}$$

Their Hilbert product with Schur-orthonormalized spherical harmonics $\mathrm{y}^L_b=i^L\sqrt{\frac{4\pi}{1+2L}}\mathrm{Y}^L_b$,

$$\begin{array}{rl}\langle\mathbf{\Phi}^\star(\vec x_2)|\mathbf{\Phi}(\vec x_1)\rangle &= j_{2J}(Q|\vec x|)\mathrm{y}^{2J}_b(\vec\omega)=\{\vec x|Q^2,2J;b\},\quad \vec x=\vec x_1-\vec x_2,\\ \{\vec x|Q^2,0\} &= \int\frac{d^3q}{2\pi Q}\ \delta(\vec q^{\,2}-Q^2)e^{-i\vec q\vec x}\ \ =\int d^2\omega\ \frac{e^{-iQ\vec\omega\vec x}+e^{iQ\vec\omega\vec x}}{2},\\ \{\vec x|Q^2,1;b\} &= \int\frac{d^3q}{2\pi Q}\ \frac{\vec q}{|\vec q|}\delta(\vec q^{\,2}-Q^2)e^{-i\vec q\vec x}=\int d^2\omega\ u(\vec\omega)\circ\sigma^3\frac{e^{-iQ\vec\omega\vec x}+e^{iQ\vec\omega\vec x}}{2}\circ u^\star(\vec\omega),\end{array}$$

gives coefficients of infinite-dimensional irreducible representations $|Q^2,L\}$ of $\mathbf{SO}(3)\ \vec\times\,\mathbb{R}^3$. They are matching products of spherical harmonics with spherical Bessel functions that depend on the symmetric space $\mathbf{SO}(3)\ \vec\times\,\mathbb{R}^3/\mathbf{SO}(3)\cong\mathbb{R}^3$:

$$\begin{array}{l}\vec x\longmapsto\ \ \{\vec x|Q^2,L;b\}=j_L(Qr)\mathrm{y}^L_b(\vec\omega)=\frac{j_L(Qr)}{r^L}(i\vec x)^L_b\in\mathbb{C},\\ \text{e.g., } L=0:\ \vec x\longmapsto\left\{\begin{array}{rll}\{\vec x|Q^2,0\} &= \int d^2\omega\ e^{-iQ\vec\omega\vec x} &= \frac{\sin Qr}{Qr}\\ &= j_0(Qr), & \\ \{\vec x|Q^2,1;b\} &= \int d^2\omega\ \vec\omega\ e^{-iQ\vec\omega\vec x} &= i\frac{\vec x}{r}d_{Qr}j_0(Qr)\\ &= \frac{j_1(Qr)}{r}i\vec x. & \end{array}\right.\end{array}$$

Representation coefficients with nontrivial rotation properties can be obtained by derivatives $\frac{\partial}{\partial\vec x}$.

There is the distributive Schur orthogonality for the representation coefficients: Hilbert spaces for different rotation- or translation-invariant $\{Q^2,J\}$ are orthogonal:

$$\begin{array}{l}\int d^3x\ \{Q'^2,L';b'|\vec x\}\{\vec x|Q^2,L;b\}\\ \quad=\int_0^\infty r^2dr\int d^2\omega\ j_{L'}(Q'r)\ \overline{\mathrm{y}^{L'}_{b'}(\vec\omega)}\mathrm{y}^L_b(\vec\omega)j_L(Qr)\\ \quad=\frac{2\pi^2}{Q^2}\delta(Q-Q')\frac{1}{1+2L}\delta^{LL'}(\mathbf{1}_{1+2L})_{bb'},\\ \text{e.g., }\int d^3x\int\frac{d^3q'}{2\pi Q'}\delta(\vec q'^2-Q'^2)e^{i\vec q'\vec x}\int\frac{d^3q}{2\pi Q}\delta(\vec q^{\,2}-Q^2)e^{-i\vec q\vec x}\\ \quad=2\pi\int\frac{d^3q}{Q'Q}\delta(\vec q^{\,2}-Q'^2)\delta(\vec q^{\,2}-Q^2)=\frac{2\pi^2}{Q^2}\delta(Q-Q').\end{array}$$

The decomposition of the identity for the square integrable $\mathbf{SO}(3)$-compatible $\mathbb{R}^3$-functions with the orthogonal canonical and a harmonic basis (distributive completeness)

$$\mathrm{id}_{L^2(\frac{\mathbf{SO}(3)\ \vec\times\ \mathbb{R}^3}{\mathbf{SO}(3)})}\cong{}^{\oplus}\!\int d^3x\ |\vec x\}\{\vec x|={}^{\oplus}\!\int_0^\infty\frac{Q^2dQ}{2\pi^2}\bigoplus_{L=0}^{\infty}(1+2L)\bigoplus_{b=-L}^{L}|Q^2,L;b\}\{Q^2,L;b|$$

displays the Plancherel measure of the irreducible representation classes

$$\int_0^\infty \frac{Q^2 dQ}{2\pi^2}\sum_{L=0}^\infty (1+2L)$$

and describes the harmonic Fourier-Bessel analysis

$$\begin{aligned}\{\vec{x}|f\} &= f(\vec{x}) &&= \int_0^\infty \frac{Q^2 dQ}{2\pi^2}\sum_{L=0}^\infty (1+2L) \sum_{b=-L}^{L} \frac{j_L(Qr)}{r^L}(i\vec{x})^L_b \; \tilde{f}^L_b(Q^2),\\ \{Q^2, L; b|f\} &= \tilde{f}^L_b(Q^2) &&= \int_0^\infty r^2 dr \int d^2\omega \; \frac{j_L(Qr)}{r^L}\overline{(i\vec{x})^L_b} \; f(\vec{x}).\end{aligned}$$

7.10.3 Particle Representations of Poincaré Groups

For inducing representations of the rank-2 Poincaré group $\mathbf{SO}_0(1,3) \,\vec{\times}\, \mathbb{R}^4$, in a real 5-dimensional indefinite unitary faithful representation

$$\mathbf{SO}_0(1,3) \,\vec{\times}\, \check{\mathbb{R}}^4 = \{\begin{pmatrix} \mathbf{SO}_0(1,3) & q \\ 0 & 1 \end{pmatrix} \mid q \in \check{\mathbb{R}}^4\},$$

there are four types of spacetime translation characters (energy-momenta) with fixgroups in the Lorentz group:

$$\underset{q=0}{\begin{pmatrix} \mathbf{SO}_0(1,3) & 0 \\ 0 & 1 \end{pmatrix}}, \quad \underset{\substack{q^2=-Q^2<0\\ Q\in\mathbb{R}}}{\begin{pmatrix} \mathbf{SO}_0(1,2) & 0 & 0 \\ 0 & 1 & Q \\ 0 & 0 & 1 \end{pmatrix}}, \quad \underset{\substack{q^2=m^2>0\\ m\in\mathbb{R}}}{\begin{pmatrix} 1 & 0 & m \\ 0 & \mathbf{SO}(3) & 0 \\ 0 & 0 & 1 \end{pmatrix}}, \quad \underset{\substack{q^2=0,\; q\neq 0\\ \pm|\vec{q}|\in\mathbb{R}}}{\begin{pmatrix} 1 & 0 & 0 & \pm|\vec{q}| \\ 0 & \mathbf{SO}(2) & Q & 0 \\ 0 & 0 & 1 & \pm|\vec{q}| \\ 0 & 0 & 0 & 1 \end{pmatrix}}.$$

For the first two types one has to know the representations of the noncompact Lorentz groups $\mathbf{SO}_0(1,3) \sim \mathbf{SL}(\mathbb{C}^2)$ (given by Naimark and sketched below) and $\mathbf{SO}_0(1,2) \sim \mathbf{SU}(1,1) \cong \mathbf{SL}(\mathbb{R}^2)$ (given by Bargmann). The fixgroup for lightlike energy-momenta is a semidirect product $\mathbf{SO}(2) \,\vec{\times}\, \mathbb{R}^2$ with boosts $\mathbb{R}^2$. The Hilbert representations of the 2-dimensional Euclidean group are induced with fixgroups $\mathbf{SO}(2)$ and $\{1\}$ as given above:

$$\text{for } \mathbf{SO}(2) \,\vec{\times}\, \mathbb{R}^2: \quad \underset{Q=0}{\begin{pmatrix} 1 & 0 & 0 & \pm|\vec{q}| \\ 0 & \mathbf{SO}(2) & 0 & 0 \\ 0 & 0 & 1 & \pm|\vec{q}| \\ 0 & 0 & 0 & 1 \end{pmatrix}}, \quad \underset{Q\neq 0}{\begin{pmatrix} 1 & 0 & 0 & \pm|\vec{q}| \\ 0 & \mathbf{1}_2 & Q & 0 \\ 0 & 0 & 1 & \pm|\vec{q}| \\ 0 & 0 & 0 & 1 \end{pmatrix}}$$

Therefore, the irreducible Poincaré group Hilbert representations are induced as follows:

$$\begin{aligned} & \mathbf{irrep}_+ \; \mathbf{SO}_0(1,3) \\ \uplus \; & \mathbf{irrep}_+ \; [\mathbf{SO}_0(1,2) \times \mathbb{R}^4] \\ \uplus \; & \mathbf{irrep}_+ \; [\mathbf{SO}(3) \times \mathbb{R}^4] \\ \uplus \; & \mathbf{irrep}_+ \; [\mathbf{SO}(2) \times \mathbb{R}^4] \uplus \mathbf{irrep}_+ \; [\mathbb{R}^2 \times \mathbb{R}^4] \overset{\text{ind}}{\longmapsto} \mathbf{irrep}_+ \; [\mathbf{SO}_0(1,3) \,\vec{\times}\, \mathbb{R}^4]. \end{aligned}$$

With respect to the characteristic two invariants, rational and continuous, the representations of the Cartan Lie algebras are relevant.

Minkowski ("linear") spacetime and noninteracting ("free") matter are "unified" by Wigner's definition: Massive and massless particle fields are spin $\mathbf{SU}(2)$ and axial rotation $\mathbf{SO}(2)$-intertwiners, constructed with $\mathbf{SL}(\mathbb{C}^2)/\mathbf{SU}(2)$ and $\mathbf{SL}(\mathbb{C}^2)/\mathbf{SO}(2)$-transmutators, and acted on by irreducible Hilbert representations of the Poincaré group.

As seen from experiments, stable particles use only representations with *causal translation-invariant* $q^2 = m^2 \geq 0$ which come with a nontrivial *compact stabilgroup* for rotation properties. To include half-integer spin, "double valued" representations of the Lorentz group $\mathbf{SO}_0(1,3) \cong \mathbf{SL}(\mathbb{C}^2)/\mathbb{I}(2)$ are admitted, i.e., particles are induced representations of $\mathbf{SL}(\mathbb{C}^2) \vec{\times} \mathbb{R}^4$. With complex representation spaces, a nontrivial particle-antiparticle number can be included by $\mathbf{U}(1)$-representations

$$\begin{aligned}\mathbf{irrep}_+ \, [\mathbf{U}(1) \times \mathbf{SU}(2) \times \mathbb{R}^4] \uplus \mathbf{irrep}_+ \, [\mathbf{U}(1) \times \mathbf{SO}(2) \times \mathbb{R}^4]&\\ \overset{\text{ind}}{\longmapsto} \mathbf{irrep}_+ \, [\mathbf{U}(1) \times \mathbf{SL}(\mathbb{C}^2) \vec{\times} \mathbb{R}^4].&\end{aligned}$$

Free particle fields, in the following only for massive particles $m > 0$, embed Lorentz compatibly representations of time translations $\mathbb{R}$ and of the Euclidean scattering group $\mathbf{SU}(2) \vec{\times} \mathbb{R}^3$ in Poincaré group representations. The Hilbert space with $\mathbf{U}(1) \times \mathbf{SL}(\mathbb{C}^2) \vec{\times} \mathbb{R}^4$-representation, induced by a Hilbert representation of the charge-spin-translation group on $W \cong \mathbb{C}^{1+2J}$,

$$\mathbf{U}(1) \times \mathbf{SU}(2) \times \mathbb{R}^4 \ni (e^{i\alpha}, u, x) \longmapsto e^{iZ\alpha} 2J(u) e^{iqx} \in \mathbf{U}(1+2J) \text{ with } q^2 = m^2,$$

consists of intertwiners that map the spin orbits in the Lorentz group into $\mathbf{SU}(2)$-orbits in a representation space W. The space $W(\vec{q}) = W \times \{\vec{q}\}$ for each momentum of the energy-momentum hyperboloid is spanned by creation operators:

$$\begin{aligned}\text{distributive basis:} \quad & \{|m^2, J, Z; \vec{q}, a\rangle \,\big|\, \vec{q} \in \mathbb{R}^3, \; a = 1, \ldots, 1+2J\},\\ & \mathrm{u}(\vec{q})^a|0\rangle = |m^2, J, Z; \vec{q}, a\rangle = |m^2, J, Z; \mathrm{y}, a\rangle.\end{aligned}$$

The elements of the distributive basis $|m^2, J, Z; \vec{q}, a\rangle$ (not Hilbert vectors) are called "eigenstates" for an (anti)particle with the invariants $\mathbf{U}(1)$-charge Z, mass m, spin J, and eigenvalue momentum $\vec{q}$ and third spin direction a. Hyperbolic coordinates y, appropriate for the Lorentz group action, are equivalent to the more familiar momentum coordinates $\vec{q}$:

$$\begin{aligned}\mathcal{Y}^3 \ni \;\; \mathrm{y} = \vartheta(q^2)\vartheta(q_0)\tfrac{q}{|q|} &= \begin{pmatrix} \cosh\psi \\ \frac{\vec{q}}{|\vec{q}|}\sinh\psi \end{pmatrix} = \tfrac{1}{|q|}\begin{pmatrix} \vartheta(q_0)q_0 \\ \vec{q} \end{pmatrix},\\ \delta(\mathrm{y}) &= \tfrac{1}{\sinh^2\psi}\delta(\psi)\delta(\vec{\omega}), \quad \int d^3\mathrm{y} = \int_0^\infty \sinh^2\psi \, d\psi \int d^2\omega,\\ \int_0^\infty \sinh^2\psi \, d\psi &= \int_0^\infty \tfrac{q^2 dq}{q_0}\Big|_{q_0=\sqrt{m^2+\vec{q}^2}} = \int_m^\infty dq_0 \, |\vec{q}|\Big|_{|\vec{q}|=\sqrt{q_0^2-m^2}},\\ \vartheta(q^2)e^{iqx} &= e^{\epsilon(q_0)i|q|\mathrm{y}x}, \quad \vartheta(q^2)\cos qx = \cos|q|\mathrm{y}x.\end{aligned}$$

The Hilbert product distribution comes with the Dirac distribution on the energy-momentum hyperboloid (the $\mathbf{U}(1)$-invariant Z is omitted)

$$\begin{aligned}\langle \mathrm{u}^\star_{a_2}\mathrm{u}^{a_1}\rangle_{\mathrm{F}} = \delta^{a_1}_{a_2} &\stackrel{\mathrm{ind}}{\longmapsto} \langle \mathrm{u}^\star(\vec q_2)_{a_2}\mathrm{u}(\vec q_1)^{a_1}\rangle_{\mathrm{F}}\\ &= \langle m^2, J; \vec q_2, a_2|m^2, J; \vec q_1, a_1\rangle &= \delta^{a_1}_{a_2} 4\pi q_0\, \delta(\vec q_1 - \vec q_2)\\ &= \langle m^2, J; \mathrm{y}_2, a_2|m^2, J; \mathrm{y}_1, a_1\rangle &= \delta^{a_1}_{a_2} 4\pi\delta(\mathrm{y}_1 - \mathrm{y}_2)\end{aligned}$$

with the completeness

$$\begin{aligned}{}^{\oplus}\!\!\int \tfrac{d^3q}{4\pi q_0}|m^2, J; \vec q, a\rangle\langle m^2, J; \vec q, a| &= {}^{\oplus}\!\!\int \tfrac{d^3\mathrm{y}}{4\pi}|m^2, J; \mathrm{y}, a\rangle\langle m^2, J; \mathrm{y}, a|\\ &\cong \mathrm{id}_{L^2(\mathcal{Y}^3, \mathbb{C}^{1+2J})} = \mathbf{1}_{1+2J}\, \mathrm{id}_{L^2(\mathcal{Y}^3)}.\end{aligned}$$

Hilbert spaces with different $\mathbf{U}(1)$-charge $Z \neq Z'$, e.g., for particle-antiparticle, translation- or rotation-invariant, are orthogonal.

The representation space is the Hilbert space of the square integrable mappings on the energy-momentum 3-hyperboloid $\mathcal{Y}^3$. The Hilbert vectors $|m^2, J; w\rangle$ use an expansion with the hyperbolic "directions" for the distributive basis (canonical expansion with $\vec q$-indexed Fock spaces)

$$w \in L^2(\mathcal{Y}^3, \mathbb{C}^{1+2J}) : \left\{\begin{aligned} |m^2, J; w\rangle &= {}^{\oplus}\!\!\int \tfrac{d^3q}{4\pi q_0}\, w(\vec q)_a|m^2, J; \vec q, a\rangle\\ &= {}^{\oplus}\!\!\int \tfrac{d^3\mathrm{y}}{4\pi}\, w(\mathrm{y})_a|m^2, J; \mathrm{y}, a\rangle,\\ \langle m^2, J; w_2|m^2, J; w_1\rangle &= \int \tfrac{d^3q}{4\pi q_0}\, \overline{w_2(\vec q)_a} w_1(\vec q)_a\\ &= \int \tfrac{d^3\mathrm{y}}{4\pi}\, \overline{w_2(\mathrm{y})_a} w_1(\mathrm{y})_a.\end{aligned}\right.$$

The spacetime translation action on the Hilbert vectors is built by that on the distributive basis.

The normalization of the integral of the distributive basis goes with the hyperboloid volume, i.e., it is not an element of $L^2(\mathcal{Y}^3, \mathbb{C}^{1+2J})$,

$$\begin{aligned}|m^2, J; 1, a\rangle &= {}^{\oplus}\!\!\int \tfrac{d^3\mathrm{y}}{4\pi}|m^2, J; \mathrm{y}, a\rangle \notin L^2(\mathcal{Y}^3, \mathbb{C}^{1+2J}),\\ \langle m^2, J; 1, a|m^2, J; 1, b\rangle &= \delta_{ab}{}^{\oplus}\!\!\int \tfrac{d^3\mathrm{y}}{4\pi} = \delta_{ab}\int \tfrac{d^4q}{2\pi}\vartheta(q_0)\delta(q^2 - 1).\end{aligned}$$

The integral of the scalar product distribution is a scalar $\mathbf{SO}_0(1,3) \vec{\times} \mathbb{R}^4$-representation coefficient, not square integrable,

$$\begin{aligned}\int \tfrac{d^3\mathrm{y}_1 d^3\mathrm{y}_2}{(4\pi)^2}\langle m^2, J; \mathrm{y}_2, a_2|\cos qx|m^2, J; \mathrm{y}_1, a_1\rangle &= \delta_{a_1a_2}\int \tfrac{d^3\mathrm{y}}{4\pi}\cos m\mathrm{y}x\\ = \delta_{a_1a_2}\int \tfrac{d^4q}{4\pi}\delta(q^2-1)e^{iqx} &= \delta_{a_1a_2}\tfrac{-\vartheta(x^2)\pi\mathcal{N}_{-1}(|x|)+\vartheta(-x^2)2\mathcal{K}_1(|x|)}{2|x|}.\end{aligned}$$

Particle fields (chapter "Particle Quantum Fields") use transmutators (boost representations) that relate to each other irreducible spin and finite-dimensional Lorentz group representations

$$D(\tfrac{q}{m}) \stackrel{\mathbf{SU}(2)}{=} \bigoplus_\iota D^\iota(\tfrac{q}{m}) \Rightarrow \left\{\begin{aligned}\mathbf{D}^\iota(m^2)^j &= {}^{\oplus}\!\!\int \tfrac{d^3q}{4\pi q_0}\, \mathrm{u}(\vec q)^a D^\iota(\tfrac{q}{m})^j_a \in V \subset W^{\mathcal{Y}^3},\\ \mathbf{D}^\iota_\lambda(m^2)^j &= \mathbf{D}^\iota(m^2)^k D(\lambda)^j_k \text{ for } \lambda \in \mathbf{SL}(\mathbb{C}^2).\end{aligned}\right.$$

The transmutators are not $L^2(\mathcal{Y}^3, \mathbb{C}^{1+2J})$ square integrable. Transmutators pair the spin group representation Hilbert spaces with the minimal Lorentz representation spaces $\mathbf{SU}(2) \bullet W \subseteq \mathbf{SL}(\mathbb{C}^2) \bullet V$, e.g., $1, \Lambda(\frac{q}{m}), s(\frac{q}{m})$ for the scalar

$\boldsymbol{\Phi}$, vector $\mathbf{Z}$, and Dirac spinor $\binom{\mathbf{r}}{\mathbf{l}}$ fields $V \otimes W^T \cong \mathbb{C} \otimes \mathbb{C}, \mathbb{C}^4 \otimes \mathbb{C}^3, \mathbb{C}^2 \otimes \mathbb{C}^2$ respectively, all with mass $m > 0$ and a transmutator notation

$$\begin{pmatrix} \boldsymbol{\Phi} \\ \mathbf{Z}^j \\ \mathbf{r}^C \\ \mathbf{l}^{\dot{C}} \end{pmatrix}(0) \cong \begin{pmatrix} \mathbf{E} \\ \boldsymbol{\Lambda}^j \\ \mathbf{s}^C \\ \hat{\mathbf{s}}^{\dot{C}} \end{pmatrix}(m^2) = {}^{\oplus}\!\!\int \frac{d^3q}{4\pi q_0} \begin{pmatrix} \frac{u(\vec{q})+\mathrm{a}^\star(\vec{q})}{\sqrt{2}} & \\ \frac{\mathrm{u}(\vec{q})^a+\mathrm{u}^\star(\vec{q})^a}{\sqrt{2}} & \Lambda(\frac{q}{m})^j_a \\ \frac{u(\vec{q})^A+\mathrm{a}^\star(\vec{q})^A}{\sqrt{2}} & s(\frac{q}{m})^C_A \\ \frac{u(\vec{q})^A-\mathrm{a}^\star(\vec{q})^A}{\sqrt{2}} & \hat{s}(\frac{q}{m})^{\dot{C}}_A \end{pmatrix}$$

$$\text{with} \quad \begin{cases} \mathrm{u} \in W,\ \mathrm{u}^\star \in W^T,\ \mathrm{a} \in \overline{W}^T,\ \mathrm{a}^\star \in \overline{W}, \\ W \oplus \overline{W}^T = W_{\text{doub}},\quad W_{\text{doub}} \oplus W^T_{\text{doub}} = \mathbf{W}. \end{cases}$$

The Lorentz group action is accompanied by a Wigner rotation

$$\lambda \in \mathbf{SL}(\mathbb{C}^2): \quad u(\lambda, \tfrac{q}{m}) = s(\Lambda.\tfrac{q}{m})^{-1} \circ \lambda \circ s(\tfrac{q}{m}) \in \mathbf{SU}(2).$$

The particle fields are Lorentz irreducible spacetime translation orbits

$$\begin{pmatrix} \boldsymbol{\Phi} \\ \mathbf{Z}^j \\ \mathbf{r}^C \\ \mathbf{l}^{\dot{C}} \end{pmatrix}(x) \cong \begin{pmatrix} \mathbf{E} \\ \mathbf{A}^j \\ \mathbf{s}^C \\ \hat{\mathbf{s}}^{\dot{C}} \end{pmatrix}(m^2|x) = {}^{\oplus}\!\!\int \frac{d^3q}{4\pi q_0} \begin{pmatrix} \frac{e^{iqx}u(\vec{q})+e^{-iqx}\mathrm{a}^\star(\vec{q})}{\sqrt{2}} & \\ \frac{e^{iqx}\mathrm{u}(\vec{q})^a+e^{-iqx}\mathrm{u}^\star(\vec{q})^a}{\sqrt{2}} & \Lambda(\frac{q}{m})^j_a \\ \frac{e^{iqx}u(\vec{q})^A+e^{-iqx}\mathrm{a}^\star(\vec{q})^A}{\sqrt{2}} & s(\frac{q}{m})^C_A \\ \frac{e^{iqx}u(\vec{q})^A-e^{-iqx}\mathrm{a}^\star(\vec{q})^A}{\sqrt{2}} & \hat{s}(\frac{q}{m})^{\dot{C}}_A \end{pmatrix},$$

$$\begin{aligned} \mathbf{D}^\iota(m^2|x)^j &= {}^{\oplus}\!\!\int \frac{d^3q}{4\pi q_0}\, e^{iqx}\mathrm{u}(\vec{q})^a D^\iota(\tfrac{q}{m})^j_a, \\ \text{with } \mathbf{D}^\iota_\lambda(m^2|x)^j &= \mathbf{D}^\iota(m^2|\Lambda^{-1}.x)^k D(\lambda)^j_k. \end{aligned}$$

For a spacetime field, the canonical boost expansion ${}^{\oplus}\!\!\int \frac{d^3q}{4\pi q_0}$ for the induced $\mathbf{SL}(\mathbb{C}^2)$-representation is simultaneously the harmonic expansion (Fourier analysis) with respect to the translation representations.

The Hilbert product of the fields give coefficients of infinite-dimensional representations of $\mathbf{SO}_0(1,3) \vec{\times} \mathbb{R}^4$, which depend on the symmetric space $\mathbf{SO}_0(1,3) \vec{\times} \mathbb{R}^4/\mathbf{SO}_0(1,3) \cong \mathbb{R}^4$ (chapter "Propagators"):

$$\begin{aligned} x \longmapsto \langle\{\boldsymbol{\Phi}(x_2), \boldsymbol{\Phi}(x_1)\}\rangle &\cong \{x|m^2, [0|0]\} \\ = \int \frac{d^4q}{4\pi m^2}\, \delta(q^2 - m^2) e^{iqx} &= \begin{cases} \int \frac{d^3\mathrm{y}}{4\pi} \cos my x, \\ \int \frac{d^3q}{4\pi m^2 q_0} \cos q_0 x_0\, e^{-i\vec{q}\vec{x}}\Big|_{q_0}, \\ \int_m^\infty \frac{dq_0|\vec{q}|}{m^2} \cos q_0 x_0\, j_0(|\vec{q}|r)\Big|_{|\vec{q}|}. \end{cases} \end{aligned}$$

Representation coefficients with nontrivial Lorentz properties arise by derivatives $\frac{\partial}{\partial x}$. Schur's orthogonality is established up to the divergent $\mathcal{Y}^3$-measure; the coefficients are not square integrable, e.g.,

$$\int d^4x\ \{m'^2, [0|0]|x\}\{x|m^2, [0|0]\} = \frac{\pi^2}{m^3}\delta(m - m') \int d^4q\ \delta(q^2 - 1).$$

The quantum algebra is the tensor algebra $\bigotimes \mathbf{W}^{\mathcal{Y}^3}$ modulo the quantization via the duality distributions of the basic momentum operators

$$[\mathrm{u}^\star_a, \mathrm{u}^b]_\epsilon = \delta^b_a \overset{\text{ind}}{\longmapsto} [\mathrm{u}^\star(\vec{p})_a, \mathrm{u}(\vec{q})^b]_\epsilon = \delta^b_a 4\pi q_0 \delta(\vec{q} - \vec{p}).$$

The quantization opposite (anti-) commutators implement the Lie algebra with the basic space endomorphisms $\mathbf{AL}(W) = W \otimes W^T$ for the finite-dimensional case and $\mathbf{AL}(W^{\mathcal{Y}^3})$ for the infinite-dimensional one. The Lie algebra of the charge-Poincaré group is represented in the Lie algebra of the definite unitary automorphism group

$$\log[\mathbf{U}(1) \times \mathbf{SL}(\mathbb{C}^2)] \ \vec{\oplus} \ \mathbb{R}^4 \longrightarrow \log \mathbf{U}(W^{\mathcal{Y}^3})$$

induced from $\mathbf{U}(1) \times \mathbf{SU}(2) \times \mathbb{R}^4$ representations as follows:

$$\begin{array}{rcl} \frac{[\mathrm{u}^b, \mathrm{u}_a^\star]_{-\epsilon}}{2} & \stackrel{\text{ind}}{\longmapsto} & \oplus\!\int \frac{d^3q}{4\pi q_0} \frac{[\mathrm{u}(\vec{q})^b, \mathrm{u}^\star(\vec{q})_a]_{-\epsilon}}{2}, \\ [\frac{[\mathrm{u}^b, \mathrm{u}_a^\star]_{-\epsilon}}{2}, \mathrm{u}^c] = \delta_b^c \mathrm{u}^b & \stackrel{\text{ind}}{\longmapsto} & [\oplus\!\int \frac{d^3q}{4\pi q_0} \frac{[\mathrm{u}(\vec{q})^b, \mathrm{u}^\star(\vec{q}, a]_{-\epsilon}}{2}, \mathrm{u}(\vec{p})^c] = \delta_b^c \mathrm{u}(\vec{p})^b. \end{array}$$

This has been used for the implementation of Poincaré Lie algebras in the chapter "Massive Particle Quantum Fields."

7.11 Group Representations on Homogeneous Functions

In general, the regular representation of a group on its functions is decomposable. Homogeneous functions of integer degree (harmonic polynomials) constitute finite-dimensional Hilbert representation spaces for compact groups. Finite dimensions, integer winding numbers, integer powers, and the degrees of polynomials are related to each other and characteristic for irreducible representations of compact groups as familiar from $\mathbf{SU}(2)$ with dimensions $1 + 2J$ and powers $\bigvee^{2J} u$ or with the structure of the harmonic polynomials. Simple Lie algebras have fundamental representations as $\mathbb{N}_0$-basis of the finite-dimensional irreducible representation cone, e.g., the Pauli representation for $\log \mathbf{SU}(2)$ and the two Weyl representations for $\log \mathbf{SL}(\mathbb{C}^2)$. The finite-dimensional irreducible representations act on the totally symmetric tensor products of the fundamental representations (chapters "Spin, Rotations, and Position" and "Lorentz Operations"). If a group G acts on a vector space $V \cong \mathbb{C}^d$ with basis $\{e^j\}$, the group acts on all tensor powers with the symmetry classes as invariant subspaces. The totally symmetric powers are isomorphic to the polynomials, homogeneous of degree n, with the monomials as basis:

$$\begin{aligned} &\bigvee^k V \cong \mathbb{C}[e^1, \dots, e^d]^n \ni p^n(e^1, \dots, e^d) = \sum_{n_1, \dots, n_d} \alpha_{n_1 \dots n_d} (e^1)^{n_1} \cdots (e^d)^{n_d} \\ &\quad \text{with } n_1 + \dots + n_d = n = \dim_{\mathbb{C}} \bigvee^k V = \binom{d+k-1}{k} = \deg p, \\ &\quad \gamma \in \mathbb{C} \Rightarrow p^n(\gamma e^1, \dots, \gamma e^n) = \gamma^n p(e^1, \dots, e^d), \quad n \in \mathbb{N} \end{aligned}$$

The polynomials with integer degree, related to integer invariants, can be generalized to homogeneous functions of complex degree, related to continuous complex invariants. Such an extension can be used for the infinite-dimensional representations of nonabelian noncompact groups. The homogeneity powers, characterizing irreducible representation spaces, give the invariants and eigenvalues of Cartan subgroup representations, integer $\mathbb{Z}$ for $\mathbf{U}(1)$ and $\mathbf{SO}(2)$, and continuous complex $i\mathbb{R} \oplus \mathbb{R}$ for $\mathbf{D}(1)$ and $\mathbf{SO}_0(1,1)$.

7.11.1 Homogeneous Functions on $\mathbf{SL}(\mathbb{C}^2)$-Spinors

The finite-dimensional representations of

$$\mathbf{SL}(\mathbb{C}^2) = \{s = e^{i\vec{\alpha}+\vec{\beta}} = \left(\begin{smallmatrix} a & b \\ c & d \end{smallmatrix}\right) \bigm| \det s = 1\} \cong \mathbf{SU}(2) \times \mathbf{SL}(\mathbb{C}^2)/\mathbf{SU}(2)$$

act on polynomials, homogeneous in a left und right Weyl spinor basis as indeterminates (chapter "Lorentz Operations")

$$p(z^1, z^2)^{[2L|2R]} \in \bigvee^{2L} \mathbb{C}^2 \otimes \bigvee^{2R} \mathbb{C}^2.$$

They can be written as double homogeneous polynomials with the indeterminates (z^1, z^2) having a conjugation.

As shown by Gel'fand and Naimark and sketched superficially in the following, to obtain Hilbert $\mathbf{SL}(\mathbb{C}^2)$-representations, the polynomials have to be generalized to homogeneous functions, not necessarily with integer powers for the representations of the noncompact degree of freedom $\mathbf{SO}_0(1,1) \subset \mathcal{Y}^3 \cong \mathbf{SL}(\mathbb{C}^2)/\mathbf{SU}(2)$. The representation spaces are subspaces of the vector space $\mathbb{C}(\mathbb{C}^2) = \{f : \mathbb{C}^2 \longrightarrow \mathbb{C} \bigm| \text{continuous}\}$ with functions on a spinor space $\mathbb{C}^2$ with the defining $\mathbf{SL}(\mathbb{C}^2)$-representation. The function space is acted on by the right regular representation

$$\begin{array}{ccc} \mathbb{C}^2 & \xrightarrow{s} & \mathbb{C}^2 \\ f \downarrow & & \downarrow f_s \\ \mathbb{C} & \xrightarrow[\mathrm{id}_\mathbb{C}]{} & \mathbb{C} \end{array}, \quad \begin{aligned} f_s(z^1, z^2) &= f((z^1, z^2)s), \\ (z^1, z^2)s &= (z^1, z^2)\left(\begin{smallmatrix} a & b \\ c & d \end{smallmatrix}\right) = (az^1 + cz^2, bz^1 + dz^2), \\ \mathbb{C}^2,\ \mathbb{C}(\mathbb{C}^2) &\in \underline{\mathbf{vec}}_{\mathbf{SL}(\mathbb{C}^2)}. \end{aligned}$$

The function space $\mathbb{C}(\mathbb{C}^2)$ is highly reducible. The irreducible $\mathbf{SL}(\mathbb{C}^2)$-representations are induced and characterized by the irreducible representations of a Cartan subgroup $\mathbf{SO}(\mathbb{C}^2) = \mathbf{SO}(2) \times \mathbf{SO}_0(1,1) \cong \mathbf{U}(1) \times \mathbf{D}(1) = \mathbf{GL}(\mathbb{C})$,

$$\begin{aligned} &u = e^{i\alpha_3} \in \mathbf{U}(1),\ d = e^{\beta_3} \in \mathbf{D}(1),\ \ \gamma = ud \in \mathbf{GL}(\mathbb{C}), \\ &\gamma^{\zeta_1}\overline{\gamma}^{\zeta_2} = u^{2j}d^\delta \text{ with } [\zeta_1|\zeta_2] = [\tfrac{\delta+2j}{2}|\tfrac{\delta-2j}{2}], \end{aligned}$$

acting on vector subspaces by the functions homogeneous with respect to the action of $\mathbf{U}(1) \times \mathbf{D}(1)$,

$$\mathbb{C}[z^1, z^2]^{[\zeta_1-1|\zeta_2-1]} = \{f \in \mathbb{C}(\mathbb{C}^2) \bigm| f(\gamma z^1, \gamma z^2) \ = \gamma^{\zeta_1-1}\overline{\gamma}^{\zeta_2-1} f(z^1, z^2),\ \zeta_{1,2} \in \mathbb{C}\}.$$

The functions use integer winding numbers $2j$ (spin j) for the compact group $\mathbf{SO}(2)$ and a complex boost eigenvalue δ for the noncompact group $\mathbf{SO}_0(1,1)$. The Cartan subgroup weights

$$\begin{array}{rl}(2j,\delta)\in & \operatorname{spec}\mathbf{SO}(2)\times\operatorname{spec}\mathbf{SO}_0(1,1)\cong\mathbb{Z}\times\mathbb{C},\\ & \mathbf{SO}(2)\subset\mathbf{SU}(2),\ \mathbf{SO}_0(1,1)\subset\mathbf{SL}(\mathbb{C}^2)/\mathbf{SU}(2)\end{array}$$

are the powers in its representations,

$$\mathbf{U}(1)\times\mathbf{D}(1)\ni\gamma=e^{i\alpha+\beta}\longmapsto e^{i\alpha 2j+\beta\delta}.$$

The continuous complex $[\zeta_1-1|\zeta_2-1]$-homogeneity generalizes the polynomial $[2l|2r]$-homogeneity for finite-dimensional representations.

Since the $[\zeta_1-1|\zeta_2-1]$-homogeneous functions have the orbit properties

$$f(z^1,z^2)=\frac{(z^2)^{\zeta_1}(\overline{z^2})^{\zeta_2}}{z^2\overline{z^2}}f(\tfrac{z^1}{z^2},1),$$

the group $\mathbf{SL}(\mathbb{C}^2)$ acts on the corresponding vector space containing the "start functions" that depend only on one complex variable $F(z)=f(z,1)$ in the following way:

$$\begin{array}{rl}s & \longmapsto[2j,\delta](s),\quad F\longmapsto[2j,\delta](s)(F)=F_s,\\ F_s(z) & =(bz+d)^{\zeta_1-1}(\overline{bz+d})^{\zeta_2-1}F(\frac{az+c}{bz+d})=(\frac{bz+d}{\overline{bz+d}})^j|bz+d|^{\delta-2}F(\frac{az+c}{bz+d}),\end{array}$$

involving the linear fractional transformation of the closed complex plane

$$\overline{\mathbb{C}}\ni z\longmapsto\frac{az+c}{bz+d}\in\overline{\mathbb{C}},\quad\det\begin{pmatrix}a&b\\c&d\end{pmatrix}=1.$$

7.11.2 Principal and Supplementary $\mathbf{SL}(\mathbb{C}^2)$-Representations

In the irreducible $\mathbf{SL}(\mathbb{C}^2)$-representations a Cartan subgroup comes with a power for both real subgroups,

$$\mathbf{SO}(\mathbb{C}^2)\ni e^{(i\alpha+\beta)\sigma^3}\longmapsto e^{(i\alpha 2j+\beta\delta)\sigma^3},\quad(2j,\delta)\in\mathbb{Z}\times\mathbb{C}.$$

The finite-dimensional representations (chapter "Lorentz Operations") are characterized by an integer $\delta=2d\in\mathbb{Z}$ where d and spin j have an integer sum, i.e., both are integer or both are half-integer. In this case, the spin-boost pair can be replaced by a left-right winding number pair:

$$\begin{array}{l}(2j,\delta)=(2j,2d)\in\mathbb{Z}\times\mathbb{Z}\text{ and }j+d\in\mathbb{Z},\\ [\zeta_1|\zeta_2]=[d+j|d-j]=[2l|2r].\end{array}$$

For the unitary irreducible $\mathbf{SL}(\mathbb{C}^2)$-representations a Cartan subgroup has to be represented in one of the two unitary group types in 2-dimensions:

$$\mathbf{SO}(\mathbb{C}^2)\ni e^{(i\alpha+\beta)\sigma^3}\longmapsto e^{(i\alpha 2j+\beta\delta)\sigma^3}\in\left\{\begin{array}{c}\mathbf{SU}(2),\\ \mathbf{SU}(1,1).\end{array}\right.$$

The $\mathbf{U}(2)$-unitary $\mathbf{SO}(\mathbb{C}^2)$-representations have to satisfy

$$\begin{aligned}&[(i\alpha 2j+\beta\delta)\sigma^3]^\star = (-i\alpha 2j+\beta\overline{\delta})\sigma^3 = -(i\alpha 2j+\beta\delta)\sigma^3\\ &\Rightarrow \begin{cases} 2j \in \mathbb{Z},\\ \delta = -\overline{\delta} = iQ \in i\mathbb{R}.\end{cases}\end{aligned}$$

This induces the *principal series*

$$e^{(i\alpha+\beta)\sigma^3} \longmapsto \begin{pmatrix} e^{i(\alpha 2j+\beta Q)} & 0\\ 0 & e^{-i(\alpha 2j+\beta Q)}\end{pmatrix} \in \mathbf{SO}(2) \subset \mathbf{SU}(2),$$

characterized by an integer winding number $2j$ and an imaginary boost eigenvalue iQ,

$$\mathbf{weights}\,^{(2,0)}\mathbf{SL}(\mathbb{C}^2) = \{(2j, iQ)\} = \mathbb{Z}\times i\mathbb{R}.$$

The equivalence classes of the irreducible representations take into account the self-duality of $\mathbf{SO}(\mathbb{C}^2)$, i.e., $(j,Q)\cong -(j,Q)$. This leads to two principal series:

$$\mathbf{irrep}\,^{(2,0)}\mathbf{SL}(\mathbb{C}^2) = \mathbf{irrep}_+^{(2,0)}\mathbf{SL}(\mathbb{C}^2) \cong \mathbb{N}_0\times\mathbb{R}_+ \ni (2J, Q_\pm^2).$$

The principal representations act on an L^2-Hilbert space with product

$$F \longmapsto [2j, iQ](s)(F): \quad \langle F|F\rangle = \int dz d\overline{z}\; \overline{F(z)}F(z), \quad \begin{cases} z = x+iy,\\ dz\, d\overline{z} = dx\, dy.\end{cases}$$

The indefinite unitary group $\mathbf{U}(1,1)$ with conjugation $\times$,

$$\begin{pmatrix}\alpha & \beta\\ \gamma & \delta\end{pmatrix}^\times = \begin{pmatrix}0 & 1\\ 1 & 0\end{pmatrix}\begin{pmatrix}\alpha & \beta\\ \gamma & \delta\end{pmatrix}^\star\begin{pmatrix}0 & 1\\ 1 & 0\end{pmatrix} = \begin{pmatrix}\overline{\delta} & \overline{\beta}\\ \overline{\gamma} & \overline{\alpha}\end{pmatrix},$$

gives as condition for the boost eigenvalue and for the spin winding number

$$\begin{aligned}&[(i\alpha 2j+\beta\delta)\sigma^3]^\times = -(-i\alpha 2j+\beta\overline{\delta})\sigma^3 = -(i\alpha 2j+\beta\delta)\sigma^3\\ &\Rightarrow \begin{cases} 2j = -2j = 0 \in \mathbb{Z},\\ \delta = \overline{\delta} = \kappa \in \mathbb{R}.\end{cases}\end{aligned}$$

This induces the *supplementary series*

$$e^{(i\alpha+\beta)\sigma^3} \longmapsto \begin{pmatrix} e^{\beta\kappa} & 0\\ 0 & e^{-\beta\kappa}\end{pmatrix} \in \mathbf{SO}_0(1,1) \subset \mathbf{SU}(1,1),$$

characterized by trivial spin and a real boost eigenvalue κ,

$$\mathbf{weights}\,^{(1,1)}\mathbf{SL}(\mathbb{C}^2) = \{(0,\kappa)\} = \{0\}\times\mathbb{R},$$

and for the irreducible representation classes with the self-duality of $\mathbf{SO}_0(1,1)$,

$$\mathbf{irrep}\,^{(1,1)}\mathbf{SL}(\mathbb{C}^2) \cong \{0\}\times\mathbb{R}^- \ni (0, -\kappa^2).$$

For a certain range of the boost eigenvalue, the supplementary representations act on a Hilbert space whose scalar product involves a positive-type function:

$$\begin{aligned} F &\longmapsto [0,\kappa](s)(F) \text{ with } 0 \le \kappa^2 \le 1, \\ \langle F|F\rangle &= \langle \hat{F} * F\rangle_d = -\tfrac{1}{4\Gamma(-\kappa)} \int dz_1 d\overline{z}_1 dz_2 d\overline{z}_2\, \overline{F(z_1)}\, \tfrac{1}{|z_1-z_2|^{2+2\kappa}}\, F(z_2), \end{aligned}$$

unitary representation	weights $\mathbf{SL}(\mathbb{C}^2)$	$=$	$\mathbb{Z} \times i\mathbb{R} \uplus \{0\} \times \mathbb{R}$
$\mathbf{SO}(\mathbb{C}^2) \longrightarrow \mathbf{SU}(2)$ (principal, Hilbert)	$\mathbf{irrep}_+^{(2,0)}\mathbf{SL}(\mathbb{C}^2)$	$\cong$	$\mathbb{N}_0 \times \mathbb{R}_+ \ni (2J, Q^2_\pm)$
$\mathbf{SO}(\mathbb{C}^2) \longrightarrow \mathbf{SU}(1,1)$ (supplementary, partly Hilbert)	$\mathbf{irrep}^{(1,1)}\mathbf{SL}(\mathbb{C}^2)$	$\cong$	$\{0\} \times \mathbb{R}^- \ni (0, -\kappa^2)$

unitary $\mathbf{SL}(\mathbb{C}^2)$-representations

7.11.3 Continuous Quantum Numbers

To generalize the $\mathbf{SL}(\mathbb{C}^2)$-structures: The irreducible $\mathbf{SL}(\mathbb{C}^{1+r})$-representations, $r \ge 1$, are characterizable by the representations of a Cartan subgroup $\mathbf{SO}(\mathbb{C}^2)^r$ with r integer $\mathbf{SO}(2)$-winding numbers and r $\mathbf{SO}_0(1,1)$-weights as homogeneity powers, expressible by $1+r$ numbers with trivial sum (traceless Lie algebra):

$$(Z_0, \dots, Z_r, \delta_0, \dots, \delta_r)_0 \in (\mathbb{Z} \times \mathbb{C})^r \text{ with } \sum_{k=0}^r Z_k = 0 = \sum_{k=0}^r \delta_k.$$

The unitary irreducible representations represent a Cartan subgroup with self-dual 1-dimensional subgroups in the possible $1+r$ types of unitary subgroups:

$$\begin{aligned} \mathbf{SL}(\mathbb{C}^{1+r}) \supset \mathbf{SO}(\mathbb{C}^2)^r \longrightarrow\ & \mathbf{SU}(1+s, r-s) \subset \mathbf{SL}(\mathbb{C}^{1+r}), \\ & s = r, \dots, 0. \end{aligned}$$

For the *principal* series one has an $\mathbf{SU}(1+r)$-representation of a Cartan subgroup:

$$\mathbf{weights}^{(1+r,0)}\mathbf{SL}(\mathbb{C}^{1+r}) \ = \{(Z_0, \dots, Z_r, iQ_0, \dots, iQ_r)_0\} = (\mathbb{Z} \times i\mathbb{R})^r.$$

The *supplementary* series involve some equal $\mathbf{SO}(2)$-winding number pairs (Z, Z) and the same number of corresponding "mixed" $\mathbf{SO}_0(1,1)$-weight pairs $(iQ+\kappa, iQ-\kappa)$, e.g., one pair

$$\begin{aligned} &(Z, Z, Z_2, \dots, Z_r, iQ+\kappa, iQ-\kappa, iQ_2, \dots, iQ_r)_0 \in \mathbf{weights}^{(r,1)}\mathbf{SL}(\mathbb{C}^{1+r}) \\ &\text{with } 2Z + \sum_{k=2}^r Z_k = 0 \text{ and } 2Q + \sum_{k=2}^r Q_k = 0, \\ &\mathbf{weights}^{(r,1)}\mathbf{SL}(\mathbb{C}^{1+r}) = (\mathbb{Z} \times i\mathbb{R})^{r-1} \times (\{0\} \times \mathbb{R}). \end{aligned}$$

For $r \geq 2$ there exist nontrivial *degenerate principal* representations with $\mathbf{SU}(1+s)$-unitarity, $s = 1, \ldots, r-1$, for the represented Cartan subgroup.

Altogether, one obtains with respect to the different unitary signatures one principal nondegenerate series, r supplementary ones, and $r-1$ principal degenerate series:

$\mathbf{SU}(1+s, r-s)$-unitary $s = r$: principal nondegenerate $s = r-1, \ldots, 0$: supplementary	$\textbf{weights}\ ^{(1+s,r-s)}\mathbf{SL}(\mathbb{C}^{1+r}) \cong (\mathbb{Z} \times i\mathbb{R})^s \times \mathbb{R}^{r-s}$
$\mathbf{SU}(1+s)$-unitary $s = 1, \ldots, r-1$: principal degenerate	$\textbf{weights}\ ^{(1+s,0)}\mathbf{SL}(\mathbb{C}^{1+r}) \cong (\mathbb{Z} \times i\mathbb{R})^s$

	general	finite-dimensional
general	$(\mathbb{Z} \times \mathbb{C})^r$	$(\mathbb{Z} \times \mathbb{Z})^r$
$\mathbf{SU}(1+s, r-s)$ unitary	$(\mathbb{Z} \times i\mathbb{R})^s \times \mathbb{R}^{r-s}$ $s = 0, \ldots, r$	$\mathbb{Z}^r$ $s = 0$

weights of irreducible $\mathbf{SL}(\mathbb{C}^{1+r})$-representations

With respect to the irreducible representation classes one has to take into account equivalences.

7.12 Harmonic Analysis of Hyperboloids

The infinite-dimensional Hilbert representations of the Lorentz groups are used for the harmonic analysis of functions on the nonabelian hyperboloids [6]:

$$s = 2, 3, \cdots : \mathcal{Y}^s \cong \mathbf{SO}_0(1, s)/\mathbf{SO}(s).$$

With the Cartan decomposition

$$\mathbf{SO}_0(1, s) = \mathbf{SO}(s) \circ \mathbf{SO}_0(1, 1) \circ \mathbf{SO}(s)$$

the hyperboloids have real rank 1 and imaginary rank R for $s = 2R, 1+2R$ with Cartan tori $\mathbf{SO}(2)^R$. The minimal nonabelian cases $s = 2, 3$ are characteristic for the even- and odd-dimensional nonabelian hyperboloids.

The harmonic analysis of the maximal noncompact abelian group $\mathbb{R} \cong \mathbf{SO}_0(1, 1) \cong \mathcal{Y}^1$ can be formulated with hyperbolic concepts: Cartan spacetime $\mathbb{R}^2$ embeds both the hyperbolas $\mathcal{Y}^1$ and all their tangents $\log \mathbf{SO}_0(1, 1) \cong \mathbb{R}$. With the forward-backward momenta $\epsilon|p|$, $\epsilon = \pm 1 \in \Omega^0$, as invariant linear forms (eigenvalues), the defining $\mathbf{SO}_0(1, 1)$-representation is the Lorentz product qy with a lightlike energy-momentum q, i.e., $q^2 = 0$, normalized with the momentum invariant $|p|$,

$$\mathcal{Y}^1 \ni \mathrm{y} = \begin{pmatrix} \cosh \psi \\ \sinh \psi \end{pmatrix}, \quad q = |p| \begin{pmatrix} 1 \\ \epsilon \end{pmatrix} \in \mathbb{R}_+ \times \Omega^0, \quad e^{\epsilon\psi} = \cosh \psi + \epsilon \sinh \psi = \frac{q\mathrm{y}}{|p|}.$$

The harmonic analysis with the irreducible Hilbert representations

$$\mathcal{Y}^1 \ni y \longmapsto e^{ip\psi} = e^{i|p|\epsilon\psi} = \left(\frac{q\mathrm{y}}{|p|}\right)^{i|p|}, \quad p = \epsilon|p|$$

looks in this parametrization as follows, with forward and backward separation:

$$f(\psi) = \int_0^\infty \frac{dp}{2\pi} \left(\frac{q\mathrm{y}}{|p|}\right)^{i|p|} \tilde{f}(p) \leftrightarrow \tilde{f}(p) = \int d\mathcal{Y}^1(\psi) \left(\frac{q\mathrm{y}}{|p|}\right)^{-i|p|} f(\psi).$$

The $\mathcal{Y}^1$-measure has the corresponding parametrization $\int d\mathcal{Y}^1(\psi) = \int_0^\infty d\psi$.

The Lorentz compatible formalism with the powers of the defining $\mathbf{SO}_0(1,1)$-representations $(e^{\epsilon\psi})^{i|p|} = (\frac{q\mathrm{y}}{|p|})^{i|p|}$ as irreducible representations is generalizable to the nonabelian cases $\mathcal{Y}^s$, $s \geq 2$, with $\mathbf{SO}_0(1,s)$-invariant products: With $\mathcal{Y}^s$ and its tangent spaces $\mathbb{R}^s$ embedded in $\mathbb{R}^{1+s}$, the harmonic analysis of the $\mathcal{Y}^s$-functions use the tangent space forms $\vec{p} \in \mathbb{R}^s$ (momenta). The future lightcone V_0^s with $\{0\}$ in $\mathbb{R}^{1+s}$ is isomorphic to the $\mathcal{Y}^s$-tangent space

$$\log \mathcal{Y}^s \cong \mathrm{V}_0^s \cong \mathbb{R}^s \cong \mathbb{R}_+ \times \Omega^{s-1}.$$

The momenta $\vec{p}$ can be embedded as 1-dimensional lightlike energy-momenta:

$$\begin{array}{rlcrl} \mathbb{R}^{1+s} \supset \mathcal{Y}^s \ni \mathrm{y} & = \begin{pmatrix} \cosh\psi \\ \frac{\vec{x}}{r}\sinh\psi \end{pmatrix} & \leftrightarrow & \mathbb{R}^{1+s} \supset \mathrm{V}_0^s \ni q & = |\vec{p}| \begin{pmatrix} 1 \\ \frac{\vec{p}}{|\vec{p}|} \end{pmatrix}, \\ \mathrm{y}^2 & = 1 & & q^2 & = 0, \\ \text{e.g., } \mathbb{R}^3 \supset \mathcal{Y}^2 \ni \mathrm{y} & = \begin{pmatrix} \cosh\psi \\ \sinh\psi\cos\theta \\ \sinh\psi\sin\theta \end{pmatrix} & \leftrightarrow & \mathbb{R}_+ \times \Omega^1 \cong \mathrm{V}_0^2 \ni q & = |\vec{p}| \begin{pmatrix} 1 \\ 1 \\ 0 \end{pmatrix}. \end{array}$$

The hyperboloid $\mathcal{Y}^s$ (nonlinear s-dimensional position) embeds hyperbolas. The normalized rotation $\mathbf{SO}(s)$-invariant product contains the defining representations of hyperbolas $\mathbf{SO}_0(1,1)$ in $\mathcal{Y}^s$, related to each other by rotations $\mathbf{SO}(s)$ and indexed by an Ω^1-orientation angle θ. It is the nonabelian extension of the abelian exponent:

$$\begin{array}{llll} s = 1: & \frac{q\mathrm{y}}{|p|} = \cosh\psi + \epsilon\sinh\psi & = e^{\epsilon\psi}, & \epsilon = \pm 1, \\ s \geq 2: & \frac{q\mathrm{y}}{|\vec{p}|} = \cosh\psi + \cos\theta\sinh\psi & = \cos^2\frac{\theta}{2}\, e^{\psi} + \sin^2\frac{\theta}{2}\, e^{-\psi}, & \cos\theta = \frac{\vec{p}\vec{x}}{|\vec{p}|r}. \end{array}$$

The minimal nonabelian case $\mathcal{Y}^2$ is characteristic since the eigenvalues q can always be written in this 2-component form. The additional degrees of freedom are spherical. The defining $\mathcal{Y}^1$-representations for the characteristic case $\mathcal{Y}^2$ are related to each other by axial rotations $\mathbf{SO}(2)$. $|\vec{p}|$ is the invariant.

It is useful to compare with the compact spheres $\Omega^s \cong \mathbf{SO}(1+s)/\mathbf{SO}(s)$: There is the $L^2(\mathcal{Y}^s)$-analogous treatment for the harmonic analysis [11] of the square integrable functions $L^2(\Omega^s)$. With the hyperboloid-sphere transition $\psi \leftrightarrow i\chi$ to

$$\begin{array}{llcll} s = 1: & \mathbb{R}^2 \supset \Omega^1 \ni \vec{\omega} = \begin{pmatrix} \cos\chi \\ i\sin\chi \end{pmatrix} & \leftrightarrow & \mathbb{N} \times \Omega^0 \ni \vec{q} & = L\begin{pmatrix} 1 \\ i\epsilon \end{pmatrix}, \\ s \geq 2: & \mathbb{R}^{1+s} \supset \Omega^s \ni \vec{\omega} = \begin{pmatrix} \cos\chi \\ i\frac{\vec{x}}{r}\sin\chi \end{pmatrix} & \leftrightarrow & \mathbb{R}^{1+s} \ni \vec{q} & = L\begin{pmatrix} 1 \\ i\frac{\vec{L}}{L} \end{pmatrix}, \quad L = |\vec{L}|, \end{array}$$

one uses embedded subgroups $\mathbf{SO}(2)$:

$$\begin{array}{llll} s = 1: & \frac{\vec{q}\vec{\omega}}{L} = \cos\chi + i\epsilon\sin\chi & = e^{i\epsilon\chi}, & \epsilon = \pm 1, \\ s \geq 2: & \frac{\vec{q}\vec{\omega}}{L} = \cos\chi + i\cos\theta\sin\chi & = \cos^2\frac{\theta}{2}\, e^{i\chi} + \sin^2\frac{\theta}{2}\, e^{-i\chi}, & \cos\theta = \frac{\vec{L}\vec{x}}{Lr}. \end{array}$$

The functions for the rotated hyperbolas and circles are diagonal matrix elements of the nonabelian groups in a corresponding Cartan and Euler parametrization:

$$\begin{array}{lll}
\mathbf{SO}_0(1,2) & \sim \mathbf{SU}(1,1) & \ni \begin{pmatrix} \cosh\psi+\cos\theta\sinh\psi & -i\sin\theta\sinh\psi \\ i\sin\theta\sinh\psi & \cosh\psi-\cos\theta\sinh\psi \end{pmatrix} = v\circ e^{\sigma_3\psi}\circ v^\star, \\
\mathbf{SO}(3) & \sim \mathbf{SU}(2) & \ni \begin{pmatrix} \cos\chi+i\cos\theta\sin\chi & \sin\theta\sin\chi \\ -\sin\theta\sin\chi & \cos\chi-i\cos\theta\sin\chi \end{pmatrix} = v\circ e^{i\sigma_3\chi}\circ v^\star, \\
e^{\sigma_3\psi} & \in \mathbf{SO}_0(1,1), & \\
e^{i\sigma_3\chi} & \in \mathbf{SO}(2), & v = e^{i\sigma_1\frac{\theta}{2}} = \begin{pmatrix} \cos\frac{\theta}{2} & i\sin\frac{\theta}{2} \\ i\sin\frac{\theta}{2} & \cos\frac{\theta}{2} \end{pmatrix} \in \mathbf{SO}(2).
\end{array}$$

The irreducible representations for the noncompact nonabelian subgroups $\mathbf{SO}_0(1,2)$ have an imaginary power $i|\vec{p}|$ with the invariant $|\vec{p}|=m$:

$$s\geq 2: \quad \cosh\psi+\cos\theta\sinh\psi \longmapsto (\cosh\psi+\cos\theta\sinh\psi)^{i|\vec{p}|} = (\tfrac{qy}{|\vec{p}|})^{i|\vec{p}|}.$$

Therefore, the hyperboloid functions $L^2(\mathcal{Y}^s)$ have the harmonic analysis

$$\begin{aligned}
f(\vec{\psi}) &= \int d\check{\mathcal{Y}}^s(\vec{p})\frac{(\cosh\psi+\cos\theta\sinh\psi)^{i|\vec{p}|}}{(\cosh\psi+\cos\theta\sinh\psi)^{\frac{s-1}{2}}}\tilde{f}(\vec{p}) \\
&\leftrightarrow \tilde{f}(\vec{p}) = \int d\mathcal{Y}^s(\vec{\psi})\frac{(\cosh\psi+\cos\theta\sinh\psi)^{-i|\vec{p}|}}{(\cosh\psi+\cos\theta\sinh\psi)^{\frac{s-1}{2}}}f(\vec{\psi}).
\end{aligned}$$

The irreducible representations are normalized with a length factor $(\frac{qy}{|\vec{p}|})^{\frac{s-1}{2}}$ for the corresponding sphere Ω^{s-1}.

The invariant for spheres are integer powers $L=0,1,2,\dots$ (angular momenta):

$$\begin{aligned}
s=1: &\quad e^{i\epsilon\chi}=\cos\chi+i\epsilon\sin\chi &&\longmapsto e^{iL\epsilon\chi}=(\tfrac{q\vec{\omega}}{L})^L, \\
s\geq 2: &\quad \cos\chi+i\cos\theta\sin\chi &&\longmapsto (\cos\chi+i\cos\theta\sin\chi)^L=(\tfrac{q\vec{\omega}}{L})^L.
\end{aligned}$$

There is an expression for the harmonic analysis of $L^2(\Omega^s)$, analogous to that for $L^2(\mathcal{Y}^s)$, now with discrete L-summation.

The circle-integration of the integer powers of the diagonal matrix elements for $\mathbf{SO}(3)$ gives the Legendre polynomials with integer index (invariant). They constitute a Hilbert basis of $L^2(\mathbf{SO}(2))\cong L^2(-1,1)$ (chapter "Quantum Probability"):

$$\begin{aligned}
L=0,1,\cdots: &\quad \mathrm{P}^L(\cos\chi)=\int_0^{2\pi}\frac{d\theta}{2\pi}(\cos\chi+i\cos\theta\sin\chi)^L, \\
&\quad \frac{d^2\mathrm{P}^L}{d\chi^2}+\frac{\cos\chi}{\sin\chi}\frac{d\mathrm{P}^L}{d\chi}=\frac{1}{\sin\chi}\frac{d}{d\chi}\sin\chi\frac{d\mathrm{P}^L}{d\chi}=-L(1+L)\mathrm{P}^L, \\
&\quad \mathbb{R}\ni\xi\longmapsto\mathrm{P}^L(\xi) \text{ with } \frac{d}{d\xi}(1-\xi^2)\frac{d\mathrm{P}^L}{d\xi}=-L(1+L)\mathrm{P}^L, \\
\int_{-1}^1\frac{d\xi}{2}\,\mathrm{P}^L(\xi)\,\mathrm{P}^{L'}(\xi) &= \frac{1}{1+2L}\delta_{LL'}, \quad \sum_{L=0}^\infty(1+2L)\mathrm{P}^L(\xi)\,\mathrm{P}^L(\xi')=\delta(\tfrac{\xi-\xi'}{2}).
\end{aligned}$$

Their hyperbolic partners arising as circle-integrated complex powers of the diagonal matrix elements for $\mathbf{SO}_0(1,2)$ are the *Legendre functions* with continuous complex index (invariant) μ:

$$\begin{aligned}
\mu\in\mathbb{C}: &\quad \mathrm{P}^\mu(\cosh\psi)=\int_0^{2\pi}\frac{d\theta}{2\pi}(\cosh\psi+\cos\theta\sinh\psi)^\mu, \\
&\quad \frac{d^2\mathrm{P}^\mu}{d\psi^2}+\frac{\cosh\psi}{\sinh\psi}\frac{d\mathrm{P}^\mu}{d\psi}=\frac{1}{\sinh\psi}\frac{d}{d\psi}\sinh\psi\frac{d\mathrm{P}^\mu}{d\psi}=\mu(1+\mu)\mathrm{P}^\mu, \quad \psi>0, \\
&\quad \mathbb{R}\ni\zeta\longmapsto\mathrm{P}^\mu(\zeta) \text{ with } \frac{d}{d\zeta}(\zeta^2-1)\frac{d\mathrm{P}^\mu}{d\zeta}=\mu(1+\mu)\mathrm{P}^\mu.
\end{aligned}$$

The Legendre functions for $\mu = im - \frac{1}{2}$, called *cone functions*, are coefficients of the irreducible principal $\mathbf{SL}(\mathbb{R}^2)$-representations. They can be used as orthonormal distributive basis for the Hilbert space $L^2(\mathbf{SO}_0(1,1) \cong L^2(1,\infty)$

$$\begin{array}{rl} \int_1^\infty d\zeta \ \mathrm{P}^{im-\frac{1}{2}}(\zeta)\ \mathrm{P}^{-im'-\frac{1}{2}}(\zeta) & = \frac{\pi}{m \tanh \pi m}\delta(m-m'), \\ \int_0^\infty \frac{m \tanh \pi m}{\pi} dm \ \mathrm{P}^{im-\frac{1}{2}}(\zeta)\ \mathrm{P}^{-im-\frac{1}{2}}(\zeta') & = \delta(\zeta - \zeta'). \end{array}$$

There occurs the Plancherel measure $\frac{1}{\pi}\Pi^s(m^2)dm$ for the invariants $m^2 = \vec{p}^2$ of the arising irreducible $\mathbf{SO}_0(1,s)$-representations, different for even and odd space dimensions $s = 2R, 1+2R$:

$$\begin{array}{rl} \int d\check{\mathcal{Y}}^s(\vec{p}) & = \int_0^\infty \frac{1}{\pi}\Pi^s(m^2)\ dm \int \frac{d^{s-1}\omega}{(2\pi)^{s-1}} \end{array}$$

$$\Pi^s(m^2) = \left|\frac{\Gamma(im+\frac{s-1}{2})}{\Gamma(im)}\right|^2 = \left\{\begin{array}{ll} \Gamma(R-\frac{1}{2})^2 \times m^2 \frac{\tanh \pi m}{\pi m} \prod_{k=1}^{R-1}(1+\frac{4m^2}{(2k-1)^2}), & s = 2R = 2,4,6,\ldots, \\ \Gamma(R)^2 \times m^2 \prod_{k=1}^{R-1}(1+\frac{m^2}{k^2}), & s = 1+2R = 3,5,7,\ldots. \end{array}\right.$$

It is computed with the properties of the Γ-function

$$\mu \in \mathbb{C}:\ \Gamma(\overline{\mu}) = \overline{\Gamma(\mu)},\quad \Gamma(\mu)\Gamma(1-\mu) = \frac{\pi}{\sin \pi\mu},\quad |\Gamma(i\mu)|^2 = \frac{\pi}{\mu \sinh \pi\mu},$$

$$\left|\frac{\Gamma(i\mu+\frac{s-1}{2})}{\Gamma(\frac{s-1}{2})}\right|^2 = \left\{\begin{array}{ll} \frac{1}{\cosh \pi\mu}\prod_{k=1}^{R-1}(1+\frac{4\mu^2}{(2k-1)^2}), & s = 2R = 2,4,6,\ldots, \\ \frac{\pi\mu}{\sinh \pi\mu}\prod_{k=1}^{R-1}(1+\frac{\mu^2}{k^2}), & s = 1+2R = 3,5,7,\ldots. \end{array}\right.$$

The harmonic analysis of the functions on $\mathcal{Y}^2 \cong \mathbf{SL}(\mathbb{R}^2)/\mathbf{SO}(2)$ employs the principal series representations of the group $\mathbf{SL}(\mathbb{R}^2) \sim \mathbf{SO}_0(1,2)$. The Plancherel measure $\frac{m}{\pi}\tanh \pi m\ dm = \frac{m}{\pi^2}\frac{1}{\cosh \pi m}\ d\cosh \pi m$ contains a hyperbolic function. Also, the harmonic analysis of the functions on the hyperboloid (nonlinear 3-dimensional position) $\mathcal{Y}^3 \cong \mathbf{SL}(\mathbb{C}^2)/\mathbf{SU}(2)$ employs the principal series representations of the Lorentz group $\mathbf{SL}(\mathbb{C}^2) \sim \mathbf{SO}_0(1,3)$. The Plancherel measure $\frac{m^2}{\pi}\ dm$ contains the dilation structure.

The additional product factor $\prod_{k=1}^{R-1}$ in the Plancherel measures is nontrivial for additional tori $\mathbf{SO}(2)^{R-1}$. The full products are self-dual hyperbolic functions, the noncompact partners of the self-dual spherical cosine $\cos \pi m$ and sine $\frac{\sin \pi m}{\pi m}$ (product theorem of Weierstrass [1])

$$\cosh \pi m = \prod_{k=1}^{\infty}(1+\frac{4m^2}{(2k-1)^2}),\quad \frac{\sinh \pi m}{\pi m} = \prod_{k=1}^{\infty}(1+\frac{m^2}{k^2}).$$

In the Plancherel measures there arise the "R-tails" $\prod_{k=R}^{\infty}$.

MATHEMATICAL TOOLS

7.13 Convolutions

The Cartesian product $\Phi_1 \diamond \Phi_2$ of two disjoint additive mappings (chapter "The Kepler Factor") into a unital algebra $\Phi_i : \mathcal{S}_i \longrightarrow A$, $i = 1, 2$, e.g., the numbers $A = \mathbb{K}$, formulated with generalized mappings (distributions, measures)

$$\begin{array}{rll} \mathbf{\Phi}_i : & S_i \longrightarrow A, & \Phi_i \cong d\mu_i(x_i)\, \mathbf{\Phi}_i(x_i), \\ \mathbf{\Phi}_1 \diamond \mathbf{\Phi}_2 : & S_1 \times S_2 \longrightarrow A, & \Phi_1 \diamond \Phi_2 \cong d\mu_1(x_1) d\mu_2(x_2)\, \mathbf{\Phi}_1(x_1) \diamond \mathbf{\Phi}_2(x_2), \end{array}$$

can be transformed by a measurable mapping $h \in \mathbf{mes}(S_1 \times S_2, S)$,

$$h : S_1 \times S_2 \longrightarrow S,$$

into a disjoint additive mapping on S, the h-image, called the *h-convoluted* mapping (distribution, measure), the *convolution product*

$$\begin{array}{rl} \Phi_1 *_h \Phi_2 : \mathcal{S} \longrightarrow A, & \Phi_1 *_h \Phi_2(X) = h \bullet (\Phi_1 \diamond \Phi_2)(X) = \Phi_1 \diamond \Phi_2(h^{-1}[X]) \\ & = \int_{h^{-1}[X] \subseteq S_1 \times S_2} d\mu_1(x_1) d\mu_2(x_2)\, \mathbf{\Phi}_1(x_1) \diamond \mathbf{\Phi}_2(x_2), \\ \Phi_1 *_h \Phi_2 & \cong d\mu(x)\, (\mathbf{\Phi}_1 *_h \mathbf{\Phi}_2)(x), \\ (\mathbf{\Phi}_1 *_h \mathbf{\Phi}_2)(x) & = \int_{S_1 \times S_2} d\mu_1(x_1) d\mu_2(x_2) \mathbf{\Phi}_1(x_1) \diamond \mathbf{\Phi}_2(x_2) \delta(h(x_1, x_2), x). \end{array}$$

Analogous definitions apply for products with more factors. The mapping h, if obvious, will be omitted.

The existence of the convolution product depends on the properties of the generalized mappings $\mathbf{\Phi}$ and the composition function h, which has to be discussed carefully.

Of importance is the convolution of mappings (distributions, measures) of a *locally compact group and a symmetric space* $G \times S \longrightarrow S$, $h(k, y) = k \bullet y$ where the group (unimodular) carries a Haar measure $d^G k$ and the G-set a left-invariant measure $d^S(k \bullet y) = d^S y$,

$$\begin{array}{rl} (\mathbf{\Phi}_G * \mathbf{\Phi}_S)(x) & = \int_{G \times S} d^G k\, d^S y\, \mathbf{\Phi}_G(k) \diamond \mathbf{\Phi}_S(y)\, \delta(k \bullet y, x) \\ & = \int_G d^G k\, \mathbf{\Phi}_G(k) \diamond \mathbf{\Phi}_S(k^{-1} \bullet x), \end{array}$$

especially for a *group acting on a vector space* $G \times V \longrightarrow V$, $k \bullet y = D(k).y$, or a *group acting on a coset* $G \times G/H \longrightarrow G/H$, $k \bullet y = kgH$, e.g., the *group acting on itself* with the group law $G \times G \longrightarrow G$,

$$\begin{array}{rl} (\mathbf{\Phi}_1 * \mathbf{\Phi}_2)(g) & = \int_{G \times G} d^G k_1\, d^G k_2\, \mathbf{\Phi}_1(k_1) \diamond \mathbf{\Phi}_2(k_2)\, \delta(k_1 k_2, g) \\ & = \int_G d^G k\, \mathbf{\Phi}_1(k) \diamond \mathbf{\Phi}_2(k^{-1} g) \\ & = \int_G d^G k\, \mathbf{\Phi}_1(g k^{-1}) \diamond \mathbf{\Phi}_2(k). \end{array}$$

7.14 Abelian Convolution of Functions and Distributions

In the following, complex functions or distributions on the abelian group $\mathbb{R}^d$ with Haar-Lebesgue measures are denoted by $L^p_{dq}(\mathbb{R}^d, \mathbb{C}) = L^p$, etc.

Continuous functions with compact support $\mathcal{C}_c$ are closed under the associative abelian convolution for the additive group $q \in \mathbb{R}^d$:

$$\begin{array}{lll} \mathcal{C}_c \in \star\underline{\mathbf{aag}}_{\mathbb{C}} : & \mathcal{C}_c * \mathcal{C}_c & \longrightarrow \mathcal{C}_c, \\ & (f_1 * f_2)(q) & = \int dq_1 dq_2 f_1(q_1)\ \delta(q_1 + q_2 - q)\ f_2(q_2). \end{array}$$

$\mathcal{C}_c$ is dense in all Lebesgue spaces L^p. The continuously differentiable functions are stable under L^1-convolution:

$$L^1 \in \star\underline{\mathbf{n\hat{a}ag}}_{\mathbb{C}},\quad L^p, \mathcal{C}^n \in \star\underline{\mathbf{mod}}_{L^1}.$$

For functions and distributions

$$\begin{array}{ccccc} L^\infty \supset \mathcal{C}_c \supset \mathcal{C}_c^\infty & \subset & \mathcal{S} & \subset & \mathcal{C}^\infty \\ \cap & & \cap & & \cap \\ \mathcal{D}'_c & \subset & \mathcal{S}' & \subset & \mathcal{D}' \supset \mathcal{M} \supset L^1 \end{array}$$

the convolution is defined if one factor comes with compact support,

$$\begin{array}{llll} \mathcal{C}_c^\infty & * & \mathcal{D}' & \longrightarrow & \mathcal{C}^\infty, \\ \mathcal{D}'_c & * & (\mathcal{C}_c^\infty, \mathcal{S}, \mathcal{C}^\infty, \mathcal{D}'_c, \mathcal{S}', \mathcal{D}') & \longrightarrow & (\mathcal{C}_c^\infty, \mathcal{S}, \mathcal{C}^\infty, \mathcal{D}'_c, \mathcal{S}', \mathcal{D}'). \end{array}$$

The distributions with compact support constitute a commutative unital algebra with convolution product $*$ and the Dirac distribution δ_0 as unit. $\mathcal{C}_c^\infty$ is a nonunital subalgebra. All spaces involved are stable under convolution with $\mathcal{D}'_c$ and thus corresponding modules

$$\begin{array}{l} \mathcal{D}'_c \in \star\underline{\mathbf{aag}}_{\mathbb{C}},\quad (\mathcal{C}_c^\infty, \mathcal{S}, \mathcal{C}^\infty, \mathcal{D}'_c, \mathcal{S}', \mathcal{D}') \in \underline{\mathbf{mod}}_{\mathcal{D}'_c}, \\ \text{unit } \delta_0 = dq\ \delta(q). \end{array}$$

Distributions with support in a cone, stable under addition $q_1 + q_2$, are a convolution algebra,

$$\mathcal{D}'(\mathbb{R}^d_+) \in \star\underline{\mathbf{aag}}_{\mathbb{C}}.$$

The convolution of two functions can be rolled over to the convolution of one reflected function with a distribution wherever defined,

$$\langle \mu, f * g\rangle = \langle f_- * \mu, g\rangle \text{ with } f_-(q) = f(-q).$$

The tempered distributions $\mathcal{S}'$, like Feynman particle propagators for Minkowski space, do not constitute a convolution algebra. To find their convolutors, one starts from the *Fourier compatibility between multiplication* $\bullet$ *and* *convolution* $*$: The product of Fourier transforms is the Fourier transform of the convolution product, wherever defined:

$$\mathcal{V} * \mathcal{W} \longrightarrow \mathcal{U}, \quad \mathbf{F}.\mathcal{V} \bullet \mathbf{F}.\mathcal{W} \longrightarrow \mathbf{F}.\mathcal{U}, \quad \tilde{\mu} \bullet \tilde{\nu} = \widetilde{\mu * \nu}.$$

The multiplication, where one partner has to be a function, goes pointwise for two functions and by dual rollover for distributions:

$$(f \bullet g)(x) = f(x)g(x), \quad \langle f \bullet \mu, g \rangle = \langle \mu, f \bullet g \rangle.$$

For *functions* $\mathcal{S}$ *rapidly decreasing* at infinity and the *slowly increasing (tempered) distributions* $\mathcal{S}'$ with the Fourier isomorphisms

$$\left(\mathcal{S} \subset L^2 \cong [L^2]' \subset \mathcal{S}'\right) \overset{\mathbf{F}}{\cong} \left(\mathcal{S} \subset L^2 \cong [L^2]' \subset \mathcal{S}'\right),$$

the *multiplication operators* $\mathcal{S}_\bullet$ in $\mathcal{C}^\infty$,

$$\begin{aligned} &\mathcal{S}_\bullet \bullet (\mathcal{S}, \mathcal{S}_\bullet, \mathcal{S}') \longrightarrow (\mathcal{S}, \mathcal{S}_\bullet, \mathcal{S}') \\ &\mathcal{S}_\bullet \in \star\underline{\mathbf{aag}}_{\mathbb{C}}, \quad (\mathcal{S}, \mathcal{S}_\bullet, \mathcal{S}') \in \underline{\mathbf{mod}}_{\mathcal{S}_\bullet} \text{ unital with multiplication } \bullet, \end{aligned}$$

are given by the *slowly increasing (tempered) functions* (increase at infinity is limited by polynomials). By Fourier transformation, they are related bijectively to the *rapidly decreasing distributions* $\mathcal{S}'_*$,

$$\mathcal{S}_\bullet \overset{\mathbf{F}}{\cong} \mathcal{S}'_*,$$

which are called *convolution operators*, since

$$\begin{aligned} &\mathcal{S}'_* * (\mathcal{S}, \mathcal{S}'_*, \mathcal{S}') \longrightarrow (\mathcal{S}, \mathcal{S}'_*, \mathcal{S}'), \\ &\mathcal{S}'_* \in \star\underline{\mathbf{aag}}_{\mathbb{C}}, \quad (\mathcal{S}, \mathcal{S}'_*, \mathcal{S}') \in \underline{\mathbf{mod}}_{\mathcal{S}'_*} \text{ unital with convolution } * . \end{aligned}$$

$\mathcal{S}$ is a nonunital subalgebra. The Fourier transform is an algebra isomorphism from the unital multiplication algebra $\mathcal{S}_\bullet$ to the unital convolution algebra $\mathcal{S}'_*$,

$$\mathbf{F} \in \overset{\circ}{\mathbf{aag}}_{\mathbb{C}} (\mathcal{S}_\bullet, \mathcal{S}'_\star), \quad \widetilde{f \bullet g} = \tilde{f} * \tilde{g}.$$

$\mathcal{S}'_*$ can be locally characterized, rather complicatedly, by finite sums of derivatives of continuous functions rapidly decreasing at infinity. For any integer n there is such a sum with the derivatives depending on n. In addition there is a related fall-off property.

Multiplication functions and convolution distributions (not topologically dual) are related to the other functions and distributions as follows:

$$\begin{array}{ccccccc} \mathcal{C}_c^\infty & \subset & \mathcal{S} & \subset & \mathcal{S}_\bullet & \subset & \mathcal{C}^\infty \\ \cap & & \cap & & \cap & & \cap \\ \mathcal{D}'_c & \subset & \mathcal{S}'_* & \subset & \mathcal{S}' & \subset & \mathcal{D}' \end{array}$$

With $\mathcal{S}' \bullet \mathcal{S} \longrightarrow \mathcal{S}'_*$ and $\mathcal{S}' * \mathcal{S} \longrightarrow \mathcal{S}_\bullet$ multiplication and convolution for the dual pair with the rapidly decreasing functions and slowly increasing distributions are related to each other in the commutative diagram with the Fourier isomorphisms

$$\begin{array}{ccc} \mathcal{S}' \times \mathcal{S} & \stackrel{\bullet}{\longrightarrow} & \mathcal{S}'_* \\ {\scriptstyle \mathbf{F}\times\mathbf{F},\cong}\downarrow & & \downarrow{\scriptstyle \mathbf{F},\cong} \\ \mathcal{S}' \times \mathcal{S} & \underset{*}{\longrightarrow} & \mathcal{S}_\bullet \end{array}, \quad \begin{array}{ccc} (\mu(x), f(x)) & \longmapsto & (\mu \bullet f)(x) = \mu(x)f(x) \\ \downarrow & & \downarrow \\ (\tilde{\mu}(q), \tilde{f}(q)) & \longmapsto & (\tilde{\mu} * \tilde{f})(q) = \widetilde{\mu \bullet f}(q) \end{array}.$$

Therefore the dual product can be obtained either by integration of the x-dependent product or by the convolution product value for $q = 0$:

$$\left.\begin{array}{rl} (\mu, f)(x) & = \int \frac{dq}{(2\pi)^d}\, e^{iqx} (\tilde{\mu}, \tilde{f})(q), \\ (\tilde{\mu}, \tilde{f})(q) & = \int dx\, e^{-iqx} (\mu, f)(x), \end{array}\right\} \quad \mu(x)f(x) = \int \frac{dq}{(2\pi)^d}\, e^{iqx} \tilde{\mu} * \tilde{f}(q),$$
$$\langle \mu, f \rangle = \int dx\, \mu(x)f(x) = \tilde{\mu} * \tilde{f}(0) = \int dq\, \tilde{\mu}(-q)\tilde{f}(q).$$

7.15 Parabolic Subgroups

The induction of representations of affine groups is a good introduction to the more general parabolic induction for representations of a semisimple Lie group G (also for reductive Lie group with suitable interpretations), especially for noncompact ones. The abelian subgroups, i.e., the maximal compact Cartan tori $(\exp i\mathbb{R})^{r_c}$ and, especially, the maximal noncompact Cartan planes $(\exp \mathbb{R})^{r_{nc}}$, with their irreducible Hilbert representations $e^{i\alpha} \longmapsto e^{Zi\alpha}$, $\mathbb{Z} \in \mathbb{Z}^{r_c}$ and $e^{\beta} \longmapsto e^{iq\beta}$, $q \in \mathbb{R}^{r_{nc}}$, respectively, are the subgroups for the inducing procedure.

With an Iwasawa factorization (chapter "Spin, Rotations, and Position") $G = \mathcal{K} \circ \mathcal{A} \circ \mathcal{N}$ into maximal compact $\mathcal{K}$ and triagonal $\mathcal{A} \circ \mathcal{N}$ with maximal noncompact abelian $\mathcal{A}$ one can define a *minimal parabolic subgroup* $\mathcal{S} = \mathcal{M} \circ \mathcal{A} \circ \mathcal{N}$ by extension of $\mathcal{A}$ with its centralizer $Z_G(\mathcal{A}) = \mathcal{M}$. Explicit examples for Lorentz groups in three and four dimensions are

$$\begin{array}{lll} \mathbf{SO}_0(1,2) \sim & \mathbf{SL}(\mathbb{R}^2) & = \mathbf{SO}(2) \circ \quad [\mathbb{I}(2) \circ \mathbf{SO}_0(1,1) \circ e^{\mathbb{R}}], \\ \mathbf{SO}_0(1,3) \sim & \mathbf{SL}(\mathbb{C}^2) & = \mathbf{SU}(2) \circ [\mathbf{SO}(2) \circ \mathbf{SO}_0(1,1) \circ e^{\mathbb{C}}]. \end{array}$$

A minimal parabolic subgroup is used in the *Bruhat decomposition* into double cosets $\mathcal{S}\backslash G/\mathcal{S}$.

A *parabolic subgroup* of G is a closed subgroup containing a minimal parabolic subgroup $\mathcal{S}$. A parabolic subgroup has the *Langlands factorization (decomposition)*

$$S = M \circ A \circ N, \quad \log S = \log M \oplus \log A \oplus \log N \text{ (as vector space)}$$

with the properties for the mutually orthogonal Lie subalgebras:

$$\begin{aligned} \text{noncompact abelian: } & \log A, \quad \text{nilpotent: } \log N, \\ A\text{-centralizer: } & \log M \oplus \log A = \{l \in \log G \mid [l, \log A] = \{0\}\}, \\ N\text{-normalizing: } & [\log M \oplus \log A, \log N] \subseteq \log N. \end{aligned}$$

The adjoint action of $\log A$ decomposes the nilpotent $\log N$ into an orthogonal sum of common eigenspaces. One can visualize parabolic subgroups as block-triagonal, e.g., in $G = \mathbf{SL}(\mathbb{R}^n)$:

$$\left(\begin{array}{cc|ccc|c} \times & \times & n & n & n & n \\ \times & \times & n & n & n & n \\ \hline 0 & 0 & \times & \times & \times & n \\ 0 & 0 & \times & \times & \times & n \\ 0 & 0 & \times & \times & \times & n \\ \hline 0 & 0 & 0 & 0 & 0 & \times \end{array}\right),$$

with $M \circ A$ block-diagonal $\times$ and N with diagonal $\mathbf{1}$ and entries n strictly above and to the right of the blocks.

One has with a parabolic subgroup an

$$\text{Iwazawa-Langlands factorization } G = \mathcal{K} \circ S = \mathcal{K} \circ M \circ A \circ N.$$

In comparing parabolic subgroups with affine groups $H \vec{\times} \mathbb{R}^n$, $M \circ A$ are the anlaogue to the direct product subgroups $H_0 \times \mathbb{R}^n$ with fixgroups H_0.

Bibliography

[1] V. Bargmann, Irreducible representations of the Lorentz group, *Ann. of Math.*, 48 (1947), 568-640.

[2] H. Behnke, F. Sommer, *Theorie der analytischen Funktionen einer komplexen Veränderlichen* (1962), Springer-Verlag, Berlin, Göttingen, Heidelberg.

[3] S. Bochner, Monotone Funktionen, Stieltjessche Integrale und harmonische Analyse, *Math. Ann.*, 108 (1933), 378-410.

[4] N. Bourbaki, *Intégration, Chapitres 1-8* (1952-1959), Hermann, Paris.

[5] J. Dixmier, *C^*-Algebras* (1969, English Translation 1977), North-Holland Publishing Company, Amsterdam, New York, Oxford.

[6] G.B. Folland, *A Course in Abstract Harmonic Analysis* (1995), CRC Press, Boca Raton, Ann Arbor, London, Tokyo.

[7] W. Fulton, J. Harris, *Representation Theory* (1991), Springer.

[8] I.M. Gelfand, M.A. Neumark, *Unitäre Darstellungen der klassischen Gruppen* (1950, German Translation 1957), Akademie Verlag, Berlin.

[9] I.M. Gelfand and D.A. Raikov, Irreducible unitary Representations of locally bicompact groups, *Mat. Sb* 13 (55) (1942), 301-316

[10] I.M. Gel'fand, G.E. Shilov, *Generalized Functions I (Properties and Operations)* (1958, English translation 1963), Academic Press, New York and London.

[11] I.M. Gel'fand, M.I. Graev, N.Ya. Vilenkin, *Generalized Functions V (Integral Geometry and Representation Theory)* (1962, English translation 1966), Academic Press, New York and London.

[12] S. Helgason, *Groups and Geometric Analysis* (1984), Academic Press, New York. London, Sydney, Tokyo, Toronto, etc.

[13] A.A. Kirillov, *Elements of the Theory of Representations* (1976), Springer-Verlag, Berlin, Heidelberg, New York.

[14] A. Knapp, *Representation Theory of Semisimple Groups* (1986), Princeton University Press, Princeton.

[15] G.W. Mackey, On induced representations of groups (1951), Amer. J. Math. 73, 576-592.

[16] G.W. Mackey, *Induced Representations of Groups and Quantum Mechanics* (1968), W.A. Benjamin, New York, Amsterdam.

[17] M.A. Neumark, *Lineare Darstellungen der Lorentzgruppe* (1958, German Translation 1963), VEB Deutscher Verlag der Wissenschaften, Berlin.

[18] F. Peter, H. Weyl, Die Vollständigkeit der primitiven Darstellungen einer geschlossenen kontinuierlichen Gruppe, *Math. Ann.* 97 (1927), 737-755.

[19] T.O. Sherman, Fourier analysis on the sphere, *Trans. Amer. Math. Soc.* **209**, 1-31 (1975).

[20] R.S. Strichartz, Harmonic analysis on hyperboloids, *Journal of Functional Analysis* **12**, 341-383 (1973).

[21] F. Treves, *Topological Vector Spaces, Distributions and Kernels* (1967), Academic Press, New York, London.

[22] E. P. Wigner, On unitary representations of the inhomogeneous Lorentz group, *Annals of Mathematics* 40 (1939), 149-204.

8

RESIDUAL SPACETIME REPRESENTATIONS

In Feynman propagators (chapter "Propagators") with energy-momentum poles $\frac{i}{\pi}\frac{1}{q^2+io-m^2}$, the Fourier transform of the real part, i.e., of the distribution $\delta(q^2-m^2)$, supported by the energy-momentum hyperboloid ("on-shell"), represents free particles with real momenta $\vec{q}^2 = q_0^2 - m^2 > 0$ by coefficients of the translations in the Poincaré group, e.g., $e^{iq_0t}\frac{\sin|\vec{q}|r}{r}$. The imaginary "momenta" $q_0^2 - m^2 = -Q^2 < 0$ in the principal value "off-shell" distribution $\frac{1}{q_{\rm P}^2-m^2}$ leads to interactions, e.g., to Yukawa interactions in $e^{iq_0t}\frac{e^{-|Q|r}}{r}$. Spacetime interactions are supported by the causal bicone. The harmonic analysis of the future cone $\mathbf{D}(2) \cong \mathbf{GL}(\mathbb{C}^2)/\mathbf{U}(2)$ (unitary relativity) as a nonlinear homogeneous spacetime model, i.e., of the mappings $W^{\mathbf{D}(2)}$ of the full linear group, constant on $\mathbf{U}(2)$, into $\mathbf{U}(2)$-representation spaces $W \cong \mathbb{C}^{1+2J}$, involves the representations of the acting extended Lorentz group $\mathbf{GL}(\mathbb{C}^2)$, which have to be used for spacetime interactions. Free particle fields are not complete for the harmonic analysis of nonlinear spacetime, genuine interaction fields are necessary [4, 14]. Interactions cannot be expanded completely with free particles.

Representations of linear and nonlinear spacetime embed time and position representations. Representation coefficients of 3-dimensional hyperbolic position $\mathcal{Y}^3$ as symmetric space for Lorentz operations $\mathbf{SO}_0(1,3)$ can be written with Fourier transformed 3-sphere momentum measures (chapter "The Kepler Factor") as seen in Hilbert-space-valued Schrödinger-bound state functions, e.g., for the hydrogen ground state $e^{-|m|r} = \int \frac{d^3q}{2\pi^2}\frac{2|m|}{(\vec{q}^2+m^2)^2}e^{-i\vec{q}\vec{x}}$. These representations of nonlinear position $\mathcal{Y}^3$ with a dipole singularity sphere for imaginary momenta $\vec{q}^2 = -m^2$ have to be embedded into causally supported representation coefficients of nonlinear spacetime $\mathbf{D}(2) \cong \mathbf{D}(1) \times \mathcal{Y}^3$. The embedding energy-momentum distributions do not describe free particles: The Lorentz invariant mass for the representation of the position degree of freedom comes as a singularity in a higher-order pole, starting with a dipole distribution $\frac{d^4q}{(q^2-m^2)^2}$, as required by Lorentz compatible embedding of the 3-sphere measures $\frac{d^3q}{(\vec{q}^2+Q^2)^2}\frac{2|Q|}{}$ with energy-dependent invariant $Q^2 = m^2 - q_0^2$.

Multipole energy-momentum distributions lead, via their Fourier transforms with appropriate integration contours, to residual representations of symmetric spaces. Representations of time (harmonic oscillator), of position (scattering and bound waves), of spacetime translations (on-shell part of Feynman propagators), and of nonlinear spacetime (spheres, hyperboloids, multipole interactions) will be formulated in the language of residual representations with their characterizing invariant singularities.

8.1 Linear and Nonlinear Spacetime

Minkowski translations $\mathbb{R}^4$ with position translations $\mathbb{R}^3$ contain as substructure Cartan translations $\mathbb{R}^2$ with time and 1-dimensional position translations $\mathbb{R}$ (chapter "Spacetime as Unitary Operation Classes"). Cartan and Minkowski translations are parametrizable by Hermitian (2×2) matrices, diagonal for Cartan spacetime,

$$\begin{array}{ccccc} t \in \mathbb{R} & \longrightarrow & x^0 + \sigma_3 x^3 \in \mathbb{R}^2 & \longrightarrow & x = x^0 + \vec{x} = \begin{pmatrix} x^0 + x^3 & x^1 - ix^2 \\ x^1 + ix^2 & x^0 - x^3 \end{pmatrix} \in \mathbb{R}^4. \\ & \nearrow & & \nearrow & \\ x^3 \in \mathbb{R} & \longrightarrow & \vec{x} \in \mathbb{R}^3 & & \end{array}$$

A group $\mathbf{D}(1) = \exp \mathbb{R}$ acts both on the time and the 1-dimensional position translations. For Cartan translations, it is rearranged from $\mathbf{D}(1) \times \mathbf{D}(1)$ to $\mathbf{D}(\mathbf{1}_2) \times \mathbf{SO}_0(1,1)$ with the rotation free orthochronous Lorentz group (self-dual dilations). Together with the position rotations $\mathbf{SO}(3)$, it is embedded into the $\mathbf{D}(\mathbf{1}_4)$-extended Poincaré group,

$$\begin{array}{ccccc} \mathbf{D}(1) \,\vec{\times}\, \mathbb{R} & \longrightarrow & [\mathbf{D}(1) \times \mathbf{SO}_0(1,1)] \,\vec{\times}\, \mathbb{R}^2 & \longrightarrow & [\mathbf{D}(1) \times \mathbf{SO}_0(1,3)] \,\vec{\times}\, \mathbb{R}^4. \\ & \nearrow & & \nearrow & \\ \mathbf{D}(1) \,\vec{\times}\, \mathbb{R} & \longrightarrow & [\mathbf{D}(1) \times \mathbf{SO}(3)] \,\vec{\times}\, \mathbb{R}^3 & & \end{array}$$

Spacetime has an order structure: Time future is embedded into Cartan and Minkowski future,

$$\begin{array}{rcll} \mathbb{R}_+ \ni t = \vartheta(t)t & \hookrightarrow & \vartheta(x^2)\vartheta(x^0)(x^0 + \sigma_3 x^3) & = x \in \mathbb{R}^2_+, \\ & \hookrightarrow & \vartheta(x^2)\vartheta(x^0)(x^0 + \vec{x}) & = x \in \mathbb{R}^4_+. \end{array}$$

The futures are used as noncompact spaces (open cones without "skin"), i.e., without the strict presence $x = 0$ and without lightlike translations for nontrivial position $s = 1, 3$,

$$x \in \mathbb{R}^{1+s}_+ \Rightarrow x^2 > 0, \; s = 0, 1, 3.$$

Time future is the causal group $\mathbf{D}(1) = \exp \mathbb{R}$,

$$\mathbb{R}_+ \ni t \;= e^{\psi_0} \in \mathbf{D}(1) \cong \mathbf{GL}(\mathbb{C})/\mathbf{U}(1).$$

Cartan future is the direct product of causal group and abelian Lorentz group,

$$x = e^{\psi_0 + \sigma_3 \psi} \in \mathbb{R}^2_+ \cong \mathbf{D}(1) \times \mathbf{SO}_0(1,1).$$

The action of the full linear group, called the extended Lorentz group, on Minkowski translations

$$\mathbf{GL}(\mathbb{C}^2) \times \mathbb{R}^4 \longrightarrow \mathbb{R}^4, \quad g \bullet x = g \circ x \circ g^\star$$

leaves the future invariant. The future has an orbit parametrization with Lie algebra coefficients:

$$\begin{array}{rl} x = e^{\psi_0 + \vec{\psi}} = & u(\frac{\vec{\psi}}{\psi}) \circ e^{\psi_0 + \sigma_3 \psi} \circ u(\frac{\vec{\psi}}{\psi})^\star \in \mathbb{R}^4_+ \cong \mathbf{GL}(\mathbb{C}^2)/\mathbf{U}(2), \\ \text{with} & \left\{ \begin{array}{ll} e^{\psi_0} & = \sqrt{x^2} \in \mathbf{D}(1), \\ e^{\pm\psi} & = \sqrt{\frac{x_0 \pm r}{x_0 \mp r}} \in \mathbf{SO}_0(1,1), \quad |\vec{\psi}| = \psi, \quad \frac{\vec{\psi}}{\psi} = \frac{\vec{x}}{r}. \end{array} \right. \end{array}$$

It involves two rotation degrees of freedom for the 2-sphere $\Omega^2 \cong \mathbf{SU}(2)/\mathbf{SO}(2)$,

$$u(\tfrac{\vec{\psi}}{\psi}) = \begin{pmatrix} \cos\frac{\theta}{2} & ie^{-i\varphi}\sin\frac{\theta}{2} \\ ie^{i\varphi}\sin\frac{\theta}{2} & \cos\frac{\theta}{2} \end{pmatrix} = \frac{1}{\sqrt{2r(r+x_3)}} \begin{pmatrix} r + x_3 & -ix_1 - x_2 \\ ix_1 - x_2 & r + x_3 \end{pmatrix} \in \mathbf{SU}(2)/\mathbf{SO}(2).$$

The future of Minkowski spacetime $\mathbb{R}^{1+s}$ is an orbit of $\mathbf{D}(1) \times \mathbf{SO}_0(1,s)$. One- and four-dimensional future are the first two entries in the symmetric space chain $\mathbf{GL}(\mathbb{C}^n)/\mathbf{U}(n)$, $n = 1, 2, \ldots$, which are the manifolds of the unitary groups in the general linear group, canonically parametrized in the polar decomposition $g = u \circ |g|$ with the real n^2-dimensional ordered absolute values $x = |g| = \sqrt{g^\star \circ g} \in \mathbb{R}^{n^2}_+$ of the general linear group. They are the positive cone of the ordered C*-algebras with the complex $n \times n$ matrices (chapter "Spacetime as Unitary Operation Classes").

Nonlinear spacetimes are symmetric spaces, in general not vector spaces. They are parametrizable by the future of its tangent linear spacetime. At each point, the tangent space of nonlinear spacetime is a full translation space:

$$\mathbb{R}^{1+s} \cong \left\{ \begin{array}{ll} \log \mathbf{D}(1), & s = 0, \\ \log \mathbf{D}(1) \;\oplus\; \log \mathbf{SO}_0(1,1), & s = 1, \\ \log \mathbf{GL}(\mathbb{C}^2)/\mathbf{U}(2), & s = 3. \end{array} \right.$$

The Lie algebra coefficients $(\psi_0, \vec{\psi}) \in \mathbb{R} \times \mathbb{R}^3$ are related to tangent time and position as follows:

$$\begin{array}{rl} x = x_0 + \vec{x} & = e^{\psi_0}(\cosh\psi + \frac{\vec{\psi}}{\psi}\sinh\psi) = 1 + \psi_0 + \vec{\psi} + \cdots, \\ x_0 \pm |\vec{x}| & = e^{\psi_0 \pm |\psi|}. \end{array}$$

Nonlinear Minkowski spacetime contains many familiar homogeneous subspaces in the manifold decomposition

$$\mathbf{D}(2) \cong \mathbf{D}(1) \times \mathbf{SL}(\mathbb{C}^2)/\mathbf{SU}(2).$$

One-dimensional future $\mathbf{D}(1)$ is multiplied by the 3-dimensional Lobachevsky space, i.e., by the 3-hyperboloid (nonlinear position),

$$\frac{x}{\sqrt{x^2}} = e^{\vec{\psi}} \in \mathcal{Y}^3 \cong \mathbf{SL}(\mathbb{C}^2)/\mathbf{SU}(2) \cong \mathbf{SO}_0(1,3)/\mathbf{SO}(3),$$

which contains 2-dimensional non-Euclidean planes (2-hyperboloids) $\mathcal{Y}^2 \cong \mathbf{SO}_0(1,2)/\mathbf{SO}(2)$ and 1-dimensional hyperboloids (abelian Lorentz groups) $\mathcal{Y}^1 \cong \mathbf{SO}_0(1,1)$.

Harmonic analysis of the tangent groups for linear spacetime

$$\begin{array}{ccccccc} \mathbb{R} & \longrightarrow & \mathbf{SO}_0(1,1)\,\vec{\times}\,\mathbb{R}^2 & \longrightarrow & \mathbf{SO}_0(1,2)\,\vec{\times}\,\mathbb{R}^3 & \longrightarrow & \mathbf{SO}_0(1,3)\,\vec{\times}\,\mathbb{R}^4 \\ & \nearrow & & \nearrow & & \nearrow & \\ \mathbb{R} & \longrightarrow & \mathbf{SO}(2)\,\vec{\times}\,\mathbb{R}^2 & \longrightarrow & \mathbf{SO}(3)\,\vec{\times}\,\mathbb{R}^3 & & \end{array}$$

involves the irreducible Hilbert representations for free scattering waves and free particles (Wigner), i.e., of the functions on affine groups (tangent groups) $\mathbf{SO}(s)\,\vec{\times}\,\mathbb{R}^s$ (Euclidean groups) and $\mathbf{SO}_0(1,s)\,\vec{\times}\,\mathbb{R}^{1+s}$ (Poincaré groups).

Harmonic analysis of nonlinear spacetime

$$\begin{array}{ccccccc} \mathbf{D}(1) & \longrightarrow & \mathbf{D}(1)\times\mathcal{Y}^1 & \longrightarrow & \mathbf{D}(1)\times\mathcal{Y}^2 & \longrightarrow & \mathbf{D}(2)\cong\mathbf{D}(1)\times\mathcal{Y}^3 \\ & \nearrow & & \nearrow & & \nearrow & \\ \mathbf{SO}_0(1,1) & \longrightarrow & \mathcal{Y}^2 & \longrightarrow & \mathcal{Y}^3 & & \end{array}$$

embeds the harmonic analysis of the flat, spherical, and hyperbolic symmetric subspaces.

8.2 Residual Representations

The method of residual representations with (energy-)momentum distributions is intended to generalize, especially to nonabelian noncompact operations, the cyclic Hilbert representations of translations via positive (energy-)momentum measures. It uses the Fourier transformed Radon measures $\mathcal{M}(\check{\mathbb{R}}^n)$ which are essentially bounded function $L^\infty(\mathbb{R}^n)$. The form of residual representation leads to a generalization of the Feynman propagators as used in canonical quantum field theory.

The goal of the residual representation method is to translate the relevant representation structures of homogeneous spaces (real Lie groups) and its tangent translations (Lie algebras) – invariants, normalizations, product representations, etc. – into the language of rational complex (energy-)momentum functions with its poles, residues, and convolution products.

8.2.1 Residual Representations of Symmetric Spaces

Harmonic analysis of a symmetric space G/H with real Lie groups $G \supseteq H$ analyzes complex G/H-mappings with respect to irreducible G-representations with the related invariants. The eigenvalues (weights) of the group G-representations are a subset of the linear Lie algebra forms $(\log G)^T$. For translations all linear forms are weights, the (energy-)momenta. For simple groups, the weights constitute a subset of the weight space W^T (linear forms of a Cartan Lie algebra W) with the dimension the Lie algebra rank, $\dim_{\mathbb{R}} W^T = \mathrm{rank}_{\mathbb{R}} \log G$. The weights are discrete for a compact group. The Lie algebra is acted on by the adjoint representation of the group in the affine group $G\,\vec{\times}\,\log G$, its forms by the coadjoint (dual) one. Invariants are multilinear Lie algebra forms, e.g., linear for abelian groups or the bilinear Killing form for semisimple groups.

The tangent spaces of G/H are isomorphic to the corresponding Lie algebra classes, denoted by $\log G/H = \log G/\log H$ with $\dim_{\mathbb{R}} \log G/H = \dim_{\mathbb{R}} G - \dim_{\mathbb{R}} H$. It inherits the adjoint action of the group G, the linear forms the coadjoint one.

Now the definition of residual representations: (Generalized) functions (representation coefficients) of a symmetric space G/H,

$$\mu : (G/H)_{\text{repr}} \longrightarrow \mathbb{C}, \quad x \longmapsto \mu(x),$$

are assumed to be parametrizable by vectors $x \in V$ (translations) of an orbit in a real vector space with fixgroup H,

$$x \in G \bullet x_0 \cong G/H, \quad G \bullet x_0 \subseteq V \cong \mathbb{R}^n,$$

e.g., a group G by its Lie algebra $\log G$ (canonical coordinates) such as $\mathbf{SU}(2) \cong \{e^{i\vec{\sigma}\vec{x}} \mid \vec{x} \in \mathbb{R}^3\}$ or the hyperboloid $\mathbf{SO}_0(1,3)/\mathbf{SO}(3) \cong \{x \in \mathbb{R}^4 \mid x^2 = \ell^2 > 0,\ x_0 > 0\}$ by the vectors of a timelike orbit. With the dual space $q \in V^T \cong \mathbb{R}^n$ (by abuse of language called (energy-)momenta, also in the general case), e.g., the dual Lie algebra, the representations of G/H are characterizable by G-invariants $\{I_1, \dots, I_R\}$, with rational values for a compact group and rational or continuous values for a noncompact group. The invariants are given by q-polynomials and can be built by multilinear invariants, $q = m$ for an abelian group, quadratic invariants $q^2 = \pm m^2$, e.g., Killing form invariants.

If there exists a distribution of the (energy-)momenta, especially a Radon measure $\tilde{\mu} \in \mathcal{M}(\check{\mathbb{R}}^n)$, whose Fourier transformation gives functions μ of the symmetric space and if the generalized function $\tilde{\mu}$ comes as a quotient of two polynomials, where the invariant zeros of the denominator polynomial $Q(q)$ characterize a G-representation

$$\tilde{\mu}(q) \cong \frac{P(q)}{Q(q)} \text{ with } Q(q)\text{-factors } \{(q-m)^n, (q^2 \pm m^2)^n, (q^k \pm m^k)^n\},\ m \in \mathbb{R},$$

then μ is called a residual representation of G/H.

A representation of a symmetric space G/H contains representations of subspaces K, e.g., of subgroups $\mathbf{SO}(2) \subset \mathbf{SU}(2)$ or $\mathbf{SO}_0(1,1) \subset \mathbf{SL}(\mathbb{C}^2)/\mathbf{SU}(2)$. A residual G/H-representation with canonical tangent space parameters $x = (x_K, x_\perp)$ has a projection to a residual K-representation by integration $\int d^{n-s}x_\perp$ over the complementary space $\frac{\log G/H}{\log K} \cong \mathbb{R}^{n-s}$,

$$K \longrightarrow \mathbb{C}, \quad x_K \longmapsto \mu(x_K, 0) = \int \frac{d^{n-s}x_\perp}{(2\pi)^{n-s}} \mu(x) = \int d^s q_K\ \tilde{\mu}(q_K, 0) e^{iq_K x_K}.$$

The integration picks up the Fourier components for trivial tangent space forms (energy-momenta) $q_\perp = 0$ of $\frac{\log G/H}{\log K}$.

A Fourier integral involves irreducible representations $x \longmapsto e^{iqx}$ of the underlying translations $x \in V \cong \mathbb{R}^n$. With that, residual representations with positive distributions $\tilde{\mu}$ of the (energy-)momenta (characters) $q \in V^T$ give cyclic translation representation coefficients.

With velocities and actions measured in units $(c, \hbar)$ all energy and momentum invariants can be measured in mass units. Nontrivial invariants $m \neq 0$ can be used as intrinsic units by a rescaling of translations $x \longmapsto \frac{x}{|m|}$ and (energy-) momenta $q \longmapsto |m|q$ to obtain dimensionless Lie parameters and eigenvalues. To include the trivial case $m = 0$, invariants will be kept in most cases, and somewhat inconsequentially, in the dimensional form.

8.2.2 Spherical, Hyperbolic, Feynman, and Causal Distributions

(Energy-)momentum measures are used in the definition of free particle representations. The Lebesgue measure $\frac{d^n q}{(2\pi)^n}$ is the Plancherel measure for the irreducible translation representations $\mathbb{R}^n \ni x \longmapsto e^{iqx} \in \mathbf{U}(1)$ and Haar measure $d^n x$. For irreducible representations of affine groups $G \vec{\times} \mathbb{R}^n$ it is modified by Dirac distributions of (energy-)momenta on homogeneous spaces G/H. They describe interaction-free structures with cyclic translation representations.

For the circle one has different parametrizations, e.g.,

$$\Omega^1 \ni \begin{pmatrix} q_0 \\ iq \end{pmatrix}, \text{ for semi-circle: } \begin{pmatrix} \cos\chi \\ i\sin\chi \end{pmatrix}_{-\frac{\pi}{2}}^{\frac{\pi}{2}} = \frac{1}{\sqrt{1+p^2}} \begin{pmatrix} 1 \\ ip \end{pmatrix}_{-\infty}^{\infty} = \frac{1}{1+v^2} \begin{pmatrix} 1-v^2 \\ 2iv \end{pmatrix}_{-1}^{1}.$$

Therefore, the Euclidean group relevant measure (chapter "Propagators") for the momentum direction sphere, i.e., for the compact classes of orthogonal groups $\mathbf{SO}(1+s)/\mathbf{SO}(s) \cong \Omega^s$, has the parametrizations, also for $s = 0$ where applicable,

$$\begin{aligned}
&|\Omega^s| = \int d^s\omega = \int d^{1-s}q \; 2\delta(q_0^2 + \vec{q}^2 - 1) = \int_0^\pi (\sin\chi)^{s-1} d\chi \int d^{s-1}\omega = \int \frac{2d^s p}{(\vec{p}^2+1)^{\frac{1+s}{2}}},\\
&\text{polar decomposition: } q = |q|\vec{\omega} \text{ with } |q|^2 = q_0^2 + \vec{q}^2,\ \vec{\omega} \in \Omega^s,\\
&\text{with } |\Omega^s| = \frac{2\pi^{\frac{1+s}{2}}}{\Gamma(\frac{1+s}{2})} = 2, 2\pi, 4\pi, 2\pi^2, \ldots,\ \frac{|\Omega^{s-2}|}{|\Omega^s|} = \frac{s-1}{2\pi},\ \frac{\Gamma(\frac{1}{2}+R)}{\Gamma(\frac{1}{2})} = 2^{1-2R}\frac{\Gamma(2R)}{\Gamma(R)}.
\end{aligned}$$

For noncompact classes of orthogonal groups there is the Poincaré-group-relevant measure of the one-shell positive energylike hyperboloid $\mathbf{SO}_0(1,s)/\mathbf{SO}(s) \cong \mathcal{Y}^s$ whose parametrizations can be obtained with the spherical-hyperbolic transition $(i\vec{q}, i\chi, i\vec{p}, i\vec{v}) \rightarrow (\vec{q}, \psi, \vec{p}, \vec{v})$,

$$\begin{aligned}
&\int d^s \mathrm{y} = \int d^{1+s}q \; 2\vartheta(q_0)\delta(q_0^2 - \vec{q}^2 - 1) = \int_0^\infty (\sinh\psi)^{s-1} d\psi \int d^{s-1}\omega = \int \frac{d^s q}{\sqrt{\vec{q}^2+1}},\\
&\text{"polar" decomposition: } q = |q|\mathrm{y} \text{ with } |q|^2 = q_0^2 - \vec{q}^2,\ \mathrm{y} \in \mathcal{Y}^s.
\end{aligned}$$

Finally, there is the measure of the momentumlike hyperboloid, the noncompact classes of noncompact groups $\mathbf{SO}_0(1,s)/\mathbf{SO}_0(1,s-1) \cong \mathcal{Y}^{(1,s-1)}$:

$$\begin{aligned}
&\int d^s \mathrm{s} = \int d^{1+s}q \; 2\delta(q_0^2 - \vec{q}^2 + 1) = 2\int_{\vec{q}^2 \geq 1} \frac{d^s q}{\sqrt{\vec{q}^2 - 1}} = \int_{-\infty}^{\infty} (\cosh\psi)^{s-1} d\psi \int d^{s-1}\omega,\\
&\text{"polar" decomposition: } q = |q|\mathrm{s} \text{ with } |q|^2 = -q_0^2 + \vec{q}^2,\ \mathrm{s} \in \mathcal{Y}^{(1,s-1)}.
\end{aligned}$$

The Dirac "on-shell" and the principal value (with $q_{\rm P}^2$) "off-shell" distributions are imaginary and real part of the (anti-) Feynman distributions

$$\begin{array}{rl}\log(q^2 \mp io - \mu^2) & = \log|q^2 - \mu^2| \mp i\pi\vartheta(\mu^2 - q^2)\\ \frac{\Gamma(1+N)}{(q^2\mp io-\mu^2)^{1+N}} & = -(-\frac{\partial}{\partial q^2})^{1+N}\log(q^2 \mp io - \mu^2)\\ = (-\frac{\partial}{\partial q^2})^N \frac{1}{q^2\mp io-\mu^2} & = \frac{\Gamma(1+N)}{(q_{\rm P}^2-\mu^2)^{1+N}} \pm i\pi\delta^{(N)}(\mu^2 - q^2)\\ \text{for } \mu^2 \in \mathbb{R} & \text{and } N = 0, 1, \ldots .\end{array}$$

Feynman distributions are possible for any signature $\mathbf{O}(t, s)$ with positive or negative invariant μ^2.

Characteristic for and compatible only with the orthochronous Lorentz group $\mathbf{SO}_0(1, s)$ are the advanced (future) and retarded (past) causal energy-momentum distributions with positive invariant m^2 only. They are distinguished by their energy q_0 behavior:

$$\begin{array}{rl}\log((q \mp io)^2 - m^2) & = \log|q^2 - m^2| \mp i\pi\epsilon(q_0)\vartheta(m^2 - q^2)\\ \frac{\Gamma(1+N)}{((q\mp io)^2-m^2)^{1+N}} & = -(-\frac{\partial}{\partial q^2})^{1+N}\log((q \mp io)^2 - m^2)\\ = (-\frac{\partial}{\partial q^2})^N \frac{1}{(q\mp io)^2-m^2} & = \frac{\Gamma(1+N)}{(q_{\rm P}^2-m^2)^{1+N}} \pm i\pi\epsilon(q_0)\delta^{(N)}(m^2 - q^2)\\ \text{for } m^2 \geq 0 & \text{and } (q \mp io)^2 = (q_0 \mp io)^2 - \vec q^2.\end{array}$$

8.2.3 Residual Distributions

In this subsection all coefficients for noncompact operations are given [13] that will be relevant for the spacetime representations in the following. They can be obtained as residues in the form of Fourier transformed measures and involve Bessel, Neumann, and Macdonald functions (chapter "Propagators").

The causal structure of the reals and its unitary representations occur in the Fourier transformed causal measures,

$$m, \nu \in \mathbb{R}: \qquad \int \frac{dq}{2i\pi} \frac{\Gamma(1-\nu)}{(q-io-m)^{1-\nu}} e^{iqx} = \vartheta(x)\frac{e^{imx}}{(ix)^\nu}.$$

Here and in the following the integrals hold wherever the Γ-functions are defined.

The scalar distributions for the definite orthogonal groups in general dimension with real and imaginary singularities on spheres Ω^{s-1} with $\vec q^2 = \pm 1$ give Macdonald functions $\mathcal{K}_\nu$ and Hankel (Bessel with Neumann) functions $\mathcal{H}_\nu^{1,2} = \mathcal{J}_\nu \pm i\mathcal{N}_\nu$

$$\begin{array}{l}\mathbf{O}(s),\ s = 1, 2, 3, \ldots,\\ r = \sqrt{\vec x^2},\quad \nu \in \mathbb{R},\end{array}\left\{\begin{array}{rl}\int \frac{d^s q}{\pi^{\frac{s}{2}}} \frac{\Gamma(\frac{s}{2}-\nu)}{(\vec q^2)^{\frac{s}{2}-\nu}} e^{i\vec q\vec x} & = \frac{\Gamma(\nu)}{(\frac{r^2}{4})^\nu},\\ \int \frac{d^s q}{\pi^{\frac{s}{2}}} \frac{\Gamma(\frac{s}{2}-\nu)}{(\vec q^2+1)^{\frac{s}{2}-\nu}} e^{i\vec q\vec x} & = \frac{2\mathcal{K}_\nu(r)}{(\frac{r}{2})^\nu} = \frac{\pi(i\mathcal{J}_\nu-\mathcal{N}_\nu)(ir)}{(\frac{r}{2})^\nu},\\ \int \frac{d^s q}{\pi^{\frac{s}{2}}} \frac{\Gamma(\frac{s}{2}-\nu)}{(\vec q^2-io-1)^{\frac{s}{2}-\nu}} e^{i\vec q\vec x} & = \frac{\pi(i\mathcal{J}_\nu-\mathcal{N}_\nu)(r)}{(\frac{r}{2})^\nu} = \frac{2\mathcal{K}_\nu(-ir)}{(\frac{r}{2})^\nu}.\end{array}\right.$$

The angle integration is different for even and odd dimensions,

$$s \ge 2: \ \int \frac{d^s q}{|\Omega^{s-2}|} \mu(\vec{q}^2) e^{i\vec{q}\vec{x}} = \int_0^\infty q^{s-1} dq\ \mu(q^2) \int_0^\pi (\sin\chi)^{s-2} d\chi\ e^{iqr\cos\chi}$$
$$= \begin{cases} \int_0^\infty q^{1+2R} dq\ \mu(q^2) & \int_0^\pi (1-\cos^2\chi)^R d\chi\ e^{iqr\cos\chi}, & s = 2+2R, \\ -\int_0^\infty q^{2+2R} dq\ \mu(q^2) & \int_{-1}^1 (1-\zeta^2)^R d\zeta\ e^{iqr\zeta}, & s = 3+2R, \end{cases}$$

The integrals can be obtained by 2-sphere spread from the values for $R = 0$.

All (half)integer index functions arise by derivation $\frac{d}{d\frac{r^2}{4\pi}} = \frac{2\pi}{r}\frac{d}{dr}$:

$$\mathbb{R}_+ \ni r \longmapsto \frac{(2\mathcal{K}_\nu,\ \pi\mathcal{J}_\nu,\ \pi\mathcal{N}_\nu)(r)}{(\frac{r}{2\pi})^\nu} = \begin{cases} \left(-\frac{d}{d\frac{r^2}{4\pi}}\right)^N (e^{-r},\ \cos r,\ \sin r), \\ \qquad \nu + \frac{1}{2} = N = 0, 1, 2, \dots, \\ \left(-\frac{d}{d\frac{r^2}{4\pi}}\right)^N \Big(2\mathcal{K}_0(r),\ \pi\mathcal{J}_0(r),\ \pi\mathcal{N}_0(r)\Big), \\ \qquad \nu = N = 0, 1, 2, \dots. \end{cases}$$

The half-integer index functions start from the exponentials. The noncompact and compact self-dual representations of the reals come with imaginary and real poles in the complex plane,

$$\mathbf{O}(1): \quad \int \frac{dq}{\pi} \frac{1}{q^2 - io \pm 1} e^{iqx} = \begin{cases} e^{-r}, & \text{poles at } q = \pm i, \\ ie^{ir}, & \text{poles at } q = \pm 1, \end{cases}$$

They involve the positive-type function for the basic self-dual spherical representation $\mathbb{R} \ni x \longmapsto \cos x$ and the basic self-dual hyperbolic one $\mathbb{R} \ni x \longmapsto e^{-|x|}$.

The integer index functions begin with 2-dimensional (energy-)momentum integrals, which integrate over the $\mathbb{R}$-representation coefficients,

$$\mathbf{O}(2): \begin{cases} \int \frac{d^2q}{\pi} \frac{1}{\vec{q}^2+1} e^{i\vec{q}\vec{x}} & = 2\mathcal{K}_0(r) & = \int d\psi\ e^{-r\cosh\psi}, \\ \int \frac{d^2q}{\pi} \frac{1}{\vec{q}^2 - io - 1} e^{i\vec{q}\vec{x}} & = -\pi[\mathcal{N}_0 - i\mathcal{J}_0](r) & = \int d\psi\ e^{ir\cosh\psi}, \\ \int \frac{d^2q}{\pi} \frac{1}{\vec{q}^2_{\mathrm{P}} - 1} e^{i\vec{q}\vec{x}} & = -\pi\mathcal{N}_0(r) & = \int d\psi\ \cos(r\cosh\psi), \\ \int d^2q\ \delta(\vec{q}^2 - 1) e^{i\vec{q}\vec{x}} & = \pi\mathcal{J}_0(r) & = \int d\psi\ \sin(r\cosh\psi) \\ & & = \int_{-\pi}^{\pi} d\chi\ \cos(r\cos\chi), \end{cases}$$

$$\begin{matrix}\mathbf{O}(1,1): \\ |x| = \sqrt{|x^2|}\end{matrix} \begin{cases} \int \frac{d^2q}{i\pi} \frac{1}{q^2 - io + 1} e^{iqx} & = \vartheta(x^2) 2\mathcal{K}_0(|x|) - \vartheta(-x^2)\pi[\mathcal{N}_0 + i\mathcal{J}_0](|x|), \\ \int \frac{d^2q}{i\pi} \frac{1}{q^2 - io - 1} e^{iqx} & = \vartheta(-x^2) 2\mathcal{K}_0(|x|) - \vartheta(x^2)\pi[\mathcal{N}_0 - i\mathcal{J}_0](|x|), \\ \int d^2q\ \delta(q^2 - 1) e^{iqx} & = \vartheta(-x^2) 2\mathcal{K}_0(|x|) - \vartheta(x^2)\pi\mathcal{N}_0(|x|), \\ -\int \frac{d^2q}{\pi} \frac{1}{q^2_{\mathrm{P}} - 1} e^{iqx} & = \vartheta(x^2)\pi\mathcal{J}_0(|x|). \end{cases}$$

For $\nu \ge 0$, only the Bessel functions are regular at $r = 0$,

$$\begin{aligned} \begin{pmatrix} 2\mathcal{K}_0 \\ -\pi\mathcal{N}_0 \end{pmatrix}(r) &= -\sum_{k=0}^\infty \frac{(\pm\frac{r^2}{4})^k}{(k!)^2} [\log\tfrac{r^2}{4} - 2\Gamma'(1) - 2\varphi(k)], \\ \varphi(0) &= 0,\ \ \varphi(k) = 1 + \tfrac{1}{2} + \ldots + \tfrac{1}{k},\ \ k = 1, 2, \ldots, \\ -\Gamma'(1) &= \lim_{k\to\infty}[\varphi(k) - \log k] = 0.5772\ldots \text{ (Euler's constant)}, \\ \mathcal{J}_0(r) &= \sum_{k=0}^\infty \frac{(-\frac{r^2}{4})^k}{(k!)^2}. \end{aligned}$$

The half-integer index functions for $\nu > 0$ arise by derivation with respect to the group parameter from the rotation free case with $\mathbb{R}$-representations for $\nu = -\frac{1}{2}$, starting with $\nu = \frac{1}{2}$, whereas the integer index functions start from $\nu = 0$ which requires a finite integration of $\mathbb{R}$-representations:

$$\text{for } \nu = -\tfrac{1}{2}: \ \cos r \ \begin{array}{ll} \nearrow & 2\frac{\sin r}{r} = \int_{-1}^{1} d\zeta \ \cos \zeta r = -\frac{d}{d\frac{r^2}{4}} \cos r \quad \text{for } \nu = \frac{1}{2}, \\ \searrow & \pi \mathcal{J}_0(r) = \int_{-1}^{1} \frac{d\zeta}{\sqrt{1-\zeta^2}} \cos \zeta r \qquad \text{for } \nu = 0. \end{array}$$

The Bessel, Neumann and Macdonald functions with half-integer index, e.g. $\{\cos, \sin, \exp\}$, are used for odd dimensions, e.g. for the groups $\mathbf{D}(1), \mathbf{SO}(1+2R), \mathbf{SO}_0(1,2R)$, $R = 1, 2, \ldots$. The corresponding functions with integer index, e.g. $\{\pi\mathcal{J}_0, \pi\mathcal{N}_0, 2\mathcal{K}_0\}$, are used for even dimensions, e.g. for the groups $\mathbf{D}(1) \times \mathbf{SO}_0(1, 2R-1), \mathbf{SO}(2+2R)$,

$$\mathbf{SO}_0(t,s) \text{ with } t+s = \left\{ \begin{array}{llll} \nu + \frac{3}{2} & = 1+N & = 1+2R & = 1,3,5,\ldots \\ \nu + 2 & = 2+N & = 2R & = 2,4,6,\ldots \end{array} \right.$$

As familiar from the local isomorphisms $\mathbf{SU}(2) \sim \mathbf{SO}(3)$ and $\mathbf{SL}(\mathbb{R}^2) \sim \mathbf{SO}_0(1,2)$, on the one hand, and $\mathbf{SL}(\mathbb{C}^2) \sim \mathbf{SO}_0(1,3) \sim \mathbf{SO}(\mathbb{C}^3)$ (as real Lie groups), on the other hand, the difference between odd and even dimensions is also related [7] to the transition from real $\mathbb{R}$ to complex $\mathbb{C} = \mathbb{R} \oplus i\mathbb{R}$ (doubled reals).

By analytic continuation one obtains for indefinite orthogonal groups

$$\begin{array}{c} \mathbf{O}(t,s): \\ t \geq 1, \ s \geq 1, \\ d = t+s = 2,3,\ldots, \\ |x| = \sqrt{|x^2|}, \\ x^2 = \vec{x}_t^2 - \vec{x}_s^2, \end{array} \left\{ \begin{array}{ll} \int \frac{d^d q}{i^s \pi^{\frac{d}{2}}} \frac{\Gamma(\frac{d}{2}-\nu)}{(q^2-io)^{\frac{d}{2}-\nu}} e^{iqx} & = \frac{\Gamma(\nu)}{\left(\frac{x^2+io}{4}\right)^{\nu}}, \\ \int \frac{d^d q}{i^s \pi^{\frac{d}{2}}} \frac{\Gamma(\frac{d}{2}-\nu)}{(q^2-io+1)^{\frac{d}{2}-\nu}} e^{iqx} & = \frac{\vartheta(x^2) 2\mathcal{K}_\nu(|x|) - \vartheta(-x^2) i\pi \mathcal{H}^2_{-\nu}(|x|)}{|\frac{x}{2}|^\nu} \\ & \quad -\delta_\nu^N i\pi \sum_{k=1}^{N} \frac{1}{(N-k)!} \delta^{(k-1)}(-\frac{x^2}{4}). \end{array} \right.$$

For integer $N = 1, 2, \ldots$, there arise, via the phase of the logarithm. $x^2 = 0$ supported Dirac distributions

$$\begin{aligned} \log(-x^2 - io) &= \log|x^2| - i\pi\vartheta(x^2), \\ \left(-\frac{\partial}{\partial \frac{x^2}{4}}\right)^k \vartheta(x^2) &= \delta^{(k-1)}(-\tfrac{x^2}{4}), \ k = 1, 2, \ldots . \end{aligned}$$

The residual normalizations for positive and negative invariants a are

$$\mathbf{O}(t,s), \ t+s = d = 1,2,\ldots, \ a \in \mathbb{R}: \ \int \frac{d^d q}{i^s \pi^{\frac{d}{2}}} \frac{\Gamma(\frac{d}{2}-\nu)}{(q^2-io+a)^{\frac{d}{2}-\nu}} = \frac{\Gamma(-\nu)}{(a-io)^{-\nu}}.$$

Orthogonally invariant distributions are embedded in hyperbolically invariant ones, e.g., for $(1,s)$-spacetime with the general Lorentz groups

$$\begin{array}{l}\mathbf{O}(1,s),\\ s=1,2,\dots,\\ N=1,2,\dots,\end{array}\left\{\begin{array}{rl}\int\frac{d^dq}{i^s\pi^{\frac{d}{2}}}\frac{\Gamma(\frac{d}{2}-\nu)}{(q^2-io)^{\frac{d}{2}-\nu}}e^{iqx} &= \frac{\Gamma(\nu)}{\left(\frac{x^2+io}{4}\right)^{\nu}},\\ \int\frac{d^dq}{i^s\pi^{\frac{d}{2}}}\frac{\Gamma(\frac{d}{2}-\nu)}{(q^2-io+1)^{\frac{d}{2}-\nu}}e^{iqx} &= \frac{\vartheta(x^2)2\mathcal{K}_\nu(|x|)-\vartheta(-x^2)\pi[\mathcal{N}_{-\nu}+i\mathcal{J}_{-\nu}](|x|)}{|\frac{x}{2}|^\nu}\\ & \quad -\delta_\nu^N i\pi\sum_{k=1}^N\frac{1}{(N-k)!}\delta^{(k-1)}(-\frac{x^2}{4}),\\ \int\frac{d^dq}{i^s\pi^{\frac{d}{2}}}\frac{e^{i\nu\pi}\Gamma(\frac{d}{2}-\nu)}{(q^2-io-1)^{\frac{d}{2}-\nu}}e^{iqx} &= \frac{\vartheta(-x^2)2\mathcal{K}_\nu(|x|)-\vartheta(x^2)\pi[\mathcal{N}_{-\nu}-i\mathcal{J}_{-\nu}](|x|)}{|\frac{x}{2}|^\nu}\\ & \quad +\delta_\nu^N i\pi\sum_{k=1}^N\frac{1}{(N-k)!}\delta^{(k-1)}(-\frac{x^2}{4}).\end{array}\right.$$

With respect to hyperbolic differential equations with $\mathbf{SO}_0(1,s)$, Huygens principle with spherical $\mathbf{SO}(s)$-boundary conditions holds for odd dimensions $1+s$, not, however, for even spacetime dimensions [5].

Now special cases to be used below: For $\nu=-\frac{1}{2}$ there are *no singularities*:

$$\mathbf{O}(s):\left\{\begin{array}{rl}\int\frac{d^sq}{|\Omega^s|}\frac{2}{(\vec q^2)^{\frac{1+s}{2}}}e^{i\vec q\vec x} &= -r,\\ \int\frac{d^sq}{|\Omega^s|}\frac{2}{(\vec q^2+1)^{\frac{1+s}{2}}}e^{i\vec q\vec x} &= e^{-r},\\ \int\frac{d^sq}{|\Omega^s|}\frac{2}{(\vec q^2-io-1)^{\frac{1+s}{2}}}e^{i\vec q\vec x} &= ie^{ir},\end{array}\right.$$

$$\mathbf{O}(1,s):\left\{\begin{array}{rl}\int\frac{d^{1+s}q}{i^s|\Omega^{1+s}|}\frac{2}{(q^2-io)^{\frac{2+s}{2}}}e^{iqx} &= -|x|[\vartheta(x^2)+i\vartheta(-x^2)],\\ \int\frac{d^{1+s}q}{i^s|\Omega^{1+s}|}\frac{2}{(q^2-io+1)^{\frac{2+s}{2}}}e^{iqx} &= \vartheta(x^2)e^{-|x|}+\vartheta(-x^2)e^{-i|x|},\\ \int\frac{d^{1+s}q}{i^s|\Omega^{1+s}|}\frac{2}{(q^2-io-1)^{\frac{2+s}{2}}}e^{iqx} &= i\vartheta(-x^2)e^{-|x|}+i\vartheta(x^2)e^{i|x|}.\end{array}\right.$$

For $\nu=0$ there is a *logarithmic singularity* in $\mathcal{K}_0$ and $\mathcal{N}_0$:

$$\mathbf{O}(s):\left\{\begin{array}{rl}\int\frac{d^sq}{|\Omega^{s-1}|}\frac{2}{(\vec q^2+1)^{\frac{s}{2}}}e^{i\vec q\vec x} &= 2\mathcal{K}_0(r),\\ \int\frac{d^sq}{|\Omega^{s-1}|}\frac{2}{(\vec q^2-io-1)^{\frac{s}{2}}}e^{i\vec q\vec x} &= -\pi[\mathcal{N}_0-i\mathcal{J}_0](r),\end{array}\right.$$

$$\mathbf{O}(1,s):\left\{\begin{array}{rl}\int\frac{d^{1+s}q}{i^s|\Omega^s|}\frac{2}{(q^2-io+1)^{\frac{1+s}{2}}}e^{iqx} &= \vartheta(x^2)2\mathcal{K}_0(|x|)-\vartheta(-x^2)\pi[\mathcal{N}_0+i\mathcal{J}_0](|x|),\\ \int\frac{d^{1+s}q}{i^s|\Omega^s|}\frac{2}{(q^2-io-1)^{\frac{1+s}{2}}}e^{iqx} &= \vartheta(-x^2)2\mathcal{K}_0(|x|)-\vartheta(x^2)\pi[\mathcal{N}_0-i\mathcal{J}_0](|x|).\end{array}\right.$$

The Fourier transformed *simple poles* are used for representations of the affine groups $\mathbf{SO}(s)\vec\times\mathbb{R}^s$ and $\mathbf{SO}_0(1,s)\vec\times\mathbb{R}^{1+s}$ (more below):

$$\mathbf{O}(s):\left\{\begin{array}{rl}\int\frac{d^sq}{\pi^{\frac{s}{2}}}\frac{1}{\vec q^2}e^{i\vec q\vec x} &= \frac{\Gamma(\frac{s-2}{2})}{(\frac{r}{2})^{s-2}},\\ \int\frac{d^sq}{\pi^{\frac{s}{2}}}\frac{1}{\vec q^2+1}e^{i\vec q\vec x} &= \frac{2\mathcal{K}_{\frac{s-2}{2}}(r)}{(\frac{r}{2})^{\frac{s-2}{2}}},\\ \int\frac{d^sq}{\pi^{\frac{s}{2}}}\frac{1}{\vec q^2-io-1}e^{i\vec q\vec x} &= -\frac{\pi[\mathcal{N}_{\frac{s-2}{2}}-i\mathcal{J}_{\frac{s-2}{2}}](r)}{(\frac{r}{2})^{\frac{s-2}{2}}},\end{array}\right.$$

$$\mathbf{O}(1,s):\left\{\begin{array}{lcl}\int \frac{d^{1+s}q}{i^s\pi^{\frac{1+s}{2}}}\,\frac{1}{q^2-io}e^{iqx} &=& \frac{\Gamma(\frac{s-1}{2})}{\left(\frac{x^2+io}{4}\right)^{\frac{s-1}{2}}}:\\[2ex] \int \frac{d^{1+s}q}{i^s\pi^{\frac{1+s}{2}}}\,\frac{1}{q^2-io+1}e^{iqx} &=& \frac{\vartheta(x^2)2\mathcal{K}_{\frac{s-1}{2}}(|x|)-\vartheta(-x^2)\pi[\mathcal{N}_{-\frac{s-1}{2}}+i\mathcal{J}_{-\frac{s-1}{2}}](|x|)}{|\frac{x}{2}|^{\frac{s-1}{2}}}\\ && -\delta^N_{\frac{s-1}{2}}i\pi\sum_{k=1}^{N}\frac{1}{(N-k)!}\delta^{(k-1)}(-\frac{x^2}{4}),\\[2ex] \int \frac{d^{1+s}q}{i\pi^{\frac{1+s}{2}}}\,\frac{1}{q^2-io-1}e^{iqx} &=& \frac{\vartheta(-x^2)2\mathcal{K}_{\frac{s-1}{2}}(|x|)-\vartheta(x^2)\pi[\mathcal{N}_{-\frac{s-1}{2}}-i\mathcal{J}_{-\frac{s-1}{2}}](|x|)}{|\frac{x}{2}|^{\frac{s-1}{2}}}\\ && +\delta^N_{\frac{s-1}{2}}i\pi\sum_{k=1}^{N}\frac{1}{(N-k)!}\delta^{(k-1)}(-\frac{x^2}{4}).\end{array}\right.$$

The lightcone-supported Dirac distributions arise for even-dimensional spacetime with nonflat position, i.e., for $(1,s)=(1,3),(1,5),\ldots$.

The one-dimensional pole integrals are spread to odd dimensions starting with $1+s=3$ and a singularity at $|x|=0$:

$$\mathbf{O}(3):\left\{\begin{array}{rcll}\int \frac{d^3q}{\pi^2}\,\frac{1}{\vec q^2}e^{i\vec q\vec x} &=& \frac{2}{r}, &\\ \int \frac{d^3q}{\pi^2}\,\frac{1}{\vec q^2+1}e^{i\vec q\vec x} &=& -\frac{\partial}{\partial\frac{r^2}{4}}e^{-r} &= 2\frac{e^{-r}}{r},\\ \int \frac{d^3q}{\pi^2}\,\frac{1}{\vec q^2-io-1}e^{i\vec q\vec x} &=& -\frac{\partial}{\partial\frac{r^2}{4}}ie^{ir} &= 2\frac{e^{ir}}{r},\end{array}\right.$$

$$\mathbf{O}(1,2):\left\{\begin{array}{rcl}-\int \frac{d^3q}{\pi^2}\,\frac{1}{q^2-io}e^{iqx} &=& 2\frac{\vartheta(x^2)-i\vartheta(-x^2)}{|x|},\\ -\int \frac{d^3q}{\pi^2}\,\frac{1}{q^2-io+1}e^{iqx} &=& 2\frac{\vartheta(x^2)e^{-|x|}-\vartheta(-x^2)ie^{-i|x|}}{|x|},\\ -\int \frac{d^3q}{\pi^2}\,\frac{1}{q^2-io-1}e^{iqx} &=& 2\frac{\vartheta(x^2)e^{i|x|}-i\vartheta(-x^2)e^{-|x|}}{|x|}.\end{array}\right.$$

The *dipoles in three dimensions* are without singularity:

$$\mathbf{O}(3):\left\{\begin{array}{rcl}\int \frac{d^3q}{\pi^2}\,\frac{1}{(\vec q^2)^2}e^{i\vec q\vec x} &=& -r,\\ \int \frac{d^3q}{\pi^2}\,\frac{1}{(\vec q^2+1)^2}e^{i\vec q\vec x} &=& e^{-r},\\ \int \frac{d^3q}{\pi^2}\,\frac{1}{(\vec q^2-io-1)^2}e^{i\vec q\vec x} &=& ie^{ir},\end{array}\right.$$

$$\mathbf{O}(1,2):\left\{\begin{array}{rcl}-\int \frac{d^3q}{\pi^2}\,\frac{1}{(q^2-io)^2}e^{iqx} &=& |x|[\vartheta(x^2)-i\vartheta(-x^2)],\\ -\int \frac{d^3q}{\pi^2}\,\frac{1}{(q^2-io+1)^2}e^{iqx} &=& \vartheta(x^2)e^{-|x|}+\vartheta(-x^2)e^{-i|x|},\\ -\int \frac{d^3q}{\pi^2}\,\frac{1}{(q^2-io-1)^2}e^{iqx} &=& i\vartheta(x^2)e^{i|x|}+i\vartheta(-x^2)e^{-|x|}.\end{array}\right.$$

The 2-dimensional integrals are spread to even dimensions, starting with

$1 + s = 4$:

$$\mathbf{O}(4): \quad \begin{cases} \int \frac{d^4q}{\pi^2} \frac{1}{\vec{q}^2} e^{i\vec{q}\vec{x}} &= \frac{4}{r^2}, \\ \int \frac{d^4q}{\pi^2} \frac{1}{\vec{q}^2+1} e^{i\vec{q}\vec{x}} &= -\frac{\partial}{\partial \frac{r^2}{4}} 2\mathcal{K}_0(r) &= \frac{2\mathcal{K}_1(r)}{\frac{r}{2}}, \\ \int \frac{d^4q}{\pi^2} \frac{1}{\vec{q}^2-io-1} e^{i\vec{q}\vec{x}} &= \frac{\partial}{\partial \frac{r^2}{4}} \pi[\mathcal{N}_0 - i\mathcal{J}_0](r) &= -\frac{\pi[\mathcal{N}_1 - i\mathcal{J}_1](r)}{\frac{r}{2}}, \end{cases}$$

$$\mathbf{O}(1,3): \quad \begin{cases} -\int \frac{d^4q}{i\pi^2} \frac{1}{q^2-io} e^{iqx} &= \frac{4}{x^2+io}, \\ -\int \frac{d^4q}{i\pi^2} \frac{1}{q^2-io+1} e^{iqx} &= \frac{\vartheta(x^2)2\mathcal{K}_1(|x|)-\vartheta(-x^2)\pi[\mathcal{N}_{-1}+i\mathcal{J}_{-1}](|x|)}{\frac{|x|}{2}} - i\pi\delta(\frac{x^2}{4}), \\ \int \frac{d^4q}{i\pi^2} \frac{1}{q^2-io-1} e^{iqx} &= \frac{\vartheta(-x^2)2\mathcal{K}_1(|x|)-\vartheta(x^2)\pi[\mathcal{N}_{-1}-i\mathcal{J}_{-1}](|x|)}{\frac{|x|}{2}} + i\pi\delta(\frac{x^2}{4}). \end{cases}$$

Dipoles for four spacetime dimensions lead to maximally logarithmic singularities.

8.3 Residual Representations of the Reals

The simplest case of residual representations is for time and 1-dimensional position with energy and momentum distributions respectively. The representations yield, for a linear invariant, matrix elements of the real 1-dimensional compact and noncompact groups $\mathbf{U}(1) = \exp i\mathbb{R}$ and $\mathbf{D}(1) = \exp \mathbb{R}$ respectively and, for dual invariants, of their self-dual spherical and hyperbolic doublings $\mathbf{SO}(2)$ and $\mathbf{SO}_0(1,1)$ respectively.

8.3.1 Rational Complex Representation Functions

An irreducible $\mathbb{R}$-representation is the residue of a rational complex energy function, or, equivalently, a Fourier transformed Dirac distribution supported by the linear invariant energy $m \in \mathbb{R}$:

$$\mathbb{R} \ni t \longmapsto e^{imt} = \oint \frac{dq}{2i\pi} \frac{1}{q-m} e^{iqt} = \int dq\ \delta(q-m) e^{iqt} \in \mathbf{U}(1).$$

This gives the prototype of a residual representation. The integral $\oint$ circles the singularity in the mathematically positive direction.

For the abelian group $\mathbf{D}(1) \cong \mathbb{R}$, where the dimension coincides with the rank and where the eigenvalues q are the group invariants m, the transition to the residual form is a trivial transcription to the singularity $q = m$. This will be different for nonabelian groups with dimension strictly larger than the rank, e.g., for the rotations $\mathbf{SO}(3)$, with dimension 3 and rank 1, with the invariant a square $\vec{q}^2 = m^2$ of the three $\mathbb{R}^3$-eigenvalues $\vec{q}$.

In the Fourier transformations of the future and past distributions the real-imaginary decomposition into Dirac and principal value distributions goes with the order function decomposition $\vartheta(\pm t) = \frac{1 \pm \epsilon(t)}{2}$ in the functions on future $\mathbb{R}_+$ and past $\mathbb{R}_-$,

$$\text{causal: } \mathbb{R}_\pm \ni \vartheta(\pm t)t \longmapsto \pm \int \frac{dq}{2i\pi} \frac{1}{q \mp io - m} e^{iqt} = \vartheta(\pm t) e^{imt}.$$

All those distributions originate from the same representation functions with one pole in the compactified complex plane:

$$\mathbb{C} \ni q \longmapsto \frac{1}{q-m} \in \overline{\mathbb{C}}, \quad m \in \mathbb{R}.$$

The position $q = m$ of the singularity is related to the continuous invariant. The Fourier transforms with different contours around the pole represent via $\vartheta(\pm t)$ the causal structure of the reals.

A representation distribution with nontrivial residue can be normalized,

$$\mathbb{R} \ni 0 \longmapsto 1 = \oint \frac{dq}{2i\pi} \frac{1}{q-m} = \text{res}_m \frac{1}{q-m} = \langle m|m\rangle.$$

The residual normalization gives, simultaneously, both the normalization of the unit $t = 0$ representation $t \longmapsto e^{imt}$ (pure state) and the scalar product of the normalized eigenvector (pure cyclic vector) $|m\rangle$.

8.3.2 Compact and Noncompact Dual Invariants

Poles at dual compact representation invariants $q^2 = m^2$ can be combined from linear poles at $q = \pm|m|$, the invariants for the dual irreducible subrepresentations.

The Fourier transforms of the causal and (anti-)Feynman energy distributions are functions on the cones, the bicone, and the group with $\mathbf{SO}(2)$ matrix elements:

$$\begin{array}{llll} \text{causal:} & \mathbb{R}_\pm \ni \vartheta(\pm t)t \longmapsto & \pm\int \frac{dq}{i\pi} \frac{\binom{q}{|m|}}{(q\mp io)^2-m^2} e^{iqt} & = \vartheta(\pm t) 2\binom{\cos mt}{i \sin |m|t}, \\ \text{bicone:} & \mathbb{R}_+ \uplus \mathbb{R}_- \ni t \longmapsto & \pm\int \frac{dq}{i\pi} \frac{\binom{|m|}{q}}{q^2\mp io-m^2} e^{iqt} & = \binom{1}{\pm\epsilon(t)} e^{\pm i|mt|}, \\ \text{group:} & \mathbb{R} \ni t \longmapsto & \int dq \binom{|m|}{q} \delta(q^2 - m^2) e^{iqt} & = \binom{\cos mt}{i \sin |m|t}. \end{array}$$

The normalization for $t = 0$ uses different matrix elements for the causal residues with two poles with equal imaginary part and for the Feynman residues with two poles with opposite imaginary part:

$$\begin{array}{ll} \text{causal:} & \mathbb{C} \ni q \longmapsto \frac{2q}{q^2-m^2} \in \overline{\mathbb{C}}, \quad \sum \oint_{\pm|m|} \frac{dq}{2i\pi} \frac{2q}{q^2-m^2} = \sum \text{res}_{\pm|m|} \frac{2q}{q^2-m^2} = 2, \\ \text{Feynman:} & \mathbb{C} \ni q \longmapsto \frac{2|m|}{q^2-m^2} \in \overline{\mathbb{C}}, \quad \pm \oint_{\pm|m|} \frac{dq}{2i\pi} \frac{2|m|}{q^2-m^2} = \text{res}_{\pm|m|} \frac{\pm 2|m|}{q^2-m^2} = 1. \end{array}$$

The functions with noncompact dual representation invariants $q^2 = -m^2$ give, as Fourier transformed Ω^1-measure, noncompact matrix elements of faithful cyclic $\mathbf{D}(1)$-representations, not irreducible:

$$\begin{aligned} \mathbf{SO}_0(1,1) \cong \mathbb{R} \ni x \quad & \longmapsto \int \frac{dq}{\pi} \frac{|m|}{q^2+m^2} e^{iqx} \\ & = \oint \frac{dq}{2i\pi} \left[\frac{\vartheta(-x)}{q-i|m|} - \frac{\vartheta(x)}{q+i|m|}\right] e^{iqx} = e^{-|mx|}. \end{aligned}$$

The representation-relevant residues are taken at imaginary "momenta" $q = \pm i|m|$ in the complex momentum plane:

$$\mathbb{C} \ni q \longmapsto \frac{2|m|}{q^2+m^2} \in \overline{\mathbb{C}}, \quad \oint_{\pm i|m|} \frac{dq}{2\pi} \frac{2|m|}{q^2+m^2} = \text{res}_{\pm i|m|} \frac{\pm 2i|m|}{q^2+m^2} = 1.$$

8.4 Residual Representations of Tangent Groups

Cyclic Hilbert representations of the abelian translations are used for affine groups $G \vec{\times} \mathbb{R}^n$. The representation collects, via direct integrals over the orbits of the homogeneous group, the translation representations with Dirac distributions of the (energy-)momentum eigenvalues. This will be given for Euclidean and Poincaré groups with their orthogonal homogeneous groups. Physically relevant examples for the Hilbert spaces with square integrable functions on momentum spheres (nonrelativistic scattering) and energy-momentum hyperboloids (free particles) have been given in the chapter "Harmonic Analysis."

8.4.1 Euclidean Groups

The irreducible infinite-dimensional Hilbert representations of the Euclidean groups $\mathbf{SO}(s) \vec{\times} \mathbb{R}^s$, $s \geq 1$, are, for nontrivial translation invariant and $s \geq 2$, inducible with fixgroup $\mathbf{SO}(s-1)$ (chapter "Harmonic Analysis"). The scalar representation coefficients for the Euclidean spaces $\mathbf{SO}(s) \vec{\times} \mathbb{R}^s / \mathbf{SO}(s) \cong \mathbb{R}^s$,

$$\begin{aligned} m^2 > 0: \quad \mathbb{R}^s \ni \vec{x} \longmapsto \quad & D^s(m^2 r^2) = \int d^s q \; \delta(\vec{q}^2 - m^2) e^{-i\vec{q}\vec{x}} \\ &= \tfrac{|m|^{s-2}}{2} \int d^{s-1}\omega \; e^{-i|m|\vec{\omega}\vec{x}} = \tfrac{|m|^{s-2}}{2} \int d^{s-1}\omega \; \cos |m|\vec{\omega}\vec{x}, \end{aligned}$$

use the Fourier transformed measure of the momentum direction sphere $\vec{\omega} = \frac{\vec{q}}{|\vec{q}|} \in \Omega^{s-1}$. The scalar representation coefficient can be normalized as a positive function for a cyclic translation representation where the momentum sphere has the invariant $|m|$ as intrinsic unit:

$$\begin{aligned} \text{state: } d^s(m^2 r^2) &= \int \tfrac{d^s q}{|m|^{s-1}|\Omega^{s-1}|} \; 2|m| \delta(\vec{q}^2 - m^2) e^{-i\vec{q}\vec{x}} \\ &= \int \tfrac{d^{s-1}\omega}{|\Omega^{s-1}|} \; \cos |m|\vec{\omega}\vec{x}, \quad d^s(0) = 1, \; |\Omega^{s-1}| = \tfrac{2\pi^{\frac{s}{2}}}{\Gamma(\frac{s}{2})}. \end{aligned}$$

For $s \geq 3$, the scalar representation coefficients arise by a 2-sphere spread via derivations $-\frac{d}{d\frac{r^2}{4\pi}}$,

$$D^s(r^2) = -\frac{d}{d\frac{r^2}{4\pi}} D^{s-2}(r^2) = \frac{\pi \mathcal{J}_{\frac{s-2}{2}}(r)}{(\frac{r}{2\pi})^{\frac{s-2}{2}}},$$

and embed the self-dual $\mathbb{R}$-representation matrix element

$$\mathbb{R} \ni x \longmapsto D^1(r^2) = \int dq \; \delta(q^2 - 1) e^{-iqx} = \cos r.$$

For odd-dimensional spaces with $\mathbf{SO}(1 + 2R)$, they involve half-integer-index (spherical) Bessel functions, whereas integer-index Bessel functions are used for even-dimensional spaces with $\mathbf{SO}(2R)$, both with rank R

$$D^s(r^2) = \begin{cases} 2^{2R-1} \frac{\pi j_{R-1}(r)}{(\frac{r}{2\pi})^{R-1}} = \left(-\frac{d}{d\frac{r^2}{4\pi}}\right)^R \cos r, & \text{for } s = 1 + 2R = 1, 3, \ldots, \\ \frac{\pi \mathcal{J}_{R-1}(r)}{(\frac{r}{2\pi})^{R-1}} = \left(-\frac{d}{d\frac{r^2}{4\pi}}\right)^{R-1} \pi \mathcal{J}_0(r), & \text{for } s = 2R = 2, 4, \ldots. \end{cases}$$

The integrals sum over the embedded $\mathbb{R}$-representation coefficients

$$\begin{aligned} D^2(r^2) &= \int_0^\pi d\chi \ \cos(r\cos\chi) &&= \pi\mathcal{J}_0(r), \\ D^3(r^2) &= \pi\int_{-1}^1 d\zeta \ \cos(r\zeta) &&= 2\pi j_0(r) = -\frac{d}{d\frac{r^2}{4\pi}}\cos r. \end{aligned}$$

$\mathbf{SO}(s)$-nontrivial degrees of freedom $\mathbb{R}^s \cong \mathbb{R}_+ \times \Omega^{s-1}$ use derivations

$$(\vec{q})^L \sim (i\frac{\partial}{\partial\vec{x}})^L, \quad L = 0, 1, \ldots, \quad \frac{\partial}{\partial\vec{x}} = \frac{\vec{x}}{r}\frac{\partial}{\partial r} = 2\vec{x}\frac{\partial}{\partial r^2}.$$

8.4.2 Poincaré Groups

The irreducible Hilbert representations of the Poincaré group $\mathbf{SO}_0(1,s) \vec{\times} \mathbb{R}^{1+s}$, $s \geq 1$, for positive translation–invariant, e.g., massive particle representations for $m^2 > 0$, are inducible with fixgroup $\mathbf{SO}(s)$ (chapter "Harmonic Analysis"). They come as the Fourier transformed infinite measure of the energy-momentum hyperboloid with the directions as eigenvalues $\frac{q}{|q|} = \pm \mathrm{y} \in \mathcal{Y}^s_\pm \cong \mathbf{SO}_0(1,s)/\mathbf{SO}(s)$. The scalar representation coefficients for the hyperbolic spaces $\mathbf{SO}_0(1,s) \vec{\times} \mathbb{R}^{1+s}/\mathbf{SO}_0(1,s) \cong \mathbb{R}^{1+s}$ read with $|x| = \sqrt{|x^2|}$,

$$\begin{aligned} m^2 > 0: \quad \mathbb{R}^{1+s} \ni x \longmapsto D^{(1,s)}(m^2x^2) &= \int d^{1+s}q \ \delta(q^2 - m^2)e^{iqx} \\ &= |m|^{s-1}\int d^s\mathrm{y} \ \cos|m|\mathrm{y}x, \\ D^{(1,s)}(x^2) = \frac{d}{d\frac{x^2}{4\pi}} D^{(1,s-2)}(x^2) &= \frac{-\vartheta(x^2)\pi\mathcal{N}_{-\frac{s-1}{2}}(|x|) + \vartheta(-x^2)2\mathcal{K}_{\frac{s-1}{2}}(|x|)}{|\frac{x}{2\pi}|^{\frac{s-1}{2}}}. \end{aligned}$$

The hyperbolic invariant $|m|$ is used as intrinsic unit in the coefficient, renormalized with the $(s-1)$ spherical degrees of freedom

$$d^{(1,s)}(m^2x^2) = \int \frac{d^{1+s}q}{|m|^{s-1}|\Omega^{s-1}|} \ 2\delta(q^2 - m^2)e^{iqx} = \int \frac{d^s\mathrm{y}}{|\Omega^{s-1}|} \ 2\cos|m|\mathrm{y}x.$$

It is not finite for $x = 0$.

The representation coefficients embed time and 1-position representations

$$\begin{aligned} \mathbb{R} \ni t &\longmapsto \int dq \ \delta(q^2 - 1)e^{iqt} &&= \cos t, \\ \mathbb{R} \ni z &\longmapsto \int \frac{dq}{\pi} \ \frac{1}{q^2+1}e^{-iqz} &&= e^{-|z|}, \end{aligned}$$

for $s \geq 2$ as 2-sphere spreads. For odd dimension and $\mathbf{SO}_0(1,2R)$, they involve half-integer index functions, hyperbolic Macdonald and spherical Bessel functions

$$\begin{aligned} D^{(1,2)}(x^2) &= 2\pi\frac{-\vartheta(x^2)\sin|x| + \vartheta(-x^2)e^{-|x|}}{|x|}, \\ D^{(1,2R)}(x^2) &= \left(\frac{\partial}{\partial\frac{x^2}{4\pi}}\right)^R [\vartheta(x^2)\cos|x| + \vartheta(-x^2)e^{-|x|}] \\ &= \frac{\vartheta(x^2)(-1)^R\pi\mathcal{J}_{R-\frac{1}{2}}(|x|) + \vartheta(-x^2)2\mathcal{K}_{R-\frac{1}{2}}(|x|)}{|\frac{x}{2\pi}|^{R-\frac{1}{2}}} \text{ for } 1+s = 1+2R = 3, 5, \ldots. \end{aligned}$$

For even spacetime dimension and $\mathbf{SO}_0(1,2R-1)$ they start with the rank $R=1$ Poincaré group by integrating $\mathbb{R}$-representation coefficients on a hyperbola:

$$\begin{aligned} D^{(1,1)}(x^2) &= \int d\psi\ [\vartheta(x^2)\cos(|x|\cosh\psi) + \vartheta(-x^2)e^{-|x|\cosh\psi}] \\ &= -\vartheta(x^2)\pi\mathcal{N}_0(|x|) + \vartheta(-x^2)2\mathcal{K}_0(|x|). \end{aligned}$$

A 2-sphere spread gives the Hilbert representation of the rank-2 Poincaré group with Minkowski translations and, in general, the integer index functions

$$\begin{aligned} D^{(1,2R-1)}(x^2) &= \left(\frac{\partial}{\partial\frac{x^2}{4\pi}}\right)^{R-1} [-\vartheta(x^2)\pi\mathcal{N}_0(|x|) + \vartheta(-x^2)2\mathcal{K}_0(|x|)] \\ &= \frac{\vartheta(x^2)(-1)^R\pi\mathcal{N}_{R-1}(|x|) + \vartheta(-x^2)2\mathcal{K}_{R-1}(|x|)}{|\frac{x}{2\pi}|^{R-1}} \text{ for } 1+s=2R=2,4,\ldots. \end{aligned}$$

$\mathbf{SO}_0(1,s)$-nontrivial coefficients use derivations

$$(q)^L \sim (-i\tfrac{\partial}{\partial x})^L,\ L=0,1,\ldots,\ \ \frac{\partial}{\partial x} = 2x\frac{\partial}{\partial x^2}.$$

8.5 Residual Representations of Position

Self-dual spherical $\mathbf{SO}(2)$-coefficients of translation representations are positive-type functions in $L^\infty(\mathbb{R})_+$, with Dirac measures in $\mathcal{M}(\mathbb{R})_+$

$$\mathbb{R} \ni t \longmapsto 2\cos\omega t = \int dq\ 2|q|\delta(q^2-\omega^2)e^{iqt}.$$

Bound waves and interactions are characterizable by self-dual hyperbolic $\mathbf{SO}_0(1,1)$-coefficients that are square integrable functions in $L^\infty(\mathbb{R})_+ \cap L^2(\mathbb{R})$ and have a rational function as positive Radon measure,

$$\mathbb{R} \ni z \longmapsto e^{-|Qz|} = \int \frac{dq}{\pi}\frac{Q|}{q^2+Q^2}e^{-iqz}.$$

The spherical and hyperbolic invariants come from a real and "imaginary" momentum pair as poles in the complex momentum plane, i.e., from $q=\pm\omega$ and $q=\pm i|Q|$ respectively. In contrast to the hyperbolic state, the spherical state is decomposable as a direct sum.

The 1-dimensional quantum-mechanical example is given by the Schrödinger functions of the harmonic oscillator. They are position representation coefficients with the representation invariant the inverse intrinsic length $Q^2 = \frac{1}{\ell^4} = kM$. Here the hyperbolic state $z \longmapsto e^{-|Qz|}$ with positive definite coordinate shows up in a reparametrization with the square of the usual position parameter $z = \frac{x^2}{2}$ (chapter "The Kepler Factor"):

$$[\ell^2\frac{p^2}{2} + \frac{x^2}{2\ell^2}]\psi(x) = \frac{E}{\omega}\psi(x) \Rightarrow \psi_0(x) = e^{-\frac{x^2}{2\ell^2}} = e^{-|Qz|} \text{ with } \frac{2E_0}{\omega} = 1.$$

In contrast to scattering waves, bound waves for nonabelian position groups use higher-order momentum poles, where the order depends on the position space dimension. This will be exemplified by the nonrelativistic hydrogen atom bound waves, which represent the noncompact nonabelian group $\mathbf{SO}_0(1,3)$ and start with momentum dipoles.

8.5.1 The Multipoles of the Hydrogen Atom

The hyberbolic structure of a nonrelativistic dynamics with the Coulomb-Kepler potential $\frac{1}{r}$ and the invariance of the Lenz-Runge "perihelion" vector has been exploited quantum-mechanically by Fock [4]. With the additional rotation invariance the bound state vectors come in irreducible k^2-dimensional representations of the group $\mathbf{SO}(4) = \frac{\mathbf{SU}(2)\times\mathbf{SU}(2)}{\mathbb{I}(2)}$, centrally correlating two $\mathbf{SU}(2)$'s, with the integer invariant $k = 1 + 2J = 1, 2, \ldots$.

The measure of the 3-sphere as the manifold of the orientations of the rotation group $\mathbf{SO}(3)$ in the invariance group $\mathbf{SO}(4)$ has a momentum parametrization by a dipole (chapter "The Kepler Factor"),

$$\Omega^3 \cong \mathbf{SO}(4)/\mathbf{SO}(3), \quad \frac{1}{\sqrt{\vec{q}^2+1}}\begin{pmatrix}1\\ i\vec{q}\end{pmatrix} \in \Omega^3 \subset \mathbb{R}^4 \Rightarrow |\Omega^3| = \int d^3\omega = \int \frac{2d^3q}{(\vec{q}^2+1)^2} = 2\pi^2.$$

Ω^3-integration of the pure translation states $\mathbb{R}^3 \ni \vec{x} \longmapsto e^{-i\vec{q}\vec{x}} \in \mathbf{U}(1)$, i.e., the Fourier transformed Ω^3-measure, gives the hydrogen ground state function as a scalar representation coefficient of 3-position space:

$$\mathcal{Y}^3 \cong \mathbf{SO}_0(1,3)/\mathbf{SO}(3) \cong \mathbf{SL}(\mathbb{C}^2)/\mathbf{SU}(2) \ni \vec{x} \longmapsto \int \frac{d^3q}{\pi^2}\frac{|Q|}{(\vec{q}^2+Q^2)^2}e^{-i\vec{q}\vec{x}} = e^{-|Q|r}.$$

In the bound waves, 3-position space is represented as a 3-hyperboloid with a continuous invariant Q^2 for the imaginary "momenta" $\vec{q}^2 = -Q^2$ on a 2-sphere Ω^2 and a discrete rotation invariant $2J \in \mathbb{N}$.

The bound waves are matrix elements of infinite-dimensional cyclic principal $\mathbf{SL}(\mathbb{C}^2)$-representations, where,with the Cartan subgroups $\mathbf{SO}(2)\times\mathbf{SO}_0(1,1)$, the irreducible ones are characterized by one integer and one continous invariant. In the language of induced representations, the bound waves of the hydrogen atom are rotation $\mathbf{SO}(3)$-intertwiners on the group $\mathbf{SO}_0(1,3)$ ($\mathcal{Y}^3$-functions) with values in Hilbert spaces with $\mathbf{SO}(3)$-representations in $(1+2J)^2$-dimensional $\mathbf{SO}(4)$-representations.

For the nonrelativistic hydrogen atom, the rotation dependence $\vec{x}$ is effected by momentum derivation of the Ω^3-measure,

$$\vec{x}e^{-r} = \int \frac{d^3q}{\pi^2}\frac{4i\vec{q}}{(1+\vec{q}^2)^3}e^{-i\vec{q}\vec{x}} \text{ with } \frac{4\vec{q}}{(1+\vec{q}^2)^3} = -\frac{\partial}{\partial\vec{q}}\frac{1}{(1+\vec{q}^2)^2}.$$

The 3-vector factor $\frac{2\vec{q}}{1+\vec{q}^2}$ is uniquely supplemented to a normalized 4-vector on the 3-sphere, a parametrization of the sphere

$$\frac{1}{1+\vec{q}^2}\begin{pmatrix}\vec{q}^2-1\\ 2i\vec{q}\end{pmatrix} = \begin{pmatrix}\cos\chi\\ \frac{\vec{q}}{|\vec{q}|}i\sin\chi\end{pmatrix} = \begin{pmatrix}p_0\\ i\vec{p}\end{pmatrix} = p \in \Omega^3 \subset \mathbb{R}^4, \quad p_0^2 + \vec{p}^2 = 1.$$

The totally symmetric traceless products of the normalized 4-vector build the higher-order Ω^3-harmonics $\mathrm{Y}^{(2J,2J)}(p) \sim (p)^{2J}$.

The Kepler bound waves in $(1+2J)^2$-multiplets come with $2J$-dependent multipoles:

$$\mathcal{Y}^3 \ni \vec{x} \longmapsto \int \frac{d^3q}{\pi^2}\frac{1}{(1+\vec{q}^2)^2}(p)^{2J}e^{-i\vec{q}Q\vec{x}} \text{ with } \begin{cases} p &= \frac{1}{1+\vec{q}^2}\begin{pmatrix}\vec{q}^2-1\\ 2i\vec{q}\end{pmatrix}, \\ Q &= \frac{1}{1+2J}. \end{cases}$$

As illustration, the $k = 2$ bound state quartet with tripole vector and Laguerre polynomials

$$\begin{aligned} Q = \tfrac{1}{2}: \quad \int \tfrac{d^3q}{\pi^2} \tfrac{1}{(1+\vec{q}^2)^2} \begin{pmatrix} p_0 \\ i\vec{p} \end{pmatrix} e^{-i\vec{q}Q\vec{x}} &= \int \tfrac{d^3q}{\pi^2} \tfrac{1}{(1+\vec{q}^2)^3} \begin{pmatrix} \vec{q}^2 - 1 \\ 2i\vec{q} \end{pmatrix} e^{-i\vec{q}Q\vec{x}} \\ = \begin{pmatrix} \frac{1-Qr}{2} \\ \frac{Q\vec{x}}{2} \end{pmatrix} e^{-Qr} &= \begin{pmatrix} \frac{1}{4}\mathrm{L}_1^1(2Qr) \\ \frac{Q\vec{x}}{2}\mathrm{L}_2^0(2Qr) \end{pmatrix} \end{aligned}$$

and the $k = 3$ bound state nonet with quadrupole tensor

$$\begin{aligned} Q = \tfrac{1}{3}: \quad \int \tfrac{d^3q}{\pi^2} \tfrac{1}{(1+\vec{q}^2)^2} \begin{pmatrix} 3p_0^2 - \vec{p}^2 \\ ip_0\vec{p} \\ 3\vec{p} \otimes \vec{p} - \mathbf{1}_3\vec{p}^2 \end{pmatrix} e^{-i\vec{q}Q\vec{x}} &= \int \tfrac{d^3q}{\pi^2} \tfrac{4}{(1+\vec{q}^2)^4} \begin{pmatrix} 3(\frac{\vec{q}^2-1}{2})^2 - \vec{q}^2 \\ i\vec{q}\,\frac{\vec{q}^2-1}{2} \\ 3\vec{q} \otimes \vec{q} - \mathbf{1}_3\vec{q}^2 \end{pmatrix} e^{-i\vec{q}Q\vec{x}} \\ = \begin{pmatrix} 1 - 2Qr + \frac{2Q^2r^2}{3} \\ \frac{2-Qr}{3}\frac{Q\vec{x}}{2} \\ \frac{Q^2}{2}(\mathbf{1}_3\frac{r^2}{3} - \vec{x} \otimes \vec{x}) \end{pmatrix} e^{-Qr} &= \begin{pmatrix} \frac{1}{3}\mathrm{L}_1^2(2Qr) \\ \frac{Q\vec{x}}{2}\,\frac{1}{6}\mathrm{L}_3^1(2Qr) \\ \frac{Q^2}{2}(\mathbf{1}_3\frac{r^2}{3} - \vec{x} \otimes \vec{x})\mathrm{L}_5^0(2Qr) \end{pmatrix}. \end{aligned}$$

8.5.2 Residual Representations of Hyperbolic Positions

Distributions of s-dimensional momenta $\vec{q} \in \mathbb{R}^s$ with the action of the rotation group $\mathbf{SO}(s)$ are used for representations[6, 11] of the hyperboloids $\mathcal{Y}^s$ and spheres Ω^s. For $s = 1$, "flat" and "hyberbolic" are isomorphic. The residual representations of nonabelian noncompact hyperboloids and compact spheres with $s \geq 2$ have to embed the nontrivial representations of the abelian groups with continuous and integer dual invariants respectively:

$$\begin{aligned} \mathbf{SO}_0(1,1) \cong \mathcal{Y}^1 \ni x &\longmapsto \int \tfrac{dq}{\pi} \tfrac{|m|}{q^2+m^2} e^{iqx} &&= e^{-|mx|}, \quad m^2 > 0, \\ \mathbf{SO}(2) \cong \Omega^1 \ni e^{ix} &\longmapsto \pm \int \tfrac{dq}{i\pi} \tfrac{|m|}{q^2 \mp io - m^2} e^{iqx} &&= e^{\pm i|mx|}, \quad |m| = 1, 2, \ldots. \end{aligned}$$

The pole invariants $\{\pm i|m|\}$ and $\{\pm|m|\}$ on the discrete sphere $\Omega^0 = \{\pm 1\}$ are embedded, for the nonabelian case, into singularity spheres Ω^{s-1} that arise in the Cartan factorization

$$\begin{aligned} \mathbf{SO}_0(1,s)/\mathbf{SO}(s) &\cong \mathcal{Y}^s \cong \mathbf{SO}_0(1,1) \circ \Omega^{s-1}, \\ \mathbf{SO}(1+s)/\mathbf{SO}(s) &\cong \Omega^s \cong \mathbf{SO}(2) \circ \Omega^{s-1}. \end{aligned}$$

The rank of the orthogonal groups gives the real (noncompact) rank 1 for the odd-dimensional hyperboloids, i.e., one continuous noncompact invariant,

$$\begin{aligned} \mathrm{rank}_{\mathbb{R}}\mathbf{SO}_0(t,s) = R \text{ for } t+s = 2R \text{ and } t+s &= 2R+1, \\ \mathrm{rank}_{\mathbb{R}}\mathbf{SO}_0(1, 2R-1) - \mathrm{rank}_{\mathbb{R}}\mathbf{SO}(2R-1) &= 1, \\ \mathrm{rank}_{\mathbb{R}}\mathbf{SO}_0(1, 2R) - \mathrm{rank}_{\mathbb{R}}\mathbf{SO}(2R) &= 0. \end{aligned}$$

Odd-dimensional hyperboloids and spheres, $\mathcal{Y}^s$ and Ω^s with $s = 2R - 1$, will be considered as a generalization of the minimal and characteristic nonabelian case $s = 3$ with nontrivial rotations for the nonrelativistic hydrogen atom above.

The coefficients of residual representations of hyperboloids $\mathcal{Y}^{2R-1}$ use the Fourier transformed measure of the momentum sphere Ω^{2R-1} with singularity sphere Ω^{2R-2} for imaginary "momenta" as eigenvalues with continuous non-compact invariant $\vec{q}^2 = -m^2 < 0$. $\mathbf{SO}(2R)$-multiplets arise via the sphere parametrization $\frac{1}{\vec{q}^2+m^2}\begin{pmatrix} \vec{q}^2 - m^2 \\ 2i|m|\vec{q} \end{pmatrix} \in \Omega^{2R-1} \subset \mathbb{R}^{2R}$ with $\frac{1}{|\Omega^{2R-1}|} = \frac{\Gamma(R)}{2\pi^R}$,

$$\text{for } \mathcal{Y}^{2R-1},\ R = 1, 2, \ldots,$$
$$\vec{x} \longmapsto \begin{cases} \int \frac{d^{2R-1}q}{|\Omega^{2R-1}|} \frac{2|m|}{(\vec{q}^2+m^2)^R} e^{-i\vec{q}\vec{x}} & = e^{-|m|r}, \\ \int \frac{d^{2R-1}q}{|\Omega^{2R-1}|} \frac{2R|m|}{(\vec{q}^2+m^2)^{1+R}} \begin{pmatrix} \vec{q}^2 - m^2 \\ 2i|m|\vec{q} \end{pmatrix} e^{-i\vec{q}\vec{x}} & = \begin{pmatrix} R-1-|m|r \\ \vec{x} \end{pmatrix} e^{-|m|r}. \end{cases}$$

Each state $\{\vec{x} \longmapsto e^{-|m|r}\} \in L^\infty(\mathbf{SO}_0(1, 2R-1))_+$ with $m^2 > 0$, characterizes an infinite-dimensional Hilbert space with a faithful cyclic representation of $\mathbf{SO}_0(1, 2R-1)$ as familiar for $R = 2$ from the principal series representations of the Lorentz group $\mathbf{SO}_0(1,3)$. The positive type-function defines the Hilbert product:

$$\begin{array}{rl} \text{distributive basis:} & \{|m^2; \vec{q}\rangle \mid \vec{q} \in \mathbb{R}^{2R-1}\}, \\ \text{scalar product distribution:} & \langle m^2; \vec{q}\,'|m^2; \vec{q}\rangle = \frac{2|m|}{(\vec{q}^2+m^2)^R} |\Omega^{2R-1}| \delta(\vec{q} - \vec{q}\,'), \\ \text{Hilbert vectors:} & |m^2; f\rangle = \int \frac{d^{2R-1}q}{|\Omega^{2R-1}|} f(\vec{q}) |m^2; \vec{q}\rangle, \\ & \langle m^2; f|m^2; f'\rangle = \int \frac{d^{2R-1}q}{|\Omega^{2R-1}|} \overline{f(\vec{q})} \frac{2|m|}{(\vec{q}^2+m^2)^R} f'(\vec{q}). \end{array}$$

There is a representation of each abelian noncompact subgroup in the Cartan decomposition $\mathcal{Y}^{2R-1} \cong \mathbf{SO}_0(1,1) \circ \Omega^{2R-2}$ with the action on a distributive basis and hence on the Hilbert vectors:

$$\begin{array}{rl} \mathbf{SO}_0(1,1)\text{-representations for all } \vec{\omega} \in \Omega^{2R-2} : e^{-\vec{\omega}\vec{x}} & \longmapsto e^{-i|\vec{q}|\vec{\omega}\vec{x}} = e^{-i\vec{q}\vec{x}} \in \mathbf{U}(1), \\ \text{action of all } \mathbf{SO}_0(1,1) : \ |m^2; \vec{q}\rangle & \longmapsto e^{-i\vec{q}\vec{x}} |m^2; \vec{q}\rangle, \end{array}$$
$$\begin{array}{rl} \text{cyclic vector:} & |m^2; 1\rangle = \int \frac{d^{2R-1}q}{|\Omega^{2R-1}|} |m^2; \vec{q}\rangle \\ \text{with} & \int \frac{d^{2R-1}q\, d^{2R-1}q'}{|\Omega^{2R-1}|^2} \langle m^2; \vec{q}\,'|e^{-i\vec{q}\vec{x}}|m^2; \vec{q}\rangle = e^{-|m|r}. \end{array}$$

The scalar product is written with the positive-type function, e.g., for 3-dimensional position $R = 2$ with intrinsic unit

$$\begin{array}{rl} \langle f|f'\rangle = \int \frac{d^3q}{2\pi^2} \overline{f(\vec{q})} \frac{2}{(\vec{q}^2+1)^2} f'(\vec{q}) & = \int d^3x_1 d^2x_2\, \overline{\tilde{f}(\vec{x}_2)}\, e^{-|\vec{x}_1 - \vec{x}_2|} \tilde{f}'(\vec{x}_1) \\ \text{with } f(\vec{q}) & = \int d^3x\, \tilde{f}(\vec{x}) e^{i\vec{q}\vec{x}}. \end{array}$$

It can be brought in the form of square integrability $L^2(\mathbb{R}^3)$ by absorption of the square root,

$$\psi(\vec{q}) = \frac{\sqrt{8\pi}}{\vec{q}^2+1} f(\vec{q}) \Rightarrow \langle f|f'\rangle = \int \frac{d^3q}{(2\pi)^3} \overline{\psi(\vec{q})}\, \psi'(\vec{q}) = \int d^3x\, \overline{\tilde{\psi}(\vec{x})}\, \tilde{\psi}'(\vec{x}).$$

Therefore, all infinite-dimensional Hilbert spaces for different continuous invariants $m^2 > 0$ are subspaces of one Hilbert space $L^2(\mathcal{Y}^{2R-1})$ with $\mathcal{Y}^{2R-1} \cong \mathbb{R}^{2R-1}$. States with different invariants are not orthogonal, i.e., , they have a nontrivial Schur scalar product

$$\{d^{m_1^2}|d^{m_2^2}\} = \int d^{2R-1}x\, e^{-|m_1|r} e^{-|m_2|r} = |\Omega^{2R-2}| \frac{\Gamma(2R)}{(|m_1|+|m_2|)^{2R-1}}.$$

The orthogonality of the $\mathcal{Y}^3$-representation coefficients with different invariant $m^2 = \frac{1}{(1+2J)^2}$ in the hydrogen atom is a consequence of the different rotation invariants J.

The corresponding matrix elements of representations of odd-dimensional spheres are obtained by real-imaginary transition. They involve Feynman distributions with supporting singularity sphere Ω^{2R-2} for real momenta with integer compact invariant $\vec{q}^2 = m^2$, $|m| = 1, 2, \dots$,

$$\begin{array}{ll}\text{for } \Omega^{2R-1}: & \vec{x} \longmapsto \left\{ \begin{array}{rl} \pm\int \frac{d^{2R-1}q}{i|\Omega^{2R-1}|} \frac{2|m|}{(\vec{q}^2 \mp io - m^2)^R} e^{-i\vec{q}\vec{x}} & = e^{\pm i|m|r}, \\ \int \frac{d^{2R-1}q}{\pi^{R-1}} |m| \delta^{(R-1)}(m^2 - \vec{q}^2) e^{-i\vec{q}\vec{x}} & = \cos|m|r. \end{array}\right. \\ R = 1, 2, \dots & \end{array}$$

The irreducible representation spaces are finite dimensional, e.g., for $R = 2$ isomorphic to $\mathbb{C}^{1+2L}$. The irreducible spaces for different discrete invariants, e.g., $L = 0, 1, \dots$, are Schur-orthogonal subspaces of the infinite-dimensional Hilbert space $L^2(\Omega^{2R-1})$.

The residual normalization for complex representation functions

$$\mathbb{R} \times \Omega^{2R-1} \hookrightarrow \mathbb{C} \times \Omega^{2R-1} \ni \vec{q} = |\vec{q}|\,\frac{\vec{q}}{|\vec{q}|} \longmapsto \frac{2\mu}{(\vec{q}^2+\mu^2)^R} \in \overline{\mathbb{C}}, \quad \mu \in \mathbb{C}$$

has to take into account the sphere degrees of freedom in $\mathbb{C} \times \Omega^{2R-1}$, e.g., for $\mathcal{Y}^{2R-1}$,

$$\mathrm{res}_{\pm i|m|} \frac{2i|m|}{(\vec{q}^2+m^2)^R} = \int \frac{d^{2R-1}q}{i|\Omega^{2R-1}|} \frac{2i|m|}{(\vec{q}^2+m^2)^R} = \oint_{\pm i|m|} \frac{dq}{2i\pi} \frac{(q^2)^{R-1} 2i|m|}{(q^2+m^2)^R} = 1.$$

The higher-order q^2-power is compensated by the q^2-power of the measure. Nonscalar functions have trivial residue.

The tangent translations for the nonabelian Lie algebras $\log \mathbf{SO}(1, 2R-1)$ for the hyperboloids and $\log \mathbf{SO}(2R)$ for the spheres are represented by Yukawa potentials and spherical waves (half-integer index Macdonald and Hankel functions respectively), which arise by 2-sphere spread of the states

$$\begin{array}{rll} R = 2, 3, \dots, \text{ for } \mathcal{Y}^{2R-1}: & \vec{x} \longmapsto \int \frac{d^{2R-1}q}{|\Omega^{2R-1}|} \frac{2}{(\vec{q}^2+m^2)^{R-1}} e^{-i\vec{q}\vec{x}} & = 2\frac{e^{-|m|r}}{r}, \\ \text{for } \Omega^{2R-1}: & \vec{x} \longmapsto \pm\int \frac{d^{2R-1}q}{i|\Omega^{2R-1}|} \frac{2}{(\vec{q}^2 \mp io - m^2)^{R-1}} e^{-i\vec{q}\vec{x}} & = 2\frac{e^{\pm i|m|r}}{r}. \end{array}$$

8.6 Residual Representations of Causal Spacetime

Faithful Hilbert representations of unitary relativity $\mathbf{GL}(\mathbb{C}^2)/\mathbf{U}(2)$ are infinite-dimensional. The hyperisospin $\mathbf{U}(2)$-induced representations of the extended Lorentz group $\mathbf{GL}(\mathbb{C}^2)$ are subrepresentations of the two-sided regular representation of $\mathbf{GL}(\mathbb{C}^2) \times \mathbf{GL}(\mathbb{C}^2)$. The two-sided dichotomous action group is realized in the electroweak standard model (chapter "Gauge Interactions") by external and internal transformations, faithful for $\mathbf{SL}(\mathbb{C}^2) \times \mathbf{U}(2)$ acting on the left-handed isodoublet lepton field, for $\mathbf{SO}_0(1,3) \times \mathbf{SO}(3)$ on the isospin gauge vector field, for $\mathbf{SL}(\mathbb{C}^2) \times \mathbf{U}(1)$ on the right-handed isosinglet lepton field, and for $\mathbf{SO}_0(1,3)\times\{1\}$ on the hypercharge gauge vector field. With the notable exception of the Higgs field, all isospin $\mathbf{SU}(2)$-representations of the standard model fields are isomorphic to subrepresentations of their Lorentz group $\mathbf{SL}(\mathbb{C}^2)$-representations.

The manifold of unitary relativity is parametrizable by the future cone of linear spacetime. It is called causal spacetime

$$\mathbf{GL}(\mathbb{C}^2)/\mathbf{U}(2)\cong\mathbf{D}(1)\times\mathbf{SO}_0(1,3)/\mathbf{SO}(3)\cong\mathcal{D}^4\cong\mathbb{R}^4_+.$$

The acting product group $\mathbf{D}(1)\times\mathbf{SO}_0(1,3)$ contains, in addition to the Lorentz group, a dilation group $\mathbf{D}(1)$, which, in a relativistic framework, is related to the causal group for strictly positive "eigentime" $\sqrt{\vartheta(x_0)\vartheta(x^2)x^2}\in\mathbf{D}(1)$.

The induced $\mathbf{GL}(\mathbb{C}^2)$-representations act on mappings $w:\mathbb{R}^4_+\longrightarrow W$ that relate future spacetime points to hyperisospin $\mathbf{U}(2)$-orbits in a vector space W. The complex functions of the causal cone

$$\mathbb{R}^4_+\ni x=\vartheta(x_0)\vartheta(x^2)x\longmapsto\vartheta(x_0)\vartheta(x^2)f(x)\in\mathbb{C},$$

are the coefficients of unitary relativity with all representations of $\mathbf{GL}(\mathbb{C}^2)$ that contain trivial hyperisospin $\mathbf{U}(2)$-representations.

8.6.1 Harmonic Analysis of the Causal Cartan Plane

Causal spacetime $\mathcal{D}^4$ has real rank 2 as dimension of a Cartan plane $\mathcal{D}^2=\mathbf{D}(1)\times\mathbf{SO}_0(1,1)$ as maximal noncompact abelian group, which is the diagonal part in its Cartan decomposition

$$\begin{aligned}\mathcal{D}^4\ni x=\begin{pmatrix}x_0+x_3&x_1-ix_2\\x_1+ix_2&x_0-x_3\end{pmatrix}&=u(\tfrac{\vec{x}}{r})\circ\begin{pmatrix}x_0+r&0\\0&x_0-r\end{pmatrix}\circ u^\star(\tfrac{\vec{x}}{r}),\\ \Omega^2\ni\tfrac{\vec{x}}{r}\longmapsto u(\tfrac{\vec{x}}{r})&\in\mathbf{SU}(2),\\ \begin{pmatrix}x_0+r&0\\0&x_0-r\end{pmatrix}&=\sqrt{x_0^2-r^2}\begin{pmatrix}\sqrt{\frac{x_0+r}{x_0-r}}&0\\0&\sqrt{\frac{x_0-r}{x_0+r}}\end{pmatrix}\in\mathbf{D}(1)\times\mathbf{SO}_0(1,1).\end{aligned}$$

The parametrization of the group by future cone translations differs from the exponential Lie algebra parametrization, $\mathbf{D}(1)\ni e^{\psi_0}=\vartheta(x_0)x_0$. The neutral group element of $\mathcal{D}^2$ distinguishes a hyberboloid that can be used to introduce an intrinsic length or mass scale $|m|$ for translations $\underline{x}$ with a length dimension

$$\mathcal{D}^2\ni e^\psi=m\underline{x}\Rightarrow\left\{\begin{aligned}\psi&=\psi_0\mathbf{1}_2+\psi_3\sigma_3,\ \tanh\psi_3=\tfrac{x_3}{x_0},\\ \psi_0&=\log|m\underline{x}|,\ |x|=\sqrt{x^2},\\ \psi_0=0&\iff m^2\underline{x}^2=1.\end{aligned}\right.$$

The irreducible unitary representations of the product group are:

$$\begin{aligned}\mathbf{D}(1)\times\mathbf{SO}_0(1,1)&\longmapsto\mathbf{U}(1),\\ e^{\psi_0\mathbf{1}_2+\psi_3\sigma_3}=\mathbf{1}_2x_0+\sigma_3x_3&\longmapsto e^{i(q_0\psi_0+q_3\psi_3)}=|x|^{iq_0}\left(\tfrac{x_0+x_3}{x_0-x_3}\right)^{iq_3}\\ \text{with}&\ \ (q_0,q_3)\in\mathbb{R}^2.\end{aligned}$$

The characteristic function and the defining translation parametrization of the causal Cartan line and plane can be written with advanced energy-momentum measures with a pole at a trivial invariant $q = 0$ and $q^2 = 0$:

$$\text{for } \mathbf{D}(1) \cong \mathbf{SO}_0(1,1): \quad \left\{ \begin{array}{lcl} \vartheta(x_0) & = & \int \frac{dq_0}{2i\pi} \frac{1}{q_0 - io} e^{iq_0 x_0}, \\ \vartheta(x_0)x_0 & = & -\int \frac{dq_0}{2\pi} \frac{1}{(q_0-io)^2} e^{iq_0 x_0}, \end{array} \right.$$

$$\text{for } \mathbf{D}(1) \times \mathbf{SO}_0(1,1): \quad \left\{ \begin{array}{lcl} \vartheta(x_0)\vartheta(x^2)\pi & = & -\int \frac{d^2q}{2\pi} \frac{1}{(q-io)^2} e^{iqx}, \\ \vartheta(x_0)\vartheta(x^2)x\pi & = & \int \frac{d^2q}{2\pi} \frac{2iq}{[(q-io)^2]^2} e^{iqx}, \end{array} \right.$$
$$\text{with } x^2 = x_0^2 - x_3^2,\ (q-io)^2 = (q_0 - io)^2 - q_3^2.$$

The Fourier transform of an energy-momentum function holomorphic in the lower complex energy q_0 plane is valued in the future cone, i.e., supported by causal line and plane. In the Lorentz compatible translation parametrization, the harmonic analysis of the Cartan plane uses $\mathbf{SO}_0(1,1)$-invariants $q^2 = m^2$ as energy-momentum singularities:

$$\mathbf{D}(1) \cong \mathbf{SO}_0(1,1): \quad \left\{ \begin{array}{rl} \int \frac{dq_0}{2i\pi} \frac{1}{q_0 - io - m} e^{iq_0 x_0} & = \vartheta(x_0) e^{imx_0}, \\ \int \frac{dq_0}{2\pi} \frac{iq}{-(q_0-io)^2 + m^2} e^{iq_0 x_0} & = \vartheta(x_0) \cos mx_0, \\ \int \frac{dq_0}{2\pi} \frac{1}{-(q_0-io)^2 + m^2} e^{iq_0 x_0} & = \vartheta(x_0) \frac{\sin mx_0}{m}, \end{array} \right.$$

$$\mathbf{D}(1) \times \mathbf{SO}_0(1,1): \quad \left\{ \begin{array}{rl} \int \frac{d^2q}{2\pi} \frac{1}{-(q-io)^2 + m^2} e^{iqx} & = \vartheta(x_0)\vartheta(x^2)\pi \mathcal{J}_0(|mx|), \\ \int \frac{d^2q}{2\pi} \frac{2iq}{[-(q-io)^2 + m^2]^2} e^{iqx} & = \vartheta(x_0)\vartheta(x^2)x\pi \mathcal{J}_0(|mx|). \end{array} \right.$$

The line $\mathbf{D}(1)$ has rank 1 with one representation characterizing invariant pole in self-dual form $q^2 = m^2$. The two continuous invariants for the rank-2 causal plane are implemented, in a residual representation, by two poles in the complex energy-momentum plane

$$\mathbf{D}(1) \times \mathbf{SO}_0(1,1) \ni \vartheta(x_0)\vartheta(x^2)x \longmapsto \int \frac{d^2q}{2i\pi} \frac{2q}{[-(q-io)^2 + m_\kappa^2][(q-io)^2 - m_0^2]} e^{iqx}.$$

Below, the product of the two energy-momentum poles will be related to the product structure of the represented group. The two poles can be taken as the endpoints of a finite $\mathbf{SO}_0(1,1)$-invariant singularity line $q^2 \in \{\zeta m_0^2 + (1 - \zeta)m_\kappa^2 \mid \zeta \in [0,1]\}$, which gives the Fourier transformation of energy-momentum logarithms,

$$\begin{aligned} \int \frac{d^2q}{2\pi} \frac{2iq}{[-(q-io)^2 + m_\kappa^2][-(q-io)^2 + m_0^2]} e^{iqx} &= \int \frac{d^2q}{2\pi} \int_0^1 d\zeta \frac{2iq}{[-(q-io)^2 + \zeta m_0^2 + (1-\zeta)m_\kappa^2]^2} e^{iqx} \\ &= x \int \frac{d^2q}{2\pi(m_0^2 - m_\kappa^2)} \log \frac{(q-io)^2 - m_0^2}{(q-io)^2 - m_\kappa^2} e^{iqx} \\ &= \int \frac{d^2q}{2\pi} \int_{m_\kappa^2}^{m_0^2} \frac{dM^2}{m_0^2 - m_\kappa^2} \frac{2iq}{[-(q-io)^2 + M^2]^2} e^{iqx} \\ &= \vartheta(x_0)\vartheta(x^2)x \int_{m_\kappa^2}^{m_0^2} \frac{dM^2}{m_0^2 - m_\kappa^2} \pi \mathcal{J}_0(|Mx|) \\ &= \vartheta(x_0)\vartheta(x^2) \frac{x}{2} \frac{\partial}{\partial \frac{x^2}{4}} \frac{\pi \mathcal{J}_0(|m_0 x|) - \pi \mathcal{J}_0(|m_\kappa x|)}{m_0^2 - m_\kappa^2}. \end{aligned}$$

In the not-Lorentz compatible direct product form the residual representations of the group for Cartan spacetime read

$$\begin{array}{rl}\mathbf{D}(1)\times\mathbf{SO}_0(1,1)\ni(\vartheta(x_0)x_0,x_3)\longmapsto & \int\frac{dq_0dq_3}{(2\pi)^2}\frac{2iq_0}{(q_3^2+m_\kappa^2)[-(q_0-io)^2+m_0^2]}e^{i(q_0x_0-q_3x_3)}\\ & =\vartheta(x_0)\cos m_0x_0\,\frac{e^{-|m_\kappa x_3|}}{|m_\kappa|}\end{array}$$

8.6.2 Harmonic Analysis of Causal Spacetimes

A causal Cartan plane is a maximal noncompact abelian group in the product of the causal group for eigentime and the Lorentz group for general nontrivial position dimension:

$$\mathbf{D}(1)\times\mathbf{SO}_0(1,1)\subseteq\mathbf{D}(1)\times\mathbf{SO}_0(1,s),\quad s=1,2,\dots.$$

The Cartan decomposition [5] ($G=KAK$) of the generalized Lorentz group uses a maximal compact subgroup and a maximal noncompact abelian subgroup:

$$\mathbf{SO}_0(1,s)=\mathbf{SO}(s)\circ\mathbf{SO}_0(1,1)\circ\mathbf{SO}(s).$$

As familiar from Cartan's B and D-series, the orthogonal groups come in two basically different types, those for odd and those for even dimensions. The causal scalar functions and the defining tangent vector parametrizations with $\vartheta(x)=\vartheta(x_0)\vartheta(x^2)$ have the harmonic analysis with $\mathbb{R}^{1+s}$-characters and an advanced energy-momentum measure $(q-io)^2=(q_0-io)^2-\vec{q}^2$:

$$\mathbf{SO}_0(1,s):\left\{\begin{array}{rcl}\vartheta(x)\begin{pmatrix}\frac{x}{|x|}\\|x|\end{pmatrix}&=&\int\frac{d^{1+2R}q}{|\Omega^{1+2R}|}\frac{1}{[-(q-io)^2]^{1+R}}\begin{pmatrix}iq\\1\end{pmatrix}e^{iqx},\\ && 1+s=1-2R=1,3,\dots,\\ \vartheta(x)\begin{pmatrix}1\\x\end{pmatrix}\pi&=&\int\frac{d^{2R}q}{|\Omega^{2R-1}|}\frac{1}{[-(q-io)^2]^{R}}\begin{pmatrix}1\\\frac{2iqR}{-(q-io)^2}\end{pmatrix}e^{iqx},\\ && 1+s=2R=2,4,\dots.\end{array}\right.$$

The spherical degrees of freedom show up in the order of the pole and in the normalization of the integration $d^{1+s}q$ with measures of the n-dimensional unit spheres $|\Omega^n|=\frac{2\pi^{\frac{1+n}{2}}}{\Gamma(\frac{1+n}{2})}$: for odd spacetime dimensions $1+2R$, the equally dimensioned sphere $\Omega^{1+s}=\Omega^{1+2R}$ is used in contrast to even dimensions $2R$ with Ω^{2R-1} for the position degrees of freedom. The causal spacetime coefficients with one invariant are

$$\begin{array}{rcl}\int\frac{d^{1+2R}q}{|\Omega^{1+2R}|}\frac{1}{[-(q-io)^2+m^2]^{1+R}}\begin{pmatrix}iq\\1\end{pmatrix}e^{iqx}&=&\vartheta(x)\begin{pmatrix}\frac{x}{|x|}\cos m|x|\\\frac{\sin m|x|}{m}\end{pmatrix},\\ \int\frac{d^{2R}q}{|\Omega^{2R-1}|}\frac{1}{[-(q-io)^2+m^2]^{R}}\begin{pmatrix}1\\\frac{2iqR}{-(q-io)^2+m^2}\end{pmatrix}e^{iqx}&=&\vartheta(x)\begin{pmatrix}1\\x\end{pmatrix}\pi\mathcal{J}_0(|mx|).\end{array}$$

The classes of even-dimensional Lorentz groups with respect to their maximal compact rotation groups constitute a hyperboloid $\mathcal{Y}^{2R-1}$ (rotation relativity) as position manifold in causal spacetime manifolds with the causal group $\mathbf{D}(1)$ ("eigentime"). A causal manifold can be Lorentz compatibly parametrized by the open future cone of corresponding flat Poincaré spacetime $\mathbf{SO}_0(1,2R-1) \vec{\times} \mathbb{R}^{2R}$:

$$\begin{aligned}\mathbf{D}(1) \times \mathbf{SO}_0(1,2R-1)/\mathbf{SO}(2R-1) &\cong \mathcal{D}^{2R} = \mathbf{D}(1) \times \mathcal{Y}^{2R-1} \\ &\cong \mathbb{R}^{2R}_+ = \{x \in \mathbb{R}^{2R} \mid x = \vartheta(x)x\}.\end{aligned}$$

Even-dimensional spacetime $\mathcal{D}^{2R}$ with two continuous invariants for a Cartan plane $\mathcal{D}^2$ (real rank 2) is represented as the Fourier transformed product of two energy-momentum distributions, one with a simple pole and the other with a pole of order $R = 1, 2, \ldots$:

$$\begin{aligned}\mathcal{D}^{2R} \ni \vartheta(x)x &\longmapsto \int \frac{d^{2R}q}{i|\Omega^{2R-1}|} \frac{2q}{[-(q-io)^2+m_\kappa^2]^R[(q-io)^2-m_0^2]} e^{iqx} \\ &= \int \frac{d^{2R}q}{|\Omega^{2R-1}|} \int_0^1 d\zeta \frac{2iqR(1-\zeta)^{R-1}}{[-(q-io)^2+\zeta m_0^2+(1-\zeta)m_\kappa^2]^{1+R}} e^{iqx} \\ &= x \int \frac{d^{2R}q}{|\Omega^{2R-1}|} \int_0^1 d\zeta \frac{(1-\zeta)^{R-1}}{[-(q-io)^2+\zeta m_0^2+(1-\zeta)m_\kappa^2]^{R}} e^{iqx}.\end{aligned}$$

The simple pole embeds the representations of time translations $\mathbf{D}(1) \cong \mathbb{R}$, the order-$R$ pole those of the Lorentz group $\mathbf{SO}_0(1,2R-1)$ for hyperbolic position $\mathcal{Y}^{2R-1} \cong \mathbb{R}^{2R-1}$ (more below). Together with $R-1$ discrete invariants (imaginary rank) of the orthogonal group $\mathbf{SO}(2R-1)$ for a Cartan torus $\mathbf{SO}(2)^{R-1}$ the acting group has rank $1+R$.

Unitary relativity $\mathcal{D}^4 = \mathbf{D}(1) \times \mathcal{Y}^3$ as the minimal nonabelian case with three space dimensions $s = 3$ has real rank 2 and imaginary rank 1. The $\mathbf{GL}(\mathbb{R}^4)$-invariant measure $\frac{d^4q}{[(q-io)^2]^2}$ has a dipole [4]. The $\mathcal{D}^4$-representation coefficients with two invariants are:

$$\begin{aligned}\mathbf{GL}(\mathbb{C}^2)/\mathbf{U}(2) \cong \mathbb{R}^4_+ \ni \vartheta(x)x &\longmapsto \int \frac{d^4q}{2i\pi^2} \frac{2q}{[-(q-io)^2+m_\kappa^2]^2[(q-io)^2-m_0^2]} e^{iqx} \\ &= -x \int \frac{d^4q}{2\pi^2(m_0^2-m_\kappa^2)^2} [\log \frac{(q-io)^2-m_0^2}{(q-io)^2-m_\kappa^2} + \frac{m_0^2-m_\kappa^2}{(q-io)^2-m_\kappa^2}] e^{iqx} \\ &= \vartheta(x) \frac{x}{2} \frac{\partial}{\partial \frac{x^2}{4}} \left[\frac{\partial}{\partial \frac{x^2}{4}} \frac{\pi \mathcal{J}_0(|m_0 x|) - \pi \mathcal{J}_0(|m_\kappa x|)}{(m_0^2-m_\kappa^2)^2} - \frac{\pi \mathcal{J}_0(|m_\kappa x|)}{m_0^2-m_\kappa^2} \right].\end{aligned}$$

8.7 Time and Position Subgroup Representations

A group representation represents all subgroups. The projections of representations of unitary relativity $\mathbf{GL}(\mathbb{C}^2)/\mathbf{U}(2)$, i.e., of causal spacetime $\mathcal{D}^4 = \mathbf{D}(1) \times \mathcal{Y}^3$ and $\mathbf{D}(1) \times \mathbf{SO}_0(1,3)$, to those of the factors causal group for eigentime and hyperbolic position lead respectively to representations for free states und to interactions. Free particles are related to representations of the causal group $\mathbf{D}(1) \cong \mathbb{R}$ (translations), Lorentz compatibly embedded in representations of causal spacetime $\mathcal{D}^4$, whereas interactions are related to embedded representations of hyperbolic position $\mathcal{Y}^3$, a symmetric space of the Lorentz group $\mathbf{SO}_0(1,3)$. In the future cone, the action of the causal group $\mathbf{D}(1)$ may be called 'hyperbolic hopping', from hyperboloid to hyperboloid, and the action of the dilative Lorentz group $\mathbf{SO}_0(1,1)$ 'hyperbolic stretching', inside one hyperboloid.

8.7.1 Projection to and Embedding of Particles and Interactions

The representation coefficients of the acting product group for even-dimensional causal spacetime $\mathcal{D}^{2R}$, $R = 1, 2, \dots$

$$\begin{array}{l}\mathbf{D}(1) \times \mathbf{SO}_0(1, 2R-1) \text{ for } \left\{\begin{array}{ll} \mathbf{D}(1) \times \mathcal{Y}^{2R-1} & = \mathcal{D}^{2R}, \\ \mathcal{M}(\mathbb{R}^{2R}_+) * L^\infty(\mathbb{R}^{2R}_+) & = L^\infty(\mathbb{R}^{2R}_+) \end{array}\right. \\ \text{with } \mathbf{D}(1) \cong \mathbb{R}, \ \ \mathcal{Y}^{2R-1} \cong \mathbf{SO}_0(1, 2R-1)/\mathbf{SO}(2R-1) \cong \mathbb{R}^{2R-1}\end{array}$$

are the convolution product of a causally embedded coefficient $d^{2R}_{m^2}$ for the Lorentz group $\mathbf{SO}_0(1, 2R-1)$ with a bounded function $|\mathcal{J}_0(r)| \le \mathcal{J}_0(0) = 1$ and a Lorentz compatibly embedded Radon measure ω_{m^2} for the causal group (eigentime) $\mathbf{D}(1)$

$$\begin{pmatrix} d^{2R}_{m^2}(x) \\ \omega_{m^2}(x) \\ d^{2R}_{m^2} * \omega_{m^2}(x) \end{pmatrix} \sim \begin{pmatrix} 1 \\ x\Gamma(R)(\frac{\partial}{\partial \frac{x^2}{4}})^R \\ x \end{pmatrix} \vartheta(x)\pi\mathcal{J}_0(|mx|) \ = \int \frac{d^{2R}q}{|\Omega^{2R-1}|} \begin{pmatrix} \frac{1}{[-(q-io)^2+m^2]^R} \\ \frac{2iq}{-(q-io)^2+m^2} \\ \frac{2iqR}{[-(q-io)^2+m^2]^{1+R}} \end{pmatrix} e^{iqx}$$

with the explicit form of the Radon distribution for $N = 0, 1, \dots,$

$$(\frac{\partial}{\partial \frac{x^2}{4}})^{1+N}\vartheta(x^2)\mathcal{J}_0(|mx|) = \sum_{k=-N}^{0} \frac{(m^2)^k}{k!}\delta^{(N+k)}(-\frac{x^2}{4}) + \vartheta(x^2)(\frac{\partial}{\partial \frac{x^2}{4}})^{1+N}\mathcal{J}_0(|mx|).$$

The pointwise multiplied harmonic components come with a pole of order R, 1 and $R+1$.

The Lorentz compatible (generalized) functions on causal spacetime can be projected to (generalized) functions on time translations and on hyperbolic position,

$$\text{for } \mathcal{D}^{2R}: \ \ \vartheta(x)\mu(x) \longmapsto \left\{\begin{array}{lll} \int d^{2R-1}x & \vartheta(x)\mu(x) & \text{for } \mathbf{D}(1), \\ \int dx_0 & \vartheta(x)\mu(x) & \text{for } \mathcal{Y}^{2R-1}. \end{array}\right.$$

The time projections by integration over position, i.e., for trivial momenta $\vec{q} = 0$,

$$\begin{aligned}\text{for } \mathbf{D}(1): \ & \int \frac{|\Omega^{2R-1}| d^{2R-1}x}{(2\pi)^{2R}} \begin{pmatrix} 1 \\ x\Gamma(R)(\frac{\partial}{\partial \frac{x^2}{4}})^R \end{pmatrix} \vartheta(x)\pi\mathcal{J}_0(|mx|) \\ & = \int \frac{dq_0}{2\pi} \begin{pmatrix} \frac{1}{[-(q_0-io)^2+m^2]^R} \\ \frac{2iq_0}{-(q_0-io)^2+m^2} \end{pmatrix} e^{iq_0x_0} = \vartheta(x_0) \begin{pmatrix} \frac{1}{\Gamma(R)}(-\frac{\partial}{\partial m^2})^{R-1}\frac{\sin mx_0}{m} \\ 2\cos mx_0 \end{pmatrix}\end{aligned}$$

contain in the lower component a positive-type function $\cos \in L^\infty(\mathbb{R})_+$ for time translations with real invariants $|q_0| = \pm|m|$,

$$\begin{array}{lll}\mathbf{D}(1) \ni \vartheta(x_0)x_0 \longmapsto & 2\vartheta(x_0)\cos mx_0 & = \int \frac{dq_0}{2\pi} \frac{2iq_0}{-(q_0-io)^2+m^2} e^{iq_0x_0} \\ \hookrightarrow & x\Gamma(R)(\frac{\partial}{\partial \frac{x^2}{4}})^R\vartheta(x)\pi\mathcal{J}_0(|mx|) & = \int \frac{d^{2R}q}{|\Omega^{2R-1}|} \frac{2iq}{-(q-io)^2+m^2} e^{iqx}\end{array}$$

which can be used for free particles as translation invariant $q^2 = m^2$ in a Poincaré group representation. For $R \ge 2$, the upper component displays matrix elements of indefinite unitary faithful translation representations [3] (halfinteger index spherical Bessel functions [13]),

$$\begin{aligned}(-\frac{\partial}{\partial m^2})^{R-1}\frac{\sin mx_0}{m} &= (-\frac{\partial}{\partial m^2})^{R-2}\frac{\sin mx_0 - mx_0\cos mx_0}{2m^3}\\ &= x_0^{2R-1}(-\frac{\partial}{\partial t^2})^{R-1}\frac{\sin t}{t},\ t=|m|x_0.\end{aligned}$$

The projections on the position hyperboloid with the $\mathbf{SO}_0(1,2R-1)$-coefficients by integration over time, i.e., for trivial energy $q_0=0$,

$$\begin{aligned}\text{for } \mathcal{Y}^{2R-1}:\ &\int \frac{dx_0}{2\pi}\begin{pmatrix}1\\ x\Gamma(R)(\frac{\partial}{\partial\frac{x^2}{4}})^R\end{pmatrix}\vartheta(x)\pi\mathcal{J}_0(|mx|)\\ &= \int \frac{d^{2R-1}q}{|\Omega^{2R-1}|}\begin{pmatrix}\frac{1}{(\vec q^2+m^2)^R}\\ \frac{2i\vec q}{\vec q^2+m^2}\end{pmatrix}e^{-i\vec q\vec x} = \begin{pmatrix}\frac{e^{-|m|r}}{2|m|}\\ \Gamma(R)(\frac{4m^2}{r^2})^{R-1}|m|(-\frac{\partial}{\partial m^2})^{R-1}\frac{\vec x}{r}\frac{e^{-|m|r}}{|m|}\end{pmatrix},\end{aligned}$$

give, in the upper component, a positive-type function $\exp\in L^\infty(\mathcal{Y}^{2R-1})_+$ for hyperbolic position with an imaginary invariant $|\vec q| = \pm i|m|$,

$$\begin{aligned}\mathcal{Y}^{2R-1}\ni\vec x\longmapsto\ &\frac{e^{-|m|r}}{2|m|} &&= \int\frac{d^{2R-1}q}{\Omega^{2R-1}|}\frac{1}{(\vec q^2+m^2)^R}e^{-i\vec q\vec x}\\ \hookrightarrow\ &\vartheta(x)\pi\mathcal{J}_0(|mx|) &&= \int\frac{d^{2R}q}{\Omega^{2R-1}|}\frac{1}{[-(q-io)^2+m^2]^R}e^{iqx}.\end{aligned}$$

The lower component involves generalized Yukawa forces (halfinteger index hyperbolic Macdonald functions [13]), i.e., exponential forces $\epsilon(x)e^{-|mx|}$ for Cartan spacetime $R=1$ and Yukawa forces proper $\frac{\vec x}{r}\frac{1+|m|r}{2r^2}e^{-|m|r}$ for Minkowski spacetime $R=2$ with the inverse invariant of the $\mathbf{SO}_0(1,1)$-representations as characteristic range $\frac{1}{|m|}$,

$$\begin{aligned}|m|\vec x(\frac{m^2}{r^2})^{R-1}(-\frac{\partial}{\partial m^2})^{R-1}\frac{e^{-|m|r}}{|m|r} &= |m|^3\vec x(\frac{m^2}{r^2})^{R-2}(-\frac{\partial}{\partial m^2})^{R-2}\frac{1+|m|r}{2|m|^3r^3}e^{-|m|r}\\ &= |m|^{2R-1}\vec x(-\frac{\partial}{\partial\rho^2})^{R-1}\frac{e^{-\rho}}{\rho},\quad \rho=|m|r.\end{aligned}$$

The harmonic components of cyclic representations of nonlinear 3-dimensional position as hyperboloid $\mathcal{Y}^3\cong\mathbf{SO}_0(1,3)/\mathbf{SO}(3)$ are positive momentum measures with dipoles $R=2$. The $\mathcal{Y}^3$-functions arise as bound state wave functions of the nonrelativistic hydrogen atom, e.g., the positive-type function $\vec x\longmapsto e^{-r}=\int\frac{d^3q}{\pi^2}\frac{1}{(\vec q^2+1)^2}e^{i\vec q\vec x}$ as Schrödinger function for the ground state.

The imaginary eigenvalues in the dipole $\frac{1}{[-(q-io)^2+m^2]^2}$ for the representation of hyperbolic position $\mathcal{Y}^3$ in unitary relativity $\mathcal{D}^4$ cannot be used for translations. The mass $\vec q^2=-m^2$ for $q_0^2=0$ characterizes an interaction, not a free particle.

The projections of representations of Cartan spacetime $\mathcal{D}^2$ on representations of the causal group $\mathbf{D}(1)$ and of position $\mathcal{Y}^1$ are

$$\begin{aligned}\text{time:}\quad &\mathbb{R}_+\ni\vartheta(t)t\longmapsto &&\int\frac{dz}{2\pi}\int\frac{d^2q}{2i\pi[-(q-io)^2+m_\kappa^2]}\frac{2q}{(q-io)^2-m_0^2}e^{iqx}\\ &&&=\vartheta(t)2\frac{\cos m_0t-\cos m_\kappa t}{m_0^2-m_\kappa^2},\\ \text{position:}\quad &\mathbf{SO}_0(1,1)\cong\mathbb{R}\ni z\longmapsto &&\int\frac{dt}{2\pi}\int\frac{d^2q}{2i\pi[-(q-io)^2+m_\kappa^2]}\frac{2q}{(q-io)^2-m_0^2}e^{iqx}\\ &&&= -\frac{2}{m_0^2-m_\kappa^2}\epsilon(z)\frac{\partial}{\partial|z|}V(|z|).\end{aligned}$$

The position projection displays exponential interactions

$$V(|z|) = \frac{e^{-|m_\kappa z|}}{|m_\kappa|} - \frac{e^{-|m_0 z|}}{|m_0|}, \quad \frac{\partial}{\partial|z|}V(|z|) = e^{-|m_\kappa z|} - e^{-|m_0 z|}.$$

Correspondingly, 4-dimensional spacetime with a dipole for the Lorentz scalar future measure is projected to representation coefficients of time future and of 3-dimensional hyperbolic position:

$$\begin{array}{rrl} \text{time:} & \mathbb{R}_+ \ni \vartheta(t)t \longmapsto & \int \frac{d^3x}{8\pi^2} \int \frac{d^4q}{2i\pi^2[-(q-io)^2+m_\kappa^2]^2} \frac{2q}{(q-io)^2-m_0^2} e^{iqx} \\ & & = \vartheta(t)\frac{2}{m_0^2-m_\kappa^2}\left(\frac{\cos m_0 t-\cos m_\kappa t}{m_0^2-m_\kappa^2} + \frac{m_\kappa t \sin m_\kappa t}{2m_\kappa^2}\right), \\ \text{position:} & \mathcal{Y}^3 \ni \vec{x} \longmapsto & \int \frac{dt}{2\pi} \int \frac{d^4q}{2i\pi^2[-(q-io)^2+m_\kappa^2]^2} \frac{2q}{(q-io)^2-m_0^2} e^{iqx} \\ & & = \frac{2}{(m_0^2-m_\kappa^2)^2} \frac{\vec{x}}{r}\frac{\partial}{\partial r}V_3(r). \end{array}$$

There arise Yukawa and exponential interactions

$$\begin{array}{rl} V_1(r) & = \frac{e^{-|m_\kappa|r}}{|m_\kappa|} - \frac{e^{-|m_0|r}}{|m_0|} - \frac{m_0^2-m_\kappa^2}{2|m_\kappa|^3}(1+|m_\kappa|r)e^{-|m_\kappa|r} \\ V_3(r) & = \frac{\partial}{\partial r^2}V_1(r) = \frac{e^{-|m_0|r}-e^{-|m_\kappa|r}}{r} - \frac{m_0^2-m_\kappa^2}{2|m_\kappa|}e^{-|m_\kappa|r}. \end{array}$$

An exponential interaction is the 2-sphere spread of a 1-dimensional position representation $\frac{1}{2}e^{-r} = -\frac{\partial}{\partial \frac{r^2}{4}}(1+r)e^{-r}$ with an r-proportional contribution from a dipole.

Bibliography

[1] V. Bargmann, Irreducible representations of the Lorentz group, *Ann. of Math.*, 48 (1947), 568-640.

[2] H. Behnke, F. Sommer, *Theorie der analytischen Funktionen einer komplexen Veränderlichen* (1962), Springer-Verlag, Berlin, Göttingen Heidelberg.

[3] H. Boerner, *Darstellungen von Gruppen* (1955), Springer, Berlin, Göttingen, Heidelberg.

[4] V. Fock, Zur Theorie des Wasserstoffatoms, *Zeitschrift für Physik* 98 (1935), 145-154.

[5] G.B. Folland, *A Course in Abstract Harmonic Analysis* (1995), CRC Press, Boca Raton, Ann Arbor, London, Tokyo.

[6] I.M. Gel'fand, G.E. Shilov, *Generalized Functions I (Properties and Operations)* (1958, English translation 1963), Academic Press, New York and London.

[7] I.M. Gel'fand, M.I. Graev, N.Ya. Vilenkin, *Generalized Functions V (Integral Geometry and Representation Theory)* (1962, English translation 1966), Academic Press, New York and London.

[8] W. Heisenberg, *Einführung in die einheitliche Feldtheorie der Elementarteilchen* (1967), Hirzel, Stuttgart.

[9] S. Helgason, *Groups and Geometric Analysis* (1984), Academic Press, New York. London, Sydney, Tokyo, Toronto, etc.

[10] A. Knapp, Representation Theory of Semisimple Groups (1986), Princeton University Press, Princeton.

[11] T.O. Sherman, Fourier analysis on the sphere, *Trans. Amer. Math. Soc.* **209**, 1-31 (1975)

[12] R.S. Strichartz, Harmonic analysis on hyperboloids, *Journal of Functional Analysis* **12**, 341-383 (1973).

[13] N.Ja. Vilenkin, A.U. Klimyk, *Representations of Lie Groups and Special Functions* (1991), Kluwer Academic Publishers, Dordrecht, Boston, London.

[14] H. Saller, On the non-particle structure of gauge interactions, *Nuovo Cimento* 105A (1992), 455-491.

[15] H. Saller, Realizations of causal manifolds by quantum fields, *International Journal of Theoretical Physics* **36**, 2783-2826 (1997).

[16] H. Saller, Representations of spacetime as unitary operation classes; or against the monoculture of particle fields, *International Journal of Theoretical Physics* **38**, 1697-1733 (1999).

[17] H. Saller, Residual representations of spacetime, *International Journal of Theoretical Physics* **40**, 1209-1248 (2001).

9

SPECTRUM OF SPACETIME

In Wigner's classification, linear spacetime and free particles originate from one operational concept and its representations, from an affine subgroup with Lorentz transformations acting on translations. Why the free particles have the characteristic invariants, i.e., the observed masses m^2, spins J, and, for the additional internal $\mathbf{U}(1)$-operations, charge numbers z, is not explained by classifying the irreducible Hilbert representations of the Poincaré group. The actual spectrum of matter $(m^2, 2J, z) \in \mathbb{R}_+ \times \mathbb{N} \times \mathbb{Z}$ together with the normalization of particles and the coupling constants of interactions has to be understood by additional structures, e.g., by representations of a nonlinear spacetime model.

The multilinear algebra structure of quantum operations involves typical ensembles of representations ("towers of bound states"), which are products of one basic representation, defining the relevant operation group. Characteristic examples are the free states of translations, which are familiar from the equidistant linear spectrum of the harmonic oscillator; representations of time translations $\mathbb{R} \cong \mathbf{D}(1)$; and the bound states of the nonrelativistic hydrogen atom as representations of hyperbolic 3-dimensional position $\mathcal{Y}^3 \cong \mathbf{SO}_0(1,3)/\mathbf{SO}(3) \cong \mathbb{R}^3$ with the inverse squared energy spectrum.

A pointwise product of positive-type functions $d \in L^\infty(G)_+$ of a real Lie group is a positive-type function for the product representation,

$$d_1 \cdot d_2(g) = \langle a_1|D_1(g)|a_1\rangle\langle a_2|D_2(g)|a_2\rangle = \langle a_1, a_2|D_1 \otimes D_2(g)|a_1, a_2\rangle.$$

For the harmonic components, one has to use the convolution $\tilde{d}_1 * \tilde{d}_2$.

The characters (representation classes, dual group) as eigenvalues of the additive group $\check{\mathbb{R}}^d$—energies for time translations $\mathbb{R}$ and momenta for position translations $\mathbb{R}^3$—give rise to convolution algebras of the corresponding distributions (functions, measures). Nonlinear spacetime $\mathbf{D}(2) \cong \mathbf{GL}(\mathbb{C}^2)/\mathbf{U}(2)$ as a homogeneous space of the extended Lorentz group $\mathbf{GL}(\mathbb{C}^2)$ with tangent Minkowski translations $x \in \mathbb{R}^4$ can be represented by residues of Fourier transformed energy–momentum $q \in \check{\mathbb{R}}^4$ functions (chapter "Residual Spacetime Representations"). The representation-characterizing invariants arise as poles in the complex energy and momentum plane.

Product representations come with the product of representation coefficients, i.e., in a residual formulation with the convolution $*$ of (energy–)momentum distributions. The convolution itself picks up a residue,

$$* \sim \delta(q_1 + q_2 - q) \sim \mathrm{res}_{q_1+q_2=q}.$$

The convolution adds (energy–)momenta of singularity manifolds as imaginary and real eigenvalues for compact and noncompact representation invariants. The Radon (energy–)momentum measures are a convolution algebra, which reflects the pointwise multiplication property of the essentially bounded function classes:

$$\mathcal{M}(\check{\mathbb{R}}^n) * \mathcal{M}(\check{\mathbb{R}}^n) \subseteq \mathcal{M}(\check{\mathbb{R}}^n), \quad L^\infty(\mathbb{R}^n) \cdot L^\infty(\mathbb{R}^n) \subseteq L^\infty(\mathbb{R}^n).$$

In the Feynman integrals of special relativistic quantum field theory as convolutions of energy–momentum distributions, the on-shell parts for translation representations give product representation coefficients of the Poincaré group, i.e., energy–momentum distributions for free states (multiparticle measures, below). The off-shell interaction contributions are not convolutable. This is the origin of the "divergence" problem in quantum field theories with interactions. With respect to Poincaré group representations, the convolution of Feynman propagators makes no sense.

In this chapter the convolution structure of time, position, and spacetime representations is considered. In the end an attempt is made to determine, from eigenvalue equations, the spectrum of spacetime $\mathbf{D}(2) \cong \mathbb{R}^4_+ = \mathbf{D}(1) \times \mathcal{Y}^3$, i.e., invariant masses and normalizations of energy-momentum poles for the representations of the causal group, Lorentz compatibly embedded into nonlinear Minkowski spacetime. Perhaps one can characterize this as an attempt to find a Lorentz compatible solution of the bound state problem in a potential $V_3(r)$ with exponential and Yukawa contributions which has been given above (chapter "Residual Spacetime Representations") as the projection of the representation of nonlinear spacetime to representations of hyperbolic 3-position.

Only some illustrations of an explicit calculation are given for the determination of the particle properties as spectrum of a homogeneous spacetime model. If the proposal for the solution of such a difficult problem really goes in the right direction, both the qualitative foundations and the concrete realization require more work.

9.1 Convolutions for Abelian Groups

Product representations of translations $\mathbb{R}^n$ with sum and difference of the energy–momentum invariants arise as the pointwise product of positive-type functions $L^\infty(\mathbb{R}^n)_+$ or the convolution of positive energy-momentum Radon measures $\mathcal{M}(\check{\mathbb{R}}^n)_+$.

The simplest case is given for 1-dimensional translations, e.g., for time translations $t \in \mathbb{R}$ with an addition of the energy invariants in the irreducible and self-dual representation,

$$\mathcal{C}_b(\mathbb{R})_+ \cdot \mathcal{C}_b(\mathbb{R})_+ = \mathcal{C}_b(\mathbb{R})_+ \quad \begin{cases} e^{im_1t} \cdot e^{im_2t} &= e^{im_+t}, \\ 2\cos m_1t \cdot 2\cos m_2t &= 2\cos m_+t + 2\cos m_-t \end{cases}$$
$$\text{with} \quad m_\pm = m_1 \pm m_2,$$
$$\mathcal{M}(\check{\mathbb{R}})_+ * \mathcal{M}(\check{\mathbb{R}})_+ = \mathcal{M}(\check{\mathbb{R}})_+ \quad \begin{cases} \delta(q-m_1) * \delta(q-m_2) = \delta(q-m_+) \\ 2|q|\delta(q^2-m_1^2) * 2|q|\delta(q^2-m_2^2) \\ = 2|q|\delta(q^2-m_+^2) + 2|q|\delta(q^2-m_-^2). \end{cases}$$

9.1.1 Convolutions with Linear Invariants

The residual product for the two causal function algebras, conjugate and orthogonal to each other, and the Dirac convolution algebra is summarized with the residually normalized representation functions and the integration contours:

$$\vartheta(\pm t)e^{imt} = \pm \int \frac{dq}{2i\pi} \frac{1}{q \mp io - m} e^{iqt}.$$

causal time $\mathbf{D}(1)$ and energies $\mathbb{R}$
$(\overset{1}{*}, q) = (\pm \frac{*}{2i\pi}, q \mp io)$ causal, orthogonal
$\frac{1}{q-m_1} \overset{1}{*} \frac{1}{q-m_2} = \frac{1}{q-m_+}$
$\delta(q-m_1) * \delta(q-m_2) = \delta(q-m_+)$

The normalization factor for the residual product is the 1-sphere measure as used in the residue,

$$\oint \frac{dq}{2i\pi} = \text{res}, \quad \frac{*}{2\pi} \cong \frac{1}{|\Omega^1|}\delta(q_1 + q_2 - q).$$

There is the more general convolution

$$\frac{\Gamma(1+\nu_1)}{(q-m_1)^{1+\nu_1}} \overset{1}{*} \frac{\Gamma(1+\nu_2)}{(q-m_2)^{1+\nu_2}} = \frac{\Gamma(1+\nu_1+\nu_2)}{[q-m_+])^{1+\nu_1+\nu_2}},$$

which generalizes the integer–power derivatives $(\frac{\partial}{\partial m})^N$ for nontrivial nildimensions $N = 1, 2, \ldots$ to real powers $\nu \in \mathbb{R}$ wherever the Γ-functions are defined.

9.1.2 Convolutions with Self-Dual Invariants

The causal distributions with compact dual invariants

$$\pm \frac{1}{i\pi} \frac{q}{(q \mp io)^2 - m^2} = |m|\delta(q^2-m^2) \pm \frac{1}{i\pi} \frac{q}{q_{\mathrm{P}}^2 - m^2} = \pm \frac{1}{2i\pi}\left(\frac{1}{q \mp io - |m|} + \frac{1}{q \mp io + |m|}\right)$$

keep the property of constituting orthogonal convolution algebras, conjugate to each other:

$$\vartheta(\pm t) 2\cos mt = \pm \int \frac{dq}{i\pi} \frac{q}{(q \mp io)^2 - m^2} e^{iqt}.$$

causal time $\mathbf{D}(1)$ and energies $\mathbb{R}$
$(\overset{1}{*}, q^2) = (\pm \frac{*}{2i\pi}, (q \mp io)^2)$ causal, orthogonal
$\frac{2q}{q^2-m_1^2} \overset{1}{*} \frac{2q}{q^2-m_2^2} = \frac{2q}{q^2-m_+^2} + \frac{2q}{q^2-m_-^2}$

Since the Feynman energy distributions combine advanced and retarded distributions,

$$\pm\frac{1}{i\pi}\frac{|m|}{q^2\mp io-m^2} = |m|\delta(q^2-m^2)\pm\frac{1}{i\pi}\frac{|m|}{q_{\mathrm{P}}^2-m^2} = \pm\frac{1}{2i\pi}\left(\frac{1}{q\mp io-|m|}-\frac{1}{q\pm io+|m|}\right),$$

they constitute convolution algebras, conjugate to each other, however not orthogonal, $e^{+i|m_1 t|}\cdot e^{-i|m_2 t|}\neq 0$,

$$e^{\pm i|mt|} = \pm\int\frac{dq}{i\pi}\frac{|m|}{q^2\mp io-m^2}e^{iqt}.$$

bicone time $\mathbb{R}_+ \uplus \mathbb{R}_-$ and energies $\mathbb{R}$
$(\overset{1}{*},q^2)=(\pm\frac{*}{2i\pi},q^2\mp io)$ Feynman, not orthogonal
$\frac{2\vert m_1\vert}{q^2-m_1^2}\overset{1}{*}\frac{2\vert m_2\vert}{q^2-m_2^2}=\frac{2\vert m_+\vert}{q^2-m_+^2}$

The faithful Hilbert representations of $\mathcal{Y}^1\cong\mathbf{SO}_0(1,1)\cong\mathbb{R}$ (1-dimensional abelian position) with Fourier transformed Ω^1-measures and noncompact dual invariants constitute a convolution algebra,

$$e^{-|mz|} = \int\frac{dq}{\pi}\frac{|m|}{q^2+m^2}e^{-iqz}.$$

position $\mathcal{Y}^1$ and "momenta" $\mathbb{R}$
$\Omega^1\vert=2\pi,\quad \overset{1}{*}=\frac{*}{2\pi}$
$\frac{2\vert m_1\vert}{q^2+m_1^2}\overset{1}{*}\frac{2\vert m_2\vert}{q^2+m_2^2}=\frac{2\vert m_+\vert}{q^2+m_+^2}$

9.2 Convolutions for Position Representations

Residual representations of Euclidean, spherical, and hyperbolic spaces are characterized by *singularity spheres* with real momenta (imaginary eigenvalues) for scattering structures and imaginary "momenta" (real eigenvalues) for bound structures. The convolution of the related "momentum" functions reflect pointwise multiplications of Bessel, Neumann, and Macdonald functions,

$$\begin{array}{rll}
\int\frac{d^sq}{\pi^{\frac{s}{2}}}\,\delta(\vec{q}^{\,2}-1)e^{-i\vec{q}\vec{x}} &= \frac{\mathcal{J}_{\frac{s-2}{2}}(r)}{(\frac{r}{2})^{\frac{s-2}{2}}}, & \text{Euclidean},\\
\int\frac{d^sq}{\pi^{\frac{s}{2}}}\,\frac{\Gamma(\frac{s}{2}-\nu)}{(\vec{q}^{\,2}-io-1)^{\frac{s}{2}-\nu}}e^{-i\vec{q}\vec{x}} &= \frac{i\pi\mathcal{H}^{(1)}_\nu(r)}{(\frac{r}{2})^\nu} = -\frac{\pi[\mathcal{N}_\nu-i\mathcal{J}_\nu](r)}{(\frac{r}{2})^\nu}, & \text{spherical},\\
\int\frac{d^sq}{\pi^{\frac{s}{2}}}\,\frac{\Gamma(\frac{s}{2}-\nu)}{(\vec{q}^{\,2}+1)^{\frac{s}{2}-\nu}}e^{-i\vec{q}\vec{x}} &= \frac{2\mathcal{K}_\nu(r)}{(\frac{r}{2})^\nu}, & \text{hyperbolic}.
\end{array}$$

9.2.1 Convolutions for Euclidean Scattering

Interaction-free product structures convolute Dirac distributions for cyclic translation representations. In contrast to the convolution of Dirac distributions for self-dual invariants with basic spherically self-dual 2-dimensional representations,

$$\begin{array}{rl}
\text{abelian }\mathbb{R}: & 2P_1\delta(q^2-P_1^2)*2P_2\delta(q^2-P_2^2)\\
& \quad = 2P_-\delta(q^2-P_-^2)+2P_+\delta(q^2-P_+^2)\\
\text{with} & P_{1,2}>0,\;\; P_\pm=|P_1\pm P_2|,
\end{array}$$

the convolution of Dirac distributions for the infinite-dimensional representations of the Euclidean groups, $s\geq 2$, with the sphere radii as momentum invariants $\vec q^2 = P^2 > 0$ leads to position translation representations with the momentum sphere radii between the invariants, $P_-^2 \le \vec q^2 \le P_+^2$,

$$\begin{array}{l}\mathbf{SO}(s)\ \vec\times\,\mathbb{R}^s: \quad \delta(\vec q^2 - P_1^2) \underset{2|\Omega^{s-2}|}{*} \delta(\vec q^2 - P_2^2) = \frac{|Q|^{s-3}}{|\vec q|}\vartheta(P_+^2 - \vec q^2)\vartheta(\vec q^2 - P_-^2).\\ s = 2, 3, \ldots .\end{array}$$

The convolution product is normalized with the $(s-2)$-sphere. There arises a momentum-dependent normalization factor $|Q|$, which contains the characteristic two-particle convolution function,

$$Q^2 = \frac{-\Delta(\vec q^2)}{4\vec q^2}, \quad \Delta(\vec q^2) = \Delta(\vec q^2, P_1^2, P_2^2) = (\vec q^2 - P_+^2)(\vec q^2 - P_-^2).$$

It is symmetric in the three invariants involved:

$$\Delta(a, b, c) = a^2 + b^2 + c^2 - 2(ab + ac + bc) = (a + b - c)^2 - 4ab.$$

The minimal cases $s = 2, 3$ are characteristic for even- and odd-dimensional position, for scattering representations in three dimensions:

$$\begin{array}{rl}\mathcal{C}_b(\mathbb{R}^3)_+ \cdot \mathcal{C}_b(\mathbb{R}^3)_+ = \mathcal{C}_b(\mathbb{R}^3)_+: & \frac{\sin P_1 r}{r} \cdot \frac{\sin P_2 r}{r} = \frac{\cos P_- r - \cos P_+ r}{2r^2},\\ \mathcal{M}(\check{\mathbb{R}}^3)_+ * \mathcal{M}(\check{\mathbb{R}}^3)_+ = \mathcal{M}(\check{\mathbb{R}}^3)_+: & \delta(\vec q^2 - P_1^2) \underset{4\pi}{*} \delta(\vec q^2 - P_2^2)\\ & = \frac{1}{|\vec q|}\vartheta(P_+^2 - \vec q^2)\vartheta(\vec q^2 - P_-^2).\end{array}$$

The square of a representation is a normalized positive-type function,

$$\left(\frac{\sin Pr}{Pr}\right)^2 = \frac{1 - \cos 2Pr}{2(Pr)^2}, \quad \delta(\vec q^2 - P^2) \underset{4\pi}{*} \delta(\vec q^2 - P^2) = \frac{1}{|\vec q|}\vartheta(4P^2 - \vec q^2).$$

9.2.2 Convolutions for Hyperboloids

Cyclic representations of the general Lorentz group $\mathbf{SO}_0(1, 2R-1)$ for the hyperboloid $\mathcal{Y}^{2R-1} \cong \mathbf{SO}_0(1, 2R-1)/\mathbf{SO}(2R-1)$ with real rank 1 and noncompact invariant $\vec q^2 = -m^2 < 0$ are characterized by continuous functions of positive type $\mathcal{C}_b(\mathcal{Y}^{2R-1})_+ \subset L^\infty(\mathcal{Y}^{2R-1})_+$,

$$\mathcal{Y}^{2R-1} \cong \mathbb{R}^{2R-1} \ni \vec x \longmapsto \int \frac{d^{2R-1}q}{|\Omega^{2R-1}|} \frac{2|m|}{(\vec q^2 + m^2)^R} e^{-i\vec q\vec x} = e^{-|m|r}.$$

Nontrivial properties for the maximal compact group, the rotations $\mathbf{SO}(2R-1)$, $R \ge 2$, arise by derivations $\frac{\partial}{\partial \vec x} \sim -i\vec q$. These Lorentz group representations start from the characteristic hyperbolic exponentials for the maximal noncompact abelian subgroup with imaginary singularities $q = \pm im$,

$$\mathbf{SO}_0(1, 1) \cong \mathbb{R} \ni x \longmapsto \int \frac{dq}{\pi} \frac{|m|}{q^2 + m^2} e^{-iqx} = e^{-|mx|}.$$

The representations are faithful and cyclic, but not irreducible. They are square integrable and not Schur-orthogonal for different invariants $m_1^2 \neq m_2^2$.

Product representations $e^{-|m_1|r} \cdot e^{-|m_2|r} = e^{-|m_+|r}$ convolute the positive momentum measures. The measure of the associated compact unit sphere

Ω^{2R-1} is used for the residual normalization (more on the normalization below). The representations of 3-dimensional hyperbolic position $\mathcal{Y}^3$ use the Fourier transformed Ω^3-measure, familiar from the nonrelativistic hydrogen Schrödinger functions. The radii of the "momentum" spheres as invariants are added up in the convolution

$$e^{-|m|r} = \int \frac{d^3q}{\pi^2} \frac{|m|}{(\vec q^2+m^2)^2} e^{-i\vec q\vec x}.$$

position $\mathcal{Y}^3 \cong \mathbf{SO}_0(1,3)/\mathbf{SO}(3)$ and "momenta" $\mathbb{R}^3$ with $\mathbf{SO}(3)$
$\lvert\Omega^3\rvert = 2\pi^2, \quad \overset{3}{*} = \frac{*}{2\pi^2}$
$\frac{2\lvert m_1\rvert}{(\vec q^2+m_1^2)^2} \overset{3}{*} \frac{2\lvert m_2\rvert}{(\vec q^2+m_2^2)^2} = \frac{2\lvert m_+\rvert}{(\vec q^2+m_+^2)^2}$

In general, the representations of odd-dimensional hyperboloids $\mathcal{Y}^{2R-1}$ come with Fourier transformed Ω^{2R-1}-measures and imaginary singularity sphere Ω^{2R-2} for the "momentum" eigenvalues. The sphere measures can be obtained by invariant momentum derivatives

$$(-\frac{\partial}{\partial \vec q^2})^{R-1} \frac{|m|}{\vec q^2+m^2} = \Gamma(R)\frac{|m|}{(\vec q^2+m^2)^R}, \quad R = 1,2,\dots .$$

Product representations arise by the convolution with the sphere volume as residual normalization

$$e^{-|m|r} = \int \frac{d^{2R-1}q}{|\Omega^{2R-1}|} \frac{2|m|}{(\vec q^2+m^2)^R} e^{-i\vec q\vec x}.$$

position $\mathcal{Y}^{2R-1} \cong \mathbf{SO}_0(1,2R-1)/\mathbf{SO}(2R-1),\ R=1,2,\dots$ and "momenta" $\mathbb{R}^{2R-1}$ with $\mathbf{SO}(2R-1)$
$\lvert\Omega^{2R-1}\rvert = \frac{2\pi^R}{\Gamma(R)}, \quad \overset{2R-1}{*} = \frac{*}{\lvert\Omega^{2R-1}\rvert}$
$(\frac{\partial}{\partial \vec q})^{L_1} \frac{2\lvert m_1\rvert}{(\vec q^2+m_1^2)^R} \overset{2R-1}{*} (\frac{\partial}{\partial \vec q})^{L_2} \frac{2\lvert m_2\rvert}{(\vec q^2+m_2^2)^R} = (\frac{\partial}{\partial \vec q})^{L_1+L_2} \frac{2\lvert m_+\rvert}{(\vec q^2+m_+^2)^R}$ for $L=0,1,\dots$

The convolution may involve tensor products for $\mathbf{SO}(2R-1)$-representations. In general, nontrivial $\mathbf{O}(t,s)$-properties are effected by the convolution-compatible (energy–)momentum derivatives

$$\begin{aligned} \frac{\partial}{\partial q} = 2q\frac{\partial}{\partial q^2}, \quad \frac{\partial}{\partial q}\otimes q &= \mathbf{1}_{t+s} + q\otimes q\, 2\frac{\partial}{\partial q^2}, \\ \frac{\partial}{\partial q}\otimes \frac{\partial}{\partial q} &= (\mathbf{1}_{t+s} + q\otimes q\, 2\frac{\partial}{\partial q^2})2\frac{\partial}{\partial q^2}, \quad \dots, \end{aligned}$$

which, acting on multipoles, raise the pole order,

$$-\frac{\partial}{\partial q}\frac{\Gamma(R)}{(q^2+\mu^2)^R} = \frac{2q\,\Gamma(1+R)}{(q^2+\mu^2)^{1+R}}.$$

9.2.3 Convolutions for Spheres

The representations of odd-dimensional spheres use a singularity sphere Ω^{2R-2} with real momentum eigenvalues in the convolutions

$$e^{\pm i|m|r} = \pm\int \frac{d^{2R-1}q}{i|\Omega^{2R-1}|} \frac{2|m|}{(\vec q^2 \mp io - m^2)^R} e^{-i\vec q\vec x}.$$

sphere $\Omega^{2R-1} \cong \mathbf{SO}(2R)/\mathbf{SO}(2R-1),\ R=1,2,\dots$ and momenta $\mathbb{R}^{2R-1}$ with $\mathbf{SO}(2R-1)$
$\lvert\Omega^{2R-1}\rvert = \frac{2\pi^R}{\Gamma(R)}, \quad (\overset{2R-1}{*}, \vec q^2) = (\pm\frac{*}{i\lvert\Omega^{2R-1}\rvert}, \vec q^2 \mp io)$ not orthogonal
$(\frac{\partial}{\partial \vec q})^{L_1} \frac{2\lvert m_1\rvert}{(\vec q^2-m_1^2)^R} \overset{2R-1}{*} (\frac{\partial}{\partial \vec q})^{L_2} \frac{2\lvert m_2\rvert}{(\vec q^2-m_2^2)^R} = (\frac{\partial}{\partial \vec q})^{L_1+L_2} \frac{2\lvert m_+\rvert}{(\vec q^2-m_+^2)^R}$ for $L=0,1,\dots$

9.2.4 Residual Normalization

Above, the abelian convolutions $* \cong \delta(q_1+q_2-q)$ of the $\mathbb{R}^n$-Lebesgue measures $d^n q$ of energies and momenta are "rationalized" with respect to the product representations of a noncompact Cartan line in such a way that the compact spherical degrees do not show up (no π's). The normalization results from spheres in the definition of higher-dimensional residues with a characteristic difference for odd- and even-dimensional spaces.

Since the convolution with a Dirac distribution amounts to a residue $f(q) = \int dp\ \delta(p-q)f(p) = \oint \frac{dp}{2i\pi}\frac{f(p)}{p-q}$, the convolution normalization for the time representation coefficients is given by the normalization of the residue for the irreducible $\mathbf{D}(1)$-representation coefficient with real pole $q = m$,

$$\tilde{\mathcal{D}}^1: \quad \frac{*}{2i\pi} \text{ from } \int \frac{dq}{2i\pi}\frac{1}{q-io-m}e^{iqx_0} = \vartheta(x_0)e^{imx_0};$$

2π is the length of the circle $\Omega^1 \cong \mathbf{U}(1)$, the compact representation image of $\mathbf{D}(1)$. It normalizes the energy Plancherel measure $\frac{dq}{2\pi}$ for the time translation Haar measure dt.

The convolution normalization for $\mathbf{SO}_0(1,1)$ is determined by the residual normalization in the faithful cyclic representation coefficient whose self-duality causes the factor 2 for the two imaginary poles $q = \pm i|m|$,

$$\tilde{\mathcal{Y}}^1: \quad \frac{*}{2\pi} \text{ from } \int \frac{dq}{2\pi}\frac{2|m|}{q^2+m^2}e^{-iqx} = e^{-|mx|}.$$

In general for odd-dimensional hyperboloids, the residual normalization of the rotation scalar functions of positive type uses the measures of the compact partner spheres,

$$\tilde{\mathcal{Y}}^{2R-1}: \quad \overset{2R-1}{*} = \frac{*}{|\Omega^{2R-1}|} \text{ from } \int \frac{d^{2R-1}q}{|\Omega^{2R-1}|}\frac{2|m|}{(\vec{q}^2+m^2)^R}e^{-i\vec{q}\vec{x}} = e^{-|m|r}.$$

The momentum eigenvalues lie on a sphere $\{\vec{q} \in \check{\mathbb{R}}^{2R-1} \mid \vec{q}^2 = -m^2\} \cong \Omega^{2R-2}$. For the nonabelian case $R \geq 2$, the sphere normalization $\frac{1}{|\Omega^{2R-1}|} = \frac{\Gamma(R)}{2\pi^R}$ differs from the "flat" normalization $\frac{1}{|\Omega^1|^{2R-1}} = \frac{1}{(2\pi)^{2R-1}}$ for self-dually represented translations $\mathbb{R}^{2R-1} \longrightarrow \mathbf{SO}(2)^{2R-1}$.

9.3 Convolution of Singularity Hyperboloids

The convolution via the integration prescription $\int d^d q_1 d^d q_2 \delta(q_1+q_2-q)$ involves the Dirac distribution for the linear combination of real (energy–)momenta. This does not determine completely the complex integration contour for distributions with squared invariants in more than 1 dimension (singularity surfaces),

$$\mathbf{O}(t,s),\quad t+s \geq 2: \quad q \in \mathbb{R}^{t+s},\quad q^2 - m^2 = 0,\quad m^2 \in \mathbb{R}.$$

(energy–)momenta eigenvalues $\{q \in \mathbb{R}^2 | \, q^2 - m^2 = 0\}$ constitute, for the nontrivial case, a 1-dimensional invariant manifold, either a circle Ω^1 or a two-branch hyperbola $\mathcal{Y}^1_+ \uplus \mathcal{Y}^1_-$ and $\mathbb{R} \uplus \mathbb{R}$. This is in contrast to dual invariants, where the 0-dimensional invariance manifold $\{q \in \mathbb{R} \mid q^2 - m^2 = 0\}$ consists of discrete points.

The convolution can be performed by joining (interpolating) first the invariant-determining denominator polynomials of the (energy–)momentum distributions where the singularity orders are added,

$$\begin{aligned} \frac{\Gamma(\nu_1)\cdots\Gamma(\nu_k)}{R_1^{\nu_1}\cdots R_k^{\nu_k}} &= \int_0^1 d\zeta_1 \cdots \int_0^1 d\zeta_k \delta(\zeta_1 + \cdots + \zeta_k - 1) \frac{\zeta_1^{\nu_1 - 1}\cdots\zeta_k^{\nu_k - 1}\Gamma(\nu_1 + \cdots + \nu_k)}{(\zeta_1 R_1 + \cdots + \zeta_k R_k)^{\nu_1 + \cdots + \nu_k}}, \\ & \nu_j \in \mathbb{R}, \quad \nu_j \neq 0, -1, -2, \ldots, \end{aligned}$$

e.g., for the fundamental distribution for nonlinear spacetime with two continuous real invariants

$$\frac{2q}{q^2 - m_\zeta^2}\frac{1}{(q^2 - m_s^2)^R} = -\frac{\partial}{\partial q}\int_0^1 d\zeta \, \frac{\zeta^{R-1}}{[q^2 - \zeta m_s^2 - (1-\zeta)m_0^2]^R}.$$

The product of two distributions with real invariants $m^2 \in \mathbb{R}$,

$$\frac{\Gamma(1+\nu_1)}{(q_1^2 - m_1^2)^{1+\nu_1}} \times \frac{\Gamma(1+\nu_2)}{(q_2^2 - m_2^2)^{1+\nu_2}} = \int_0^1 d\zeta_{1,2}\delta(\zeta_1 + \zeta_2 - 1)\frac{\zeta_1^{\nu_1}\zeta_2^{\nu_2}\Gamma(2+\nu_1+\nu_2)}{[\zeta_1(q_1^2 - m_1^2) + \zeta_2(q_2^2 - m_2^2)]^{2+\nu_1+\nu_2}},$$

can be written with center of mass (energy–)momentum q and relative (energy–)momentum p,

$$\zeta_1 q_1^2 + \zeta_2 q_2^2 = (p + \tfrac{\zeta_1 - \zeta_2}{2}q)^2 + \zeta_1\zeta_2 q^2 \text{ with } \left\{ \begin{array}{ll} q_1 + q_2 &= q, \\ q_1 - q_2 &= 2p, \\ d^d q_1 d^d q_2 &= d^d q \, d^d p. \end{array} \right.$$

For two equal-type Feynman distributions $\frac{1}{q^2 \mp io - m^2}$ the product inherits this Feynman type both for the center of mass and the relative (energy–)momentum distributions,

$$q_1^2 \pm io \text{ with } q_2^2 \pm io \Rightarrow \zeta_1 q_1^2 + \zeta_2 q_2^2 \pm io.$$

That is different for two equal type causal distributions $\frac{1}{(q \mp io)^2 - m^2}$ with $(q \mp io)^2 = (q_0 - io)^2 - \vec{q}^2$, for $\mathbf{SO}_0(1, s)$ and positive invariants $m^2 \geq 0$: The product gives an equal-type causal distribution for center of mass energy–momenta q, but with an indefinite $\zeta_1 - \zeta_2 \cong 2\zeta_1 - 1 \in [-1, 1]$, both an advanced and a retarded distribution for the relative energy–momenta p,

$$(q_{1,2} \pm io)^2 \Rightarrow \left\{ \begin{array}{l} \zeta_1 q_1^2 + \zeta_2 q_2^2 \pm 2io(\zeta_1 q_1^0 + \zeta_2 q_2^0) \\ = (p + \frac{\zeta_1 - \zeta_2}{2}q)^2 \pm 2io\epsilon(\zeta_1 - \zeta_2)p^0 + \zeta_1\zeta_2 q^2 \pm ioq^0. \end{array} \right.$$

For a complete definition of the convolution of causal distributions the prescription $\delta(q_1 + q_2 - q)$ for the real part is supplemented by the prescription to use, for the imaginary part, the equal-type causal distribution also for the relative energy–momenta.

Summarizing: The *convolution for $\mathbf{O}(t,s)$-distributions on $\mathbb{R}^{t+s}$, $t+s\ge 2$*, for two equal-type distributions with real invariants is defined to yield the same distribution type:

$$\begin{array}{rl}\frac{\Gamma(1+\nu_1)}{(q_1^2-m_1^2)^{1+\nu_1}} * \frac{\Gamma(1+\nu_2)}{(q_2^2-m_2^2)^{1+\nu_2}} & = \int_0^1 d\zeta_{1,2}\delta(\zeta_1+\zeta_2-1)\int d^dp \frac{\zeta_1^{\nu_1}\zeta_2^{\nu_2}\Gamma(2+\nu_1+\nu_2)}{(p^2+\zeta_1\zeta_2q^2+\zeta_1m_1^2+\zeta_2m_2^2)^{2+\nu_1+\nu_2}}\\ & = i^s\pi^{\frac{d}{2}}\int_0^1 d\zeta_{1,2}\delta(\zeta_1+\zeta_2-1)\frac{\zeta_1^{\nu_1}\zeta_2^{\nu_2}\Gamma(\frac{4-d}{2}+\nu_1+\nu_2)}{(\zeta_1\zeta_2q^2+\zeta_1m_1^2+\zeta_2m_2^2)^{\frac{4-d}{2}+\nu_1+\nu_2}},\\ \text{with equal type} & Q^2+io,\ Q^2-io,\ (Q-io)^2,\ (Q+io)^2,\\ \text{for all} & Q\in\{q_1,q_2,q,p\}.\end{array}$$

Advanced and retarded measures for the orthochronous group $\mathbf{SO}_0(1,s)$ are convolution-orthogonal to each other. This is in contrast to the conjugated Feynman measures

$$\mathbf{O}(t,s),\ t\ge 1,\ m^2\ge 0: (F_{m^2},F^\star_{m^2}) = \pm\frac{1}{i\pi}\frac{1}{q^2\mp io-m^2} = \delta_{m^2}\pm i\mathrm{P}_{m^2},$$

which in general are not convolution-orthogonal to each other. ¿From their individual convolution,

$$\begin{pmatrix} F_{m_1^2} * F_{m_2^2}\\ F^\star_{m_1^2} * F^\star_{m_2^2}\end{pmatrix} = \delta_{1*2}\pm i\mathrm{P}_{1*2},$$

one obtains the Dirac and principal value contribution

$$\begin{array}{rll}\delta_{1*2} & = \frac{1}{2}(F_{m_1^2} * F_{m_2^2} + F^\star_{m_1^2} * F^\star_{m_2^2}) & = \delta_{m_1^2} * \delta_{m_2^2} - \mathrm{P}_{m_1^2} * \mathrm{P}_{m_2^2},\\ \mathrm{P}_{1*2} & = \frac{1}{2i}(F_{m_1^2} * F_{m_2^2} - F^\star_{m_1^2} * F^\star_{m_2^2}) & = \delta_{m_1^2} * \mathrm{P}_{m_2^2} + \mathrm{P}_{m_1^2} * \delta_{m_2^2}.\end{array}$$

9.4 Convolutions for Spacetime

Residual spacetime representations are characterized by energy–momentum singularity hyperboloids.

The convolution of energy–momentum Feynman distributions for ($d = 1+s$)-dimensional spacetime reflects pointwise multiplications of Hankel and Macdonald functions

$$\int \frac{d^dq}{i^s\pi^{\frac{d}{2}}}\frac{e^{i\nu\pi}\Gamma(\frac{d}{2}-\nu)}{(q^2-io-1)^{\frac{d}{2}-\nu}}e^{iqx} = \frac{\vartheta(-x^2)2\mathcal{K}_\nu(|x|)+\vartheta(x^2)i\pi\mathcal{H}^{(1)}_{-\nu}(|x|)}{|\frac{x}{2}|^\nu} + \delta_\nu^N i\pi\sum_{k=1}^N\frac{1}{(N-k)!}\delta^{(k-1)}(-\frac{x^2}{4}).$$

The on-shell Dirac distribution (real part) with the Neumann functions gives coefficents of Poincaré group representations:

$$\int \frac{d^dq}{\pi^{\frac{d}{2}-1}}\ \delta(q^2-1)e^{iqx} = \frac{\vartheta(-x^2)2\mathcal{K}_{\frac{d-2}{2}}(|x|)-\vartheta(x^2)\pi\mathcal{N}_{-\frac{d-2}{2}}(|x|)}{|\frac{x}{2}|^{\frac{d-2}{2}}}.$$

Particle propagators have poles, for even-dimensional spacetimes $\mathbf{SO}_0(1,2R-1)\ \vec{\times}\ \mathbb{R}^{2R}$ (Cartan, Minkowski, ... for $2R=2,4,\dots$),

$$\int \frac{d^{2R}q}{i\pi}\frac{1}{q^2-io-1}e^{iqx} = \left(\frac{\partial}{\partial\frac{x^2}{4\pi}}\right)^{R-1}[\vartheta(-x^2)2\mathcal{K}_0(|x|)+\vartheta(x^2)i\pi\mathcal{H}^{(1)}_0(|x|)].$$

Feynman integrals in perturbation theory involve convolutions of energy-momentum distributions for pointwise products of spacetime distributions. In general, they do not make sense since $\mathcal{S}'(\mathbb{R}^d)$ is not a convolution algebra. For Minkowski spacetime, there arise undefined products ("divergences") of generalized functions from the energy–momentum principal value for the causally supported off-shell part $\frac{1}{q_{\rm P}^2-m^2} \overset{\mathbb{R}^4}{\sim} \delta(x^2)+\cdots$:

$$\begin{array}{cccc} & [\delta(q^2-m_1^2)+\frac{1}{i\pi}\frac{1}{q_{\rm P}^2-m_1^2}] & * & [\delta(q^2-m_2^2)+\frac{1}{i\pi}\frac{1}{q_{\rm P}^2-m_2^2}] \\ \overset{\mathbb{R}^4}{\sim} & [\frac{1}{x_{\rm P}^2}+\frac{1}{i\pi}\delta(x^2)+\cdots] & \cdot & [\frac{1}{x_{\rm P}^2}+\frac{1}{i\pi}\delta(x^2)+\cdots]. \end{array}$$

The convolution of energy-momentum distributions adds the spacetime translation eigenvalues to the eigenvalue q of the product representation, e.g., for scalar multipole distributions

$$\begin{aligned} &\pm\frac{1}{i\pi}\frac{\Gamma(1+n_1)}{(q^2\mp io-m_1^2)^{1+n_1}} * \cdots * \pm\frac{1}{i\pi}\frac{\Gamma(1+n_k)}{(q^2\mp io-m_k^2)^{1+n_k}} \\ &\quad = (\pm\tfrac{1}{i\pi})^k \int d^{1+s}q_1\cdots d^{1+s}q_k\delta(\sum_{j=1}^k q_j - q)\prod_{j=1}^k \frac{\Gamma(1+n_j)}{(q_j^2\mp io-m_j^2)^{1+n_j}}. \end{aligned}$$

The convoluted distributions have to be all of the same type, either all Feynman q^2-io or all anti-Feynman q^2+io.

Two distributions in Cartan spacetime have the product

$$\begin{aligned} \text{for } \mathbb{R}^2:\ &\pm\frac{1}{i\pi}\frac{\Gamma(1+\nu_1)}{(q^2\mp io-m_1^2)^{1+\nu_1}} * \pm\frac{1}{i\pi}\frac{\Gamma(1+\nu_2)}{(q^2\mp io-m_2^2)^{1+\nu_2}} \\ &= \frac{1}{i\pi}\int_0^1 d\zeta_{1,2}\delta(\zeta_1+\zeta_2-1)\int\frac{d^2p}{i\pi}\frac{\zeta_1^{\nu_1}\zeta_2^{\nu_2}\Gamma(2+\nu_1+\nu_2)}{[p^2\mp io+\zeta_1\zeta_2q^2-\zeta_1m_1^2-\zeta_2m_2^2]^{2+\nu_1+\nu_2}}. \end{aligned}$$

The convolution is the *residue with respect to the relative energy–momentum* $p=\frac{q_1-q_2}{2}$ *dependence*,

$$\text{for } \mathbb{R}^2:\quad \pm\int\frac{d^2p}{i\pi}\frac{\Gamma(2+\nu)}{(p^2\mp io+a)^{2+\nu}} = \frac{\Gamma(1+\nu)}{(\mp io+a)^{1+\nu}}.$$

The result depends on the center of mass energy–momentum $q=q_1+q_2$:

$$\begin{aligned} \text{for } \mathbb{R}^2:\ &\pm\frac{1}{i\pi}\frac{\Gamma(1+\nu_1)}{(q^2\mp io-m_1^2)^{1+\nu_1}} * \pm\frac{1}{i\pi}\frac{\Gamma(1+\nu_2)}{(q^2\mp io-m_2^2)^{1+\nu_2}} \\ &= \pm\frac{1}{i\pi}\int_0^1 d\zeta\frac{\zeta^{\nu_1}(1-\zeta)^{\nu_2}\Gamma(1+\nu_1+\nu_2)}{[\zeta(1-\zeta)(q^2\mp io)-\zeta m_1^2-(1-\zeta)m_2^2]^{1+\nu_1+\nu_2}}. \end{aligned}$$

Here and in the following the convolutions exist only for pole orders where the involved Γ-functions are defined. Elsewhere, there arise "divergences."

The convolution of two energy–momentum distributions in Minkowski spacetime,

$$\begin{aligned} \text{for } \mathbb{R}^4:\ &\mp\frac{1}{i\pi^2}\frac{\Gamma(2+\nu_1)}{(q^2\mp io-m_1^2)^{2+\nu_1}} * \mp\frac{1}{i\pi^2}\frac{\Gamma(2+\nu_2)}{(q^2\mp io-m_1^2)^{2+\nu_2}} \\ &= \frac{1}{i\pi^2}\int_0^1 d\zeta_{1,2}\delta(\zeta_1+\zeta_2-1)\int\frac{d^4p}{i\pi^2}\frac{\zeta_1^{1+\nu_1}\zeta_2^{1+\nu_2}\Gamma(4+\nu_1+\nu_2)}{[p^2\mp io+\zeta_1\zeta_2q^2-\zeta_1m_1^2-\zeta_2m_2^2]^{4+\nu_1+\nu_2}}, \end{aligned}$$

leads with

$$\text{for } \mathbb{R}^4:\quad \mp\int\frac{d^4p}{i\pi^2}\frac{\Gamma(3+\nu)}{(p^2\mp io+a)^{3+\nu}} = \frac{\Gamma(1+\nu)}{(\mp io+a)^{1+\nu}}$$

to the same characteristic integral as for Cartan spacetime:

$$\begin{aligned}\text{for } \mathbb{R}^4: \quad &\mp\frac{1}{i\pi^2}\frac{\Gamma(2+\nu_1)}{(q^2\mp io-m_1^2)^{2+\nu_1}} * \mp\frac{1}{i\pi^2}\frac{\Gamma(2+\nu_2)}{(q^2\mp io-m_2^2)^{2+\nu_2}}\\ &= \mp\frac{1}{i\pi^2}\int_0^1 d\zeta \frac{\zeta^{1+\nu_1}(1-\zeta)^{1+\nu_2}\Gamma(2+\nu_1+\nu_2)}{[\zeta(1-\zeta)(q^2\mp io)-\zeta m_1^2-(1-\zeta)m_2^2]^{2+\nu_1+\nu_2}}.\end{aligned}$$

9.4.1 Convolutions for Free Particles

The Fourier transformed principal part of a Feynman distribution for an orthochronous group $\mathbf{SO}_0(1,s)$ can be written as an order function times an on-shell part:

$$F_{m^2} = \frac{1}{i\pi}\frac{1}{q^2-io-m^2} = \delta_{m^2} + i\mathrm{P}_{m^2}: \quad \begin{cases} \delta_{m^2} &= \frac{\delta_{|m|}+\delta_{-|m|}}{2},\\ \mathrm{P}_{m^2} &\sim i\epsilon(x_0)\epsilon_{m^2},\\ \epsilon_{m^2} &= \frac{\delta_{|m|}-\delta_{-|m|}}{2}.\end{cases}$$

In the principal value convolution contribution of two Feynman propagators for spacetime $\mathbb{R}^{1+s}$ the order function drops out $\epsilon(x_0)^2 = 1$:

$$\begin{aligned}\pm\frac{1}{i\pi}\frac{1}{q^2\mp io-m_1^2} * \pm\frac{1}{i\pi}\frac{1}{q^2\mp io-m_2^2} = 2\Big[&\vartheta(+q_0)\delta(q^2-m_1^2) * \vartheta(+q_0)\delta(q^2-m_2^2)\\ &+ \vartheta(-q_0)\delta(q^2-m_1^2) * \vartheta(-q_0)\delta(q^2-m_2^2)\Big]\\ &\pm \frac{1}{i\pi}\Big[\delta(q^2-m_1^2) * \frac{1}{q_{\mathrm{P}}^2-m_2^2} + \frac{1}{q_{\mathrm{P}}^2-m_1^2} * \delta(q^2-m_2^2)\Big].\end{aligned}$$

The principal value square is also an on-shell convolution only. The convolution of translation representation coefficients from the real part of the propagator (free particles) gives corresponding coefficients for product representations (product of free particles):

$$\begin{aligned}\delta_{1*2} &= \tfrac{1}{2}(F_{m_1^2} * F_{m_2^2} + F^\star_{m_1^2} * F^\star_{m_2^2}) &&= \delta_{m_1^2} * \delta_{m_2^2} - \mathrm{P}_{m_1^2} * \mathrm{P}_{m_2^2}\\ &= \delta_{m_1^2} * \delta_{m_2^2} + \epsilon_{m_1^2} * \epsilon_{m_2^2} &&= \frac{\delta_{|m_1|}*\delta_{|m_2|}+\delta_{-|m_1|}*\delta_{-|m_2|}}{2}.\end{aligned}$$

The set with all $(1+s)$-dimensional "filled up" forward (backward) energy–momentum hyperboloids is an additive cone. Therefore, the distributions supported by positive and negative energy-momentum are convolution algebras, however, not orthogonal to each other:

$$\begin{aligned}&\{q \succeq |m_1|\} + \{q \succeq |m_2|\} = \{q \succeq |m_+|\}\\ &\text{with } \delta_{\pm|m|} \sim 2|m|\vartheta(\pm q_0)\delta(q^2-m^2) \in \mathcal{D}'(\mathbb{R}^{1+s}_\pm) \in \underline{\mathbf{aag}}_{\mathbb{C}} \text{ (convolution product)}.\end{aligned}$$

The convolution for abelian time with self-dual invariants $m_{1,2}^2 > 0$,

$$\begin{aligned}\text{abelian } \mathbb{R}: \quad &|m_1|\vartheta(\pm q_0)\delta(q^2-m_1^2) * |m_2|\vartheta(\pm q_0)\delta(q^2-m_2^2)\\ &= |m_+|\vartheta(\pm q_0)\delta(q_0^2-m_+^2),\end{aligned}$$

is embedded into the convolution of nonabelian hyperboloids for product representations of the translations (the real part δ_{1*2} for simple pole Feynman propagators). With the hyberboloid "radii" as energy–momentum invariants $q^2 = m^2 \geq 0$,

$$\mathbf{SO}_0(1,s) \vec{\times} \mathbb{R}^{1+s}: \quad \vartheta(\pm q_0)\delta(q^2 - m_1^2) \underset{2|\Omega^{s-1}|}{*} \vartheta(\pm q_0)\delta(q^2 - m_2^2) \\ = \frac{|Q|^{s-2}}{|q|}\vartheta(\pm q_0)\vartheta(q^2 - m_+^2), \quad s = 1, 2, 3, \ldots,$$

they involve the two-particle threshold factor

$$Q^2 = \frac{\Delta(q^2)}{4|q|^2}, \quad \Delta(q^2) = \Delta(q^2, m_1^2, m_2^2) = (q^2 - m_+^2)(q^2 - m_-^2).$$

For nontrivial position, the convolution (phase space integral) of s-dimensional on-shell hyperboloids (particle measures) does not lead to s-dimensional on-shell hyperboloids $\delta(q^2 - m_+^2)$. It leads to translation representations with energy–momenta over the free particle threshold at the mass sum $q^2 \geq m_+^2$, i.e., $\vec{q}^2 = q_0^2 - m_+^2 \geq 0$, with $m_\pm = |m_1| \pm |m_2|$. Here, the energy is enough to produce two free particles with masses $m_{1,2}$ and momentum $(\vec{q}_1 + \vec{q}_2)^2 \geq 0$.

The minimal cases $s = 1, 2$ are characteristic for even- and odd-dimensional spacetime. The Poincaré group $\mathbf{SO}_0(1,3) \vec{\times} \mathbb{R}^4$ is the minimal case with rotations:

$$\begin{aligned} \mathcal{M}(\check{\mathbb{R}}^4)_+ \; * \; \mathcal{M}(\check{\mathbb{R}}^4)_+ = \mathcal{M}(\check{\mathbb{R}}^4)_+ : \quad & \vartheta(q_0)\delta(q^2 - m_1^2) \underset{4\pi}{*} \vartheta(q_0)\delta(q^2 - m_2^2) \\ &= \sqrt{\frac{(q^2 - m_+^2)(q^2 - m_-^2)}{(q^2)^2}}\, \vartheta(q_0)\vartheta(q^2 - m_+^2) \\ \text{for } m_1^2 = m_2^2 = m^2: \quad &= \sqrt{\frac{q^2 - 4m^2}{q^2}}\, \vartheta(q_0)\vartheta(q^2 - 4m^2). \end{aligned}$$

Such convolutions arise, e.g., as the nondivergent on-shell contribution in the quantum electrodynamical vacuum polarization by electron-positron pairs.

9.4.2 Off-Shell Convolution Contributions

Energy-momentum convolutions combine the points on the hyperbolic-spherical singularity surfaces for particle-interaction structures, determined by the invariants. The characteristic new feature is the on-shell off-shell convolution, i.e., of compact with noncompact invariants. The convolution contribution in the mixed terms is not for product representations of the spacetime translations,

$$\mathrm{P}_{1*2} = \frac{1}{2i}(F_{m_1^2} * F_{m_2^2} - F^\star_{m_1^2} * F^\star_{m_2^2}) = \delta_{m_1^2} * \mathrm{P}_{m_2^2} + \mathrm{P}_{m_1^2} * \delta_{m_2^2}.$$

The divergences in Minkowski space arise from the mixed terms (mathematically meaningless),

$$\delta(q^2 - m_1^2) * \frac{1}{q_\mathrm{P}^2 - m_2^2} \overset{\mathbb{R}^4}{\sim} \frac{1}{x_\mathrm{P}^2} \cdot \delta(x^2) + \cdots .$$

Only for trivial position does the principal value part also add the invariant poles,

$$s = 0: \quad \mathrm{P}_{1*2} \sim i\epsilon(t)\frac{\delta_{|m_1|} * \delta_{|m_2|} - \delta_{-|m_1|} * \delta_{-|m_2|}}{2} \sim \mathrm{P}_{m_+^2}.$$

The characteristic effect of a convolution of noncompact with compact invariant comes in the principal value part for nontrivial position degrees of freedom:

$$\begin{array}{rcccc}\delta(q^2-m^2) & \sim & \vartheta(q^2-m^2), & & \\ \frac{1}{q_{\rm P}^2-m^2} & \sim & \vartheta(q^2-m^2) & + & \vartheta(-q^2+m^2) \\ & & \cup & & \cup \\ & & \text{compact (free)} & + & \text{noncompact.} \\ & & e^{imt} & & e^{-|mz|}\end{array}$$

The denominator polynomial in the convolution square above has two energy-momentum-dependent zeros,

$$\begin{array}{l}-P(\zeta)=\zeta(1-\zeta)q^2-\zeta m_1^2-(1-\zeta)m_2^2=-q^2[\zeta-\zeta_1(q^2)][\zeta-\zeta_2(q^2)],\\ \zeta_{1,2}(q^2)=\frac{q^2-m_+m_-\pm\sqrt{\Delta(q^2)}}{2q^2}\text{ with } m_\pm=|m_1|\pm|m_2|,\end{array}$$

which are either both real or complex conjugate to each other according to the sign of the discriminant $\Delta(q^2)$ (two-particle threshold factor):

$$\begin{array}{rl}\vartheta(\Delta(q^2))= & \vartheta(q^2-m_+^2)+\vartheta(m_-^2-q^2),\\ \vartheta(-\Delta(q^2))= & \vartheta(m_+^2-q^2)\vartheta(q^2-m_-^2).\end{array}$$

Furthermore, real zeros, in the case of $\Delta(q^2)\geq 0$, are in the integration ζ-interval $[0,1]$ only for energy–momenta over the threshold $\vartheta(q^2-m_+^2)$:

$$\begin{array}{rcll}\zeta_{1,2}(m_+^2) & = & \frac{|m_2|}{|m_1|+|m_2|} & \in [0,1],\\ \zeta_{1,2}(m_-^2) & = & \frac{-|m_2|}{|m_1|-|m_2|} & \notin [0,1],\end{array}$$

and graphically

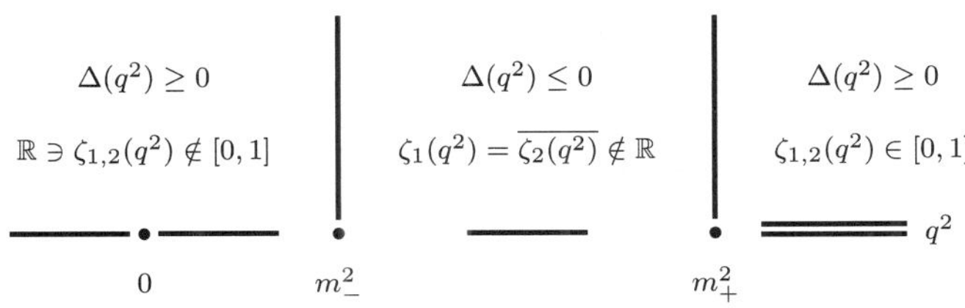

Therefore, the convolution of two energy–momentum pole distributions contains as relevant contribution

$$\int_0^1 d\zeta\,\frac{1}{\zeta(1-\zeta)(q^2\mp io)-\zeta m_1^2-(1-\zeta)m_2^2}$$

$$=\int_0^1 d\zeta\left[\frac{1}{\zeta(1-\zeta)q_{\rm P}^2-\zeta m_1^2-(1-\zeta)m_2^2}\pm i\pi\delta(\zeta(1-\zeta)q^2-\zeta m_1^2-(1-\zeta)m_2^2)\right]$$

$$=-\frac{2}{\sqrt{|\Delta(q^2)|}}\left[\vartheta(\Delta(q^2))\log\left|\frac{\Sigma(q^2)-2\sqrt{\Delta(q^2)}}{m_+^2-m_-^2}\right|+\vartheta(-\Delta(q^2))\arctan\frac{2\sqrt{-\Delta(q^2)}}{\Sigma(q^2)}\right]$$

$$\pm\frac{2i\pi}{\sqrt{\Delta(q^2)}}\vartheta(q^2-m_+^2)$$

with $\Delta(q^2)=(q^2-m_+^2)(q^2-m_-^2),\ \ \Sigma(q^2)=(q^2-m_+^2)+(q^2-m_-^2).$

The convolution product depends on the two variables $\{q^2 - m_+^2, q^2 - m_-^2\}$. The spacetime original convolution of compact with noncompact invariants shows up for energy–momenta under the threshold $\vartheta(m_+^2 - q^2)$, illustrated in one more example:

$$\int_0^1 d\zeta \, \frac{1}{\zeta(q^2 \mp io) - m^2} = \frac{1}{q^2}\left[\log\left|\frac{q^2-m^2}{m^2}\right| \pm i\pi\vartheta(q^2 - m^2)\right].$$

In the convolution of distributions of odd-dimensional spaces, e.g., for energy $q \in \mathbb{R}$, the integral compensates the m_-^2-pole from the discriminant,

$$\begin{aligned}
\text{for } \mathbb{R}: \quad \frac{|m_1|}{q^2 \mp io - m_1^2}\left(\pm \frac{*}{i\pi}\right)\frac{|m_2|}{q^2 \mp io - m_2^2} &= \int_0^1 d\zeta \, \frac{|m_1 m_2|}{[-\zeta(1-\zeta)(q^2 \mp io) + \zeta m_1^2 + (1-\zeta) m_2^2]^{\frac{3}{2}}} \\
&= \frac{|m_+|}{q^2 \mp io - m_+^2} \\
\text{with } \frac{1}{P(\zeta)^{\frac{3}{2}}} &= -\frac{4}{(q^2 - m_+^2)(q^2 - m_-^2)} \frac{d^2\sqrt{P(\zeta)}}{d\zeta^2}.
\end{aligned}$$

9.4.3 Residual Products for Spacetime

The convolution product for even-dimensional causal spacetimes $\mathcal{D}^{2R} = \mathbf{D}(1) \times \mathcal{Y}^{2R-1}$ with real rank 2 involves convolutions for the causal group (eigentime) $\mathbf{D}(1) \cong \mathbb{R}$ and for the Lorentz group for hyperbolic position $\mathcal{Y}^{2R-1} \cong \mathbf{SO}_0(1, 2R-1)/\mathbf{SO}(2R-1)$.

The convolutions of Cartan energy-momentum pole distributions are

spacetime $\mathcal{D}^2 = \mathbf{D}(1) \times \mathcal{Y}^1$ with $\mathbf{SO}_0(1,1)$, $\lvert\Omega^1\rvert = 2\pi$
$(\overset{2}{*}, q^2) = \begin{cases} (\mp \frac{*}{2i\pi}, (q \mp io)^2), & \text{causal, orthogonal} \\ (\mp \frac{*}{i\pi}, q^2 \mp io), & \text{Feynman, not orthogonal} \end{cases}$
$\frac{1}{-q^2+m_1^2} \overset{2}{*} \frac{1}{-q^2+m_2^2} = \int_0^1 d\zeta \frac{1}{-\zeta(1-\zeta)q^2 + \zeta m_1^2 + (1-\zeta) m_2^2}$

The different factor 2 for Feynman and causal measures originates from the different residual structure in the complex plane: for causal measures both poles are in the same half plane, for Feynman measures one in the upper and one in the lower half plane.

In contrast to the factors for time and position, both odd-dimensional with real rank 1, the convolutions for minimal 2-dimensional Cartan spacetime with rank 2 do not produce invariant pole singularities (0-dimensional) for product representations. The residual products of even-dimensional spaces display pole distributions only before the finite ζ-integration over an invariant line singularity (1-dimensional). The pole distributions can be written with spectral functions, e.g., for one vanishing mass,

$$\begin{aligned}
\int_0^1 d\zeta \, \frac{1}{-\zeta q^2 + m^2} &= \int_{m^2}^\infty \frac{dM^2}{M^2} \frac{1}{-q^2 + M^2}, \\
\int_0^1 d\zeta \, \frac{1-\zeta}{-\zeta q^2 + m^2} &= \int_{m^2}^\infty \frac{dM^2}{M^2} \frac{M^2 - m^2}{M^2} \frac{1}{-q^2 + M^2}.
\end{aligned}$$

After ζ-integration, there arise logarithms and no energy–momentum poles. The logarithm is typical for a finite integration [1], e.g., for a function holomorphic on the integration curve (where defined):

$$\int_\beta^\alpha dz\ f(z) = \sum \mathrm{res}[f(z)\log\tfrac{z-\alpha}{z-\beta}],\quad \begin{cases} \int_\beta^\infty dz\ f(z) & = -\sum \mathrm{res}[f(z)\log(z-\beta)], \\ \int_{-\infty}^\infty dz\ f(z) & = 2i\pi\sum \mathrm{res} f(z), \end{cases}$$

with the sum of all residues in the closed complex plane, cut along the integration curve, here,

$$\begin{aligned} \int_0^1 \tfrac{d\zeta}{-\zeta q^2+m^2} &= \sum \mathrm{res}[\tfrac{1}{-\zeta q^2+m^2}\log\tfrac{\zeta-1}{\zeta}] &&= \frac{\log(1-\frac{q^2}{m^2})}{-q^2}, \\ \int_0^1 d\zeta\ \tfrac{1-\zeta}{-\zeta q^2+m^2} &= \sum \mathrm{res}\tfrac{1-\zeta}{-\zeta q^2+m^2}\log\tfrac{\zeta-1}{\zeta} &&= \frac{(1-\frac{m^2}{q^2})\log(1-\frac{q^2}{m^2})-1}{-q^2}. \end{aligned}$$

In the second case there is a nontrivial residue at the holomorphic point $\zeta = \infty$.

The corresponding structures for Minkowski spacetime as minimal case with nontrivial rotation degrees of freedom are as follows:

$$\boxed{\begin{array}{c} \text{spacetime } \mathcal{D}^4 = \mathbf{D}(2) = \mathbf{D}(1) \times \mathcal{Y}^3 \\ \text{with } \mathbf{SO}_0(1,3),\ \ |\Omega^3| = 2\pi^2 \end{array}}$$

$$\boxed{(\overset{4}{*}, q^2) = \begin{cases} (\mp\frac{*}{2i\pi^2}, (q\mp io)^2), & \text{causal, orthogonal} \\ (\mp\frac{*}{i\pi^2}, q^2\mp io), & \text{Feynman, not orthogonal} \end{cases}}$$

$$\boxed{\begin{array}{cccl} \frac{\partial}{\partial q^2}\frac{1}{-q^2+m_1^2} & \overset{4}{*} & \frac{\partial}{\partial q^2}\frac{1}{-q^2+m_2^2} & = \frac{\partial}{\partial q^2}\int_0^1 d\zeta \frac{1}{-\zeta(1-\zeta)q^2+\zeta m_1^2+(1-\zeta)m_2^2} \\ = \frac{1}{(-q^2+m_1^2)^2} & \overset{4}{*} & \frac{1}{(-q^2+m_2^2)^2} & = \int_0^1 d\zeta \frac{\zeta(1-\zeta)}{[-\zeta(1-\zeta)q^2+\zeta m_1^2+(1-\zeta)m_2^2]^2} \end{array}}$$

In general, one obtains the even-dimensional spacetime $\mathcal{D}^{2R}$ distributions of energy–momenta by relativistically compatible 2-sphere spread. Measures for higher-dimensional spacetimes are obtained by derivation of order $R-1$, the rank of the maximal compact group $\mathbf{SO}(2R-1)$,

$$R = 1, 2, \ldots:\quad \frac{1}{\Gamma(R)}\left(\frac{\partial}{\partial q^2}\right)^{R-1}\frac{1}{-q^2+m^2} = \frac{1}{(-q^2+m^2)^R},$$

with the convolution

$$\boxed{\begin{array}{c} \text{spacetime } \mathcal{D}^{2R} = \mathbf{D}(1) \times \mathcal{Y}^{2R-1},\ \ R = 1, 2, \ldots \\ \text{with } \mathbf{SO}_0(1, 2R-1),\ \ |\Omega^{2R-1}| = \frac{2\pi^R}{\Gamma(R)} \end{array}}$$

$$\boxed{(\overset{2R}{*}, q^2) = \begin{cases} (\mp\frac{*}{i|\Omega^{2R-1}|}, (q\mp io)^2), & \text{causal, orthogonal} \\ (\mp\frac{*\,2}{i|\Omega^{2R-1}|}, q^2\mp io), & \text{Feynman, not orthogonal} \end{cases}}$$

$$\boxed{(\tfrac{\partial}{\partial q})^{L_1}\frac{1}{(-q^2+m_1^2)^R} \overset{2R}{*} (\tfrac{\partial}{\partial q})^{L_2}\frac{1}{(-q^2+m_2^2)^R} = (\tfrac{\partial}{\partial q})^{L_1+L_2}\frac{1}{\Gamma(R)}(\tfrac{\partial}{\partial q^2})^{R-1}\int_0^1 d\zeta\frac{1}{-\zeta(1-\zeta)q^2+\zeta m_1^2+(1-\zeta)m_2^2}}$$

With the remaining finite 1-dimensional integration $\int_0^1 d\zeta$, the residual normalization for even-dimensional spacetime $\mathcal{D}^{2R}$ with a Cartan plane uses the volume of the odd-dimensional unit sphere Ω^{2R-1}, the compact partner of the embedded position hyperboloid $\mathcal{Y}^{2R-1}$,

$$\overset{2R}{*} = -\frac{*}{i|\Omega^{2R-1}|} \text{ from } \int \frac{d^{2R}q}{|\Omega^{2R-1}|}\frac{1}{[-(q-io)^2+m^2]^R}e^{iqx} = \vartheta(x)\pi\mathcal{J}_0(|mx|).$$

Coefficients of nontrivial representations of the Lorentz group $\mathbf{SO}_0(1,2R-1)$ are effected by energy–momentum derivatives $(\frac{\partial}{\partial q})^L$, $L=1,2,\ldots$, with the examples, wherever the Γ-functions are defined for $\nu\in\mathbb{R}$,

$$\begin{aligned}
\frac{\Gamma(R+\nu_1)}{(-q^2+m_1^2)^{R+\nu_1}} \overset{2R}{*} \frac{\Gamma(R+\nu_2)}{\Gamma(R)\,(-q^2+m_2^2)^{R+\nu_2}} &= (\tfrac{\partial}{\partial q^2})^{R-1}[\nu_1,\nu_2](q^2)\\
&= \int_0^1 d\zeta \frac{\zeta^{R-1+\nu_1}(1-\zeta)^{R-1+\nu_2}\Gamma(R+\nu_1+\nu_2)}{[-\zeta(1-\zeta)q^2+\zeta m_1^2+(1-\zeta)m_2^2]^{R+\nu_1+\nu_2}},\\
\frac{2q\ \Gamma(1+R+\nu_1)}{(-q^2+m_1^2)^{1+R+\nu_1}} \overset{2R}{*} \frac{\Gamma(R+\nu_2)}{\Gamma(R)\,(-q^2+m_2^2)^{R+\nu_2}} &= 2q(\tfrac{\partial}{\partial q^2})^{R}[\nu_1,\nu_2](q^2)\\
&= 2q\int_0^1 d\zeta \frac{\zeta^{R+\nu_1}(1-\zeta)^{R+\nu_2}\Gamma(1+R+\nu_1+\nu_2)}{[-\zeta(1-\zeta)q^2+\zeta m_1^2+(1-\zeta)m_2^2]^{1+R+\nu_1+\nu_2}},\\
\frac{2q\ \Gamma(1+R+\nu_1)}{(-q^2+m_1^2)^{1+R+\nu_1}} \overset{2R}{*} \frac{2q\ \Gamma(1+R+\nu_2)}{\Gamma(R)\,(-q^2+m_2^2)^{1+R+\nu_2}} &= 2\tfrac{\partial}{\partial q}\otimes q\ (\tfrac{\partial}{\partial q^2})^{R}[\nu_1,\nu_2](q^2).
\end{aligned}$$

The convolution of Feynman propagators in 4-dimensional flat spacetime $R=2$, e.g., $\frac{1}{q^2+io-m_1^2} * \frac{1}{q^2+io-m_2^2}$ as arising in perturbation expansions for flat spacetime quantum field theories, is not defined. This shows up in the singularities of the Γ-functions. The divergent parts $\frac{1}{q_{\rm P}^2-m_1^2} * \frac{1}{q_{\rm P}^2-m_2^2}$ and $\delta(q^2-m_1^2) * \frac{1}{q_{\rm P}^2-m_2^2}$ involve the off-shell interaction contribution. They are not coefficients of translation product representations like the meaningful on-shell convolution above $\delta(q^2-m_1^2) * \delta(q^2-m_2^2)$.

9.5 Tangent Structures for Spacetime

The dual of a Lie algebra $L=\log G\cong\mathbb{R}^n$, i.e., its linear forms L^T, is easier accessible than the group dual $\check{G}$, which in general is no group, but, e.g., a cone or a direct sum of cones. L^T contains all eigenvalues of the Lie algebra action. Multilinear forms of the eigenvalues give the invariants that characterize the group dual. The Lie group acts on itself $G\times G\longrightarrow G$ by left and right multiplication, on its Lie algebra $G\times L\longrightarrow L$, and on the linear forms $G\times L^T\longrightarrow L^T$ by adoint and coadjoint action respectively.

9.5.1 Differential Operators and Lie Algebra Kernels

In a representation framework, bound state vectors (particles) and interactions have a close connection, expressible with the transition from Lie group representation coefficients, by derivation, to those of Lie algebra representations. This derivative transition is illustrated for 3-dimensional position, where a third-order derivative of the ground state wave function $\vec{x}\longmapsto e^{-r}$ of the nonrelativistic hydrogen atom, a positive-type $L^\infty(\mathbb{R}^3)$-function, leads via the Yukawa potential to the Yukawa force,

$$\frac{\partial}{\partial\vec{x}}\frac{\partial}{\partial r^2}e^{-r} = -\frac{\partial}{\partial\vec{x}}\frac{e^{-r}}{2r} = \frac{\vec{x}}{r}\frac{1+r}{2r^2}e^{-r}.$$

As will be discussed below, it is group–theoretically interpretable that the pointwise product $\frac{1}{r}\cdot e^{-mr}$ of the Coulomb potential with a bound state wave function gives a Yukawa potential.

The invariant differential operators [5] of functions on a group $L^\infty(G) = L^1(G)'$ and on its symmetric spaces G/H are related to the center of the enveloping algebra for the Lie algebra $\log G$. For semisimple Lie groups the Killing form related Laplace-Beltrami operator with its eigenfunctions and eigendistributions plays the essential role.

The "inverses" of invariant derivatives D with respect to the Dirac distribution $D\gamma^{-1} = \delta_0$ define fundamental kernels γ^{-1}, familiar from Green functions (distributions) of differential equations of motion [7]. The image of group functions $d \in \check{\mathcal{G}}$ under the linear transformation $\gamma^{-1} \cdot d$ with a fundamental kernel gives the associated $\check{\mathcal{G}}$-module with the Lie algebra kernels, denoted by $\log \check{\mathcal{G}}$, and called, for physics, G-induced interactions

$$\gamma^{-1} \in \mathcal{M}(G), \quad L^\infty(G) \supseteq \check{\mathcal{G}} \longrightarrow \log \check{\mathcal{G}} = \gamma^{-1} \cdot \check{\mathcal{G}} \subseteq \mathcal{M}(G).$$

It is a submodule of the Radon distribution $\mathcal{M}(G) \cdot L^\infty(G) \subseteq \mathcal{M}(G)$ with the convolution and pointwise product for functions and Fourier transforms in the parametrization by tangent translations

$$\begin{array}{c|c|c} * & \log \check{\mathcal{G}} & \check{\mathcal{G}} \\ \hline \log \check{\mathcal{G}} & \log \check{\mathcal{G}} & \check{\mathcal{G}} \\ \check{\mathcal{G}} & \check{\mathcal{G}} & - \end{array} \qquad \check{\mathcal{G}} \qquad \begin{array}{c|c|c} \cdot & \log \check{\mathcal{G}} & \check{\mathcal{G}} \\ \hline \log \check{\mathcal{G}} & - & \log \check{\mathcal{G}} \\ \check{\mathcal{G}} & \log \check{\mathcal{G}} & \check{\mathcal{G}} \end{array}, \quad \log \check{\mathcal{G}} = \gamma^{-1} \cdot \check{\mathcal{G}}$$

$$\updownarrow \qquad \text{Fourier} \qquad \updownarrow$$

$$\begin{array}{c|c|c} \cdot & \log \tilde{\mathcal{G}} & \tilde{\mathcal{G}} \\ \hline \log \tilde{\mathcal{G}} & \log \tilde{\mathcal{G}} & \tilde{\mathcal{G}} \\ \tilde{\mathcal{G}} & \tilde{\mathcal{G}} & - \end{array} \qquad \tilde{\mathcal{G}} \qquad \begin{array}{c|c|c} * & \log \tilde{\mathcal{G}} & \tilde{\mathcal{G}} \\ \hline \log \tilde{\mathcal{G}} & - & \log \tilde{\mathcal{G}} \\ \tilde{\mathcal{G}} & \log \tilde{\mathcal{G}} & \tilde{\mathcal{G}} \end{array}, \quad \log \tilde{\mathcal{G}} = \tilde{\gamma}^{-1} * \tilde{\mathcal{G}}$$

from group product for product representations

For time, position, and spacetime the "inverse derivative" distributions have poles at the trivial invariant and the Fourier transforms:

$$\left.\begin{array}{c} \mathbb{R}: \\ N = 0, 1, \dots, \end{array}\right\} \quad \int \frac{dq}{2i\pi} \frac{\Gamma(1+N)}{(q-io)^{1+N}} e^{iqx} = \vartheta(x)(ix)^N,$$

$$\left.\begin{array}{c} \mathbf{SO}(2R-1): \\ R = 1, 2, \dots, \\ \frac{\Gamma(N+\frac{1}{2})}{\Gamma(\frac{1}{2})} = \frac{2\Gamma(2N)}{4^N\Gamma(N)}, \end{array}\right\} \quad \int \frac{d^{2R-1}q}{2\pi^R} \begin{pmatrix} \frac{\Gamma(R-1-N)}{(\vec{q}^2)^{R-1-N}} \\ \frac{2i\vec{q}\,\Gamma(R-N)}{(\vec{q}^2)^{R-N}} \end{pmatrix} e^{-i\vec{q}\vec{x}} = \frac{\Gamma(2N)}{\Gamma(N)} \frac{2}{r^{1+2N}} \begin{pmatrix} 1 \\ \vec{x} \end{pmatrix},$$

$$\mathbf{SO}_0(1, 2R-1): \quad \int \frac{d^{2R}q}{2\pi^R} \begin{pmatrix} \frac{\Gamma(R)}{[-(q-io)^2]^R} \\ \frac{\Gamma(R-1-N)}{[-(q-io)^2]^{R-1-N}} \\ \frac{2iq\,\Gamma(1+R)}{[-(q-io)^2]^{1+R}} \\ \frac{2iq\,\Gamma(R-N)}{[-(q-io)^2]^{R-N}} \end{pmatrix} e^{iqx} = \pi\vartheta(x_0) \begin{pmatrix} \vartheta(x^2) \\ \delta^{(N)}(-\frac{x^2}{4}) \\ x\vartheta(x^2) \\ x\delta^{(N)}(-\frac{x^2}{4}) \end{pmatrix}.$$

9.5.2 Fundamental Interactions of Spacetime

The abelian time group $\mathbf{D}(1) \cong \mathbb{R}$ is isomorphic to the additive group structure of its Lie algebra. This is reflected by the coincidence of the fundamental kernel for the causal group with the trivial group representation. The fundamental kernel inverting the invariant derivative $\frac{d}{dx_0}$ acts as identity; the abelian causal group induces only trivial interactions

$$\begin{aligned} \gamma^{-1}(x_0) &= \vartheta(x_0), \ \frac{d}{dx_0}\vartheta(x_0) = \delta(x_0), \\ \vartheta(x_0)\vartheta(x_0)e^{imx_0} &= \vartheta(x_0)e^{imx_0} = \int \frac{dq}{2i\pi} \, \frac{1}{q-io-m} e^{iqx_0}. \end{aligned}$$

$\frac{1}{q-io} \ * \ \tilde{\mathcal{D}}^1 = \log \tilde{\mathcal{D}}^1$				
$\frac{1}{q-io}$	$\frac{*}{2i\pi}$	$\frac{1}{q-io-m}$	$=$	$\frac{1}{q-io-m}$
$\frac{1}{q-io-m_1}$	$\frac{*}{2i\pi}$	$\frac{1}{q-io-m_2}$	$=$	$\frac{1}{q-io-m_+}$

The inverse of the invariant derivative $\frac{d}{dx}$ is the sign function as fundamental kernel for 1-dimensional position, acting on the selfdual representation coefficients of the hyperboloid $\mathbf{SO}_0(1,1) \cong \mathbb{R}$,

$$\begin{aligned} \gamma^{-1}(x) &= \frac{\epsilon(x)}{2}, \ \frac{d}{dx}\frac{\epsilon(x)}{2} = \delta(x), \\ \frac{\epsilon(x)}{2} e^{-|mx|} &= \int \frac{dq}{2\pi} \, \frac{iq}{q^2+m^2} e^{-iqx}. \end{aligned}$$

$\frac{i}{q_{\mathrm{P}}} \ * \ \tilde{\mathcal{Y}}^1 = \log \tilde{\mathcal{Y}}^1$				
$\frac{i}{q_{\mathrm{P}}}$	$\frac{*}{2\pi}$	$\frac{2iq}{q^2+m^2}$	$=$	$\frac{\lvert m\rvert}{q^2+m^2}$
$\frac{2iq}{q^2+m_1^2}$	$\frac{*}{2\pi}$	$\frac{2iq}{q^2+m_2^2}$	$=$	$\frac{2\lvert m_+\rvert}{q^2+m_+^2}$
$\frac{2iq}{q^2+m_1^2}$	$\frac{*}{2\pi}$	$\frac{2\lvert m_2\rvert}{q^2+m_2^2}$	$=$	$\frac{2iq}{q^2+m_+^2}$

with the principal value $\frac{i}{q_{\mathrm{P}}} = \frac{iq}{q^2+o^2}$.

For hyperbolic position $\mathcal{Y}^{2R-1} \cong \mathbf{SO}_0(1, 2R-1)/\mathbf{SO}(2R-1)$ with nonabelian degrees of freedom $R \geq 2$ and nontrivial Cartan torus $\mathbf{SO}(2)^{R-1}$, the fundamental kernel (interaction) as inverse of the invariant generalized Laplacian $(\vec{\partial}^2)^{R-1}$ is the Coulomb potential $\frac{1}{r}$. The action on the functions of positive type $\vec{x} \longmapsto e^{-|m|r}$ gives Yukawa potentials as position kernels,

$$\begin{aligned} R \geq 2: \ \gamma^{-1}(\vec{x}) &= \frac{1}{r}, \ -(\partial^2)^{R-1}\frac{1}{r} = \frac{(2\pi)^{2R-1}}{|\Omega^{2R-1}|}\delta(\vec{x}); \ R = 2: -\partial^2\frac{1}{r} = 4\pi\delta(\vec{x}), \\ \frac{1}{r}e^{-|m|r} &= \int \frac{d^{2R-1}q}{|\Omega^{2R-1}|} \, \frac{1}{(\vec{q}^2+m^2)^{R-1}} e^{-i\vec{q}\vec{x}}. \end{aligned}$$

$\frac{1}{(\vec{q}^2)^{R-1}} \ * \ \tilde{\mathcal{Y}}^{2R-1} = \log \tilde{\mathcal{Y}}^{2R-1}, \quad R = 2, 3, \ldots$				
$\frac{1}{(\vec{q}^2)^{R-1}}$	$\overset{2R-1}{*}$	$\frac{2\lvert m\rvert}{(\vec{q}^2+m^2)^R}$	$=$	$\frac{1}{(\vec{q}^2+m^2)^{R-1}}$
$\frac{1}{(\vec{q}^2+m_1^2)^{R-1}}$	$\overset{2R-1}{*}$	$\frac{2\lvert m_2\rvert}{(\vec{q}^2+m_2^2)^R}$	$=$	$\frac{1}{(\vec{q}^2+m_+^2)^{R-1}}$
$\frac{2i\vec{q}}{(\vec{q}^2+m_1^2)^R}$	$\overset{2R-1}{*}$	$\frac{2\lvert m_2\rvert}{(\vec{q}^2+m_2^2)^R}$	$=$	$\frac{2i\vec{q}}{(\vec{q}^2+m_+^2)^R}$

Causal spacetime $\mathcal{D}^{2R} = \mathbf{D}(1) \times \mathcal{Y}^{2R-1}$ with real rank 2 has two fundamental kernels, the Lorentz compatibly embedded kernels of the causal group (eigentime) and of hyperbolic position. Both are future lightcone supported Dirac distributions,

$$\begin{aligned} \int \frac{d^{2R}q}{i|\Omega^{2R-1}|}\frac{q}{(q-io)^2}e^{iqx} &= \Gamma(R)\frac{x}{2}\pi\vartheta(x_0)\delta^{(R-1)}(-\frac{x^2}{4}), \\ R \geq 2: \quad \int \frac{d^{2R}q}{|\Omega^{2R-1}|}\frac{1}{[-(q-io)^2]^{R-1}}e^{iqx} &= (R-1)\pi\vartheta(x_0)\delta(\frac{x^2}{4}), \\ (\partial^2)^{R-1}\int \frac{d^{2R}q}{|\Omega^{2R-1}|}\frac{1}{[-(q-io)^2]^{R-1}}e^{iqx} &= \partial \int \frac{d^{2R}q}{i|\Omega^{2R-1}|}\frac{q}{(q-io)^2}e^{iqx} = \frac{(2\pi)^{2R}}{|\Omega^{2R-1}|}\delta(x). \end{aligned}$$

The time and position projections lead back to the embedded kernels, i.e., to the characteristic function $\vartheta(x_0)$ for time and for $R = 1$ position to $\frac{\epsilon(z)}{2} = \frac{x}{2r}$, and for $R \geq 2$ to the Coulomb potential $\frac{1}{r}$,

$$\begin{pmatrix} \int \frac{|\Omega^{2R-1}|\, d^{2R-1}x}{(2\pi)^{2R}} \\ \int \frac{dx_0}{2\pi} \end{pmatrix} \int \frac{d^{2R}q}{i|\Omega^{2R-1}|} \, \frac{q}{(q-io)^2} e^{iqx} = \begin{pmatrix} \vartheta(x_0) \\ \frac{\vec{x}}{r} \frac{\Gamma(2R-1)}{2(r^2)^{R-1}} \end{pmatrix},$$

$$R \geq 2: \quad \begin{pmatrix} \int \frac{|\Omega^{2R-1}|\, d^{2R-1}x}{(2\pi)^{2R}} \\ \int \frac{dx_0}{2\pi} \end{pmatrix} \int \frac{d^{2R}q}{|\Omega^{2R-1}|} \, \frac{1}{[-(q-io)^2]^{R-1}} e^{iqx} = \begin{pmatrix} \vartheta(x_0) \frac{x_0^{2R-3}}{\Gamma(2R-2)} \\ \frac{1}{r} \end{pmatrix}.$$

The spacetime kernels are computed with the following convolutions:

$\frac{q}{q^2} \ast \tilde{\mathcal{D}}^{2R} \cup \frac{1}{(q^2)^{R-1}} \ast \tilde{\mathcal{D}}^{2R} = \log \tilde{\mathcal{D}}^{2R},\ R = 1, 2, \ldots$				
	$\frac{q}{-q^2}$	$\overset{2R}{\ast}$	$\frac{1}{(-q^2+m^2)^R}$	$= \int_0^1 d\zeta \, \frac{(1-\zeta)^{R-1} q}{-\zeta q^2 + m^2}$
	$\frac{q}{-q^2+m_1^2}$	$\overset{2R}{\ast}$	$\frac{1}{(-q^2+m_2^2)^R}$	$= \int_0^1 d\zeta \, \frac{(1-\zeta)^R q}{-\zeta(1-\zeta)q^2 + \zeta m_1^2 + (1-\zeta) m_2^2}$
$R \geq 2:$	$\frac{1}{(-q^2)^{R-1}}$	$\overset{2R}{\ast}$	$\frac{1}{(-q^2+m^2)^R}$	$= \int_0^1 d\zeta \, \frac{\zeta^{R-2}}{(-\zeta q^2 + m^2)^{R-1}}$
	$\frac{1}{(-q^2+m_1^2)^{R-1}}$	$\overset{2R}{\ast}$	$\frac{1}{(-q^2+m_2^2)^R}$	$= \int_0^1 d\zeta \, \frac{\zeta^{R-2}(1-\zeta)^{R-1}}{[-\zeta(1-\zeta)q^2 + \zeta m_1^2 + (1-\zeta) m_2^2]^{R-1}}$

9.5.3 Feynman Propagators

A Feynman propagator for flat spacetime, e.g., for a massive scalar field,

$$\langle \{\mathbf{\Phi}(y), \mathbf{\Phi}(x)\} + i\epsilon(x_0 - y_0) i[\mathbf{\Phi}(y), \mathbf{\Phi}(x)] \rangle = \frac{i}{\pi} \int d^4q \, \frac{1}{q^2 + io - m^2} e^{iq(x-y)},$$

is a combination of a representation coefficient $\langle \{\mathbf{\Phi}(y), \mathbf{\Phi}(x)\} \rangle$ of the Poincaré group $\mathbf{SO}_0(1,3) \vec{\times} \mathbb{R}^4$ for particles and an embedded kernel $\epsilon(x_0 - y_0) i[\mathbf{\Phi}(y), \mathbf{\Phi}(x)]$ of hyperbolic position for interactions. The Fourier transformed energy-momentum distributions in Feynman propagators give spacetime functions (Macdonald $\mathcal{K}_0$ and Neumann $\mathcal{N}_0$) for the real part (on-shell) and Radon distributions with Bessel functions $\mathcal{J}_0$ for the imaginary part (off-shell),

$$\begin{aligned} \frac{i}{\pi} \int d^{2R}q \, \frac{1}{q^2 + io - 1} e^{iqx} &= \left(\frac{\partial}{\partial \frac{x^2}{4\pi}}\right)^{R-1} [\vartheta(-x^2) 2\mathcal{K}_0(|x|) - \vartheta(x^2)\pi(\mathcal{N}_0 + i\mathcal{J}_0)(|x|)] \\ &= \left(\frac{\partial}{\partial \frac{x^2}{4\pi}}\right)^{R-1} \int d\psi \, [\vartheta(-x^2) e^{-|x|\cosh\psi} + \vartheta(x^2) e^{-i|x|\cosh\psi}]. \end{aligned}$$

The projection of a Feynman propagator $\frac{i}{\pi}\frac{1}{q^2 + io - m^2}$ to time and position by position and time integration, respectively, displays a translation representation coefficient $\cos m x_0$ only for the on-shell part $\delta(q^2 - m^2)$, whereas the principal value off-shell part $\frac{i}{\pi}\frac{1}{q_{\mathrm{P}}^2 - m^2}$ with causal support in spacetime is position-projected to the exponential potential $e^{-|m|r}$ for 2-dimensional spacetime and to the Yukawa potential $\frac{e^{-|m|r}}{r}$ for 4-dimensional spacetime,

$$\begin{aligned} \begin{pmatrix} \int \frac{d^{2R-1}x}{(2\pi)^{2R-1}} \\ \int \frac{dx_0}{2\pi} \end{pmatrix} \frac{i}{\pi} \int d^{2R}q \, \frac{1}{q^2 + io - 1} e^{iqx} &= \begin{pmatrix} -\int \frac{dq_0}{i\pi} \, \frac{1}{q_0^2 + io - 1} e^{iq_0 x_0} \\ \int \frac{d^{2R-1}q}{i\pi} \, \frac{1}{\vec{q}^2 + 1} e^{-i\vec{q}\vec{x}} \end{pmatrix} \\ &= \begin{pmatrix} \cos x_0 \\ 0 \end{pmatrix} - i \begin{pmatrix} \epsilon(x_0) \sin x_0 \\ (-\frac{\partial}{\partial \frac{r^2}{4\pi}})^{R-1} e^{-r} \end{pmatrix}. \end{aligned}$$

The off-shell contributions are no coefficients of Poincaré group representations,

$$\begin{aligned}
\int \frac{d^{2R}q}{\pi} \frac{1}{q_{\rm P}^2-1} e^{iqx} &= i\epsilon(x_0) \int d^{2R}q \; \epsilon(q_0)\delta(q^2-1)e^{iqx} \\
&= -\left(\frac{\partial}{\partial \frac{x^2}{4}}\right)^{R-1} \vartheta(x^2)\pi\mathcal{J}_0(|x|), \\
(\partial^2+1)\int \frac{d^{2R}q}{\pi} \frac{1}{q_{\rm P}^2-1} e^{iqx} &= -\frac{(2\pi)^R}{\pi}\delta(x).
\end{aligned}$$

The advanced and retarded energy-momentum distributions are off-shell:

$$\begin{aligned}
\int \frac{d^{2R}q}{2\pi} \frac{1}{(q\mp io)^2-1} e^{iqx} &= \vartheta(\pm x_0) \int \frac{d^{2R}q}{\pi} \frac{1}{q_{\rm P}^2-1} e^{iqx} \\
&= \pm i\vartheta(\pm x_0) \int d^{2R}q \; \epsilon(q_0)\delta(q^2-1)e^{iqx}.
\end{aligned}$$

9.5.4 Duality for Group Representation and Fundamental Kernel

In residual representations the characterizing invariants of time $\mathbf{D}(1)$ and of hyperbolic position $\mathcal{Y}^{2R-1}$, both with real rank 1, arise as singularities of the harmonic components,

$$\tilde{d}_m(q) = \begin{cases} \frac{1}{q-io-m} & \text{for } \tilde{\mathcal{D}}^1, \\ \frac{2iq}{q^2+m^2} & \text{for } \tilde{\mathcal{Y}}^1, \\ \frac{2|m|}{(\vec{q}^2+m^2)^R} & \text{for } \tilde{\mathcal{Y}}^{2R-1}, \; R \geq 2. \end{cases}$$

The invariants are the intrinsic units for energy and momentum. The denominator energy-momentum polynomials define eigenvalue equations for the invariants:

$$\frac{1}{\tilde{d}_m(q)} = 0, \quad \begin{cases} q-m &= 0, \text{ linear}, \\ \vec{q}^2+m^2 &= 0, \text{ self-dual}. \end{cases}$$

Since the invariants of product representations of even-dimensional causal spacetimes with real rank 2 do not arise as pole structures in the convolution products, there has to be found a spacetime generalizable equivalent formulation for the characterization and determination of the invariants. Such a formulation uses the energy-momentum functions for the tangent Lie algebra kernels associated to the group representations,

$$\tilde{\mathcal{G}} \ni \tilde{d}_m(q) \longmapsto \tilde{\gamma}^{-1} * \tilde{d}_m(q) \in \log \tilde{\mathcal{G}}.$$

The harmonic components of the fundamental kernels can be normalized with the representation invariant (intrinsic unit) to a dimensionless "inverse derivative,"

$$\tilde{\gamma}^{-1}(q) = \begin{cases} \frac{M}{q-io} & \text{for } \tilde{\mathcal{D}}^1 \longrightarrow \log \tilde{\mathcal{D}}^1, \\ \frac{iqM}{q^2+o^2} & \text{for } \tilde{\mathcal{Y}}^1 \longrightarrow \log \tilde{\mathcal{Y}}^1, \\ (\frac{M^2}{\vec{q}^2})^{R-1} & \text{for } \tilde{\mathcal{Y}}^{2R-1} \longrightarrow \log \tilde{\mathcal{Y}}^{2R-1}, \; R \geq 2. \end{cases}$$

The invariants of group and fundamental kernel coincide, $m = M$, if their representation coefficients are dual to each other as seen in the dual product, which can be written as the convolution at trivial energy and momenta $q = 0$,

$$1 = \langle \gamma^{-1}, d_m \rangle = \tilde{\tilde{\gamma}}^{-1} * \tilde{d}_m(0),$$

$$\left.\begin{array}{llllll} \tilde{\mathcal{D}}^1: & \frac{-M}{q-io} & \underset{2i\pi}{*} & \frac{1}{q-io-m} & = \frac{-M}{q-io-m} \\ \tilde{\mathcal{Y}}^1: & \frac{iq|M|}{q^2+o^2} & \underset{2\pi}{*} & \frac{2iq}{q^2+m^2} & = \frac{|Mm|}{q^2+m^2} \\ \tilde{\mathcal{Y}}^{2R-1}: & (\frac{M^2}{\vec{q}^2})^{R-1} & \overset{2R-1}{*} & \frac{2|m|}{(\vec{q}^2+m^2)^R} & = \frac{(M^2)^{R-1}}{(\vec{q}^2+m^2)^{R-1}} \end{array}\right\} \overset{q=0}{=} 1 \text{ for } M^2 = m^2.$$

9.5.5 The Relative Time-Position Normalization of Causal Spacetime

The representations of even-dimensional spacetime as a product of homogeneous spaces for time and hyperbolic position $\mathcal{D}^{2R} = \mathbf{D}(1) \times \mathcal{Y}^{2R-1}$,

$$\mathcal{D}^{2R} \ni \vartheta(x)x \longmapsto \int \tilde{d}^{2R}_{\kappa^2}(q) \frac{d^{2R}q}{i|\Omega^{2R-1}|} \frac{2q|m_0|}{(q-io)^2 - m_0^2} e^{iqx} = d^{2R}_{\kappa^2} * \omega_{m_0^2}(x),$$

can be considered as Lorentz-compatibly embedded representations of the causal group (eigentime) $\mathbf{D}(1)$. The factor representing the Lorentz group $\mathbf{SO}_0(1, 2R-1)$ is the Bessel function, a metric–inducing function of positive type,

$$\int \tilde{d}^{2R}_{\kappa^2}(q) \frac{d^{2R}q}{|\Omega^{2R-1}|} e^{iqx} = \vartheta(x)\pi \mathcal{J}_0(|m_\kappa x|) = d^{2R}_{\kappa^2}(x).$$

The dimensionless hyperbolic energy-momentum measure with one position characterizing invariant m_κ^2 is the derivative of a logarithm:

$$\tilde{d}^{2R}_{\kappa^2}(q) \frac{d^{2R}q}{|\Omega^{2R-1}|} = \frac{d^{2R}q}{2\pi^R} \frac{\Gamma(R)}{[-(q-io)^2 + m_\kappa^2]^R} = \frac{d^{2R}q}{2\pi^R} \left(\frac{\partial}{\partial q^2}\right)^R \log[-(q-io)^2 + m_\kappa^2].$$

In the *hyperbolic convolution product* $\overset{2R}{*_\kappa}$, defined with the hyperbolic measure $\tilde{d}^{2R}_{\kappa^2}(q) \frac{d^{2R}q}{|\Omega^{2R-1}|}$, the right factor is multiplied with the representation of hyperbolic position

$$\overset{2R}{*_\kappa} \cong \overset{2R}{*} \tilde{d}^{2R}_{\kappa^2}(q) = * \frac{i}{|\Omega^{2R-1}|[-(q-io)^2 + m_\kappa^2]^R}.$$

The hyperbolic convolution describes, in a Lorentz compatible embedding, the action of the Lie algebra kernels of the causal group $\mathbf{D}(1)$ on the group $\mathbf{D}(1)$-representations,

$$\begin{array}{llll} \tilde{\gamma}_{2R}^{-1} \overset{2R}{*_\kappa} \tilde{\mathcal{D}}^1 & = \tilde{\gamma}_{2R}^{-1} * \tilde{\mathcal{D}}^{2R}, & \tilde{\gamma}_{2R}^{-1} \overset{2R}{*_\kappa} \tilde{\omega} & = \tilde{\gamma}_{2R}^{-1} \overset{2R}{*} [\tilde{d}^{2R}_{\kappa^2} \cdot \tilde{\omega}] \\ \text{Fourier} \updownarrow & & & \\ \gamma_{2R}^{-1} \overset{2R}{\cdot_\kappa} \mathcal{M}(\mathbb{R}^{2R}_+) & = \gamma_{2R}^{-1} \cdot L^\infty(\mathbb{R}^{2R}_+), & \gamma_{2R}^{-1} \overset{2R}{\cdot_\kappa} \omega & = \gamma_{2R}^{-1} \cdot [d^{2R}_{\kappa^2} * \omega]. \end{array}$$

For $2R$-dimensional causal spacetime, the duality condition between the embedded $\mathbf{D}(1)$-representation and the embedded fundamental $\mathbf{D}(1)$-kernel with intrinsic unit $|m_0|$,

$$\tilde{\gamma}_{2R}^{-1}(q) = \frac{q|m_0|}{q^2} \quad \text{for } \tilde{\mathcal{D}}^{2R} \xrightarrow{\tilde{\gamma}_{2R}^{-1} *} \log \tilde{\mathcal{D}}^{2R},$$

determines the ratio $\kappa^2 = \frac{m_\kappa^2}{m_0^2}$ of the invariants. The dual product of the embedded $\mathbf{D}(1)$-Radon distributions is defined by the embedded metric inducing function of hyperbolic position. For duality, it has to give the unit

$$\begin{aligned} \langle \gamma_{2R}^{-1}, \omega_{m_0^2} \rangle_{d_{\kappa^2}^{2R}} = \langle \gamma_{2R}^{-1}, d_{\kappa^2}^{2R} * \omega_{m_0^2} \rangle &= \frac{q|m_0|}{q^2} \overset{2R}{*_\kappa} \frac{2q|m_0|}{q^2 - m_0^2} \Big|_{q=0} \\ &= -\frac{\mathbf{1}_{2R}}{R} \log_R \kappa^2 = \mathbf{1}_{2R}. \end{aligned}$$

The hyperbolic convolution in the duality condition

$$\begin{aligned} \frac{q}{q^2} \overset{2R}{*_\kappa} \frac{2q}{q^2-1} &= \frac{q}{q^2} \overset{2R}{*} \frac{1}{(-q^2+\kappa^2)^R} \frac{2q}{q^2-1} = \frac{q}{q^2} \overset{2R}{*} \int_0^1 d\xi \, \frac{(1-\xi)^{R-1}(-2qR)}{[-q^2+\xi+(1-\xi)\kappa^2]^{1+R}} \\ &= \frac{\partial}{\partial q} \otimes q \int_0^1 d\zeta \int_0^1 d\xi \frac{(1-\zeta)^{R-1}(1-\xi)^{R-1}}{-\zeta q^2+\xi+(1-\xi)\kappa^2}, \end{aligned}$$

is computed with

$$\begin{aligned} \frac{q}{q^2-m_1^2} \overset{2R}{*} \frac{-2qR}{(-q^2+m_2^2)^{1+R}} &= \frac{q}{q^2-m_1^2} \overset{2R}{*} -\frac{\partial}{\partial q} \frac{1}{(-q^2+m_2^2)^R} \\ &= \frac{\partial}{\partial q} \otimes q \int_0^1 d\zeta \frac{(1-\zeta)^R}{-\zeta(1-\zeta)q^2+\zeta m_1^2+(1-\zeta)m_2^2}. \end{aligned}$$

The duality condition can also be formulated as a normalization condition for the $\mathbf{D}(1)$-representing Lorentz scalar Radon measure with the harmonic components $\tilde{d}_{m_0^2}(q) = \frac{2m_0^2}{q^2-m_0^2}$,

$$\begin{aligned} R = -\log_R \kappa^2 &= \int \tilde{d}_{\kappa^2}^{2R}(q) \frac{d^{2R}q}{i|\Omega^{2R-1}|} \, \frac{2m_0^2}{(q-io)^2-m_0^2} = \langle d_{m_0^2} \rangle_{d_{\kappa^2}^{2R}} \\ &= \int_0^1 d\zeta \frac{\zeta^{R-1}}{1-\zeta(1-\kappa^2)} = \frac{1}{1-\kappa^2} \int_{\kappa^2}^1 \frac{dM^2}{M^2} \, \left(\frac{1-M^2}{1-\kappa^2}\right)^{R-1} \\ &= -\frac{1}{(1-\kappa^2)^R}[\log \kappa^2 + \sum_{k=1}^{R-1} \frac{(1-\kappa^2)^k}{k}] = \sum_{k=R}^{\infty} \frac{(1-\kappa^2)^{k-R}}{k}. \end{aligned}$$

Functions like the "R-tail" of the logarithm $\log_R \kappa^2$ are typical for the normalization of hyperbolic representations, e.g., the Plancherel measure of the irreducible representations for the harmonic analysis [6] of functions on nonabelian odd-dimensional hyperboloids $L^2(\mathcal{Y}^{2R-1})$, $R = 2, 3, \ldots$, which is given by $m_R^2 = m^2$ for $R = 2$ and by $m_R^2 = m^2 \prod_{k=1}^{R-2}(1 + \frac{m^2}{k^2})$ for $R \geq 3$ where the full product is the hyperbolic Macdonald function $\prod_{k=1}^{\infty}(1 + \frac{m^2}{k^2}) = \frac{\sinh \pi m}{\pi m}$.

The invariant ratio κ^2 is the most important characteristic number for the residual representations of even-dimensional causal spacetime $\mathcal{D}^{2R}$ with real rank 2. It relates the two Lorentz invariants to each other as intrinsic units for the embedded representations of the causal group (eigentime) $\mathbf{D}(1)$ with $q_0 = \pm|m_0|$ (translation–invariant m_0^2) and position $\mathcal{Y}^{2R-1}$ with $|\vec{q}| = \pm i|m_\kappa|$ (interaction–invariant m_κ^2).

For abelian Cartan spacetime $\mathcal{D}^2$ with one time and one position dimension $\mathbf{D}(1) \cong \mathbb{R} \cong \mathcal{Y}^1$, the invariants are equal: The advanced energy-momentum measure is a dipole $\frac{2q|m_0|}{(-q^2+m_\kappa^2)(q^2-m_0^2)} \to -\frac{2q|m|}{(-q^2+m^2)^2}$. For nonabelian spacetime $R = 2, 3, \ldots$ the integration $\frac{dM^2}{M^2} = d\log M^2$ comes with a nontrivial factor $(\frac{1-M^2}{1-\kappa^2})^{R-1}$ from the Cartan torus $\mathbf{SO}(2)^{R-1}$. The logarithm of the mass ratio goes with the rank $-\log\kappa^2 \sim 1 + R = 3, 4, \ldots$,

$$-R = \log_R \kappa^2 = \begin{cases} \frac{1}{1-\kappa^2}\log\kappa^2 & \Rightarrow \kappa^2 = 1, \qquad R = 1, \\ \frac{1}{(1-\kappa^2)^2}(\log\kappa^2 + 1 - \kappa^2) & \Rightarrow \kappa^2 \sim e^{-3} \sim \frac{1}{20.1}, \qquad R = 2. \end{cases}$$

For unitary relativity $\mathcal{D}^4 \cong \mathbf{GL}(\mathbb{C}^2)\mathbf{U}(2)$, the ratio $\kappa^2 = \frac{m_\kappa^2}{m_0^2}$ of the interaction invariant to the translation invariant determines the coupling constants of the lightcone–supported massless gauge fields [9].

9.6 Translation Invariants as Particle Masses

The time $\mathbf{D}(1) \cong \mathbb{R}$ invariants of the powers of one defining representation of causal spacetime $\mathcal{D}^4 \cong \mathbf{D}(1) \times \mathcal{Y}^3 \cong \mathbb{R}^4_+$ are proposed to determine the mass spectrum of relativistic particles. Since the causal spacetime group $\mathbf{GL}(\mathbb{C}^2)$ has real rank 2, i.e., two characterizing continuous invariants $\{m_0^2, m_\kappa^2\}$, the invariants for the products are related to both the embedded causal group $\mathbf{D}(1)$ and Lorentz group $\mathbf{SO}_0(1,3)$-representations of 3-position $\mathcal{Y}^3$.

In contrast to the linear spacing for the time $\mathbf{D}(1)$-invariants (harmonic oscillator),

$$\begin{array}{l} \mathbf{D}(1) \cong \mathbb{R} \ni t \longmapsto e^{imt} = \int dq\, \delta(q-m)e^{iqt} \longmapsto (e^{imt})^k \in L^\infty(\mathbb{R})_+, \\ \{q = E_k = km \mid k = 0, 1, 2, \ldots\}, \end{array}$$

and the "squared spacing" for the continuous $\mathbf{SO}_0(1,3)$-invariants as visible in the bound waves of 3-position $\mathcal{Y}^3$ (nonrelativistic hydrogen atom and periodic system of atoms),

$$\begin{array}{l} \mathcal{Y}^3 \ni \vec{x} \longmapsto e^{-|m|r} = \int \frac{d^3q}{\pi^2}\frac{|m|}{(\vec{q}^2+m^2)^2}e^{-i\vec{q}\vec{x}} \longmapsto (e^{-|m|r})^k \in L^\infty(\mathcal{Y}^3)_+, \\ \{\vec{q}^2 = \frac{1}{2E_k} = -k^2m^2 \mid k = 1 + 2J, J = 0, \frac{1}{2}, 1, \frac{3}{2}, \ldots\}, \end{array}$$

both groups with real rank 1 and one continuous invariant, there is no such simple regularity for the masses of spacetime particles. The simple energy-momentum pole structure in the products for causal group and hyperbolic position is a peculiarity of real rank 1.

9.6.1 Eigenvalues from Causal Group Kernels

For the causal group $\mathbf{D}(1)$, the convolution powers of the energy distribution for the defining representation with intrinsic unit $\underline{q} = qm$

$$\tilde{d}(q) = \frac{1}{q-1}, \quad (\tilde{d})^{*1+k} = \underbrace{\tilde{d} \overset{1}{*} \tilde{d} \overset{1}{*} \cdots \overset{1}{*} \tilde{d}}_{(1+k)\text{ times}} \quad \text{with } (\overset{1}{*}, q) = (\tfrac{*}{2i\pi}, q - io),$$

are acted on by the fundamental kernel,

$$\begin{array}{rl} \tilde{\gamma}^{-1} \overset{1}{*} \tilde{\mathcal{D}}^1 & = \log \tilde{\mathcal{D}}^1 \text{ with } \gamma^{-1} = -\frac{1}{q}, \\ \frac{-1}{q} \overset{1}{*} (\frac{1}{q-1})^{*1+k} = \frac{-1}{q} \overset{1}{*} \frac{1}{q-(1+k)} & = \frac{-1}{q-(1+k)} = \tilde{\gamma}^k(q). \end{array}$$

The eigenvalue of a $\mathbf{D}(1)$-product representation $(\tilde{d})^{*k}(q) = \frac{1}{q-k}$ is given by the singularity $q = k$ with the eigenvalue equation $\frac{1}{(\tilde{d})^{*k}(q)} = q - k = 0$. With tangent kernels and $\log \tilde{\mathcal{D}}^1$-functions $\frac{-1}{q-(1+k)}$, the eigenvalue is not given by the singularity $q = 1+k$ of $\tilde{\gamma}^k$, but by the condition that, there, the residue arising by the convolution is equal to 1,

$$\tilde{\gamma}^{-1} \overset{1}{*} (\tilde{d})^{*1+k}(q) = \tilde{\gamma}^k(q) = 1 \Rightarrow q = k = 0, \pm 1, \pm 2, \ldots$$

At the eigenvalue $q = k$, the fundamental kernel is in duality with the representation measure.

9.6.2 Mass Zero as Spacetime Translation Invariant

To obtain the $\mathbf{D}(1)$-eigenvalue equations as embedded in even-dimensional causal spacetime $\mathcal{D}^{2R}$, the convolution powers of the energy-momentum distribution for the defining spacetime representation with intrinsic unit $\underline{q}^2 = m_0^2 q^2$ are convoluted with the embedded fundamental kernel for $\mathbf{D}(1)$ to yield the $\mathbf{D}(1)$-kernels of the spacetime tangent module:

$$\begin{array}{rl} \tilde{\gamma}_{2R}^{-1} \overset{2R}{*_\kappa} \tilde{\mathcal{D}}^1 & = \tilde{\gamma}_{2R}^{-1} \overset{2R}{*} \tilde{\mathcal{D}}^{2R} \subseteq \log \tilde{\mathcal{D}}^{2R} \\ \text{with } \tilde{\gamma}_{2R}^{-1}(q) & = \frac{q}{q^2}, \quad (\overset{2R}{*}, q^2) = (-\frac{*}{i|\Omega^{2R-1}|}, (q - io)^2), \\ \frac{q}{q^2} \overset{2R}{*} \left(\frac{2q}{q^2-1}\right)^{*\kappa, 1+k} & = \frac{q}{q^2} \overset{2R}{*} \left(\frac{1}{(-q^2+\kappa^2)^R} \frac{2q}{q^2-1}\right)^{*1+k} = \tilde{\gamma}_{2R}^k(q). \end{array}$$

The translation invariants q^2 are given by those energy-momenta where fundamental kernel and spacetime representation are in duality, i.e., where the tangent residues are projectors,

$$\tilde{\gamma}_{2R}^k(q) = \mathbf{1} = \sum_\iota \mathcal{P}_\iota,$$

with a decomposition into nondecomposable projectors on the right–hand side.

The simplest nontrivial eigenvalue equation,

$$\begin{array}{rl} \frac{q}{q^2} \overset{2R}{*_\kappa} \frac{2q}{q^2-1} = \tilde{\gamma}_{2R}^0(q) & = \frac{\partial}{\partial q} \otimes q \int_0^1 d\xi \int_0^1 d\zeta \frac{(1-\xi)^{R-1}(1-\zeta)^{R-1}}{-\xi q^2 + \zeta(1-\kappa^2) + \kappa^2} \\ & = \mathcal{P}_1 r_{2R}^1(q^2, \kappa^2) + \mathcal{P}_0 r_{2R}^0(q^2, \kappa^2), \end{array}$$

is decomposed with two projectors

$$\frac{\partial}{\partial q} \otimes q = \mathbf{1}_{2R} + 2q \otimes q \frac{\partial}{\partial q^2} = \mathcal{P}_1 + \mathcal{P}_0 (1 + 2q^2 \frac{\partial}{\partial q^2}), \quad \left\{ \begin{array}{rl} \mathcal{P}_1 & = \mathbf{1}_{2R} - \frac{q \otimes q}{q^2}, \\ \mathcal{P}_0 & = \frac{q \otimes q}{q^2}. \end{array} \right.$$

The q^2-dependent residual functions for the two related eigenvalue equations

$$r_{2R}^1(q^2, \kappa^2) = 1, \quad r_{2R}^0(q^2, \kappa^2) = 1$$

arise by integrations over singularities:

$$\begin{aligned} r^1_{2R}(q^2,\kappa^2) &= \int_0^1 d\xi \int_0^1 d\zeta \tfrac{(1-\xi)^{R-1}(1-\zeta)^{R-1}}{-\xi q^2+\zeta(1-\kappa^2)+\kappa^2} \\ &= \int_0^{q^2} \tfrac{dQ^2}{(q^2)^R} \int_{\kappa^2}^1 \tfrac{dM^2}{(1-\kappa^2)^R} \tfrac{(1-Q^2)^{R-1}(1-M^2)^{R-1}}{-Q^2+M^2} \\ &= \tfrac{1}{(q^2)^R(\kappa^2-1)^R} \int_0^{q^2} dz_1 \int_0^{\kappa^2-1} dz_2 \tfrac{z_1^{R-1} z_2^{R-1}}{z_1+z_2+1-q^2}, \\ r^0_{2R}(q^2,\kappa^2) &= (1+2q^2\tfrac{\partial}{\partial q^2}) r^1_{2R}(q^2,\kappa^2) \\ &= (1-2R) r^1_{2R}(q^2,\kappa^2) + 2(\tfrac{1-q^2}{q^2})^{R-1} \int_{\kappa^2}^1 \tfrac{dM^2}{(1-\kappa^2)^R} \tfrac{(1-M^2)^{R-1}}{-q^2+M^2}. \end{aligned}$$

The duality condition above for the embedded $\mathbf{D}(1)$-representation and its fundamental kernel, leading to the determination of the ratio for interaction and translation invariant, can be read as an eigenvalue solution at $q^2 = 0$, i.e., for Poincaré group representations with mass zero,

$$q^2 = 0 \Rightarrow r^1_{2R}(0,\kappa^2) = r^0_{2R}(0,\kappa^2) = 1.$$

With a positive residual normalization the transversal components with projector $\mathcal{P}_1$ can be related to the nontrivially polarized particle modes in the massless gauge fields [9].

9.6.3 Eigenvalue Equations with Logarithmic Residues

The real rank 2 of even-dimensional spacetimes leads to integrations over singularity lines with characteristic logarithmic residues, which involve a real and an imaginary part. They have their origin in the integration for a finite path with two endpoints [1], e.g., where defined:

$$\begin{aligned} \int_\beta^\alpha \tfrac{dz}{z-\delta} &= \mathrm{res}[\tfrac{1}{z-\delta}\log\tfrac{z-\alpha}{z-\beta}] = \log\tfrac{\delta-\alpha}{\delta-\beta} = \log\left|\tfrac{\delta-\alpha}{\delta-\beta}\right| + i\arg\tfrac{\delta-\alpha}{\delta-\beta} \\ \tfrac{1}{2i\pi}\oint \tfrac{dz}{z-\delta} &= \mathrm{res}\tfrac{1}{z-\delta} = 1. \end{aligned}$$

This leads to complicated looking formulas in contrast to the odd-dimensional integrations without generic logarithms for real rank 1 time and hyperbolic positions.

The logarithmic residues involve integer powers $z^k_{\log}$ with logarithms and the harmonic series $\varphi(k)$,

$$\begin{aligned} \tfrac{1}{z} &= \tfrac{d}{dz}\log z = (\tfrac{d}{dz})^{1+k}\tfrac{z^k_{\log}}{k!}, \quad k=0,1,\dots, \\ z^k_{\log} &= z^k[\log z - \varphi(k)] \text{ with } \begin{cases} \varphi(0) &= 0, \\ \varphi(k) &= 1+\dots+\tfrac{1}{k} = 1, \tfrac{3}{2}, \tfrac{11}{3!}, \dots. \end{cases} \end{aligned}$$

The residual functions are computed systematically via integration by parts,

$$\begin{aligned} &\int_0^{\alpha_{1,2}} dz_{1,2} \tfrac{z_1^{k_1} z_2^{k_2}}{z_1+z_2+\gamma} = \int_0^{\alpha_{1,2}} dz_{1,2} z_1^{k_1} z_2^{k_2} \partial_1\partial_2 (z_1+z_2+\gamma)^1_{\log} \\ &= \underline{z}_1^{k_1} \underline{z}_2^{k_2} (\underline{z}_1+\underline{z}_2+\gamma)^1_{\log} + k_1 k_2 \int_0^{\alpha_{1,2}} dz_{1,2} z_1^{k_1-1} z_2^{k_2-1} (z_1+z_2+\gamma)^1_{\log} \\ &\quad -\underline{z}_1^{k_1} k_2 \int_0^{\alpha_2} dz_2 z_2^{k_2-1} (\underline{z}_1+z_2+\gamma)^1_{\log} - \underline{z}_2^{k_2} k_1 \int_0^{\alpha_1} dz_1 z_1^{k_1-1} (z_1+\underline{z}_2+\gamma)^1_{\log}. \end{aligned}$$

An underlining denotes the prescription to take the difference at the integration limits,

$$r(\underline{z}) = r(\alpha) - r(0).$$

With the parameters

$$\alpha_1 = q^2,\ \alpha_2 = \kappa^2 - 1,\ \alpha_1 + \alpha_2 = q^2 + \kappa^2 - 1, \quad \left\{ \begin{array}{rl} \gamma & = 1 - q^2, \\ \alpha_1 + \gamma & = 1, \\ \alpha_2 + \gamma & = \kappa^2 - q^2, \\ \alpha_1 + \alpha_2 + \gamma & = \kappa^2, \end{array} \right.$$

the resdidual functions above for abelian 2-dimensional Cartan spacetime require two integrations for $\frac{1}{z} = (\frac{d}{dz})^2 z^1_{\log}$:

$$\begin{aligned}
\boxed{r_2^1(q^2, \kappa^2)} &= \frac{1}{\alpha_1\alpha_2} \int_0^{\alpha_{1,2}} dz_{1,2} \frac{1}{z_1+z_2+\gamma} \\
&= \frac{(\underline{z}_1+\underline{z}_2+\gamma)^1_{\log}}{\alpha_1\alpha_2} = \frac{(\alpha_1+\alpha_2+\gamma)^1_{\log}+(\gamma)^1_{\log}-(\alpha_1+\gamma)^1_{\log}-(\alpha_2+\gamma)^1_{\log}}{\alpha_1\alpha_2} \\
&= \int_0^{q^2} \frac{dQ^2}{q^2} \int_{\kappa^2}^1 \frac{dM^2}{1-\kappa^2} \frac{1}{-Q^2+M^2} = \frac{(\kappa^2)^1_{\log}+(1-q^2)^1_{\log}-(1)^1_{\log}-(\kappa^2-q^2)^1_{\log}}{q^2(\kappa^2-1)} \\
&= \boxed{-\frac{(1-q^2)\log(1-q^2)-(\kappa^2-q^2)\log(\kappa^2-q^2)+\kappa^2\log\kappa^2}{q^2(1-\kappa^2)}}, \\
\boxed{r_2^0(q^2, \kappa^2)} &= -r_2^1(q^2,\kappa^2) + 2\int_{\kappa^2}^1 \frac{dM^2}{1-\kappa^2}\frac{1}{-q^2+M^2} = -r_2^1(q^2,\kappa^2) + \frac{2}{1-\kappa^2}\log\frac{1-q^2}{\kappa^2-q^2} \\
&= \boxed{\frac{(1+q^2)\log(1-q^2)-(\kappa^2+q^2)\log(\kappa^2-q^2)+\kappa^2\log\kappa^2}{q^2(1-\kappa^2)}},
\end{aligned}$$

with the special cases

$$\begin{aligned}
r_2^1(q^2, 1) &= -\frac{1}{q^2}\log(1-q^2), \\
r_2^1(q^2, 0) &= -\frac{1-q^2}{q^2}\log(1-q^2) - \log(-q^2).
\end{aligned}$$

$2R$-dimensional spacetime requires up to $2R$-integrations for $\frac{1}{z} = (\frac{d}{dz})^{2R}\frac{z^{2R-1}_{\log}}{(2R-1)!}$, e.g. for 4-dimensional Minkowski spacetime:

$$\begin{aligned}
\boxed{r_4^1(q^2, \kappa^2)} = &\ \frac{1}{\alpha_1^2\alpha_2^2}\int_0^{\alpha_{1,2}} dz_{1,2}\frac{z_1 z_2}{z_1+z_2+\gamma} \\
= &\ \frac{(\alpha_1+\alpha_2+\gamma)^1_{\log}}{\alpha_1\alpha_2} + \int_0^{\alpha_{1,2}} dz_{1,2}\partial_1\partial_2\frac{(z_1+z_2+\gamma)^3_{\log}}{3!\alpha_1^2\alpha_2^2} \\
&-\alpha_1\int_0^{\alpha_2} dz_2\partial_2\frac{(\alpha_1+z_2+\gamma)^2_{\log}}{2\alpha_1^2\alpha_2^2} - \alpha_2\int_0^{\alpha_1} dz_1\partial_1\frac{(z_1+\alpha_2+\gamma)^2_{\log}}{2\alpha_1^2\alpha_2^2} \\
= &\ \frac{(\alpha_1+\alpha_2+\gamma)^1_{\log}}{\alpha_1\alpha_2} + \frac{(\alpha_1+\alpha_2+\gamma)^3_{\log}+(\gamma)^3_{\log}-(\alpha_1+\gamma)^3_{\log}-(\alpha_2+\gamma)^3_{\log}}{3!\alpha_1^2\alpha_2^2} \\
&+\frac{\alpha_1(\alpha_1+\gamma)^2_{\log}+\alpha_2(\alpha_2+\gamma)^2_{\log}-(\alpha_1+\alpha_2)(\alpha_1+\alpha_2+\gamma)^2_{\log}}{2\alpha_1^2\alpha_2^2} \\
= &\ \int_0^{q^2}\frac{dQ^2}{(q^2)^2}\int_{\kappa^2}^1\frac{dM^2}{(1-\kappa^2)^2}\frac{(1-Q^2)(1-M^2)}{-Q^2+M^2} \\
= &\ \frac{(\kappa^2)^1_{\log}}{q^2(\kappa^2-1)} + \frac{(\kappa^2)^3_{\log}+(1-q^2)^3_{\log}-(1)^3_{\log}-(\kappa^2-q^2)^3_{\log}}{3!(q^2)^2(\kappa^2-1)^2} + \frac{q^2(1)^2_{\log}+(\kappa^2-1)(\kappa^2-q^2)^2_{\log}-(q^2+\kappa^2-1)(\kappa^2)^2_{\log}}{2(q^2)^2(\kappa^2-1)^2} \\
= &\ \boxed{\frac{(1-q^2)^3\log(1-q^2)+q^2(1-\kappa^2)(1-2\kappa^2-q^2)-(3-2\kappa^2-q^2)(\kappa^2-q^2)^2\log(\kappa^2-q^2)+\kappa^2(3\kappa^2-2\kappa^4-6q^2+3q^2\kappa^2)\log\kappa^2}{3!(q^2)^2(1-\kappa^2)^2}}.
\end{aligned}$$

It is needed for $\kappa^2 \ll q^2$:

$$\begin{array}{rl}\boxed{r_4^1(q^2,0)} &= \int_0^{q^2}\frac{dQ^2}{(q^2)^2}\int_0^1 dM^2\frac{(1-Q^2)(1-M^2)}{-Q^2+M^2}\\ &= \boxed{\frac{(1-q^2)^3\log(1-q^2)}{6(q^2)^2}+\frac{1-q^2}{6q^2}-\frac{(3-q^2)\log(-q^2)}{6}},\\ \boxed{r_4^0(q^2,0)} &= -3r_4^1(q^2,0)+2\int_0^1 dM^2\frac{(1-q^2)(1-M^2)}{q^2(-q^2+M^2)}\\ &= -3r_4^1(q^2,0)+2\frac{(1-q^2)^2}{q^2}\log\frac{1-q^2}{-q^2}-2\frac{1-q^2}{q^2}\\ &= \boxed{2\frac{(1-q^2)^2}{q^2}\log\frac{1-q^2}{-q^2}-2\frac{1-q^2}{q^2}-\frac{(1-q^2)^3\log(1-q^2)}{2(q^2)^2}-\frac{1-q^2}{2q^2}+\frac{(3-q^2)\log(-q^2)}{2}}.\end{array}$$

9.7 Normalization of Translation Representations

Starting from a generating fundamental representation, the residue of a product representation defines its normalization. For spacetime, the determination of the residues requires the transition from inverse derivative energy–momentum distributions (kernels) to the associated distributions for the representations of the spacetime translations.

The exponential from the Lie algebra $\mathbb{R}$ (time translations) to the group $\exp\mathbb{R} = \mathbf{D}(1)$ can be reformulated in the language of residual representations with energy functions by a geometric series

$$e^{imt} = \sum_{k=0}^{\infty}\frac{(imt)^k}{k!} = \oint\frac{dq}{2i\pi}\frac{1}{q-m}e^{iqt} = \oint\frac{dq}{2i\pi}\frac{1}{q}\sum_{k=0}^{\infty}\left(\frac{m}{q}\right)^k e^{iqt};$$

where $-\frac{m}{q}$ is the inverse derivative energy function for the representation function $\frac{1}{q-m}$.

9.7.1 Geometric Transformation and Mittag-Leffler Sum

Translation representations are characterized by (energy–)momentum distributions with simple poles. Meromorphic complex functions have only pole singularities. In the compactified complex plane $\overline{\mathbb{C}}$ they constitute the field of rational functions. The representation distributions for one dimension (pole functions) have negative degree:

$$\overline{\mathbb{C}} \ni q \longmapsto \rho(q) = \frac{P^n(q)}{P^k(q)} = \frac{\alpha_0+\alpha_1 q+\cdots+\alpha_n q^n}{\gamma_0+\gamma_1 q+\cdots+\gamma_k q^k} \in \overline{\mathbb{C}},\quad \alpha_j,\gamma_j\in\mathbb{C},\ \gamma_k\neq 0,\quad k\le n.$$

The geometric transformations for $\mathbf{D}(1)$ (time) with $z = \frac{q}{m}$,

$$\frac{1}{z} = \tilde{\gamma} \longmapsto \frac{\tilde{\gamma}}{1-\tilde{\gamma}} = \frac{1}{z-1},$$

are elements of the broken rational (conformal) bijective transformations of the closed complex plane

$$\overline{\mathbb{C}} \ni \zeta \longmapsto \frac{a\zeta+c}{b\zeta+d} \in \overline{\mathbb{C}}$$

with real coefficients as a group isomorphic to

$$g = \begin{pmatrix} a & b \\ c & d \end{pmatrix} \in \mathbf{SL}(\mathbb{R}^2) \cong \mathbf{SU}(1,1) \sim \mathbf{SO}_0(1,2).$$

The group transforms circles and lines into circles and lines (as a set, not individually). Upper and lower half-planes $x \pm io$ remain stable. The eigenvalue $\tilde{\gamma} = 1$ becomes a pole:

$$\begin{pmatrix} a & b \\ c & d \end{pmatrix} = \begin{pmatrix} 1 & -1 \\ 0 & 1 \end{pmatrix}: \ \tilde{\gamma} \longmapsto \frac{\tilde{\gamma}}{1-\tilde{\gamma}}, \ (1,0) \longmapsto (\infty, 0).$$

With one fixpoint $\tilde{\gamma} = 0$ the transformation is parabolic, i.e., an element of the $\mathbb{R}$-isomorphic subgroup $\begin{pmatrix} 1 & b \\ 0 & 1 \end{pmatrix}$.

The geometric transformation will be generalized in order to associate functions with pole singularities to the complex eigenvalue functions $\tilde{\gamma}(z)$ for spacetime,

$$\tilde{\gamma}(z) \longmapsto \frac{\tilde{\gamma}(z)}{1-\tilde{\gamma}(z)}.$$

An eigenvalue $z_0 \in \{z \mid \tilde{\gamma}(z) = 1\}$ gives a pole. If the zero z_0 is simple with $\tilde{\gamma}$ holomorphic there, it defines, by geometric transformation of its Taylor series, a Laurent series [1] and a residue

$$\begin{aligned} \tilde{\gamma}(z) &= 1 + (z - z_0)\tilde{\gamma}'(z_0) + \sum_{k=2}^{\infty} \frac{(z-z_0)^k}{k!} \tilde{\gamma}^{(k)}(z_0), \\ \frac{\tilde{\gamma}(z)}{1-\tilde{\gamma}(z)} &= \frac{\operatorname{res}(z_0)}{z-z_0} + \sum_{k=0}^{\infty} (z - z_0)^k a_k(z_0), \\ \operatorname{res}(z_0) &= -\frac{1}{\tilde{\gamma}'(z_0)}. \end{aligned}$$

Each eigenvalue $\{z_k \mid \tilde{\gamma}(z_k) = 1\}$ has its own principal part. Their sum, called a Mittag-Leffler sum, replaces the simple pole for $\mathbf{D}(1)$:

$$\begin{aligned} \tilde{\gamma}(z) &\longmapsto \frac{\tilde{\gamma}(z)}{1-\tilde{\gamma}(z)} = \sum_{z_k} \frac{a_{-1}(z_k)}{z-z_k} + \cdots, \\ \text{generalizing } \frac{1}{z} &\longmapsto \frac{\frac{1}{z}}{1-\frac{1}{z}} = \frac{1}{z-1}. \end{aligned}$$

Therefore, one obtains for an eigenvalue function for spacetime $\mathcal{D}^{2R}$ and its projectors at the invariant solutions

$$\tilde{\gamma}(q^2) = \tilde{\gamma}^{-1} * \tilde{d}(q^2) = \mathbf{1} \Rightarrow q^2 \in \{m^2\}$$

the transition to complex representation functions $\tilde{\mathcal{G}}_0$, assumed with simple poles,

$$\tilde{\mathcal{G}} \longrightarrow \log \tilde{\mathcal{G}} \longrightarrow \tilde{\mathcal{G}}_0, \quad \tilde{d} \longmapsto \tilde{\gamma}(q^2) \longmapsto \frac{\tilde{\gamma}(q^2)}{\mathbf{1}-\tilde{\gamma}(q^2)} = \sum_{m^2} \frac{\operatorname{res}(m^2)}{q^2-m^2} + \cdots.$$

The residue is the negative inverse of the derivative of the energy–momentum tangent function at the invariant

$$\tilde{\gamma}(q^2) = \mathbf{1} + (q^2 - m^2)\frac{\partial \tilde{\gamma}}{\partial q^2}(m^2) + \cdots \Rightarrow \operatorname{res}(m^2) = -\frac{1}{\frac{\partial \tilde{\gamma}}{\partial q^2}(m^2)}.$$

The simple poles can be used for the representation of the Poincaré group $\mathbf{SO}_0(1, 2R-1) \vec{\times} \mathbb{R}^{2R}$. The residue gives the normalization of the associated representation

$$\text{on-shell part of } \frac{i}{\pi} \frac{\mathrm{res}(m^2)}{q^2 + io - m^2} = \mathrm{res}(m^2)\delta(q^2 - m^2).$$

9.7.2 Gauge-Coupling Constants as Residues at Mass Zero

In the residual product of the fundamental spacetime representation with the dual inverse derivative

$$\frac{q}{q^2} \overset{2R}{*} \frac{1}{(-q^2+\kappa^2)^R} \frac{2q}{q^2 - 1} = [\mathbf{1}_{2R} + q \otimes q \, 2\frac{\partial}{\partial q^2}] \, r^1_{2R}(q^2, \kappa^2),$$

the residual normalization $\mathrm{res}(0, \kappa^2)$ for the massless solution $r^1_{2R}(0, \kappa^2) = 1$ is given by the inverse of the negative derivative of the eigenvalue function there:

$$\begin{aligned} -\frac{1}{\mathrm{res}(0,\kappa^2)} = \frac{\partial r^1_{2R}}{\partial q^2}(0, \kappa^2) &= \frac{1}{R(1+R)(1-\kappa^2)^R} \int_{\kappa^2}^1 dM^2 \frac{(1-M^2)^{R-1}}{M^4} \\ &= \frac{1}{R(1+R)} [\frac{1}{\kappa^2} + (R-1) \log_R \kappa^2] \\ &= \frac{1 - R(R-1)\kappa^2}{R(1+R)\kappa^2} \quad \text{with } -\frac{1}{R} \log_R \kappa^2 = 1. \end{aligned}$$

One has the numerical values for Cartan and Minkowski spacetime

$$-\mathrm{res}(0, \kappa^2) = \begin{cases} 2\kappa^2 = 2, & R = 1, \\ \frac{6\kappa^2}{1-2\kappa^2} \sim \frac{6}{e^3 - 2} \sim \frac{1}{3}, & R = 2. \end{cases}$$

With the geometric transformation the Laurent series gives an energy-momentum distribution for a spacetime translation representation with invariant zero and residual normalization. With appropriate integration contour, it can be used as propagator for a mass-zero spacetime vector field with coupling constant $-\mathrm{res}(0, \kappa^2)$:

$$\mathbf{SO}_0(1, 2R-1) \vec{\times} \mathbb{R}^{2R} : \text{ on-shell part of } \frac{i}{\pi} \frac{\eta_{jk}\mathrm{res}(0,\kappa^2)}{q^2 + io} = \eta_{jk}\mathrm{res}(0, \kappa^2)\delta(q^2).$$

This vector field has, in addition to an $\mathbf{SO}_0(1,1)$-related pair with neutral signature, $2R-2$ particle-interpretable degrees of freedom (chapter "Massless Quantum Fields"), which are related to the spherical degrees of freedom $\Omega^{2R-2} \subset \mathcal{D}^{2R}$ and the compact fixgroup $\mathbf{SO}(2R-2)$ in the massless particle fixgroup $\mathbf{SO}(2R-2) \vec{\times} \mathbb{R}^{2R-2}$. Those degrees of freedom have a positive scalar product:

$$-\eta_{jk} = \begin{cases} \begin{pmatrix} -1 & 0 \\ 0 & \mathbf{1}_{2R-1} \end{pmatrix} \cong \begin{pmatrix} 0 & 0 & 1 \\ 0 & \mathbf{1}_{2R-2} & 0 \\ 1 & 0 & 0 \end{pmatrix} & \text{for } \mathbf{SO}_0(1, 2R-1) \vec{\times} \mathbb{R}^{2R}, \\ \begin{pmatrix} -1 & 0 \\ 0 & 1 \end{pmatrix} \cong \begin{pmatrix} 0 & 1 \\ 1 & 0 \end{pmatrix} & \text{for } \mathbf{SO}_0(1,1) \vec{\times} \mathbb{R}^2, \\ \begin{pmatrix} -1 & 0 \\ 0 & \mathbf{1}_3 \end{pmatrix} \cong \begin{pmatrix} 0 & 0 & 1 \\ 0 & \mathbf{1}_2 & 0 \\ 1 & 0 & 0 \end{pmatrix} & \text{for } \mathbf{SO}_0(1,3) \vec{\times} \mathbb{R}^4. \end{cases}$$

The particle-interpretable degrees of freedom start with rank 2 Minkowski spacetime. There, the two degrees of freedom with a positive scalar product have left and right polarization for the axial $\mathbf{SO}(2)$-rotations.

If adjoint representations of compact internal degrees of freedom, e.g., of $\mathbf{U}(2)$ hypercharge-isospin, are included, the accordingly normalized residues of the arising mass-zero solutions in 4-dimensional spacetime may be compared with the coupling constants in the propagators of massless gauge fields as in the standard model of electroweak interactions (chapter "Gauge Interactions")

$$\mathbf{U}(2) \times [\mathbf{SO}_0(1,3) \vec{\times} \mathbb{R}^4] : \ \tfrac{-\eta^{jk} G^2}{q^2 + io} \text{ with } G^2 \in (g_1^2, g_2^2 | g^2, \gamma^2) \sim (\tfrac{1}{8.4}, \tfrac{1}{2.5} | \tfrac{1}{10.9}, \tfrac{1}{1.9}).$$

Without the introduction of the internal degrees of freedom only the order of magnitude of the normalizations G^2 can be compared with the residues above for the simple massless poles from representations of spacetime $\mathbf{D}(2) \cong \mathbf{GL}(\mathbb{C}^2)/\mathbf{U}(2)$,

$$\text{for } \log \mathbf{D}(2) \cong \mathbb{R}^4 : \quad G^2 \leftrightarrow -\text{res}(0, \kappa^2) \sim \tfrac{1}{3}.$$

MATHEMATICAL TOOLS

9.8 Divergences in Feynman Integrals

The following convolution products are valid only where the Γ-functions are defined:

$$\mathbf{SO}_0(1,3) \vec{\times} \mathbb{R}^4$$

$\overset{4}{*}$	$\frac{\Gamma(2+n)}{(q^2-m^2)^{2+n}}$	$\frac{2q\ \Gamma(3+n)}{(q^2-m^2)^{3+n}}$
$\frac{1}{q^2}$	$I_0^{1+n}(q^2)$	$2qI_2^{2+n}(q^2)$
$\frac{q}{q^2}$	$qI_1^{1+n}(q^2)$	$\frac{\partial}{\partial q} \otimes q\ I_1^{1+n}(q^2)$
$\frac{q}{(q^2)^2}$	$qI_2^{2+n}(q^2)$	$\frac{\partial}{\partial q} \otimes q\ I_2^{2+n}(q^2)$

$$(\overset{4}{*}, q^2) = (\mp \tfrac{2*}{i|\Omega^3|}, q^2 \mp io) = (\mp \tfrac{*}{i\pi^2}, q^2 \mp io), \text{ Feynman,}$$
$$q \cong q^j, \quad \tfrac{\partial}{\partial q} \otimes q = \mathbf{1}_4 + 2q \otimes q \tfrac{\partial}{\partial q^2} \cong \delta_k^j + 2q^j q_k \tfrac{\partial}{\partial q^2}.$$

The integrals for natural $n = 0, 1, \ldots,$

$$\begin{aligned}
I_0^{1+n}(q^2) &= \int_0^1 d\zeta \tfrac{\Gamma(1+n)}{(q^2\zeta - m^2)^{1+n}} &&= \tfrac{1}{q^2}(\tfrac{\partial}{\partial m^2})^n \log(1 - \tfrac{q^2}{m^2}),\\
I_1^{1+n}(q^2) &= \int_0^1 d\zeta \tfrac{(1-\zeta)\Gamma(1+n)}{(q^2\zeta - m^2)^{1+n}} &&= \tfrac{1}{q^2}(\tfrac{\partial}{\partial m^2})^n[-1 + (1 - \tfrac{m^2}{q^2}) \log(1 - \tfrac{q^2}{m^2})],\\
I_2^{1+n}(q^2) &= \int_0^1 d\zeta \tfrac{\zeta\ \Gamma(1+n)}{(q^2\zeta - m^2)^{1+n}} &&= \tfrac{1}{q^2}(\tfrac{\partial}{\partial m^2})^n[1 + \tfrac{m^2}{q^2} \log(1 - \tfrac{q^2}{m^2})]\\
&= -\tfrac{\partial}{\partial q^2} I_0^n(q^2) &&= I_0^{1+n}(q^2) - I_1^{1+n}(q^2),
\end{aligned}$$

can be determined with derivatives

$$\begin{aligned}
\frac{\Gamma(1+n)}{z^{1+n}} &= (-\tfrac{d}{dz})^{1+n}\log(-z),\\
\log(-z) &= (\tfrac{d}{dz})^n \frac{z^n}{\Gamma(1+n)}[\log(-z)-\varphi(n)],\\
z^n\log(-z) &= \frac{d}{dz}\frac{z^{1+n}}{1+n}[\log(-z)-\tfrac{1}{1+n}].
\end{aligned}$$

Convergent integrals have to vanish for $q^2 \to \infty$. The integrals are logarithmically divergent with $\log q^2$ for $\Gamma(0)$ and quadratically divergent with q^2 for $\Gamma(-1)$, which exemplifies the general divergences with $(q^2)^n$ for $\Gamma(-n)$. Such integrals make no sense. The divergent integrals can be "obtained" up to q^2-polynomials by deriving with respect to the invariant q^2 and then taking appropriate values for n. This limit, in turn, can be written as a q^2-derivation

$$\begin{aligned}
\frac{\partial}{\partial q^2}\int_0^1 d\zeta \frac{\Gamma(1+n)}{(q^2\zeta-m^2)^{1+n}} &= -\int_0^1 d\zeta \frac{\zeta\,\Gamma(2+n)}{(q^2\zeta-m^2)^{2+n}},\\
\text{for } n\to -1: \quad &\to -\int_0^1 d\zeta \frac{\zeta}{q^2\zeta-m^2} = -\frac{\partial}{\partial q^2}\int_0^1 d\zeta \log(q^2\zeta-m^2),\\
(\frac{\partial}{\partial q^2})^2\int_0^1 d\zeta \frac{(1-\zeta)\Gamma(1+n)}{(q^2\zeta-m^2)^{1+n}} &= \int_0^1 d\zeta \frac{\zeta^2(1-\zeta)\Gamma(3+n)}{(q^2\zeta-m^2)^{3+n}},\\
\text{for } n\to -2: \quad &\to \int_0^1 d\zeta \frac{\zeta^2(1-\zeta)}{q^2\zeta-m^2} = \frac{\partial}{\partial q^2}\int_0^1 d\zeta\; \zeta(1-\zeta)\log(q^2\zeta-m^2)\\
&= (\frac{\partial}{\partial q^2})^2\int_0^1 d\zeta\;(1-\zeta)(q^2\zeta-m^2)[\log(q^2\zeta-m^2)-1],\\
\frac{\partial}{\partial q^2}\int_0^1 d\zeta \frac{\zeta\,\Gamma(2+n)}{(q^2\zeta-m^2)^{2+n}} &= -\int_0^1 d\zeta \frac{\zeta^2\Gamma(3+n)}{(q^2\zeta-m^2)^{3+n}},\\
\text{for } n\to -2: \quad &\to -\int_0^1 d\zeta \frac{\zeta^2}{q^2\zeta-m^2} = -\frac{\partial}{\partial q^2}\int_0^1 d\zeta\;\zeta\log(q^2\zeta-m^2).
\end{aligned}$$

For polynomials of finite degree one may try to connect the unknown coefficients with experimental numbers.

If the divergent convolutions with negative n arise as part of a well-defined convolution (no "divergences"), one can combine integrals with compatible regularizations:

$$\begin{aligned}
\frac{1}{q^2-m^2} \overset{4}{*}_{\rm reg} \frac{1}{q^2} &= \qquad -\int_0^1 d\zeta \log(q^2\zeta-m^2),\\
\frac{2q}{q^2-m^2} \overset{4}{*}_{\rm reg} \frac{q}{q^2} &= \frac{\partial}{\partial q}\otimes q \int_0^1 d\zeta\;(1-\zeta)(q^2\zeta-m^2)[\log(q^2\zeta-m^2)-1],\\
\frac{2q}{(q^2-m^2)^2} \overset{4}{*}_{\rm reg} \frac{q}{q^2} &= \frac{\partial}{\partial q}\otimes q \int_0^1 d\zeta\;(\zeta-1)\log(q^2\zeta-m^2),\\
\frac{2q}{q^2-m^2} \overset{4}{*}_{\rm reg} \frac{q}{(q^2)^2} &= \frac{\partial}{\partial q}\otimes q \int_0^1 d\zeta\;(-\zeta)\log(q^2\zeta-m^2).
\end{aligned}$$

From this the higher-order poles, including the convergent convolutions, can be computed by m^2-derivations, as seen in the transition from the second to the third integral. For $m^2=0$ the third integral coincides with the fourth one up to a constant, which is irrelevant on the logarithmic divergence level.

The remaining integrals with the massless pole convolution characteristic logarithm $\log(q^2\zeta-m^2)$ can be combined from

$$\begin{aligned}
(m^2,q^2)^n &= \int_0^1 d\zeta\;(m^2-q^2\zeta)^n\log(q^2\zeta-m^2)\\
&= \frac{(m^2-q^2)^{1+n}-(m^2)^{1+n}}{(1+n)^2q^2} + \frac{(m^2)^{1+n}\log(-m^2)-(m^2-q^2)^{1+n}\log(q^2-m^2)}{(1+n)q^2},\\
(0,q^2)^n &= \frac{(-q^2)^n}{1+n}(-\frac{1}{1+n}+\log q^2),\quad n=0,1,\ldots,
\end{aligned}$$

with the lowest powers

$$(m^2,q^2)^n = \left\{\begin{array}{ll}
-1+\frac{m^2\log(-m^2)-(m^2-q^2)\log(q^2-m^2)}{q^2}, & n=0,\\
\frac{-2m^2+q^2}{4}+\frac{m^4\log(-m^2)-(m^2-q^2)^2\log(q^2-m^2)}{2q^2}, & n=1,\\
\frac{-3m^4+3q^2m^2-(q^2)^2}{9}+\frac{m^6\log(-m^2)-(m^2-q^2)^3\log(q^2-m^2)}{3q^2}, & n=2,
\end{array}\right.$$

used in the integrals above:

$$
\begin{aligned}
-\int_0^1 d\zeta \log(q^2\zeta - m^2) &= -(m^2, q^2)^0 = 1 - \frac{m^2 \log(-m^2) - (m^2 - q^2)\log(q^2 - m^2)}{q^2}, \\
\int_0^1 d\zeta\, (1-\zeta)(q^2\zeta - m^2)[\log(q^2\zeta - m^2) - 1] &= \left(\frac{3m^2 - q^2}{6}\right) + \frac{(m^2-q^2)(m^2,q^2)^1 - (m^2,q^2)^2}{q^2} \\
&= -\frac{11}{36}q^2 + \frac{11}{12}m^2 - \frac{m^4}{6q^2} + \frac{m^4(m^2-3q^2)\log(-m^2) - (m^2-q^2)^3\log(q^2-m^2)}{6(q^2)^2}, \\
\int_0^1 d\zeta\, (\zeta - 1)\log(q^2\zeta - m^2) &= \frac{(m^2-q^2)(m^2,q^2)^0 - (m^2,q^2)^1}{q^2} \\
&= \frac{3}{4} - \frac{m^2}{2q^2} + \frac{m^2(m^2-2q^2)\log(-m^2) - (m^2-q^2)^2\log(q^2-m^2)}{2(q^2)^2},
\end{aligned}
$$

$$
\begin{aligned}
\int_0^1 d\zeta(-\zeta)\log(q^2\zeta - m^2) &= \frac{-m^2(m^2,q^2)^0 + (m^2,q^2)^1}{q^2} \\
&= \frac{1}{4} + \frac{m^2}{2q^2} + \frac{-m^4\log(-m^2) + (m^2+q^2)(m^2-q^2)\log(q^2-m^2)}{2(q^2)^2}.
\end{aligned}
$$

For the first integral the q^2-linear and constant terms are irrelevant, for the second and third one the constant terms.

Bibliography

[1] H. Behnke, F. Sommer, *Theorie der analytischen Funktionen einer komplexen Veränderlichen* (1962), Springer-Verlag, Berlin, Göttingen, Heidelberg.

[2] S. Bochner, Monotone Funktionen, Stieltjessche Integrale und harmonische Analyse, Math. Ann., 108 (1933), 378-410.

[3] G.B. Folland, A Course in Abstract Harmonic Analysis (1995), CRC Press, Boca Raton, Ann Arbor, London, Tokyo.

[4] W. Heisenberg, *Einführung in die einheitliche Feldtheorie der Elementarteilchen* (1967), Hirzel, Stuttgart.

[5] S. Helgason, *Groups and Geometric Analysis* (1984), Academic Press, New York. London, Sydney, Tokyo, Toronto, etc..

[6] R.S. Strichartz, Harmonic analysis on hyperboloids, *Journal of Functional Analysis* 12, 341-383 (1973).

[7] F. Treves, *Topological Vector Spaces, Distributions and Kernels* (1967), Academic Press, New York, London.

[8] H. Saller, Matter as spectrum of spacetime representations, hep-th/0304034.

[9] H. Saller, Gauge coupling constants as residues of spacetime representations, hep-th/0505059.

Index